Practical Handbook of

Physical Properties of Rocks and Minerals

Edited by

Robert S. Carmichael

Head, Geophysics Program
Department of Geology
University of Iowa
Iowa City, Iowa

CRC Press, Inc.
Boca Raton, Florida

Library of Congress Cataloging-in-Publication Data

CRC practical handbook of physical properties of rocks and minerals / edited by Robert S. Carmichael.
p.cm.
Bibliography: p.
Includes index.
ISBN 0-8493-3703-8
1. Rocks—Handbooks, manuals, etct. 2. Mineralogy—Handbooks, manuals, etc. I. Carmichael, Robert S. II. Title: Practical handbook of physical properties of rocks and minerals.
QE431.6.P5C73 1989 88-21004
552—dc19 CIP

Direct all inquiries to CRC Press, Inc., 2000 Corporate Blvd., N. W., Boca Raton, Florida, 33431.

International Standard Book Number 0-8493-3703-8

Library of Congress Card Number 88-21004
Printed in the United States

PREFACE

This *Practical Handbook* has been constructed to serve as a convenient, compact, yet comprehensive source of basic information. The technical data have been compiled and selectively edited to provide an organized and definitive presentation of the physical properties of rocks and their constituent minerals. The formal is primarily tabular and graphical for easy reference and comparisons. There is also instructive textual material to present, explain, and clarify the data.

Physical properties of rocks and minerals are of necessary interest and utility in many fields of work—including geology, petrophysics, geophysics, materials science, geochemistry, and geotechnical engineering. Assembled and collated data sets are finding ever-increasing application in interdisciplinary studies and projects. The scale of investigation ranges from the molecular and crystalline, up to terrestrial studies of Earth and other planetary bodies. Geologists are interested in age-dating rocks to reconstruct the origin of mineral deposits; seismologists in prospective earthquake prediction using premonitory physical changes; crystallographers in the synthesis of minerals with special optical or physical properties; exploration geophysicists in the variation of physical properties of subsurface rocks to permit detection of natural resources such as oil and gas, geothermal energy, and ores of metals; geotechnical engineers in the nature and behavior of the materials on or in or of which such structures as buildings, dams, tunnels, bridges, or underground storage vaults are to be constructed; solid-state physicists in the magnetic, electrical, and mechanical properties of materials for electronic devices, computer components, or high-performance ceramics; petroleum reservoir engineers in the response on well logs or in drilling in deep rock at elevated temperature and pressure.

The various physical properties are addressed here in Sections, which have been prepared by recognized authorities who are among the leaders of their specialties. These contributing authors are drawn from leading university, governmental research, and industry establishments. They assisted with the precursory presentation in CRC Press' *Handbook of Physical Properties of Rocks,* published in three volumes in 1982 to 1984. That material has been re-assembled, edited and abridged, and updated for the present work.

The initial section sets the stage for presenting and assessing physical properties, by outlining the "Mineral composition of rocks". This treats the chemical composition and physical characteristics of rock types (igneous, sedimentary, metamorphic), of pore fluids (including geothermal fluids), of economic ores and fuels (coal, petroleum, oil shale and tar sands, radioactive minerals), and of marine sediments. Properties of minerals and crystals are presented, including petrographic characteristics. Compositions are also given for the Earth's crust and mantle, and for meteorites and Moon rock. The Section is assembled by Kenneth F. Clark, Ph.D., Professor in the Department of Geological Sciences at the University of Texas at El Paso, Texas.

Subsequent sections deal with the various individual physical properties. The first, "Densities of rocks and minerals", presents a massive data set on this fundamental and diagnostic property by succinct use of histograms and statistical analysis of density ranges. Densities are included for minerals, rocks, and soils. Outline is given of the determination of density and porosity by calculation and by *in situ* methods. The Section is contributed by Gary R. Olhoeft, Ph.D., of the Petrophysics and Remote Sensing Branch, U.S. Geological Survey, Denver, Colorado, and Gordon R. Johnson of the same agency.

The Section on "Inelastic properties" addresses strength and rheology of rocks and minerals. It presents a compilation of experimental data on the results of laboratory tests in rock mechanics; stress-strain relations; the effects of pore fluids, time and stress rate, and temperature; and rock friction. It is prepared by Stephen H. Kirby, Ph.D., of the Office of Earthquake Studies, U.S. Geological Survey, Menlo Park California, and John W. McCormick, now with the State University of New York in Plattsburgh, New York.

''Magnetic properties of minerals and rocks'' presents the magnetic and crystalline properties of rock-forming magnetic minerals and analogous materials. Included are the various types of remanent magnetization, and the magnetic properties susceptibility, coercive field, Curie temperature, anisotropy, and saturation magnetization. The variation of properties with chemical composition, grain size and shape, and temperature and pressure, is given. This Section is compiled by the coordinating author and editor of this volume, Robert S. Carmichael, Ph.D.

''Electrical properties'' summarizes the conductivity/resistivity and dielectric constants of minerals and dry rocks; along with the variation of electrical properties with temperature, pressure, the frequency at which measurement is made, and lithology and porosity. Also given are: induced polarization; the resistivity of brine and of water-bearing rocks; electrical properties and electric logs of sedimentary rocks, of *in situ* sequences of rocks; and properties of coal, permafrost, and the Earth's interior. The Section is assembled by George V. Keller, Ph.D., of the Department of Geophysics at Colorado School of Mines, Golden Colorado.

The Section on ''Seismic velocities'' gives compressional and shear wave velocities for various rocks, minerals, marine sediments and water, aggregates and glasses, the Earth's crust and mantle, glaciers and permafrost. Laboratory and *in situ* measurements are included, along with the variation of seismic velocity with degree of fluid saturation, and with pressure and temperature. The material is compiled by Nikolas I. Christensen, Ph.D., now of the Department of Earth and Atmospheric Sciences at Purdue University, West Lafayette Indiana. In a companion Section on ''Seismic attenuation'', attention is focussed on methods of laboratory and seismological determination of attenuation, having application to oil exploration and terrestrial studies. Data are given for p- and s-wave attenuation for minerals, for sedimentary and crystalline rocks, and for the Earth, and the invluence of strain amplitude, pressure, frequency, and fluid saturation. The work is presented by Marius Vassiliou, Ph.D., now of Rockwell International Science Center, Thousand Oaks, California and by Bernhard R. Tittmann, Ph.D., and Carlos A. Salvado of the Earth and Planetary Sciences Group at Rockwell International Science Center in Thousand Oaks, California.

''Radioactivity properties of minerals and rocks'' reviews the radioactive isotope systems used in geochronology (age dating), and the decay constants, radiogenic heat production in rocks, and the radioactive minerals. This is given by W. Randall Van Schmus, Ph.D., of the Department of Geology, University of Kansas, Lawrence Kansas.

The Section on ''Spectroscopic properties of rocks and minerals'' is a comprehensive presentation of the interaction of geoscience materials with electromagnetic radiation in the visible and infrared range. The properties of absorption/transmission, reflection and emission are given, along with the spectral characteristics of minerals and rocks. The material was diligently compiled by Graham R. Hunt, Ph.D., then of the Petrophysics and Remote Sensing Branch of the U.S. Geological Survey, Denver Colorado, shortly before his untimely death.

''Engineering properties of rock'' reviews factors and tests relating to rock appraisal, characterization, and the assessment of properties such as strength, hardness, elastic constants, and deformation that would be of special interest in geotechnical engineering. Data for various engineering properties is given, including the effects of pore-water pressure. The Section is compiled by Allen W. Hatheway, Ph.D., in the Department of Geological Engineering, University of Missouri, Rolla Missouri, and George A. Kiersch, Ph.D., emeritus professor from Cornell University, Ithaca, New York.

My thanks are extended to all who have contributed to the formulation and execution of this work. The editorial function at CRC Press was performed by Sandy Pearlman, Amy Skallerup, and Paul Gottehrer. The University of Iowa provided partial release time in the form of a Developmental Leave. Final work was completed at the facilities of Meiji University—at the School of Science and Engineering and with assistance of Dr. Takeo Hamamoto of the Office of International Programs—and at the Tokyo Institute of Technology—with opportunity and support there provided by Dr. Akira Sawaoka of the Research Lab of Enginnering Materials.

Robert S. Carmichael
Iowa City Iowa, and Tokyo Japan
April 1988

THE AUTHOR

Robert S. Carmichael, Ph.D., is Professor of Geophysics and Geology in the Department of Geology, University of Iowa, Iowa City. He graduated from the University of Toronto with a B.A.Sc. degree in geophysics/engineering physics, and then earned M.S. and Ph.D. degrees in Earth and Planetary Science from the University of Pittsburgh. His thesis specialties were in seismology and rock magnetism, and while there, he was an Andrew Mellon University Fellow.

After graduation in 1967, he spent a year at Osaka University in Japan as a postdoctoral Research Fellow of the Japan Society for Promotion of Science, working in high-pressure geophysics. Upon return, he joined Shell Oil's Research Center in Houston as a research geophysicist in petroleum exploration. Now at the University of Iowa, Dr. Carmichael has research interests in rock properties, exploration geophysics, high-pressure geophysics and magnetics, and earthquakes in the central Midcontinent region. In the past few years he has been faculty geoscientist on a Semester-at-Sea circumglobal voyage, exchange visitor at the University of Iceland in Reykjavik, U.S. organizer for the Fulbright Summer Geology Program in Iceland, and visiting professor for a year at Meiji University in Tokyo.

He has authored numerous scientific articles, and done consulting for geotechnical and seismic problems. He is a member of the American Geophysical Union, Society of Exploration Geophysicists, Iowa Academy of Science, and Society of Terrestrial Magnetism and Electricity.

CONTRIBUTORS

Nikolas I. Christensen, Ph.D., earned his degrees at the University of Wisconsin. He has worked in geology and geophysics at the Universities of Southern California and Washington. Dr. Christensen is now a Professor of Geoscience at Purdue University. His research centers on elastic properties of rocks and minerals, crystal physics, and applications to the crust of the Earth.

Kenneth F. Clark, Ph.D., has degrees from the University of Durham (United Kingdom) and New Mexico. He has worked as a geologist with Anglo-American Corporation/South Africa and with Cornell University. From 1971 to 1980 he was at the University of Iowa as Professor of Geology. His research has been in economic geology, mineral deposits, and tectonism and mineralization. Dr. Clark is now with the Department of Geological Sciences at the University of Texas at El Paso.

Allen W. Hatheway, Ph.D., has degrees from the Universities of California/Los Angeles and Arizona. He is a registered geologist, engineering geologist, and civil engineer in several states and has worked in consulting geotechnical engineering for LeRoy Crandall & Associates, Woodward-Clyde Consultants, Shannon and Wilson, Inc. and Haley & Aldrich, Inc. He is now with the University of Missouri at Rolla as Professor of Geological Engineering. His technical interests include engineering geology and engineering properties of rocks.

Graham R. Hunt, Ph.D., D.Sc., earned degrees from the University of Sydney in Australia and worked in spectroscopy at Tufts University, M.I.T., and the Air Force Cambridge Research Labs. At the time of writing his contribution, he was a Senior Research Scientist with the U.S. Geological Survey as Chief of the Petrophysics and Remote Sensing Branch in Denver. Dr. Hunt's research has been in spectroscopy and physical chemistry, molecular structure, and the remote sensing of the composition of terrestrial and extraterrestrial surfaces. He has since died after a brief illness.

Gordon R. Johnson graduated from Colorado State University and has since worked for the U.S. Geological Survey in geophysical exploration for minerals and in physical properties of rocks. He is with the Petrophysics and Remote Sensing Branch in Denver.

George V. Keller, Ph.D., graduated from Pennsylvania State University and then worked for the U.S. Geological Survey. He is now a Professor of Geophysics at the Colorado School of Mines and former Head of the Department. He is co-author of the book, *Electrical Methods in Geophysical Prospecting*. Dr. Keller's research includes electrical prospecting, geothermal resources, physical rock properties, and the Earth's crust.

George A. Kiersch, Ph.D., graduated from the Colorado School of Mines and the University of Arizona. He worked as a geologist with the Army Corps of Engineers and directed exploration programs for the University of Arizona and Southern Pacific Company before joining Cornell University in 1960 as Professor of Engineering Geology. He served as Chairman of Geological Sciences there from 1965 to 1971. Dr. Kiersch's interests have been in engineering geology, mineral deposits, and geomechanics. He is now Emeritus Professor from Cornell.

Stephen H. Kirby, Ph.D., has degrees from the University of Illinois and the University of California at Los Angeles. He has worked for the U.S. Geological Survey in their Heavy Metals Branch in Denver, and is now a geophysicist in their Office of Earthquakes, Volcanoes and Engineering in Menlo Park, California. The physical properties of rocks and minerals are his primary research interest.

John W. McCormick, Ph.D., obtained his degrees at the Pennsylvania State University and University of California at Los Angeles. He worked as a geophysicist at the U.S. Geological Survey's Office of Earthquake Studies in Menlo Park, California and is now in the Department of Computer Sciences at the State University of New York at Plattsburgh.

Gary R. Olhoeft, Ph.D., has degrees from Massachusetts Institute of Technology and the University of Toronto. He has worked for Kennecott Copper Corporations, NASA, and Lockheed Electronics Company, and is now a research geophysicist for the U.S. Geological Survey in Denver. His research interests are in the physical and chemical properties of rocks and minerals.

Carlos A. Salvado graduated from the University of Rhode Island and then earned M.S. degrees in physics from the University of Colorado and in geophysics from the California Institute of Technology. He has worked for IBM and Sierra Geophysics, Inc., and as a consultant for the Marine Research Laboratory of Maine, and was a Senior Research Associate at Rockwell International Science Center in Thousand Oaks, California from 1980 to 1984. Since then he has been an independent consultant and entrepreneur. He is currently President of Applied Sciences Corporation in Carlsbad, California, a company which develops instrumentation for the characterization of fiber-reinforced composite materials.

W. Randall Van Schmus, Ph.D., earned degrees from California Institute of Technology and the University of California at Los Angeles. He has worked for the Lunar-Planetary Research Branch of the Air Force Cambridge Research Labs, and since 1967 for the department of Geology at the University of Kansas where he is a professor, his research interests are in the geology, geochronology, and geophysics of Precambrian rocks and the mineralogy and petrology of meteorites.

Bernhard R. Tittmann, Ph.D., earned his degrees at George Washington University and the University of California at Los Angeles. He has worked for Hughes Aircraft, Inc. and North American Aviation/Rockwell International, now being Manager of the Earth and Planetary Sciences group at the latter's Science Center in Thousand Oaks, California. He has spent a year as a faculty physicist at the University of California at Los Angeles, and a year as Visiting Professor at the University of Paris. His technical interests center on seismic and mechanical properties, solid state physics, and ultrasonics.

Marius S. Vassiliou, Ph.D., received his doctorate in geophysics and electrical engineering from the California Institute of Technology in 1983. He also holds a Master's degree in Computer Science from the University of Southern California, and a Bachelor's degree from Harvard University. He has consulted to Rockwell and TRW, and worked at ARCO Oil and Gas Co.'s research and development laboratory. Since 1985 he has been a Member of the Technical Staff at the Rockwell International Science Center in Thousand Oaks, California. His technical interests are in geophysics, materials science, and numerical computation.

TABLE OF CONTENTS

SECTION IV

SECTION V

SECTION VI

Nikolas I. Christensen

SECTION VII

Marius Vassiliou, Carlos A. Salvado, and Bernhard R. Tittmann

SECTION VIII

SECTION IX

SECTION X

Section I
Mineral Composition of Rocks

By
Kenneth F. Clark

INTRODUCTION

The composition of naturally occurring assemblages of minerals has a fundamental bearing on the physical properties of rocks of which they are constituents. Considerations of rock density, elasticity, seismic, thermal, electrical, magnetic, and radioactive properties, for example, are largely dictated by aggregate properties of individual minerals. Thus, any fundamental consideration of the physical properties of rocks must be prefaced by a scrutiny of the chemical and physical properties of minerals. This leads to a consideration of crystal structures, which in turn is governed by atomic characteristics and chemical bonding.[278]

In the following pages, the properties of minerals and rocks of the Earth have been compiled, taking into account their terrestrial distribution that is governed by geologic processes. Relevant data on extraterrestrial materials is indicated, as are naturally occurring substances that do not strictly fit the definition of mineral or rock. Included here are several chemically and physically contrasting fluids located on and below the surface of the Earth. Last, the mineral compositions of earth materials that have historically proved to be useful to man are surveyed, including organic and radioactive substances.

COSMIC ABUNDANCES

Earth materials are considered to be a sample of cosmic matter. Extraterrestrial matter (solid, liquid, gaseous) has been grouped[301] into stellar matter, interstellar matter, and matter of our solar system. Our own solar system including sun, planets, matter of comets and meteorites, and gaseous atmosphere of the sun and planets is assumed to have a common origin. Knowledge of the chemical composition of cosmic matter allows recognition of relationships between solar systems and the origin and development of terrestrial matter.

As noted earlier,[301] the chemical composition of extraterrestrial matter can be studied by spectral analyses of luminous matter, by examining meteorites, or by examining lunar rock samples. Table 1 compares several compilations of elemental cosmic abundances. Tables 2 to 9 show the classification of meteorites plus their mean chemical composition and mineralogy. Tables 10 to 12 provide data on averages of major elemental analyses of basaltic lunar rocks. In Figure 1, weight percent CaO is plotted against weight percent Al_2O_3 in glasses from the Apollo 14 landing site and compared to typical values in meteorites. Major elemental compositions of some ANT-Suite (anorthositic-noritic-troctolitic) rocks[281] are given in Table 12 and their compositional fields and nomenclature are plotted on the 0l-An-SiO_2 pseudoternary diagram in Figure 2.

EARTH'S CRUST

Several theories suggest that the Earth was formed by the cold accretion of particles of metal, troilite, and silicates with bulk composition approximated by chondrites.[397] The chondritic meteorites are made up of three different phases or groups of phases: nickel-iron, iron sulfide, and silicate minerals. Within the Earth, elements are distributed between phases that can be formed according to their relative affinity for metal or for silicate. While gravity controls the relative positions of the phases, the distribution of elements within these phases depends upon chemical potentials.[221] Whatever the origin and chemical differentiation that took place in the history of the Earth, a plausible hypothesis about the chemistry of the interior can be made by distributing the metal and silicates of meteorites so that they satisfy the density and elastic proper-

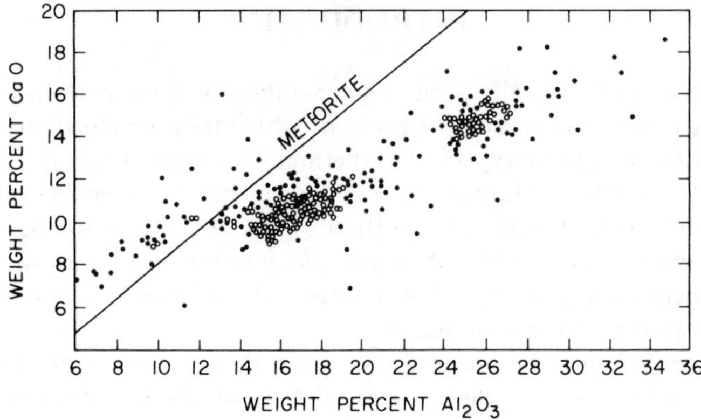

FIGURE 1. Plot of weight percent CaO versus weight percent Al_2O_3 in glasses from Apollo 14 landing site regolith.

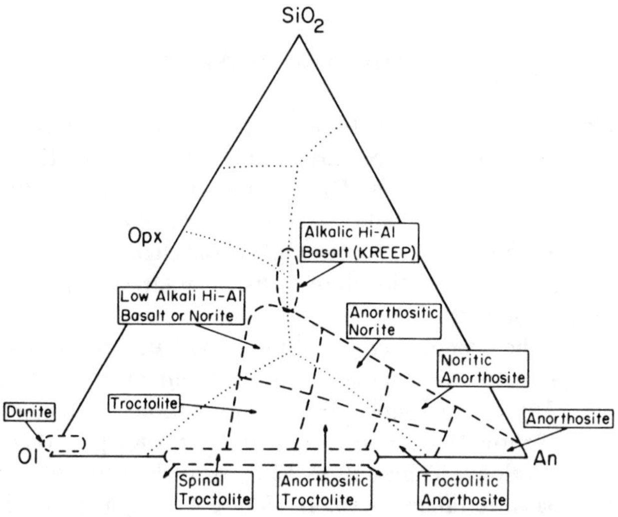

FIGURE 2. Compositional fields and nomenclature of ANT-Suite Rocks plotted on the 01-An-SiO_2 pseudoternary diagram. (High alumina (KREEP) basal fields are also included for reference.)

ties deduced from geophysical measurements. This gives rise to a generally accepted model of a core consisting largely of molten iron surrounded by a mantle made up mostly of silicates of iron and magnesium[188] and a surficial crust. This differentiation may be the most significant event in the Earth's history; it led to the formation of the crust and eventually to continents and probably initiated escape of gases from the interior that resulted in formation of the atmosphere and oceans.[278]

The distribution of rock types in large structural units of the crust and crustal layers is given in Tables 13 and 14. Elemental data for types of crust are given in Tables 15 to 17.

MANTLE AND CORE

The core and mantle together essentially determine the bulk composition of the Earth. The mantle constitutes 67.2% of the Earth's total mass and 90% of its volume.[397] The composition of the upper mantle may be obtained directly by the study of rocks that have been directly derived from the mantle and emplaced in the crust by tectonic processes.[294] Some of these ultramiafic intrusions are documented in Table 18, whereas xenoliths of mantle origin are shown by mineralogy and/or major element analysis in Tables 19 to 21. The term pyrolite, as shown in Table 22, is the inferred parental material that gave rise to mantle-derived ultramafic rocks that remained after a basaltic component had been extracted,[294] and is consistent with an Earth model derived from material resembling carbonaceous chondrites.

CRYSTALS AND MINERALS

A crystal is a solid body bounded by plane natural surfaces that are an expression of the regular internal arrangement of constitutent atoms or ions.[222] The classification of crystals based on planes, axes, and centers of symmetry using the Hermann-Maugin sumbols is shown in Table 23. Ionic radii and electronegativities are displayed in Table 24. Chemical classification of minerals, isomorphism, and structural classification of silicates are identified in Tables 25 and 26. The predominant rare earth elements as fixed in minerals are listed in Table 27. For X-ray powder diffraction identification of minerals, see Reference 28.

PETROGRAPHIC CHARACTERISTICS

Petrography is the systematic description and classification of rocks.[243] The polarizing microscope is specifically designed for the study of minerals and rocks whereby distinctive optical properties can be detected. Some minerals are opaque (Table 28) and can only be evaluated in reflected light, whereas others are transparent or translucent and can be viewed in transmitted light.

MINERALOGY OF ROCK KINDREDS

The main groups of rocks are termed igneous and sedimentary and their metamorphic equivalents. Igneous rocks are the products of crystallization of naturally occurring silicate melts, whereas sedimentary rocks result from accumulation of materials at the surface of the earth by various processes under the influence of agents such as wind, water, and ice. Metamorphic rocks result from recrystallization of igneous and sedimentary rocks at relatively high temperatures and pressures. The various mineralogic associations that constitute igneous, sedimentary, and metamorphic rocks are shown in Table 29.

IGNEOUS ROCKS

A classification of igneous rocks based on relative mineral abundances is portrayed in Figure 6. Included here are volcanic and plutonic rocks whose essential and accessory minerals are shown in Table 30. Compositions of average igneous rock are given by several investigators in Table 31 and norms of granitic, intermediate, gabbroic-basaltic, peridotitic, anorthositic, and alkalic rocks are included in Tables 32 to 36. In the calculation of the norm the various oxides, determined by chemical analysis, are combined sequentially to form the normative mineral components.[52] The Cross, Iddings, Pirsson, and Washington method (CIPW) is shown for several igneous rocks and is described in standard works.[177]

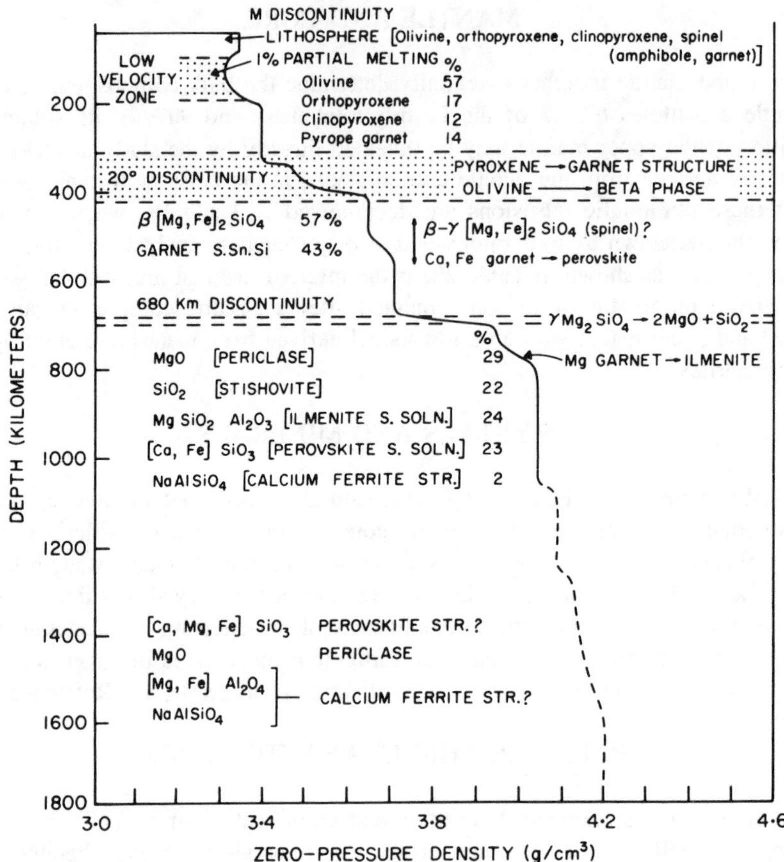

FIGURE 3. Possible mineral assemblages and corresponding zero-pressure densities for a model mantle of pyrolite composition.

Chemical analyses of igneous rock suites that characterize particular segments of the earth, for example, oceanic basins, continental crust, or tectonic belts, or that evolved through various evolutionary processes are given in Tables 37 to 57. In addition to major rock-forming elements and corresponding normative mineralogy, trace elemental and isotopic data are included in several instances. Additionally, rare earth element (RRE) contents of various rock kindreds[124] are cited in Tables 53 to 55. Finally analyses of gases from fumaroles and volcanoes are shown in Tables 56[52] and 57.[327]

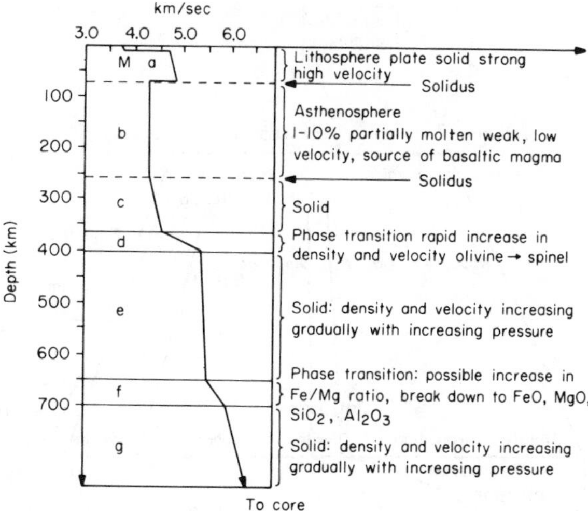

FIGURE 4. A modern view of the structure outermost 700 km of the Earth is illustrated by a plot of S-wave velocity against depth.

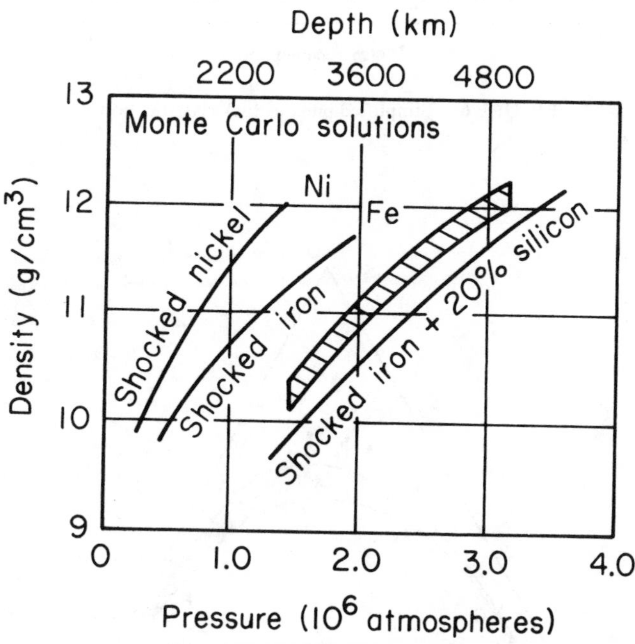

FIGURE 5. Density in the Earth's fluid core plotted against depth below the surface and against pressure.

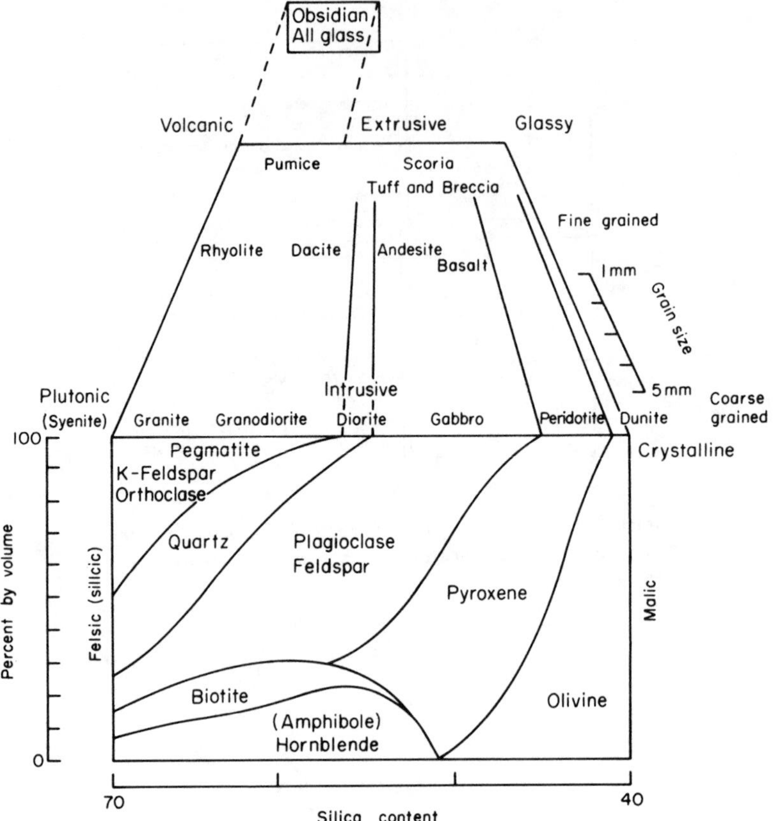

FIGURE 6. Simplified igneous rock classification.

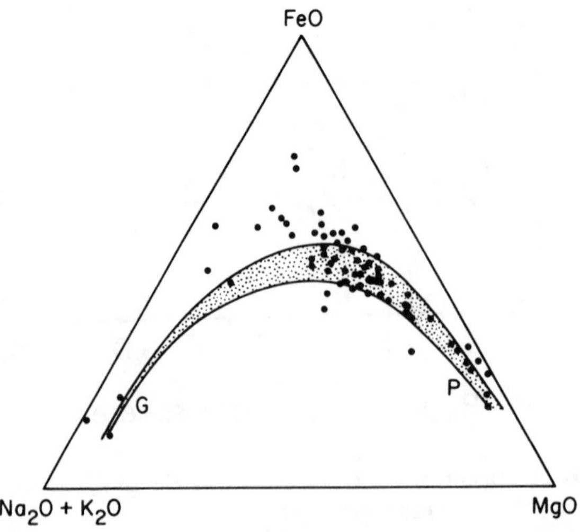

FIGURE 7. Plot of analyses of Karroo diabases and associated picrites and granophyres.

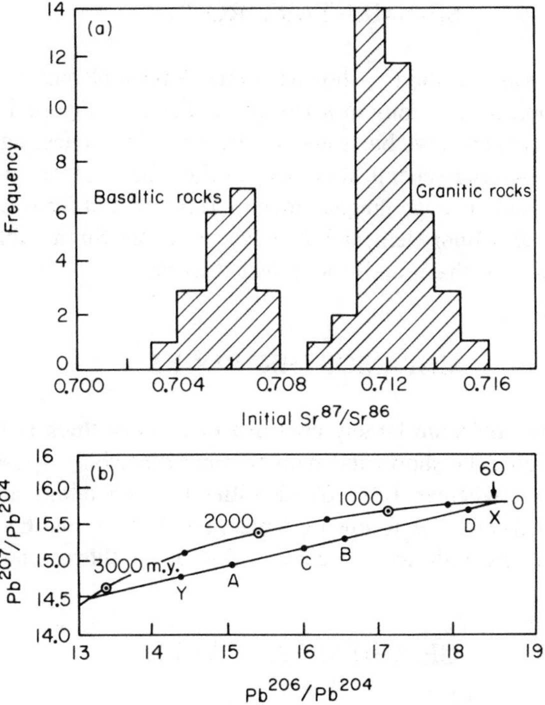

FIGURE 8. Isotopic data for strontium and lead, Hebridean Tertiary volcanic province.

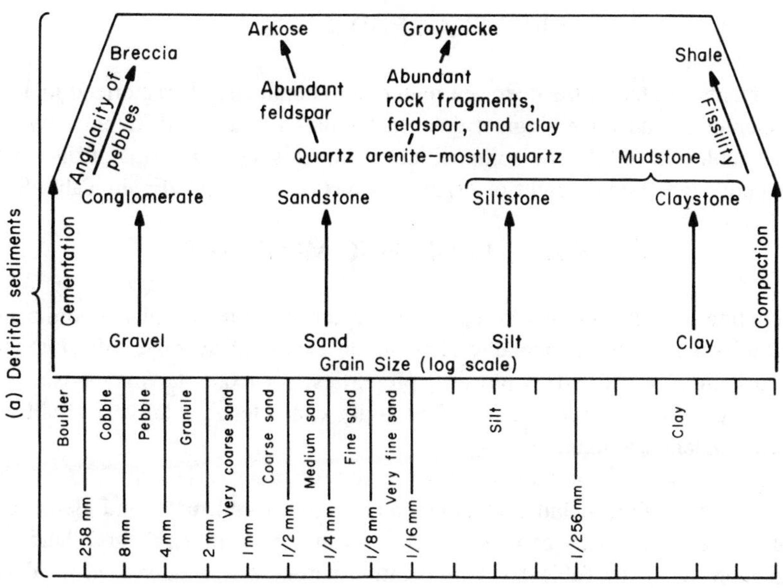

FIGURE 9. Sedimentary rock classification.

SEDIMENTARY ROCKS

The sedimentary rock classification [278] shown in Figure 9 is supplemented by characterizing common sediment chemical variations in a triangular diagram, Figure 10. Shales are not markedly differentiated chemically, but some sandstones, limestones, and saline residues are abnormally pure.[272] Average chemical composition of shale, sandstone, limestone, and sediment are compared with average igneous rock in Table 58 and for selected sedimentary rocks in Tables 59 and 60. Minor elemental abundances are shown in Table 61 and average mineralogical constitution for shales and other selected sedimentary rocks are given in Tables 90 and 91, respectively.

METAMORPHIC ROCKS

Regional metamorphic rocks are largely confined to areas of thick sedimentation at the continental margins. Figure 11 shows the metamorphic assemblages (facies) in terms of temperature and pressure, whereas Table 65 identifies the mineralogy and major element constitution of different facies. The following tables, numbers 66 and 67 compare average composition of pelitic rocks with average composition of phyllites, schists, gneisses, and amphibolites.

GEOCHEMICAL CYCLE

The migration of elements within the crust, probably supplemented by addition of primary material from the mantle, provides the concept of the geochemical cycle[221] (Figure 12). The conversion of igneous rock to various sedimentary rocks is given in Table 68. Table 69 shows major oxide analyses of silt, siltstone, and loess. The ionic potential and behavior of elements in sedimentary processes is given in Figure 13. Tables 70 and 71 compare the composition of seawater with that of fresh water and also that of the crust, granite, basalt, and shale.

DEEP SEA SEDIMENTS

Those sediments located in the deep sea region are chemically characterized in Tables 72 to 74. The elemental and mineralogical data on nodules formed in different environments are shown in Tables 75 to 82. The composition of the biosphere in terms of distribution of elements as percentage body weight of organisms has been compiled[221] in Table 83.

ORES AND ECONOMIC MINERALS

The concluding sections of this compilation document minerals of economic interest. Because of the concentration of metallic elements in excess of average crustal abundances (Clarke of concentration[221]) certain physical properties, for example, magnetism, density, conductivity of electrical currents, and radioactivity characterize relatively small areas of the crust where metals accumulate.

Table 84 is one form of commercial classification of mineral deposits[98,333] by element, mineral, rock, or fluid that includes metallic, nonmetallic fossil fuels, and elements found in waters and gases. The metal content of the most important minerals and their densities are given in more detail in Table 85. Comparative hardness is widely used in identifying ore minerals and is shown in Table 86.

Detailed mineralogic and optical properties including solubility of evaporite minerals [33] have been compiled in Table 87.

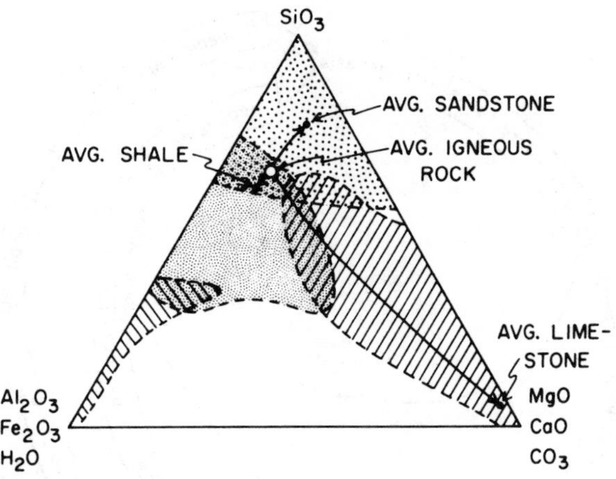

FIGURE 10. Triangle diagram showing range in composition of the common sediments.

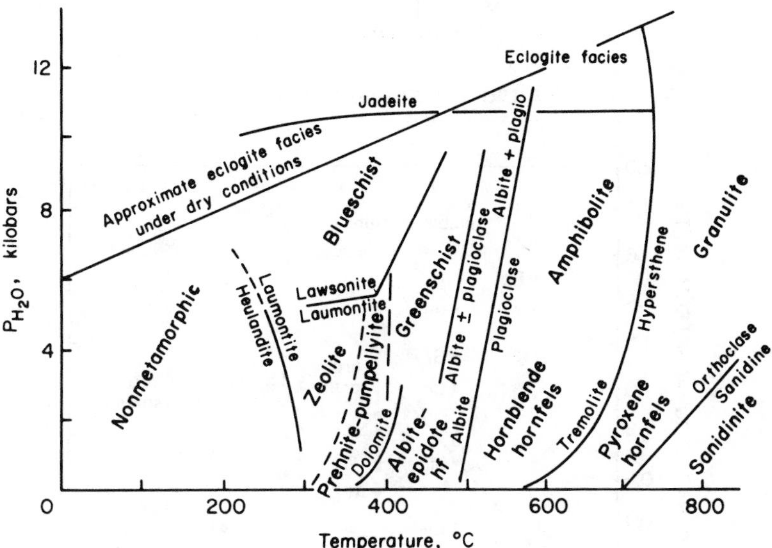

FIGURE 11. Metamorphic facies boundaries.

COAL

The average composition and relative proportions of constituent elements in wood, peat, and various ranks of coal are listed in Tables 88 and 89, respectively. The type classification of coal, in which components are amenable to optical microscope identification, is given in Tables 90 and 91.

Proximate and ultimate (major element) analyses including calorific value for a number of coals are shown in more detail in Tables 92 and 93.

The correlation of nomenclature of the transmitted light to the reflected light[178] and other microscopic characteristics are shown in Tables 94 and 95.

Mineral inclusions in coal are given in Table 96, whereas mean analytical values of elements for coals in the eastern and western U.S., are listed in Tables 97 and 98, respectively.

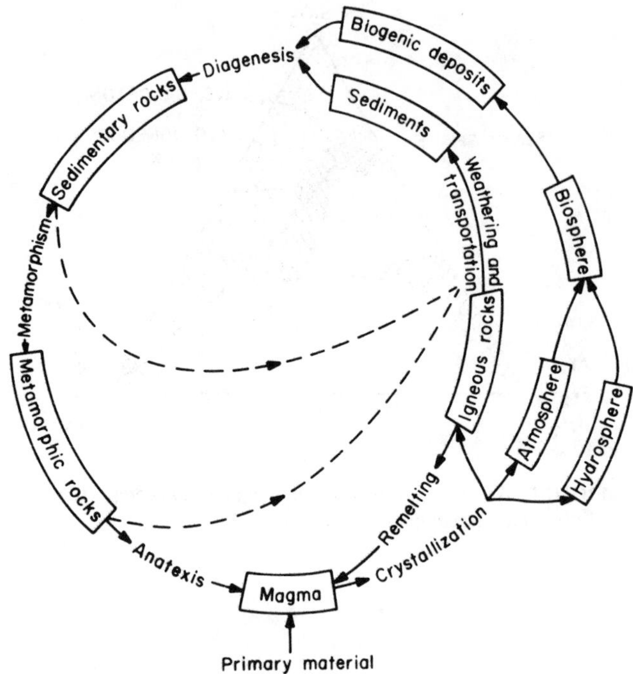

FIGURE 12. The geochemical cycle.

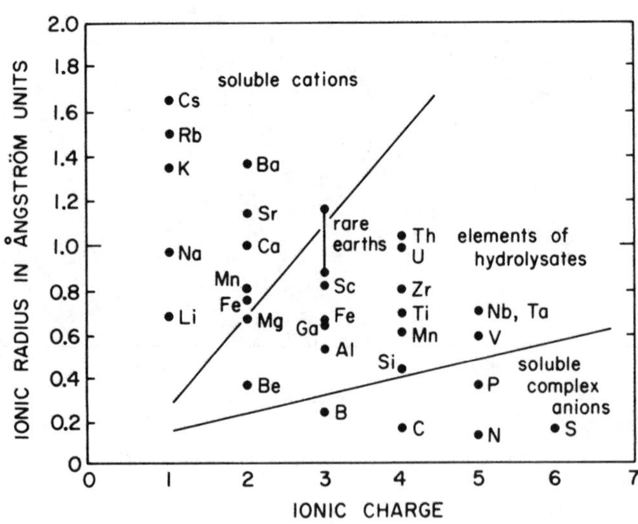

FIGURE 13. Ionic potential and the behavior of elements in sedimentary processes.

PETROLEUM, INCLUDING OIL SHALE AND TAR SAND

Elemental analyses of oils in terms of carbon, hydrogen and oxygen, specific gravity, and gravity Beaumé (a specific gravity of 1 compared with water, is 10° on the Beaumé scale[292]) are given in Tables 99 and 100, plus color and distillation parameters of naturally occurring crude oils from selected wells in Table 101. The composition of some crude oils in terms of the main hydrocarbon groups, resins and asphaltenes, is shown in Table 102. Natural gas analyses are compiled in Table 103.

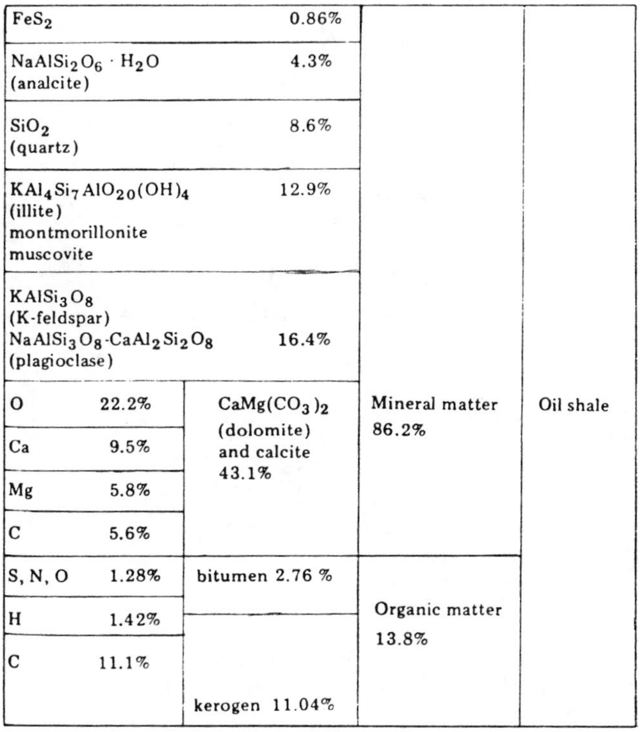

Component	%	Group	Matter	
FeS_2	0.86%			
$NaAlSi_2O_6 \cdot H_2O$ (analcite)	4.3%			
SiO_2 (quartz)	8.6%			
$KAl_4Si_7AlO_{20}(OH)_4$ (illite) montmorillonite muscovite	12.9%			
$KAlSi_3O_8$ (K-feldspar) $NaAlSi_3O_8$-$CaAl_2Si_2O_8$ (plagioclase)	16.4%			
O	22.2%	$CaMg(CO_3)_2$ (dolomite) and calcite 43.1%	Mineral matter 86.2%	Oil shale
Ca	9.5%			
Mg	5.8%			
C	5.6%			
S, N, O	1.28%	bitumen 2.76 %	Organic matter 13.8%	
H	1.42%			
C	11.1%	kerogen 11.04%		

FIGURE 14. General scheme of the oil shale components.

Oil shale research has received considerable interest in the last few years and oil shale elemental constitution and mineralogical characteristics are compared in Tables 104, 105, and Figure 14.

Comparison of various age conventional pooled oils in relation to heavy oils extracted from tar sands are distinct as plotted in Figure 15. The oil extracts are higher in resin and asphaltene but lower in hydrocarbon content than the pooled oils.[77] Table 106 compares several parameters of bitumen and synthetic crude from tar sands.

RADIOACTIVE MINERALS

Average analyses of uranium source rocks and surface ground waters by element in relation to uranium are given in Tables 107 and 108, respectively.

GEOTHERMAL FLUIDS

Equations for geothermometers are detailed in Table 109. A schematic representation of variations in approximate resistivity in areas of moderate increase and also rapid increase of temperature with depth, where a gas or dry steam layer might occur, is given in Figure 16. Figure 17 schematically relates resistivity and temperature gradient to subsurface rocks in terms of their geothermal potential, whereas Figure 18 shows a similar relationship for resistivity plotted against salinity.

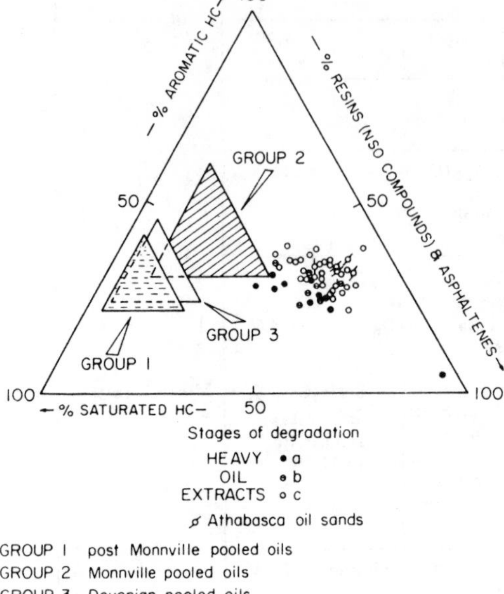

FIGURE 15. Composition diagram showing relationship of heavy oils extracted from cores to conventional pooled oils.

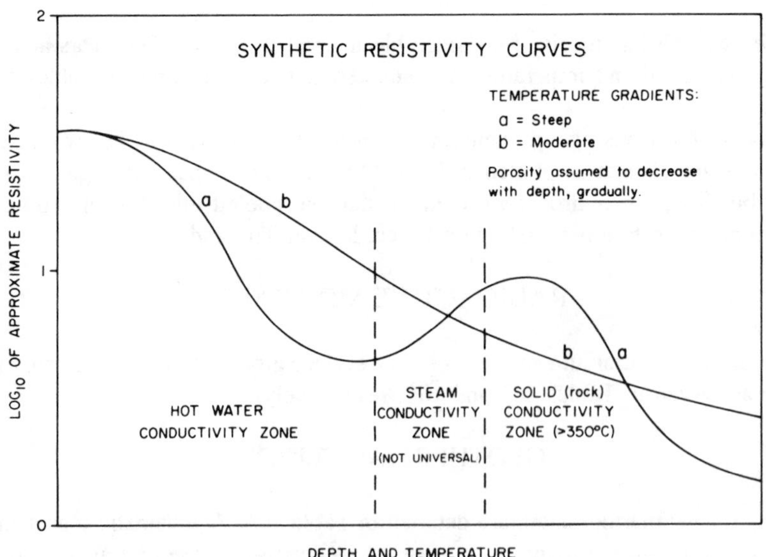

FIGURE 16. Synthetic resistivity depth curves in geothermal exploration techniques.

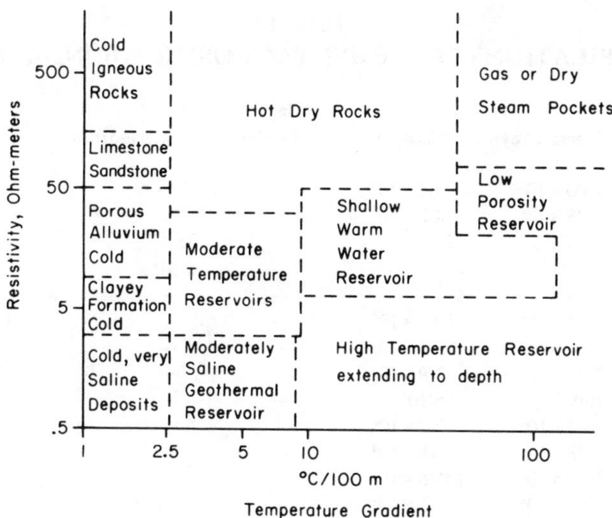

FIGURE 17. Resistivity and temperature gradient to subsurface rocks in terms of their geothermal potential.

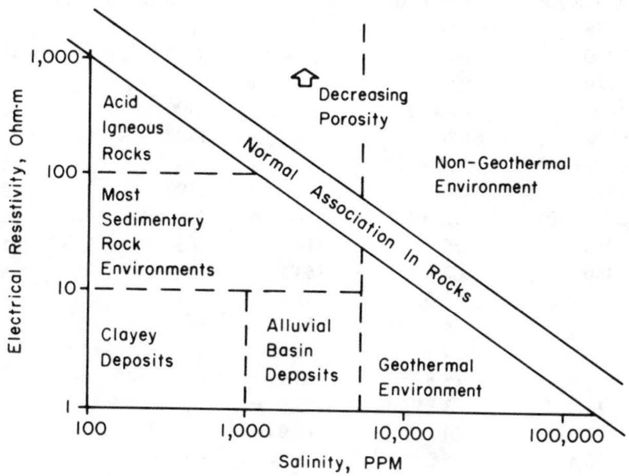

FIGURE 18. Resistivity plotted against salinity.

Table 1
COMPILATIONS OF ELEMENTAL COSMIC ABUNDANCES[49]

Atomic no.	Element	Suess-Urey	Cameron	Clayton-Fowler	Chondrites	Sun	B stars
1	H	4.00×10^{10}	3.2×10^{10}	—	—	3.2×10^{10}	3.6×10^{10}
2	He	3.08×10^{9}	2.6×10^{9}	—	—	—	5.7×10^{9}
3	Li	100	38	—	38	0.29	—
4	Be	20	7	—	0.64[a]	7.2	—
5	B	24	6	—	—	—	—
6	C	3.5×10^{6}	1.66×10^{7}	—	—	1.66×10^{7}	7.1×10^{6}
7	N	6.6×10^{6}	3.0×10^{6}	—	—	3.0×10^{6}	5.4×10^{6}
8	O	2.15×10^{7}	2.9×10^{7}	—	—	2.9×10^{7}	2.1×10^{7}
9	F	1600	$\sim 10^{3}$	—	—	—	1.2×10^{5}
10	Ne	8.6×10^{6}	2.9×10^{6}	—	—	—	1.9×10^{7}
11	Na	4.38×10^{4}	4.18×10^{4}	—	—	6.3×10^{4}	—
12	Mg	9.12×10^{5}	1.046×10^{6}	—	—	7.9×10^{5}	3.1×10^{6}
13	Al	9.48×10^{4}	8.93×10^{4}	—	—	5.0×10^{4}	5.8×10^{4}
14	Si	1.00×10^{6}	1.00×10^{6}	—	—	1.00×10^{6}	1.00×10^{6}
15	P	1.00×10^{4}	9320	—	—	6900	1.08×10^{4}
16	S	3.75×10^{5}	6.0×10^{5}	—	—	6.3×10^{5}	1.08×10^{6}
17	Cl	8850	1836	—	—	—	5.8×10^{4}
18	Ar	1.5×10^{5}	2.4×10^{5}	—	—	—	2.9×10^{5}
19	K	3160	2970	—	3290	1580	—
20	Ca	4.90×10^{4}	7.28×10^{4}	—	4.5×10^{4}	4.5×10^{4}	—
21	Sc	28	29	—	32	21	—
22	Ti	2440	3140	—	2090	1510	—
23	V	220	590	—	—	158	—
24	Cr	7800	1.20×10^{4}	—	6400	5000	—
25	Mn	6850	6320	—	7200	2500	—
26	Fe	6.00×10^{5}	8.50×10^{4}	—	—	1.17×10^{5}	—
27	Co	1800	750	—	1190	1380	—
28	Ni	2.74×10^{4}	1.5×10^{4}	—	2.48×10^{4}	2.6×10^{4}	—
29	Cu	212	39	316	186	3500	—
30	Zn	486	202	360	—	800	—
31	Ga	11.4	9.05	32	—	7.2	—
32	Ge	50.5	134	49.1	18.7	62	—
33	As	4.0	4.4	2.18	—	—	—
34	Se	67.6	18.8	30.9	18.8	—	—
35	Br	13.4	3.95	6.9	—	—	—
36	Kr	51.3	20	17.9	—	—	—
37	Rb	6.5	5.0	4.24	4.6—5.8	9.5	—
38	Sr	18.9	21	17.55	—	13.5	—
39	Y	8.9	3.6	11.35	—	5.6	—
40	Zr	54.5	23	39.8	—	54	—
41	Nb	1.00	0.81	3.5	—	2.8	—
42	Mo	2.42	2.42	2.77	—	2.5	—
44	Ru	1.49	1.58	0.83	1.3[b]	0.93	
45	Rh	0.214	0.26	0.13	0.27	0.19	
46	Pd	0.675	1.00	0.601	—	0.51	
47	Ag	0.26	0.26	0.166	0.131	0.044	
48	Cd	0.89	0.89	0.804	—	0.91	
49	In	0.11	0.11	0.071	0.0013	0.46	
50	Sn	1.33	1.33	1.88	—	1.10	
51	Sb	0.246	0.15	0.09	—	2.8	
52	Te	4.67	3.00	1.86	5—2[c]	—	
53	I	0.80	0.46	0.21	0.02—0.3[c]	—	
54	Xe	4.0	3.15	2.61	—	—	
55	Cs	0.456	0.25	0.13	0.10,0.14	—	
56	Ba	3.66	4.0	3.68	3.96,6.44	4.0	
57	La	2.00	0.38	0.70	0.40	—	

Table 1 (continued)
COMPILATIONS OF ELEMENTAL COSMIC ABUNDANCES[49]

Atomic no.	Element	Suess-Urey	Cameron	Clayton-Fowler	Chondrites	Sun	B stars
58	Ce	2.26	1.08	1.17	0.62	—	
59	Pr	0.40	0.16	0.176	0.15	—	
60	Nd	1.44	0.69	0.777	0.74	—	
62	Sm	0.664	0.24	0.595	0.25	—	
63	Eu	0.187	0.083	0.149	0.078,0.097	—	
64	Gd	0.684	0.33	0.410	0.36	—	
65	Tb	0.0956	0.054	0.083	0.056	—	
66	Dy	0.556	0.33	0.449	0.39	—	
67	Ho	0.118	0.076	0.084	0.078	—	
68	Er	0.316	0.21	0.359	0.21	—	
69	Tm	0.0318	0.032	0.060	0.039	—	
70	Yb	0.220	0.18	0.387	0.19	1.07	
71	Lu	0.050	0.031	0.035	0.036	—	
72	Hf	0.438	0.16	0.236	—	—	
73	Ta	0.065	0.021	0.030	0.017,0.019	—	
74	W	0.49	0.11	0.184	0.11	—	
75	Re	0.135	0.054	0.052	—	—	
76	Os	1.00	0.73	0.511	0.73[b]	—	
77	Ir	0.821	0.500	0.39	0.38	—	
78	Pt	1.625	1.157	0.80	—	—	
79	Au	0.145	0.13	0.13	0.13	—	
80	Hg	0.284	0.27	0.62	.04—8.4	—	
81	Tl	0.108	0.11	0.74	$(3.8—10) \times 10^{4}$	—	
82	Pb	0.47	2.2	6.5	0.05—28	2.5[d]	
83	Bi	0.144	0.14	0.92	0.0016	—	
90	Th	—	0.069	—	0.026	—	
92	U	—	0.042	—	$(7.2—7.9) \times 10^{-3}$	—	

[a] Sill and Willis, *Geochim. Cosmochim. Acta,* 26, 1209, 1962.
[b] Bate and Huizenga, *Geochim. Cosmochim. Acta,* 27, 345, 1963.
[c] Goles and Anders, *J. Geophys. Res.,* 66, 3075, 1961.
[d] Helliwell, 1961, *Astrophys. J.,* 133, 566, 1961.

Table 2
MEAN CHEMICAL COMPOSITION OF METEORITES COMPARED WITH THAT OF ERUPTIVE ROCKS[301]

Element	Iron meteorites	Stony meteorites	Chondrites	Eruptive rocks
0	—	35.71	34.84	46.60
Fe	89.70	23.31	25.07	5.00
Si	—	18.07	17.78	27.72
Mg	—	13.67	14.38	2.00
S	0.08	1.80	2.09	0.05
Ca	—	1.73	1.39	3.63
Ni	9.10	1.53	1.34	0.08
Al	—	1.52	1.32	8.13
Na	—	0.65	0.68	2.83
Cr	—	0.32	0.25	0.02
K	—	0.17	0.084	2.59
C	0.12	0.15	0.1	0.03
Co	0.62	0.12	0.08	0.002
P	0.18	0.11	0.05	0.12
Ti	—	0.11	0.066	0.44

Table 3
DISTRIBUTION OF ELEMENTS IN METEORITES AND METEORITIC PHASES [ppm][301]

Symbol	Nickel-iron (Heide)	Troilite (Heide)	Chondrite (Vinogradov)	Silicate (Heide)	Meteorites (total) (Heide)	Meteorites (total) (Rankama and Sahama)
Ag	4	18	0.094	—	2	2.0
Al	40	—	13,000	17,400 ± 1,400	13,391	13,800
Ar	—	—	—	present	—	present
As	360	1,020	0.3	20	149	present
Au	4	0.5	0.17	—	0.65	0.65
B	—	—	2	3	2.3	1.5
Ba	—	—	6	8	6.15	6.9
Be	—	—	3.6	1	0.8	1
Bi	0.5	2	0.003	0.02	0.25	present
Br	1	—	0.5	25	19.4	20
C	1,200	—	400	400	492	300
Ca	500	—	14,000	19,700 ± 2,000	15,231	13,300
Cd	8	30	0.1	1.6	4.8	present
Ce	—	—	0.5	2.5	1.9	1.77
Cl	Present	—	70	900	692	1,000 ± 1,500
Co	6,300 ± 200	100	800	206 ± 86	1,135	1,200
Cr	300	1,200	2,500	3,450 ± 590	2,792	3,340
Cs	—	—	0.1	0.1	0.08	0.08
Cu	310	500	100	1.6	87.4	170
Dy	—	—	0.35	2.5	1.9	1.80
Er	—	—	0.2	2.1	1.6	1.48
Eu	—	—	0.08	0.33	0.25	0.25
F	—	—	28	40	30.8	28
Fe	907,800 — 2,600	611,000	250,000	156,400 ± 5,400	306,969	288,000
Ga	36	0.5	3	0.5	5.96	4.2
Gd	—	—	0.4	2	1.5	1.42
Ge	500	30	10	5	83	79
H	—	—	—	630 ± 150	485	present
He	present	—	—	present	present	present
Hf	—	—	0.5	1	0.8	1.6
Hg	—	0.2	3	<0.01	0.02	present
Ho	—	—	0.07	0.72	0.55	0.51
I	0.6	—	0.04	1.25	1.1	1
In	0.5	0.5	0.001	—	0.1	0.15
Ir	4	0.4	0.48	—	0.65	0.65
K	—	—	850	1,990 ± 200	1,531	1,540
La	—	—	0.3	2.2	1.7	1.58
Li	—	—	3	3	2.3	4
Lu	—	—	0.035	0.65	0.5	0.46
Mg	320	—	140,000	158,200 ± 3,000	121,742	123,000
Mn	300	460	2,000	2,960 ± 670	2,359	2,080
Mo	10.0	11	0.6	2.5	5.3	5.3
N	present	—	1	0.9	0.7	present
Na	—	—	7,000	7,790 ± 530	5,992	5,950
Nb	0.2	—	0.3	0.5	0.4	0.41
Nd	—	—	0.6	3.7	2.85	2.59
Ni	85,900 ± 2,400	1,000	13,500	1,380 ± 240	14,354	15,680
O	—	—	350,000	410,200 ± 9,500	315,538	323,000
Os	8	9	0.5	—	1.9	1.92
P	1,800	3,000?	500	700	1,046	1,050
Pb	56	20	0.2	2	11.7	11
Pd	9	2	1	—	1.5	1.54
Pr	—	—	0.1	1	0.8	0.75
Pt	20	2	2	0.083	3.3	3.25

Table 3 (continued)
DISTRIBUTION OF ELEMENTS IN METEORITES AND METEORITIC PHASES [PPM][301]

Symbol	Nickel-iron (Heide)	Troilite (Heide)	Chondrite (Vinogradov)	Silicate (Heide)	Meteorites (total) (Heide)	Meteorites (total) (Rankama and Sahama)
Rb	—	—	5	4.5	3.5	3.5
Re	0.008,2	0.001	8×10^{-4}	0.008	0.007,5	0.002,0
Rh	5	0.4	0.19	—	0.8	0.80
Ru	20	9	1	—	2.2	2.23
S	present	347,633	20,000	—	26,740	21,200
Sb	2	7.8	0.1	0.1	0.98	present
Se	—	—	6	5	3.8	4
Se	3	100	10	13	18	7
Si	40	—	180,000	205,700 ± 1,800	158,237	163,000
Sm	—	—	0.2	1.3	1	0.95
Sn	100	15	1	5	20.4	20
Sr	—	—	10	26	20	20
Ta	<0.06	—	0.02	<0.38	0.3	0.30
Tb	—	—	0.05	0.64	0.5	0.45
Te	—	1.3	0.5	—	0.1	0.1
Th	0.04	—	0.04	2	1.54	0.8
Ti	100	—	500	1,800	1,400	1,320
Tl	—	0.3	0.001	0.15	0.15	Present
Tm	—	—	0.04	0.38	0.3	0.26
U	0.007	—	0.015	0.4	0.31	0.36
V	6	—	70	50	39.4	39
W	8.1	—	0.15	18	15.1	15
Y	—	—	0.8	7	5.4	4.72
Yb	—	—	0.2	2	1.5	1.42
Zn	115	1,520	50	3.4	137	138
Zr	8	—	30	100	78.2	73

Table 4
GASES IN METEORITES[301]

	CO_2	CO	CH_4	H_2	N_2	Total
Stony meteorites	3.77	0.24	0.20	0.50	0.09	4.80
Iron meteorites	0.21	0.67	0.02	1.67	0.24	2.81

Table 5
CHEMICAL COMPOSITION OF TEKTITES (TAYLOR AND SACHS, 1964), CHONDRITES AND ACHONDRITES (VINOGRADOV, 1958) AND MOON BASALT [MASS %][301]

	Tektites (Australite)	Chondrites (Silicate phase)	Achondrites (Free from feldspar)	Moon Basalt (Mean Comp. Apollo 11)
SiO_2	73.31	47.0	52.56	41
TiO_2	0.41	0.14	0.12	11
Al_2O_3	11.68	3.09	1.09	9
Fe_2O_3	0.60 }	15.40	11.45	19
FeO	4.03 }			
MgO	2.02	29.48	30.47	8
CaO	3.55	2.41	1.20	11
Na_2O	1.25	1.2	0.36	0.5
K_2O	2.26	0.21	0.11	

Table 6
COMPOSITIONS[a] OF CARBONACEOUS CHONDRITES (ALL VALUES IN WEIGHT PERCENT)[185]

Component	Meteorite name										
	Orguell	Ivuna	Mighel	Nawapali	Haripura	Santa Cruz	Murray	Ornams	Cold Bokkeveld	Lance	Mokoia
Fe	0.00	0.00	0.00	0.00	0.00	0.00	0.00	0.72	0.00	2.19	0.00
Ni	0.00	0.00	0.00	0.00	0.00	0.00	0.00	—	0.00	1.50	0.00
Co	0.00	0.00	0.00	0.00	0.00	0.00	0.00	—	0.00	0.07	0.00
FeS	15.07	18.38	10.05	7.67	14.93	9.04	7.67	6.43	8.16	6.49	6.74
SiO_2	22.56	22.71	27.81	27.08	26.39	29.36	28.69	33.52	27.33	33.23	33.40
TiO_2	0.07	0.07	0.08	0.09	0.08	0.12	0.09	—	0.08	0.13	0.10
Al_2O_3	1.65	1.62	2.15	2.09	2.27	2.19	2.19	—	2.29	2.93	2.51
MnO	0.19	0.23	0.21	0.17	0.19	0.19	0.21	0.18	0.19	0.20	0.19
FeO	11.39	9.45	19.13	20.76	15.12	22.34	21.08	—	20.17	24.80	25.43
MgO	15.81	16.10	19.46	17.89	18.04	21.16	19.77	23.87	18.73	23.54	23.98
CaO	1.22	1.89	1.66	2.20	2.01	2.30	1.92	1.99	1.56	2.64	2.56
Na_2O	0.74	0.75	0.63	0.54	0.70	0.50	0.22	0.55	0.61	0.58	0.51
K_2O	0.07	0.07	0.05	0.16	0.07	0.12	0.04	0.17	0.05	0.14	0.04
P_2O_5	0.28	0.41	0.30	0.26	0.23	0.32	0.32	0.35	0.30	0.32	0.38
H_2O^+						9.23	9.98	0.25			
H_2O^-	19.89	18.68	12.86	16.41	13.70	1.10	2.44	0.18	15.17	1.40	2.07
Cr_2O_3	0.36	0.33	0.36	0.38	0.45	0.39	0.44	0.50	0.42	0.49	0.52
NiO	1.23	1.34	1.53	1.54	1.71	1.64	1.50	0.00	1.49	0.00	1.64
CoO	0.06	0.06	0.07	0.06	0.08	0.08	0.08	0.00	0.08	0.00	0.08
C	3.10	4.83	2.48	2.50	4.00	2.54	2.78	—	1.30	0.46	0.47
Loss on ignition	6.96	4.10	0.36	2.03	—	—	0.62	—	2.23	—	—
Sum	100.65	101.02	99.19	101.83	99.97	102.62	100.04	—	100.16	101.11	100.62

[a] Analyses by H. B. Wiik (13).

Table 7
MINERALS FOUND IN METEORITES (ACCORDING TO MASON, FROM MOORE, 1962, AND SUPPLEMENTS)[301]

Mineral	Formula	Mineral	Formula
Nickel-iron	(Fe, Ni)	Calcite	$CaCO_2$
Kamacite	(With about 6% Ni)	Ilmenite	$FeTiO_3$
Taenite	(With about 13—48% Ni)	Rutile	TiO_2
Copper	Cu	Magnetite	Fe_3O_4
Gold	Au	Chromite	$FeCr_2O_4$
Diamond	C	Spinel	$MgAl_2O_4$
Graphite	C	Quartz	SiO_2
Cliftonite	C	Tridymite	SiO_3
Sulfur	S	Cristobalite	SiO_2
Moissanite	SiC	Apatite	$Ca_3(PO_4)_2Cl$
Cohenite	Fe_2C	Merrillite*	$Na_2Ca_2(PO_4)_2O$
Schreibersite*	$(Fe, Ni)_3P$	Farringtonite*	$Mg_2(PO_4)_2$
Osbornite*	TiN	Stanfieldite	P-mineral
Troilite	FeS	Epsomite	$MgSO_4$-$7H_2O$
Oldhamite	CaS	Zircon	$ZrSiO_4$
Alabandite	MnS	Olivine	$(Mg, Fe)_2SiO_4$
Sphalerite	ZnS	Orthopyroxene	$(Mg, Fe)SiO_3$
Chalcopyrrhotite	$CuFeS_2$	Clinopyroxene	$(Ca, Mg, Fe)SiO_3$
Djerfisherite	Alk-Cu-Fe-Sulfide	Ureyite	$NaCrSi_2O_6$
Pentlandite	$(Fe, Ni)_8S_8$	Merrihueite	$(K, Na)_2(Mg, Fe)_5Si_{12}O_{30}$
Daubreelithe*	$FeCr_2S_4$	Richterite	Na-tremolite
Lawrencite*	$FeCl_2$	Plagioclase	$(Na, Ca)(Al, Si)_4O_8$
Magnesite	$MgCO_3$	Serpentine	$Mg_6Si_4O_{10}(OH)_8$

Note: Minerals marked by an asterisk have not been found on the Earth.

Table 8
CLASSIFICATION OF METEORITES (ADAPTED FROM KEIL)[386]

Group	Class	Major minerals
	Stones	
Chondrites	Enstatite ch.	(Clino-)enstatite, kamacite, troilite, plagioclase
	Bronzite-olivine high-iron group ch.	Olivine, bronzite, clinopyroxene, plagioclase, kamacite, taenite, troilite
	Hypersthene-olivine ch., low-iron group	Olivine, hypersthene (bronzite), clinopyroxene, plagioclase, kamacite, taenite, troilite
	Low-iron-low-metal group ch.	Olivine, bronzite, plagioclase, kamacite
	High-iron-low-metal group ch., (olivine-pigeonite ch. and so on.)	Olivine, pigeonite, plagioclase, kamacite
	Carbonaceous ch.	Chlorite (?), epsomite, magnetite, troilite, olivine, pyroxene
Achondrites	Enstatite ach., aubrites	Enstatite, olivine, plagioclase
Ca-poor	Bronzite ach., diogenites	Bronzite, clinopyroxene
	Olivine ach., chassignites	Olivine, chromite
	Olivine-pigeonite ach., ureilites	Olivine, pigeonite, kamacite
Ca-rich	Augite ach., angrites	Augite, olivine, troilite
	Diopside-olivine ach., nakhalites	Diopside, olivine, plagioclase
	Orthopyroxene-pigeonite-plagioclase ach., eu-eucrites	Orthopyroxene, pigeonite, plagioclase
	Howardites	Pigeonite, orthopyroxene, augite, plagioclase
	Stony-Irons	
(Siderolites, Lithosiderites)	Pallasites or olivine st.-i	Olivine, kamacite, taenite
	Siderophyres or bronzite-tridymite st.-i.	Bronzite, kamacite, taenite, tridymite
	Lodranites or bronzite-olivine st.-i.	Bronzite, olivine, kamacite, taenite, troilite
	Mesosiderites or pyroxene-plagioclase st.-.i	Orthopyroxene, plagioclase, kamacite, taenite, troilite
	Irons	
Siderite s	Hexahedrites	Kamacite
	Octahedrites:	Kamacite, taenite, troilite
	coarsest O.	
	coarse O.	
	medium O.	
	fine O.	
	finest O.	
	Ataxites (Ni-rich)	Taenite, kamacite

Table 9
MODAL CONTENT (IMPORTANT MINERALS) OF A FEW TYPES OF METEORITES [VOLUME-%] (FROM HEIDE, 1957, AND OTHER AUTHORS)

					Achondrites	
Minerals	Meteoritic iron	Pallasites	Mesosiderites	Chondrites	Free from feldspar	Containing feldspar
Nickel-iron	98.3	50	45	9	0.6	(0.5)
Olivine	—	48	1.5	45	15	3
Pyroxene	—		31	25—30	70	45
Feldspar	—		16	11	10	50
Troilite	0.12	0.3	3	6	1	0.6
Schreibersite	1.12	0.2	2.5	(—)	(—)	(—)

Table 10
EXAMPLES[a] OF MAJOR ELEMENT ANALYSES AND AVERAGES OF BASALTIC LUNAR ROCKS WITH IGNEOUS TEXTURES[185]

	Average Compositions[b]				
Oxide	Apollo 11	Apollo 12	Apollo 14	15085,34[c]	15555,157[c]
SiO_2	40.10	47.10	47.70	46.39	44.75
Al_2O_3	8.60	12.80	21.44	5.79	9.85
Fe_2O_3	0.00	0.00	0.00	—	—
FeO	18.90	17.40	7.78	26.75	23.40
MgO	7.74	6.80	7.29	8.20	8.03
CaO	10.70	11.40	13.05	9.12	10.72
Na_2O	0.46	0.64	0.70	0.21	0.30
K_2O	0.30	0.07	0.48	0.07	0.09
TiO_2	12.20	3.17	1.16	3.07	2.64
P_2O_5	0.2	0.17	0.42	0.09	0.07
MnO	0.25	0.24	0.11	0.37	0.32
Cr_2O_3	0.37	0.31	0.25	0.67	0.77

[a] All values in weight percent.
[b] From Rose et al.(56).
[c] From Mason et al.(56).

Table 11
MINERALOGICAL[a] MODAL ANALYSES OF SOME MARE ROCKS WITH IGNEOUS TEXTURES[185]

Mineral component	Rock no. 10017	10045	10058
Clinopyroxene[b]	59.4	53.2	45.7
Plagioclase	25.1	26.9	37.1
Olivine	—	3.1	—
Cristobalite	tr.	1.8	5.1
Fe-Ti oxides (mostly ilmenite)	14.5	11.3	10.5
Troilite	0.36	1.3	0.27
Iron	0.04	0.18	0.03

[a] From Brown et al. (1970) Mineralogical, chemical and petrological features of Apollo 11 rocks and their relationship to igneous processes: Proc. Apollo 11 Lunar Sci. Conf., Geochim. et Cosmochim. Acta, Suppl. 1, Vol. 1, A. A. Levinson, Ed., p. 197, Pergamon Press. All values in volume percent.

[b] Includes pyroxferroite (9.3% in 10045 and present in all). Counts made in transmitted light and corrected after counting in reflected light (approx. 2300 counts in each sample from approx. 3 cm^2).

Table 12
MAJOR[a] ELEMENT COMPOSITIONS OF SOME ANT-SUITE ROCKS (IN WEIGHT PERCENT)[282]

	1	2	3	4	5	6a	6b	7	8	9	10
SiO_2	39.93	37.5	41.1	43.2	52	49.8	49.5	44.08	41.94	44.09	45.0
TiO_2	0.03	0.05	0.07	0.03	0.3	0.08	0.16	0.02	0.02	0.41	0.37
Al_2O_3	1.53	15.9	29.9	19.1	15.5	18.4	20.87	35.49	28.23	26.59	24.4
Cr_2O_3	0.34	0.49	0.18	0.3	0.16	0.31	0.23	0.00	0.16	0.13	0.12
FeO	11.34	5.8	3.7	5.4	7.4	6.02	5.05	0.23	7.40	6.19	5.4
MnO	0.13	0.16	0.03	0.04	0.12	0.10	0.08	0.00	0.09	0.08	0.08
MgO	43.61	33.7	9.6	21.0	15.9	14.5	11.76	0.09	6.34	6.06	9.5
CaO	1.14	6.2	14.8	10.1	9.1	10.5	11.71	19.68	15.74	15.43	14.3
Na_2O	<0.02	0.14	0.29	0.2	0.33	0.30	0.35	0.34	0.20	0.30	0.41
K_2O	0.00	0.04	0.04	0.05	0.08	0.05	0.06	<0.01	0.01	0.06	0.09
P_2O_5	0.04	0.02	0.04	n.d	n.d	n.d	0.04	0.01	n.d	0.03	0.08
Total	98.12	100.00	99.76	99.42	100.89	99.96	99.81	99.94	100.14	99.52	99.75

Note:
1. Dunite 72415.2. LSPET (1973) Includes 0.01% S.
2. Spinel troctolite 67435. Prinz et al. (1973a).
3. Spinel troctolite 65785.1-1. Dowty et al. (1974a). Includes 0.01% ZrO_2.
4. Troctolite 76535. Gooley et al. (1974).
5. "Civet-Cataclast" norite 72255.42. Haskin et al. (1974).
6. Norite 78235. (a) Dymek et al. (1975), (b) Winzer et al. (1975).
7. Anorthosite 15415. LSPET (1972).
8. Troctolitic anorthosite 62237. Dymek et al. (1975).
9. Anorthositic gabbro 77017.2. LSPET (1973). Includes 0.15% S.
10. Average highland rocks, Luna 20 site. Prinz et al. (1973b).

n.d. Not determined.

[a] Although dunite is not a member of the ANT-suite according to the original definition (Prinz et al., 1971, 1973b), it is included here because it appears to be genetically linked to the ANT rocks and is an early, presumably cumulate rock from the lunar highlands.

Table 13
DISTRIBUTION OF ROCK TYPES IN LARGE STRUCTURAL UNITS OF THE CRUST AND THE CRUSTAL LAYERS. (DATA FROM RONOV AND YAROSHEVSKY, 1969)[397]

Crustal layers in different crustal units	Average thickness (km)	Volume (km^3)	Mass (10^{24}g)	Types of rocks and abundances. Percent volume of layer, except percent area for oceanic Layer 1
		Continental Platform		
Sedimentary	1.8	135	0.35	Sands, 23.6. Clays, 49.5. Carbonates, 21.0. Evaporites, 2.0. Basalts, 3.9
		Continental Geosynclinal Folded Belts		
Sedimentary	10.0	365	0.94	Sands, 18.7. Clays and Shales, 39.4. Carbonates, 16.3. Evaporites, 0.3. Basalts, 12.6. Andesites, 10.2. Rhyolites, 2.5.
		Subcontinental, Shelf and Slope		
Sedimentary	2.9	190	0.48	Similar to above groups
		Continental and Subcontinental		
Granitic		3590	9.81	Granites, 18.1. Granodiorites, 19.9.
continental	20.1			Syenites, 0.3. Gabbro, 3.7.
subcontinental	9.1			Peridotites, 0.1. Gneisses, 37.6. Schists, 9.0. Marbles, 1.5. Amphibolites, 9.8
Basaltic		3760	10.91	Acid igneous and metamorphic
continental	20.1			rocks, 50.0. Basic igneous and
subcontinental	11.7			metamorphic rocks, 50.0
		Oceanic		
Layer 1 Sedimentary	0.4	120	0.19	Terrigenous, 7.3. Calcareous, 41.5 Siliceous, 17.0. Red Clays, 31.2
Layer 2	0.6	175	0.44	Sediments, 50.0.
	0.6	175	0.52	Basalts, 50.0
Layer 3	5.7	1700	4.92	Oceanic tholeiitic basalts, 99.0. Alkaline differentiates, 1.0

Table 14
ABUNDANCES OF MAIN ROCK TYPES AND MINERALS IN THE CRUST (DATA FROM RONOV AND YAROSHEVSKY, 1969)[397]

Rocks	% Volume of Crust	Minerals	% Volume of Crust
	Sedimentary		
		Quartz	12
Sands	1.7	Alkali feldspar	12
Clays and shales	4.2	Plagioclase	39
Carbonates (including salt-bearing deposits)	2.0	Micas	5
		Amphiboles	5
	Igneous		
		Pyroxenes	11
Granites	10.4	Olivines	3
Granodiorites, diorites	11.2	Clay minerals (+ chlorites)	4.6
Syenites	0.4	Calcite (+ aragonite)	1.5
Basalts, gabbros, amphibolites, eclogites	42.5	Dolomite	0.5
		Magnetite (+ titanomagnetite)	1.5
Dunites, periodotites	0.2	Others (garnets, kyanite, andalusite, sillimanite, apatite, etc.)	4.9
	Metamorphic		
Gneisses	21.4		
Schists	5.1		
Marbles	0.9		
Totals		**Totals**	
Sedimentary	7.9	Quartz + feldspar	63
Igneous	64.7	Pyroxene + olivine	14
Metamorphic	27.4	Hydrated silicates	14.6
		Carbonates	2.0
		Others	6.4

Table 15
EARTH'S CRUST GEOCHEMISTRY[299]

Volumes, Masses, and Average Chemical Composition of the Crust

Types of crust	Shell	Volumes (10^6km^3)	Average thickness (km)	Mass (10^{24}g)	Components, wt % and mass (10^{24}g)[a] SiO_2	TiO_2	Al_2O_3	Fe_2O_3	FeO	MnO	MgO	CaO	Na_2O	K_2O	P_2O_5	Corg	CO_2	S[b]	Cl	H_2O^+
Conti-	Sedimentary	500	3.4	1.29	49.90	0.65	12.97	2.99	2.80	0.11	3.06	11.70	1.70	2.04	0.16	0.48	8.20	0.18	0.21	2.90
nental					0.645	0.008	0.167	0.039	0.036	0.001	0.039	0.151	0.022	0.026	0.002	0.006	0.106	0.002	0.003	0.037
	"Granitic"	3000	20.1	8.20	63.94	0.57	15.18	2.00	2.86	0.10	2.21	3.98	3.06	3.29	0.20	0.17	0.84	0.04	0.05	1.53
					5.243	0.047	1.245	0.164	0.234	0.008	0.181	0.326	0.251	0.270	0.016	0.014	0.069	0.003	0.004	0.125
	"Basaltic"	3000	20.1	8.70	58.23	0.90	15.49	2.86	4.78	0.19	3.85	6.05	3.10	2.58	0.30	0.11	0.51	0.03	0.03	1.00
					5.066	0.078	1.348	0.249	0.416	0.016	0.335	0.526	0.270	0.224	0.026	0.009	0.044	0.003	0.003	0.087
	Total	6500	43.6	18.19	60.22	0.73	15.18	2.48	3.77	0.14	3.05	5.51	2.99	2.86	0.24	0.16	1.20	0.05	0.06	1.37
	Continental				10.954	0.133	2.760	0.452	0.686	0.025	0.555	1.003	0.543	0.520	0.044	0.029	0.219	0.008	0.010	0.249
Subcon-	Sedimentary	210	3.2	0.52	49.90	0.65	12.97	2.99	2.80	0.11	3.06	11.70	1.70	2.04	0.16	0.48	8.20	0.18	0.21	2.90
tinental					0.258	0.003	0.067	0.016	0.015	0.001	0.016	0.061	0.009	0.010	0.001	0.002	0.043	0.001	0.001	0.015
	"Granitic"	590	9.1	1.61	63.94	0.57	15.18	2.00	2.86	0.10	2.21	3.98	3.06	3.29	0.20	0.17	0.84	0.04	0.05	1.53
					1.029	0.009	0.244	0.032	0.046	0.002	0.036	0.064	0.049	0.053	0.003	0.003	0.013	0.001	0.001	0.025
	"Balsaltic"	740	11.4	2.14	58.23	0.90	15.49	2.86	4.78	0.19	3.85	6.05	3.10	2.58	0.30	0.11	0.51	0.03	0.03	1.00
					1.247	0.019	0.332	0.061	0.102	0.004	0.082	0.130	0.066	0.055	0.006	0.002	0.011	0.001	0.001	0.021
	Total Sub-	1540	23.7	4.27	59.35	0.73	15.07	2.55	3.82	0.16	3.14	5.97	2.90	2.79	0.23	0.16	1.57	0.07	0.07	1.43
	continental				2.534	0.032	0.643	0.109	0.163	0.007	0.134	0.254	0.124	0.119	0.010	0.007	0.067	0.003	0.003	0.061
Oceanic	Sedimentary	120	0.4	0.19	40.73	0.61	11.45	4.60	0.97	0.47	2.94	16.29	1.13	2.01	0.15	0.26	13.27	—	—	5.17
	(layer I)				0.077	0.001	0.022	0.009	0.002	0.001	0.006	0.031	0.002	0.004	0.003	0.005	0.025	—	—	0.010
	Volcanic sedimen- tary	530	1.8	1.45	45.62	1.02	14.24	3.37	4.23	0.31	5.43	13.67	2.04	1.03	0.14	0.12	6.04	—	—	2.74
	(layer II)				0.661	0.015	0.206	0.049	0.061	0.0045	0.079	0.198	0.030	0.015	0.002	0.002	0.088	—	—	0.040
	"Basaltic"	1520	5.1	4.40	49.72	1.37	16.57	2.35	6.85	0.17	7.52	11.48	2.80	0.21	0.13	0.01	—	0.03	0.03	0.66
					2.188	0.060	0.729	0.103	0.306	0.0075	0.331	0.505	0.123	0.009	0.006	0.0004	—	0.001	0.001	0.029
	Total	2170	7.3	6.04	48.44	1.26	15.85	2.67	6.11	0.21	6.89	12.15	2.57	0.46	0.13	0.04	1.87	0.02	0.02	1.31
Oceanic					2.926	0.076	0.957	0.161	0.369	0.013	0.416	0.734	0.155	0.028	0.008	0.003	0.113	0.001	0.001	0.079
	Total Crust	10,210	20.0	28.50	57.60	0.84	15.30	2.53	4.27	0.16	3.88	6.99	2.88	2.34	0.22	0.14	1.40	0.04	0.05	1.37
					16.414	0.240	4.360	0.722	1.218	0.045	1.105	1.992	0.822	0.667	0.062	0.040	0.399	0.012	0.014	0.389

[a] For each shell the upper line is weight percent; the lower line is mass (10^{24}g).

[b] The mean composition of the sedimentary shell differs from that of Table 1 by recalculation of SO_2 to S.

Table 16
ABUNDANCES OF CHEMICAL ELEMENTS IN THE EARTH'S CRUST AND CHONDRITES[a] (IN PPM) (RECALCULATION OF THE DATA OF A. VINOGRADOV, 1962, AND S. TAYLOR, 1964)[299]

Element	Continental crust	Crust as a whole	Chondrites
H	(1,530)	(1,520)	—
Li	20	18	3
Be	3	2	3.6
B	10	9	2
C	200(5,070)	180(5,220)	400
N	19	19	1
O	461,000(472,650)	456,000(468,780)	350,000
F	585	544	28
Na	23,550(22,060)	22,700(21,370)	7,000
Mg	23,300(18,500)	27,640(23,400)	140,000
Al	82,300(80,210)	83,600(80,950)	13,000
Si	281,500(279,380)	273,000(277,970)	180,000
P	1,050(1,040)	1,120(960)	500
S	350(530)	340(400)	20,000
Cl	145(620)	126(500)	70
K	20,850(23,620)	18,400(19,420)	850
Ca	41,500(40,000)	46,600(49,960)	14,000
Sc	22	25	6
Ti	5,650(4,380)	6,320(5,040)	500
V	120	136	70
Cr	102	122	2,500
Mn	950(1,110)	1,060(1,240)	2,000
Fe	56,300(46,810)	62,200(50,880)	250,000
Co	25	29	800
Ni	84	99	13,500
Cu	60	68	100
Zn	70	76	50
Ga	19	19	3
Ge	1.5	1.5	10
As	1.8	1.8	0.3
Se	0.05	0.05	10
Br	2.4	2.5	0.5
Rb	90	78	5
Sr	370	384	10
Y	33	31	1
Zr	165	162	30
Nb	20	20	0.3
Mo	1.2	1.2	0.6
Ru	—	—	—
Rb	—	—	—
Pd	0.015	0.015	1
Ag	0.075	0.08	0.1
Cd	0.15	0.16	0.1
In	0.25	0.24	0.01
Sn	2.3	2.1	1
Sb	0.2	0.2	0.1
Te	—	—	0.5
J	0.45	0.46	0.04
Cs	3	2.6	0.1

Table 16 (continued)
ABUNDANCES OF CHEMICAL ELEMENTS IN THE EARTH'S CRUST AND CHONDRITES[a] (IN PPM) (RECALCULATION OF THE DATA OF A. VINOGRADOV, 1962, AND S. TAYLOR, 1964)[299]

Element	Continental crust	Crust as a whole	Chondrites
Ba	425	390	6
La	39	34.6	0.3
Ce	66.5	66.4	0.5
Pr	9.2	9.1	0.1
Nd	41.5	39.6	0.8
Sm	7.05	7.02	0.2
Eu	2.0	2.14	0.08
Gd	6.2	6.14	0.4
Tb	1.2	1.18	0.05
Dy	—	—	0.35
Ho	1.3	1.26	0.07
Er	3.5	3.46	0.2
Tu	0.52	0.5	0.04
Yb	3.2	3.1	0.2
Lu	—	—	0.03
Hf	3.0	2.8	0.5
Ta	2.0	1.7	0.02
W	1.25	1.2	0.15
Re	0.0007	0.0007	0.0008
Os	—	—	0.5
Ir	—	—	0.5
Pt	—	—	2
Au	0.004	0.004	0.17
Hg	0.085	0.086	3
Tl	0.85	0.72	0.001
Pb	14	13	0.2
Bi	0.0085	0.0082	0.003
Th	9.6	8.1	0.04
U	2.7	2.3	0.015

[a] In parentheses are shown the abundances of the main elements obtained by a quantitative method based on the volume measurements.

Table 17
THE MAJOR CHEMICAL ELEMENTS IN THE EARTH'S CRUST[222]

	Weight percent	Atom percent	Ionic radius (Å)	Volume percent
O	46.40	62.19	1.40	94.04
Si	28.15	21.49	0.42	0.88
Al	8.23	6.54	0.51	0.48
Fe	5.63	2.16	0.74	0.49
Mg	2.33	2.05	0.66	0.33
Ca	4.15	2.22	0.99	1.18
Na	2.36	2.20	0.97	1.11
K	2.09	1.15	1.33	1.49

Table 18
CALCULATED VOURINOS BULK COMPOSITION COMPARED WITH COMPOSITIONS OF LIZARD, TINAQUILLO, AND ST. PAUL ROCKS HIGH-TEMPERATURE PERIDOTITES[294]

	Vourinos[a] (A)	(B)	Lizard[b]	Tinaquillo[c]	St. Paul's Rocks[d]
SiO_2	46.3	44.9	44.8	44.9	43.6
TiO_2	0.2	0.1	0.2	0.1	0.3
Al_2O_2	4.7	3.0	4.2	3.2	3.7
FeO[e]	5.2	5.1	8.2	7.6	8.0
MnO	0.1	0.1	0.1	0.1	0.1
MgO	35.5	39.9	39.2	40.0	38.5
CaO	3.5	2.5	2.4	3.0	2.6
Na_2O	0.6	0.4	0.2	0.2	0.3
K_2O	0.1	0.1	0.05	0.02	0.1
Cr_2O_2	0.2	0.3	0.4	0.5	0.5
NiO	0.2	0.3	0.2	0.3	0.3
P_2O_3	0.04	0.02	0.01	—	0.1

[a] Moores (1970) A and B analyses represent use of different methods in arriving at bulk composition.
[b] Green (1964).
[c] Green (1963).
[d] Melson et al. (1970).
[e] All iron is calculated as FeO.

Table 19
PROPORTIONS OF ROCK TYPES FOUND IN LARGE SUITES OF ULTRAMAFIC AND MAFIC XENOLITHS FROM SOUTH AFRICAN KIMBERLITE PIPES (AFTER MATHIAS, SIEBERT, AND RICKWOOD, 1970, AND MacGREGOR, PERSONAL COMMUNICATION, 1969)[294]

Rock	Mineralogy	MSR %[a] (1970)	McGregor %[b] (1969)
	Peridotite-Pyroxenite Association		
Dunite	Ol,	0.3	0.5
Harzburgite	O1, Opx ± Sp	16	26
Lherzolites	Ol, Opx, Cpx ± Sp	14	11
Garnet harzburgite	Ol, Opx, Ga	18	21
Garnet lherzolite	Ol, Opx, Cpx, Ga	43	39
Pyroxenites	Ol, Cpx ± Ga	6	2.5
Others		3	
	Eclogitic Association[c]		
Eclogite	Ga, Cpx	63	
2-px Eclogite	Ga, Cpx, Opx	2	
Kyanite eclogite	Ga, Cpx, Ky	8	
Corundum eclogite	Ga, Cor, Cpx	6	
Quartz eclogite	Ga, Cpx, Qz	0.6	
Plagioclase eclogite[d]	Ga, Cpx, Plag	8	
Garnet granulite[d]	Ga, Cpx, Plag, Qz	3	
Others		9	

Note: Abbreviations: Ol = olivine, Opx = orthopyroxene, Cpx = Ca-rich clinopyroxene, Ga = pyrope-rich garnet, Ky = kyanite, Cor = corundum, Qz = quartz. Plag = plagioclase, Sp = spinel.

[a] Population of 295 xenoliths.
[b] Population of about 200 xenoliths.
[c] Population of 171 xenoliths.
[d] Possibly granulites of crustal origin.

Table 20
COMPOSITIONS OF 15 GARNET PERIDOTITE XENOLITHS FROM SOUTH AFRICAN DIAMOND PIPES. (AFTER CARSWELL AND DAWSON, 1970)[294]

	(1)	(2)	(3)
SiO_2	46.5	45.5	44.5 — 47.9
TiO_2	0.3	0.2	0.02 — 2.3
Al_2O_3	1.8	2.7	1.1 — 3.3
Cr_2O_3	0.4	0.3	0.2 — 0.5
FeO	6.7	7.0	5.9 — 8.4
MnO	0.1	0.1	0.1 — 0.2
NiO	0.3	0.3	0.25 — 0.4
MgO	42.0	41.9	37.7 — 45.7
CaO	1.5	1.9	0.9 — 3.5
Na_2O	0.2	0.2	0.06 — 0.4
K_2O	0.2	0.1	0.0 — 0.4
P_2O_5	0.02	0.03	0 — 0.05

Note: (1) Mean of 9 analyses by Carswell and Dawson. (2) Mean of 6 analyses collected from previous literature. (3) Composition range in all 15 analyzed garnet peridotites.

Table 21
COMPOSITIONS OF COLLECTIONS OF SPINEL PERIDOTITE NODULES OCCURRING IN THE ALKALI BASALT SUITE[294]

	(1)	(2)	(3)	(4)
SiO_2	41.10	44.4	44.5	45.0
TiO_2	0.08	0.04	0.07	0.07
Al_2O_3	0.56	1.66	2.69	3.01
Cr_2O_3	0.35	0.47	0.43	0.41
Fe_2O_3	1.24	1.41	1.46	1.28
FeO	9.31	7.50	6.71	6.70
NiO	0.44	0.27	0.26	0.25
MnO	0.15	0.13	0.11	0.11
MgO	46.33	42.3	40.9	39.7
CaO	0.17	1.64	2.61	3.15
Na_2O	—	0.11	0.22	0.24
K_2O	—	0.04	0.01	0.04

Note: (1) Mean of three analyses of nodules with Al_2O_3 contents of less than 1% (Harris, Reay, and White, 1967). (2) Mean of analyses of 27 nodules from Puy Beanite, west of Riom, Puy-de-Dome (Hutchison, 1970). (3) Mean of analyses of 42 nodules from Le Puy, Haute-Loire (Hutchison, 1970). (4) Mean of analyses of 20 nodules from the basalte de Rocher du Lion, (Vilminot, 1965).

Table 22
PYROLITE COMPOSITIONS[294]

	For models					Limiting cases		Average mantle pyrolite[h]
Constituent	a	b	c	d	e	f	g	
SiO_2	45.2	44.9	46.1	45.6	42.9	44.9	46.5	45.1
TiO_2	0.7	0.24	0.2	0.2	0.2	0.1	0.2	0.2
Al_2O_3	3.5	4.3	4.3	3.9	5.8	3.2	3.6	4.6
Cr_2O_3	0.4	0.4	—	0.4	0.2	0.5	0.4	0.3
Fe_2O_3	0.5	—	—	—	0.3	—	1.0	0.3
FeO	8.0	8.2	8.2	5.2	8.9	7.6	9.4	7.6
MnO	0.14	0.1	—	0.1	0.14	0.1	0.2	0.1
NiO	0.2	0.2	—	0.3	0.2	0.3	—	0.2
MgO	37.5	38.9	37.6	37.7	37.2	40.0	33.0	38.1
CaO	3.1	2.5	3.1	3.0	3.7	3.0	5.1	3.1
Na_2O	0.6	0.23	0.4	0.5	0.4	0.2	0.5	0.4
K_2O	0.13	0.02	0.03	0.1	0.003	0.0006	0.2	0.02
P_2O_5	0.06	0.02	—	0.03	—	—	0.01	0.02
								100.0

[a] 3:1 Peridotite-basalt mix model pyrolite (Ringwood, 1966).
[b] 99% Lizard peridotite (Green, 1964) + 1% nephelinite.
[c] 83% Residual harzburgite + 17% primitive oceanic tholeiite.
[d] Vourinos ophiolite complex.
[e] For upper mantle model (Carter, 1970).
[f] Least-fractionated alpine ultramafic (smallest degree of partial melting of pyrolite).
[g] Ultramafic liquid (highest degree of partial melting of pyrolite).
[h] Mean of columns b-e.

Table 23
THE THIRTY-TWO CRYSTAL CLASSES[222]

1	Axes: 2 Fold	3 Fold	4 Fold	6 fold	Planes	Center	Hermann-Maugin Symbols	Class name	System
1	—	—	—	—	—	—	1	Pedial	Triclinic
2	—	—	—	—	—	yes	1	Pinacoidal	
3	—	—	—	—	1	—	m	Domatic	
4	1	—	—	—	—	—	2	Sphenoidal	Monoclinic
5	1	—	—	—	1	yes	2/m	Prismatic	
6	1	—	—	—	2	—	mm2	Orthorhombic pyramidal	
7	3	—	—	—	—	—	222	Orthorhombic disphenoidal	Orthorhombic
8	3	—	—	—	3	yes	2/m 2/m 2/m	Orthorhombic dipyramidal	
9	—	1	—	—	—	—	3	Trigonalpyramidal	
10	—	1	—	—	—	yes	3	Rhombohedral	
11	—	1	—	—	3	—	3m	Ditrigonal pyramidal	Trigonal
12	3	1	—	—	—	—	32	Trigonal trapezohedral	
13	3	1	—	—	3	yes	32/m	Hexagonal scalenohedral	
14	—	1	—	—	1	—	6	Trigonal dipyramidal	
15	—	—	—	1	—	—	6	Hexagonal pyramidal	
16	—	—	—	1	1	yes	6/m	Hexagonal dipyramidal	
17	3	1	—	—	4	—	6m2	Ditrigonal dipyramidal	Hexagonal
18	—	—	—	1	6	—	6mm	Dihexagonal pyramidal	
19	6	—	—	1	—	—	622	Hexagonal trapezohedral	
20	6	—	—	1	7	yes	6/m 2/m 2/m	Dihexagonal dipyramidal	
21	1	—	—	—	—	—	4	Tetragonal disphenoidal	
22	—	—	1	—	—	—	4	Tetragonal pyramidal	
23	—	—	1	—	1	yes	4/m	Tetragonal dipyramidal	
24	3	—	—	—	2	—	42m	Tetragonal scalenohedral	Tetragonal
25	—	—	—	—	4	—	4mm	Ditetragonal pyramidal	
26	4	—	1	—	—	—	422	Tetragonal trapezohedral	
27	4	—	1	—	5	yes	4/m 2/m 2/m	Ditetragonal dipyramidal	
28	3	4	—	—	—	—	23	Tetartoidal	
29	3	4	—	—	3	yes	2/m3	Diploidal	
30	3	4	—	—	6	—	43m	Hextetrahedral	Isometric
31	6	4	3	—	—	—	432	Gyroidal	
32	6	4	3	—	9	yes	4/m 3 2/m	Hexoctahedral	

Table 24
IONIC RADII AND ELECTRONEGATIVITIES[188]

Element	Ion[a]	Radius for 6-coordination ("octahedral"), Å[b]	Observed coordination numbers[c]	Electronegativity[d]	Approx. ionic character of bond with oxygen, %[e]
Aluminum	Al^{3+}	0.51	4, 6	1.5	60
Antimony	Sb^{3+}	0.76	6		66
	Sb^{5+}	0.62	4, 6	1.9	48
Arsenic	As^{3+}	0.58	4, 6		60
	As^{5+}	0.46	4	2.0	38
Barium	Ba^{++}	1.34	8—12	0.9	84
Beryllium	Be^{++}	0.35	4	1.5	63
Bismuth	Bi^{3+}	0.96	6, 8	1.9	66
Boron	B^{3+}	0.23 (0.21, 0.22)	3, 4	2.0	43
Bromine	Br^{-}	1.96		2.8	
Cadmium	Cd^{++}	0.97	6, 8	1.7	66
Calcium	Ca^{++}	0.99	6, 8	1.0	79
Carbon	C^{4+}	0.16(0.15)	3	2.5	23
Cerium	Ce^{3+}	1.07	6, 8	1.1	74
Cesium	C^{3+}	1.67 (1.82)	12	0.7	89
Chlorine	Cl^{-}	1.81		3.0	
Chromium	Cr^{3+}	0.63	6	1.6	53
	Cr^{6+}	0.52	4		23
Cobalt	Co^{++}	0.72	6	1.8	65
Copper	Cu^{+}	0.96	6, 8	1.9	71
	Cu^{++}	0.72	6	2.0	57
Fluorine	F^{-}	1.36		4.0	
Gallium	Ga^{3+}	0.62	4, 6	1.6	57
Germanium	Ge^{4+}	0.53	4	1.8	49
Gold	Au^{+}	1.37	8—12	2.4	62
Hafnium	Hf^{4+}	0.78	6	1.3	70
Indium	In^{3+}	0.81	6	1.7	62
Iodine	I^{-}	2.20		2.5	
	I^{5+}	0.62	6		54
Iron	Fe^{++}	0.74	6	1.8	69
	Fe^{3+}	0.64	6	1.9	54
Lead	Pb^{++}	1.20	6—10	1.8	72
Lithium	Li^{+}	0.68	6	1.0	82
Magnesium	Mg^{++}	0.66	6	1.2	71
Manganese	Mn^{++}	0.80	6	1.5	72
	Mn^{3+}	0.66	6		51
	Mn^{4+}	0.60	4, 6		38
Mercury	Hg^{++}	1.10	8	1.9	62
Molybdenum	Mo^{4+}	0.70	6		58
	Mo^{6+}	0.62	4, 6	1.8	47
Nickel	Ni^{++}	0.69	6	1.8	60
Niobium	Nb^{5+}	0.69	6	1.6	56
Nitrogen	N^{5+}	0.13 (0.12)	3	3.0	9
Oxygen	O^{-}	1.40		3.5	
Palladium	Pd^{++}	0.80	6	2.2	61
Phosphorus	P^{5+}	0.35	4	2.1	35
Platinum	Pt^{++}	0.80	6	2.2	65
Potassium	K^{+}	1.33	8—12	0.8	87
Radium	Ra^{++}	1.43	8—12	0.9	83
Rare-earth metals	La^{3+}	1.14	8	1.1	77

Table 24 (continued)
IONIC RADII AND ELECTRONEGATIVITIES[188]

Element	Ion[a]	Radius for 6-coordination ("octahedral"), Å[b]	Observed coordination numbers[c]	Electronegativity[d]	Approx. ionic character of bond with oxygen, %[e]
	Ce^{3+}—Ho^{3+}	1.07—0.91	6, 8	1.1—1.2	73—75
	Er^{3+}—Lu^{3+}	0.89—0.85	6	1.2	76
Rhenium	Re^{4+}	0.72	6		63
	Re^{7+}	0.56	4, 6		51
Rubidium	Rb^{+}	1.47	8—12	0.8	87
Scandium	Sc^{3+}	0.81	6	1.3	65
Selenium	Se^{-}	[2.00]		2.4	
	Se^{6+}	0.42	4		26
Silicon	Si^{4+}	0.42	4	1.8	48
Silver	Ag^{+}	1.26	8, 10	1.9	71
Sodium	Na^{+}	0.97	6, 8	0.9	83
Strontium	Sr^{++}	1.12	8	1.0	82
Sulfur	S^{-}	1.85		2.5	
	S^{6+}	0.30 (0.29)	4		20
Tantalum	Ta^{5+}	0.68	6	1.5	63
Tellurium	Te^{-}	[2.24]		2.1	
	Te^{6+}	0.56	4, 6		36
Thallium	Tl^{+}	1.47	8—12		79
	Tl^{3+}	0.95	6, 8	1.8	58
Thorium	Th^{4+}	1.02	6, 8	1.3	72
Tin	Sn^{++}	0.93	6	1.8	73
	Sn^{4+}	0.71	6	1.9	57
Titanium	Ti^{3+}	0.76	6		60
	Ti^{4+}	0.68	6	1.5	51
Tungsten	W^{6+}	0.62	4, 6	1.7	57
Uranium	U^{4+}	0.97	6, 8		68
	U^{6+}	0.80	6	1.7	62
Vanadium	V^{3+}	0.74	6	1.6	57
	V^{4+}	[0.65]	6		45
	V^{5+}	0.59	4, 6		36
Yttrium	Y^{3+}	0.92	6	1.2	74
Zinc	Zn^{++}	0.74	4, 6	1.7	63
Zirconium	Zr^{4+}	0.79	6	1.4	65

[a] Only ions commonly found in naturally occurring minerals are listed.

[b] Sources J. Green, *Geol. Soc. America Bull.*, Vol. 70, pp. 1127—1184, 1959, and L. H. Ahrens, *Geochim. et Cosmochim. Acta*, 2, 155, 1952, except for three values (enclosed in square brackets) estimated by F. G. Smith, in *Physical Geochemistry*, Addison-Wesley Publishing Company, Inc., Reading, Mass., 1963. To find radii for other kinds of coordination from the octahedral radii, the following rules may be used: For coordination number 4, subtract 0.03 from the octahedral radius. For coordination number 8, add 0.04 to the octahedral radius. For coordination number 10, add 0.09 to the octahedral radius. For coordination number 12, add 0.13 to the octahedral radius. These rules give radii within 0.01 Å of the best value in all except five cases. The values for these five are given in parentheses.

[c] Source: G. Smith, op. cit.

[d] Source: L. Pauling, *The Nature of the Chemical Bond*, Cornell University Press, Ithaca, N. Y., 1960. The numbers are in arbitrary units, ranging from 0.7 for Cs to 4.0 for F.

[e] Source: Smith, op. cit., calculated by Smith from electronegativity values estimated by A. S. Povarennykh, *Dokl. Akad. Nauk SSSR*, 109, 993, 1956.

Table 25
CHEMICAL[a] CLASSES OF MINERALS[278]

Class	Defining anions	Example
Native elements	None: no charged ions	Copper, Cu
Sulfides and similar compounds	Sulfide: S^- and similar anions	Pyrite, FeS_2
Oxides and	O^-	Hematite, Fe_2O_3
Hydroxides	OH^-	Brucite, $Mg(OH)_2$
Halides	Cl^-, F^-, Br^-, I^-	Halite, NaCl
Carbonates and similar compounds	CO_3^-	Calcite, $CaCO_3$
Sulfates and similar compounds	SO_4 -- and similar anions	Barite, $BaSO_4$
Phosphates and similar compounds	PO_4^{-3} and similar anions	Apatite, $Ca_5F(PO_4)_3$
Silicates	SiO_3	Pyroxene, $MgSiO_3$

[a] This classification, derived originally by Berzelius in the nineteenth century and used extensively by Dana, is a simplified form of the scheme used by Berry and Mason in *Elements of Mineralogy,* W. H. Freeman and Company, 1968.

Table 26
STRUCTURAL CLASSIFICATION OF THE SILICATES[222]

Classification	Structural arrangement	Si:O	Examples
Nesosilicates	Independent tetrahedra	1:4	Forsterite, Mg_2SiO_4
Sorosilicates	Two tetrahedra sharing one oxygen	2:7	Hemimorphite, $Zn_4Si_2O_7(OH)_2 \cdot H_2O$
Cyclosilicates	Closed rings of tetrahedra each sharing two oxygens	1:3	Beryl, $Be_3Al_2Si_6O_{18}$
Inosilicates	Continuous single chains of tetrahedra each sharing two oxygens	1:3	Enstatite, $MgSiO_3$
	Continuous double chains of tetrahedra sharing alternately two and three oxygens	4:11	Anthophyllite, $Mg_7(Si_4O_{11})_7(OH)_2$ Talc, $Mg_2Si_40_{10}(OH)_2$
Phyllosilicates	Continuous sheets of tetrahedra each sharing three oxygens	2:5	Phlogopite, $KMg_3(AlSi_3O_{10})(OH)_2$
Tektosilicates	Continuous framework of tetrahedra each sharing all four oxygens	1:2	Quartz, SiO_2; Nepheline, $NaAlSiO_4$

Table 27
THE PREDOMINANT RARE EARTH MINERALS

Mineral	Formula of mineral	Contents %			Genetic type of formation
		$(Car)_2O_3$[a]	$(yttr)_2O_3$[b]	$(R.E.E.)_2O_3$	
		Complex oxides			
Knopite	(Ca, Ce) TiO_3	6.81	—	—	Contacto-metasomatic, magmatic
Loparite	$(Na, Ca, Ce)_2 (Ti, Nb)_2O_6F$	31—33	—	—	Magmatic
Pyrochlore	$NaCaNb_2O_6F$	4.36—5.90	0.46	4.36—6.36	Pegmatitic
Koppite	$NaCaNb_2O_6F$	9.83	—	—	Contacto-metasomatic
Fergusonite	(Y,Fe,Ce) $(Nb,Ta,Ti)O_4$	0.2—4.0	28—40	31—41	Pegmatitic (granites)
Euxenite	(Y,Ca,Ce,U,Th)	0.4—2.4	24—28	25—30	Pegmatitic (granites)
Polycrase	(Y,Ca,Ce,U,Th) $(Ti,Nb,Ta)_2O_6$	0.6—2.6	25.27	26.29	
Eshinite	(Ca,Ce,Fe",Th) $(Ti,Nb)_2O_6$	19.50	4.53	25.0	Pegmatitic (nephetine syenites)
Priorite	(Y,Er,Cat, Fe",Th) $(Ti,Nb)_2O_6$	2—4.3	17—29	21—30	Pegmatitic (granites)
Lamarshite	$(Y,Er)(Nb,Ta)_2O_6$	0.9—4.2	8—17	10—19	
Khlopinite	(Y,U,Th)(Nb,Ti,Fe) O_3	—	17.65	—	
Viikite	$(Y,U,Fe)(Nb,Ta,Ti)_2 (O,OH)_6$	0.5—8.6	0.8—29	3—33	Pegmatitic (granites)
Braunerite	$(U,Ca,Fe)TiO_6$	0.3—7.3	1.8—4.3	07.35	Pegmatitic contacto-metasomatic
Uranitite	$1(U,Th)O_2mUO_3 \cdot n$ PbO	—	—	04.4	
a) Braggerite	var. forms of uranitite (Cont. Th. R)			6.16	Pegmatitic (granites)
b) Cleveite	Var. forms of uranitite (cont. R, Th)	—	—	15.0	
		Carbonates			
Parisite	$(Ca,La)_2Ca(CO_3)_3F_2$	55—61	0.0—7.86	5.5—61.0	Samarskite, hydrothermal. Contactometasomatic hydrothermal
Bartnasite	$(Ce, La) (CO)_3F$	73—76			Contacto-metasomatic hydrothermal
		Phosphates			
Xenotime	YPO_4	0.9—2.1	57—68	57—68	Pegmatitic
Monazite	$(Ce,La,Y,Th) PO_4$	52—74	1.1—5.0	56—75	Alluvial
Rhabdophanite	$(Ce,Y)(PO)_4$	55—62	0.0—8.5	62—64	Hypergenic
Apatite	$Ca_5(PO_4)_3$ (F,Cl,OH)	0.7—4.9	—	—	Magmatic
		Silicates			
Yttrialite	$(Y,Th,U,Fe)_2 Si_2O_7$	3.3—8.2	43.4—49.3	49—51	Pegmatitic (granites)
Orthite	$(Ca,Ce)_2 (Al,Fe)_3Si_3O_{12}$ (O,OH)	11—22.5	0.1—6.1	11—23.3	Magmatic, pegmatitic sedimentary metamorphic
Cyrtolite	$ZrSiO_4 \cdot nH_2O$	—1.16	8.93	10.1	Pegmatitic
Rinkolite	$Na_2Ca_4CeTi (Si_4O_{15}) (F,OH_3)$	13.7—14.4	0.9—1.8	15.5—19.1	Pegmatitic (nepheline syenites)
Lovchorrite	$Na_2Ca_4CeTi (Si_4O_{15})(F,OH_3)$	11—15	1.3—3.4	4.14—17	
Gadolinite	$Y,Fe Be_2Si_2O_{10}$	5—32	22—50	—	Pegmatitic (granites)

[a] Oxides of R.E.E. of the ceric subgroup.

[b] Oxides of R.E.E. of the yttric subgroup.

Table 28
OPAQUE MINERALS

Mineral	Composition	Color	Form
Graphite	C	Grey to silvery in reflected light	Flakes, dust, sometimes hexagonal plates
Magnetite	Fe_3O_4	Grey, generally dull in reflected light	Anhedral grains, square or rectangular cross sections; reticulating plates in triangular or rhombic pattern with leucoxene after ilmenite-magnetite intergrowths
Chromite	$FeO \cdot Cr_2O_3$	Dull grey in reflected light; thin edges brown, isotropic	Often crystals showing rectangular or square cross sections; also grains and aggregates
Pyrite	FeS_2	Pale brassy yellow	Usually crystals give square or rectangular cross sections; also grains and aggregates
Pyrrhotite	Fe_nS_{n+1}	Bronze-yellow	Usually anhedral grains and aggregates
Chalcopyrite	$CuFeS_2$	Bright brassy yellow	Usually anhedral grains, aggregates, and stringers
Arsenopyrite	FeAsS	Silvery white	Anhedral grains, aggregates and axe-shaped or needle-like crystals
Galena	PbS	Silvery white	Usually anhedral grains and masses, commonly showing triangular pits due to plucking out of cleavage pieces; often as stage is rotated, reflection will momentarily be given by cleavage surfaces
Ilmenite	$FeO \cdot TiO_2$	Grey, indistinguishable from magnetite	Frequently intergrown with magnetite, and often altered to leucoxene; best distinguished from magnetite by lack of magnetism; tabular crystals
Leucoxene	(Secondary Ti minerals)	White, yellowish, brownish, porcelain-like luster	Often intergrown with magnetite; as grains and aggregates in altered biotite; may often be recognized as fine-grained sphene or rutile when examined under high power

Note: The common opaque minerals, except graphite and leuxocene, are metallic by reflected light. Certain sections of some common nonmetallic minerals, such as basal sections of strongly colored tourmaline and biotite, may appear opaque, especially if the section is a little thick. Examination of thin edges of the grains and comparison with more favorably oriented grains of the same mineral in the section will assist in the identification.

Table 29
THE MINERALOGY OF DIFFERENT ROCK ASSOCIATIONS[222]

The Magmatic Environment

1. Igneous Rocks
   a. Silicates: quartz (tridymite, cristobalite), feldspars, feldpathoids, analcime, olivine, enstatic-hypersthene, augite, aegirine, hornblende, biotite, muscovite, zircon, sphene.
   b. Phosphates: apatite, monazite.
   c. Oxides: magnetite, ilmenite, chromite.
   d. Sulfides: pyrite, pyrrhotite.
   e. Elements: platinum group, diamond (in ultrabasic rocks).
2. Pegmatites
   a. Silicates: quartz, feldspars (especially microcline and albite), feldspathoids, aegirine, hornblende, biotite, muscovite, phlogopite, tourmaline, spodumene, lepidolite, zircon, thorite, allanite, spessartine, beryl, topaz, scapolite.
   b. Phosphates: apatite, amblygonite, monazite.
   c. Oxides: magnetite, ilmenite, hematite, cassiterite, uraninite, column-bite-tantalite, corundum.
   d. Halides: fluorite, cryolite.
   e. Sulfides: arsenopyrite, stibnite, bismuthinite, molybdenite.
   f. Elements: antimony, bismuth.
3. Hydrothermal Deposits
   a. Silicates: quartz, feldspars, muscovite, chlorite, epidote, hornblende, tourmaline, zeolites, topaz, apophyllite, rhodonite, datolite, axinite, pectolite.
   b. Sulfates: barite.
   c. Carbonates: calcite, dolomite, ankerite, magnesite, rhodochrosite, witherite.
   d. Oxides: cassiterite, magnetite, hematite, thnenite, uraninite, rutile, anatase, brookite.
   e. Halides: fluorite.
   f. Sulfides: chalcopyrite, bornite, chalcocite, enargite, sphalerite, galena, tetrahedrite, pyrite, marcasite, pyrrhotite, pyrargyrite, proustite, arsenopyrite, stibnite, cinnabar, and many others.
   g. Elements: gold, silver, arsenic, antimony, bismuth.

3A. Secondary Alteration Products of Ore Minerals
   a. Silicates: chrysocolla, hemimorphite.
   b. Sulfates, etc.: anglesite, crocoite, wulfenite.
   c. Phosphates, vanadates: carnotite, tyuyamunite, vanadinite, pyromorphite, mimetite.
   d. Carbonates: malachite, azurite, smithsonite, cerussite.
   e. Oxides: cuprite, hematite, goethite
   f. Elements: silver, copper.

4. Fumarolic and Hot Springs Deposits
   a. Silicates: quartz (chalcedony), opal, zeolites.
   b. Sulfates: gypsum, alunite, and many others.
   c. Oxides: hematite, magnetite.
   d. Halides: halite, sylvite, salmmoniac (NH_4Cl), and many others.
   e. Sulfides: pyrite, cinnabar, stibnite, covellite.
   f. Elements: sulfur.

The Sedimentary Environment

5. Resistates
   a. Silicates: quartz, feldspars, muscovite, biotite, garnet, tourmaline, staurolite, zircon, thorite, topaz, kyanite, andalusite.
   b. Phosphates: monazite.
   c. Oxides: magnetite, ilmenite, corundum, columbite-tantalite, cassiterite, rutile, spinel, chromite.
   d. Elements: gold, platinum metals, diamond.
6. Hydrolysates
   a. Silicates: quartz (chalcedony), opal, clay minerals, glauconite, chamosite.
   b. Oxides: bauxite.
7. Oxidates
   a. Oxides: limonite, hematite, pyrolusite, psilomelane.
8. Reduzates
   a. Carbonates: siderite.
   b. Sulfides: pyrite, marcasite.
   c. Elements: sulfur.

Table 29 (continued)
THE MINERALOGY OF DIFFERENT ROCK ASSOCIATIONS[222]

9. Precipitates
   a. Phosphates: apatite (phosphorite).
   b. Carbonates: calcite, aragonite, dolomite.
10. Evaporites
   a. Sulfates: gypsum, anhydrite, many others.
   b. Carbonates: calcite, aragonite, dolomite, sodium carbonates.
   c. Borates: kernite, borax.
   d. Nitrates: soda-niter.
   e. Halides: halite, sylvite, carnallite, many others.

The Metamorphic Environment

11. Low Grade
   a. Silicates: quartz, albite, talc, serpentine, chlorite, tremolite-actinolite, epidote, muscovite, sphene, prehnite, tourmaline, pyrophyllite, spessartine.
   b. Carbonates: calcite, dolomite, magnesite, siderite.
   c. Oxides: rutile, anatase, brookite, magnetite, hematite, brucite.
   d. Sulfides: pyrite, pyrrhotite.
   e. Elements: graphite.
12. Medium Grade
   a. Silicates: quartz, plagioclase, microcline, orthoclase, kyanite, andalusite, staurolite, serpentine, forsterite, anthophyllite, cummingtonite, cordierite, garnet, hornblende, epidote, muscovite, biotite, tourmaline, scapolite, idocrase.
   b. Carbonates: calcite, dolomite.
   c. Oxides: rutile, magnetite, hematite, ilmenite, corundum, spinel.
   d. Sulfides: pyrite, pyrrhotite.
   e. Elements: graphite.
13. High Grade
   a. Silicates: quartz, plagioclase, orthoclase, microcline, andalusite, sillimanite, forsterite, pyroxenes, cordierite, garnet, wollastonite, hornblende, scapolite, tourmaline, sphene.
   b. Carbonates: calcite.
   c. Oxides: magnetite, hematite, ilmenite, corundum, spinel, rutile.
   d. Sulfides: pyrite, pyrrhotite.
   e. Elements: graphite.

Meteorites

14. Iron Meteorites
   a. Elements: nickel-iron, graphite, diamond.
   b. Sulfides, etc.: troilite (FeS), schreibersite $(Fe,Ni,Co)_3P$, cohenite (Fe_3C).
15. Stony Meteorites
   a. Silicates: olivine, enstatite, hypersthene, diopside, plagioclase.
   b. Sulfides: troilite (FeS).
   c. Elements: nickel-iron, graphite.

Table 30
MINERALOGY OF THE IGNEOUS ROCKS[222]

	Leucocratic	Kelanocratic
Essential	SiO_2	
	Quartz; tridymite and/or cristobalite in some volcanic rocks	Olivine
	Feldspars	Pyroxenes
	K-feldspars: Orthoclase, microcline	Enstatite
	Perthite: K-feldspar—albite intergrowth	Hypersthene
	Na-Ca feldspars: Plagioclase	Augite
		Aergirine
	Feldspathoids	
	Nepheline	Hornblende
	Leucite	
	Sodalite	Biotite
	Cancrinite	
Accessory	Apatite	Ilmenite
	Muscovite	Magnetite
	Corundum	Pyrite
	Sphene	Pyrrhotite
	Fluorite	
	Zircon	

Table 31
COMPOSITION OF THE "AVERAGE IGNEOUS ROCK"[385]

	Clarke, Washington (1924): arithmetic mean of 5159 analyses of magmatic rocks (%)	Sederholm (1925): average of rocks from Finland by area (%)	Grout (1938): average of rocks from the Canadian Shield by area (%)	Shaw et al. (1967): average of statistically weighted rocks from the Canadian Precambrian Shield (%)	Average by area computed from Table 44 and rock analyses reported by Nockolds (1954) (%)
SiO_2	59.12	67.58	63.08	64.93	66.4
TiO_2	1.05	0.41	0.81	0.52	0.7
Al_2O_3	15.34	14.66	16.75	14.63	14.9
Fe_2O_3	3.08	1.27	2.38	1.36	1.5
FeO	3.80	3.29	2.91	2.75	3.0
MnO	0.12	0.04	0.02	0.068	0.08
MgO	3.49	1.69	1.78	2.24	2.2
CaO	5.08	3.40	4.07	4.12	3.8
Na_2O	3.84	3.07	3.64	3.46	3.6
K_2O	3.13	3.56	3.07	3.10	3.3
H_2O^+	1.15	0.79	0.79	0.79	0.6
P_2O_5	0.30	0.11	0.22	0.15	0.18
CO_2	0.10	0.12	0.39	1.28	n.d.

Table 32
AVERAGE CHEMICAL COMPOSITION AND NORMATIVE MINERALS OF GRANITIC ROCKS

	Alkali granites [1a] (48)	Alkali rhyolites [1b] (21)	Granites [2a] (72)	Rhyolites [2b] (22)	Quartz monzonites [5a] (121)	Quartz latites [5b] (58)	Granodiorites [3a] (137)	Rhyodacites [3b] (115)	Quartz diorites [4a] (58)	Dacites [4b] (50)
SiO_2	73.86	74.57	72.08	73.66	69.15	70.15	66.88	66.27	66.15	63.58
TiO_2	0.20	0.17	0.37	0.22	0.56	0.42	0.57	0.66	0.62	0.64
Al_2O_3	13.75	12.58	13.86	13.45	14.63	14.41	15.66	15.39	15.56	16.67
Fe_2O_3	0.78	1.30	0.86	1.25	1.22	1.68	1.33	2.14	1.36	2.24
FeO	1.13	1.02	1.67	0.75	2.27	1.55	2.59	2.23	3.42	3.00
MnO	0.05	0.05	0.06	0.03	0.06	0.06	0.07	0.07	0.08	0.11
MgO	0.26	0.11	0.52	0.32	0.99	0.63	1.57	1.57	1.94	2.12
CaO	0.72	0.61	1.33	1.13	2.45	2.15	3.56	3.68	4.65	5.53
Na_2O	3.51[a]	4.13	3.08	2.99	3.35	3.65	3.84	4.13	3.90	3.98
K_2O	5.13[a]	4.73	5.46	5.35	4.58	4.50	3.07	3.01	1.42	1.40
H_2O^+	0.47	0.66	0.53	0.78	0.54	0.68	0.65	0.68	0.69	0.56
P_2O_5	0.14	0.07	0.18	0.07	0.20	0.12	0.21	0.17	0.21	0.17
qz[b]	32.2	31.1	29.2	33.2	24.8	26.1	21.9	20.8	24.1	19.6
or	30.0	27.8	32.2	31.7	27.2	26.7	18.3	17.8	8.3	8.3
ab	29.3	35.1	26.2	25.1	28.3	30.9	32.5	35.1	33.0	34.1
an	2.8	2.0	5.6	5.0	11.1	9.5	16.4	14.5	20.8	23.3
c	1.4	—	0.8	0.9	—	—	—	—	—	—
$CaSiO_3$	—	0.1	—	—	—	0.2	—	1.3	0.3	1.3
$MgSiO_3$	0.6	0.3	1.3	0.8	2.5	1.6	3.9	3.9	4.9	5.3
$FeSiO_3$	1.1	0.6	1.7	—	2.2	0.8	2.9	1.3	4.1	2.8
ac	—	—	—	—	—	—	—	—	—	—
mt	1.2	1.9	1.4	1.9	1.9	2.5	1.9	3.0	2.1	3.3
il	0.5	0.3	0.8	0.5	1.1	0.8	1.1	1.4	1.2	1.2
ap	0.3	0.2	0.4	0.2	0.5	0.3	0.5	0.3	0.5	0.3

Note: Effusive rocks include obsidians; rock number in brackets; number of analyses used for average in parentheses.

[a] Pegmatites mainly differ in averages by their higher potassium and slightly lower sodium contents (~ 6.3% K_2O) etc.

[b] The following abbreviations are used for normative minerals: qz = quartz; or = K-feldspar; ab = albite; an = anorthite; c = corundum; lc = leucite; ne = nepheline; ac = acmite; mt = magnetite; il = ilmenite; ap = apatite; cc = calcite.

Table 33
AVERAGE CHEMICAL COMPOSITION AND NORMATIVE MINERALS OF INTERMEDIATE ROCKS

	Alkali syenites [6a] (25)	Alkali trachytes [6b] (15)	Syenites [7a] (18)	Trachytes [7b] (24)	Monzonites [8a] (46)	Latites [8b] (42)	Monzodiorites [9a] (56)	Latite andesites [9b] (38)	Diorites [11a] (50)	Andesites [11b] (49)
SiO_2	61.86	61.95	59.41	58.31	55.36	54.02	54.66	56.00	51.86	54.20
TiO_2	0.58	0.73	0.83	0.66	1.12	1.18	1.09	1.29	1.50	1.31
Al_2O_3	16.91	18.03	17.12	18.05	16.58	17.22	16.98	16.81	16.40	17.17
Fe_2O_3	2.32	2.33	2.19	2.54	2.57	3.83	3.26	3.74	2.73	3.48
FeO	2.63	1.51	2.83	2.02	4.58	3.98	5.38	4.36	6.97	5.49
MnO	0.11	0.13	0.08	0.14	0.13	0.12	0.14	0.13	0.18	0.15
MgO	0.96	0.63	2.02	2.07	3.67	3.87	3.95	3.39	6.12	4.36
CaO	2.54	1.89	4.06	4.25	6.76	6.76	6.99	6.87	8.40	7.92
Na_2O	5.46	6.55	3.92	3.85	3.51	3.32	3.76	3.56	3.36	3.67
K_2O	5.91	5.53	6.53	7.38	4.68	4.43	2.76	2.60	1.33	1.11
H_2O+	0.53	0.54	0.63	0.53	0.60	0.78	0.60	0.92	0.80	0.86
P_2O_5	0.19	0.18	0.38	0.20	0.44	0.49	0.43	0.33	0.35	0.28
qz	1.7	—	2.0	—	—	0.5	2.0	7.2	0.3	5.7
or	35.0	32.8	38.4	43.9	27.8	26.1	16.7	15.6	7.8	6.7
ab	46.1	54.0	33.0	28.8	29.3	27.8	31.9	29.9	28.3	30.9
an	4.2	3.3	10.0	9.7	15.8	19.2	21.1	22.2	25.8	27.2
ne	—	0.6	—	2.0	—	—	—	—	—	—
$CaSiO_3$	3.0	2.1	3.0	4.2	6.3	4.5	4.5	4.1	5.6	4.2
$MgSiO_3$	2.4	1.6	5.0	3.2	8.0	9.7	9.9	8.5	15.3	10.9
$FeSiO_3$	2.1	—	2.1	0.5	4.1	2.4	5.4	3.0	8.5	5.3
Mg_2SiO_4	—	—	—	1.4	0.8	—	—	—	—	—
Fe_2SiO_4	—	—	—	0.2	0.4	—	—	—	—	—
mt	3.3	3.3	3.3	3.7	3.7	5.6	4.9	5.3	3.9	5.1
il	1.2	1.4	1.5	1.2	2.1	2.3	2.1	2.4	2.9	2.4
ap	0.5	0.4	1.0	0.5	1.0	1.2	1.0	0.8	0.8	0.7

Note: Rock number in brackets; number of analyses used for averages in parentheses.

Table 34
AVERAGE CHEMICAL COMPOSITION AND NORMATIVE MINERALS OF GABBROIC-BASALTIC ROCKS

	Gabbros [12a] (160)	Tholeiitic basalts [12b] (137)	Alkali olivine basalts [12b] (96)
SiO_2	48.36	50.83	45.78[a]
TiO_2	1.32	2.03	2.63
Al_2O_3	16.81	14.07	14.64
Fe_2O_3	2.55	2.88	3.16
FeO	7.92	9.00	8.73
MnO	0.18	0.18	0.20
MgO	8.06	6.34	9.39[a]
CaO	11.07	10.42	10.74
Na_2O	2.26	2.23	2.63[a]
K_2O	0.56	0.82	0.95[a]
H_2O+	0.64	0.91	0.76
P_2O_5	0.24	0.23	0.39
qz	—	3.5	—
or	3.3	5.0	6.1
ab	18.9	18.9	18.3
an	34.2	25.9	24.7
ne	—	—	2.3
$CaSiO_3$	8.0	10.3	10.8
$MgSiO_3$	14.0	15.8	7.1
$FeSiO_3$	7.4	11.2	2.9
Mg_2SiO_4	4.3	—	11.5
Fe_2SiO_4	2.5	—	5.0
mt	3.7	4.2	4.6
il	2.4	3.8	5.0
ap	0.6	0.5	1.0

[a] The mean of literature data from Turner, Verhoogen (1960) is higher in SiO_2 (48.4), Na_2O (3.2), K_2O (1.3), mainly compensated by lower MgO.

Table 35
AVERAGE CHEMICAL COMPOSITION AND NORMATIVE MINERALS OF PERIDOTITIC AND ANORTHOSITIC ROCKS

	Peridotite (23)	Anorthosite [12a] (9)		Peridotite (23)	Anorthosite [12a] (9)
SiO_2	43.54	54.54	qz	—	1.4
TiO_2	0.81	0.52	or	1.7	6.7
Al_2O_3	3.99	25.72	ab	4.7	39.3
Fe_2O_3	2.51	0.83	an	7.5	45.9
FeO	9.84	1.46	ne	—	—
MnO	0.21	0.02	$CaSiO_3$	3.9	0.3
MgO	34.02	0.83	$MgSiO_3$	14.8	2.1
CaO	3.46	9.62	$FeSiO_3$	2.6	1.2
Na_2O	0.56	4.66	Mg_2SiO_4	49.1	—
K_2O	0.25	1.06	Fe_2SiO_4	9.6	—
H_2O+	0.76	0.63	mt	3.7	1.2
P_2O_5	0.05	0.11	il	1.5	0.9
			ap	0.1	0.3

Note: Rock number in brackets; number of analyses used for averages in parentheses.

Table 36
AVERAGE CHEMICAL COMPOSITION AND NORMATIVE MINERALS OF ALKALIC ROCKS

	Nepheline syenites [13a] (80)	Phonolites [13b] (47)	Essexites [14a] (15)	Nepheline tephrites [14] (8)	Leucite tephrites [14b] (31)	Ijolites [15a] (11)	Olivine nephelinites [15b] (21)	Olivine leucitites [15b] (11)	Olivine melilitites [15] (10)
SiO_2	55.38	56.90	46.88	44.82	47.05	42.58	40.29	43.64	37.08
TiO_2	0.66	0.59	2.81	2.65	1.54	1.41	2.90	2.54	3.31
Al_2O_3	21.30	20.17	17.07	15.42	16.05	18.46	11.32	10.82	8.08
Fe_2O_3	2.42	2.26	3.62	4.28	3.49	4.01	4.87	5.11	5.12
FeO	2.00	1.85	5.94	6.61	5.78	4.19	7.69	5,89	7.23
MnO	0.19	0.19	0.16	0.16	0.17	0.20	0.22	0.15	0.18
MgO	0.57	0.58	4.85	7.27	6.20	3.22	13.28	13.86	16.19
CaO	1.98	1.88	9.49	10.32	10.80	11.38	12.99	10.66	16.30
Na_2O	8.84	8.72	5.09	5.30	2.35	9.55	3.14	2.16	2.30
K_2O	5.34	5.42	2.64	1.26	5.38	2.55	1.44	4.09	1.36
H_2O+	0.96	0.96	0.97	1.56	0.60	0.55	1.08	0.72	1.89
P_2O_5	0.19	0.17	0.48	0.35	0.59	1.52	0.78	0.63	0.96
CO_2	0.17	—	—	—	—	0.38	—	—	—
Cl	—	0.23	—	—	—	—	—	—	—
SO_3	—	0.13	—	—	—	—	—	—	—
or	31.1	31.7	15.6	7.8	22.2	10.0	—	6.9	—
ab	32.0	36.2	14.7	12.6	—	—	—	—	—
an	2.8	1.7	16.1	14.5	17.5	—	12.8	6.1	7.5
lc	—	—	—	—	7.4	3.9	6.5	13.8	6.5
ne	23.3	18.7	15.3	17.3	10.8	43.7	14.2	9.9	10.5
Ca_2SiO_4	—	—	—	—	—	—	1.6	—	12.8
$CaSiO_3$	2.1	2.9	11.6	14.3	13.6	18.5	17.2	17.8	10.7
$MgSiO_3$	1.2	1.4	8.2	10.4	9.3	8.0	13.1	14.5	8.6
$FeSiO_3$	0.8	0.9	2.4	2.5	3.2	2.4	2.2	1.1	0.8
Mg_2SiO_4	0.1	—	2.8	5.5	4.3	—	14.1	14.1	22.3
Fe_2SiO_4	0.1	—	0.8	1.5	1.7	—	2.7	1.2	2.5
mt	3.5	3.3	5.3	6.3	5.1	5.8	7.2	7.4	7.4
il	1.4	1.2	5.3	5.0	2.9	2.7	5.5	4.9	6.2
ap	0.4	0.3	1.2	0.8	1.3	3.6	1.8	1.5	2.3
cc	0.4	—	—	—	—	0.9	—	—	—

Note: Rock number in brackets; number of analyses used for averages in parentheses.

Table 37
CHEMICAL COMPOSITIONS (OXIDES, WT%) AND CIPW NORMS OF ROCKS FROM MID-ATLANTIC RIDGE[52]

	1	2	3	4	5	6	7	8	9	10
SiO_2	49.20	49.02	49.27	47.94	49.00	47.50	49.70	48.65	43.15	48.56
TiO_2	2.03	1.46	1.26	0.75	1.46	1.83	1.49	1.44	2.70	0.24
Al_2O_3	16.09	18.04	15.91	17.45	15.50	16.00	14.85	15.99	13.46	18.69
Fe_2O_3	2.72	1.58	2.76	1.21			2.16	2.18	4.52	2.27
FeO	7.77	6.22	7.60	8.47	9.77	12.20	8.27	6.19	8.22	4.30
MnO	0.18	0.13	0.13	0.13	—	—	0.18	0.15	0.11	0.11
MgO	6.44	7.85	8.49	10.19	8.00	5.37	8.56	9.66	10.80	9.26
CaO	10.46	11.51	11.26	11.26	10.80	11.40	11.17	11.52	9.80	12.67
Na_2O	3.01	2.92	2.58	2.37	2.90	2.57	2.69	2.71	3.47	1.88
K_2O	0.14	0.08	0.19	0.09	0.21	0.49	0.15	0.57	1.63	0.07
P_2O_5	0.23	0.12	0.13	0.08	—	—	0.13	0.21	0.75	0.02
H_2O^+	0.70	0.64	0.35	0.23			0.61	0.75	1.21	1.72
H_2O^-	0.95	0.57	0.51	0.15	1.19	3.28	0.16	0.30	0.15	0.17
Total	99.92	100.14	100.44	100.32	98.83	100.64	100.12	100.32	99.97	99.96
Q	0.3									
or	0.8	0.5	1.1	0.6	1.27	2.96	1.11	3.34	9.63	0.56
ab	25.7	24.4	21.8	20.0	25.03	22.25	23.06	23.06	9.67	15.72
an	29.8	36.3	31.2	36.7	29.23	31.38	27.52	29.75	16.34	42.46
ne									10.67	
di	17.4	16.6	19.2	15.2	16.19	15.42	21.81	21.10	22.23	15.96
hy	16.2	7.7	13.6	4.5	9.14	6.63	13.19	1.26		14.07
ol		9.0	5.9	19.7	10.08	9.48	6.33	14.45	16.76	4.95
mt	4.0	2.3	4.0	1.8	2.22	2.22	3.25	3.25	6.55	3.25
il	3.8	2.7	2.4	1.4	2.83	3.56	2.89	2.74	5.13	0.46
ap	0.5	0.3	0.3	0.2	0.33	0.34	0.34	0.48	1.64	0.05
Total	98.5	99.8	99.5	100.1	96.32	94.24	99.50	99.43	98.62	97.48

Note: Explanation of column headings (1) Oceanic tholeiite, depth 2910 m; 20°40′S, 13°16′W (Engle and Engel, 1964a, D2-1). (2) Oceanic tholeiite (diabase), depth 2388 m; 9°39′N, 40°27′W (Engel and Engel, 1964a, D5-5). (3) Oceanic tholeiite, depth 3566 m, rift floor; 28°53′N, 43°20′W (G. D. Nicholls, 1965, Table 1, analysis 1). (4) High-alumina basalt, some locality as 3 (G. D. Nicholls, 1965, Table 2, analysis 2). (5) Oceanic tholeiite, depth 4200 m, rift floor; 30°08′N, 43°37′W (Kay et al., 1970, analysis A150-21-1C). (6) Basalt, depth 3700 m; 31°49′N, 42°25′W (Kay et al., 1970, analysis GE160). (7) Basalt, depth 3700 m; 31°49′N, 42°25′W (Muir and Tilley, 1966, p. 195, analysis 3). (8) Basalt, depth 3600 m, rift floor; 45°44′N, 27°44′W (Muir and Tilley, 1964b, Table 1, analysis 5). (9) Alkali olivine basalt, depth between 2000 and 3000 m; a few kilometers northeast of St. Paul's Rocks; 1°1′N, 29°21′W (Melson et al., 1967). (10) Laminated gabbro, depth 4000 to 5000 m, Romanche trench, 0°14′N, 17°7′W (Melson and Thompson, 1970).

Table 38
ATOMIC[a] ABUNDANCES (PPM) AND ABUNDANCE RATIOS OF TRACE ELEMENTS AND ISOTOPES, LAVAS OF MID-ATLANTIC RIDGE[52]

	1	1a[b]	2	2a[c]	5	6	8	9	10
Ba	5		5			11.7	16	300	10
Co	32		26				38		
Cr	220		280				700	250	900
Ni	87		78		190	100	220	270	200
Pb		1.29		0.56					2
Rb	1.14	1.42	<10		1.90	12.9	22		<20
Sr	190(134)		90		90	105	320	500	110
Th	0.15			0.13					
U	0.16			0.09					
V	280		240				350		110
Zr	160		62				45	200	10
K/Rb	1020	950			700	366	230		
Rb/Sr	∿0.01		<0.1		0.02	0.12	0.07		<0.2
Th/U	0.9			1.4					
Sr^{87}/Sr^{86}		0.7032[c]							
Pb^{206}/Pb^{204}		18.471		18.816					
Pb^{207}/Pb^{204}		15.54		15.68					
Pb^{208}/Pb^{204}		38.01		38.65					
U^{238}/Pb^{204}		7.9		10.5					

[a] Column headings numbered as in Table 50.
[b] Values reported by Gast (1967) and Tatsumoto (1966).
[c] Normalized to 0.7080 for standard E and A $SrCO_3$ (Hedge and Peterman, 1970, p. 119).

Table 39
CHEMICAL COMPOSITIONS (OXIDES, WT %). CIPW NORMS[a] AND ATOMIC ABUNDANCES AND ABUNDANCE RATIOS OF TRACE ELEMENTS AND ISOTOPES OF OCEAN-FLOOR LAVAS, RISES OF EAST PACIFIC OCEAN[52]

	1	2	3	4	5	6		1a	3a	4a
SiO_2	49.80	49.13	48.30	59.00	49.90	50.10	Ba	25	19.4	54.8
TiO_2	2.02	1.23	2.19	1.75	1.08	2.18	Ce		16.5	75
Al_2O_3	14.88	14.97	14.30	12.60	17.30	13.80	Cs		0.074	0.082
Fe_2O_3	1.55	3.28					Co	35		
FeO	10.24	5.72	11.70	12.00	7.60	12.30	Cr	160		
MnO	0.21	0.16					Ni	58	58	10
MgO	6.74	7.68	6.70	1.70	7.08	6.11	Pb	0.49		
CaO	10.72	12.68	10.10	5.60	12.78	10.90	Rb	1.06	5	7.25
Na_2O	2.91	2.37	2.75	4.25	2.45	2.83	Sr	110(86)	107	105
K_2O	0.24	0.16	0.18	0.65	0.18	0.16	Th	0.21		
P_2O_5	0.28	0.15					U	0.09		
H_2O^+	0.54	1.06					V	400		
H_2O^-	0.06	1.25	1.29	1.78	0.80		Zr	150		
							K/Rb	1890	310	770
Total	100.19	99.84	97.51	99.33	99.17	98.38	Cs/Rb		0.014	0.011
							Th/U	2.3		
Q		0.79		13.66		0.04	Sr^{87}/Sr^{86}	0.7025		
or-	1.1	0.89	1.10	3.92	1.08	0.96	Pb^{206}/Pb^{204}	18.24		
ab	24.6	20.01	24.08	36.70	20.99	24.25	Pb^{207}/Pb^{204}	15.53		
an	26.7	29.75	27.06	13.66	36.12	24.79	Pb^{208}/Pb^{204}	38.03		
di	22.0	25.73	19.59	11.84	21.90	23.96	U^{238}/Pb^{204}	6.4		
hy	13.7	12.79	17.73	14.27	13.16	19.27				
ol	5.7		3.54		2.12					
mt	2.3	4.76	2.25	2.22	2.20	2.20				
il	3.7	2.36	4.30	3.39	2.08	4.19				
ap	0.6	0.35	0.34	0.33	0.33	0.33				
Total	100.4	97.43	99.99	99.99	99.98	99.99				

Note: Explanation of column headings (1) Glassy basalt, East Pacific rise, depth 2300 m; 12°52′S, 110°57′W (Engel et al., 1965, table 1, analysis PVD-3). (1a) Trace-element and isotopic data for analysis 1 (Engel et al., 1965; Tatsumoto et al., 1965; Tatsumoto, 1966). (2) Basalt, Mohole drill core, depth 3746 m; off Guadalupe Island (East Pacific rise), 28°59′N, 117°30′W (Engel and Engel, 1961, p. 1799, analysis 1). (3,3a) Basalt, East Pacific rise, depth 3120 m; 7°08′N, 103°15′W (Kay et al., 1970, p. 1593, analysis V2023). (4,4a) "Andesite" glass, East Pacific rise, depth 3182 m; 5°31′S, 106°46′W (Kay et al., 1970, p. 1593, analysis V2140). (5) Basalt, Gordo rise, depth 2500 m; 41°15′N, 127°28′W (Kay et al., 1970, p. 1592, analysis 13E). (6) Basalt, Juan de Fuca rise, depth 2502 m; 44°36′N, 130°19′W (Kay et al., 1970, p. 1591, analysis 2C).

[a] For analyses 3—6, norms are calculated (Kay et al., 1970) assuming a low state of oxidation of Fe (Fe_2O_3 = 1.50%) and reasonable values of MnO (0.18%) and P_2O_3 (0.15%).

Table 40
CHEMICAL COMPOSITIONS (OXIDES, WT%) AND CIPW NORMS OF ROCKS IN DIFFERENTIATED DIABASE SHEETS[52]

	1	2	3	4	5	6	7	8	9	10
SiO_2	47.41	51.98	52.67	58.59	53.32	53.18	53.20	55.08	60.37	67.62
TiO_2	0.89	1.21	2.76	1.56	0.70	0.65	0.72	1.52	1.13	0.84
Al_2O_3	8.66	14.48	11.94	11.26	14.18	15.37	15.59	15.36	13.77	12.31
Fe_2O_3	2.81	1.37	3.50	3.00	0.97	0.76	0.83	3.43	2.00	5.66
FeO	11.15	8.92	11.08	10.80	8.54	8.33	8.54	9.06	8.20	3.04
MnO	0.20	0.16	0.19	0.19	0.18	0.15	0.16	0.16	0.17	0.11
MgO	19.29	7.59	3.94	1.35	7.23	6.71	6.11	1.82	0.83	0.32
CaO	6.76	10.33	8.06	3.66	11.22	11.04	11.03	7.64	6.14	2.52
Na_2O	1.35	2.04	2.78	3.68	1.38	1.65	1.62	2.29	2.82	2.92
K_2O	0.43	0.84	1.29	2.52	0.87	1.03	1.14	1.84	2.42	3.38
P_2O_5	0.10	0.14	0.31	0.85	0.20	0.08	0.15	0.19	0.33	0.24
H_2O^+	1.45	0.88	1.02	1.52	1.00	0.67	1.09	1.42	1.26	0.68
H_2O^-	0.11	0.16	0.20	0.43	0.64	0.45	0.27	0.50	0.52	0.62
Total	100.61	100.10	100.14	99.41	100.43	100.07	100.45	100.31	99.96	100.26
Q		2.80	8.09	14.33	6.58	4.80	5.39	13.20	18.18	32.46
or	2.39	4.73	7.62	15.12	5.14	6.09	6.74	10.56	14.46	20.02
ab	11.53	16.92	23.52	31.22	11.68	13.96	13.71	19.39	23.58	24.63
an	16.21	28.23	16.29	6.61	29.93	31.49	31.90	26.13	17.79	10.56
di	13.14	18.19	18.10	5.15	20.05	18.71	18.02	9.47	9.44	0.86
hy	25.91	23.60	13.69	15.78	22.21	21.37	20.42	11.28	8.92	0.40
ol	23.77									
mt	4.18	2.03	5.65	4.41	1.41	1.10	1.20	4.87	3.02	7.98
il	1.65	2.28	5.24	2.98	1.33	1.23	1.37	2.89	2.13	1.52
ap	0.23	0.32	0.73	2.02	0.47	0.19	0.35	0.44	0.76	0.55
Total	99.01	99.10	98.93	97.62	98.80	98.94	99.10	98.23	98.28	98.98

Note: Explanation of column headings (1) Olivine diabase, 20 m above base, Palisades sill (300 m thick), New Jersey; cf. Table 9-6, analysis 4 (K. R. Walker, 1969a, Table 8, analysis W-824-60). (2) Chilled diabase, base of Palisades sill (300 m thick), New Jersey (K. R. Walker, 1969a, Table 8, analysis W-899 LC-60). (3) Hypersthene diabase, 210 m above base of Palisades sill, New Jersey (K. R. Walker, 1969a, Table 8, analysis W-J-60). (4) Granophyric diabase 240 m above base of Palisades sill, New Jersey (K. R. Walker, 1969a, Table 8, analysis W-F-60). (5) Chilled diabase, upper contact Great Lake sheet, Tasmania (McDougall, 1964a, p. 123, analysis 21). (6) Chilled diabase, average of 13 analyses, Tasmanian Jurassic diabases (McDougall, 1962, p. 294). (7) Diabase near middle of Great Lake sheet (530 m thick), Tasmania; cf. Table 9-7, analysis 2 (McDougall, 1964a, analysis 5123-700). (8) Diabase transitional to granophyre, Red Hill dike, Tasmania; cf. Table 9-7, analysis 7 (McDougall, 1962, analysis M-395). (9) Fayalite granophyre, Red Hill dike, Tasmania (McDougall, 1962, analysis M-12). (10) Granophyre, Red Hill dike, Tasmania (McDougall, 1962, analysis M-8).

Table 41
CHEMICAL COMPOSITIONS (MAJOR OXIDES, WT %) AND CIPW NORMS OF BASALTIC ROCKS OF "PLATEAU MAGMA TYPE" AND ASSOCIATED LAVAS, HEBRIDEAN PROVINCE, NORTHWEST BRITAIN[52]

	1	2	3	4	5	6	7
SiO_2	45.34	45.52	46.12	46.97	47.90	49.68	66.13
TiO_2	1.13	1.48	1.81	1.59	1.57	2.13	0.61
Al_2O_3	14.67	17.58	13.94	15.00	15.28	16.99	16.03
Fe_2O_3	2.40	4.17	1.95	1.71	1.70	3.45	3.17
FeO	9.15	8.14	10.46	8.94	9.10	8.99	0.70
MnO	0.22	0.17	0.18	0.37	0.17	0.27	0.10
MgO	13.32	8.46	11.08	10.52	7.30	2.79	0.84
CaO	9.12	9.72	9.05	10.70	12.07	5.46	1.45
Na_2O	1.86	1.86	3.11	2.18	2.81	5.78	5.34
K_2O	0.24	0.62	0.57	0.63	0.53	1.90	4.82
P_2O_5	0.09	0.07	0.23	0.12	0.16	0.48	0.08
H_2O^+	1.69	1.85	1.49	0.38	1.27	1.77	0.36
H_2O^-	1.05	0.32	0.40	0.63	0.46	0.34	0.43
Total	100.28	99.96	100.39	99.74	100.32	100.03	100.06
Q							12.12
or	1.45	3.8	3.34	3.3	2.78	11.12	28.73
ab	15.72	15.6	20.96	17.8	22.27	38.90	45.06
an	30.86	37.7	22.38	29.8	27.80	14.46	5.28
ne			2.84		0.71	5.61	
di	11.61	7.1	17.89	18.2	25.44	6.53	0.22
hy	9.24	18.6		5.9			2.26
ol	22.72	5.9	24.33	17.8	13.62	10.67	
mt	3.48	6.0	2.90	2.6	2.55	4.98	1.37
il	2.13	2.9	3.50	3.0	3.04	4.10	0.67
ap	0.20	0.2	0.34	0.3	0.34	1.34	0.18
Total	97.41	97.8	98.48	98.7	98.55	97.71	99.09[a]

Note: Explanation of column headings (1) Olivine basalt, upper lava series, Antrim (Patterson, 1951, p. 286, analysis 5). (2) Olivine basalt, Ard Bheinn, Arran (King, 1955, p. 328, analysis 1). (3) Olivine basalt, Skye (Tilley and Muir, 1962, p. 212, analysis 1). (4) Diabase ("normal dolerite"), northern Skye (Anderson and Dunham, 1966, p. 147, analysis Q). (5) Olivine basalt, Fingal's Cave, Staffa (Tilley and Muir, 1962, analysis 2). (6) Mugearite, type locality, Skye (Muir and Tilley, 1961, p. 190, Tables 4,5; analysis 1). (7) Trachyte, Skye (Anderson and Dunham, 1966, p. 118, analysis XI).

[a] Including ac 3.20.

Table 42
CHEMICAL ANALYSES (MAJOR OXIDES, WT %) AND CIPW NORMS OF Q-NORMATIVE BASIC AND INTERMEDIATE ROCKS, HEBRIDEAN PROVINCE, NORTHERN BRITAIN[52]

	1	2	3	4	5	6
SiO_2	50.41	50.36	53.92	57.09	54.18	58.34
TiO_2	1.30	1.06	0.79	0.80	1.97	1.08
Al_2O_3	15.14	14.51	15.81	13.76	13.74	14.09
Fe_2O_3	2.71	2.61	3.05	2.74	1.88	2.34
FeO	7.95	8.09	5.06	4.98	10.79	7.78
MnO	0.17	0.12	0.11	0.12	0.30	0.17
MgO	6.57	6.26	5.32	4.29	2.42	4.24
CaO	11.30	10.77	10.22	8.12	6.34	6.96
Na_2O	2.29	2.48	2.09	2.51	3.46	2.66
K_2O	0.82	0.99	1.46	1.95	1.85	1.30
P_2O_5	0.15	0.45	0.12	0.13	1.30	0.06
H_2O^+	1.01	1.10	1.01	1.43	1.40	0.85
H_2O^-	0.72	1.27	1.19	1.45	0.26	0.27
Total	100.54	100.07	100.15	99.37	99.89	100.14
Q	1.43	1.46	8.90	14.47	7.8	14.3
or	4.84	5.84	8.62	11.52	10.9	7.8
ab	19.35	20.97	17.67	20.50	29.3	22.5
an	28.70	25.53	29.59	21.05	16.5	22.5
di	21.27	20.27	15.17	11.25	5.5	9.6
hy	16.37	16.63	11.46	10.91	18.7	16.5
mt	3.94	3.89	4.42	3.96	2.7	3.5
il	2.46	2.02	1.50	1.52	3.8	2.1
ap	0.37	0.99	0.27	0.30	3.1	0.1
Total	98.73	97.60	97.60	95.48	98.3	98.9

Note: Explanation of column headings (1) Tholeiitic diabase, Kielderhead dike, Northumberland (Holmes and Harwood, 1929, p. 15). (2) Basalt of Giant's Causeway, Nudale lava series, Antrim (Holmes, 1936, p. 91). (3) Tholeiitic diabase, Coley Hill dike, near Newcastle-upon-Tyne (Holmes and Harwood, 1929, p. 32). (4) Tholeiitic diabase, Hebburn dike, Durham (Holmes and Harwood, 1929, p. 35). (5) Ferrodiorite, associated with marscoite, Skye (Wager et al., 1965, p. 283, analysis 6A). (6) "Andesite", Arran (King, 1955, p. 328, analysis 4).

[a] Samples 1 to 4 have been equated with the nonporphyrite central magma type.

Table 43
CHEMICAL COMPOSITIONS (OXIDES, WT %) AND CIPW NORMS OF CHILLED BORDER GABBROS (PROBABLE LIQUID COMPOSITIONS) OF LAYERED BASIC INTRUSIONS[52]

	1	2	3	4	5		6
SiO_2	50.68	48.08	50.68	49.86	52.55	Ba	25
TiO_2	1.06	1.17	0.45	2.02	0.30	Co	55
Al_2O_3	13.55	17.22	17.64	11.17	15.38	Cr	170
Fe_2O_3	1.17	1.32	0.26	2.37	0.70	Ni	180
FeO	9.08	8.44	9.88	14.70	10.26	Rb	4
MnO	0.18	0.16	0.15	0.28	0.21	Sr	267
MgO	9.70	8.62	7.67	3.44	6.84	V	190
CaO	11.22	11.38	10.47	10.59	10.30	Zr	50
Na_2O	1.79	2.37	1.87	2.46	2.70		
K_2O	0.63	0.25	0.24	0.91	0.22		
P_2O_5	0.10	0.10	0.09	1.20	0.01		
H_2O^+	0.53	1.01	0.42	0.50	0.41		
H_2O^-	0.06	0.05	0.06	0.08	0.06		
Total	99.75	100.17	99.88	99.58	99.94		
Q				3.48			
or	3.4	1.48	1.39	5.56	2.22		
ab	15.1	20.05	15.72	20.96	22.79		
an	26.9	35.62	39.06	16.40	28.77		
di	23.1	16.43	9.95	24.21	18.56		
hy	23.3	7.63	31.82	18.17	26.00		
ol	2.7	13.54					
mt	1.6	1.91	0.46	3.48	0.93		
il	2.0	2.22	0.91	3.80	0.61		
ap	0.2	0.24	0.20	2.85			
Total	98.3	99.12	99.51	98.91	99.88		

Note: Explanation of column headings. Analyses 1—3 represent less differentiated, analyses 4 and 5 more differentiated magmatic liquids. (1) Muskox intrusion (average of two analyses; C. H. Smith and H. Kapp, 1963, p. 33). (2) Skaergaard marginal gabbro (Wager and Brown, 1967, Table 7, analysis 4507). (3) Base of Stillwater (H. H. Hess, 1960, p. 162, analysis G70). (4) Roof of Bushveld (H. H. Hess, ibid., BV52). (5) Roof of Great "Dike" (H. H. Hess, ibid., GD29). (6) Atomic abundances of trace elements in chilled border gabbros, Skaergaard (Wager and Brown, 1967, pp. 193—201).

Table 44
SOME GEOCHEMICAL DATA FOR CONTINENTAL THOLEIITIC ROCKS COMPARED WITH OCEANIC THOLEIITES[52]

	1	2	3	4	5	6	7	8	9	10
K/Na	0.63	0.26—0.48	0.23—0.55	0.70	0.15—0.41	0.17	0.10	0.1—0.3	0.10—0.30	0.05
K/Rb	200—220	200—270	400—425	200—270		350	500	280	400—500	1300—1500
Rb/Sr	0.25	0.12—0.4	0.03—0.07	0.32		0.028	0.015	0.058	0.28	0.01
$U/K \times 10^4$	1.25	1.4—3.2	0.4—0.77		0.38—0.64			1.2—1.4	0.3—1.0	0.75
$Th/K \times 10^4$	5	5—7.5	2.7—3.0		1.8—4.2			3.5—3.9	2.18—2.8	1.5
Sr^{87}/Sr^{86} (initial)	0.7115	0.712	0.7057	0.706—0.710	0.704—0.706	0.7029[b]	0.7065	0.7078	0.7039—0.7040	0.7026
Sr (ppm)	130	100—140	180—400	180, 230	120—180	13	267	260	250—800[a]	100—200
Ba (ppm)		160—430	150—350	700, 750	200		25		150[a]	14
U (ppm)	0.9	0.6—1.6	0.2—0.4		0.35			1.9	0.2	0.1
Th (ppm)	3.3	2.2—5.4	1.3—1.5		1.8			0.53, 0.78	0.7	0.2

Note: Explanation of column headings (1) Diabases of Tasmania (143—167 m.y.). Data from Heier et al. (1965); Compston et al. (1968). (2) Diabases (Ferrar dolerites) of Antarctica (147—163 m.y.). Cf. Tables 9-10, 9-11. Data from Gunn (1966) and Compston et al. (1968). (3) Diabases of Karroo, South Africa (154—190 m.y.). Cf. Tables 9-2, 9-3, 9-5. Data from Compston et al. (1968). (4) Two tholeiitic basalts, Nuanetsi region, Karroo province, South Africa. Cf. Tables 9-4, 9-5. (5) Triassic diabases, New Jersey. Cf. Tables 9-7 to 9-10. (6) Peridotite, ultramafic zone, Stillwater complex, Montana. Data from Stueber and Murthy (1966, pp. 1247, 1249, 1252). (7) Marginal gabbro, Skaergaard intrusion, Greenland. Data from Wager and Brown (1967; cf. Table 9-16). (8) Tertiary tholeiitic basalts, Brighton, Tasmania. Data from Edwards (1950) and Compston et al. (1968, pp. 140—141). (9) Hawaiian tholeiitic basalts. Data from Faure and Hurley (1963, p. 38); G. A. Macdonald and Katsura (1964); Heier et al. (1965); and Hedge and Peterman (1970). (10) Some abyssal tholeiites of the ocean. Data from Engel et al. (1965) and Hedge and Peterman (1970).

[a] Tholeiitic basalt, Kilauea (Wager and Mitchell, 1953; cf. Table 8-13).
[b] Present ratio 0.7063, recalculated assuming age 2700 m.y.

Table 45
CHEMICAL COMPOSITIONS OF SOME AFRICAN CARBONATITES (LAVAS), TRACHYTES, FENITES, AND KIMBERLITES[52]

	1	2	3	4	5	6	7
SiO_2	12.99	Tr.	58.43	62.73	55.50	36.58	27.93
TiO_2	1.74	0.10	0.34			2.67	2.73
Al_2O_3	3.03	0.08	17.84		13.77	7.15	4.47
Fe_2O_3	7.93	0.26	5.09			6.69	7.04
FeO	4.44		0.00			4.99	5.12
MnO	0.40	0.04	0.42			0.34	0.23
MgO	8.55	0.49	0.43			22.55	25.42
CaO	35.97	12.74	0.80			6.05	10.01
SrO	0.63	1.24					
BaO	0.15	0.95	0.18				
Na_2O	0.73	29.53	0.38	0.38	3.61	0.28	0.21
K_2O	0.20	7.58	13.90	14.97	9.20	0.47	1.18
P_2O_5	3.32	0.83	0.35		0.16	0.38	1.07
H_2O^+	3.45	8.59	1.05			8.51	7.89
H_2O^-	1.65		0.11			3.31	0.68
Co_2	14.79	31.75					5.61
SO_2	0.35	2.00					
Cl		3.86					
F	0.30	2.69					0.89
Total	100.62[a]	100.73[b]	99.32			99.97	100.56[c]

Note: Explanation of column headings. (1) Vesicular carbonatitic lava, Kalyango volcano, Fort Portal area, Toro-Ankole province, Uganda (Von Knorring and Du Bois, 1961). (2) Carbonatite lava (pahoehoe), extruded from Oldoinyo Lengai, Tanzania (Dawson, 1964b, p. 106, no. BD 114). (3) Trachyte, Toror Hills, Uganda (Sutherland, 1965, p. 370, no. 3). (4) Feldspathic fenite, Toror Hills, Uganda (Sutherland, 1965, p. 371, no. 15); partial analysis. (5) Fenitized granitic basement, Toror Hills, Uganda (Sutherland, 1965, p. 371, no. 16); partial analysis. (6) Kimberlite, Lesotho (Dawson, 1962b, p. 551, analysis 1). (7) Micaceous kimberlite, Lesotho (Dawson, 1962b, p. 551, analysis 6).

[a] S = 0.35.
[b] Total less O = 2.00 for F and Cl.
[c] Including Cr_2O_3 = 0.08.

Table 46
REPRESENTATIVE ANALYSES AND TRACE-ELEMENT ABUNDANCES (PPM) OF LAVAS FROM THE NEW GUINEA-NEW BRITAIN ACTIVE VOLCANIC ARC[52]

	New Guinea			Talasea				Rabaul							
	1	2	3	4	5	6	7	8	9	10		4	5	6	7
SiO_2	53.05	60.10	55.60	51.57	58.60	66.34	75.33	50.11	55.27	64.95	Zr	30	75	115	150
TiO_2	0.35	0.56	0.81	0.80	0.89	0.71	0.27	0.95	0.86	0.84	Sr	355	395	270	200
Al_2O_3	13.90	16.40	16.50	15.91	15.38	14.63	12.58	18.89	17.40	15.25	Rb	5	15	25	55
Fe_2O_3	6.07	4.25	2.90	2.74	2.22	1.63	1.58	3.67	3.80	2.04	Ni	100	70	75	40
FeO	3.99	5.30	8.20	7.04	6.71	3.77	0.88	5.80	4.17	3.00	V	275	235	40	35
MnO	0.16	0.12	0.21	0.17	0.18	0.14	0.07	0.18	0.15	0.24	Cr	125	15	5	5
MgO	8.60	3.10	3.10	6.73	3.22	1.48	0.24	4.64	3.80	1.42	Ba	150	345	350	645
CaO	10.68	7.30	7.30	11.74	7.02	4.04	1.25	10.86	8.22	3.64	K/Rb	740	807	712	576
Na_2O	2.57	1.75	3.13	2.41	3.84	4.50	4.02	2.76	3.42	4.88	Sr/Rb	0.014	0.038	0.093	0.275
K_2O	0.90	0.84	1.93	0.44	1.46	2.14	3.82	0.94	1.53	2.89	Sr^{87}/Sr^{86}[a]	0.7035	0.7036	0.7036	0.7035
P_2O_5	0.25	0.15	0.36	0.11	0.25	0.21	0.02	0.19	0.23	0.30					
H_2O^+	Nil	0.11	0.13	0.35	0.30	0.35	0.31	0.79	0.79	0.55					
H_2O^-	Nil	0.16	0.23	0.10	0.07	0.12	0.09	0.20	0.23	0.04					
Total	100.52	100.14	100.40	100.11	100.14	100.06	100.46	99.98	99.87	100.04					

Note: Explanation of column headings (1)Manam Island, basaltic-andesite flow from main crater, 1957—1958 eruption (W. R. Morgan, 1966). (2) Karkar Island, pyroxene-olivine andesite lave (W. R. Morgan, 1966). (3) Long Island, pyroxene-andesite lave (W. R. Morgan, 1966). (4) Basalt, Lake Dakataua, Talasca (Lowder and Carmichael, 1970, no. 311). (5) Pyroxene andesite, historic flow (1890s) of Mt. Makalia, Talasca (Lowder and Carmichael, 1970, no. 114). (6) Dacite from Lake Dakataua, Talasca (Lowder and Carmichael, 1970, no. 306). (7) Rhyolite obsidian, Volupai road, Talasca (Lowder and Carmichael, 1970, no. 343). (8) Basalt, Rabaul caldera (Heming, in press). (9) Basaltic andesite, Rabaul caldera (Heming, in press). (10) Bomb from 1937 eruption of Vulcan, Rabaul caldera (Heming, in press).

[a] Strontium isotopic data taken from Peterman, et al. (1970a).

Table 47
AVERAGE[a] INITIAL STRONTIUM ISOTOPIC VALUES (Sr^{87}/Sr^{86}) FOR SELECTED ROCK TYPES OF ACTIVE VOLCANIC ARCS AND CONTINENTAL MARGINS (ALL DATA ADJUSTED TO A VALUE OF 0.7089 FOR MIT $SrCo_3$)[52]

	Basalt (<52% SiO_2)	Basaltic Andesite (52—55% SiO_2)	Andesite (55—63% SiO_2)	Dacite (63—68% SiO_2)	Rhyolite (>68% SiO_2)
New Britain	0.7035	0.7036	0.7036	0.7036	0.7035
Tonga	—	0.7037	0.7042	0.7043	
Marianas	0.7042	—	0.7042	0.7038	—
Izu Islands	0.7036	—	0.7040	—	0.7034
Caribbean					
St. Kitts	0.7036	0.7040	0.7038	—	—
St. Vincent	0.7042	0.7040	0.7039	—	—
Carriacou	0.7052	—	0.7054	—	—
North Japan	0.7043	—	0.7041	—	—
California					
Mt. Shasta	—	0.7039	0.7030	0.7032	—
Mt. Lassen	0.7039	0.7032	0.7040	—	—
Medicine Lake	0.7034	0.7037	—	—	0.7040
Central America	0.7035	0.7042	0.7036	0.7036	0.7042
New Zealand					
Taupo	0.7042	—	0.7055	0.7051	0.7053
Average[b]	0.7040	0.7038	0.7040	0.7037	0.7038

[a] Data taken from Pushkar (1968), Peterman et al. (1970a, 1970b), Hedge (1966), Hedge and Knight (1969), Hedge and Lewis (1971), Ewart and Stipp (1968) Oversby and Ewart (1972), Ewart et al. (1973).

[b] Excludes New Zealand average andesite, dacite, and rhyolite.

Table 48
AVERAGE[a] ISOTOPIC COMPOSITION OF LEAD AND AVERAGE CONCENTRATIONS OF Pb, U, AND Th (ppm) IN SELECTED VOLCANIC ROCKS OF ACTIVE VOLCANIC ARCS AND CONTINENTAL MARGINS[52]

	Pb	U	Th	Pb^{206}/Pb^{204}	Pb^{207}/Pb^{204}	Pb^{208}/Pb^{204}
New Zealand (Taupo)						
Basalt	—	—	—	18.749	15.618	38.62
Andesite	—	—	—	18.773	15.604	38.611
Tonga						
Andesite	1.83	0.19	—	18.528	15.549	38.118
Dacite	3.85	0.37	—	18.525	15.544	38.131
Kermadec						
Basalt	1.88	0.18	—	18.552	15.560	38.436
Caribbean (St. Kitts)						
Basalt	—	—	—	18.868	15.640	38.629
Andesite	—	—	—	18.952	15.658	38.711
Japan (tholeiitic)						
Basalt	3.22	0.15	0.26	18.507	15.68	38.75
Andesite	4.04	0.38	0.38	18.385	15.66	38.65
Japan (high Al_2O_3)						
Basalt	4.32	0.38	1.04	18.384	15.67	38.67
Andesite	5.07	0.59	1.69	18.364	15.68	38.67
Dacite	6.02	1.46	4.11	18.346	15.67	38.65
Japan (alkali)						
Olivine basalt	3.35	0.68	3.34	18.08	15.56	38.60
New Zealand (Auckland)						
Alkali olivine basalt	—	—	—	19.130	15.580	38.764

[a] Data taken from Armstrong and Cooper (1971), Tatsumoto and Knight (1969), Oversby and Ewart (1972).

Table 49
CHEMICAL COMPOSITIONS (OXIDES, WT %), CIPW NORMS AND TRACE-ELEMENT ABUNDANCES (ppm) OF TYPICAL ROCKS, MESOZOIC BATHOLITHS, CALIFORNIA[52]

	1	2	3	4	5		3	4
SiO_2	50.78	62.2	66.92	71.42	75.4	Ba	1000	1000
TiO_2	0.77	0.7	0.47	0.36	0.1	Co	10	5
Al_2O_3	20.40	16.6	15.19	14.03	13.3	Cr	10	6
Fe_2O_3	1.75	1.4	1.45	0.89	0.3	Ni	5	3
FeO	6.20	4.5	2.52	1.63	0.74	Pb	20	20
MnO	0.09	0.06	0.08	0.05	0.08	Sr	700	300
MgO	6.49	2.7	1.74	0.70	0.12	V	60	40
CaO	10.24	5.7	3.79	1.91	0.48	Zr	100	500
Na_2O	2.20	3.4	3.16	2.86	4.1			
K_2O	0.45	1.6	3.82	5.35	4.5			
P_2O_5	0.05	0.09	0.18	0.09	0.01			
H_2O^+			0.06	0.08	0.46			
H_2O^-	0.65	0.6	0.48	0.35				
Total	100.07	99.55	99.86	99.72	99.59			
Q	3.18	12.90	22.62	28.38	32.58			
or	2.22	8.90	22.76	31.68	26.69			
ab	18.34	28.82	26.72	24.11	34.58			
an	41.98	28.91	15.84	9.46	2.50			
di	7.20	2.07	1.83					
hy	21.42	13.57	6.18	3.55	1.36			
mt	2.55	2.09	2.09	1.39	0.46			
il	1.52	1.67	0.90	0.65				
ap	0.33	0.34	0.35					
c				0.10	0.71			
Total	98.74	99.27	99.29	99.32	98.88			

Note: Explanation of column headings (1) Average composition of San Marcos gabbro, southern California batholith (Larsen, 1948, p. 50, analysis B). (2) Average composition of Bonsall quartz diorite, southern California batholith (Larsen, 1948, p. 67, analysis G). (3) Lamarck granodiorite, east central Sierra Nevada (Bateman et al., 1963, p. 29, analysis 12). (4) Tungsten Hills quartz monzonite, east central Sierra Nevada (Bateman et al., 1963, p. 29, analysis 20). (5) Quartz monzonite, Cathedral Peaks type, east central Sierra Nevada (Bateman et al., 1963, p. 29, analysis 25).

Table 50

ATOMIC ABUNDANCES (ppm), ABUNDANCE RATIOS, AND ISOTOPIC ABUNDANCES OF RADIOACTIVE AND RELATED ELEMENTS. MESOZOIC GRANITIC BATHOLITHS, WESTERN UNITED STATES[52]

	1	2	3	4	5	6	7	8
$K \times 10^{-3}$	28	25	34	45	3.1	14.9	28.4	
Rb	83	64	129	218				
Sr	431	610	52	108				
Th	11	7.3	16.2	36.3				
U	3.4	1.5	4	9.2	0.22	2	1.5	
K/Rb	340	390	260	205				
Rb/Sr	0.191	0.15	0.28	2.01				
Th/U[a]	4	4.8	4	4.9				
$U/K \times 10^4$	1.2	0.6	1.2	2.0	0.7	0.3	0.6	
$Th/K \times 10^4$	4	2.9	4.8	8				
Sr^{87}/Sr^{86} (initial)	0.7078	0.7063	0.7073	0.7081				
Pb^{206}/Pb^{204}	17.98	16.94	17.96	18.14				18.95
Pb^{208}/Pb^{204}	38.23	37.68	38.23	38.41				38.52

Note: Explanation of column headings. 1-4. Boulder batholith (Doe et al., 1968; Tilling and Gottfried, 1969); lead values in determined potassium feldspars. 1. Mafic granodiorite (Unionville pluton); 2 Felsic granodiorite (Rader Creek pluton); 3 Butte quartz monzonite (75% of exposed batholith); 4 Alaskite. 5-7. Southern California batholith (Larsen and Gottfriend, 1961, p. 73). 5 Hornblende gabbro, San Marcos pluton; SiO_2 = 48.16; 6 Bonsall quartz diorite; SiO_2 = 62.28; 7 Woodson Mountain granodiorite; SiO_2 = 72.55; 8 Southern California batholith; leucogranite (cited by Doe, 1967, p. 53).

[a] Computed from individual Th/U ratios.

Table 51
CHEMICAL ANALYSES (OXIDES WT %) AND CIPW NORMS OF SOME REPRESENTATIVE POTASSIC PRECAMBRIAN GRANITES[52]

	1	2	3	4	5
SiO_2	71.32	73.84	69.50	77.04	70.11
TiO_2	0.27	0.22	0.45	0.05	0.42
Al_2O_3	14.31	13.09	12.34	12.39	14.11
Fe_2O_3	0.88	0.87	2.02	0.15	1.14
FeO	1.18	1.79	3.97	0.27	2.62
MnO	0.07	0.03	0.06	0.01	0.06
MgO	0.52	0.23	0.35	0.17	0.24
CaO	1.05	1.12	2.33	0.22	1.66
Na_2O	2.62	2.32	2.50	1.96	3.03
K_2O	6.87	6.24	5.80	6.91	6.03
P_2O_5	0.03	0.05	0.08	0.12	0.09
H_2O^+	0.65	0.45	0.43	0.41	0.23
H_2O^-	0.15	0.07	0.10	0.06	0.07
F		0.17	0.11		0.09
Total	99.92	100.49	100.04	99.76	99.90
Q	26.27	33.61	26.68	38.67	23.19
C	0.51	1.05		1.57	
or	40.49	36.74	34.52	40.83	35.58
ab	22.01	19.41	20.98	16.58	28.03
an	5.56	3.98	5.29	0.31	5.70
di			2.87		1.02
hy	2.49	2.84	4.88	0.73	3.40
mt	1.39	1.16	3.01	0.22	1.79
il	0.46	0.46	0.91	0.10	0.76
ap		0.14	0.34	0.28	0.20
Total	99.10	99.39	99.48	99.29	99.67

Note: Explanation of column headings. (1) Charnockite, Madras, India (Subramaniam, 1959, p. 348, analysis 5). (2) Biotite rapakivi, Ahvenisto massif, Finland (Savolahti, 1956, p. 77, analysis 5). (3) Hornblende rapakivi, Ahvenisto massif, Finland (Savolahti, 1956, p. 77, analysis 10). (4) Potassic granite, Helsinki, Finland (Harme, 1965, p. 12). (5) Hornblende granite, Adirondacks, New York (Buddington, 1957, p. 293, analysis 1).

Table 52
TRACE ELEMENT CONTENT OF SOME COMMON ROCKS[122]

Sample	Concentration, ppm								
	U	Ba	Sr	Pb	Tl	Zn	Cd	K	Rb
Composite "granite"[a] > 70% SiO_2 (221)	4.2	—	132	28.7	—	—	—	3.70[b]	186
Composite "granodiorite"[a] 60 to 70% SiO_2 (191)	2.60	—	371	19.2	—	—	—	3.11	140
Composite "diorite"[a] < 60% SiO_2 (85)	1.63	—	635	14.6	—	—	—	2.41	93
Composite granite I[c] (27)	3.14	570	348	22.2	—	—	—	3.29	133
Composite granite II[d] (50)	2.3	685	283	—	—	—	—	3.04	125
Composite basalt[a] < 55% SiO_2 (282)	1.07	—	526	12.7	—	—	—	1.07	34
Composite basalt[d] (250)	1.07	334	461	6.59	—	—	—	0.95	29
Granulities[e]	0.5	—	—	—	—	—	—	2.5	(40)
High-calcium granite[f]	3.0	420	440	15	0.72	60	0.13	2.52	110
Oceanic tholeiites[g]	0.09	14	130	0.75	—	—	—	0.12	1.0
Basalt[f]	1.0	330	465	6	0.21	105	0.22	0.83	30
Granodiorite[h]	2.5	500	400	15	0.7	60	0.1	2.5	100
Diorite-granulite[i]	0.6	(100)	500	8	(0.1)	60	(0.1)	1.5	20
Abyssal basalt	0.1	15	130	0.75	(0.1)	100	0.2	0.12	1.0

[a] Composite of specimens analyzed in the Rock Analyses Laboratory, University of Minnesota.
[b] This value measured in percent.
[c] Composite of 27 Finnish Precambrian granites.
[d] Composite of basalts, analyzed by Turekian and Kulp (1956).
[e] Lambert and Heier (1966, 1967); rubidium content based on K/Rb ratios reported orally at American Geophysical Union meetings.
[f] Turekian and Wedepohl (1961).
[g] Engel et al. (1965); U and Pb content, Tatsumoto (1966).
[h] Parentheses indicate very uncertain or arbitrary estimates.
[i] Composite of 50 Precambrian granites from the western United States.

Table 53
ABUNDANCES OF THE RARE-EARTH ELEMENTS IN THE EARTH'S CRUST[134]

		According to A. E. Fersman		According to A. P. Vinogradov[a]		According to V. M. Goldschmidt	
Atomic number	Symbol of the element	Weight %	% of total	Weight %	% of total	g per ton	% of total
39	Y	$5 \cdot 10^{-3}$	33.70	$2.8 \cdot 10^{-3}$	17.78	31	20.66
57	La	$6.5 \cdot 10^{-4}$	4.38	$1.8 \cdot 10^{-3}$	11.43	19	12.65
58	Ce	$2.9 \cdot 10^{-3}$	19.55	$4.5 \cdot 10^{-3}$	28.59	44	29.33
59	Pr	$4.5 \cdot 10^{-4}$	3.03	$7 \cdot 10^{-4}$	4.44	5.6	3.72
60	Nd	$1.7 \cdot 10^{-3}$	11.45	$2.5 \cdot 10^{-3}$	15.87	24	15.99
61	Pm	?	—	?	—	?	—
62	Sm	$7 \cdot 10^{-4}$	4.71	$7 \cdot 10^{-4}$	4.44	6.5	4.33
63	Eu	$2 \cdot 10^{-5}$	0.13	$1.2 \cdot 10^{-4}$	0.76	1.0	0.66
64	Gd	$7.5 \cdot 10^{-4}$	5.05	$1 \cdot 10^{-4}$	6.35	6.3	4.19
65	Tb	$1 \cdot 10^{-4}$	0.67	$1.5 \cdot 10^{-4}$	0.95	1.0	
66	Dy	$7.5 \cdot 10^{-4}$	5.05	$4.5 \cdot 10^{-4}$	2.85	4.3	2.86
67	Ho	$1 \cdot 10^{-4}$	0.67	$1.3 \cdot 10^{-4}$	0.83	1.2	0.80
68	Er	$6.5 \cdot 10^{-4}$	4.38	$4 \cdot 10^{-4}$	2.55	2.4	1.60
69	Tu	$1 \cdot 10^{-4}$	0.67	$8 \cdot 10^{-5}$	0.51	0.3	0.26
70	Yb	$8 \cdot 10^{-4}$	5.41	$3 \cdot 10^{-4}$	1.91	2.6	1.73
71	Lu	$1.7 \cdot 10^{-4}$	1.14	$1 \cdot 10^{-4}$	0.64	0.7	0.46
Total		0.0146%	100%	0.0158%	100%	149.9	100%

[a] A. P. Vinogradov's data ignores the contents of R.E.E. in the hydrosphere and atmosphere.

Table 54
MEAN CONTENT OF R.E.E. IN THE MAIN TYPES OF ROCK, % (ACCORDING TO A. P. VINOGRADOV'S DATA)[124]

Element	Ultra-basic rocks (dunites, peridotites, pyroxenites)	Basic rocks (basalts, gabbro, norites, diabases and others)	Intermediate rocks (diorites and andesites)	Acid rocks, (granites, liparites, riolites and others)	Sedimentary rocks (clays and schists)	Earth's crust (2 parts of acid rocks + 1 part of basic rocks) %	Schists[a]
Y	$4.5 \cdot 10^{-4}$	$1.8 \cdot 10^{-3}$	$3 \cdot 10^{-3}$	$2 \cdot 10^{-3}$	$3.3 \cdot 10^{-3}$	$2 \cdot 10^{-3}$	$3.1 \cdot 10^{-3}$
La	—	$2 \cdot 10^{-3}$	$4 \cdot 10^{-3}$	$4.6 \cdot 10^{-3}$	$4 \cdot 10^{-3}$	$4 \cdot 10^{-3}$	$1.9 \cdot 10^{-3}$
Ce	—	10^{-3}	$3 \cdot 10^{-3}$	$6 \cdot 10^{-3}$	$3 \cdot 10^{-3}$	$4 \cdot 10^{-3}$	$4.4 \cdot 10^{-3}$
Pr	—	$1.3 \cdot 10^{-4}$	—	$1 \cdot 10^{-3}$	$5 \cdot 10^{-4}$	$7 \cdot 10^{-4}$	$5.6 \cdot 10^{-4}$
Nd	—	$(1 \cdot 10^{-3})$	$2 \cdot 10^{-3}$	$4 \cdot 10^{-3}$	$1.8 \cdot 10^{-3}$	$3 \cdot 10^{-3}$	$2.4 \cdot 10^{-3}$
Pm	—	?	—	?	?	?	—
Sm	—	$(1.5 \cdot 10^{-4})$	—	$6 \cdot 10^{-4}$	$5 \cdot 10^{-4}$	$(4 \cdot 10^{-4}$	$6.5 \cdot 10^{-4}$
Eu	—	—	—	$(1.7 \cdot 10^{-4})$	$(1 \cdot 10^{-4})$	(-10^{-4})	$1 \cdot 10^{-4}$
Gd	—	$(2 \cdot 10^{-4})$	—	$(1 \cdot 10^{-3})$	$5 \cdot 10^{-4}$	$(7 \cdot 10^{-4})$	$6.3 \cdot 10^{-4}$
Tb	—	—	—	$(2.5 \cdot 10^{-4})$	$9 \cdot 10^{-4}$	(10^{-4})	$1 \cdot 10^{-4}$
Dy	—	$(1.5 \cdot 10^{-4})$	—	$(5 \cdot 10^{-4})$	$4 \cdot 10^{-4}$	$(4 \cdot 10^{-4})$	$4.3 \cdot 10^{-4}$
Ho	—	—	—	—	$(1 \cdot 10^{-4})$	?	$1.2 \cdot 10^{-4}$
Er	—	$1 \cdot 10^{-4}$	—	$(2.5 \cdot 10^{-4})$	$2.5 \cdot 10^{-4}$	$(2 \cdot 10^{-4})$	$2.4 \cdot 10^{-4}$
Tu	—	—	—	$(2 \cdot 10^{-4})$	(10^{-5})	$(1 \cdot 10^{-4})$	$3 \cdot 10^{-5}$
Yb	—	$(1 \cdot 10^{-4})$	—	$(2 \cdot 10^{-4})$	$2.2 \cdot 10^{-4}$	$(2 \cdot 10^{-4})$	$2.6 \cdot 10^{-4}$
Lu	—	—	—	$(2 \cdot 10^{-4})$	$(2 \cdot 10^{-5})$	$(1 \cdot 10^{-4})$	$7 \cdot 10^{-5}$

[a] The data on the schists are taken from Minam's investigation.

Table 55
MINERALS OF THE NEPHELINIC SYENITES CONTAINING R.E.E.[124]

Mineral	Composition of mineral		
	$(R.E.E.)_2O_3$	CaO	SrO
Eudialite	1.7—2.6	7.11	up to 1.42
Loparite	31—33	3.18—5.46	2.0—3.4
Rinkite	22—23	23	—
Rinkolite	16—19	24—26	1.62—3.30
Lorchorrite	14—17	24—27	1.12—3.56
Nordite	19.25	4.46	7.40
Stenstrupine	25—38	1.85—4.55	—
Apatite	0.7—4.9	51—54	1.75—11.42
Belovite	24	5.23	33.60
Erikite	40—56	1.57—1.81	0.90
Britholite	60.54	11.28	—
Ankilite	46	1.52	21.03
Synchysite	52—55	12—17	—
Cordylite	49	1.91	—

Table 56
ANALYSES[a] OF GASES COLLECTED FROM FUMAROLES AT SHOWA-SHINZAN, USU VOLCANO, JAPAN, DOME OF HYPERSTHENE DACITE EXTRUDED IN 1944—1945[52]

Fumarole no.	Temp. (°C)	CO_2	CH_4	NH_3	H_2	HCl	HF	H_2S	SO_2	P_4O_{10}[b]	(%)	(%)
A-1(9081)	750	4700	5.8	4.3	1808	389	199	7.2	120	0.0016	0.723	99.25
A-6a	700	3620	8.3	0.4	1450	510	209	37	89	0.0019	0.592	99.39
C-2	645	3660	8.0	0.6	1210	490	200	30	91	0.0012	0.569	99.41
B-4a	464	7825	12.0	8.6	440	130	76	9.2	10	0.0019	0.859	99.10
A-4a	460	3550	3.2	1.1	811	569	306	15	10	0.0022	0.537	99.24
C-4	430	3230	6.2	—	400	669	230	45	89	0.0053	0.567	99.41
B-1	328	8485	14.2	0.7	660	140	62	100	13	0.0030	0.948	99.00
B-4b	300	7820	12.8	0.2	380	150	45	87	36	0.0011	0.853	99.07
A-6b	203	3900	Nil	0.5	Nil	748	83	52	70	0.0008	0.486	99.1
C-3(9063)	194	1970	4.1	0.26	351	120	11	110	13	—	0.258	99.72

Components of low volatility in fumarole A-1 (9081) but collected almost 5 years before the samples above.

Values in ppm

SiO_2	253	Cu	0.03
Al	15	Zn	0.5
Fe	1.3	As	0.7
Ca	4.6	Ag	0.003
Mg	32	Sn	0.03
Na	22	Pb	0.03
K	15		

[a] Results given in volume percent ($\times 10^4$) except for water. Data taken from D. E. White and Waring (1963).

[b] Reported as P in mg per 1000 liters; recalculated to P_4O_{10} gas in volume percent ($\times 10^4$).

Table 57
ANALYSES OF GASES FROM KILAUEA[327] GIVEN IN VOLUME PERCENT. ALL CHLORINE CALCULATED AS Cl_2.

	Sample					
	J8	J11	J13	J16	S3	S9
CO_2	47.68	20.93	16.96	18.03	33.48	8.32
CO	1.46	0.59	0.58	0.56	1.42	0.82
H_2	0.48	0.32	0.96	0.67	1.56	1.82
N_2	2.41	4.13	3.35	3.11	12.88	8.92
Ar	0.14	0.31	0.66	0.08	0.45	0.29
SO_2	11.15	11.42	7.91	8.53	29.83	16.80
S_2	0.04	0.25	0.09	0.15	1.79	2.48
SO_2	0.42	0.55	2.46	2.53		
Cl_2	0.04	0.00	0.10	0.08	0.17	1.01
H_2O	36.18	61.56	67.52	66.25	17.97	59.97

Table 58
CHEMICAL COMPOSITION OF AVERAGE ROCKS (AFTER CLARKE)[272]

Constituent	Igneous rock	Shale	Sandstone	Limestone	Sediment[a]
SiO_2	59.14	58.10	78.33	5.19	57.95
TiO_2	1.05	0.65	0.25	0.06	0.57
Al_2O_3	15.34	15.40	4.77	0.81	13.39
Fe_2O_3	3.08	4.02	1.07	0.54	3.47
FeO	3.80	2.45	0.30	—	2.08
MgO	3.49	2.44	1.16	7.89	2.65
CaO	5.08	3.11	5.50	42.57	5.89
Na_2O	3.84	1.30	0.45	0.05	1.13
K_2O	3.13	3.24	1.31	0.33	2.86
H_2O	1.15	5.00	1.63	0.77	3.23
P_2O_5	0.30	0.17	0.08	0.04	0.13
CO_2	0.10	2.63	5.03	41.54	5.38
SO_2	—	0.64	0.07	0.05	0.54
BaO	0.06	0.05	0.05	—	—
C	—	0.80	—	—	0.66
	99.56	100.00	100.00	99.84	99.93

[a] Shale 82%, sandstone 12%, limestone 6% (after Leith and Mead, 1915).

Table 59
AVERAGE CHEMICAL COMPOSITION OF SELECTED SEDIMENTARY ROCKS[385]

	Sandstones (253)[a] Clarke (1924) (%)	Sandstones[b] (from platforms) (3,700)[a] Vinogradov, Ronov (1956) (%)	Greywackes (61)[c] Pettijohn (1963) (%)	Shales (mainly from geosynclines) (277)[a] Clarke (1924), Goldschmidt (1933), Minami (1935), Shaw (1956) (%)	Shales (from platforms) (6,800)[a] Vinogradov, Ronov (1956) (%)	Tillites (68)[d] Goldschmidt (1933) (%)	Limestones (93)[a] Wedepohl (unpubl.) (%)	Carbonate rocks (from platforms) (1,500-8,300)[d], Vinogradov, Ronov (1956) (%)	Cherts (10)[d] Cressmann (1962) (%)	Pelagic clays (430)[e] Landergren (1964) (%)
SiO_2	78.7	70.0	66.7	58.9	50.7	58.9	6.9	8.2	89.9	54.9
TiO_2	0.25	0.58	0.6	0.78	0.78	0.79	0.05	n.d.	0.2	0.78
Al_2O_2	4.8	8.2	13.5	16.7	15.1	15.9	1.7	2.2	3.7	16.6
Fe_2O_2	1.1	2.5[a]	1.6	2.8	4.4[a]	3.3	0.98	1.0[a]	2.3	7.7
FeO	0.3	1.5[a]	3.5	3.7	2.1[a]	3.7	1.3	0.68[a]	n.d.	
MnO	0.03[a]	0.06[a]	0.1	0.09	0.08	0.10	0.08	0.07[a]	0.1	(2.0)
MgO	1.2	1.9	2.1	2.6	3.3	3.3	0.97	7.7	0.5	3.4
CaO	5.5	4.3	2.5	2.2	7.2	3.2	47.6	40.5	0.3	0.72
Na_2O	0.45	0.58	2.9	1.6	0.8	2.1	0.08	n.d.	0.7	1.3
K_2O	.3	2.1	2.0	3.6	3.5	3.9	0.57	n.d.	0.7	2.7
H_2O+	1.3	3.0	2.4	5.0	5.0	3.0	0.84	n.d.	1.2	(9.2)
P_2O_2	0.08	0.10[a]	0.2	0.16	0.10[a]	0.21	0.16	0.07[a]	0.9	(0.72)
C/CO_2	n.d./5.0	0.26[a]/3.9	0.1/1.2	0.6[a]/1.3	0.67[a]/6.1	n.d./0.6	38.3	0.23[a]/35.5	(0.3)/n.d.	computed carbonate free
S/SO_2	n.d./0.07	n.d./0.7	0.1/0.3	0.24[a]/n.d.	n.d./0.6	0.08/0.09	0.11/0.02	n.d./3.1	n.d.	n.d.

[a] Data are from literature compilations different from those listed at head of the column.
[b] Psammitic rocks in general.
[c] Similar to an average of 70 greywackes from the Harz mountains (Germany), the type locality.
[d] Number of samples used for average in parentheses.
[e] Weighted average for the three oceans, sea salt free samples (35 samples of sea salt containing Pacific clays, reported by Goldberg and Arrhenius, 1958, contain about 6 to 7% NaCl. This is a high value compared with 3—4% NaCl in Atlantic clays, analyzed by Behne, 1953.).

Table 60
CHEMICAL COMPOSITION OF SHALES; AVERAGE OF 277 SAMPLES FROM THE LITERATURE

CLARKE (1922), SHAW (1956), MINAMI (1935), GOLDSCHMIDT (1933)

	%
SiO_2	58.9
TiO_2	0.78
Al_2O_2	16.7
Fe_2O_2	2.8
FeO	3.7
MnO	0.09
MgO	2.6
CaO	2.2
Na_2O	1.6
K_2O	3.6
H_2O^+	5.0
P_2O_2	0.16
CO_2	1.3

Table 61
AVERAGE MINOR ELEMENT CONTENT ABUNDANCES IN PELITIC ROCK[386]

Li	66	Cu	45	Rh	0.000x	Sm	7.3	Ir	0.000x
Be	3	Zn	95	Pd	0.00x	Eu	1.6	Pt	0.00x
B	100	Ga	19	Ag	0.07	Gd	7.0	Au	0.005
F	740	Ge	1.6	Cd	0.8	Tb	1.2	Hg	0.4
P	700[a]	As	10	In	0.1	Dy	5.5	Tl	1.4
S	2400	Se	0.6	Sn	6.0	Ho	1.6	Pb	20
Cl	180[a]	Br	4	Sb	1.5	Er	3.9	Bi	0.1
Sc	13	Rb	140	J	2.2	Tm	0.6	Th	12
Ti	4600	Sr	300	Cs	5.5	Yb	3.7	U	3.7
V	130	Y	41	Ba	580	Lu	0.7		
Cr	90	Zr	160	La	40	Hf	2.8		
Mn	850	Nb	18	Ce	95	Ta	2.0		
Co	19	Mo	2.6	Pr	9.7	W	1.8		
Ni	68	Ru	0.000x	Nd	59	Os	0.000x		

[a] Insoluble by washing (~ 6000 ppm soluble Cl).

Table 62
AVERAGE MINERAL COMPOSITION OF SHALES ON THE BASIS OF > 70,000 X-RAY ANALYSES OF AMERICAN AND EUROPEAN SHALES[386]

	%
Illite (and expanded clay minerals)	45—55
Quartz	20
Feldspar (Plagioclase > K-feldspar)	10—15
Chlorite, Kaolinite	14
Calcite, Dolomite	3

Table 63
AVERAGE MINERAL COMPOSITION OF SELECTED SEDIMENTARY ROCK TYPES, IN PERCENTS

	Sandstones Huckenholz (1963a)	Greywackes Huckenholz (1963a)	Shales[a] Wedepohl (1967)	Limestones[b]
Quartz	82	37	20	
Plagioclase Potassium feldspar	5	28	10—15	
Muscovite (M)/illite (I)	8 (M)		45—55 (I)	14
Chlorite (C) Kaolinite		29 (C)	14	
Calcite Dolomite	3	3	3	84
Accessories	2	3	1	2

Note: The important mineral of a group is indicated in parentheses abbreviated.

[a] Data on the clay minerals are mainly averages of Weaver's (1967) 70,000 X-ray analyses of American shales, which are comparable with the author's averages of European Paleozoic and Mesozoic shales. The quartz content is only from European shales, which contain on the average more plagioclase than K-feldspar. The figure for illite includes expanded clay minerals.

[b] By computing the mineral constituents from the analysis of 345 limestones which Clarke (1924) has published (7.5% silicates, 36% dolomite), one realises that a large number of dolomites has been included in the composite of American limestones. The average carbonate rocks from the Russian platform contain about 12% silicates and 35% dolomite. It seems to be reasonable to assume that between a quarter and a third of all carbonate rocks consists of dolomites.

Table 64
SUMMARY OF METAMORPHIC FACIES[396]

I. Shallow Contact Metamorphism

A. Very low fluid pressures of generally less than 1500 bars; sequence of facies in response to rising temperature

(1) Albite—epidote—hornfels facies
(2) Hornblende—hornfels facies
(3) K—feldspar—cordierite—hornfels facies (formerly called pyroxene—hornfels facies)
(4) Sanidinite facies

II. Regional Dynamothermal Metamorphism

A. Low and intermediate pressures[a] of about 2000 to 5000—6000 bars, $P_f \simeq P_s$; sequence of facies in response to rising temperature

(1) Greenschist facies
(2) Cordierite—amphibolite facies

B. High and very high pressures in excess of 5000—6000 bars, $P_f \simeq P_s$; sequence of facies in response to rising temperature

(1) Greenschist facies (in some cases glaucophanitic greenschist facies)
(2) Almandine—amphibolite facies
(3) Special facies at high temperatures and $P_{H_2O} << P_s$: Granulite facies

III. Burial Metamorphism

A. Low temperatures; intermediate pressures, $P_f \simeq P_s$; sequence of facies in response to slightly rising temperature

(1) Laumontite—prehnite—quartz facies (formerly called zeolitic facies)
(2) Pumpellyite—prehnite—quartz facies (formerly called prehnite—pumpellyite— metagreywacke facies)

B. Low temperatures; high and very high pressures, $P_f \simeq P_s$; sequence of facies in response to rising pressure

(1) Lawsonite—albite facies
(2) Lawsonite—glaucophane facies (formerly called glaucophane—schist facies) As the temperature increases, there is a transition to the glaucophanitic greenschist facies.

[a] These conditions may also be realized at localized deep-seated contact metamorphism as contrasted to shallow contact metamorphism.

Table 65
THE MINERALOGY OF THE DIFFERENT FACIES[221]

Facies	Si	Al, Si	K, Al, Si	Na, Al, Si	Ca, Al, Si	Ca, Si, (CO_2)	Si, Fe, (Mg), (CO_2)	Si, Mg, (Fe), (CO_2)	Si, Mg, Ca, (CO_2)	Ca, Al, Mg, Si	Mg, (Fe), Al, Si	Fe, (Mg), Al, Si	K, Mg, Fe, Al, Si
Greenschist	Quartz		Muscovite Microcline	Albite	Zoisite	Quartz + calcite	Biderite + quartz	Magnesite + quartz Talc	Dolomite + quartz Talc + calcite Tremolite		Chlorite	Chloritoid	Muscovite + chlorite
Epidote-amphibolite	Quartz	Kyanite	Muscovite Microcline	Albite	Zoisite	Quartz + calcite	Cummingtonite	Talc Serpentine Anthophyllite	Tremolite	Blue-green hornblende	Chlorite	Chloritoid Almandite	Biotite
Amphibolite	Quartz	Kyanite Billinman-ite	Muscovite Microcline		Zoisite Plagioclase Grossularite	Wollastonite Quartz + calcite	Cummingtonite	Authophyllite Forsterite	Tremolite Diopside	Green hornblende	Cordierite	Almandite Staurolite	Biotite
Granulite	Quartz	Kyanite Sillimanite	Orthoclase		Plagioclase	Wollastonite Quartz + calcite	Hypersthene	Enstatite Forsterite	Diopside	Augite	Pyrope-almandite		Orthoclase + pyroxene
Pyroxene hornfels	Quartz	Andalusite Sillimanite	Orthoclase		Grossularite Plagioclase	Wollastonite	Hypersthene	Enstatite Forstarite	Dioside	Brown hornblende Augite	Cordierite Pyrope-almandite		Biotite
Sanidinite	Tridym-ite	Mullite	Sanidine		Plagioclase	Wollastonite (Pseudowollaston-ite) Larnite Rankinite	Clinohyper-sthene	Forsterite Pigeonite Clinoenatatite	Diopside Melilite Merwinite	Augite	Cordierite		Orthoclase + pyrox-ene

Table 66
AVERAGE COMPOSITIONS OF PELITIC ROCKS OF THE WORLD[239]

	Clays, shales, and slates (85 analyses)		Phyllites, schists, and gneisses (70 analyses)	
	Average	Standard deviation	Average	Standard deviation
SiO_2	59.93	6.33	63.51	8.94
TiO_2	0.85	0.57	0.79	0.67
Al_2O_3	16.62	3.33	17.35	5.08
Fe_2O_3	3.03	2.08	2.00	1.66
FeO	3.18	1.84	4.71	2.44
MgO	2.63	1.98	2.31	1.82
CaO	2.18	2.54	1.24	0.92
Na_2O	1.73	1.27	1.96	1.06
K_2O	3.54	1.33	3.35	1.31
H_2O	4.34	2.38	2.42	1.53
CO_2	2.31[a]	2.60	0.22[b]	0.22

[a] Determined in only 43 analyses.
[b] Determined in only 19 analyses.

Table 67
AVERAGE CHEMICAL COMPOSITIONS OF METAMORPHIC ROCKS (ANHYDROUS BASIS)[239]

	1 Phyllites	2 Mica schists	3 Two-mica gneisses	4 Quartzo-feldspathic gneisses	5 Amphibolites
SiO_2	60.0	64.3	67.7	70.7	50.3
TiO_2	1.1	1.0	—	0.5	1.6
Al_2O_3	20.7	17.5	16.6	14.5	15.7
Fe_2O_3	3.0	2.1	1.9	1.6	3.6
FeO	4.8	4.6	3.4	2.0	7.8
MnO	0.1	0.1	—	0.1	0.2
MgO	2.9	2.7	1.8	1.2	7.0
CaO	1.2	1.9	2.0	2.2	9.5
Na_2O	2.0	1.9	3.1	3.2	2.9
K_2O	4.0	3.7	3.5	3.8	1.1
P_2O_5	0.2	0.2	—	0.2	0.3
	100.0	100.0	100.0	100.0	100.0

Note: 1, 2, 4, 5 after Poldervaart (1955); 3 after Lapadu-Hargues (1945).

Table 68
CONVERSION OF IGNEOUS TO SEDIMENTARY ROCKS (ANALYSES IN GRAMS)[120]

	1 Average igneous rock (after Brotzen, 1966) (gm/kg)	2 Average limestone (assuming 70% of CaO of igneous rock)	3 Remainder	4 Average shale (assuming 97% of K_2O in igneous remainder)	5 Remainder	6 Average sandstone	7 Remainder, found in ocean, pore waters, evaporites
SiO_2	635	4	631	535	96	100% → 96	0
Al_2O_3	159	1	158	146	12	9	3
Fe_2O_3	29	—	29	36	−7	2	−9
FeO	33	—	33	26	7	2	5
MgO	29	6	23	21	2	1	1
CaO	49	70% → 34	15	13	2	4	−2
Na_2O	33	—	33	9	24	1	23
K_2O	33	—	33	97% → 32	1	2	−1
CO_2	—	34	−34	13	−47	3	−50
H_2O	—	—	—	34	−34	2	−36
Total	1,000	79		865		122	

Note: HC1 required to convert oxides to chlorides, 27g; CO_2 required for carbonate minerals, 50g.

Table 69
CHEMICAL COMPOSITION OF SILT AND LOESS[274]

Constituent	Loess near Galena, Ill.[a]	Loess Vicksburg, Miss.[b]	Loess Kansas City, Mo.[c]	Loess Kansu, China[d]	Summer silt Leppakosi, Finland[e]	Siltstone Iron River, Mich.[f]
SiO_2	64.61	60.69	74.46	59.30	59.20	59.19
TiO_2	0.40	0.52	0.14	0.60	1.20	1.45
Al_2O_3	10.64	7.95	12.26	11.45	16.14	14.61
Fe_2O_3	2.61	2.61	3.25	2.32	4.36	1.51
FeO	0.51	0.67	0.12	1.55	3.24	11.28
MnO	0.05	0.12	0.02	—	0.09	0.10
MgO	3.69	4.56	1.12	2.29	3.14	2.94
CaO	5.41	8.96	1.69	9.78	2.52	0.09
Na_2O	1.35	1.17	1.43	1.80	3.82	0.12
K_2O	2.06	1.08	1.83	2.17	1.97	2.38
H_2O^+	2.05 (combined H_2O^+ and H_2O^-)	1.14 (combined)	2.70 (combined)	0.96 (combined)	1.16	4.69
H_2O^-					1.15	0.07
P_2O_5	0.06	0.13	0.09	0.20	0.17	0.01
CO_2	6.31	9.63	0.49	7.41	—	1.25
SO_3	0.11	0.12	0.06	—	—	0.08*
C(organic)	0.13	0.19	0.12	—	1.94	0.25
Cl	0.07	0.08	0.05	—	—	—
Total	100.06	99.62	99.83	99.83	100.10	100.02

* Sulfide sulfur.

[a-c] R. B. Riggs analyst (Clarke, 1924, p. 514).
[d] Barbour, 1927, p. 283.
[e] Late-glacial varved sediment; L. Lokka analyst (Eskola, 1932).
[f] Dunn Creek slate (Precambrian); drill core, Homer Mine; an iron-rich siltstone. C. Warshaw, analyst (James et al., 1968, Table 4).

Table 70
CHEMICAL COMPARISON OF SEA WATER AND FRESH WATER[386]

Element	Seawater (S) ppm	Author	Freshwater (F) ppm	Author	S/F
Cl	18980		6.7	Li, C	2.8×10^3
Na	10560		6.0	Li, C	1.8×10^3
SO_4^{2-}	2650		11.6	Li,C	2.4×10^2
Mg	1270		3.8	Li,C	3.4×10^2
Ca	400		17.5	Li,C	2.3×10
K	380		2.2	Li,C	1.8×10
HCO_3^-	140		47	Li,C	3
Br	65	G	0.014	H.u.K,Be,K	7×10^3
Sr	8	G	0.057	Su,Du,O	1.4×10^2
B	4.8	G	0.020	H.u.T,Du,K	2.4×10^2
Si	3	G	17	Da	1.8×10^{-1}
F	1.3	G	0.20	H.u.K,Su,F.u.R.	6.5
PO_4^{3-}	0.21	G	0.06(?)	Si,Li	3.5(?)
Li	0.2	G	0.0016	Du, Y An	1.3×10^2
Rb	0.12	B	0.0014	Y,Du,Bo,Kr	8.6×10
I	0.05	B	0.002	H,Su	2.5×10
Mo	0.013	Is,G	0. 001	Du,Su,Kh,Kr	2.6×10
Ba	0.013	B	0.04	Du,L	3.2×10^{-1}
Zn	0.010	G	0.010	Si,Su,K,Hu,Kh,Li	1
Fe	0.007	F,G	0.23	Si,Su,Du,Kh	3×10^{-2}
Ni	0.005	Sch	0.004	(T)Du,N.Kh	1.2
Mn	0.003	F,G	0.009	Si,Du,N,T,Kh	3×10^{-1}
As	0.003		G 0.002	Si,Su	1.5
U	0.003	G	≤0.001	Si,R,A,Fi,I,Ko,H.u.P	≥3
Cu	0.002	F,G	0.007	Z,Si,Su,N,H,Du,T,K,Kh	3×10^{-1}
V	0.002	G	0.001	N,Su,Kh	2
Ti	0.001	G	0.005(?)	T,Du,Kh	2×10^{-1}(?)
Al	0.001	S	0.30	Su,Du,Kh	2.8×10^{-3}
Ni	0.005	Sch	0.004	(T)Du,N,Kh	1.2
Mn	0.003	F,G	0.009	Si,Du,N,T,Kh	3×10^{-1}
As	0.003	G	0.002	Si,Su	1.5
U	0.003	G	≤ 0.001	Si,R,A,Fi,I,Ko,H.u.P	≥3
Cu	0.002	F,G	0.007	Z,Si,Su,N,H,Du,T,K,Kh	3×10^{-1}
V	0.002	G	0.001	N,Su,Kh	2
Ti	0.001	G	0.005(?)	T,Du,Kh	2×10^{-1}(?)
Al	0.001	S	0.30	Su,Du,Kh	2.8×10^{-3}

Table 71
AVERAGE ABUNDANCES OF ELEMENTS IN THE EARTH'S CRUST, IN THREE COMMON ROCKS, AND IN SEAWATER (ppm)

Element	Crust[347]	Granite[347]	Basalt[347]	Shale[373]	Seawater[133]
O	46.4×10^4				857,000
Si	28.2×10^4	32.3×10^4	24.0×10^4	23.8×10^4	3.0
Al	8.2×10^4	7.7×10^4	8.8×10^4	8.0×10^4[361]	0.01
Fe	5.6×10^4	2.7×10^4	8.6×10^4	4.7×10^4[361]	0.01
Ca	4.1×10^4	1.6×10^4	6.7×10^4	2.5×10^4	400
Na	2.4×10^4	2.8×10^4	1.9×10^4	0.66×10^4	10,500
Mg	2.3×10^4	0.16×10^4	4.5×10^4	1.34×10^4	1,350
K	2.1×10^4	3.3×10^4	0.83×10^4	2.3×10^4	380
Ti	5,700	2,300	9,000	4,500	0.001
H	1,400				108,000
P	1,050	700	1,400	770	0.07
Mn	950	400	1,500	850[361]	0.002
F	625	850	400	500	1.3
Ba	425	600	250	580[361]	0.03
Sr	375	285	465	450	8.0
S	260	270	250	220	885
C	200	300	100	1,000	28
Zr	165	180	150	200	
V	135	20	250	130	0.002
Cl	130	200	60	160	19,000
Cr	100	4	200	100	0.00005
Rb	90	150	30	140[361]	0.12
Ni	75	0.5	150	95	0.002
Zn	70	40	100	80	0.01
Ce	67[361]	87[361]	48[361]	50	5.2×10^{-6}
Cu	55	10	100	57	0.003
Y	33	40	25	30	0.0003
Nd	28	35[361]	20	23	9.2×10^{-6}
La	25	40	10	40	1.2×10^{-5}
Co	25	1	48	20	0.0001
Sc	22	5	38	10	0.00004
Li	20	30	10	60	0.17
N	20	20	20	60	0.5
Nb	20	20	20	20	0.00001
Ga	15	18	12	19[361]	0.00003
Pb	12.5	20	5	20	0.00003
B	10	15	5	100	4.6
Th	9.6	17	2.2	11	0.00005
Sm	7.3[361]	9.4[361]	5.3[361]	5.5	1.7×10^{-6}
Gd	7.3[361]	9.4[361]	5.3[361]	6.5	2.4×10^{-6}
Pr	6.5[361]	8.3[361]	4.6[361]	5	2.6×10^{-6}
Dy	5.2[361]	6.7[361]	3.8[361]	4.5	2.9×10^{-6}
Yb	3	3.8[361]	2.1[361]	3	2.0×10^{-6}
Hf	3	4	2	6	
Cs	3	5	1	5[361]	0.0005
Be	2.8	5	0.5	3	6×10^{-7}
Er	2.8	3.8[361]	2.1[361]	2.5	2.4×10^{-6}
U	2.7	4.8	0.6	3.2	0.003
Br	2.5	1.3	3.6	6	65
Sn	2	3	1	6[361]	0.0008
As	1.8	1.5	2	6.6	0.003
Ge	1.5	1.5	1.5	2	0.00006
Mo	1.5	2	1	2	0.01
W	1.5	2	1	2	0.0001
Ho	1.5[361]	1.9[361]	1.1[361]	1	8.8×10^{-7}

Table 71 (continued)
AVERAGE ABUNDANCES OF ELEMENTS IN THE EARTH'S CRUST, IN THREE COMMON ROCKS, AND IN SEAWATER (ppm)

Element	Crust[347]	Granite[347]	Basalt[347]	Shale[373]	Seawater[133]
Eu	1.2	1.5[361]	0.8[361]	1	4.6×10^{-7}
Tb	1.1[361]	1.5[361]	0.8[361]	0.9	
Lu	0.8[361]	1.1[361]	0.6[361]	0.7	4.8×10^{-7}
Tm	0.25[361]	0.3[361]	0.2[361]	0.25	5.2×10^{-7}
I	0.5	0.5	0.5	1	0.06
Tl	0.45	0.75	0.1	1	<0.00001
Cd	0.2	0.2	0.2	0.3	0.00011
Sb	0.2	0.2	0.2	1.5[361]	0.0005
Bi	0.17	0.18	0.15	0.01	0.00002
In	0.1	0.1	0.1	0.05	<0.02
Hg	0.08	0.08	0.08	0.4	0.00003
Ag	0.07	0.04	0.1	0.1	0.00004
Se	0.05	0.05	0.05	0.6	0.0004

Note: Au, Pt metals, Re, and Te are less than 0.05 ppm in rocks and less than 0.00001 ppm in seawater. Concentrations of inert gases in seawater: He, 5×10^{-6} ppm; Ne, 0.0001 ppm; Ar, 0.6 ppm; Kr, 0.0003 ppm; Xe, 0.0001 ppm.

The heading "crust" means the continental crust only, and this part of the crust is assumed to be made up of roughly equal parts of basalt and granite. "Shale" includes recent clays as well as shales, but not the fine-grained sediments of the deep sea. "Seawater" is normal surface water with a chlorinity of 19‰.

Table 72
AVERAGE CONTENT OF ELEMENTS IN ALL DEEP-SEA SEDIMENTS[191]

	%		%
Si	23.0	Ni	0.032
Al	9.2	Cu	0.074
Fe	6.5	Cr	0.0093
Ti	0.73	V	0.045
Mg	2.1	Pb	0.015
Ca	2.9	Mo	0.0045
Na	4.0	Zr	0.018
K	2.5	Yb	0.0021
Sr	0.071	Y	0.015
Ba	0.39	La	0.014
B	0.03	Sc	0.025
Mn	1.25	Co	0.016
Ga	0.0019		

Table 73
AVERAGE CHEMICAL COMPOSITION OF MAIN TYPES OF DEEP-SEA SEDIMENTS[191]

	Red clay %	Radiolarian ooze %	Diatom ooze %	Globigerina ooze %	Picropod ooze %
SiO_2	62.10	56.00	67.92	31.71	3.65
Al_2O_3	16.06	10.52	0.55	11.10	0.80
Fe_2O_3	11.83	14.99	0.39	7.03	3.06
MnO_2	0.55	3.23		tr	
CaO	0.28	0.39			
MgO	0.50	0.25			
$Ca_3P_2O_5$	0.19	1.39	0.41	2.80	2.44

Table 74
CONTENT OF TRACE ELEMENTS IN DEEP-SEA SEDIMENTS[191] (ppm)

	Hemipelagic sediments	Globigerina ooze	Radiolarian ooze	Mn nodules	Diatom ooze	Brown clay
Li				55		78
Rb						391
Ca						13
Sr						60
Ba		180	180			200
B	16—155	155				
Sc		4.6	3	3		4
Zr						140
Mn						1,770
Ni						253
Cu				3,000		160
Y	0	8	0	8		8
Ag				0.2		
Ga_2O_3		50	100	50	50	1,000
GeO_2						50
Se				19		
Cd		4	4	51—84	4	4

Table 75
CHEMICAL ANALYSIS OF MARINE STANDARD NODULE GRLD—126[113]

	Acid-soluble Fraction of Sample								Total Sample					
Element	1	2	3	4	5	6	7	8	9	10	11	12	13	14
Fe (wt. %)	8.66 (0.12)	9.71 (0.30)	10.8	9.94	9.5	10.4	9.06	9.55 (0.15)	9.77 (0.23)	10.6	10.37 (0.44)	10.32	10.6 (0.5)	9.4
Mn (wt. %)	23.6 (1.5)	22.88 (0.34)	23.7	23.68	22.9	25.1	23.5	22.65 (0.23)	24.7 (0.7)	25.2	23.09 (0.22)	22.44	25.5 (1.0)	24.4
Cu (wt. %)	0.592 (0.018)	0.641 (0.009)	0.59	0.626	0.65	0.62	0.67	0.609 (0.014)	0.630 (0.060)	0.669	0.670 (0.030)	0.605	1.02 (0.25)	0.66
Ni (wt. %)	0.760 (0.011)	0.941 (0.019)	0.85	0.993	0.92	1.03	0.59	0.929 (0.087)	1.00 (0.04)	1.05	0.988 (0.070)	1.09	1.25 (0.25)	1.01
Co (ppm)	1350 (29)	1487 (30)	1400	1550	9300	1130	1800	1524 (43)	1380 (50)	1600	1437 (26)	2408	NA	1578
Zn (ppm)	1217 (28)	1274 (15)	1330	NA	1415	NA	1630	1220 (16)	3700 (400)	NA	1276 (22)	NA	1400 (500)	NA
Ca (wt.%)	1.38 (0.02)	1.36 (0.01)	1.3	1.43	1.36	NA	NA	1.28 (0.02)	NA	1.58	1.29 (0.001)	1.46	1.42 (0.11)	NA
Mg (wt. %)	1.33 (0.02)	1.41 (0.03)	1.3	1.53	1.67	NA	NA	1.53 (0.02)	NA	1.80	1.51 (0.024)	0.87	0.72	NA
Na (wt. %)	2.81 (0.11)	1.71 (0.06)	1.9	1.94	1.66	NA	NA	NA	1.84 (0.06)	2.15	NA	1.93	NA	NA
K (wt. %)	0.82 (0.01)	0.59 (0.02)	0.9	0.99	0.76	NA	NA	NA	NA	1.04	NA	0.93	1.55 (0.11)	NA
Si (wt. %)	NA	NA	NA	NA	NA	NA	NA	NA	NA	NA	NA	NA	9.78 (0.97)	NA
Al (wt. %)	NA	NA	NA	NA	NA	NA	NA	NA	NA	NA	NA	NA	2.50 (0.35)	2.5
Ti (wt. %)	NA	NA	NA	NA	NA	NA	NA	NA	NA	NA	NA	NA	0.32 (0.08)	NA

Note: Values in parentheses are standard deviations; NA = not analyzed.

1 —Michigan-1973 (Callender/Roberts), 10% HCl-30% H_2O_2 soluble; atomic absorption.
2 —Michigan-1975 (Callender/Shedlock); 10% HCl-30% H_2O_2 soluble; atomic absorption.
3 —Miami (Joensuu); spectrograph.
4 —Kennecott; acid soluble; atomic absorption.
5 —Syracuse (Dean); concentrated HCl soluble; atomic absorption.
6 —Wisconsin-1973 (Morgan); concentrated HCl-$HClO_4$ soluble; atomic absorption.
7 —Washington (Piper); acid soluble; neutron activation/atomic absorption.
8 —Wisconsin-1975 (Bowser); 10% HCl-30% H_2O_2, 25% hydroxylamine-hydrochloride dithionate-citrate; atomic absorption.
9 —Battelle NW (Rancitelli); total sample; neutron activation.
10 —Kennecott; HF, HNO_3, HCl, $HClO_4$ digestion; atomic absorption.
11 —Wisconsin-1975 (Bowser); HF-boric acid digestion; atomic absorption.
12 —Hawaii (Fein); Li-tetraborate fusion, HF-HNO_3 digestion; atomic absorption/X-ray fluorescence.
13 —Hawaii (Dugolinsky); total sample; X-ray fluorescence.
14 —Hawaii (Frank); HF-HNO_3, $HClO_4$ digestion; atomic absorption.

Table 76
MINERALOGY OF HAWAIIAN MARINE MANGANESE DEPOSITS BASED ON X-RAY DIFFRACTION ANALYSES[113]

Ferromanganese Crust

Dominant — X-ray amorphous Fe, Mn oxides and hydroxides
Common — δ-MnO_2, substrate minerals
Rare — Todorokite

Substrate

Dominant — Plagioclase feldspar, pyroxene
Common — Calcite, phillipsite, montmorillonite, quartz, volcanic glass (?), olivine, illite
Rare — Aragonite, goethite, maghemite and/or magnetite, kaolinite, ilmenite (?), hematite (?)

Table 77
AVERAGE ABUNDANCES OF ELEMENTS IN MANGANESE NODULES AND OTHER FERROMANGANESE OXIDE DEPOSITS FROM THE WORLD OCEAN (wt. %)[70]

Elements	World average	Elements	World average
B	0.0277	Sr	0.0825
Na	1.9409	Y	0.031
Mg	1.8234	Zr	0.0648
Al	3.0981	Mo	0.0412
Si	8.624	Pd	$0.553 \cdot 10^{-4}$
P	0.2244	Ag	0.0006
K	0.6427	Cd	0.00079
Ca	2.5348	Sn	0.00027
Sc	0.00097	Te	0.0050
Ti	0.6424	Ba	0.2012
V	0.0558	La	0.016
Cr	0.0014	Yb	0.0031
Mn	16.174	W	0.006
Fe	15.608	Ir	$0.935 \cdot 10^{-4}$
Co	0.2987	Au	$0.248 \cdot 10^{-4}$
Ni	0.4888	Hg	$0.50 \cdot 10^{-4}$
Cu	0.2561	Tl	0.0129
Zn	0.0710	Pb	0.0867
Ga	0.001	Bi	0.0008

Table 78
AVERAGE ABUNDANCES OF Mn, Fe, Ni, Co AND Cu IN MANGANESE NODULES AND ENCRUSTATIONS FROM DIFFERENT ENVIRONMENTS (wt. %)[70]

	Sea-mounts	Plateaus	Active ridges	Other ridges	Continental borderlands	Marginal seamounts and banks	Abyssal nodules
Mn	14.62	17.17	15.51	19.74	38.69	15.65	16.78
Fe	15.81	11.81	19.15	20.08	1.34	19.32	17.27
Ni	0.351	0.641	0.306	0.336	0.121	0.296	0.540
Co	1.15	0.347	0.400	0.570	0.011	0.419	0.256
Cu	0.058	0.087	0.081	0.052	0.082	0.078	0.370
Mn/Fe	0.92	1.53	0.80	0.98	28.8	0.81	0.97
Depth (m)	1,872	945	2,870	1,678	3,547	1,694	4,460

Table 79
CHEMICAL[a] COMPOSITION OF FERROMANGANESE CONCRETIONS FROM SHALLOW MARINE ENVIRONMENTS[47]

	1	2	3	4	5	6	7	8
Si	6.30	7.66	9.15	10.90	5.56	3.55	6.41	6.32
Al	2.02	1.54	1.82	1.89	1.65	0.64	1.71	2.28
Ti		0.13	0.28	0.29	0.10	0.10	0.11	0.21
Fe[b]	14.67	22.47	19.68	22.78	26.54	18.20	6.16	3.92
Ca	1.48	1.22	1.70	1.36	4.45	10.26	1.32	5.57
Mg	0.72	0.58	0.21	0.43	1.04	0.90	1.82	1.87
K	0.07	0.75	1.43	1.45		0.41	0.99	1.03
Na	0.09	0.35	1.05	0.78		0.19	0.95	
P	1.46	0.69	1.24	0.69	1.14	0.84	0.34	0.35
Mn	28.20	14.03	13.54	9.90	6.79	14.10	32.76	30.19
S		0.08					0.07	
C		2.50	1.29	1.06	0.67			
CO_2	5.52	0.76	2.73	2.39	5.50	11.20	0.56	11.88
As					687			245
Ba		2500					2857	3090
Co		160	96	64	84	30	157	230
Cr		10	17	23	16	100	7	
Cu		48	9	17	37	10	67	17
Mo		130			18	30	231	55
Ni		750	35	47	281	100	314	77
Pb		38	9	24			nd	43
Rb								40
Sr								770
U		10						
V		150	68	98	186	10	157	
Y								28
Zn		80	113	135			23	60
Zr					42			55

Note: (1) Barents Sea.[308] Single analysis of a flat concretion. (2) Baltic Sea.[215] Composite analysis from the circum-Gotland region. (3) Gulf of Finland.[368] Mean of 9 analyses. (4) Gulf of Riga.[368] Mean of 19 analyses. (5) Black Sea.[322] Mean of: 15 analyses for Fe, Ti, Mn, P, Ni, Co, Cu, Mo, V, and Cr; 8 analyses for Si, Al, Ca, Mg, C, and CO_2; and 4 analyses for Zr. (6) Black Sea.[123] Single analysis of a nodule containing the least amount of shell material. (7) Jervis Inlet, British Columbia.[144] Mean of 2 analyses calculated on a total sample basis, except for S, Co, Cu, Pb, Ni, V, and Zn, which are for HCl-soluble fractions only. (8) Loch Fyne, Scotland.[46] Mean of 2 analyses of composite nodule samples.

[a] Major elements as wt. %, minor elements as ppm. [b] Total Fe.

Table 80
MAJOR AND MINOR ELEMENT COMPOSITION OF FERROMANGANESE CONCRETIONS AND ASSOCIATED SEDIMENTS[47]

	Black Sea[322]		Loch Fyne[45]	
	Concretions	Sediment	Concretions	Sediment
Si	5.56	17.83	6.32	22.60
Al	1.65	6.18	2.28	8.96
Ti	0.10	0.32	0.21	0.55
Fe	26.54	5.11	3.92	4.73
Ca	4.45	8.86	5.57	3.95
Mg	1.04	1.17	1.87	2.29
K			1.03	2.99
P	1.14	0.16	0.35	0.08
Mn	6.79	0.35	30.19	0.22
CO_2	5.50	9.62	11.88	4.02
As	687[321]	35[321]	245	15
Ba			3,090	590
Co	84	14	120	
Cr	16	46		
Cu	37	30	17	22
Mo	18	2	55	5
Ni	281	40	77	70
Pb			43	30
Rb			40	160
Sr			770	250
V	186	93		
Y			28	35
Zn			60	125
Zr	42	107	55	160

Table 81

YTTRIUM[a] AND RARE EARTH CONTENTS OF FERROMANGANESE CONCENTRATIONS AND ASSOCIATED SEDIMENTS FROM SHALLOW MARINE ENVIRONMENTS[47]

	1	2	3	4
La	6.2	17.0	18.2	5.85
Ce	12.0	28.2	29.7	11.36
Pr	2.5	6.7	7.0	2.62
Nd	8.2	12.5	12.0	6.01
Sm	2.5	2.7	3.7	0.96
Gd	2.5	2.7	3.7	0.89
Eu				0.19
Tb	14.7	12.7	11.0	0.24
Y				
Cy	0.5	0.5	0.4	0.94
Ho				0.08
Er	0.7	0.8	0.7	0.30
Yb	0.6	0.8	0.7	
Ce/La	1.93	1.66	1.63	1.94

Note: (1) Black Sea concretions, mean of 4 samples.[109] (2) Black Sea surface oxidized sediment, mean of 4 samples.[109] (3)Black Sea subsurface reduced sediment, mean of 4 samples.[109] (4) Loch Fyne concretions.[127]

[a] All values in ppm.

Table 82
CHEMICAL[a] COMPOSITION OF LACUSTRINE FERROMANGANESE CONCRETIONS[47]

	1	2	3	4	5	6	7	8	9	10	11	12
Si	7.85	6.86										4.89
Al	1.32	1.11										1.15
Ti	0.09	0.08	0.30									0.44
Fe[b]	35.63	38.15	15.14	22.97	19.95	20.76	17.2	16.0	40.0	20.0	33.52	23.33
Ca	1.21	0.37		1.18		1.17				1.3		0.97
Mg	0.45	0.23		0.35		0.25				0.9		0.06
Na	0.08		0.64	0.49		0.03				0.08		0.33
K	0.17	0.06	1.62	0.63		0.14				0.24		0.33
P	0.29	0.22	0.28									0.001
Mn	4.73	2.86	7.25	6.94	8.75	9.15	33.8	27.0	15.7	20.5	3.57	21.94
S	0.03	0.03	0.26									
C	1.4	1.5	0.76	1.05								1.07
CO_2	0.28			1.57								2.39
As					136	519						
Ba	1,000		2,912		8,115	10,326						10,300
Cd	10				4							
Ce					161							
Co	80	130	34	198	69	116	198	222	135	305	220	110
Cr	10	10	34			24						2
Cu	40		12	55	9	26	12	8	10	90	1,314	4
La			29		51							
Mo	30	50	10			36						
Ni	40	40	26	240		239	272	136	95	725	702	12
Pb	27		2,551				25	21	24			4
Rb			72									
Sr	300		59			148						
V	10	10	58	1	1							
Y			34									T
Zn	50		1,112	263	205	324	1,633	511	250	460	1,177	181
Zr			69									

Note: 1: Swedish lakes, estimated average composition.[215] 2: Karelian — Finnish lakes, estimated average composition.[215] 3: Oxidate crusts from lakes Windermere and Ullswater.[139] Mean of 8 analyses, except for Cu(6) and Mo(4). 4: Green Bay, Lake Michigan.[303] Mean of 23 analyses of HNO_3—H_2O_2 extracts. 5: Green Bay, Lake Michigan.[90] Mean of 6 analyses of total samples. 6: Green Bay, Lake Michigan.[302] Mean of 52 analyses of HCl—H_2O_2 extracts. 7: Grand Lake, Nova Scotia.[155] Mean of 14 analyses for Mn and Fe and 12 analyses for the minor elements. 8: Ship Harbour Lake, Nova Scotia.[55] Mean of 5 analyses. 9: Mosque Lake, Ontario.[55] Mean of 2 analyses for Mn and Fe. One analysis only for minor elements. 10: Lake Ontario.[71] Mean values from single nodule site; number analyses not specified. Values represent compositions of acid-leached (50% hot ICl) fractions. 11: Lake George, New York.[313] Mean of 7 analyses. Analytical methods not given. 12: Eningi—Lampi Lake, Central Karelia.[367] Mean of 8 analyses.

[a] Major elements as wt. %, minor elements as ppm.

[b] Total Fe.

Table 83
DISTRIBUTION[a] OF ELEMENTS AS PERCENTAGE BODY WEIGHT OF ORGANISMS[221]

Invariable			Variable		
Primary 60—1	Secondary 1—0.05	Microconstituents <0.05	Secondary	Microconstituents	Contaminants
H	Na	B	Ti	Li	He
C	Mg	Fe	V	Be	A
N	S	Si	Br	Al	Se
O	Cl	Mn		Cr	Au
P	K	Cu		F	Hg
	Ca	I		Ni	Bi
		Co		Ge	Tl
		Mo		As	
		Zn		Rb	
				Sr	
				Ag	
				Cd	
				Sn	
				Cs	
				Ba	
				Pb	
				Ra	

[a] Webb and Fearon, 1937, with additions. Webb and Fearon remark that for some of the elements the number of organisms analyzed by reliable methods is insufficient to enable to one to decide whether an element is strictly variable or invariable, and, furthermore, that the quantitative classification, though convenient, is necessarily arbitrary.

Table 84
COMMERCIAL CLASSIFICATION OF MINERAL DEPOSITS[333]

Metallic	Nonmetallic						Fossil fuels		Hydrominerals and gases
Native elements and compounds	Minerals		Crystals		Amorphous and cryptocrystalline substances	Rocks		Pools of liquids and gases	
Ores	Metallurgical and thermal insulating raw materials	Chemical and agronomic raw materials	Industrial materials and gemstones	Piezo-optical materials	Sundry materials and colored stones	Building materials and raw materials for glass and ceramics	Solid fuels and chemical raw materials	Fuels and chemical raw materials	Brines, waters and gases
Ferrous metals Fe, Ti, Cr, Mn Light metals Al, Li, Be, Mg Nonferrous metals Cu, Zn, Pb, Sb, Ni Rare and scarce metals W, Mo, Sn, Co, Hg, Bi, Zr, Cs, Nb, Ta Noble metals Au, Ag, Pt, Os, Ir Radioactive metals U, Ra, Th Dispersed elements Sc, Ga, Ge, Rb, Cd, In, Hf. Re, Te, Po, Ac Rare-earth elements La, Ce, Pr, Nd, Pm, Sm, Eu, Gd, Tb, Dy, Ho, Er, Tu, Yb, Lu	Fluxes Fluorspar Calc-spar and dolomite Feldspar and quartz Nepheline Refractories and thermal insulators Graphite Chromite Chrysotile asbestos Vermiculite Talc and talc schist Magnesite Quartzite Bauxite Highly refractory minerals Andalusite Sillimanite (fibrolite) Kyanite (disthene) Diaspore Dumortierite	Chemical raw materials Halogens (salts) Native sulfur Iron pyrite Arsenical pyrite Realgar Orpiment Fluorite Baryte Witherite Alunite (Alumstone) Celestine Strontianite Calcite Aragonite Agronomic raw materials Apatites Phosphorites Potash salts Saltpetre (nitre) Borates Datolite Tourmaline Glauconite	Dielectrids Muscovite Phlogopite Abrasives Diamond Corundum Topaz Garnet Quartz Gemstones Diamond Emerald Aquamarine Alexandrite Ruby Spinel Topaz Amethyst, etc.	Piezo crystals Piezo quartz Tourmaline Optical materials Optical fluorite Iceland spar Optical quartz	Agate Opal Obsidian Chalcedony Jasper Rhodonite Malachite Lazurite Nephrite (and jadcite) Agalmatolite Selenite Anhydrite Amber (and kauri gum)	Building materials Building stone (walling, roofing, road metal, card core) Facing stone (marble, granite, labradorite, etc.) Acid-resistant stone (andesite, felsite, etc.) Materials for artifical stone (diabase, basalt, etc.) Binding materials (marl, limestone, clay, gypsum) Fillers (gravel, sand, etc.) Hydraulic additives (trass, pumice, diatomite and tripolite, menilite shales, etc.) Pigment minerals (chald, ochre, red ochre, etc.) Raw materials for glass and ceramics Glass and Pegmatite Clây and kaolin Loess and loam	Gummites Peat Lignite Brown coal Black coal Anthracite Semisapropelites Jet Semiboghead Sapropelites Boghead Oil shales Asphalt Anthraxolite Ozokerite	Heavy naphthene petroleum Light paraffinous crude oil Fuel gas	Fresh waters (drinking, industrial) Balneological mineral waters (carbonated, hydrosulfuric, radioactive, etc.) Saline water sources Oil reservoir waters with Br, I, B, Ra, etc. Lake brines Balncological mineral silts Incombustible, inert gases He, Ne, Ar, Kr, etc.

Table 85
METAL CONTENT OF THE MOST IMPORTANT COMMERCIAL MINERALS

Element	Mineral	Formula	Mineral Content %		Density
			Theoretical	By analysis	
Aluminum	Diasporo	$HAlO_2$	47.2		3.3—3.5
	Boehmite	$AlOOH$	47.2		3
	Hydrargillite (Gibbsite)	$Al(OH)_3$	36.2		2.35
	Nepheline	$Na[AlSiO_4]$	18.9		2.6
	Leucite	$K[AlSi_2O_6]$	13.0		2.4—2.5
	Alunite	$KAl_3[SO_4]_2(OH)_6$	20.5		2.6—2.8
	Kaolinite	$Al_4[Si_4O_{10}](OH)_8$	up to 22		2.6
	Sillimanite (Fibrolite)	$Al[AlSiO_5]$	35.0		3.2
Antimony	Antimonite	Sb_2S_3	71.4	70.2—71.5	4.5—4.6
	Berthierite	$FeSb_2S_4$			4.5—4.6
	Tetrahedrite	$Cu_{12}Sb_4S_{13}$	29.2	up to 15—30	4.4—5.4
	Boulangerite	$Pb_5Sb_4S_{11}$	25.7	25.4—25.7	6.2
	Jamesonite	$Pb_4FeSb_6S_{14}$	35.39	32—34.7	5.6
	Kermesite (Red antimony)	Sb_2S_2O			4.6
(Zone of oxidation)	Oxides and hydroxides of antimony			up to 75	
Arsenic	Arsenopyrite	$FeAsS$	46.0	39.4—48.7	5.9—6.2
	Loellingite	$FeAs_2$	72.8	66.69—70.09	7—7.4
	Realgar	AsS	70.1	69.5—70.0	3.4—3.6
	Orpiment	As_2S_3	61.0	57.67—60.87	3.4—3.5
	Tennantite	$Cu_{12}As_4S_{13}$	up to 20	9.0—20.0	4.4—5.4
	Scorodite	$Fe[AsO_4]\text{-}2H_2O$	30.0		3.1—3.3
Barium	Baryte	$Ba[SO_4]$	58		4.3
	Witherite	$Ba[CO_3]$	69.5		4.5
Beryllium	Beryl	$Be_3Al_2[Si_6O_{18}]$	5.07		2.7
	Bertrandite	$Be_4[Si_2O_7](OH)_2$	56.95		2.6
	Chrysoberyl	$BeAl_2O_4$	7.15		3.7
	Helvine	$(Mn, Fe)_8[BeSiO_4]_6S_2$		2.8—5.4	3.1—3.4

Table 85 (continued)
METAL CONTENT OF THE MOST IMPORTANT COMMERCIAL MINERALS

Element	Mineral	Formula	Mineral Content %		Density
			Theoretical	By analysis	
Bismuth	Native bismuth	Bi	100	95—99	9.8
	Bismuthinite (Bismuth glance)	Bi_2S_3	81.3	72.9—82.3	6.5
	Bismutite	$Bi_2[CO_3]O_2$	87		7.0—7.4
Boron	Borax (Tincal)	$Na_2B_4O_7 \cdot 10H_2O$	11.4		1.7
	Ulexite	$NaCaB_5O_9 \cdot 8H_2O$	13.4		1.7
	Hydroboracite	$CaMgB_6O_{11} \cdot 6H_2O$	15.4		2.2
	Ascharite	$Mg[BO_2](OH)$	12.9		2.6
	Ludwigite	$(Mg, Fe^{2+})_2Fe^{3+}[BO_3]O_2$	up to 5		4.0
	Cotoit	$Mg_3[BO_3]_2$	11.4		3.1
	Datolite	$Ca_2B_2[SiO_4]_2(OH)_2$	6.8		2.9—3.0
Cadmium	Sphalerite (Zinc blende)	(Zn, Fe, Mn, Cd)S		0.05—3.2	3.5—1.2
	Smithsonite (Calamine)	$(Zn, Cd)CO_3$		0.02—0.8	4.1—4.3
Caesium	Lepidolite	$KLi_{1.5}[AlSi_3O_{10}](F, OH)_2$		0.075—0.68	2.8—2.9
	Pollucite	$(Cs, Na)[AlSi_2O_6] \cdot H_2O$	42.8		2.9
Cerium and and rare earths	Monazite	$(Ce, La)[PO_4]$	59.7	17—35	4.9—5.5
	Xenotime	$Y[PO_4]$		up to 63	4.5—4.6
	Loparite	$(Na, Ce, Ca)(Nb, Ti)O_3$		up to 34	4.7—4.9
	Orthite (Allanite)	$(Ca, Ce)_2(Fe, Fe, Mg) \cdot Al_2[SiO_4][Si_2O_7] \cdot O(OH)$		up to 23	4.1
	Parisite	$Ca(Ce, La...)_2[CO_3]_3F_2$		48—53	4.3
	Bastnaesite	$(Ce, La, Pr)[CO_3]F$		65.4	4.5—5.2
Chromium	Chromite	$FeCr_2O_4$	46.4	12—45	4—4.8
Cobalt	Smaltite (Tin white cobalt)	$CoAs_{3m2}$	28.23	13.8—21.1	6.4—6.8
	Safflorite	$CoAs_2$	28.23	6.7—23.4	7.2—7.1
	Cobaltite	(Co, Fe)AsS	35.4	26—34	6.0—6.5

Cobalt pyrites	$(Co, Ni)_3S_4$		57.96	40—53	4.8—5.8
	Cobalt-bearing pyrites			up to 13.9	
	Asbolan (Asbolite), Cobalt-bearing Psilomelane			3.15—27	2—4
	Heterogenite	$CoO \cdot 2Co_2O_3 \cdot \eta H_2O$	63.6	up to 50—60	2—4
Copper	Native copper	Cu	100		8.5—9
	Chalcocite (Copper glance, Redruthite)	Cu_2S	79.8	79.67	5.5—5.8
	Covelline (Covellite)	CuS	66.5	66.43	4.6—4.7
	Chalcopyrite	$CuFeS_2$	34.6		4.2
	Bornite	Cu_5FeS_4	up to 63.3		4.9—5.5
	Enargite	Cu_3AsS_4	48.3		4.1—5.5
	Tetrahedrite	$Cu_{12}Sb_4S_{13}$	45.77	23—45 }	
	Tennantite	$Cu_{12}As_4S_{13}$	51.57	30—53 }	4.4—5.4
	Bournonite	$PbCuSbS_3$	13.0	12.0—15.12	5.8
	Cuprite	Cu_2O	88.8		5.8—6.15
(Zone of oxidation)	Malachite	$Cu_2[CO_3](OH)_2$	57.5		4
	Azurite	$Cu_3[CO_3]_2(OH)_2$	55.3		3.8
	Chrysocolla	$CuSiO_3\ \eta H_2O$	40.4		2.0—2.3
	Brochanthite (Waringtonite, Langite)	$Cu_4[SO_4](OH)_6$	56.2		3.8—3.9
Fluorine	Fluorite	CaF_2	48.8		3.2
Gallium	Germanite	$Cu_3(Fe, Ge, Ga)S_4$		0.8 to 1.99	4.3
	Sphalerite (Zinc glance)	ZnS		traces	
Germanium	Germanite	$Cu_3(Fe, Ge, Ga)S_4$		6.2—10.19	4.3
	Renerite	$(Cu, Fe)_3(Fe, Ge)S_4$		6—7.75	4.3—4.5
Gold[a]	Native gold	(Au, Ag)		80—98	15—19
	Electrum	(Ag, Au)		50—80	12—15
	Calaverite	$AuTe_2$	43.7	41.3—42.8	9
	Sylvanite	$AuAgTe_4$	24.2	25.4—29.8	8
	Nagyagite	$Pb_5Au(Te, Sb)_4S_{5-8}$ (?)		6—13	6.8—7.5
Indium	Sphalerite (Zinc blende)	ZnS		traces	

Table 85 (continued)
METAL CONTENT OF THE MOST IMPORTANT COMMERCIAL MINERALS

Element	Mineral	Formula	Mineral Content %		Density
			Theoretical	By analysis	
Iron	Magnetite	$FeFe_2O_4$	72.35		5.2
	Haematite	Fe_2O_3	70.0		5.2
	Brown ore (Limonite)	$HFeO_2 \cdot \eta H_2O$	48—63		úp to 4
	Siderite (Chalybite)	$Fe[CO_3]$	48.21		3.0—3.8
	Chamosite	$Fe_4Al[AlSi_3O_{10}](OH)_6 \cdot \eta H_2O$		28.5—37.3	3—3.4
	Hmenite	$FeTiO_3$	36.8		4.5
Lead	Galena (Golenile, lead glance, blue lead)	PbS	86.6	82—86.6	7.5
	Bournmite (Endellianite)	$PbCuSbS_3$	42.5	40.2—43.85	5.8
	Boulangerite	$Pb_5Sb_4S_{11}$	55.4	54.7—55.6	6.2
	Jamesonite	$Pb_4FeSb_6S_{14}$	40.16	39—40	5.6
	Cerussile	$Pb[CO_3]$	77.55		6.5
(Zone of oxidation)	Anglesite	$Pb[SO_4]$	68.3		6.1—6.4
	Pyromorphite (Green lead ore)	$Pb_5[PO_4]_3Cl$	76.38		6.7—7.1
Lithium	Spodumene	$LiAl[Si_2O_6]$	3.73	1.34—3.43	3.1—3.2
	Amblygonile	$LiAl[PO_4](F, OH)$	4.7	3.3—4.67	3
	Triphyline	$LiFe[PO_4]$	4.4		3.5
	Lithiophilite	$LiMn[PO_4]$	4.4		3.5
	Lepidolite (Lithium miea)	$KLi_{1.5}[AlSi_3O_{10}] \cdot (OH, F)_2$	3.5—3.69		
	Zinnwaldite	$KLiFeAl[AlSi_3O_{10}] \cdot (OH, F)_2$	up to 2.61—2.76	1.58—1.60	2.9—3.1 2.4

	Petalite	$Li[AlSi_4O_{10}$	2.2		
Magnesium	Magnesite	$Mg[CO_3]$	28.8		3.0
	Dolomite	$CaMg[CO_3]_2$	13.2		2.9
	Carnallite	$KCl \cdot MgCl_2 \cdot 6H_2O$	8.7		1.6
	Kieserite	$Mg[SO_4] \cdot H_2O$	17.6		2.6
	Bischofite	$MgCl_2 \cdot 6H_2O$	12.0		1.6
	Olivine	$Mg_2[SiO_4]$	34.4		3.3
Manganese	Pyrolusite	MnO_2	63.2	55—63	4.7—5
	Manganite	$MnOOH$	62.5	50—62	4.2—4.3
	Psilomelane	$mMnO \cdot MnO_2 \cdot nH_2O$		40—60	4.4—4.7
	Braunite	Mn_2O_3	69.6	60—69	4.7—5
	Hausmannite	$MnMn_2O_4$	72.0	65—72	4.7—4.9
	Rhodochrosite (Dialogite)	$Mn[CO_3]$	47.8	40—45	3.5—3.7
	Rhodonite (Manganese spar)	$(Mn, Ca)[SiO_3]$	41.9	33—40	3.4—3.75
Mercury	Cinnabar	HgS	86.2		8—8.2
	Schwarzite	$(Cu, Hg)_{12}Sb_4S_{13}$		up to 17	5
Nickel	Niccolite (Kupfernickel, arsenical nickel)	$NiAs$	43.92	40.6—44.98	8
	Chloanthite	$NiAs_{3-2}$	28.14		6.2—7.2
	Pentlandite	$(Fe, Ni)_9S_8$	34.22	10—40	4.5—5
	Garnierite	$(Ni, Mg)_4[Si_4O_{10}](OH)_4 \cdot 4H_2O$		4.3—36.1	2.3—2.8
	Revdinskite (Genthite)	$(Ni, Mg)_6[Si_4O_{10}](OH)_8$		5—30	2.5—3.2
Niobium	Columbite	$(Fe, Mn)(Nb, Ta)_2O_6$		22—54.5	5.3—7.3
	Fergusonite	$Y(Nb, Ta)O_4$		20—32	4.3—6.2
	Loparite	$(Na, Ce, Ca)(Nb, Ti)O_3$		up to 8	4.7—4.9
	Pyrochlore	$(Na, Ca...)_2 \cdot (Nb, Ti)_2O_6(F, OH)$		up to 44	4—4.4
Phosphorus	Apatite	$Ca_5[PO_4]_3(F, Cl)$	41—42.3	5—35	3.2
	Phosphorite (Phosphatic deposits)	A mixture of apatite, podolite, hydroxylapatite, etc.			

Table 85 (continued)
METAL CONTENT OF THE MOST IMPORTANT COMMERCIAL MINERALS

Element	Mineral	Formula	Mineral Content %		Density
			Theoretical	By analysis	
Platinum and the platinum group	Polyxene	(Pt, Fe)	Pt 80—88	59.9—87.2	15—19
	Iridium platinum	(Ir, Pt, Fe)	Ir up to 7		
	Palladium platinum	(Pd, Pt, Fe)	Pd up to 7—40		
	Rhodium platinum	(Rh, Pt, Fe)	Rh up to 6.8		
	Platinum iridium	(Ir, Pt)	Ir up to 90	19.64—55.44	
	Osmite (Osmiridium iridosmine)	(Os, Ir)	Os up to 80		
	Palladium	Pd			11—12
	Nevyanskite	(Ir, Os)	47—77		17—21
	Sysertskite	(Os, Ir)			
	Ruthenium nevyanskite	(Ir, Os, Ru)	Ru up to 0.5		17—20
	Sperrylite	$PtAs_2$	Pt 56.6	52.57—56.4	10.5
	Cooperite	PtS	Pt 86.89	82.2—85.6	9.4
Rhenium	Molybdenite	MoS_2		Re up to 0.33	
Rubidium	Carnallite	$KCl \cdot MgCl_2 \cdot 6H_2O$		0.015—0.037	
	Lepidolite	$KLi_{1-5}[AlSi_3O_{10}](F, OH)_2$		1.19—3.46	
Selenium	Galenite	PbS		Up to 1.23	
	Pyrite (Iron pyrite)	FeS_2		traces	
	Various selenides				
Silver[b]	Native silver	Ag	100	96.78—98.45	10
	Argentite (Silver glance)	Ag_2S	87	77.58—86.71	7
	Proustite (Light red silver ore)	Ag_3AsS_3	65.4	64.5—65.37	5.6
	Pyrargyrite (Dark red silver ore)	Ag_3SbS_3	59.8	59.8	5.8
	Stephanite (Brittle silver ore)	Ag_5SbS_4	68.3	67.8—68.6	6.2

	Polybasite	$(Ag, Cu)_{16}Sb_2S_{11}$	75.5	64.3—71	6
	Pearceite	$(Ag, Cu)_{16}As_2S_{11}$	78.4	51.17—72.43	6.1
(Zone of oxidation)	Cerargyrite (Kerargyrite, horn silver)	AgCl	75.0		5.5—5.6
Sodium	Rock salt	NaCl	39.3		2.2
Strontium	Strontianite	$Sr[CO_3]$	59.3		3.7
	Celestine (Celestite)	$Sr[SO_4]$	47.7		3.9
Sulfur	Native sulfur	S	100		2
	Pyrite (Iron pyrite)	FeS_2	53.4	50.7—53.3	5.2
	Pyrrhotine (Pyrrhotite, magmetic pyrites)	$Fe_{1-2}S$	36.5	38.2—39.5	4.6
	Gypsum	$Ca[SO_4]\cdot 2H_2O$	23.2		2.3
Tantalum	Tantalite	$(Fe, Mn)(Ta, Nb)_2O_6$		43—68	6.5—8.2
	Fergusonite	$Y(Nb, Ta)O_4$		1.6—22	4.3—6.2
	Loparite	$(Na, Ce, Ca)(Nb, Ti)O_3$		up to 1	4.7—4.9
	Pyrochlore	$(Na, Ca)_2(Nb, Ta, Ti)_2O_6(OH, F)$		up to 77	
Tellurium	Calaverite	$AuTe_2$	56.4	56.9—57.87	9.0—9.4
	Sylvanite (Graphic Tellurium)	$AuAgTe_4$	62.6	60.4—62.4	7.9—8.3
	Nagyagite (Black Tellurium)	$Pb_5Au(Te, Sb)_4S_{5-8}$(?)		18—30	6.8—7.5
	Petzite	Ag_3AuTe_2	32.87	33—34.9	8.7—9.0
	Hessite	Ag_2Te	37.1	36.1—37.1	8.2—8.9
	Altaite	PbTe	38.0	30.8—38.4	8.1—8.2
Thallium	Pyrite (Iron pyrites)	FeS_2		up to 0.1—1	
	Vrbaite	$TlAs_2SbS_5$	32.16	29.52	5.3
Thorium	Thorite	$Th[SiO_4]$	71.7		4.6
	Monazite	$(Ce, La)[PO_4]$		2.02—24.1	
Tin	Cassiterite	SnO_2	78.7	69—78	6.8—7.1
	Stannine (Tin pyrites)	Cu_2FeSnS_4	27.6	25.3—27.8	4.3—4.5
	Killindrite	$Pb_3Sb_4Sb_2S_{14}$	24.8	25.38	5.5
	Tillite	$PbSnS_2$	30.51	30.02—43.4	6.4
Titanium	Rutile	TiO_2	60.0		4.2
	Hmenite	$FeTiO_3$	31.6		4.7
Uranium	Pitchblende	UO_2	33.3	up to 76.7	9.5
	Carnotite	(K, Na, Ca, Cu,	52.7	54.6	4.5

Table 85 (continued)
METAL CONTENT OF THE MOST IMPORTANT COMMERCIAL MINERALS

		$Pb)_2(UO_2)_2[VO_4]_2 \cdot 3H_2O$			
	Uranophane	$Ca(UO_2)_2[SiO_4]_2(OH_3)_2 \cdot 3H_2O$	55.6	52.4—55.7	3.8—3.9
	Brannerite	(U, Ca, Fe, Y, Th)(Ti, Fe)$_2O_6$		7.5—27.5	4.5—5.4
Tungsten	Wolframite	(Fe, Mn)[WO_4]	60.5		6.7—7.5
	Ferberite	Fe[WO_4]	60.5		7.5
	Hübnerite	Mn[WO_4]	60.7		7.1
	Scbeelite	Ca[WO_4]	63.8		6.0
Vanadium	Patronite	VS_4	28.4	28—39	
	Descloizite	Pb(Zn, Cu)[VO_4](OH)	11.6	9.8—13.7	6
	Vanadinite	$Pb_5[VO_4]_3Cl$	10.8		6.7—7.2
	Carnotite	(K, Na, Ca, Cu, Pb)$_2(UO_2)_2[VO_4]_2 \cdot 3H_2O$		11.3—12.8	7
	Roscoelite	Vanadium-containing muscovite ($KV_2[AlSi_3O_{10}](OH)_2$		4.38—16.1	2.9—3
	Vanadium titanomagnetite			0.1—0.4	
Zinc	Sphalerite (Zinc blende)	ZnS	67.1	43.6—67	3.5—4.2
	Franklinite	(Zn, Mn)Fe_2O_4		7—20.5	5—5.2
	Smithsonite	Zu[Co_3]	52.1		4.1—2.5
(Zone of oxidation)	Willemite	$Zn_2[SiO_4]$	58.6		3.9—4.2
	Electric calamine (Galmei hemimorphite)	$Zn_4[Si_2O_7][OH]_2 \cdot H_2O$	54.3		3.4—3.5
Zirconium	Zircon	Zr[SiO_4]	49.7		4.5
	Baddeleyite	ZrO_2		up to 70	4.9—5.4
	Eudialyte	(Na, Ca)$_6$Zr[Si_6O_{17}](OH, Cl)		up to 11	2.8—3.0

[a] In addition to the main ore-forming gold materials there are gold-bearing pyrites, arsenopyrites, antimonites, etc.

[b] The bulk of silver is obtained from galena, fahlerz (grey coppers), chalcocite (copper glance), and pyrites in which the silver content varies widely.

Table 86
MINERALS ARRANGED IN ORDER OF INCREASING VICKERS HARDNESS NUMBERS[36]

Mineral species	Mean	Range	Remarks
Graphite	12	12	
Molybdenite	17	16—19	⊥ to cleavage
	23	21—28	∥ to cleavage
Bismuth	18	16—19	
Tellurbismuth	21	20—21	
Argentite	24	20—30	
Hessite	33	28—41	
Orpiment	38	23—52	
Electrum	40	34—44	
Stromeyerite	41	38—44	
Altaite	51	48—57	
Gold	51	50—52	
Silver	53	48—63	
Realgar	56	53—60	
Digenite	61	56—67	
Arsenic	63	57—69	
Pyrargyrite	71	50—97	⊥ to cleavage
	106	98—126	∥ to cleavage
Covellite	72	69—78	
Galena	76	71—84	
Pyrolusite	76	76	Average hardness ⊥ to fibers
	252	252	Average hardness ∥ to fibers
	279	256—346	Isotropic sections
	292	225—405	Microcrystalline
Stibnite	77	42—109	
Chalcophanite	81	71—85	⊥ to cleavage
	124	103—165	∥ to cleavage
	133	110—178	Isotropic sections
Chalcocite	84	68—98	
Antimony	89	83—99	
Jamesonite	99	96—105	Granular allotriomorphic sections
	113	105—121	Prismatic sections
Bornite	103	97—105	
Bismuthinite	107	92—119	
Miargyrite	110	104—123	
Sylvanite	110	102—125	
Kobellite	116	69—173	
Proustite	123	109—135	
Platinum	126	125—127	
Copper	134	120—143	
Naumannite	148	115—185	
Zincite	154	150—157	⊥ to cleavage
	304	295—318	∥ to cleavage
Pearceite	160	153—164	
Enargite	160	133—185	⊥ to cleavage
	272	245—346	∥ to cleavage
Boulangerite	166	157—183	
Dyscrasite	167	162—178	
Berthierite	171	155—185	
Zinkenite	178	162—207	
Emplectite	191	168—213	∥ to elongation
	222	197—238	⊥ to elongation
Bournonite	192	185—199	
Chalcopyrite	194	186—219	
Blende	198	186—209	
Cuprite	199	192—218	

Table 86 (continued)
MINERALS ARRANGED IN ORDER OF INCREASING VICKERS HARDNESS NUMBERS[36]

Mineral species	Mean	Range	Remarks
Stannite	210	197—221	
Cubanite	213	199—228	
Pentlandite	215	202—230	
Tenorite	236	209—254	
Millerite	236	225—256	Isotropic sections
	254	235—280	∥ to elongation
	348	318—376	⊥ to elongation
Pyrrhotite	248	230—259	Anistropic sections
	303	280—318	Isotropic sections
Alabandite	251	240—266	
Coffinite	258	236—333	
Chalcostibite	276	264—285	
Niccolite	336	328—348	Anistropic sections
	446	433—455	Isotropic sections
Tennantite	338	320—361	
Freibergite	345	317—375	
Scheelite	348	285—429	
Tetrahedrite	351	328—367	
Famatinite	363	333—397	
Wolframite	373	357—394	
Manganite	410	367—459	
Carrollite	463	351—566	
Lollingite	486	421—556	⊥ to elongation
	825	739—920	∥ to elongation
Siegenite	524	503—533	
Ullmannite	525	498—542	
Betafite	525	503—560	
Ilmenite	536	519—553	Possible differences in composition
	681	659—703	
Goethite	554	525—620	Microcrystalline
	803	772—824	Coarsely crystalline
Magnetite	560	530—599	
Breithauptite	563	542—584	
Psilomelane	572	503—627	
Hausmannite	587	541—613	
Braunite	595	584—605	
Pyrochlore	613	572—665	
Hollandite	620	560—724	
Skutterudite	653	589—724	
Gersdorffite	698	665—743	
Maucherite	704	685—724	
Euxenite	707	599—782	
Rammelsbergite	712	687—778	
Brannerite	720	710—730	
Pitchblende	720	673—803	Fresh specimens, oxidation produces marked decrease in hardness
Lepidocrocite	724	690—782	
Jacobsite	734	724—745	
Davidite	745	707—803	
Hematite	755	739—822	Microcrystalline
	1,000	920—1,062	Coarsely crystalline
Pararammelsbergite	772	762—803	
Coronadite	784	767—813	
Columbite-tantalite	803	724—882	
Uraninite	808	782—839	

Table 86 (continued)
MINERALS ARRANGED IN ORDER OF INCREASING VICKERS HARDNESS NUMBERS[36]

Mineral species	Mean	Range	Remarks
Maghemite	946	894—988	
Bixbyite	1,018	1,003—1,033	
Thorianite	1,918	988—1,115	
Cassiterite	1,053	1,027—1,075	
Arsenopyrite	1,094	1,048—1,127	
Bravoite	1,097	1,003—1,288	
Marcasite	1,113	941—1,288	
Glaucodot	1,124	1,071—1,166	
Rutile	1,139	1,074—1,210	
Pyrite	1,165	1,027—1,240	
Cobaltite	1,200	1,176—1,226	
Chromite	1,206	1,195—1,210	

Table 87
PHYSICAL AND OPTICAL PROPERTIES OF THE MORE COMMON EVAPORITE MINERALS[33]

Mineral	Solubility (g/ℓ) and taste	Crystal system, space group	Unit cell dimensions a, b, c	α, β, γ	Texture and Crystal habit	Cleavage, fracture	Hardness, specific gravity	Color, luster	Optical properties	Origin and paragenesis in evaporites
Anhydrite $CaSO_4$	1.97 (20°C.) 2.10 (20°C.) 1.97 (70°C.) Tasteless	Orthorhombic Bbmm	6.24 6.99 7.00		Massive, granular. Stumpy prisms, plates.	Perfect (001) Good (010) Distinct (100) Uneven.	3.0-3.6 2.9-3.0	White, gray, bluish, brick-red. Greasy, pearly, vitreous	nα 1.570 ‖ c nβ 1.575 ‖ b nγ 1.614 ‖ a +0.44 + 2Vγ43.7° r < b weak	By metamorphism of gypsum with halite in hartsalz replaced by polyhalite hydrated to gypsum.
Astrakhanite (Blödite) $MgSO_4.Na_2SO_4 4H_2O$	High. Incongruent melting with mirabilite ppn. above 24.5°C. Slightly sharp and bitter.	Monoclinic $P(2_1/a)$	11.09 8.20 5.50	100.6°	Massive, granular, Stumpy prisms.	None. Conchoidal.	3.0-3.2 2.2	Colorless, greenish, reddish. Vitreous.	nα 1.483∧c + 41° nβ 1.486 ‖ b nγ 1.487 0.004 - 2Vα71° r > b strong	Primary in soda lakes. Secondary after kieserite in marine evaporites. Occurs with balite, löweite, kainite and Mg sulphates.
Bischofite $MgCl_2.6H_2O$	Deliquescent. 2635 (20° C) = 35.1% $MgCl_2$ soln. 2710 (60° C.) Very bitter.	Monoclinic C(2/m)	9.92 7.16 6.11	93.7°	Granular, fibrous.	Perfect (110) Distinct (001)	1.5 1.59	Colorless, white. Vitreous	nα 1.495 ‖ b nβ 1.507∧β 9.5° nγ 1.528 +0.033 + 2Vγ79° r > b weak	With tachhydrite, halite, and carnallite in end-brines. Secondary with carnallite.
Boracite $5MgO.MgCl_2 7B_2O_3$	Tasteless.	Orthorhombic			Massive, nodular. Cubes, tetrahedra and dodecahedra.	None. Conchoidal.	7.0 2.9	White, grayish, greenish, yellow Vitreous to adamantine translucent.	nα 1.662 nβ 1.667 nγ 1.673 0.011 + 2Vγ83°	Secondary nodules in gypsum and hartsalz.
Carnallite $KCl.MgCl_2.6H_2O$	Deliquescent. Incongruent melting with sylvite ppa. 1190 = 27.3 $MgCl_2$ soln. Bitter.	Orthorhombic Pseudo-trig. Pban	9.56 16.05 22.56		Massive, coarsely crystalline. Bipytamids.	None. Conchoidal.	1.0-2.5 1.6	White, red yellow. Greasy, translucent.	nα 1.466 ‖ c nβ 1.475 ‖ b nγ 1.494 ‖ a 0.028 2Vγ66° r < v weak	Primary with halite and anhydrite, rarely with sylvite, kaininte, etc. Secondary retrograde metamorphic product with halite and anhydrite.
Epsomite $MgSO_4 7H_2O$	Deliquescent. 262 (20° C.) 335 (50° C.) Sharp.	Orthorhombic $P2_1 2_1 2_1$	11.86 11.99 6.85		Botryoidal, granular, fibrous. Acicular.	Perfect (010) Distinct (011) Conchoidal.	2.0-2.5 1.75	White, reddish. Vitreous to silky, translucent.	nα 1.432 ‖ a nβ 1.455 ‖ b nγ 1.461 ‖ c 0.028 2Vγ52° r < v weak	Possible low temp. ppt. with halite, sylvite, kainite, etc. Secondary after kieserite and polyhalite.
Glaserite (Aphthitalite) $Na_2SO_4.3K_2SO_4$	145 (20° C.) Faintly salty.	Trigonal P3̄ml	5.66 7.30		Massive.	Indistinct (0001)	2.5-3.5 2.7	White, bluish, greenish. Greasy to vitreous.	nω 1.491 nε 1.499 0.008 Rarely biaxial 2V small.	Secondary in cap rocks with astrakhanite, picromerite, and mirabilite.
Glauberite $Na_2SO_4.CaSO_4$	Incongruent melting with gypsum ppn. Faintly salty.	Monoclinic C(2/c)	10-10 8.28 8.51	112.2°	Reniform. Tabular, prismatic.	Good (001) Poor (110) Conchoidal.	2.5-3.0 2.7-2.85	Pale yellow, white gray. Vitreous to pearly.	nα 1.515 nβ 1.535∧c 12° nγ 1.536 ‖ b 0.021 2Vα 7°	In marine and non-marine evaporites with halite, anhydrite, gypsum, polyhalite, thenardite, and mirabilite.

Glauber Salt	See mirabilite									
Gorgeyite $K_2Ca_5(SO_4)_6H_2O$	Incongruent melting with gypsum ppn. Bitter.	Monoclinic	17.10 6.71 18.20	113.2°	Tabular.	Good (100)	3.5 2.77	Colorless or reddish because of inclusions. Vitreous.	nα 1.560∧a 40.5° nβ 1.569‖b nγ 1.584 —— 0.024 + 2Vγ 79°	In secondary parageneses with glauberite, polyhalite, and halite.
Gypsum $CaSO_4 2H_2O$	2.0 (20° C.) Tasteless	Monoclinic A(2/a)	5.68 15.18 6.20	113.8°	Massive, granular, fibrous. Plates, needles, swallow-tail twins.	Perfect (010) Good (111), (100)	2.0 2.3-2.4	White, gray, pink. Vitreous or earthy, translucent.	nα 1.521 nβ 1.523‖b nγ 1.530∧c 52° —— 0.009 + 2Vγ 58° r > b	Primary with dolomite or halite; secondary ± halite after anhydrite or potassium salts.
Halite NaCl	264 (20. C.) 274 (60° C.) Salty.	Cubic F(4/m³) (2/m)	5.64		Massive, granular, fibrous. Cubes.	Perfect, cubic. Conchoidal.	2.5 2.1-2.2	Colorless, white, gray, red, deep blue. Vitreous.	n = 1.54	Primary or secondary in most evaporites with gypsum, anhydrite, or postassium salts.
Hartsalz	Sylvite together with anhydrite and/or kieserite, polyhalite, and halite.									
Hexahydrite $MgSO_4 6H_2O$	308 (20° C.) 355 (60° C.) Bitter.	Monoclinic C(2/c)	10.06 7.16 24.39	98.6°	Massive, fibrous. Prisms.	Good prismatic.	2.5 1.76	White, pale green. Pearly.	nα 1.426∧c – 25° nβ 1.453‖b nγ 1.456 —— 0.030 – 2Va 38°	Primary with halite and carnallite in Recent lakes. Secondary after epsomite and kieserite.
Kainite $4(KClMgSO_4).11H_2O$	Salty, slightly bitter.	Monoclinic C(2/m)	19.76 16.26 9.57	94.9°	Massive, granular. Plates, tabular.	Good (110), (100)	2.5-3.0 2.1-3.0	White, red, yellowish. Vitreous, translucent.	nα 1.494 nβ 1.505 b nγ 1.516∧c 13° —— 0.22 – 2Va almost 90° r > b weak	Primary with halite, epsomite, hexahydrite, etc. usually secondary with langbeinite after kieserite and sylvite or carnallite.
Kieserite $MgSO_4 H_2O$	Forms metastable hydrates and epsomite. 386 (80° C.) Tasteless.	Monoclinic C(2/a)	6.88 7.61 7.53	116.3°	Granular. Bipyramids.	Good (111), (113)	3.0-3.6 2.57	Colorless, white, gray, yellow. Vitreous.	nα 1.520∧c 41° nβ 1.533‖b nγ 1.584 —— 0.064 + 2Vγ 55° r > b	Secondary with sylvite, langbeinite löweite and vanthoffite.
Koeneite $2MgCl_2.3Mg(OH)_2.2Al_2O_3.3H_2O$	Slightly bitter.	Trigonal.			Massive. Plates.	Perfect (0001)	1.0-2.0 1.98	Pale yellow, pink. Silky to vitreous.	nω 1.55 nε 1.52 —— 0.03 – Ve	Secondary cavity fillings in argillaceous anhydrite.
Langbeinite $K_2SO_4.MgSO_4$	Incongruent melting below 61° C. Tasteless.	Cubic P2, 3	9.92		Massive. Modified crystals.	None, uneven.	3.0-4.0 2.8-2.83	Colorless, white, pink. Vitreous.	n = 1.534	Secondary with kieserite, sylvite, and halite or with anhydrite and polyhalite.
Leonite $K_2SO_4.4H_2O$	Incongruent melting. Slightly bitter.	Monoclinic C(2/m)	11.78 9.53 9.88	95.4°	Massive.	Prismatic.	2.5-3.0 2.2	Colorless, white, or yellowish. Vitreous to greasy.	nα 1.479 nβ 1.483‖b nγ 1.487∧a small —— 0.008 + 2Vγ almost 90°	Secondary after kainite in cap rock with halite ± sylvite, polyhalite and astrakhanite.
Löweite $6Na_2SO_4.7MgSO_4 15H_2O$	Slightly bitter.	Trigonal R3 or R3	18.96 – 13.47		Massive. Plates.	Good(001)	3.5 2.4	Pale yellow, pink. Vitreous.	nε 1.470 nω 1.490 —— 0.019 – Ve	Secondary after kieserite with vanthoffite, astrakhanite, laugbeinite, glaserite, anhydrite, etc.

Table 87 (continued)
PHYSICAL AND OPTICAL PROPERTIES OF THE MORE COMMON EVAPORITE MINERALS[33]

Mineral	Solubility (g/ℓ) and taste	Crystal system, space group	Unit cell dimensions a b c	α β γ	Texture and Crystal habit	Cleavage, fracture	Hardness, specific gravity	Color, luster	Optical properties	Origin and paragenesis in evaporites
Mirabilite $Na_2SO_4.10H_2O$	448 (20° C.)	Monoclinic $P(2_1/c)$	11.51 10.38 12.83	 107.8°	Granular, earthly, fibrous. Prismatic, tabular.	Perfect (100) Poor (001), (010)	1.5-2.0 1.48	White, colorless. Subvitreous, translucent.	nα 1.394‖b nβ 1.396 nγ 1.398Δc 31° 0.004 – 2Va 76° r < b strong.	Primary in soda lakes as winter ppt. with gypsum, halite, epsomite glauberite, glaserite, and astrakhanite. Often partly replaced by thenardite.
Picromerite $K_2SO_4.6H_2O$	Incongruent melting. Bitter.	Monoclinic $P(2_1/a)$	9.06 12.26 6.11	 104.8°	Encrustation.	Perfect (201).	2.5 2.1	Colorless, white. Vitreous.	nα 1.461Λa – 1° nβ 1.463‖b nγ 1.476 0.015 + 2Vγ 48° r > b weak.	Can be primary with sylvite at low temp., nearly always secondary after kainite and leonite with kainite, astrakhanite, and glaserite.
Polyhalite $K_2SO_4MgSO_4.2CaSO_4.SH_2O$	Incongruent melting with gypsum ppn. Tasteless.	Triclinic FT or FI	11.68 16.33 7.60°	90.6° 90.1° 91.9	Massive, fibrous, lamellar. Prismatic.	Good (100)	2.5-3.5 2.78	Brick-red, yellowish, gray. Greasy, silky.	nα 1.547Λc 12° nβ 1.560 ≈ ‖[ITO] nγ 1.567 0.020 – 2Va 64°	Secondary after anhydrite in halite Less commonly secondary in carnallite, anhydrite.
Reichardtite	See epsomite.									
Rinneite $3KCl.NaCl.FeCl_2$	Incongruent melting; astringent.	Trigonal $R\bar{3}c$	11.98 – 13.84		Coarsely granular.	Distinct (1120) Conchoidal.	3.0 2.35	Colorless, pink, yellow, violet, brown. Vitreous, when fresh but turns silky.	nω 1.589 nε 1.590 0.001 Very dependent on λ abnormal polarization colors.	Secondary after carnallite with sylvite, halite, kieserite and anhydrite.
Schönite	See picromerite.									
Sylvinite	Sylvite in combination with some halite.									
Sylvite KCl	340 (20° C.) 455 (60° C.) Salty, bitter.	Cubic $F(4/m^3)(2/m)$	6.29		Massive, granular. Cubes and octahedra.	Perfect Cubie	2.0-2.2 1.98	Colorless, white, or Yellowish red. Vitreous, greasy.	n = 1.490	Secondary after carnallite and other potash salts.
Tachhydrite $2MgCl_2CaCl_2.2H_2O$	Very hygroscopic. Sharp, bitter.	Trigonal?			Massive, granular.	Distinct (1011)	1.0-2.0 1.66	Honey-yellow, wax-yellow. Vitreous.	n'' 1.512 nœ 1.520 0.008 – Ve	Secondary with carnallite in clay Never with sylvite or kieserite.
Thenardite Na_2SO_4	Deliquescent. 388 (40° C.) 453 (60° C.) Slightly salty, bitter.	Orthorhombic Fddd	9.82 12.30 5.86		Massive. Stumpy prisms. pyramids, tabular.	Good (001)	2.7 2.68	White, pale brown. Vitreous, greasy.	nα 1.471‖c nβ 1.477‖b nγ 1.484‖a 0.013 + 2Vγ 83°	In lakes, primary with astrakhanite glauberite, epsomite. Often after mirabilite. Doubtful occurrence in marine deposits.
Vanthoffite $3Na_3SO_4.MgSO_4$	Very slightly bitter.	Monoclinic $P(2_1/a)$	9.79 9.21 8.19	 113.5°	Granular.	Subconchoidal fracture	3.5 2.69	Colorless. Vitreous.	nα 1.485 nβ 1.488‖b 1.489 0.004 – 2Va 84° r < b weak	Secondary after kieserite, löweite with halite ± langbeinite.

Note: **Other Evaporite minerals.** ***Fluorides, chlorides, and oxychlorides:*** Hydrohalite $NaCl.2H_2O$, Chlorocalcite $KCaCl_3$, Sellaite MgF_2, Zirklerite 9(Fe, Mg, Ca)$Cl_2.2Al_2O_3.3H_2O$, Douglasite $KFeCl_3.2H_2O$, and Erythrosiderite $K_2FeCl_5.H_2O$. ***Carbonates:*** **Calcite $CaCO_3$, Dolomite CaMg $(CO_3)_2$,** Ankerite (Ca, Mg, Fe) CO_3, Aragonite $CaCO_3$, and Magnesite $MgCO_3$. *Borates and borosilicates:* Ulexite $NaCaB_5O_9.8H_2O$, Proberite $NaCaB_5O_9 5H_2O$, Kaliborite $KMg_2B_{11}O_{19}.9H_2O$, Ascharite **$MgHBO_3$ (Szaibelyite),** Hydroboracite $CaMgB_6O_{11}.6H_2O$, Fabianite CaB_3O_5 (OH), Priceite $Ca_4B_{10}O_{19}.7H_2O$, Colemanite $Ca_2B_6O_{11}\cdot 5H_2O$, p-Veatchite $SrB_6O_{10}\cdot 2H_2O$, Strontioborite 4(Sr, Ca) $0.2MgO.12B_2O_3.9H_2O$, Hilgardite $Ca_2B_5O_8$ $(OH)_2$ **Cl, Sulfoborite $Mg_3SO_4(BO_2OH)4H_2O$,** Luneburgite $Mg_3(PO_4)_2B_2O$ $(OH)_2.6H_2O$; and Danburite $CaB_2Si_2O_8$. *Sulfates and phosphates:* Hemihydrite $CaSO_4.1/2H_2O$, Syngenite $K_2Ca(SO_4)_2.H_2O$, D'Ansite $Na_{21}MgCl_3(SO_4)_{10}$. Celestine $SrSO_4$, **Wagnerite Mg_2FPO_4, and Apatite Ca_5(F, OH, Cl) $(PO_4)_3$.** *Silicates:* Potash feldspar, albite, muscovite, illite, chlorites, (amesite, penninite, corrensite $NaO.12 Mg_8.25FeO.24Al_2.7Si_6.20_{20}$ $(OH)_{10}.5H_2O$ serpentine, and talc.

Table 88
AVERAGE COMPOSITION OF WOOD, PEAT, AND COALS[59]

	Carbon	Hydrogen	Nitrogen	Oxygen
Wood	49.64	6.23	0.92	43.20
Peat	55.44	6.28	1.72	38.56
Lignite	72.95	5.24	1.31	20.50
Bituminous coal	84.24	5.55	1.52	8.69
Anthracite	93.50	2.81	0.97	2.72

Table 89
RELATIVE PROPORTIONS OF CONSTITUENTS OF WOOD, PEAT, AND COALS[59]

	Carbon	Hydrogen	Nitrogen	Oxygen
Wood	100	12.5	1.8	87.0
Peat	100	11.3	3.5	64.9
Lignite	100	7.2	1.8	28.1
Bituminous coal	100	6.6	1.8	10.3
Anthracite	100	3.0	1.3	2.9

Table 90
TYPE CLASSIFICATION OF COAL, ACCORDING TO THIESSEN[266]

Types		Relative amounts of banded components	Nature of attritus	External structure and appearance of coal
	Bright	Anthraxylon more than 5% and usually predominant component Attritus usually subordinate but occasionally predominant Fusian often present but never abundant.	Translucent humic matter, thin, fibrous structure, predominant ingredient. Yellow translucent ingredients including spores, cuticles, and resin in minor amounts. Opaque matter less than 20%.	Usually coarse to fine banded but occasionally microbanded. Bright surface luster characteristic. Friable with irregular fracture.
Banded coals	Semisplint	Anthraxylon more than 5% and occasionally equal to attritus. Attritus usually predominant component. Fusain often present but usually in minor amounts.	Translucent humic matter frequently predominant ingredient. Yellow translucent ingredients including spores; cuticles, and resin frequently abundant. Opaque matter 20 to 30%. Brown matter and fusain particles in minor amounts.	Fine to microbanded. Surface luster usually bright but frequently dull. Hard with blocky fracture.
			Translucent humic matter in minor amounts.	

Table 90 (continued)
TYPE CLASSIFICATION OF COAL, ACCORDING TO THIESSEN[266]

Types		Relative amounts of banded components	Nature of attritus	External structure and appearance of coal
	Splint	Anthraxylon more than 5% but usually in minor amounts. Attritus always predominant component. Fusain often present but usually in minor amounts.	Yellow translucent ingredients abundant, particularly spores; cuticles and resin present. Opaque matter more than 30%. Brown matter frequently present. Fusain particles in minor amounts. Translucent humic matter in minor to predominant amounts.	Dull luster and grainy texture, with occasional thin bright band. Very hard, splintery fracture.
	Cannel	Anthraxylon less than 5% and usually absent. Attritus always predominant component. Fusain rare.	Yellow translucent ingredients usually predominant. Spores particularly abundant; cuticles and resin in important amounts. Opaque matter and fusain particles present. Algal remains, occasional, in minor amounts.	Smooth, uniform texture, sometimes dense grained. Dull surface luster. Conchoidal fracture.
Nonbanded coals				
	Boghead	Anthraxylon less than 5% and usually absent. Attritus always predominant. Fusain rare.	Algal remains predominant ingredient. Opaque matter present, occasionally abundant. Spores sometimes present in minor amounts.	Smooth, uniform texture, sometimes dense grained. Dull surface luster. Conchoidal fracture.

Table 91
TYPE OF COAL[178]

	Type	Main components	Opaque matter per cent	Anteraxylon per cent
Banded coal	Bright coal	Anthraxylon translucent attritus	< 20	
	Semisplint	Translucent and opaque attritus	20—30	<5
	Splint	Opaque attritus	>30	
Nonbanded coal	Cannel	Attritus with spores		<5
	Boghead	Attritus with algae		

Table 92
CHEMICAL ANALYSES OF LIVING WOOD, WOOD FROM PEAT, AND NATURAL CHARCOAL AND OF ANTHRAXYLON (An) AND FUSAIN (Fu) FROM COALS OF INCREASING RANK[a]

Substance and rank	Source	Geological age	Proximate, moisture and ash-free basis, percent				Ultimate, moisture and ash-free basis, percent										Calorific value, moisture and ash-free basis, Btu per pound		Laboratory no.	
			Volatile matter[a]		Fixed carbon		Sulfur		Hydrogen		Carbon		Nitrogen		Oxygen					
			An	Fu	An	Fu	An	Fu	An	Fu	An	FU	An	Fu	An	Fu	An	Fu	An	Fu
Cedar wood	Local	Present	89.5		10.5		0.1		6.2		51.8		0.1		41.8		8,960		B-38549	
Peat wood	Wisconsin	Pleistocene	75.2		24.8		.7		5.6		57.1		.6		36.0		9,500		B-32781	
Natural charcoal[c]	Oregon	Recent		45.2		54.8		0		3.2		64.7		0.3		31.8		11,350		A-39583
Brown coal	Washington	Unknown	67.4		32.6		1.1		5.5		57.9		.2		35.3		9,840		B-32780	
	Germany	Miocene	65.6	38.8	34.4	61.2	.5	1.3	5.8	3.7	61.1	74.5	.1	.4	32.5	20.1	10,710	11,790	E-21590	E-21591
Lignite	North Dakota	Paleocene	48.6		51.4		1.0		5.5		71.2		.6		21.7		12,390		B-40124	
	do	do	46.9	38.2	53.1	61.8	.4	.5	4.9	4.0	70.0	74.3	.9	.9	23.8	20.3	11,870	12,260	D-89080	D-89597
	do	do	45.7		54.3		1.4		4.6		73.0		.6		20.4		12,490		B-32921	
	Texas	Eocene	44.6		55.4		1.0		4.7		71.6		1.1		21.6		11,850		B-36203	
Subbituminous coal	Washington	do	40.2		59.8		.2		5.0		75.5		.4		18.9		12,950		B-22590	
	Wyoming	Paleocene	40.2		59.8		.3		5.0		75.8		1.5		17.4		12,960		B-39780	
	do	do		29.6		70.4		1.0		4.3		82.6		.5		11.6		13,870		B-44434
High-volatile bituminous coal	West Virginia	Pennsylvanian		19.8		80.2		.3		4.0		89.3		.7		5.7		14,980		A-98226
	Illinois	do	36.5		63.5		.9		5.2		81.7		1.8		10.4		14,440		B-33462	
	Ohio	do		18.9		81.1		1.9		3.5		88.1		.7		5.8		14,650		B-27623
	Pennsylvania	do	42.6	17.9	57.4	82.1	1.6	.6	5.5	3.6	83.0	89.8	1.7	.7	8.2	5.3	14,810	14,930	B-28921	A-98227
	do	do	39.2	15.2	60.8	84.8	1.1	.5	5.5	2.8	84.5	92.6	1.6	.4	7.3	3.7	15,060	14,870	B-25100	B-24646
	do	do	36.1	15.9	63.9	84.1	.9	3.6	5.4	3.0	85.3	92.6	1.5	.5	6.9	.3	15,160	15,150	B-27778	B-27779
	do	do	33.1	15.6	66.9	84.4	.7	.4	5.3	3.3	86.1	92.0	1.5	.6	6.4	3.7	15,180	15,050	B-27621	B-27622
Medium-volatile bituminous coal	do	do	27.3	9.7	72.7	90.3	1.2	3.0	5.2	2.9	88.9	93.0	1.4	.5	3.3	.6	15,600	15,230	B-29844	B-29843
	do	do	24.3		75.7		.7		5.1		89.6		1.3		3.3		15,700		B-39074	
Low-volatile bituminous coal	do	do	21.5	10.5	78.5	89.3	.7	.3	4.9	3.3	90.3	93.5	1.4	.5	2.7	2.4	15,760	15,490	B-31700	B-26584
	West Virginia	do	20.4		79.6		.6		4.8		89.9		1.8		2.9		15,650		B-26583	
	Oklahoma	do	17.9	7.3	82.1	92.7	.8	.2	4.6	2.6	88.9	94.5	1.9	.5	3.8	2.2	15,410	15,200	D-86743	D-86744
	Pennsylvania	do	17.2		82.8		.7		4.7		90.5		1.4		2.7		15,660		B-26069	
Anthracite	do	do	3.4	6.7	96.6	93.3	.6	1.2	2.5	2.5	94.9	93.4	.8	.3	1.2	2.6	15,210	14,870	B-39247	B-39464

[a] Rank of coals determined from analyses of standard face samples. Samples of anthraxylon and fusain from brown coal and anthracite were selected from gross lots of lump coal. Samples from other ranks of coal were selected from column samples collected in the same mines as the face samples.

[b] Low ash content of anthraxylon samples (less than 4%) indicated that little mineral matter was present. Ash content of fusain samples was less than 7%, with one exception. Carbon dioxide was not determined to correct for slightly high volatile matter content due to calcite.

[c] Fusain-like carbonized wood associated with volcanic ejecta at Crater Lake believed to be postglacial in age.

Table 93
SOURCE, CHEMICAL ANALYSES, AND RESULTS OF LABORATORY-SCALE HYDROGENATION TESTS OF SELECTED SAMPLES OF WOOD, ANTHRAXYLON, FUSAIN, RESINS, AND SPORES FROM COAL

State	County	Bed	Mine	Sample	Rank[a]	Laboratory no.	Condition[b]	Proximate, percent				Ultimate, percent					Calorific value Btu per lb.	Organic residues from small autoclave assays, moisture- and ash-free basis, percent	Test conditions[c]	
								Moisture	Volatile matter	Fixed carbon	Ash	Sulfur	Hydrogen	Carbon	Nitrogen	Oxygen			Temperature, °C	Pressure, lb. per sq. in.
				White cedar wood[d]		B-38549	1	1.3[e]	88.3	10.2	0.2	0.1	6.3	51.1	0.1	42.2	8,840	1.6	410	1,800
							2		89.4	10.4	.2	.1	6.2	51.8	.1	41.6	8,950			
							3		89.5	10.5		.1	6.2	51.8	.1	41.8	8,960			
Wisconsin	Manitowoc	Hawk Island Swamp		Wood	Peat	B-32781	1	16.2	60.6	20.0	3.2	.6	6.3	46.1	.5	43.3	7,660	3.7	430	1,000
							2		72.3	23.9	3.8	.7	5.3	55.0	.6	34.6	9,140			
							3		75.2	24.8		.7	5.6	57.1	.6	36.0	9,500			
Washington	King	Unnamed	Prospect pit	Anthraxylon	Br	B-32780	1	12.2	58.0	28.1	1.7	1.0	6.1	49.8	.2	41.2	8,470	6.7	430	1,000
							2		66.0	32.0	2.0	1.1	5.4	56.8	.2	31.5	9,650	1.4	430	1,500
							3		67.4	32.6		1.1	5.5	57.9	.2	36.3	9,840			
North Dakota	Mercer	do	Knife River	do	Lig	B-32921	1	32.3	29.1	34.6	4.0	.9	6.5	46.6	.4	41.6	7,970	5.6	430	1,000
							2		43.0	51.2	5.8	1.3	4.3	68.8	.6	19.2	11,770			
							3		45.7	54.3		1.4	4.6	73.0	.6	20.4	12,490			
Washington	King	No. 2	Tiger Mountain	do	Sub	B-22590	1	11.6	35.1	52.3	1.0	.2	5.7	66.0	.4	26.7	11,320	20.8	430	1,000
							2		39.7	59.2	1.1	.2	5.0	74.7	.4	18.6	12,810	4.8	430	1,700
							3		40.2	59.8		.2	5.0	75.5	.4	18.9	12,950			
Pennsylvania	Allegheny	Pittsburgh	Experimental	do	Hvab	B-25100	1	1.5	38.0	58.9	1.6	1.1	5.5	81.9	1.5	8.4	14,600	1.7	400	1,000
							2		38.6	59.8	1.6	1.1	5.4	83.1	1.6	7.2	14,820			
							3		39.2	60.8		1.1	5.5	84.5	1.6	7.3	15,060			
do	Fayette	Lower Kittanning	Indian Creek No. 1	do	Mvb	B-29844	1	.9	26.4	70.5	2.2	1.1	5.1	86.1	1.3	4.2	15,140	7.1	430	1,000
							2		26.7	71.1	2.2	1.2	5.1	86.9	1.3	3.3	15,280			
							3		27.3	72.7		1.2	5.2	88.9	1.4	3.3	15,620			
do	Cambria	do	Revloc	do	Lvb	B-31700	1	1.0	20.8	76.1	2.1	.6	4.9	87.5	1.4	3.5	15,280	18.9	430	1,000
							2		21.0	76.9	2.1	.6	4.8	88.3	1.4	2.8	15,420	10.6	440	1,800
							3		21.5	78.5		.7	4.9	90.3	1.4	2.7	15,760			
Wyoming	Campbell	Roland-Smith	Wyodak	Fusain	Sub	B-44434	1	5.9	26.1	62.1	5.9	.9	4.5	72.9	.5	15.3	12,240	44.2	430	1,500
							2		27.7	66.0	6.3	1.0	4.0	77.4	.5	10.8	13,000			
							3		29.6	70.4		1.0	4.3	82.6	.5	11.6	13,870			

Table 93 (continued)
SOURCE, CHEMICAL ANALYSES, AND RESULTS OF LABORATORY-SCALE HYDROGENATION TESTS OF SELECTED SAMPLES OF WOOD, ANTHRAXYLON, FUSAIN, RESINS, AND SPORES FROM COAL

State	County	Bed	Mine	Sample	Rank[a]	Laboratory no.	Condition[b]	Proximate, percent				Ultimate, percent					Calorific value Btu per lb.	Organic residues from small autoclave assays, moisture- and ash-free basis, percent	Test conditions[c]	
								Moisture	Volatile matter	Fixed carbon	Ash	Sulfur	Hydrogen	Carbon	Nitrogen	Oxygen			Temperature, °C	Pressure, lb. per sq. in.
Pennsylvania	Allegheny	Pittsburgh	Experimental	do	Hvab	B-24646	1	.6	13.8	77.2	8.4	.5	2.7	84.3	.4	3.7	13,530	89.0	400	1,000
							2		13.9	77.7	8.4	.5	2.6	84.8	.4	3.3	13,610	84.4	400	1,000
							3		15.2	84.8		.5	2.8	92.6	.4	3.7	14,870			
do	Fayette	Lower Kittanning	Indian Creek No. 1	do	Mvb	B-29843	1	.5	9.1	84.6	5.8	2.8	2.8	87.2	.4	1.0	14,270	83.3	430	1,000
							2		9.1	85.1	5.8	2.8	2.7	87.6	.4	.7	14,310			
							3		9.7	90.3		3.0	2.9	93.0	.5	.6	15,230			
West Virginia	Raleigh	Buckley	Winding Gulf No. 1	do	Lvb	B-26584	1	.4	10.1	86.2	3.3	.3	3.2	90.1	.5	2.6	14,920	89.6	430	1,000
							2		10.1	86.6	3.3	.3	3.2	90.4	.5	2.3	14,970	81.3	430	1,000
							3		10.5	89.5		.3	3.3	93.5	.5	2.4	15,490			
Michigan	Ingham	Unnamed	Davidson Coal Co.	Spores	Hvbb	B-22658	1	2.0	67.6	20.6	9.8	1.2	6.9	71.1	.9	10.1	13,550	4.1	400	1,000
							2		68.9	21.1	10.0	1.3	6.9	72.5	.9	8.4	13,820	1.5	430	1,800
							3		76.6	23.4		1.4	7.6	80.6	1.0	9.4	15,350			
Utah	Oarbon	King-Hiawatha	King	Resins	Hvbb	B-34925	1	.4	90.5	5.5	3.6	.4	10.1	82.2	.4	3.3	17,203	2.1	400	1,000
							2		90.9	5.5	3.6	.4	10.0	82.6	.4	3.0	17,280	1.6	430	1,800
							3		94.3	5.7		.4	10.4	85.6	.4	3.2	17,910			
Canada	British Columbia	No. 2	No. 2-8 Middlesboro Colliery	do	Hvbb	B-33460	1	.5	93.9	3.8	1.8	.2	10.1	81.7	.3	5.9	16,880	.2	430	1,000
							2		94.3	3.9	1.8	.2	10.1	82.1	.3	5.5	16,950			
							3		96.1	3.9		.2	10.2	83.6	.3	5.7	17,270			

[a] Br, brown coal; Lig, lignite; Sub, subbituminous; Lvb, low-volatile bituminous; Mvb, medium-volatile bituminous; Hvab, high-volatile B bituminous.
[b] 1, Sample as received; 2, dried at 105°C; 3, moisture- and ash-free.
[c] Time of hydrogen for all tests was 3 hr; catalyst was 1% stannous sulfide.
[d] Kiln-dried lumber.
[e] Vacuum-dried.

Table 94
COMPARISON OF STOPES-HEERLEN SYSTEM WITH SPACKMAN COAL CONSTITUENT CLASSIFICATION[178]

Stopes-Heerlen system	Classification in U.S. (Spackman/system)				
Maceral-group	Maceral-suite	Maceral-group	Range of maximum reflectance (per cent) under oil	Other distinguishing characteristics[a]	Macerals[b]
Vinitrite	Vintrinite suite	Anthrinoid group	2.50—10.00	Opaque in transmitted light. Greyish-white in reflected light.	A_{25}-A_{100}
		Vintrinoid group	0.40—2.40	Translucent in transmitted light — usually yellow, red or brown. Grey in reflected light.	V_4—V_{24}
		Xylinoid group	0.10—0.39	Translucent in transmitted light — usually buff, whitish-yellow to yellowish-brown. Dark grey in reflected light.	X_1—X_3
Exinite	Liptinite suite	Exinoid group	0.05—1.50	Coalified spore, pollen, cuticular or endodermal materials. Translucent in transmitted light — whitish-yellow, yellow, golden yellow or red. Black, dark grey to light grey in reflected light.	E_0—E_{15}
		Resinoid group	0.05—1.50	Coalified resins or other plant secretions or exudates. Translucent in transmitted light — whitish-yellow, yellow to red. Black, dark grey to light grey in reflected light.	R_0—R_{15}
Inertinite	Inertinite suite	Fusinoid group	4.00—10.00	Characteristics essentially those of fusinite (Stopes-Heerlen System).	F_{40}—F_{100}
		Semi-Fusinoid group	0.20—3.99	Characteristics essentially those of semifusinite (Stopes-Heerlen System).	SF_2—SF_{39}
		Micrinoid group	0.20—8.00	Characteristics essentially those of micrinite (Stopes-Heerlen System).	M_2—M_{80}

[a] Descriptions for transmitted light refer to thin sections of standard thickness.
[b] Provisionally differentiated into "Entity Types" and designated by type numbers (when adequately circumscribed names will be assigned ending in "-inite").

Table 95
RANK CLASSIFICATION OF COAL INCLUDING IMPORTANT MICROSCOPIC CHARACTERISTICS[178]

	Rank stages	% reflectance of vitrinite	Important microscopic characteristics	% C in vitrinite	Volatile matter % d.a.f. in vitrinite	% H_2O in situ	Calorific value of vitrite (a.f.)	Applicability of the different parameters for the determination of rank
	Peat		Large pores Details of initial plant material still recognizable Free cellulose	50		75 ~		Calorific value (a.f.) or moisture in situ (moisture-holding capacity) ↓
Brown coal	Soft brown coal		No free cellulose Plant structures still recognizable (cell cavities frequently empty)	60				
		ca. 0.3	Marked gelification and compaction takes place		ca. 53	35 ~	7200 Btu/lb (4000 kcal/kg)	
Hard brown coal	Dull brown coal		Plant structures still partly recognizable	70	ca. 49	25 ~	9900 Btu/lb (5500 kcal/kg)	Carbon (d.a.f.) ↓ Reflectance of the vitrinites ↓
	Bright brown coal	ca. 0.5	(cell cavities filled with collinite)		ca. 45	8-10 ~	12600 Btu/lb (7000 kcal/kg)	↑ Volatile matter (d.a.f.)
Hard coal	Bituminous hard coal		Exinite becomes markedly lighter in color ("Coalification jump")	80				X-ray diffraction (graphite lattice) ↓ ↑ H (d.a.f.)
		ca. 2.5	Exinite no longer distinguishable from vitrinite in reflected light	90	30			
	Anthracite		Reflectance anisotropy		10		15500 Btu/lb (8650 kcal/kg)	
	Graphite			100	0			

Table 96
PRINCIPAL MINERAL INCLUSIONS IN COAL[178]

	Intimately intergrown with the coal		Deposited in cleats and other fissures (coarsely intergrown)
	Deposited by water or windblown	Originally formed in the peat	
Clay minerals	Illite, sericite, kaolinite, leverrierite, montmorillonite, etc.		
Carbonate minerals (spars)		Accretions of siderite, dolomite, (ankerite), and calcite. $FeCO_3$ and $CaCO_3$ in fusite.	Calcite. Ankerite.
Sulfide minerals		Accretions of FeS_2. Accretions of FeS_2-$CuFeS_3$-ZnS Melnikovite.	Pyrites Marcasite Zinc blende Copper pyrites Galena
Oxides		Limonite Hematite	Gothite (needle iron ore)
Quartz	Granular quartz.	Chalcedony and quartz produced by decomposition of aluminum silicates.	Quartz Chalcedony
Salts (chlorides, sulfates, etc.)	Rock salt, thenardite, gypsum		

Table 97
MEAN ANALYTICAL VALUES FOR 23 WHOLE COAL SAMPLES FROM THE EASTERN UNITED STATES (APPALACHIAN COAL FIELDS)[132]

Element	Arithmetic mean	Geometric mean	Minimum	Maximum	Standard deviation	Number samples	Number less than values
AG	0.02ppm	0.02ppm	0.01	0.06	0.01	13	
AS	25 ppm	15 ppm	1.8	100	27	23	
B	42 ppm	28 ppm	5.0	120	32	23	
BA	200 ppm	170 ppm	72	420	110	14	
BE	1.3 ppm	1.1 ppm	0.23	2.6	0.56	23	
BR	12 ppm	8.9 ppm	0.71	26	7.6	23	
CD	0.24ppm	0.19ppm	0.10	0.60	0.18	23	23
CE	25 ppm	23 ppm	11	42	9.1	14	
CO	9.8 ppm	7.6 ppm	1.5	33	7.8	23	
CR	20 ppm	18 ppm	10	90	16	23	
CS	2.0 ppm	1.6 ppm	0.40	6.2	1.6	14	
CU	18 ppm	16 ppm	5.1	30	7.3	23	
DY	2.3 ppm	2.0 ppm	0.74	3.5	0.94	14	
EU	0.52ppm	0.47ppm	0.16	0.92	0.22	14	
F	89 ppm	84 ppm	50	150	31	23	
GA	5.7 ppm	5.2 ppm	2.9	11	2.6	23	
GE	1.6 ppm	0.87ppm	0.10	6.0	1.7	23	9
HF	1.2 ppm	1.1 ppm	0.58	2.2	0.45	14	
HG	0.20ppm	0.17ppm	0.05	0.47	0.12	23	1
I	1.7 ppm	1.4 ppm	0.33	4.9	1.1	14	1
IN	0.23ppm	0.22ppm	0.13	0.37	0.08	14	
LA	15 ppm	14 ppm	6.1	23	5.3	14	
LU	0.22ppm	0.18ppm	0.04	0.40	0.12	14	
MN	18 ppm	12 ppm	2.4	61	16	23	
MO	4.6 ppm	1.8 ppm	0.10	22	6.3	23	3
NI	15 ppm	14 ppm	6.3	28	5.7	23	
P	150 ppm	81 ppm	15	1500	300	23	
PB	5.9 ppm	4.7 ppm	1.0	18	4.0	23	3
RB	22 ppm	19 ppm	9.0	63	15	14	
SB	1.6 ppm	1.1 ppm	0.25	7.7	1.7	23	
SC	5.1 ppm	4.5 ppm	1.6	9.3	2.4	14	
SE	4.0 ppm	3.4 ppm	1.1	8.1	2.0	23	
SM	2.6 ppm	2.4 ppm	0.87	4.3	1.0	14	
SN	2.0 ppm	0.97ppm	0.20	8.0	2.4	19	7
SR	130 ppm	100 ppm	28	50	130	14	
TA	0.33ppm	0.26ppm	0.12	1.1	0.28	14	
TB	0.34ppm	0.28ppm	0.06	0.63	0.17	14	
TH	4.5 ppm	4.0 ppm	1.8	9.0	2.1	14	
TL							
U	1.5 ppm	1.3 ppm	0.40	2.9	0.73	14	
V	38 ppm	35 ppm	14	73	14	23	
W	0.69ppm	0.62ppm	0.22	1.2	0.31	14	
YB	0.83ppm	0.73ppm	0.18	1.4	0.35	14	
ZN	25 ppm	19 ppm	2.0	120	24	23	
ZR	45 ppm	41 ppm	8.0	88	18	19	
AL	1.7 %	1.6 %	1.1	3.1	0.56	23	
CA	0.47%	0.34%	0.09	2.6	0.51	23	
CL	0.17%	0.10%	0.01	0.80	0.21	23	
FE	1.5 %	1.3 %	0.50	2.6	0.69	23	
K	0.25%	0.21%	0.06	0.68	0.14	23	
MG	0.06%	0.05%	0.02	0.15	0.03	23	
NA	0.04%	0.03%	0.01	0.08	0.02	23	
SI	2.8 %	2.6 %	1.0	6.3	1.1	23	

Table 97 (continued)
MEAN ANALYTICAL VALUES FOR 23 WHOLE COAL SAMPLES FROM THE EASTERN UNITED STATES (APPALACHIAN COAL FIELDS)[132]

Element	Arithmetic mean	Geometric mean	Minimum	Maximum	Standard deviation	Number samples	Number less than values
TI	0.09%	0.09%	0.05	0.16	0.04	23	
ADL	1.2 %	0.99%	0.50	4.0	0.89	14	
MOIS	2.7 %	2.4 %	1.0	6.8	1.5	23	
VOL	33 %	32 %	17	42	8.0	23	
FIXC	55 %	54 %	45	72	7.2	23	
ASH	12 %	12 %	6.1	25	4.3	23	
Btu/lb	13111	13093	11374	13816	696	14	
C	72 %	72 %	63	80	5.3	22	
H	4.9 %	4.9 %	4.0	6.0	0.44	22	
N	1.3 %	1.3 %	0.94	1.8	0.27	22	
O	8.0 %	7.0 %	2.5	18	4.3	22	
HTA	12 %	12 %	6.2	25	4.3	23	
LTA	15 %	15 %	7.6	28	4.9	23	
ORS	0.92%	0.82%	0.35	2.5	0.48	23	
PYS	1.3 %	0.81%	0.04	2.6	0.91	23	
SUS	0.10%	0.08%	0.01	0.42	0.10	22	
TOS	2.3 %	1.9 %	0.55	5.0	1.3	23	
SXRF	2.1 %	1.8 %	0.74	4.8	1.1	23	

Note: HTA = High-temperature Ash; LTA = Low-temperature Ash; ORS = Organic Sulfur; PYS = Pyritic Sulfur; SUS = Sulfate Sulfur; TOS = Total Sulfur; SXRF = Sulfur by X-ray Fluorescence.

Table 98
MEAN ANALYTICAL VALUES FOR 28 WHOLE COAL SAMPLES FROM THE WESTERN UNITED STATES[132]

Element	Arithmetic mean	Geometric mean	Minimum	Maximum	Standard deviation	Number samples	Number less than values
AG	0.03ppm	0.02ppm	0.01	0.07	0.02	22	
AS	2.3 ppm	1.5 ppm	0.34	9.8	2.6	29	
B	56 ppm	48 ppm	16	140	32	27	
BA	500 ppm	430 ppm	160	1600	320	22	
BE	0.46ppm	0.35ppm	0.10	1.4	0.34	29	2
BR	4.7 ppm	2.1 ppm	0.50	25	7.3	29	
CD	0.18ppm	0.15ppm	0.10	0.60	0.13	29	29
CE	11 ppm	9.1 ppm	2.8	30	8.0	22	
CO	1.8 ppm	1.5 ppm	0.60	7.0	1.5	29	
CR	9.0 ppm	8.1 ppm	2.4	20	4.2	29	
CS	0.42ppm	0.16ppm	0.02	3.8	0.82	22	
CU	10 ppm	8.5 ppm	3.1	23	5.9	29	
DY	0.63ppm	0.57ppm	0.22	1.4	0.32	22	
EU	0.20ppm	0.16ppm	0.07	0.60	0.17	22	
F	62 ppm	57 ppm	19	140	26	29	
GA	2.5 ppm	2.1 ppm	0.80	6.5	1.4	29	
GE	0.91ppm	0.50ppm	0.10	3.0	0.92	29	6
HF	0.78ppm	0.70ppm	0.26	1.3	0.33	22	

Table 98 (continued)
MEAN ANALYTICAL VALUES FOR 28 WHOLE COAL SAMPLES FROM THE WESTERN UNITED STATES[132]

Element	Arithmetic mean	Geometric mean	Minimum	Maximum	Standard deviation	Number samples	Number less than values
HG	0.09ppm	0.07ppm	0.02	0.63	0.11	29	
I	0.52ppm	0.46ppm	0.20	1.0	0.25	22	11
IN	0.10ppm	0.07ppm	0.01	0.25	0.07	22	5
LA	5.2 ppm	4.5 ppm	1.8	13	3.0	22	
LU	0.07ppm	0.05ppm	0.01	0.43	0.09	22	8
MN	49 ppm	28 ppm	1.4	220	49	29	1
MO	2.1 ppm	0.59ppm	0.10	30	5.6	29	6
WI	5.0 ppm	4.4 ppm	1.5	18	3.2	29	
P	130 ppm	82 ppm	10	510	130	29	
PB	3.4 ppm	2.6 ppm	0.70	9.0	2.3	29	5
RB	4.6 ppm	2.4 ppm	0.30	29	6.6	22	6
SB	0.58ppm	0.45ppm	0.18	3.5	0.61	29	
SC	1.8 ppm	1.5 ppm	0.50	4.5	1.1	22	
SE	1.4 ppm	1.3 ppm	0.40	2.7	0.59	29	
SM	0.61ppm	0.56ppm	0.22	1.4	0.29	21	
SN	1.9 ppm	0.43ppm	0.10	15	3.8	26	21
SR	260 ppm	220 ppm	93	500	140	22	
TA	0.15ppm	0.12ppm	0.04	0.33	0.08	22	
TB	0.21ppm	0.17ppm	0.06	0.58	0.15	18	
TH	2.3 ppm	1.8 ppm	0.62	5.7	1.5	22	
TL							
U	1.2 ppm	0.99ppm	0.30	2.5	0.65	22	4
Y	14 ppm	12 ppm	4.8	43	10	29	
W	0.75ppm	0.58ppm	0.13	3.3	0.65	22	
YB	0.38ppm	0.34ppm	0.13	0.78	0.17	22	
ZN	7.0 ppm	5.0 ppm	0.30	17	4.9	29	1
ZR	33 ppm	26 ppm	12	170	31	26	
AL	1.0 %	0.88%	0.31	2.2	0.56	29	
CA	1.7 %	1.5 %	0.44	3.8	0.93	29	
CL	00.03%	0.02%	0.01	0.13	0.03	29	
FE	0.53%	0.49%	0.30	1.2	0.24	29	
K	0.05%	0.03%	0.01	0.32	0.06	29	
MG	0.14%	0.12%	0.03	0.39	0.09	29	

Table 99
ELEMENTARY ANALYSES OF PETROLEUM[292]

	Per Cent			Specific Gravity
	C	H	O	$H_2O = 1$
Heavy oil, W. Va.	83.5	13.3	3.2	0.873
Light oil, W. Va.	84.3	14.1	1.6	0.8412
Heavy oil, Pa.	84.9	13.7	1.04	0.886
Light oil, Pa.	82.0	14.8	3.2	0.816
Parma, Italy	84.0	13.4	1.8	0.786
Hanover, Germany	80.4	12.7	6.9	0.892
Galicia, Austria	82.2	12.1	5.7	0.870
Light oil, Baku, Rus.	86.3	13.6	0.1	0.884
Heavy oil, Baku, Rus.	86.6	12.3	1.1	0.938
Java	87.1	12.0	0.9	0.923
Beaumont, Texas	86.8	13.2	—	0.920

Table 100
SPECIFIC GRAVITY OF SOME AMERICAN PETROLEUMS[292]

State	Specific gravity	Gravity Baumé[1]
California (Placerita Cañon)	0.777 +	50 +
Pennsylvania	0.801—.817	46.2—42.6
Ohio	0.816—.860	42.8—32.5
Kansas	0.835—1.000	38.8—10.0
West Virginia	0.841—.873	37.6—30.0
Beaumont, Texas	0.904—.925	24.8—31.1
Wyoming	0.912—.945	23.3—11.9
California	0.920—.983	21.9—12.3

Table 101
DETERMINATIONS OF SOME PETROLEUM DISTILLATES

		Physical properties			Distillation by Engler's method 100cm³ at 60°F taken											
		Gravity at 60°F				By volume							Unsaturated hydrocarbons			
	Well depth				Begins to boil	to 150°C.		150—300°C.		Residuum		Total				
Location of well	(ft)	Specific	Degrees Baumé	Color	(°C.)	(cm³)	S.G.	(cm³)	S.G.	(cm³)	S.G.	(cm³)	Crude %	150—300°C %	Paraffin %	Asphalt %
Katalla Bay, AK[1]		0.828	39.5			21.0	.757	51.0	.820	28.0						
Oil Bay, AK[2]		0.956	16.5	Dark green	230			13.2	.878	86.8						
Contra Costa, Fresno Cty., CA[3]		0.967	14.8	Black		3.7		54.5[1]		42.8						21.0
Fresno County, Coalinga, CA[4]		0.761	54.		49	30.0[1]		60.0[2]		10.0						10.0
Oil City, Coalinga, CA[5]		0.858	33.1			28.0		50.0[1]		22.0						
Kern County, Kern River, CA[6]	1052	0.961	15.7			0		20.2	.863	79.8						21.4
McKittrick, CA[7]		0.948	17.7	Black		0		39.9[1]		60.1						14.3
Sunset, CA[8]		1.00	9.9					5.0[1]		95.0						51.0
East End Field, Los Angeles City, CA[9]	1275	0.977	13.2			0		12.0	.891	88.0						19.8
Newhall Dist., CA[10]	750	0.876	29.9			20.3	0.769	34.1	.842	45.6						10.9
Puente Dist., CA[11]	1425	0.878	29.5			23.4	0.756	33.1	.844	43.5						13.1
Santa Barbara, Lompoc, CA[12]	2500	0.957	16.2			5.2		31.5	.842	63.3						20.6
Santa Maria Field, CA[13]	1600	0.888	27.6			25.9	0.746	30.6	.847	44.5						12.0
United Oil Co., Florence, CA	2445	0.875	30.0		122	1.5		27.0	.799	70.2	.908	98.7			9.23	
Boulder Cty., CO		0.830	38.6			16.0	0.737	40.0	.800	44.0						
W. C. Jones, Robinson, IL	1140	0.849	34.9	Dark green	95	13.0	0.739	37.0	.803	50.2	.924	97.2			3.71	1.6
Spellbring Farm, sec. 7, Ohio Oil Co., IL[14]	330	0.846	31.5	Light green	63	14.0	0.775	31.0	.815							
Turner lease, sec. 33, Ohio Oil Co., IL[15]	300	0.879	29.3	Brown	65	8.0	0.701	27.0	.827							
Misner lease, Pure Oil Co., IL[16]	606	0.863	32.2	Light green	75	18.0	0.719	32.0	.807							
Brant farm, sec. 26, Ohio Oil Co., IL[17]	480	0.868	31.3	Light green	75	10.5	0.726	30.5	.814							
Chanute pool, Beach lease, Rex Oil & Gas Co., Chanute, KS	751	0.865	31.9	Dark green	109	5.0	0.735	36.0	.799	57.8	.922	98.8			4.25	1.2
Webb lease, Northland Oil & Gas Co., Erie, KS		0.874	30.2	Black	135	1.0		34.0	.800	64.1	.912	99.1			4.78	3.2
Davis lease, Dunkley & Odell, Coffeyville, KS	625	0.872	30.6	Black	100	6.0	0.729	33.0	.803	58.3	.924	97.3	24.4		5.31	0.2
Hill lease, Interstate Oil & Gas Co., Peru, KS	1070	0.865	33.6	Dark green	110	7.0	0.731	37.0	.798	63.2	.915	97.2	24.8		5.79	1.5

McKinley Crude Oil Co., Humboldt, KS		0.888	27.7	Dark green	123	1.0		29.0	.815	68.9	.925	98.9	32.8	2	3.93	2.3
Springer lease, Hardison & Streeter, Rantoul, KS	350	0.856	33.6	Black	76	11.5	0.712	29.5	.795	54.9	.927	95.9	34.0		3.45	2.3
Warren Cty. & Sunnyside, KY		0.843	36.2		67	19.0		37.0		44.0						
H. Caldwell, Monticello, KY		0.821	40.5	Dark green	67	16.0	0.715	37.5	.998	41.2	.915	94.7			4.93	.4
Caddo Oil & Mineral Co., Gilbert Well, LA		0.826	39.4	Black	136	1.5		49.0	.778	49.1	.881	99.6				.3
St. Martin's Parish, Anse-le-Butte, LA[18]	600	0.939	19.1		240			16.0		84.0						
Calcasieu Parish, Jennings, LA[19]	1000	0.909	24.0		200			41.0		59.0						5.0
St. Claire Cty., Port Huron, MI		0.833	38.0	Dark green		15.0		55.0		30.0						
Lima, OH		0.791	47.0	Dark green	23	16.0	0.700	68.0	.788	16.0						
Washington City, Macksburg, OH		0.812	42.5	Light amber		15.5		45.8[2]								
Evans lease, Julia Oil Co., Muskogee, OK	1553	0.833	38.1	Green	97	11.0	0.733	36.0	.796	52.8	.887	99.8			7.64	0.
Bartlesville pool, lot 32, Illuminating Oil Co., Bartlesville, OK	1500	0.855	33.8	Dark green	113	3.5	0.751	44.0	.787	52.4	.901	99.9			7.9	1.1
Yorgee lease, Tulsa, OK	601	0.832	38.2	Dark green	93	9.0	0.722	40.5	.781	48.5	.904	98.0	17.6	1	4.39	0.
Berryhill lease, Indiana Oil & Gas Co., Kiefer, OK	1518	0.844	35.9	Black	105	8.0	0.751	44.5	.801	48.0	.909	100.5	20.8	6	7.53	0.9
Self lease, Prairie Oil & Gas Co., Tulsa, OK	1523	0.837	37.2	Black	94	10.0	0.733	41.0	.797	47.6	.902	98.6	16.8	5	3.12	0.2
Prairie Oil & Gas Co., Morris, OK		0.846	35.5	Light green	112	3.0		34.0	.792	62.1	.887	99.1	10.0	1	11.9	0.
Buchanan lease, Burns and Caton, Morris, OK	1680	0.853	34.1	Dark green	110	5.5	0.752	36.0	.802	58.0	.900	99.5	20.0	8	3.43	0.15
Lease of Richmond Dev. Co., Muskogee, OK	1702	0.833	38.0	Green	90	11.0	0.709	37.0	.799	51.2	.886	99.2	16.4	8	1.52	0.
Prairie Oil & Gas Co., Bartlesville, OK		0.852	34.3	Dark green	103	8.0	0.738	37.0	.809	54.5	.904	99.5	24.4	4	3.75	0.2
Venango County, PA[20]		0.882	28.7	Dark brown		8.6		42.8[1]		48.7[2]						
Jefferson Cty., Beaumont, TX		0.922	21.9			1.8		17.1		81.1						7.5
Jefferson Cty., Beaumont, TX		0.921	22.1	Brown		6.5		35.0	0.872	58.6						6.3
Navarro Cty., Corsicana, TX[21]		0.860	32.7	Dark brown	77	24.8[1]	0.755	59.9[2]		15.3						
Smith Bros. & Sweeney, St. Mary's, WV	1234	0.788	47.6	Green	73	15.0	0.708	43.5	0.769	39.4	.858	97.9			9.0	0.
J. Dinsmore & Co., St. Mary's, WV	1673	0.787	47.9	Green	70	16.5	0.711	41.0	0.769	34.5	.857	92.0			5.0	0.

Table 101 (continued)
DETERMINATIONS OF SOME PETROLEUM DISTILLATES

		Physical properties			Distillation by Engler's method 100cm³ at 60°F taken											
		Gravity at 60°F				By volume							Unsaturated hydrocarbons			
						to 150°C.		150—300°C.		Residuum						
Location of well	Well depth (ft)	Specific	Degrees Baumé	Color	Begins to boil (°C.)	(cm³)	S.G.	(cm³)	S.G.	(cm³)	S.G.	Total (cm³)	Crude %	150—300°C %	Paraffin %	Asphalt %
Ohio & W. Va. Oil Co., St. Mary's, WV	590	0.804	44.1	Green	89	18.0	0.726	39.5	0.782	42.8	.865	100.3			8.9	0.
Dinsmore Oil Co., St. Mary's, WV		0.798	45.4	Green	74	8.0	0.717	43.5	0.769	42.4	.861	94.0			6.93	0.
Uinta Cty., Spring Valley, WY						28.0		24.0		48.0						
Oil Spring reserve, No. 12, Shale Spring, WY		0.857	33.3	Light green	143	tr.		51.0	0.827	48.7	.893	99.7				

Notes: 1—Sulfur trace; paraffin base 2—S, 0.098 3—[1], 150—350°C. 4—[1], 49—141°C.; [2], 141—275°C. Asphaltic base. 5—S, 0.062; [1], 150—270°C. 6—S, 0.94 7—S, 0.87; [1], 150—350°C. 8—S, 1.253; [1], 150—270°C. 9—S, 0.49 10—gravity at 15°C. 11—S, 0.36; gravity at 15°C. 12—S, 4.43; gravity at 15°C. 13—S, 1.56; gravity at 15°C. 14—S, 0.48 15—S, 0.30 16—S, 0.30; lower sand oil 17—S, 27; upper sand oil 18—S, 20; asphaltic base 19—S, 39; asphaltic base 20—[1], 150—270°C.; [2], includes loss and water 21—[1], portion of Fraction 77—203°F lost before measured; [2], 150—280°C.

Table 102
COMPOSITION OF SOME CRUDE OILS IN TERMS OF THE MAIN HYDROCARBON GROUPS, RESINS AND ASPHALTENES[169]

Type of crude	Source	Paraffins %	Naphthenes %	Aromatics %	Resins and asphaltenes %
Paraffinic		40	48	10	2
Paraffinic-naphthenic	Oklahoma City	36	45	14	5
Naphthenic	Emba-Dossor	12	75	10	3
Naphthenic-aromatic	Santa-Fe	20	45	23	12
Mixed asphaltic	Inglewood	8	42	27	23
Bermudez asphalt	Bermudez Lake	5	15	20	60

Table 103
ANALYSES OF NATURAL GAS[292]

No.	Methane, (CH_4)	Ethane (C_2H_6)	Olefine (C_2H_4)	Carbon dioxide (CO_2)	Carbon monoxide (CO)	Oxygen	Nitrogen	Hydrogen	Helium	Hydrogen sulfide
1	94.40	—	—	—	—	0.23	5.08	—	0.183	—
2	96.20	0.78	—	—	0.11	tr.	2.46	0.18	0.27	—
3	82.25	—	0.12	0.61	—	tr.	16.40	—	0.616	—
4	14.85	0.41	—	—	—	0.20	82.70	tr.	1.84	—
5	62.93	—	—	0.50	tr.	0.70	24.36	11.51	undet.	—
6		95.35		1.60	2.50	0.55	—	—	undet.	—
7	13.97	—	—	0.10	0.05	0.05	85.83	—	undet.	—
8	73.81	—	—	0.81	—	3.46	21.92	—	undet.	—
9	92.67	—	0.25	0.25	0.45	0.35	3.53	2.35	undet.	0.15
10	92.61	—	0.30	0.26	0.50	0.34	3.61	2.18	undet.	0.20
11		90.01		0.20	—	tr.	9.79	—	undet.	—
12		98.90		0.40	—	—	0.70	—	undet.	—
13	80.94	14.60	—	—	0.40	0.20	3.46	tr.	undet.	—
14	86.48	7.65	—	—	0.50	0.30	4.87	—	undet.	—
15	98.40	—	—	—	0.95	tr.	0.40	tr.	undet.	tr.
16	94.20	—	0.39	1.06	1.13	0.92	3.31	—	tr.	—
17	92.20	—	—	1.40	0.21	tr.	5.59	0.40	—	0.20
18	96.57	—	—	—	—	—	2.69	—	—	0.74
Minima	14.33	—	—	0.05	—	0.10	0.60	—	—	—
Maxima	98.30	—	—	30.40	—	9.00	85.83	—	—	—

Note: 1. Iola, Kas., 2. Buffalo, Kas., 3. Fredonia, Kas., 4. Dexter, Kas., 5. Stockton, Cal., 6. From glacial drift, Dawson, Ia., 7. Princeton, Ill., 8. Pittsfield, Ill., 9. Muncie, Ind., 10. Trenton limestone, Findlay, O., 11. Kane, McKean Co., Pa., 12. Pittsburg, Pa., 13. Big Injun sand, Shinnston, W. Va., 14. Fifty-foot sand, same locality, 15. Trenton limestone, Baldwinsville; N.Y., 16. Gas from coal mine, Scranton, Pa. 17. Kent County, Ont.; 18. Welland, Ont.

Table 104
MAIN FEATURES OF SELECTED OIL SHALES[399]

	Torbanite (N.S.W.)	Trasmanite (Tasmania)	Oil shale (Colorado)
Geological era	Permian	Permian	Eocene
Depositional conditions	Fresh water	Marine	Saline
Source	Algae	Algae?	?
Inorganic matter	Silica, clays	Mudstones	Silicates and carbonates
Analysis (% w/w) — (d.a.f. basis):			
Carbon	83.6	78.1	79.2
Hydrogen	11.3	10.2	10.5
Oxygen	3.5	6.0	6.5
Nitrogen	0.6	0.6	2.6
Sulfur	1.0	5.1	1.2
C/H wt. ratio	7.4	7.7	7.5
S.G. of pure organic matter	0.96	0.99	1.04

Table 105
MINERALOGY OF SOME OIL SHALES

Shale	Ash (%)	Mineral matter (%)	Amorphous silica and quartz (%)	Feld-spar (%)	Clay minerals (%)	Gypsum ($CaSO_4$ $2H_2O$ (%)	Pyrite (FeS_2) (%)	Calcite ($CaCO_3$) (%)	Magnesite[a] ($MgCO_3$) (%)	Siderite ($FeCO_3$) (%)
Kukersite, Estonia	36.3	47.87	9.0	6.75	13.9	1.1	4.25	56.1	—	—
Kohat, N.W.F.P. India	68.7	88.83	12.40	2.47	40.68	0.49	Trace	22.68	8.34	3.76
Broxburn, Main	67.4	76.15	16.55	11.30	45.85	0.43	1.76	2.91	2.63	11.24
Kimmeridge, Dorset	37.8	40.89	38.97	5.74	20.68	8.56	4.64	3.51	—	—
Ermelo, Transvaal	44.9	47.85	50.13	5.14	29.45	0.24	2.03	1.73	0.24	—
Tasmanite, Tasmania	79.2	82.05	56.3	6.0	23.75	1.45	1.64	—	—	—
Amherst, Burma	43.9	46.78	34.33	5.63	27.45	5.49	0.19	Trace	—	—
Boghead, Autun	65.0	79.2	32.4	n.d.	17.4	1.1	0.7	37.3	—	—
Pumpherston I	75.0	83.45	24.6	n.d.	22.9	Trace	2.35	5.8	4.15	2.15
Pumpherston II	66.3	86.77	19.3	n.d.	22.9	0.3	1.35	26.7	12.1	5.1
Middle Dunnet	77.6	84.76	26.5	n.d.	54.65	0.3	0.55	4.25	3.65	—
Newnes, N.S.W.	20.1	20.79	74.0	n.d.	17.9	0.3	0.4	—	3.3	—
Cypris shale, Brazil	65.9	69.50	66.33	n.d.	17.13	0.78	1.24	5.25	0.62	—
Massive shale, Brazil	72.8	48.5	n.d.	37.2	37.2	0.4	1.4	2.6	1.1	—

Note: n.d. = not determined

[a] Apparently Himus did not consider presence of dolomite.

Table 106
COMPARISON OF BITUMEN AND SYNTHETIC CRUDE FROM TAR SANDS[175]

	Raw bitumen	Synthetic crude product
API Gravity	9—10	35
Boiling range	400—1100°F	80—900°F
Sulfur	4.5—5.0%	0.2%
Nitrogen	0.5—1.0%	0.1%
Vanadium	150 ppm	Nil
Color	Black	Straw
Ash	1.0%	Nil

Table 107
AVERAGE ANALYSES OF URANIUM SOURCE ROCKS, HOST ROCKS, AND ORES IN SANDSTONE[119]

Quantitative and Semi-quantitative Spectrographic Analyses (Parts per Million)

Element	Granite 1	Rhyolite 2	Basalt 3	Carbonaceous shale 4	Green River oil shale 5	Average crustal sand- stone 6	Barren Salt Wash Sandstone Colorado Plateau 7	Barren Salt Wash Sandstone Uravan mineral belt 8	Mudstone Yellow Cat uranium district 9	Mineralized Salt Wash Sandstone Yellow Cat district 10	Composite uranium ore in Salt Wash Sandstone Yellow Cat district 11	Composite ore in Jurassic sandstone Colorado Plateau 12	Composite ore in Triassic sandstone Colorado Plateau 13
Si	310,000*	340,000*	>100,000			367,500							
Al	75,000	80,000	>100,000	42,000	48,000	25,300	12,000					25,000	22,000
Fe	30,000	24,000	100,000	17,000	18,800	9,900	2,400	14,000	21,500	19,000	28,200	8,700	15,000
Mg	8,200	3,600	50,000	4,000	27,000	7,100	2,300					7,600	1,700
Ca	1,600	9,400	70,000	17,000		39,500	33,000			46,000	12,280	20,000	7,000
Na	19,000	20,000	15,000	2,300	27,000	3,300	890					1,100	800
K	22,000	26,000	15,000		25,000	11,000	3,000					≈6,000	≈4,800
S						2,800	200		377	1,400	7,300		
Ti	3,200	1,400	10,000	1,900	1,500	960	510					950	1,300
Mn	550	530	1,500	88	300	Trace	220	400	470	338	292	310	240
Ba	840	840	300	250	700	170	340					750	700
Sr	440	240	300	400	1,000	<26	49					120	140
B				58	58	9-31	<10					15x	14
Cr	9.1	3.7	300	79	24	68-200	6.6	13	54	76	107	16	30
Cu	12	12	100	53	52	34	13	16	23	9	23	90	300
V	50	21	300	120	130	20	400	400	327	1,280	9,510	6,800	630
Zn	36*	89*	BLD			<20	53		≈51	228	≈228	100	310
Pb	19	28	BLD	13	24	20	<1	<4	14	11	52	96	64
Co	7.2	3	50		9.1	0	<2		16	45	≈59	11	25
Ni	≈1	<1	150	52	39	2-8	≈0.5		12	23	≈35	9.8	25
As						2.6	10	18	≈46	100	420	16.8x	200
Sb			BLD			1	≈2					≈1x	2
Th	29												
U	5					1.2	<1	11	12.8	183	3,800	1,500	
S							<2	2.3	≈9	≈51	190	11.8	6
Mo			BLD	6.5	13		<5	26	≈3	41	100	20	17
Ag			BLD			0.44		<1	≈1	<1	3.5	<1	<1

Note: BLD = Below Limit of Detection
* = Chemical analysis

Table 108
ANALYSES OF SURFACE AND GROUND WATERS[119]

Reported in parts per million

Element or radical	Well in granite Maine 1	Spring in rhyolite N. Mex. 2	Spring in salt wash sandstone Utah 3	Well in westwater sandstone N. Mex. 4	Well in Wind River sandstone Wyom. 5	Well in Rush Springs sandstone Okla. 6	Hot Spring in rhyolite tuff Cal. 7	Brine Spring in kincon shale Cal. 8
SO_4	18.0	1.8	388.0	381.0	117.0		243.0	16,800
HCO_3	74.0	41.0	202.0	314.0	185.0	216.0	432.0	1350.0
CO_3	0.0	0.0		0.0	12.0	0.0	0.0	0.0
$CaCO_3$ (hardness)	53.0	21.0		246.0	4.0	274.0	247.0	17,700
PO_4	0.0			0.0		0.0	0.1	0.2
B			0.4					
NO_3	2.0	0.2	2.6	0.0	0.0	16.0	0.0	2,500
F	0.2	1.8	0.4	0.5	0.4	0.1	5.2	31.0
Cl	17.0	2.0	13.0	7.0	9.0	48.0	66.0	1,830.0
SiO_2	53.0	50.0	11.0	18.0	11.0	20.0	45.0	11.0
Al	0.0	0.0		0.0	0.0	0.2	0.0	12.0
Fe	0.09	0.02	0.2	4.0	0.0	0.08	0.5	1.3
Ca	17.	5.4	89.0	59.0	1.5	102.0	66.0	480.0
Mg	2.5	1.8		24.0	0.1	4.9	20.0	3,980.0
Na	20.0	10.0	129.0	186.0	142.0	8.8	200.0	2,490
K	8.4	1.6	6.1	4.8	1.7	0.8	20.0	54.0
Mn	0.0	0.0	20.0	0.0	0.0		0.0	2.4
Cu							0.0	0.33
Zn	0.59						0.0	12.0
Pb			0.04					
V			0.10					
U	0.011	0.0003	0.80	0.0001	0.0001	0.12	0.0004	0.66
Se			1.0					
Solids	157.0	103.0	759.0	859.0	378.0	376.0	895.0	29,700
pH	7.0	7.8	7.9	7.6	8.9	7.1	7.4	7.8
T.	53° F	57° F		63° F	54° F	60° F	110°	54° F

Table 109
EQUATIONS FOR GEOTHERMOMETERS[358]

Silica Geothermometers (SiO_2 in ppm)[a]

Quartz, adiabatic cooling (± 2°C from 125—275°C)

$$t°C = \frac{1533.5}{5.768 - \log SiO_2} - 273.15$$

Quartz, conductive cooling (± 0.5°C from 125—250°C)

$$t°C = \frac{1315}{5.205 - \log SiO_2} - 273.15$$

Chalcedony, conductive cooling

$$t°C = \frac{1015.1}{4.655 - \log SiO_2} - 273.15$$

Na/K Geothermometers (Na, K in ppm)

White and Ellis (see text) (± 2°C from 100—275°C)

$$t°C = \frac{855.6}{\log(Na/K) + 0.8573} - 273.15$$

Fournier and Truesdell (1973)

$$t°C = \frac{777}{\log(Na/K) + 0.70} - 273.15$$

NaKCa Geothermometer (Na, K, Ca in moles/liter)

Fournier and Truesdell (1973, 1974)

$$t°C = \frac{1647}{\log (Na/K) + \beta \log (\sqrt{Ca}/Na) + 2.24} - 273.15$$

$\beta = 4/3$ for $\sqrt{Ca} / Na > 1$ and $+ < 100°C$

$\beta = 1/3$ for $\sqrt{Ca} / Na < 1$ and $+_{4/3} > 100°C$

[a] Data from Fournier (written commun., 1973).

REFERENCES

1. **Abbott, M. J.,** Petrology of the Nandewar volcano, N. S. W., Australia, *Contrib. Mineral. Petrol.,* 20, 115, 1969.
2. **Ahfeld, F.,** Mineralogy of stanite, *Neues Jahrb. Mineral. Geol. Palaeontol. Abh. Abt. A,* 68, 268, 1934.
3. **Ahrens, L. H.,** The use of ionization potentials. I. Ionic radii of the elements, *Geochim. Cosmochim. Acta,* 2, 155, 1952.
4. **Alpan, S.,** Geothermal energy explorations in Turkey, *U.N. Symp. Dev. Use of Geotherm. Res.,* 1, 25, 1975.
5. **Anderson, A. L.,** Some pseudo-eutectic textures, *Econ. Geol.,* 29, 77, 1934.
6. **Anderson, F. W. and Dunham, K. C.,** *The Geology of Northern Skye,* Mem. Geological Survey, Great Britain, 1966.
7. **Arango, E., Buitrago, A. J., Cataldi, R., Ferrara, G. C., Panichi, C., and Villegas, V. J.,** Preliminary Study on the Ruiz Geothermal Project (Columbia), Proc. U.N. Symp. Development and Utilization of Geothermal Resources, Pisa, *Geothermics,* (Special Issue 2), 2(Part 1), 43, 1970.
8. **Armbrust, G. A., Arias, J., Lahsen, A., and Trujillo, P.,** Geochemistry of the Hydrothermal Alteration at El Tatio Geothermal Field, Chile, Proc. IAVCEI Symp. Int. Volcanologia, Santiago, Chile, 1974.
9. **Armstrong, J. A. and Cooper, J. A.,** Lead isotopes in island arcs, *Bull. Volcanol.,* 35, 27, 1971.
10. **Arnason, B.,** The Hydrogen and Water Isotope Thermometer Applied to Geothermal Areas in Iceland, International Atomic Energy Agency Advisory Group Meeting on the Application of Nuclear Techniques to Geothermal Studies, Pisa, Italy, 1976.
11. **Arnórsson, S.,** Underground Temperatures in Hydrothermal Areas in Iceland as Deduced from the Silica Content of the Thermal Water, Proc. U.N. Symp. Development and Utilization of Geothermal Resources, Pisa, *Geothermics,* (Special Issue 2), 2(Part 1), 536, 1974.
12. **Arnórsson, S., Bjornsson, A., Gislason, G., and Gudmundsson, G.,** Systematic exploration of the Krisuvik High-Temperature Area, Reykjanes Peninsula, Iceland, Proc. U.N. Symp. Development and Use of Geothermal Resources, Pisa, 1975.
13. **Baba, K., Tahaki, S., Matsuo, G., and Katagiri, K.,** A Study of the Reservoir at the Matsukawa Geothermal Field, Proc. U.N. Symp. Development and Utilization of Geothermal Resources, Pisa, *Geothermics,* (Special Issue 2), 2(Part 2), 1440, 1970.
14. **Baldi, P., Ferrara, G. C., and Panichi, C.,** Geothermal research in western Campania (southern Italy): chemical and isotopic studies of thermal fluids in the Campi Flegrei, *U.N. Symp. Dev. Use Geotherm. Res.,* 1, 687, 1975.
15. **Baldi, P., Ferrara, G. C., Masselli, L., and Pieretti, G.,** Hydrogeochemistry of the region between Monte Amiata and Rome, *Geothermics,* 2, 124, 1973.
16. **Barbour, G. B.,** The loess of China, *Smithson. Inst. Annu. Rep.,* 1926, 279, 1927.
17. **Barnes, I., Hinkle, M. E., Rapp, J. B., Heropoulos, C., and Vaughn, W. W.,** Chemical composition of naturally occurring fluids in relation to mercury deposits in part of North-Central California, *U.S. Geol. Surv. Bull.,* 1382-A, A1, 1973.
18. **Bate, G. L. and Huizenga, J. R.,** Abundances of ruthenium, osmium, and uranium in some cosmic and terrestrial sources, *Geochim. Cosmochim. Acta,* 27, 345, 1963.
19. **Bateman, P. C., Clark, L. D., Huber, N. K., Moore, J. G., and Rinehart, C. D.,** The Sierra Nevada batholith, *U.S. Geol. Surv. Prof. Pap.,* 414-D, 1963.
20. **Bateman, A. M. and Lasky, S. G.,** Covellite-chalcocite solid solution and exsolution, *Econ. Geol.,* 27, 52, 1932.
21. **Beaver County News,** Phillips to Flow for Five Day Test, Milford, Utah, 76(7), 1, 1976.
22. **Bedinger, M. S., Pearson, F. J., Jr., Reed, J. E., Sniegocki, R. T., and Stone, C. G.,** The Waters of Hot Springs National Park, Arkansas — Their Origin, Nature, and Management, U.S. Geological Survey, Open File Report, 1974.
23. **Behne, W.,** Untersuchungen Zur Geochemie des Chlor und Brom, *Geochim. Cosmochim. Acta,* 3, 186, 1953.
24. **Bell, K. and Powell, J. L.,** Strontium isotope studies of alkalic rocks. The potassium rich lavas of the Birunga and Toro-Ankole regions, East and Central equatorial Africa, *J. Petrol.,* 10, 536, 1969.
25. **Berkstresser, C. F., Jr.,** Data For Springs in the Northern Coast Ranges and Klamath Mountains of California, U.S. Geological Survey, Open File Report, 1968.
26. **Berry, L. G.,** Studies of the mineral sulpho-salts, *Am. Mineral.,* 25, 726, 1940.
27. **Berry, L. G., Ed.,** Powder Diffraction File, Search Manual (mineral names), Joint Committee on Powder Diffraction Standards, 1974, 32.
28. **Berry, L. G., Ed.,** Selected Powder Diffraction Data for Minerals, Joint Committee on Powder Diffraction Standards, 1974, 833.

29. **Björnsson, S., Arnorsson, S., and Tomasson, J.**, Economic evaluation of Reykjanes thermal brine area, Iceland, *Am. Assoc. Petrol. Geol. Bull.*, 56, 2380, 1972.
30. **Blackwell, D. D. and Morgan, P.**, Geological and geophysical exploration of the Marysville geothermal area, Montana, U.S.A., *U.N. Symp. Dev. Use Geotherm. Res.*, 2, 895, 1975.
31. **Boldizsar, T. and Korim, K.**, Hydrogeology of the Pannonian geothermal basin, *U.N. Symp. Dev. Use Geotherm. Res.*, 1, 297, 1975.
32. **Borchert, H.**, Über Entmischungen in System Cu-Fe-S und ihre Bedeutung als geologische Thermometer, *Chem. Erde*, 9, 145, 1934.
33. **Borchert, H. and Muir, R. O.**, *Salt Deposits: The Origin, Metamorphism, and Deformation of Evaporites*, Van Nostrand, New York, 1964, 338.
34. **Borley, G. D.**, Potash-rich potassic rocks from southern Spain, *Mineral. Mag.*, 36, 364, 1967.
35. **Bosch, B., Deschamps, J., Leleu, M., Lopoukhine, M., Marce, A., and Vilbert, C.**, The Geothermal Zone of Lake Assal (F.T.A.I.): Geochemical and Experimental Studies, Int. Symp. Water-Rock Interactions, Prague, Czechoslovakia, 1974.
36. **Bowie, S. H. U. and Taylor, K.**, A system of ore mineral identification, *Min. Mag.*, 99, 5, 1958.
37. **Bowman, H. R., Hebert, A. J., Wollenberg, H. A., and Asaro, F.**, Trace, minor, and major elements in geothermal waters and associated rock formations (North-Central Nevada), *U.N. Symp. Dev. Use Geotherm. Res.*, 1, 699, 1975.
38. **Brondi, M., Dall' Aglio, M., and Vitrani, F.**, Lithium as a pathfinder element in the large scale hydrogeochemical exploration for hydrothermal systems, *Geothermics*, 2(3), 142, 1973.
39. **Brewer, P. G. and Spencer, D. W.**, A note on the chemical composition of the Red Sea brines, in *Hot Brines and Recent Heavy Metal Deposits in the Red Sea*, Degens, E. T. and Ross, D. A., Eds., Springer-Verlag, New York, 1969, 174.
40. **Brotzen, O.**, The average igneous rock and the geochemical balance, *Geochim. Cosmochim. Acta*, 30, 863, 1966.
41. **Brown, G. M., Emeleus, C. H., Holland, J. G., and Phillips, R.**, Mineralogical, chemical features of Apollo 11 rocks and their relationship to igneous processes, in *Proceedings of the Apollo 11 Lunar Science Conference*, Levinson, A. A., Ed., Pergamon Press, New York, 1970, 195.
42. **Buddington, A. F.**, Interrelated Precambrian granitic rocks, Northwest Adirondacks, New York, *Bull. Geol. Soc. Am.*, 68, 291, 1957.
43. **Buerger, N. W.**, The unminxing of chalcopyrite from sphalerite, *Am. Mineral.*, 19, 525, 1934.
44. **Calamai, A., Cataldi, R., Dall Aglio, M., and Ferrara, G. C.**, Preliminary report on the Cesano hot brine deposit (Northern Latium, Italy), *U.N. Symp. Dev. Use Geotherm. Res.*, 1, 305, 1975.
45. **Calvert, S. E. and Price, N. B.**, Composition of manganese nodules and manganese carbonates from Lock Fyne, Scotland, *Contrib. Mineral. Petrol.*, 29, 215, 1970.
46. **Calvert, S. E. and Price, N. B.**, Minor metal contents of recent organic rich sediments of southwest Africa, *Nature (London)*, 227, 593, 1970.
47. **Calvert, S. E. and Price, N. B.**, Shallow water, continental margin, and lacustrine nodules: distribution and geochemistry, in *Marine Manganese Deposits*, Glasby, G. P., Ed., Elsevier, New York, 1977, 45.
48. **Cameli, G. M., Rendina, M., Puxeddu, M., Rossi, A., Squarci, P., and Taffi, L.**, Geothermal research in western Campania (Southern Italy): geological and geophysical results, *U.N. Symp. Dev. Use Geotherm. Res.*, 1, 315, 1975.
49. **Cameron, A. G. W.**, Abundances of elements, in *Handbook of Physical Constants*, Clark, S. P., Ed., Geological Society of America, Mem., Boulder, Colo., 1966.
50. **Cameron, E. N.**, *Ore Microscopy*, John Wiley & Sons, New York, 1961, 293.
51. **Carmichael, I. S. E.**, The mineralogy of Thingmuli, a tertiary volcano in eastern Iceland, *Am. Mineral.*, 52, 1815, 1967.
52. **Carmichael, I. S. E., Turner, F. J., and Verhoogen, J.**, *Igneous Petrology*, McGraw-Hill, New York, 1974, 739.
53. **Carpenter, H. C. H. and Fisher, M. S.**, A metallographic investigation of native silver, *Trans. Inst. Min. Met. London*, 41, 382, 1932.
54. **Carswell, D. A.**, Possible primary upper mantle peridotite in Norwegian basal gneiss, *Lithos*, 1, 322, 1968.
55. **Carswell, D. A. and Dawson, J. B.**, Garnet peridotite xenoliths in South Africa kimberlite pipes and their petrogenesis, *Contrib. Mineral. Petrol.*, 25, 163, 1970.
56. **Carter, J. L.**, Mineralogy and chemistry of the earth's upper mantle based on the partial fusion — partial crystallization model, *Bull. Geol. Soc. Am.*, 81, 2021, 1970.
57. **Challis, G. A.**, The origin of New Zealand ultramafic intrusions, *J. Petrol.*, 6, 322, 1965.
58. **Chaturvedi, L. N. and Raymahashay, B. C.**, Geologic setting and geochemical characteristics of the Parbati Valley geothermal field, India, *U.N. Symp. Dev. Use Geotherm. Res.*, 1, 329, 1975.
59. **Clarke, F. W.**, The data of geochemistry, *U.S. Geol. Surv. Bull.*, 770, 841, 1924.

60. **Compston, W., McDougall, I., and Heier, K. S.**, Geochemical comparison of the Mesozoic basaltic rocks of Antarctica, South Africa, South America, and Tasmania, *Geochim. Cosmochim. Acta,* 32, 129, 1968.
61. **Coombs, D. S.**, Trends and affinities of basaltic magmas and pyroxenes as illustrated on the diopside-olivine-silica diagram, *Mineral. Soc. Am. Spec. Pap.*, 1, 227, 1963.
62. **Coombs, D. S. and Wilkinson, J. F. G.**, Lineages of fractionation trends in unsaturated volcanic rocks from the East Otago volcanic province (New Zealand) and related rocks, *J. Petrol.*, 10, 440, 1969.
63. **Cormy, G., Demians d'Archimbaud, J., and Surcin, J.**, Prospection Geothermique aux Antilles Francaises, Guadeloupe et Martinique, Proc. U.N. Symp. Development and Utilization of Geothermal Resources, Pisa, *Geothermics,* (Special Issue 2), 2(1), 57, 1970.
64. **Cortecci, G.**, Oxygen isotopic ratios of sulfate ions — water pairs as a possible geothermometer, *Geothermics,* 3(2), 60, 1974.
65. **Cortecci, G. and Dowgiallo, J.**, Oxygen and sulfur isotopic compositions of the sulfate ions from mineral and thermal groundwaters of Poland, *J. Hydrol.*, 27, 271, 1975.
66. **Cox, K. G. and Hornung, G.**, The petrology of the Karroo basalt of Basutoland, *Am. Mineral.*, 51, 1414, 1966.
67. **Cox, K. G., Johnson, R. L., Monkman, L. J., Stillman, C. J., Vail, J. R., and Wood, D. N.**, The geology of the Nuantesi igneous province, *Philos. Trans. R. Soc. London Ser. A*, 257, 71, 1965.
68. **Craig, H.**, Isotopic Temperatures in Geothermal Systems, Int. Atomic Energy Agency Advisory Group Meeting on the Application of Nuclear Techniques to Geothermal Studies, Pisa, 1976.
69. **Cressman, E. R.**, Nondetrital siliceous sediments, *U.S. Geol. Surv. Prof. Paper,* 440 T, 1962.
70. **Cronan, D. S.**, Environments of ferromagnesian oxide deposition, in *Marine Manganese Deposits,* Glasby, G. P., Ed., Elsevier, New York, 1976, 11.
71. **Cronan, D. S. and Thomas, R. L.**, Ferromanganese concretions in Lake Ontario, *Can. J. Earth Sci.*, 7, 1346, 1970.
72. **Cusicanqui, H., Mahon, W. A. J., and Ellis, A. J.**, The geochemistry of the El Tatio geothermal field, northern Chile, *U.N. Symp. Dev. Use Geotherm. Res.*, 1, 703, 1975.
73. **Danilchik, W.**, Dieng Geothermal Exploration Drilling Project, Indonesia, During 1972 — Interim Report, *U.S. Geol. Surv., Project Rep.*, 28, 53, 1973.
74. **Dellechaie, F.**, A hydrochemical study of the south Santa Cruz basin near Coolidge, Arizona, *U.N. Symp. Dev. Use Geotherm. Res.*, 1, 339, 1975.
75. **Demians d'Archimbaud, J. and Munier-Jolain, J. P.**, Les Progrès de l'exploration geothermique à Bouillante en Guadeloupe, *U. N. Symp. Dev. Use Geotherm. Res.*, 101, 1975.
76. **Demissie, G. and Kahsai, G. A.**, Distribution of hydrothermal areas in Ethiopia and their geothermal energy potential (Abstr.), *U.N. Symp. Dev. Use Geotherm. Res.*, 2, 1975.
77. **Deroo, G., Tissot, B., McCrossan, R. G., and Der, F.**, Geochemistry of the heavy oils of Alberta, in *Oil Sands, Fuel of the Future,* Hills, L. V., Ed., Canadian Society of Petrology and Geology Mem., Calgary, 1974, 148.
78. **Dominco, E. and Papastamatoki, A.**, Characteristics of Greek geothermal waters, *U.N. Symp. Dev. Use Geotherm. Res.*, 1, 109, 1975.
79. **Dominco, E. and Şâmilgil, E.**, The Geochemistry of the Kizildere Geothermal Field, in Framework of the Sarayköy-Denizli Geothermal Area, Proc. U.N. Symp. Development and Utilization of Geothermal Resources, Pisa, *Geothermics,* (Special Issue 2), 2(Part 1), 553, 1970.
80. **Dowty, E., Keil, K., and Prinz, M.**, Igneous Rocks from Apollo 16 Rake Samples, in Proc. 5th Lunar Science Conf., *Geochim. Cosmochim. Acta,* 1, 431, 1974.
81. **Dowgiallo, J.**, The geothermal resources of southwest Poland, *U.N. Symp. Dev. Use Geotherm. Res.*, 1, 123, 1975.
82. **Dawson, J. B.**, Basutoland Kimberlites, *Bull. Geol. Soc. Am.*, 73, 545, 1962.
83. **Dawson, J. B.**, Reactivity of the cations in carbonate magmas, *Proc. Geol. Assoc. Can.*, 15, 103, 1964.
84. **Dean, W. E.**, Fe-Mn oxydate crusts in Oneida Lake, New York, *Proc. Conf. Great Lakes Res.*, 13, 217, 1970.
85. **Doe, B. R.**, The bearing of lead isotopes on the source of granitic magma, *J. Petrol.*, 8, 51, 1967.
86. **Doe, B. R., Tilling, R. I., Hedge, C. E., and Klepper, M. R.**, Lead and strontium isotope studies of the boulder batholith, southwest Montana, *Econ. Geol.*, 63, 884, 1968.
87. **Dugolinsky, B. K.**, Chemistry and Morphology of Deep-Sea Manganese Nodules and the Significance of Associated Encrusted Protozoans on Nodule Growth, Ph.D. thesis, University of Hawaii, 1976, 288.
88. **Dymek, R. F., Albee, A. L., and Chodos, A. A.**, Comparative Petrology of Lunar Cumulate Rocks of Possible Primary Origin: Dunite 72415, troctolite 76535, Norite 78235, and Anorthosite 62237, in Proc. 6th Lunar Science Conf., *Geochim. Cosmochim. Acta,* 6(1), 301, 1975.

89. **Eckstein, Y.**, Chemical geothermometry of ground waters associated with the igneous complex of southern Sinai, *U.N. Symp. Dev. Use Geotherm. Res.*, 1, 713, 1975.
90. **Edgington, D. N. and Callender, E.**, Minor element geochemistry of Lake Michigan ferromanganese nodules, *Earth Planet. Sci. Lett.*, 8, 97, 1970.
91. **Edwards, A. B.**, Some ilmenite micro-structures and their interpretation, *Aust. Inst. Min. Met. Proc.*, 110, 39, 1938.
92. **Edwards, A. B.**, Solid solution of tetrahedrite in chalcopyrite and bornite, *Aust. Inst. Min. Met. Proc.*, 143, 1946.
93. **Edwards, A. B.**, The petrology of the Cainozoic rocks of Tasmania, *Proc. R. Soc. Victoria*, 62, 97, 1950.
94. **Edwards, A. B.**, *Textures of the Ore Minerals*, Australian Institute of Mining and Metallurgy, Melbourne, 1960, 242.
95. **Engel, A. E. J. and Engel, C. G.**, Composition of basalt cored in Mohole project (Guadalupe site), *Bull. Am. Assoc. Petrol. Geol.*, 45(1), 799, 1961.
96. **Engel, A. E. J. and Engel, C. G.**, Composition of basalts from the mid-atlantic ridge, *Science*, 144(1), 330, 1964.
97. **Engel, A. E. J., Engel, C. G., and Havens, R. G.**, Chemical characteristics of oceanic basalts and the upper mantle, *Bull. Geol. Soc. Am.*, 76, 719, 1965.
98. **Ermakov, N. P.**, *Study of Mineral-Forming Solution*, Kharkov, Iz-vo, Khark., Un-ta, 1950, 459.
99. **Esder, T. and Simsek, E.**, Geology of Izmir-Seferihisar geothermal area, western Anatolia of Turkey; determination of reservoirs by means of gradient drilling, *U.N. Symp. Dev. Use Geotherm. Res.*, 1, 349, 1975.
100. **Eskola, P.**, Conditions during the earliest geologic times, *Ann. Acad. Sci. Fenn. Ser. A*, 36, 5, 1932.
101. **Ewart, A., Bryan, W. B., and Gill, J. B.**, Mineralogy and geochemistry of the younger volcanic islands of Tonga, S. W. Pacific, *J. Petrol.*, 14, 429, 1973.
102. **Ewart, A. and Stipp, J. J.**, Petrogenesis of the volcanic rocks of the central north island, New Zealand, as indicated by a study of Sr^{87}/Sr^{86} ratio and Sr, Rb, K, U, and Th abundance, *Geochim. Cosmochim. Acta*, 32, 699, 1968.
103. **Fancelli, R. and Nuti, S.**, Locating interesting geothermal areas in the Tuscany region (Italy) by geochemical and isotopic methods, *Geothermics*, 3(4), 146, 1974.
104. **Fauve, G. and Hurley, P. M.**, The isotopic composition of strontium in oceanic and continental basalts, *J. Petrol.*, 4, 31, 1963.
105. **Ferguson, J. and Lambert, I. B.**, Volcanic exhalations and metal enrichments at Matupi Harbor, New Britain, T. P. N. G., *Econ. Geol.*, 67, 25, 1972.
106. **Ferrara, G. C., Ferrara, G., and Gonfiantini, R.**, Carbon isotopic composition of carbon dioxide and methane from steam jets of Tuscany, in Nuclear Geology of Geothermal Areas, Tongiori, E., Ed., Consiglio Nazionale Delle Ricerche Laboratorio di Geologia Nuclear, Pisa, 1963, 277.
107. **Fersman, A. E.**, K voposu o sederzhanii redkikh zemel'v apatitakh (A contribution to the problem of rare earth contents of apatites), *Dokl. Akad. Nauk. SSSR, Ser. A.*, 42, 1924.
108. **Folinsbee, R. E.**, Determination of reflectivity of ore minerals, *Econ. Geol.*, 44, 425, 1949.
109. **Fomina, L. S. and Volkov, I. I.**, Rare earths in iron-manganese concentration of the Black Sea, *Dokl. Akad. Nauk. SSSR*, 185, 188, 1969.
110. **Fouillac, C., Cailleaux, P., Michard, G., and Merlivat, L.**, Premièrs Études de Sources Thermales du Massif Central Francais au Point de vue Geothermique, *U.N. Symp. Dev. Use Geotherm. Res.*, 1, 721, 1975.
111. **Fournier, R. O. and Truesdell, A. H.**, An empirical Na-K-Ca geothermometer for natural waters, *Geochim. Chosmochim. Acta*, 37, 1255, 1973.
112. **Fournier, R. O. and Truesdell, A. H.**, Geochemical indicators of subsurface temperature. II. Estimation of temperature and fraction of hot water mixed with cold water, *U.S. Geol. Surv. J. Res.*, 2(3), 263, 1974.
113. **Frank, D. J., Meylan, M. A., Craig, J. D., and Glasby, G. P.**, Ferromanganese deposits of the Hawaiian Archipelago, *Univ. Hawaii Inst. Geophys.*, 76, 71, 1976.
114. **Franko, O. and Mucha, I.**, Geothermal resources in the central depression of the Danube lowland in Czechoslavakia (Abstr.), *U.N. Symp. Dev. Use Geotherm. Res.*, 2, 1975.
115. **Franko, O. and Racicky, M.**, The present state of development of geothermal resources in Czechoslavakia (Abstr.), *U.N. Symp. Dev. Use Geotherm. Res.*, 2, 1975.
116. **Fujii, Y. and Akeno, T.**, Chemical Prospecting of Steam and Hot Water in the Matsukawa Geothermal Area, Proc. U.N. Symp. Development and Utilization of Geothermal Resources, Pisa, *Geothermics*, (Special Issue 2), 2(Part 1), 1416, 1970.
117. **Fyfe, W. S. and Turner, F. J.**, Reappraisal of the concept of metamorphic facies, *Contrib. Mineral. Petrol.*, 12, 354, 1966.

118. **Fyfe, W. S., Turner, F. J., and Verhoogen, J.**, Metamorphic reactions and metamorphic facies, *Contrib. Mineral. Petrol.*, 12, 354, 1958.
119. **Gabelmann, J. W.**, Speculations on the uranium ore fluid, in Uranium Exploration Geology, International Energy Agency, Vienna, 1970, 315.
120. **Garrels, R. M., Mackenzie, F. T., and Siever, R.**, Sedimentary cycling in relation to the history of the continents and oceans, in *The Nature of the Solid Earth*, Robertson, E. C., Ed., McGraw-Hill, New York, 1972, 677.
121. **Gast, P. W.**, Isotope geochemistry of volcanic rocks, in *Basalts*, Vol. 1, Hess, H. H. and Poldervaart, A., Eds., John Wiley & Sons, New York, 1967, 325.
122. **Gast, P. W.**, The chemical composition of the earth, the moon, and chondritic meteorites, in *The Nature of the Solid Earth*, Robertson, E. C., Ed., McGraw-Hill, New York, 1972, 677.
123. **Geogescu, I. I. and Lupan, S.**, Contributions to the study of the ferromanganese concretions from the Black Sea, *Rev. Roum. Geol. Geophys. Geogr. Sci. Geol.*, 15, 157, 1971.
124. **Gerasimovski, V. L.**, Geochemistry of the rare-earth elements, in *Rare Earth Elements*, Academy of Science, USSR, 1959, 27.
125. **Gibson, R. E.**, The influence of pressure on the high-low inversions of quartz, *J. Phys. Chem.*, 32(1), 197, 1928.
126. **Giggenbach, W. F.**, Isotopic composition of waters of the Broadlands geothermal field, *N. Z. J. Sci.*, 14(4), 959, 1971.
127. **Glasby, G. P.**, Mechanisms of enrichment of the rarer elements in marine manganese nodules, *Mar. Chem.*, 1, 105, 1973.
128. **Glover, R. B.**, Chemical Characteristics of Water and Steam Discharges in the Rift Valley of Kenya, United Nations-Kenya, Government Geothermal Exploration Project Report, 1972, 106.
129. **Glover, R. B.**, Geothermal investigations in Kenya, *N. Z. Geochem. Group Newsletter*, 4(32), 84, 1973.
130. **Glover, R. B.**, Report on visit to Phillippines, May 1974 — Part 4, *N. Z. Dep. Sci. Ind. Res. Rep.*, 27, 1974.
131. **Glover, R. B.**, Chemical analyses of waters from Negros, Oriental, Phillippines and their geothermal significance, *N. Z. Dep. Sci. Ind. Res. Rep.*, 28, 1975.
132. **Gluskoter, H. J., Ruch, R. R., Miller, W. G., Cahill, R. A., Dreher, G. B., and Kuhn, J. K.**, Trace elements in coal: occurrence and distribution, *Ill. Geol. Surv. Circ.*, 499, 154, 1977.
133. **Goldberg, E. D.**, *Chemical Oceanography*, Riley, J. P. and Skirrow, G., Eds., Academic Press, New York, 1965, 164.
134. **Goldberg, E. D. and Arrenius, G. O. S.**, Chemistry of pacific pelagic sediments, *Geochim. Cosmochim. Acta*, 13, 153, 1958.
135. **Goldschmidt, V. M.**, Geokhimicheskie Zakony Raspredeleniya i Chastota Elementov v Kosmase (The Geochemical Laws of the Distribution and the Abundance of the Elements in the Universe,) Collective Volume; *Osnovnye idei Geokhimii (Fundamental Ideas of Geochemistry)*, Goschimtekhizdat, Leningrad, 1933.
136. **Goldschmidt, V. M.**, Grundlagen der Quantitativen Geochemie, *Fortschr. Mineral. Krist. Petrogr.*, 17, 112, 1933.
137. **Goles, G. G. and Andres, E.**, On the geochemical character of iodine in meteorites, *J. Geophys. Res.*, 66, 3075, 1961.
138. **Gonfiantini, R., Borsi, S., Ferrara, G., and Panichi, C.**, Isotopic composition of waters from the Danakil depression (Ethiopia), *Earth Planet. Sci. Lett.*, 18, 13, 1973.
139. **Gorham, E. and Swaine, D. J.**, The influence of oxidizing and reducing conditions upon the distribution of some elements in lake sediments, *Limnol. Oceanogr.*, 10, 268, 1965.
140. **Grout, F. F.**, Petrographic and chemical data on the Canadian shield, *J. Geol.*, 46, 486, 1938.
141. **Green, D. H.**, Alumina content of enstatite in a Venezuelan high-temperature peridotite, *Bull. Geol. Soc. Am.*, 74(1), 397, 1963.
142. **Green, D. H.**, The petrogenesis of the high-temperature peridotite intrusion in the Lizard area, Cornwall, *J. Petrol.*, 5, 134, 1964.
143. **Green, J.**, Geochemical table of the elements for 1959, *Bull. Geol. Soc. Am.*, 70, 1127, 1959.
144. **Grill, E. V., Murray, J. W., and MacDonald, R. D.**, Todorokite in manganese nodules from a British Colombia fjord, *Nature (London)*, 219, 358, 1968.
145. **Gringarten, A. C. and Stieltjes, L.**, Study of a Geothermal Field in the Assal Active Volcanic Rift Zone, (French Tertiary of Afars and Issaš, East Africa), Workshop on Geothermal Reservoir Engineering, Stanford University, Stanford, California, 1976.
146. **Guild, F. N.**, A microscopical study of the silver ores and their associated minerals, *Econ. Geol.*, 12, 297, 1917.
147. **Gunn, B. M.**, Differentiation in ferrar dolerites of Antarctica, *N. Z. J. Geol. Geophys.*, 5, 820, 1962.
148. **Gunn, B. M.**, Model and element variation in Antarctic tholeiites, *Geochim. Cosmochim. Acta*, 30, 881, 1966.

149. Gunter, B. D. and Musgrave, B. C., Gas chromatographic measurements of hydrothermal emanations at Yellowstone National Park, *Geochim. Cosmochim. Acta,* 30, 1175, 1966.
150. Gunter, B. D. and Musgrave, B. C., New evidence on the origin of methane in hydrothermal gases, *Geochim. Cosmochim. Acta,* 35, 113, 1971.
151. Gupta, M. L., Narain, H., and Gaur, V. K., Geothermal provinces of India as indicated by studies of thermal springs, terrestrial heat flow, and other parameters, *U.N. Symp. Dev. Use Geotherm. Res.,* 1, 387, 1975.
152. Gupta, M. L., Saxena, V. K., and Sukhija, B. S., An analysis of the hot spring activity of the Manikaran area, Himackal Pradesh, India, by geochemical studies and tritium concentrations of spring waters, *U.N. Symp. Dev. Use Geotherm. Res.,* 1, 741, 1975.
153. Härme, M., On the potassium migmatites of southern Finland, *Bull. Comm. Geol. Finlande,* 219, 1965.
154. Harris, P. G., Reay, A., and White, I. G., Chemical composition of the upper mantle, *J. Geophys. Res.,* 72, 6, 359, 1967.
155. Harriss, R. C. and Troup, A. G., Freshwater ferromanganese concentrations, chemistry and internal structure, *Science,* 166, 604, 1969.
156. Haskin, L. A., Blanchard, D. P., Korotev, R., Jacobs, J. W., Brannon, J. A., and Clark, R. S., Major and trace-element concentrations in samples from 72275 and 72255, *Smithson. Astro. Obs.,* 1, 121, 1974.
157. Hedge, C. E. and Knight, R. J., Lead and strontium isotopes in volcanic rocks from northern Honshu, Japan, *Geochim. J. Japan,* 3, 15, 1969.
158. Hedge, C. E. and Lewis, J. F., Isotope composition of strontium in three basalt-andesite centers along the lesser antilles arc, *Contrib. Mineral. Petrol.,* 32, 39, 1971.
159. Hedge, C. E. and Peterman, Z. E., The strontium isotope composition of basalts from the Gordo and Juan de Fuca rises, northeastern Pacific Ocean, *Contrib. Mineral. Petrol.,* 27, 117, 1970.
160. Heide, F., *Kleine Meteoritenkunde,* 2nd ed., Springer Verlag, Berlin, 1957.
161. Heier, K. S., Metamorphism and the chemical differentiation of the crust, *Geol. Foren. Stockholm Fork,* 87, 249, 1965.
162. Heinrich, E. W., *Mineralogy and Geology of Radioactive Raw Materials,* McGraw-Hill, New York, 1958, 645.
163. Helliwell, T. M., Oscillator strengths of lead and the lead abundance in the sun, *Astrophys. J.,* 133, 566, 1961.
164. Heming, R. F., Geology and petrology of the Rabaul Caldera, Papua, New Guinea, *Geol. Soc. Am. Bull.,* 85, 1253, 1974.
165. Hess, H. H., Stillwater igneous complex, Montana, *Geol. Soc. Am. Mem.,* 80, 225, 1960.
166. Hewitt, R. L. and Schwartz, G. M., Experiment bearing of the relation of pyrrhotite to other sulfides, *Econ. Geol.,* 32, 1070, 1937.
167. Himus, G. W., Observations on the composition of kerogen rocks and the chemical constitution of kerogen, in *Oil Shale and Cannel Coal,* Vol. 2, Institute of Petroleum, London, 1951, 112.
168. Hitosugi, T. and Yonetani, M., On the drillings of the shallow wells in Onikobe, *J. Japn. Geotherm. Energy Assoc.,* 9(1), 15, 1972.
169. Hobson, G. D. and Tiratsoo, E. N., *Introduction to Petroleum Geology,* Science Press, United Kingdom, 1975, 300.
170. Holmes, A., A record of new analyses of Tertiary igneous rocks, *R. Irish Acad. Proc.,* 43(8), 89, 1936.
171. Holmes, A., A suite of volcanic rocks from southwest Uganda containing kalsilite (a polymorph of $KAl\ SiO_4$), *Mineral. Mag.,* 26, 197, 1942.
172. Holmes, A., Petrogenesis of katungite and its associates, *Am. Mineral.,* 35, 772, 1950.
173. Holmes, A. and Harwood, H. F., The tholeiite dikes of the north of England, *Mineral. Mag.,* 22, 1, 1929.
174. Hood, D. W., Seawater chemistry, in *The Encyclopedia of Geochemistry and Environmental Sciences,* Fairbridge, R. W., Ed., Holt, Rinehart and Winston, New York, 1972, 382.
175. Humphreys, R. D., Some Engineering Aspects of the Tar Sands Project, 75th Annu. Meet. Canadian Institute of Mining, Vancouver, B. C., 1973.
176. Hutchinson, R., Paul, D. K., and Harris, P. G., Chemical composition of the upper mantle, *Mineral. Mag.,* 37, 726, 1970.
177. Hyndman, D. W., *Petrology of Igneous and Metamorphic Rocks,* McGraw-Hill, New York, 1972, 533.
178. International Committee for Coal Petrology, Type of coal, *International Handbook of Coal Petrology,* 2nd ed., Centre National de la Recherche Scientifique, Paris, 1963.
179. James, H. L., Dutton, C. E., Pettijohn, F. J., and Wier, K. L., Geology and ore deposits of the Iron River-Crystal Falls district, Iron County, Michigan, *U.S. Geol. Surv. Prof. Pap.,* 570, 184, 1968.

180. **Jangi, B. L., Prakash, G., Dua, K. J. S., Thussu, J. L., Dimri, D. B., and Pathak, C. S.,** Geothermal explorations of the Parbati Valley geothermal field, Kulu district, Himackal Pradesh, India, *U.N. Symp. Dev. Use Geotherm. Res.,* 1, 1085, 1975.
181. **Kartokusumo, W., Mahon, W. A. J., and Seal, K. E.,** Geochemistry of the Kawah Kamojang geothermal system, Indonesia, *U.N. Symp. Dev. Use Geotherm. Res.,* 1, 757, 1975.
182. **Kay, R., Hubbard, N. J., and Gast, P. W.,** Chemical characteristics and origins of the oceanic ridge volcanic rocks, *J. Geophys. Res.,* 75, 1585, 1970.
183. **King, B. C.,** The Ard Bheinn area of the central igneous complex of Arran, *Q. J. Geol. Soc. London,* 110, 323, 1955.
184. **King, B. C.,** Petrogenesis of the alkaline igneous rock sites of the volcanic and intrusive centers of eastern Uganda, *J. Petrol.,* 6, 67, 1965.
185. **King, E. A.,** *Space Geology,* John Wiley & Sons, New York, 1976, 349.
186. **Koga, A.,** Geochemistry of the Waters Discharged from Drillholes in the Otake and Hatchobaru Areas, Proc. U.N. Symp. Development and Utilization of Geothermal Resources, Pisa, *Geothermics,* (Special Issue 2), 2(Part 2), 422, 1970.
187. **Koga, A. and Noda, T.,** Geochemical prospecting in vapor-dominated fields for geothermal exploration, *U.N. Symp. Dev. Use Geotherm.Res.,* 1, 761, 1975.
188. **Krauskopf, K. B.,** *Introduction to Geochemistry,* McGraw-Hill, New York, 1967, 721.
189. **Kreiger, P.,** Bornite-klaprothlite relations, *Econ. Geol.,* 37, 387, 1940.
190. **Krishnaswamy, V. S.,** A review of Indian geothermal provinces and their potential for energy utilization, *U.N. Symp. Dev. Use Geotherm. Res.,* 1, 143, 1975.
191. **Kukal, Z.,** *Geology of Recent Sediments,* Academic Press, New York, 1971, 490.
192. **Kullerud, G.,** The FeS-ZnS system, a geological thermometer, *Norsk Geol. Tidsskr.,* 32, 61, 1953.
193. **Kusakabe, M.,** Sulphur isotopic variations in nature 10 oxygen and sulphur isotope study of Wairakei geothermal well discharges, *N. Z. J. Sci.,* 17, 183, 1974.
194. **Lahsen, A. and Trujillo, T.,** El Campo Geotermico de el Tatio, Chile, *U.N. Symp. Dev. Use Geotherm. Res.,* 157, 1975.
195. **Lambert, I. B.,** The vertical distribution of uranium, thorium, and potassium in the continental crust, *Geochim. Cosmochim. Acta,* 31, 377, 1967.
196. **Lambert, I. B. and Heier, K. S.,** The vertical distribution of thorium and uranium in the continental crust, *Am. Geophys. Union Trans.,* 47, 200, 1966.
197. **Landergren, S.,** On the geochemistry of deep-sea sediments, *Rep. Swed. Deep Sea Exped.,* 10, Spec. Inv. 5, 59, 1964.
198. **Lapadu-Hargues, P.,** Sur l'existence et la Nature de l'apport Chimique dans certains series Cristallophylliennes, *Bull. Soc. Geol. Fr.,* 5(15), 255, 1960.
199. **Larsen, E. S.,** Batholith of southern California, *Geol. Soc. Am. Mem.,* 29, 182, 1948.
200. **Larsen, E. S. and Gottfried, D.,** Distribution of uranium in rocks and minerals of the Mesozoic batholiths in western United States, *U.S. Geol. Surv. Bull.,* 1070C, 63, 1961.
201. **Longinelli, A. and Craig, H.,** Oxygen-18 variations in sulfate ions in sea water and saline lakes, *Science,* 156(3771), 56, 1967.
202. **Lowder, G. G. and Carmichael, I. S. E.,** The volcanoes of Calera of Talasea, New Britain, geology and petrology, *Bull. Geol. Soc. Am.,* 81, 17, 1970.
203. **Lund, J. W.,** Culver, G. G., and Svanevik, L. S., Utilization of intermediate-temperature geothermal water in Klamath Falls, Oregon, *U.N. Symp. Dev. Use Geotherm. Res.,* 1(2), 147, 1975.
204. **Lyon, G. L.,** Geothermal gases, in *Natural Gases in Marine Sediments,* Kaplan, I. R., Ed., Plenum Press, New York, 1974, 141.
205. **Lyon, G. L., Cox, M. A., and Hulston, J. R.,** Geothermometry in geothermal areas, *N. Z. Dept. Sci. Ind. Res. Inst. Nucl. Sci. Prog. Rep.,* 19, 51, 1973.
206. **Lyon, G. L., Cox, M. A., and Hulston, J. R.,** Gas geothermometry, *N. Z. Dept. Sci. Ind. Res. Inst. Nucl. Sci. Prog. Rep.,* 20, 41, 1973.
207. **Lyon, G. L. and Hulston, J. R.,** Recent carbon isotope and residual gas measurements in relation to geothermal temperatures, *N. Z. Inst. Nucl. Sci. Contrib.,* 407, 11, 1970.
208. **MacDonald, G. A. and Katsura, T.,** Chemical composition of Hawaiian lavas, *J. Petrol.,* 5, 82, 1964.
209. **MacDonald, R., Bailey, D. K., and Sutherland, D. S.,** Oversaturated peralkaline glassy trachytes from Kenya, *J. Petrol.,* 11, 507, 1970.
210. **MacDonald, W. J. P.,** The useful heat contained in the Broadlands geothermal field, *U.N. Symp. Dev. Use Geotherm. Res.,* 2, 1113, 1975.
211. **Mahon, W. A. J.,** The chemistry of the Orakeikorako hot spring waters, *N. Z. Geol. Surv. Bull.,* 85, 104, 1972.
212. **Mahon, W. A. J. and Finlayson, J. B.,** The chemistry of the broadlands geothermal area, New Zealand, *Am. J. Sci.,* 272, 48, 1972.

213. **Mahon, W. A. J.**, The chemical composition of natural thermal waters, in *Proc. Int. Symp. Hydrogeochemistry and Biogeochemistry,* Clark, J. W., Ed., Washington, D.C., 1976, 196.
214. **MacGregor, I. D.**, Mafic and ultramafic inclusions as indicators of the depth of origin of basaltic magmas, *J. Geophys. Res.,* 73, 3737, 1968.
215. **Manheim, F. T.**, Manganese-iron accumulations in the shallow marine environment, in *Symp. Marine Geochemistry,* Schink, D. R. and Corless, J. T., Eds., Occasional Publication Narragansett Marine Laboratory, Rhode Island, 1965, 217.
216. **Manton, W. I.**, The origin of associated basic and acid rocks in the Lebombo-Nuanetsi igneous province, South Africa, as implied by strontium isotopes, *J. Petrol.,* 9, 23, 1968.
217. **Mariner, R. H. and Willey, L. M.**, Geochemistry of thermal waters in Long Valley, California, *J. Geophys. Res.,* 81, 792, 1976.
218. **Mariner, R. H., Rapp, J. B., Willey, L. M., and Presser, T. S.**, The Chemical Composition and Estimated Minimum Thermal Reservoir Temperatures of the Principal Hot Springs of Northern and Central Nevada, U.S. Geological Survey Open File Report, 1974, 32.
219. **Mariner, R. H., Rapp, J. B., Willey, L. M., and Presser, T. S.**, The Chemical Composition and Estimated Minimum Thermal Reservoir Temperatures of Selected Hot Springs in Oregon, U.S. Geological Survey Open File Report, 1974, 27.
220. **Mason, B.**, *Principles of Geochemistry,* John Wiley & Sons, New York, 1952.
221. **Mason, B.**, *Principles of Geochemistry,* 3rd ed., John Wiley & Sons, New York, 1966, 329.
222. **Mason, B. and Berry, L. G.**, *Elements of Mineralogy,* Freeman, San Francisco, 1968, 550.
223. **Mathias, M., Siebert, J. C., and Rickwood, P. C.**, Some aspects of the mineralogy and petrology of ultramafic xenoliths in kimberlite, *Contrib. Mineral. Petrol.,* 26, 75, 1970.
224. **Matsubaya, O., Sakai, H., Kusachi, I., and Satake, H.**, Hydrogen and oxygen isotopic ratios and major element chemistry of Japanese thermal water systems, *Geochem. J.,* 7, 123, 1973.
225. **Mazor, E., Kaufman, A., and Carmi, I.**, Geochemistry of a mixed thermal spring complex, *J. Hydrol.,* 18, 289, 1973.
226. **Mazor, E.**, Atmospheric and radiogenic noble gases in thermal waters: their potential application to prospecting and steam production studies, *U.N. Symp. Dev. Use Geotherm. Res.,* 1, 793, 1975.
227. **Mazor, E., Verhagen, B. T., and Negreav, E.**, Hot springs of the igneous terrain of Swaziland, their noble gases, hydrogen, oxygen, and carbon isotopes and dissolved ions, Paper IAEA-SM-182/28, in Isotope Techniques in Groundwater Hydrology, International Atomic Energy Agency, Vienna, 1974.
228. **McConville, L. B.**, The athabasca tar sands, *Min. Eng.,* 27(1), 19, 1975.
229. **McDougall, I.**, Differentiation of the Tasmanian dolerites, Red Hill dolerite-granophyre association, *Geol. Soc. Am. Bull.,* 73, 279, 1962.
230. **McDougall, I.**, Differentiation of the Great Lake dolerite sheet, Tasmania, *J. Geol. Soc. Aust.,* 11, 107, 1964.
231. **Meidav, T. and Tonani, F.**, A critique of geothermal exploration techniques, *U.N. Symp. Dev. Use Geotherm. Res.,* 2(1), 143, 1975.
232. **Melson, W. G., Jaresewich, V. T., Cifelli, R., and Thompson, G.**, Alkali olivine basalt dredged near St. Paul's rocks, mid-Atlantic ridge, *Nature (London),* 215, 381, 1967.
233. **Melson, W. G. and Thompson, G.**, Layered basic complex in oceanic crust, Romanche fracture, *Science,* 168, 817, 1970.
234. **Mercado, S. G.**, Migración de Flúidos Geotermicos y Distribución de temperaturas en el Subsuelo del Campo Geotermico de Cerro Prieto, Baja, California, Mexico, *U.N. Symp. Dev. Use Geotherm. Res.,* 1, 487, 1975.
235. **Miller, T. P.**, Distribution and Chemical Analyses of Thermal Springs in Alaska, U.S. Geological Survey Open-File Map, 1973.
236. **Miller, T. P., Barnes, I., and Patton, W. W., Jr.**, Geologic setting and chemical characteristics of hot springs in west-central Alaska, *U.S. Geol. Surv. J. Res.,* 3(2), 149, 1975.
237. **Minami, E.**, Gehalte an Seltenen Erden in Europaischen und Japanischen Tonschiefern, *Nachr. Ges. Wiss. Göttingen Math. Phys. Kl.,* 4, 155, 1935.
238. **Minami, E.**, Selen-Gehalte von Europaischen und Japanischen Tonschiefern, *Nachr. Ges. Wiss. Göttingen Math. Phys. Kl.,* 4, 143, 1935.
239. **Miyashiro, A.**, *Metamorphism and Metamorphic Belts,* John Wiley & Sons, New York, 1973, 492.
240. **Mizutani, Y.**, Isotopic composition and underground temperature of the Otake geothermal water, Kyushu, Japan, *Geochem. J.,* 6, 67, 1972.
241. **Mizutani, Y. and Hamasuna, T.**, Origin of the Shimagamo geothermal brine, Izu, *Bull. Volcanol. Soc. Japan,* 17, 123, 1972.
242. **Moorbath, S. and Bell, J. D.**, Strontium isotope abundance studies of rubidium, strontium age determinations on Tertiary igneous rocks from the Isle of Skye, northwest Scotland, *J. Petrol.,* 6, 37, 1965.
243. **Moorehouse, W. W.**, *The Study of Rocks in Thin Section,* Harper & Row, New York, 1959, 514.

244. Morgan, W. R., A note on the petrology of some lava types from east New Guinea, *J. Geol. Soc. Aust.*, 13, 583, 1966.
245. Muir, I. D. and Tilley, C. E., Mugearites and their place in alkali igneous rocks series, *J. Geol.*, 69, 186, 1961.
246. Muir, I. D. and Tilley, C. E., Basalts from the northern part of the rift zone of the Mid-Atlantic ridge, *J. Petrol.*, 5, 409, 1964.
247. Muir, I. D. and Tilley, C. E., Basalts from the northern part of the Mid-Atlantic ridge. II, *J. Petrol.*, 7, 193, 1966.
248. Mundorff, J. C., Major thermal springs of Utah, *Utah Geol. Mineral. Surv. Water Res. Bull.*, 13, 60, 1970.
249. Nagy, B. and Colombo, U., *Fundamental Aspects of Petroleum Geochemistry*, Elsevier, New York, 1967, 388.
250. Nakamura, H., Mineral and Thermal Waters of Japan, Proc. 23rd Int. Geology Congr., Prague, 19, 45, 1969.
251. Nash, W. P. and Wilkinson, J. F. G., Shonkin sag laccolith, Montana. I. Mafic minerals and estimates of temperature, pressure, oxygen, fugacity and silica activity, *Contrib. Mineral. Petrol.*, 25, 241, 1970.
252. Nash, W. P., Carmichael, I. S. E., and Johnson, R. W., The mineralogy and petrology of Mount Suswa, Kenya, *J. Petrol.*, 10, 409, 1969.
253. Nevin, A. E. and Stauder, J., Canada-early stages of geothermal investigation in British Columbia, *U.N. Symp. Dev. Use Geotherm. Res.*, 1, 1161, 1975.
254. Newhouse, W. H., The equilibrium diagram of pyrrhotite and pentlandite and their relations in natural occurrences, *Econ. Geol.*, 22, 288, 1927.
255. Newhouse, W. H., A pyrrhotite-cubanite-chalcopyrite intergrowth from the Frood mine Sudbury, *Am. Mineral.*, 16, 334, 1931.
256. Nicholls, G. D., Basalts from the deep ocean floor, *Mineral. Mag.*, 34, 373, 1965.
257. Nininger, R. D., *Minerals for Atomic Energy*, Van Nostrand, New York, 1957, 367.
258. Nissen, A. E. and Hoyt, S. L., On the occurrence of silver in Argentiferous Galena, *Econ. Geol.*, 10, 172, 1915.
259. Noble, J. W. and Ojiambo, S. B., Geothermal explorations in Kenya, *U.N. Symp. Dev. Use Geotherm. Res.*, 1, 189, 1975.
260. Nockolds, S. R., Average chemical compositions of some igneous rocks, *Geol. Soc. Am. Bull.*, 65, 1007, 1954.
261. Nuffield, E. W., Studies of mineral sulpho-salts, *Econ. Geol.*, 42, 147, 1947.
262. Oversby, V. M. and Ewart, A., Lead isotopic compositions of Tonga-Kermadec volcanics and their petrogenetic significance, *Contrib. Mineral. Petrol.*, 37, 181, 1972.
263. Pačes, T., A systematic deviation from Na-K-Ca geothermometer below 75°C and above 10^{-4} atm $^{P}CO_2$, *Geochim. Cosmochim. Acta*, 39, 541, 1975.
264. Pačes, T. and Cermak, V., Subsurface temperatures in the Bohemian massif: geophysical measurements and geochemical estimates, *U.N. Symp. Dev. Use Geotherm. Res.*, 1, 803, 1975.
265. Panichi, C., Celati, R., Noto, P., Squarci, P., Taffi, L., and Tongiorgi, E., Oxygen and hydrogen isotope studies of the Larderello (Italy) geothermal system, Paper IAEA-SM-182/35, in Isotope Techniques in Groundwater Hydrology, International Atomic Energy Agency, Vienna, 1974.
266. Parks, B. C. and O'Donnell, H. J., Petrography of American coals, *U.S. Bur. Mines Bull.*, 550, 191, 1956.
267. Patterson, E. M., A petrochemical study of Tertiary lavas of northeast Ireland, *Geochim. Cosmochim. Acta*, 2, 283, 1951.
268. Pauling, L., *The Nature of the Chemical Bond*, Cornell University Press, Ithaca, N.Y., 1960, 644.
269. Peterman, Z. E., Carmichael, I. S. E., and Smith, A. L., Sr^{87}/Sr^{86} ratios of Quaternary lavas of the Cascade Range, northern California, *Geol. Soc. Am. Bull.*, 81, 311, 1970.
270. Peterman, Z. E., Carmichael, I. S. E., and Smith, A. L., Sr^{87}/Sr^{86} ratio in Quaternary basalts of southeastern California, *Earth Planet. Sci. Lett.*, 7, 381, 1970.
271. Petrović, A., Types of hydrogeological structures and possible hydrogeochemical provinces of thermomineral waters of Serbia, *U.N. Symp. Dev. Use Geotherm. Res.*, 1, 531, 1975.
272. Pettijohn, F. J., *Sedimentary Rocks*, Harper & Row, New York, 1957, 718.
273. Pettijohn, F. J., Chemical composition of sandstones, excluding carbonate and volcanic sands, in *Data of Geochemistry*, 6th ed., U.S. Geological Survey Professional Paper 440-S, 1963.
274. Pettijohn, F. J., *Sedimentary Rocks*, 3rd ed., Harper & Row, New York, 1975, 292.
275. Poldervaart, A., Chemistry of the earth's crust, *Geol. Soc. Am. Spec. Paper*, 62, 119, 1955.
276. Povarennykh, A. S., Quantitative evaluation of the form of chemical bonds in minerals, *Dokl. Akad. Nauk. SSSR*, 109, 993, 1956.
277. Powell, J. L. and Bell, K., Strontium isotopic studies of alkalic rocks, localities from Australia, Spain, and western United States, *Contrib. Mineral. Petrol.*, 27, 1, 1970.

278. **Press, F. and Siever, R.,** *Earth,* 2nd ed., Freeman, San Francisco, 1978, 649.
279. **Prinz, M., Bunch, T. E., and Keil, K.,** Composition and origin of lithic fragments and glasses in Apollo 11 samples, *Contrib. Mineral. Petrol.,* 32, 211, 1971.
280. **Prinz, M., Dowty, E., Keil, K., and Bunch, T. E.,** Spinel troctolite and anorthosite in Apollo 16 samples, *Science,* 179, 74, 1973.
281. **Prinz, M., Dowty, E., Keil, K., and Bunch, T. E.,** Mineralogy petrology and chemistry of lithic fragments from Luna 20 fines: origin of the cumulate ANT-suite and its relationship to high-alumina and mare basalts, *Geochim. Cosmochim. Acta,* 37, 979, 1973.
282. **Prinz, M. and Keil, K.,** Mineralogy, petrology, and chemistry of ANT-suite rocks from the Lunar Highlands, *Phys. Chem. Earth,* 10, 215, 1977.
283. **Pushker, P.,** Strontium isotope ratios in volcanic rocks of three island arc areas, *J. Geophys. Res.,* 73(2), 701, 1968.
284. **Radja, V.,** Overview of geothermal energy studies in Indonesia, *U.N. Symp. Dev. Use Geotherm. Res.,* 1, 233, 1975.
285. **Ramdohr, P.,** Neue Beobachtungen uber die Vermendbarkeit Opaker Erz als Geologische Thermometer, *Z. Prakt. Geol.,* 39(65), 89, 1931.
286. **Ramdohr, P.,** Uber Schapbachit, Matildit, und den Silber, und Wismutgehalt Mancher Bleiglanze, Sitzungberichte der Preuss, *Akad. Wiss. Phys. Math. Kl.,* 71, 1938.
287. **Rankama, K. and Sahama, T.,** *Geochemistry,* University of Chicago Press, 1950, 912.
288. **Reed, J. J.,** Chemical and modal composition of Dunite from Dun Mountain, Nelson, *N. Z. J. Geol. Geophys.,* 2, 916, 1959.
289. **Reed, M. J.,** Chemistry of Thermal Water in Selected Geothermal Areas of California, California State Division of Oil and Gas Rep. No. TR15, Sacramento, 1975, 31.
290. **Reed, M. J.,** Geology and hydrothermal metamorphism in the Cerro Prieto geothermal field, Mexico, *U.N. Symp. Dev. Use Geotherm. Res.,* 1, 539, 1975.
291. **Rich, R. A., Holland, H. D., and Petersen, U.,** *Hydrothermal Uranium Deposits,* Elsevier, New York, 1977, 264.
292. **Ries, H.,** *Economic Geology,* McGraw-Hill, New York, 1930, 74.
293. **Rightmire, C. T., Young, H. W., and Whitehead, R. L.,** Geothermal investigations in Idaho. IV. Isotopic and geochemical analyses of water from the Bruneau-Grandview and Weiser areas, southwest Idaho, *Idaho Dep. Water Res. Water Inf. Bull.,* 1976.
294. **Ringwood, A. E.,** *Composition and Petrology of the Earth's Mantle,* McGraw-Hill, New York, 1975, 188.
295. **Ringwood, A. E. and Green, D. H.,** An experimental investigation of the gabbro-eclogite transformation and some physical consequences, *Tectonophysics,* 3, 383, 1966.
296. **Ritchie, J. A.,** A determination of some base metals in Broadlands geothermal waters, *N. Z. Dept. Sci.,* 2, 209, 1973.
297. **Robertson, E. C., Fournier, R. O., and Strong, C. P.,** Hydrothermal activity in southwestern Montana, *U.N. Symp. Dev. Use Geotherm. Res.,* 1, 553, 1975.
298. **Robinson, B. W.,** Sulphur isotope equilibrium during sulphur hydrolysis at high temperatures, *Earth Planet. Sci. Lett.,* 18, 443, 1973.
299. **Ronov, A. B. and Yaroshevsky, A. A.,** Earth's crust geochemistry, in *Encyclopedia of Geochemistry and Environmental Sciences,* Fairbridge, R. W., Ed., Van Nostrand, New York, 1969, 243.
300. **Rose, H. J., Jr., Cuttitta, F., Annell, C. S., Carron, M. K., Christian, R. P., Dwornik, E., Greenland, L. P., and Lignon, D. T., Jr.,** Compositional Data for Twenty-One Fra Mauro Lunar Materials, Proc. 3rd Lunar Sci. Conf., MIT Press, Cambridge, Mass., 1972, 1215.
301. **Rösler, H. J. and Lange, H.,** *Geochemical Tables,* Elsevier, New York, 1972, 468.
302. **Rossman, R.,** Lake Michigan Ferromanganese Nodules, Ph.D. thesis, University of Michigan, Ann Arbor, 1973.
303. **Rossman, R. and Callender, E.,** Geochemistry of Lake Michigan manganese nodules, *Proc. Conf. Great Lakes Res.,* 12, 306, 1969.
304. **Ruckmick, J. C. and Noble, J. A.,** Origin of the ultramafic complex at Union Bay, southeastern Alaska, *Geol. Soc. Am. Bull.,* 70, 981, 1959.
305. **Sachanen, A. N.,** Hydrocarbons in petroleum, in *The Science of Petroleum,* Dunstan, A. E., Nash, A. W., Brooks, B. T., and Tizard, H. T., Eds., Oxford University Press, London, 1950, 55.
306. **Sahama, T. G. and Meyer, A.,** A study of the volcano Nyirangongo, Progress report, Exploration du parc national Albert, Mission d'etudes vulcanologiques, fasc., 2, 1958.
307. **Sakai, H. and Matsubaya, O.,** Isotopic geochemistry of the thermal waters of Japan and its bearing on the Kuroko ore solutions, *Econ. Geol.,* 69, 974, 1974.
308. **Samoilov, Y. V. and Titov, A. G.,** Iron-manganese rich nodules of the Black, Baltic, and Barents Seas, *Tr. Geol. Miner. Muz.,* 3, 25, 1922.
309. **Savolahti, A.,** The Ahvenisto Massif in Finland, *Bull. Comm. Geol. Finland,* 174, 1956.
310. **Schneiderhohn, H. and Ramdohr, P.,** Lehrbuch der Erzmikroskopie, 2, 1, 1931.

311. Schoen, R. and Rye, R. O., Sulphur isotope distribution in solfataras, Yellowstone National Park, *Science,* 170, 1082, 1970.
312. Schoell, M., Heating and convection within the Atlantis II deep geothermal system of the Red Sea, *U.N. Symp. Dev. Use Geotherm. Res.,* 1, 583, 1975.
313. Schoettle, M. and Friedman, G. H., Fresh water iron-manganese nodules in Lake George, New York, *Geol. Soc. Am. Bull.,* 82, 101, 1971.
314. Scholtz, D. L., The magmatic nickeliferous ore deposits of east Griqualand and Pondoland, *Trans. Geol. Soc. S. Afr.,* 39, 81, 1936.
315. Schwartz, G. M., Chalcopyrite and cubanite, *Econ. Geol.,* 22, 44, 1927.
316. Schwartz, G. M., Copper veins on Susie Island, Lake Superior, *Econ. Geol.,* 28, 762, 1928.
317. Schwartz, G. M., Intergrowths of bornite and chalcopyrite, *Econ. Geol.,* 26, 186, 1931.
318. Schwartz, G. M., Chalcocite-stromeyerite-argentite, *Econ. Geol.,* 30, 138, 1935.
319. Searle, E. J., Petrochemistry of the Aukland basalts, *N. Z. J. Geol. Geophys.,* 3, 23, 1960.
320. Sederholm, J. J., The average composition of the earth's crust in Finland, *Comm. Geol. Bull. Finlande,* 70, 3, 1925.
321. Sevast'yanov, V. F., Redistribution of arsenic during formation of iron-manganese concretions in Black Sea sediment, *Dokl. Akad. Nauk. SSSR,* 176, 191, 1967.
322. Sevast'yanov, V. F. and Volkov, I. I., Redistribution of chemical elements in the oxidised layers of the Black Sea sediments and the formation of iron-manganese nodules, *Tr. Inst. Okeanol. Akad. Nauk. SSSR,* 83, 135, 1967.
323. Seward, T. M., Equilibrium and oxidation potential in geothermal waters at Broadlands, New Zealand, *Am. J. Sci.,* 274, 190, 1974.
324. Shanker, R., Padhi, R. N., Arora, C. L., Prakash, G., Thussu, J. L., and Dua, K. J. S., Geothermal explorations of the Puga and Chumathang geothermal fields, Ladakh, India, *U.N. Symp. Dev. Use Geotherm. Res.,* 1, 245, 1975.
325. Shaw, D. M., Geochemistry of pelitic rocks. III. Major elements and general geochemistry, *Geol. Soc. Am. Bull.,* 67, 919, 1956.
326. Shaw, D. M., Reilly, G. A., Muysson, J. M., Pattenden, G., and Campbell, F. E., An estimate of the chemical composition of the Canadian precambrian shield, *Can. J. Earth Sci.,* 4, 829, 1967.
327. Shepherd, E. S., Gasses in rocks and some related problems, *Am. J. Sci.,* 325, 311, 1938.
328. Short, M. N., Microscopic determination of ore minerals, *U.S. Geol. Surv. Bull.,* 914, 314, 1940.
329. Sigvaldason, G. E. and Cuellar, G., Geochemistry of the Ahuachapan Thermal Area, El Salvador, Central America, Proc. U.N. Symp. Development and Utilization of Geothermal Resources, Pisa, *Geothermics,* (Special Issue 2), 2(Part 1), 392, 1970.
330. Sill, C. W. and Willis, C. P., The beryllium content of some meteorites, *Geochim. Cosmochim. Acta,* 26, 1209, 1962.
331. Smith, F. G., *Physical Chemistry,* Adison-Wesley, Reading, Mass., 1963, 624.
332. Smith, C. H. and Kapp, H. E., The Muskox intrusion, *Mineral. Soc. Am. Spec. Paper,* 1, 30, 1963.
333. Smirnov, V. I., *Geology of Mineral Deposits,* MIR Publishers, Moscow, 1976, 520.
334. Sorey, M. L. and Lewis, R., Discharge of hot spring systems in the Long Valley Caldera, *J. Geophys. Res.,* 81, 785, 1976.
335. Souther, J. G., Geothermal potential of western Canada, *U.N. Symp. Dev. Use Geotherm. Res.,* 1, 259, 1975.
336. Stahl, W., Aust, H., and Dounas, A., Origin of artesian and thermal waters determined by oxygen, hydrogen, and carbon isotope analyses of water samples from the Sperkhios Valley, Greece, Paper IAEA-SM-182/15, in Isotope Techniques in Groundwater Hydrology, International Atomic Energy Agency, Vienna, 1974, 317.
337. Stillwell, F. L., Observations on the mineral constitution of the Broken Hill lode, *Aust. Inst. Min. Metall. Proc.,* 64, 34, 1926.
338. Stueber, A. M. and Murthy, V. R., Strontium isotope and alkali element abundances in ultramafic rocks, *Geochim. Cosmochim. Acta,* 30, 1243, 1966.
339. Subramaniam, A. P., Charnockites of the type area near Madras, a reinterpretation, *Am. J. Sci.,* 257, 321, 1959.
340. Sumi, K. and Maeda, K., Hydrothermal alteration of main productive formation of the steam for power at Matsukawa, Japan, in Proc. Int. Symp. Hydrogeochemistry and Biogeochemistry, Clark, J. W., Ed., American Geophysical Union, Washington, D.C., 1973, 211.
341. Sutherland, D. S., Potash-trachytes and ultra potassic rocks associated with the carbonatite complex of Tororo Hills, Uganda, *Mineral. Mag.,* 35, 363, 1965.
342. Swanberg, C. A., The application of the Na-K-Ca geothermometer to thermal areas of Utah and the Imperial Valley, California, *Geothermics,* 3(2), 53, 1974.
343. Tan, E., Geothermal drilling and well testing in the Afyon area, Turkey, *U.N. Symp. Dev. Use Geotherm. Res.,* 2, 1523, 1975.

344. **Tatsumoto, M.**, Genetic relations of oceanic basalts as indicated by lead isotopes, *Science*, 153, 1094, 1966.

345. **Tatsumoto, M., Hedge, C. E., and Engel, A. E. J.**, Potassium, rubidium, strontium, thorium, and uranium in oceanic tholeiitic basalt, *Science*, 150, 886, 1965.

346. **Tatsumoto, M. and Knight, R. J.**, Isotopic composition of lead in volcanic rocks from central Honshu - with regard to basalt genesis, *Geochim. J. Jpn.*, 3, 53, 1969.

347. **Taylor, S. R.**, Abundance of chemical elements in the continental crust: a new table, *Geochim. Cosmochim. Acta*, 28(8), 1273, 1964.

348. **Taylor, S. R.**, Trace-element abundances and the chondritic earth model, *Geochim. Cosmochim. Acta*, 28, 1989, 1964.

349. **Thompson, J. M., Presser, T. S., Barnes, R. B., and Bird, D. B.**, Chemical Analysis of the Waters of Yellowstone National Park, Wyoming from 1965 to 1973, U.S. Geological Survey Open File Report 75-25, 59, 1975.

350. **Tilling, R. I. and Gotteried, T.**, Distribution of thorium, uranium, and potassium in igneous rocks of the Boulder batholith region, Montana, *U.S. Geol. Surv. Prof. Paper*, 614-E, 1969.

351. **Tilley, C. E. and Muir, I. D.**, The Hebridean plateau magma type, Edinburgh, *Geol. Soc. Trans.*, 19, 208, 1962.

352. **Tomasson, J., Fridleifsson, I. B., and Stefansson, V.**, A hydrological model for the flow of thermal water in southwestern Iceland with special reference to the Reykir and Reykjavik thermal areas, *U.N. Symp. Dev. Use Geotherm. Res.*, 1, 643, 1975.

353. **Trainer, F. W.**, Groundwater in the southwestern part of the Jemez Mountains volcanic region, *N. M. Geol. Soc. Guideb.*, 337, 1974.

354. **Truesdell, A. H.**, Geochemical Evaluation of the Dieng Mountains, Central Java, for the Production of Geothermal Energy, U.S. Geological Survey Open File Rep. IND-8, 1971, 29.

355. **Truesdell, A. H.**, Summary of section III, geochemical techniques in exploration, *U.N. Symp. Dev. Use Geotherm. Res.*, 1, 4, 1975.

356. **Truesdell, A. H.**, Chemical Evidence for Subsurface Structure and Fluid Flow in a Geothermal System, Proc. Int. Symp. Water-Rock Interactions, Prague, Czechoslavakia, 1974.

357. **Truesdell, A. H. and Fournier, R. O.**, Calculation of deep temperatures in geothermal systems from the chemistry of boiling spring waters of mixed origin, *U.N. Symp. Dev. Use Geotherm. Res.*, 1, 837, 1975.

358. **Truesdell, A. H. and Fournier, R. O.**, Procedure for estimating the temperature of a hot water component in a mixed water using a plot of dissolved silica vs. enthalpy, *U.S. Geol. Surv. Res. J.*, 1976.

359. **Truesdell, A. H. and Fournier, R. O.**, Deep Conditions in the Geothermal System of Yellowstone Park, Wyoming, From Chemical, Isotopic, and Geophysical Data, U.S. Geological Survey Open File Report, 1976.

360. **Turekian, K. K. and Kulp, J. L.**, The geochemistry of strontium, *Geochim. Cosmochim. Acta*, 10, 245, 1956.

361. **Turekian, K. K. and Wedepohl, K. H.**, Distribution of the elements in some major units of the earth's crust, *Geol. Soc. Am. Bull.*, 72, 175, 1961.

362. **Turner, F. J.**, *Metamorphic Petrology, Mineralogy, and Field Aspects*, McGraw-Hill, New York, 1968, 403.

363. **Turner, F. J. and Verhoogen, J.**, *Igneous and Metamorphic Petrology*, 2nd ed., McGraw-Hill, New York, 1960.

364. United Nation Development Programme, *Report on the Geology, Geochemistry and Hydrology of the Hot Springs of the East African Rift System in Ethiopia, Investigation of Geothermal Resources for Power Development (ETH-26)*, United Nations, New York, 1971, 433.

365. **Vakin, E. A., Polak, B. G., Sugrobov, V. M., Erlikh, E. N., Belousov, V. I., and Pilipenko, G. F.**, Recent Hydrothermal Systems of Kamchatka, Proc. U.N. Symp. Development and Utilization of Geothermal Research, Pisa, *Geothermics*, (Special Issue 2), 2(Part 2), 1116, 1970.

366. **Varne, R.**, The petrology of Moroto Mountain, eastern Uganda and the origin of nephelinites, *J. Petrol.*, 9, 169, 1968.

367. **Varentsov, I. M.**, Geochemical studies on the formation of iron-manganese nodules and crusts in recent basins. I. Eningi-Lampi Lake, Central Karelia, *Acta Mineral. Petrogr.*, 10, 363, 1972.

368. **Varentsov, I. M.**, Geochemical aspects of formation of ferromanganese ore in shelf regions of recent seas, *Acta Mineral. Petrogr.*, 21, 141, 1973.

369. **Vilminot, J. C.**, Les Enclaves de Peridotite et de Pyroxenolite a Spinelle dans le Basalt du Rocher du Lion, *Bull. Soc. Fr. Mineral. Cristallogr.*, 88, 109, 1965.

370. **Vinogradov, A. P.**, *Geokhimiya Redkikh i Rasseynnykh Khimichskikh Elementov v Pochvakh (Geochemistry of Rare and Scattered Chemical Elements in Soils)*, Publ. House Academy, Moscow, 1957.

371. **Vinogradov, A. P**, Atomic abundance of the chemical elements in the sun and in stony meteorites, *Geokhimiya*, 4, 291, 1962.

372. **Vinogradov, A. P.,** Average content of chemical elements in main types of igneous rocks of the earth's crust, *Geokhimiya,* 7, 551, 1962.
373. **Vinogradov, A. P.,** Serdniye Soderzhaniya Khimicheskikh Elementov v Glavnykh Tipakh Izerzhennykh Gornykh Porod Zemnoi Kory, *Geokhimiya,* 560, 1962.
374. **Vinogradov, A. P.,** Zakonomermosti Raspredeleniya Khimicheskikh Elementov v Zemnoi Kore (The distribution regularities of the chemical elements in the earth's crust), *Geokhimiya,* 1, 1956.
375. **Vinogradov, A. P. and Ronov, A. B.,** Composition of the sedimentary rocks of the Russian platform in relation to the history of its tectonic movements, *Geochemistry,* 6, 533, 1956.
376. **Von Knorring, O. and Du Bois, C. G. B.,** Carbonatite lava from Fort Portal area in Uganda, *Nature (London),* 192, 1064, 1961.
377. **Wager, L. R. and Brown, G. M.,** *Layered Igneous Rocks,* Freeman, San Francisco, 1967.
378. **Wager, L. R. and Mitchell, R. L.,** Trace elements in a suite of Hawaiian lavas, *Geochim. Cosmochim. Acta,* 3, 217, 1953.
379. **Wager, L. R., Vincent, E. A., Brown, G. M., and Bell, J. D.,** Marscoite and related rocks of the western Red Hills complex, Isle of Skye, *Philos. Trans. R. Soc. London Ser. A,* 257, 273, 1965.
380. **Walker, K. R.,** A mineralogical, petrological and geochechemical investigation of the Palisades sill, New Jersey, *Geol. Soc. Am. Mem.,* 115, 175, 1969.
381. **Walker, F. and Poldervaart, A.,** Karroo dolerites of the union of South Africa, *Geol. Soc. Am. Bull.,* 60, 591, 1949.
382. **Weaver, C. E.,** Potassium, illite and the ocean, *Geochim. Cosmochim. Acta,* 31, 2181, 1967.
383. **Webb, D. A. and Fearon, W. R.,** Studies on the ultimate composition of biological material. I. Aims, scope, and methods, *Sci. Proc. R. Dublin Soc.,* 21, 487, 1937.
384. **Wedepohl, K. H.,** Comparison of the deep sea and nearshore clays with reference to some minor elements, *Geochim. Cosmochim. Acta,* 14, 166, 1958.
385. **Wedepohl, K. H., Ed.,** *Handbook of Geochemistry,* Springer-Verlag, New York, 1969.
386. **Wedepohl, K. H.,** Environmental influences on the chemical composition of clays, *Phys. Chem. Earth,* 8, 305, 1971.
387. **Weigland, P. W. and Ragland, P. C.,** Geochemistry of Mesozoic dolerite dikes from eastern North America, *Contrib. Mineral. Petrol.,* 29, 195, 1970.
388. **White, D. E.,** Environments of generation of some base-metal ore deposits, *Econ. Geol.,* 63, 301, 1968.
389. **White, D. E., Barnes, I., and O'Neil, J. R.,** Thermal and mineral waters of nonmeteoritic origin, California coast ranges, *Geol. Soc. Am. Bull.,* 84, 547, 1973.
390. **White, D. E., Fournier, R. O., Muffler, L. J. R., and Truesdell, A. H.,** Physical results of research drilling in thermal areas of Yellowstone National Park, Wyoming, *U.S. Geol. Surv. Prof. Paper,* 892, 70, 1975.
391. **White, D. E., Hem, J. D., and Waring, G. A.,** Chemical composition of subsurface waters, *U.S. Geol. Surv. Prof. Paper,* 440-F, F1, 1963.
392. **White, D. E. and Truesdell, A. H.,** The geothermal resources of Taiwan, *Min. Res. Serv. Organ. Taiwan Rep.,* 105, 51, 1972.
393. **White, D. E. and Waring, G. A.,** Data of geochemistry, *U.S. Geol. Surv. Prof. Paper,* 440-K, 1, 1963.
394. **Williams, P. L., Mabey, D. R., Zohdy, A. A. R., Ackermann, H., Hoover, D. B., Pierce, K. L., and Oriel, S. S.,** Geology and geophysics of the southern Raft Rift Valley geothermal area, Idaho, USA, *U.N. Symp. Dev. Use Geotherm. Res.,* 2, 1273, 1975.
395. **Winkler, H. G. F.,** *Petrogenesis of Metamorphic Rocks,* Springer-Verlag, New York, 1957, 220.
396. **Winkler, H. G. F.,** *Petrogenesis of Metamorphic Rocks,* Springer-Verlag, New York, 1967, 237.
397. **Wyllie, P. J.,** *The Dynamic Earth,* John Wiley & Sons, New York, 1971, 416.
398. **Yamada, E.,** Geological development of the Onikobe Caldera and its hydrothermal system, *U.N. Symp. Dev. Use Geotherm. Res.,* 1, 665, 1975.
399. **Yen, T. F. and Chilingarian, G. V.,** Oil shales, *Dev. Petrol. Geol.,* 5, 292, 1976.
400. **Yoder, H. S.,** High-low quartz inversion up to 10,000 bars, *Trans. Am. Geophys. Un.,* 31, 827, 1950.
401. **Young, H. W. and Mitchell, J. C.,** Geothermal investigations in Idaho. I. Geochemistry and geologic setting of selected thermal waters, *Idaho Dep. Water Admin. Water Inf. Bull.,* 30, 43, 1973.
402. **Young, H. W. and Whitehead, R. L.,** Geothermal investigations in Idaho. II. An evaluation of thermal water in the Bruneau-Grandview area, southwest Idaho, *Idaho Dep. Water Res. Water Inf. Bull.,* 30, 125, 1975.
403. **Young, H. W. and Whitehead, R. L.,** Geothermal investigations in Idaho. III. An evaluation of thermal water in the Weiser area, Idaho, *Idaho Dep. Water Res. Water Inf. Bull.,* 30, 35, 1975.
404. **Schneiderhohn, H.,** *Met. Erz.,* 19, 523, 1922.

Section II
Densities of Rocks and Minerals

By
Gary R. Olhoeft and Gordon R. Johnson

CONCEPTS

Density is a physical property that changes significantly among various rock types owing to differences in mineralogy and porosity. If the distribution of underground rock densities is known, potentially much information can be learned about subsurface geology. Laboratory or borehole measurements of density can thus aid in the interpretation of field studies and, especially, gravity surveys.

In common usage, density is defined as the weight in air of a unit volume of an object at a specific temperature; however, in strict usage, the density of an object is defined as mass per unit volume. Weight is defined as the force that gravitation exerts on a body and thus varies with location, whereas mass is a fundamental property, a measure of the matter in a body; and mass is constant irrespective of geographic location, altitude, or barometric pressure. In many instances, such as routine density measurements of rocks, the sample weights are considered to be equivalent to their masses because the discrepancy between weight and mass will result in less error in the computed density than will experimental errors encountered in the measurement of volume. Therefore, density is often determined using weight rather than mass. Moreover, when using an equal-arm balance and standard masses to weigh an object, the effects of variations in the force of gravity are negated. The resultant measurement of apparent mass differs slightly from true mass due to the buoyant effects of air. True mass, if desired, can be computed by using a correction for the buoyant effects of air.

Specific gravity, in contrast to density, is defined as the ratio of the weight or mass in air of a unit volume of material at a stated temperature to the weight or mass in air of a unit volume gas-free distilled water at a stated temperature. Density should be reported in SI units (kg/m^3) but often is reported as g/cm^3. Specific gravity is dimensionless. Measurements of weight and volume are usually made in laboratories where normally minor variations of ambient temperature in the same laboratory as well as among laboratories have little effect on densities of rocks and minerals. For this reason the temperature at which density or specific gravity is determined is often ignored, and densities are thus commonly reported without regard to temperature. However, according to Mason,[1] ignoring the effect of temperature on determinations of density can lead to errors that are greater than the experimental error encountered while making careful routine measurements. Mason[1] discussed the concept of density in relation to temperature, errors to expect from misunderstanding of the effects of temperature, and how to apply corrections in order to minimize errors. For the sake of clarity, the expression for density should be "density at x" where x is the temperature of the material. The usage of the terms density and specific gravity are standardized in a few publications such as International Society for Rock Mechanics Committee on Laboratory Tests, Document number 2,[2] and ASTM E12-70.[3] Unfortunately, there is no standardization of these terms from one publication to another.

Table 1 summarizes the nomenclature used in this chapter. Density, as used here, refers to either bulk density in which both the volume of solid material called the grain volume and the volume of void or pore space are considered, or to grain density in which only the volume of solid material is considered. Bulk density, especially of sedimentary rocks, varies with fluid content (water) within pore spaces and can therefore be expressed in the following ways:

$$\text{Dry bulk density } (\rho_b) = \frac{W_g}{V_b} \tag{1}$$

where W_g is the weight of grains and V_b is the combined volume of grains and pore space.

Table 1
NOMENCLATURE

Weights and Volumes of Rock Constituents

W_g	Weight of grains
W_w	Weight of natural pore fluid
V_g	Volume of grains
V_p	Volume of pores
$V_{g'}$	Apparent grain volume
V_b	Bulk sample volume

Porosity and Density Terminology of Rocks

ρ_g	Grain density $= W_g/V_g$
ρ_b	Dry bulk density $= W_g/V_b$
ρ_s	Saturated bulk density $= W_g + V_p\rho_w/V_b$
ρ	Natural bulk density $= W_g + W_w/V_b$
ρ_w	Density of pore fluid
n_t	Total porosity $= (1 - V_g/V_b)\ 100$
n_a	Apparent porosity $= (V_p/V_b)\ 100$

Miscellaneous Terminology

ρ_e	Electron density
Δ_g	Vertical change in gravitational field strength

$$\text{Natural bulk density } (\rho) = \frac{W_g + W_w}{V_b} \tag{2}$$

where W_w is the weight of the natural pore water.

$$\text{Saturated bulk density } (\rho_s) = \frac{W_g + (V_p\rho_w)}{V_b} \tag{3}$$

where V_p is the volume of intercommunicating pores and ρ_w is the density of water. In contrast, grain density (ρ_g) is the ratio of the weight of the grains (W_g) to the volume of the grains (V_g).

Porosity or volume of void space is calculated from the relationship between bulk density and grain density. Total porosity is a measure of all void spaces in a porous material whether the voids are isolated or interconnected to the surface of a test sample. The formula for total porosity (n_t) expressed in percent is

$$n_t = \left(1 - \frac{V_g}{V_b}\right) 100 \tag{4}$$

Grain density may be substituted for grain volume and dry bulk density may be substituted for bulk volume yielding the equation

$$n_t = \left(1 - \frac{\rho_b}{\rho_g}\right) 100 \tag{5}$$

Apparent porosity (n_a) is a measure of the volume of those pores that intercommunicate with the surface and that are penetrable by a test fluid. The formula for apparent porosity (n_a) is

$$n_a = \left(\frac{V_p}{V_b}\right) 100 \tag{6}$$

Table 2
METHODS COMMONLY USED TO OBTAIN PORE VOLUME, BULK VOLUME, AND GRAIN VOLUME

I. Pore Volume

(A) Water absorption — several techniques
(B) Determination of volume of natural state water
(C) Absorption of an organic liquid in a vacuum
(D) Injection of mercury at high pressures into sample pores
(E) Washburn-Bunting method;[12] volume of pore air measured directly in a calibrated tube

II. Bulk Volume

(A) Calipered dimensioning of symmetrically shaped specimens
(B) Mercury displacement — several techniques
(C) Liquid displacement — on waxed specimens or impermeable material
(D) Buoyancy method — based on Archimedes' principle, requires saturated weight of specimen in air and weight of specimen suspended in liquid of known density

III. Grain Volume

(A) Liquid displacement method — on pulverized grains
(B) Boyle's law method — gas volumenometry, several variations
(C) U.S. Bureau of Mines method;[9] pressure-volume relationships of a gas system with and without sample
(D) Buoyancy method — requires weight of grains and weight of specimen suspended in liquid of known density

Some methods used to obtain apparent porosity involve a measurement of the pore volume and the determination of bulk volume but most methods involve determination of both bulk volume and apparent grain volume ($V_{g'}$) which includes both grains and isolated pores. The difference between bulk volume and apparent grain volume is equal to the volume of interconnecting pores. Therefore Equations 4 and 5 may be used to calculate apparent porosity by substituting the volume expression $V_b - V_{g'}$, for V_p into Equation 6.

METHODS FOR DETERMINING POROSITY AND DENSITY

Throughout the years many methods have been devised to determine densities and porosities of rocks and minerals. Holmes[4] discussed many of the earlier methods used to obtain specific gravities of rocks and minerals. Included in his discussion was a method used to obtain volumes of rocks and minerals at high temperatures. An extensive study of techniques and apparatus for the measurement of density was made by Hidnert and Peffer.[5] Standard porosity-density analysis procedures were categorized in API Recommended Practice.[6] Manger[7] summarized several techniques that have been used by experimenters to obtain porosity and bulk density. Among other and more recent publications that pertain to techniques of measurement are the International Society for Rock Mechanics Committee on Laboratory Tests,[2] and Muller.[8]

All methods used to obtain porosity and density involve the determination of any two of either pore volume, bulk volume, or grain volume. Table 2 summarizes the more popular methods used to obtain these three properties, each method having its advantages and limitations. The pore volume heading (Group I, Table 2) pertains to methods of obtaining the volume of interconnected pores, so these methods are thus used only to determine apparent porosity. Apparent grain volumes are determined on solid samples using either methods B, C,[9] or D listed under the grain volume heading (Group III), and true grain volumes are determined on pulverized samples using a type of Boyles Law porosimeter (Method B) or by liquid displacement (Method A). The measured values of pore, bulk, or grain volumes may differ for a given sample depending on the methods used. These variations are usually small but in some instances can significantly affect the porosity or density results. Manger[10] discussed the relationships and variations of observed values of porosity and density as a function of the methods used to obtain pore, bulk, and grain volumes. When Methods A or C are selected to determine pore volume, the sample pores are saturated with a suitable liquid using one of several techniques. Usually the volume of pore water, which is assumed to equal the pore volume, is determined by first weighing the sample saturated and again after dehydration, finding the difference between the two weights, and dividing by the density of the saturating liquid. Method B may be used to find pore volume if pores are completely saturated when samples are collected and samples are preserved in a natural-state condition until measurements are made, but usually Method B is used to determine the degree of natural-state fluid saturation within a rock. When using Method D in Group I, the volume of mercury injected into sample pores is measured directly as a function of applied pressure. Although time consuming, this method is very useful because bulk density, apparent grain density, and porosity may be calculated from the measurements and information relating to pore-size distribution, pore shape, and specific surface area of pore walls is also obtained.[11] Of course this method is destructive owing to mercury contamination of the sample pores.

The Washburn-Bunting method[12] (E) is a simple way to measure pore volumes in materials of moderate to high porosity. In principle, the air contained in the sample pores is expanded and then measured in a calibrated capillary tube. This process is accomplished by manipulating a column of mercury in such a way that a vacuum is created in the chamber surrounding the sample. The pore air then leaves the sample and is subsequently compressed at atmospheric pressure in the capillary tube. The Washburn-Bunting method is rapid but inaccurate for samples of low porosity. Unfortunately, mercury penetration into either friable samples or samples having large pore radii renders these samples useless for other tests.

The buoyancy or hydrostatic method (D, Groups II and III) is frequently used to analyze samples of rock because dry and saturated bulk density, apparent grain density, and water-accessible apparent porosity all can be calculated from the measurements. If care is used, this method will yield accurate results for most rocks owing to the fact that all measurements of volume are derived from precise measurements of weight, rather than the inexact methods generally used to determine volume. However, this method is generally not suitable for measuring friable or poorly consolidated samples unless these samples are coated with paraffin or beeswax, and then only bulk density can be determined.

When using the buoyancy method, three weighings are necessary in order to determine bulk density and apparent grain density. Samples are first dried and weighed in air (W_1). Samples are then saturated with water, weighed surface-dry in air (W_2), and finally weighed suspended by a fine-diameter wire in water (W_3). Dry bulk density is calculated using the formula

$$\rho_b = \frac{W_1}{W_2 - W_3} \times \rho_w \tag{7}$$

where ρ_w is the density of water at the particular temperature of measurement. Grain density is calculated using the formula

$$\rho_g = \frac{W_1}{W_1 - W_3} \times \rho_w \tag{8}$$

The grain densities of nonporous materials such as single crystals are often determined with considerable accuracy using the buoyancy techniques. Smakula[13] described a refined technique of hydrostatic weighing whereby grain density of crystals accurate to ±0.003% or better may be resolved. Basically the accuracy of this technique depends on precise temperature control of the immersion liquid. Another method of determining grain density by hydrostatic weighing uses the Jolly spring balance. This method, used primarily to determine the grain density of small mineral samples, is described by Holmes.[4] Athy[14] developed a method to obtain bulk density using the Jolly balance where the sample is weighed submerged in mercury.

Bulk volumes obtained using mercury displacement methods (Table 2, Group II, B) can be more or less than bulk volumes obtained by calculations from calipered dimensions of samples depending upon the textural quality of their surfaces. If samples have large surface pores or a vuggy texture, mercury will tend to fill these holes and the volume of displaced mercury will be less than the actual bulk volume; whereas if samples have relatively small pores the displaced volume of mercury may be greater than the actual sample volume owing to the negative capillarity of mercury. The mercury-displacement method is particularly useful to determine the bulk volume of irregularly shaped samples.

True grain volume and, hence, grain density and total porosity can be determined with certainty by using the liquid-displacement method for pulverized or nonporous material (Table 2, Group III, A). Most other methods yield apparent grain volume which is used to determine apparent porosity. Grain volume is determined by sequentially weighing a suitable pycnometer bottle (1) dry, (2) dry with grains, (3) with grains and filled with a liquid of known density, and (4) filled with the liquid only. The temperature of the liquid should be the same when performing Steps 3 and 4 in order to eliminate any error in the results caused by variations of density of the liquid. The grain density is calculated using the formula

$$\rho_g = \frac{b - a}{(d - a) - (c - b)}\,\rho_w \qquad (9)$$

where ρ_w is the density of displacement liquid at the specified temperature. The specific gravity of the liquid can be substituted for ρ_w if it is preferable to report results as specific gravities.

The liquid-displacement method is subject to several possible errors, the sources of which are described by Mason,[1] but when these errors are minimized the method will produce accurate results and has thus been extensively used by many workers to measure specific gravities of minerals. Semimicro- and micropycnometers are often used to determine the specific gravity of small amounts of material. Muller[8] summarized the development of the use of the micropycnometer.

Several types of porosimeters have been devised which utilize the pressure-volume relationships of Boyle's law for perfect gases. Some Boyle's law porosimeters, or gas pycnometers use air while others use a nonadsorbing gas to obtain maximum precision of measurement. A light, nonadsorbing gas such as helium will invade all void spaces except those that are completely isolated. Hence results are very nearly representative of true grain volumes; and helium-accessible porosities calculated therefrom are often equivalent to total porosities and are typically greater than water-accessible porosities. If porous samples are pulverized to the extent that isolated pores are destroyed, true grain volumes of the powdered material may then be obtained using gas pycnometry. Gas pycnometry is rapid compared to the liquid-displacement method and commonly, the accuracy of gas pycnometers compares favorably to the accuracy of liquid pycnometers. The precision of the gas pycnometer is discussed by McIntyre et al.[15] and by Zenger.[16]

DIRECT AND CALCULATED DETERMINATIONS OF DENSITY

One of the most accurate methods of determining grain density of solid material is by comparison with heavy liquids. Techniques for using this method have been extensively examined by several investigators. Pertinent references include Holmes,[4] Hidnert and Peffer,[5] and Muller.[8] Basically, the method involves immersing small fragments or powder in a liquid of approximately the same density as the solid material. The density of the liquid is adjusted until the solid material neither floats nor sinks. This is accomplished by adding miscible liquid of either greater or lesser density than the primary liquid. The final density can also be adjusted by varying the temperature.[17] The density of the liquid is subsequently accurately determined by using a standard technique such as liquid pycnometry, the Westphal balance, or by measurement of refractive index.[18]

Another method of determining grain density of solids is based on flotation of specimens in a columnar-liquid density gradient. The density gradient is prepared by mixing two miscible liquids of different densities in such a way that a continuous density gradient is formed. To calibrate the density column, index materials of known specific gravities are dropped in the column. A relationship of height vs. density can then easily be determined. A simple technique devised to accomplish this was described by Holmes.[4] Pelsmaekers and Amelinck[19] described an alternative method whereby a heat gradient was applied in a column of homogeneous liquid. ASTM method D-1505-68[20] described a density gradient column whereby an accuracy of at least 0.05% may be obtained. Commercial units are available that are in accordance with ASTM specifications.

The above flotation methods are more applicable to grain-density determinations of small mineral specimens rather than of rocks because of size limitations. However, the liquid-density gradient method may be a convenient method to simultaneously determine the grain density of various minerals in a crushed rock. Also, subtle differences in density of a single crushed mineral specimen could be easily determined by either of these methods.

With the advent of X-ray crystallography, it has become possible to determine the theoretical density of molecules in the unit cell.[1] The volume of the unit cell can be determined if the cell parameters are known.[21] Advantages of the X-ray method are that densities of minute quantities of material can be calculated in this way and that the density so determined is independent of voids or inclusions that may be in the material.[1] Also, densities obtained by the X-ray method should, with careful work, have a high degree of accuracy. Foote and Jette[22] showed that for several samples of calcite in which the lattice dimensions were precisely determined, the calculated grain densities were accurate to at least ±0.01%. If calculated X-ray densities are compared with carefully measured densities, the correctness of the chemical formula (obtained by X-ray) can perhaps be ascertained.[21] That is, if the calculated and measured densities agree, then the X-ray chemistry should be reasonably accurate.

The grain density of rocks can also be calculated from petrographic analyses.[23] If volume concentrations of the minerals and their grain densities are known, then

$$\rho_g = C_1 \cdot \rho_1 + C_2 \cdot \rho_2 + C_3\rho_3 \cdots C_n\rho_n \tag{10}$$

where C_1, C_2, etc. are the fraction of volume of the mineral components, and ρ_1, ρ_2, etc. are grain densities of the minerals. If the weight concentrations and grain densities are known, the grain density of the rock can be calculated from the equation:

$$\rho_g = \frac{1}{\dfrac{K_1}{\rho_1} + \dfrac{K_2}{\rho_2} + \dfrac{K_3}{\rho_3} \cdots \dfrac{K_n}{\rho_n}} \tag{11}$$

where K_1, K_2, etc. are the fraction by weight of the mineral components. If the chemical analysis of a rock is determined, the grain density may be calculated by substituting percentages of chemical compounds or calculated mineral constituents and naturally occurring elements and their respective grain densities into Equations 10 and 11. Similarly to calculated X-ray densities, the correctness of chemical analysis of rocks can perhaps be verified by comparison of calculated and measured grain densities. When using Equations 10 and 11, calculated density depends on (1) the accuracy of the petrographic or chemical analysis, (2) the range of density of each mineral constituent within the rock, or (3) the range of density of each of the chemical compounds that are present in the rock.

IN SITU MEASUREMENTS OF DENSITY AND POROSITY

Density Logging

Gamma rays interact with matter to produce photoelectric absorption, pair-production effects, and Compton scattering. If, as in density logging, a suitable gamma-ray source energy and detection are chosen, interactions with the rock formations can be almost exclusively limited to Compton scattering.[24] Sondes are designed in such a way that gamma rays emanating from the source penetrate the wall rock where they are "back scattered" to a detection system. If the source and detector are shielded from one another, most of the gamma rays reaching the detector will be those that have been scattered by the medium. According to theory, when Compton scattering occurs the gamma rays decrease in frequency and increase in wavelength. This scattering is proportional to the electron density ρ_e of the medium such that

$$\rho_e = \mu_o(Z/A)\rho_b \qquad (12)$$

where μ_o is Avogadro constant. Z, A, and ρ_b are, respectively, the atomic number, atomic weight, and bulk density of the material.

Because most common elements found in rocks have Z/A ratios of approximately 0.5, the electron density is mainly dependent on the ρ_b. Therefore, the response of the detector is proportional to the bulk density. Hydrogen is the most common element found in borehole logging that has a markedly different Z/A ratio (1.0). However, the presence of hydrogen in compounds generally has little effect on the Z/A ratio of the molecule. For example, the Z/A ratio of fresh water is 0.556, but water generally comprises a small percentage of the enclosing rocks. When formations containing much water or formations such as halite (Z/A = 0.4799) are logged, corrections for Z/A ratios should be applied if the true density is desired. Using the Z/A ratio of limestone (0.5000) as a standard to calibrate the gamma-gamma probe, the true density is the product of apparent density and the ratio of 0.5000 to Z/A of the material.

Although the physical principles that allow the measurement of bulk density using gamma rays have been known for several decades, it wasn't until the late 1950s before borehole logging of rock density using the gamma-gamma method was introduced. Advances in nuclear and electronic technology have made possible the use of effective sources, shields, detectors, and in-hole transistor circuitry for the development of logging techniques. Borehole logging probes capable of making quantitative density measurements were first described by Baker,[25] Caldwell,[26] Campbell and Wilson,[27] and Pickell and Heacock.[28] Basically, these probes were designed to use a collimated source and a single collimated detector. Most probes had a spring mechanism which held the source and the detector against the borehole wall in order to minimize errors due to the presence of fluid in the gap between probe and wall rock. Wahl et al.[29] described dual-spacing formation density equipment that compensate for errors due to mudcake and borehole rugosity. The probe utilizes a second detector, placed between the source and the normal detector. This short-spacing detector is especially sensitive to material near the probe. The outputs of both detectors are automatically processed by an analog computer to determine the correct formation density. More recently, Scott[30] described the development of a digital density-compensation algorithm suitable for evaluation by either an off-line computer or a small, on-site data processor. Scott's[30] results indicate that a digital procedure is more accurate than conventional analog techniques when determining formation densities.

If grain densities of the wall rocks are known, then porosities may be calculated using the values of bulk density obtained from the gamma-gamma log. With proper substitutions, Equation 2 may be rewritten as

$$\rho = \rho_g(1 - n_t) + \rho_w n_a \tag{13}$$

where n_a is the porosity expressed as a fraction. It has to be assumed here that n_a and n_t are equivalent for purposes of gamma-gamma porosity measurements. To solve for porosity

$$n = \frac{\rho_g - \rho}{\rho_g - \rho_w} \tag{14}$$

ρ_w is the density of the fluid filling the pores and is considered to be unity. If the mineralogy of sedimentary rocks is known, ρ_g can be estimated with reasonable accuracy. Davis[31] showed that, for a given sedimentary rock type, the porosity and natural bulk density are linearly related. Thus, porosity of sediments can be determined from gamma-gamma-derived densities by using either Equation 14 or by using a family of curves showing porosity vs. natural bulk density. Only one curve would probably be required for each rock type.

Porosity Logging

Neutron logging was introduced in 1941 and has since developed into an effective method of measuring formation porosity.[32] Neutron logs are made using sondes designed similar to gamma-gamma sondes. Often the same sonde with an appropriate source and detector is used to make both logs. The formation is bombarded by neutrons emanating from the source, and energy, in the form of either gamma rays, fast neutrons, or slow neutrons, is detected by a receiving element. As in gamma-gamma logs, compensated neutron logs may be obtained if two detectors located at different distances from the source are used.

Neutrons are emitted from the source at high energy and, as they repeatedly collide with and are scattered by nuclei of matter, their speed is slowed to a thermal state, at which time they move randomly about with virtually constant energy until they are captured. When neutrons are captured, their energies are conserved and emitted in the form of secondary gamma rays of capture. These high-energy gamma rays are detected by either a scintillation counter or a Geiger-Müller tube. If hydrogen nuclei are present, as in water-filled or oil-filled pores, the neutrons are slowed to a thermal state rapidly and, hence, many are captured before they travel far. Therefore, as porosity increases, fewer neutrons or gamma rays of capture reach the detector. The intensity of the detected signal is, accordingly, inversely related to the porosity of the formation.

Because water and oil contain about the same amount of hydrogen, neutron logging is considered to be an effective tool for finding formation porosity. The depth of neutron penetration into the formation depends on the porosity and fluid content. As an example, in material of low porosity in a 0.15-m diameter hole, the zone of neutron response may extend approximately 0.61 m into the wall rock, whereas in high-porosity granular rocks, such as sandstone, most of the neutron response will come from the shallow invaded zone (the portion of wall rock affected by drilling fluid).

Pirson[32] described many factors that may produce errors in the determined porosity when using neutron logs. Among these, the most serious of errors seem to be hole diameter and rugosity, presence of casing, and various problems associated with drilling fluids. Corrections can be made for most of these factors, but neutron logs are seldom used by themselves to determine *in situ* porosity.

In situ porosity may be obtained when using other nonnuclear logging methods. These methods include the short normal electric and elastic wave (sonic or acoustic) porosity logs. The short normal electric measures the resistivity of the invaded zone, and the resistivity may be used to estimate the porosity if the resistivity of the drilling fluid is known and if formations are sufficiently invaded. Acoustic log methods measure the shortest time it takes for elastic waves to travel through the formation rocks adjacent to the borehole. If the lithologies of the rocks are known, then their porosities may be determined by measuring these transit times over a short distance (usually 0.30 m).

Combinations of the above logging methods are often used to determine formation porosity. The appropriate choice of log depends on such factors as the type of drilling fluid used, borehole diameter and rugosity, and lithology. The density, acoustic, and neutron logs used singly require a knowledge of the lithology to obtain porosity, but, according to Hilchie,[33] these logs can be used in combination like simultaneous equations to solve the porosity and infer lithology without actually knowing the lithology.

Borehole Gravimetry

Borehole gravimetry has, in recent years, been developed as a reliable method of measuring *in situ* bulk density of formation rocks; however, the idea is not new. Airy[34] attempted to determine the mean density of the Earth by using a pendulum to determine the rate of change of gravity with depth in mine workings. Various investigators including Smith[35] and Hammer,[36] advanced theory and devised experimental studies, but not until the early 1950s were borehole gravimeters successfully used. Early systems had low precision (about ± 0.5 mgal or greater) and, usually, long reading times.[37-39] Howell et al.[40] described a downhole gravity meter having a precision of about ± 0.01 mgal which allowed for the determination of subsurface bulk density with an accuracy of ± 0.02 g/cm^3. McCulloh and others[41] described yet another borehole gravimeter capable of high precision (about ± 0.016 mgal) and relatively rapid reading time.

Borehole gravimetry should not be considered as an alternate to gamma-gamma density logging but usually as a complementary technique. For instance, because they integrate over a large radius of material, borehole gravity logs are compared with gamma-gamma logs to determine zones of low density, and thus high porosity, that may be remote from the borehole. Because of the sensing depth, borehole gravimeters are not affected by casing or borehole conditions such as rugosity, mudcake, or the effects of mud invasion. However, gravity readings are made at discrete intervals in boreholes, and each reading requires several minutes. Therefore, borehole gravimetry is considerably slower than is logging by conventional methods, and the logs derived from the readings are discontinuous. Two readings are made at each station with an appropriate depth interval between each reading. Simply stated, the vertical change in gravity is

$$\Delta g = (F - 4\pi G\rho)\Delta Z \tag{15}$$

where F is the free air vertical gradient, G is the universal gravitational constant, and ΔZ is the measuring interval. In terms of density, Equation 15 becomes

$$\rho = \frac{1}{4\pi G}(F - \Delta g/\Delta Z) \tag{16}$$

Thus, by measuring the change of gravity (Δg) within a measured interval (ΔZ) in a borehole, the bulk density of a considerable envelope of rock may be determined. Varying

ΔZ does not affect the radius of investigation, but the definition of the gravity field will be increased or decreased accordingly by decreasing or increasing ΔZ. The two variables, Δg and ΔZ, in Equation 16 are the major sources of error in making determinations of ρ. For example, any error of measurement resulting from inherent sensitivity limitations of the gravity meter will become larger as ΔZ decreases, and as ΔZ increases, temperature changes in the well may degrade the measurement of Δg. Other factors which may, in a smaller sense, affect the accuracy of ρ include terrain effects, borehole effects, and effects due to tides. The first two are so small that they are usually ignored, and effects of tides are always compensated for by referring to tide-correction tables. Schmoker[42] discussed the accuracy of borehole gravity data.

DATA COMPILATION

The densities of minerals are fairly well known and constant. X-ray crystallographic density determinations are precisely known to four significant figures on individual minerals.[43] However, naturally occurring minerals frequently deviate from the X-ray density by a fraction of a percent to as much as several percent (depending upon mineral type) due to deviations from stoichiometry, minor impurity contents, fluid inclusions, and related factors. Rosenholtz and Smith[44] have the most complete listing of mineral density ranges. Table 3 lists the most commonly accepted grain density of 583 minerals.

The densities of rocks are considerably more variable. Table 4 lists typical densities for various generic types of soil, but nearly any soil type can be found to fall in the range from less than 1 to nearly 2 g/cm^3. At densities higher than 2 g/cm^3, the soils generally become less friable and merge into rocks. A way to show the variations in the density of rocks to good advantage is to plot the statistical distributions. Table 5 lists the statistical properties for a variety of rock types. Sources of density data used in Table 5 include Washington[45] and Piersol and others.[46]

Figure 1 shows the histogram of grain densities for the minerals in Table 3. Across the top of the plot is a box containing the histogram counting bin width (''Div.'') in grams per cubic centimeter, the number of samples in the histogram (''No.''), the skewness (''Skew''), mean, standard deviation (''SD''), mode, and median (''Medn.''). The main plot is the histogram (raw and smoothed) as percent of total number of samples vs. density in grams per cubic centimeter. The small inset plot is the percentage of samples (vertical axis) that fall within the interval of the (mode $-$ x) to the (mode $+$ x) where x is the horizontal axis. Thus, in this plot, 72% of the samples fall in the interval from 3.03 $-$ 2.82 to 3.03 $+$ 2.82.

Figure 2 illustrates the histogram for the dry bulk density of all rocks, and, as expected, the most common (modal) value of the distribution falls at 2.65, or roughly the density of quartz. As Figure 3 illustrates, the modal value and shape of the distribution for different rocks may change considerably with rock type. This is not a fair comparison, as ''granite'' is a precise chemical or compositional term, whereas ''sandstone'' is a loose textural term, and ''basalt'' lies somewhere in between. However, there are clearly strong differences in the density histograms, even if we loosen the nomenclature just to compare intrusive, extrusive, and sedimentary rocks.

Figures 4 through 22 illustrate the histograms for various rock types where enough data were available to make the construction of a histogram meaningful. Insufficient data were available for most sedimentary rock types, metamorphic rocks, and alteration products like the clays and zeolites (see Daly et al.).[47]

Table 6 lists ranges of bulk density for various rock types that had insufficient numbers of measurements for histogram compilation. Primary sources were Daly et al.,[47] Birch,[48] Press,[49] Clark,[50] and Schock et al.[51] Mean values are not given as the modal value is more

Table 3
GRAIN DENSITIES OF MINERALS

Name/description/formula	Grain density g/cc
Acanthite/Argentite Ag_2S	7.248
Actinolite $Ca_2(Mg,Fe)_5Si_8O_{22}(OH)_2$	3.200
Aegirine $NaFeSi_2O_6$	3.550
Akermanite $MgCa_2Si_2O_7$	2.940
Alabandite MnS	4.054
Albite $NaAlSi_3O_8$	2.620
Allanite2$(Fe^{2+},Fe^{3+})Al_2[O/O/OH/SiO_4/SiO_7]$	3.800
Allemontite AsSb	6.000
Almandine $Fe_3Al_2Si_3O_{12}$	4.318
Altaite PbTe	8.246
Aluminum Al	2.698
Aluminum antimonide AlSb	4.340
Aluminum oxide-gamma Al_2O_3	3.900
Aluminum sulfate $Al_2(SO_4)_3$	2.710
Aluminum sulfide Al_2S_3	2.320
Aluminum trifluoride AlF_3	2.882
Alunite $KAl_3\ (SO_4)_2(OH)_6$	2.700
Amblygonite $(Li,Na)Al(PO)_4(F,OH)$	3.110
Amesite $(Mg,Fe;2^+)^4Al_4Si_2O_{10}(OH)_8$	2.790
Ammonia-niter $NH_4{\cdot}NO_3$	1.725
Ammonium bisulfate NH_4HSO_4	1.780
Ammonium sulfate $(NH_4)_2SO_4$	1.769
Analbite $(Na,K)AlSi_3O_8$ see Albite	2.620
Analcime $NaAlSi_2O_6{\cdot}H_2O$	2.258
Anatase TiO_2	3.840
Andalusite Al_2SiO_5	3.145
Andradite $Ca_3Fe_2Si_3O_{12}$	3.860
Anglesite $PbSO_4$	6.324
Anhydrite $CaSO_4$	2.963
Annabergite $Ni_3(AsO_4)_2{\cdot}8H_2O$	3.000
Anorthite $CaAl_2Si_2O_8$	2.760
Anthophyllite $(Mg,Fe)_7Si_8O_{22}(OH)_2$	3.000
Antigorite $Mg_3Si_2O_5(OH)_4$ Serpentine	2.600
Antimony Sb	6.698
Antlerite $Cu_3SO_4(OH)_4$	3.900
Apatite $Ca_5(PO_4)_3F$	3.180
Apophyllite $KCa_4Si_8O_{20}(F,OH){\cdot}8H_2O$	2.300
Aragonite $CaCO_3$	2.931
Arcanite K_2SO_4	2.662
Arfvedsonite $(Na,Ca)_{2\text{-}5}(Fe^{2+},Fe^{3+},Mg)_5Si_8O_{22}(OH)_2$	3.200
Arsenic As	5.780
Arsenic bromide $AsBr_3$	3.540
Arsenolite As_2O_3	3.870
Arsenopyrite FeAsS	6.162
Artinite $Mg_2CO_3(OH)_2{\cdot}3H_2O$	2.020
Atacamite $Cu_2Cl(OH)_3$	3.760
Augite $Ca(Mg,Fe^{2+},Fe^{3+},Al)(Si,Al)_2O_6C_3$	3.300
Axinite $Ca_2(Mn,Fe)Al_2BSi_4O_{15}OH$	3.300
Azurite $Cu_3(CO_3)_2(OH)_2$	3.787
Baddeleyite ZrO_2	5.826
Barite $BaSO_4$	4.480
Barium Ba	3.594
Barium chloride $BaCl_2$	3.850
Barium oxide BaO	5.992
Barium sulfide BaS	4.250

Table 3 (continued)
GRAIN DENSITIES OF MINERALS

Name/description/formula	Grain density g/cc
Barium titanate $BaTiO_3$	6.017
Barium zirconate $BaZrO_3$	5.520
Barkevikite $NaCa_2(Mg,Fe^{2+},Fe^{3+},Al)_5(Si,Al)_8O_{23}(OH)$	3.300
Beidellite $(Na,K,Mg,Ca)O{\cdot}33A_{12}(Si,Al)_4O_{10}(OH)_2{\cdot}nH_2O$	2.600
Benitoite $BaTiSi_3O_9$	3.600
Berlinite $AlPO_4$	2.618
Beryl $Be_3Al_2Si_6O_{18}$	2.641
Beryllium Be	1.847
Beryllium oxide-beta BeO	3.010
Berzelianite Cu_2Se	6.836
Bianchite $(Zn,Fe)SO_4{\cdot}6H_2O$	2.072
Biotite $K2(Mg,Fe)_{4\text{-}6}(Si,Al)_8O_{20}(OH)_4$	2.900
Bismite Bi_2O_3	9.370
Bismuth Bi	9.807
Bismuthinite Bi_2S_3	6.808
Bixbyite $(Mn,Fe)_2O_3$ Braunite	4.945
Boehmite AlO(OH)	3.071
Boracite $Mg_6B_{14}O_{26}Cl_2$	2.900
Borax $Na_2B_4O_7{\cdot}10H_2O$	1.700
Boric oxide B_2O_3	2.558
Bornite Cu_5FeS_4	5.091
Boron B	2.465
Boulangerite $Pb_5Sb_4S_{11}$	6.000
Bournonite $PbCuSbS_3$	5.800
Brochantite $Cu_4SO_4(OH)_6$	3.900
Bromargyrite AgBr	6.477
Bromellite BeO	3.010
Brucite $Mg(OH)_2$	2.368
Bunsenite NiO	6.809
CaAl-Pyroxene $CaAl_2SiO_6$	3.360
Cadmium Cd	8.643
Cadmium bromide $CdBr_2$	5.192
Cadmium stibnite CdSb	6.920
Cadmium telluride CdTe	6.200
Cadmium arsenide Cd_3As_2	6.210
Cadmoselite CdSe	5.810
Calcite $CaCO_3$	2.710
Calcium Ca	1.530
Calcium ferrite $CaFe_2O_4$	5.080
Calcium nitrate $Ca(NO_3)_2$	2.483
Calcium oxide CaO (Lime)	3.345
Calomel Hg_2Cl_2	7.166
Canorinite $(Na_2Ca)4[CO_3/(H_2O)0\text{-}3/(AlSiO_4)_6]$	2.450
Carnallite $KMgCl_3{\cdot}6H_2O$	1.610
Cassiterite SnO_2	6.993
Catophorite $(Ca,Na,K)_3(Mg,Fe^{2+},Fe^{3+},Al)_5(Si,Al)_8O_{22}(OH)_2$	3.500
Cattierite CoS_2	4.821
Celestite $SrSO_4$	3.971
Celsian $BaAl_2Si_2O_8$	3.200
Cerianite CeO_2	7.216
Cerium Ce	6.746
Cerium sesquioxide Ce_2O	6.860
Cerrusite $PbCO_3$	6.583
Cervantite Sb_2O_4	4.080
Cesium Cs	1.906

Table 3 (continued)
GRAIN DENSITIES OF MINERALS

Name/description/formula	Grain density g/cc
Cesium chloride CsCl	3.988
Cesium hydroxide CsOH	3.675
Cesium iodide CsI	4.510
Cesium monoxide Cs_2O	4.250
Chabasite $CaAl_2Si_4O_{12}{\cdot}6H_2O$	2.100
Chalcanthite $CuSO_4{\cdot}5H_2O$	2.291
Chalcocite Cu_2S	5.793
Chalcocyanite $CuSO_4$	3.603
Chalcopyrite $CuFeS_2$	4.200
Chlorargyrite AgCl	5.571
Chlorite $Mg_3(Si_4O_{10})(OH)_{12}{\cdot}Mg_3(OH)_6$	2.800
Chloromagnesite $MgCl_2$	2.333
Chromite $FeCr_2O_4$	5.086
Chromium Cr	7.187
Chrysoberyl $BeAl_2O_4$	3.650
Chrysocolla $CuSiO_3{\cdot}2H_2O$	2.200
Chrysotile $Mg_3Si_2O_5(OH)_4$ Serpentine Asbestos	2.550
Cinnabar HgS	8.187
Claudetite As_2O_3	4.186
Clausthalite PbSe	7.800
Clinoenstatite $MgSiO_3$	3.190
Clinoferrosilite $FeSiO_3$	4.005
Clinoptilolite $(Na,Ca)_{4\text{-}6}Al_6(Al,Si)Si_{19}O_{72}{\cdot}24H_2O$	2.100
Clinozoisite $Ca_2Al_3Si_3O_{12}OH$	3.290
Cobalt Co	8.836
Cobalt spinel Co_3O_4	6.070
Cobalt titanate $CoTiO_3$	4.987
Cobaltite CoAsS	6.275
Cobaltocalcite $(Ca,Co)CO_3$	4.216
Cobaltous oxide, CoO	6.438
Coccinite HgI_2	6.270
Coesite SiO_2	2.920
Cohenite Fe_3C	7.729
Colemanite $Ca_2B_6O_{11}{\cdot}5H_2O$	2.400
Columbite $(Fe,Mn)(Cb,Ta)20_6$	5.000
Platinum Cooperite PtS	10.255
Copper Cu	8.934
Copper aluminum sulfide $CuAlS_2$	3.450
Copper dichloride $CuCl_2$	3.054
Cordierite $(Mg,Fe^{2+})_2Al_4Si_5O_{18}$	2.508
Corundum Al_2O_3	3.987
Cotunnite $PbCl_2$	5.906
Covellite CuS	4.682
Cristobalite SiO_2	2.300
Crocoite $PbCrO_4$	6.120
Cryolite Na_3AlF_6	2.965
Cummingtonite $(Mg,Fe)_7[OH/Si_4O_{11}]_2$	3.211
Cuprite Cu_2O	6.105
Danburite $CaB_2Si_2O_8$	3.000
Daphnite $(Fe^{2+},Al)_6(Si,Al)_4O_{10}(OH)_8$	3.210
Datolite $CaBSiO_4OH$	2.900
Diamond C	3.515
Diaspore AlO(OH)	3.378
Dickite $Al_2Si_2O_5(OH)_4$	2.620
Digenite Cu_9S_5	5.603

Table 3 (continued)
GRAIN DENSITIES OF MINERALS

Name/description/formula	Grain density g/cc
Diopside $MgCaSi_2O_6$	3.277
Dioptase $CuSiO_2(OH)_2$	3.300
Dolomite $CaMg(CO_3)_2$	2.866
Dumortierite $(Al,Fe)_7BSi_3O_{18}$	3.350
Dysprosium Dy	8.548
Dysprosium sesquioxide Dy_2O_3	7.810
Eastonite $K_2Mg_5Al_4Si_5O_{20}(OH)_4$	2.900
Enargite Cu_3AsS_4	4.463
Enstatite $MgSiO_3$	3.209
Epidote $Ca_2(Al,Fe)_3Si_3O_{12}OH$	3.587
Epistilbite $CaAl_2Si_6O_{16}{\cdot}5H_2O$	2.250
Epsomite $MgSO_4{\cdot}7H_2O$	1.680
Erbium Er	9.006
Erbium sesquioxide Er_2O_3	8.640
Eskolaite Cr_2O_3	5.225
Europium Eu	5.245
Europium sesquioxide Eu_2O_3	7.420
Fassaite $Ca(Mg,Fe,Al)[(Si,Al)_2O_6]$	3.100
Faujasite $Na_2Ca[AlSi_2O_6]_4{\cdot}16H_2O$	1.920
Fayalite Fe_2SiO_4	4.393
Ferberite $FeWO_4$	7.521
Ferric sulfate $Fe_2(SO_4)_3$	3.097
Ferrosilite $FeSiO_3$	3.900
Ferrous oxide FeO (stoichiometric) Wuestite	5.700
Flourite CaF_2	3.179
Forsterite Mg_2SiO_4	3.213
Franklinite $(Zn,Mn,Fe^{2+})(Fe^{3+},Mn^{3+})_2O_4$	5.350
Gadolinia Gd_2O_3 Gadolinium sesquioxide	7.407
Gadolinium Gd	7.906
Gahnite $ZnAl_2O_4$	4.608
Galaxite $MnAl_2O_4$	4.078
Galena PbS	7.598
Gallium Ga	5.913
Gallium sesquioxide Ga_2O_3	6.476
Gehlenite $Ca_2Al_2SiO_7$	3.050
Geikielite $MgTiO_3$	3.895
Germanium Ge	5.326
Germanium dioxide GeO_2 (quartz type)	4.280
Gersdorffite NiAsS	5.964
Gibbsite $Al(OH)_3$	2.441
Gismodine $CaAl_2Si_2O_8{\cdot}4H_2O$	2.250
Glauconite $K_{1.5}(Fe^{3+},Mg,Al,Fe^{2+})_{4-6}(Si,Al)_8O_{20}(OH)_4$	2.300
Glaucophane $Na_2Mg_3Al_2[Si_8O_{22}](OH)_2$	3.200
Gmelinite $Na[AlSi_2O_6]{\cdot}3H_2O$	2.100
Goethite FeO(OH)	4.268
Gold Au	19.282
Goslarite $ZnSO_4{\cdot}7H_2O$	1.972
Graphite C	2.267
Greenockite CdS	4.826
Grossular $Ca_3Al_2Si_3O_{12}$	3.595
Gypsum $CaSO_4{\cdot}2H_2O$ Alabaster	2.305
Hafnia HfO_2 Hafnium dioxide	10.109
Hafnium Hf	13.242
Halite NaCl	2.163
Halloysite $Al_2Si_2O_5(OH)_4{\cdot}2H_2O$	2.550

Table 3 (continued)
GRAIN DENSITIES OF MINERALS

Name/description/formula	Grain density g/cc
Hastingsite $Na_2Ca_4(Mg,Fe)_{10}Al_4Si_{12}O_{44}(OH)_4$	3.300
Hauerite MnS_2	3.463
Hausmannite Mn_3O_4	4.856
Heazlewoodite Ni_3S_2	5.820
Hedenbergite $CaFeSi_2O_6$	3.632
Hematite Fe_2O_3	5.275
Hercynite $FeAl_2O_4$	4.265
Herzenbergite SnS	5.220
Hessite Ag_2Te	8.405
Heulandite $CaAl_2Si_7O_{18}{\cdot}6H_2O$	2.200
Holmia Ho_2O_3 Holmiun sesquioxide	8.360
Holmium Ho	8.801
Hornblende $(Ca,Na)_{2\text{-}3}(Mg,Fe^{2+},Fe^{3+},Al)_5(Si,Al)_8O_{22}(OH)_2$	3.080
Huebnerite $MnWO_4$	7.228
Humite $Mg(OH,F)_2{\cdot}3Mg_2[SiO_4]$	3.250
Huntite $CaMg_3(CO_3)_4$	2.696
Hydromagnesite $5MgO{\cdot}4CO_2{\cdot}5H_2O$	2.150
Hydrophilite $CaCl_2$	2.150
Hydroxyapatite $Ca_5(PO_4)_3OH$	3.155
Idocrase $Ca_{10}(Mg,Fe^{2+},Fe^{3+})_2Al_4Si_9O_{34}(OH)_4$ Vesuvianite	3.400
Illite $(H_{30},K)A_{18}(Si,Al)_{16}O_{40}(OH)_8$	2.660
Ilmenite $FeTiO_3$	4.788
Indium In	7.297
Iodargyrite AgI	5.684
Iridium Ir	22.564
Iron Fe	7.875
Jacobsite $MnFe_2O_4$	4.990
Jadeite $NaAlSi_2O_6$	3.400
Kaersutite $Na_2Ca_4(Mg,Fe)_6Fe_2Ti_2Al_4Si_{12}O_{44}(OH)_4$	3.200
Kainite $KMgSO_4Cl{\cdot}3H_2O$	2.130
Kaliophillite $KAlSiO_4$	2.500
Kalsilite $KAlSiO_4$	2.600
Kaolinite $Al_2Si_2O_5(OH)_4$	2.594
Karelianite V_2O_3	5.021
Kernite $Na_2B_4O_7{\cdot}4H_2O$	1.877
Kieserite $MgSO_4{\cdot}H_2O$	2.573
Klockmannite CuSe	6.121
Krennerite $AuTe_2$	8.620
Kyanite Al_2SiO_5	3.675
Labradorite $Na_2Ca_3(AlSi_3O_8)_8$	2.710
Langbeinite $K_2SO_4{\cdot}2MgSO_4$	2.830
Lantha La_2O_3 Lanthanum sesquioxide	6.510
Lanthium La	6.182
Larnite Ca_2SiO_4 Calcium Olivine	3.270
Laumontite $CaAl_2Si_4O_{12}{\cdot}4H_2O$	2.300
Laurite RuS_2	6.990
Lawrencite $FeCl_2$	3.212
Lawsonite $CaAl_2Si_2O_7(OH)_2{\cdot}H_2O$	3.090
Lead Pb	11.343
Lead nitrate $Pb(NO_3)_2$	4.530
Lead titanate $PbTiO_3$	7.940
Lead zirconate $PbZrO_3$	7.000
Lepidolite $K(Li,Al)_3(Si,Al)_4O_{10}(F,OH)_2$	2.900
Lepidomelane $K2(Fe^{3+},Fe^{2+})_{4\text{-}6}(Si,Al,Fe^{+3})_8O_{20}(OH)_4$	2.800

Table 3 (continued)
GRAIN DENSITIES OF MINERALS

Name/description/formula	Grain density g/cc
Leucite $KAlSi_2O_6$	2.469
Levyne $(Ca,Na_2)Al_2Si_4O_{12}{\cdot}6H_2O$	2.100
Lime olivine $CaSiO_4$	2.236
Limonite Amorphous Iron oxide/hydroxide	3.176
Linnaeite Co_3S_4	4.877
Litharge PbO red	9.335
Lithium Li	.533
Lithium aluminate $LiAlO_2$	2.550
Lithium chloride LiCl	2.060
Lithium hydroxide LiOH	1.460
Lithium monoxide Li_2O	2.010
Lizardite $Mg_3Si_2O_5(OH)_4$ Serpentine	2.550
Loellingite $FeAs_2$	7.477
Lutetium Lu	9.846
Lutetium sesquioxide Lu_2O_3	9.420
Maghemite Fe_2O_3 Gamma	4.880
Magnesioferrite $MgFe_2O_4$	4.487
Magnesiosilicon Mg_2Si	1.940
Magnesite $MgCO_3$	3.010
Magnesium Mg	1.737
Magnesium nitrate $Mg(NO_3)_2$	1.640
Magnesium stannide Mg_2Sn	3.591
Magnesium sulfate $MgSO_4$	2.660
Magnetite Fe_3O_4	5.200
Malachite $Cu_2(CO_3)(OH)_2$	4.031
Manganese Mn	7.470
Manganese selenide MnSe	5.590
Manganese sulfate $MnSO_4$	3.250
Manganite MnO·OH	4.330
Manganosite MnO	5.366
Marcasite FeS_2	4.870
Marshite CuI	5.710
Mascagnite $(NH_4)_2SO_4$	1.769
Massicot PbO yellow	8.000
Matlockite PbFCl	7.129
Melanterite $FeSO_4{\cdot}7H_2O$	1.898
Melilite $(Ca,Na,K)_2[(Mg,Fe,Al,Si)_3O_7]$	3.000
Mercuric chloride Hg_2Cl_2	6.470
Mercurous chloride $HgCl_2$	5.600
Merwinite $Ca_3Mg(SiO_4)_2$	3.150
Metacinnabar HgS	7.730
Microcline $KAlSi_3O_8$	2.560
Millerite NiS	5.374
Minium Pb_3O_4	8.926
Mirablite $Na_2SO_4{\cdot}10H_2O$	1.464
Molybdenite MoS_2	4.999
Molybdenum Mo	10.221
Molybdenum dioxide MoO_2	6.534
Molybdite MoO_3	4.692
Molysite $FeCl_3$	2.898
Monteponite CdO	8.239
Monticellite $CaMgSiO_4$	3.200
Montmorillonite $(Na,K,Mg,Ca)_{0.33}(Al,Mg)_2Si_4O_{10}(OH)_2{\cdot}nH_2O$	2.608
Montroycite HgO	11.211
Mordenite $Na_2(AlSi_5O_{12})_2{\cdot}6H_2O$	2.100

Table 3 (continued)
GRAIN DENSITIES OF MINERALS

Name/description/formula	Grain density g/cc
Morenosite $NiSo_4 \cdot 7H_2O$	1.948
Mullite $3Al_2O_3 \cdot 2SiO_2$	3.167
Muscovite $KAl_3Si_3O_{10}(OH)_2$	2.831
Nacrite $Al_4Si_4O_{10}(OH)_8$	2.580
Nantockite CuCl	4.139
Natrolite $Na_2Al_2Si_3O_{10} \cdot 2H_2O$	2.245
Naumanite Ag_2Se	7.863
Neodymium Nd	7.012
Neodymium sesquioxide Nd_2O_3	7.240
Nepheline $Na_3KAl_4Si_4O_{16}$	2.623
Nesquehonite $MgCO_3 \cdot 3H_2O$	1.850
Niccolite NiAs	7.776
Nickel Ni	8.910
Nickel carbonate $Ni(CO)_4$	1.320
Nickel chloride $NiCl_2$	3.550
Niobium Nb	8.580
Niobium dioxide NbO_2	5.900
Niobium monoxide NbO	7.300
Niobium pentoxide Nb_2O_5	2.845
Niter KNO_3	2.105
Nitrobarite $Ba(NO_3)_2$	3.240
Nontronite $Fe_4Si_7 \cdot 34Al_{0.66}O_{20}(OH)_2$	2.300
Oldhamite CaS	2.602
Omphacite (Ca,Na)(Mg,Fe,Al)$[Si_2O_6]$	3.300
Opal $SiO_2 \cdot nH_2O$	1.890
Orpiment As_2S_3	3.489
Orthoclase $KAlSi_3O_8$	2.570
Orthoferrosilite $FeSiO_3$	3.960
Osmium Os	22.581
Otavite $CdCO_3$	5.027
Palladium Pd	12.006
Palygorskite $(Mg,Al)_2[CH/Si_4O_{10}] \cdot 2H_2O + 2H_2O$	2.300
Paragonite $NaAl_3Si_3O_{10}(OH)_2$	2.850
Pargasite $Na_2Ca_2(Mg,Fe)_8Al_6Si_{12}O_{44}(OH)_4$	3.100
Pectolite $Ca_2NaSi_3O_8OH$	2.870
Periclase MgO	3.583
Perovskite $CaTiO_3$	4.044
Petalite $LiAlSi_4O_{10}$	2.410
Phenacite Be_2SiO_4	2.960
Phillipsite $0.5Ca_5Al_5Si_{11}O_{32} \cdot 10H_2O$	2.150
Phlogopite $KMg_3AlSi_3O_{10}(OH)_2$	2.784
Phosphorus P	1.801
Phosphorus pentoxide P_2O_5	2.390
Phosphorus trioxide P_2O_3	2.135
Picrochromite/Magnesiochromite $MgCr_2O_4$	4.414
Pigeonite (Mg,Fe,Ca)(Mg,Fe)$[Si_2O_6]$	3.400
Platinum Pt	21.460
Plattnerite PbO_2	9.375
Plutonium Pu	20.266
Polyhalite $K_2SO_4 \cdot MgSO_4 \cdot 2CaSO_4 \cdot 2H_2O$	2.780
Polymioite Ni_3S_4	4.700
Portlandite $Ca(OH)_2$	2.242
Potassium K	.862
Potassium bromate $KBrO_3$	3.270
Potassium bromide KBr	2.754

Table 3 (continued)
GRAIN DENSITIES OF MINERALS

Name/description/formula	Grain density g/cc
Potassium carbonate K_2CO_3	2.428
Potassium chlorate $KClO_3$	2.320
Potassium fluoride KF	2.505
Potassium hydroxide KOH	2.044
Potassium monoxide K_2O	2.320
Potassium orthophosphate K_3PO_4	2.564
Potassium superoxide KO_2	2.140
Powellite $CaMoO_4$	4.256
Praeseodymium sesquioxide Pr_2O_3	7.070
Praseodymium Pr	6.774
Prehnite $Ca_2Al_2Si_3O_{10}(OH)_2$	2.910
Proustite Ag_3AsS_3	5.595
Pseudobrockite Fe_2TiO_5	4.390
Pyrargyrite Ag_3SbS_3	5.851
Pyrite FeS_2	5.011
Pyrolusite MnO_2	5.234
Pyrope $Mg_3Al_2Si_3O_{12}$	3.510
Pyrophanite $MnTiO_3$	4.604
Pyrophyllite $Al_2Si_4O_{10}(OH)_2$	2.819
Pyrrhotite $Fe_{0.877}S$	4.610
Quartz SiO_2	2.648
Realgar AsS	3.590
Retgersite $NiSO_4{\cdot}6H_2O$	2.070
Rhenium Re	21.017
Rhenium dioxide ReO_2	11.400
Rhenium heptoxide Re_2O_7	6.103
Rhenium trioxide ReO_3	7.000
Rhodium Rh	12.425
Rhodochrosite $MnCO_3$	3.699
Rhodonite $MnSiO_3$	3.726
Riebeckite $Na_2Fe_2 + 3Fe_3 + 2Si_8O_{22}(OH)_4$	3.000
Rubidium Rb	1.530
Rubidium chloride RbCl	2.800
Ruthenium Ru	12.369
Rutile TiO_2	4.245
Salmiac NH_4Cl Sal Amoniac	1.527
Samarium Sm	7.528
Samarium sesquioxide Sm_2O_3	8.347
Sanadine $KAlSi_3O_8$ (high)	2.560
Sanmartinite $ZnWO_4$	7.872
Saponite $Mg_6Si_7{\cdot}34Al_{0.66}O_{20}(OH)_4$	2.350
Scacchite $MnCl_2$	2.988
Scandium Sc	2.989
Scandium sesquioxide Sc_2O_3	3.860
Scapolite $CaCO_3{\cdot}3CaAl_2Si_2O_8{\cdot}CaSO_4{\cdot}CaCl_2$	2.800
Scheelite $CaWO_4$	6.120
Scolecite $CaAl_2Si_3O_{10}{\cdot}3H_2O$	2.300
Selenium Se	4.809
Selenolite SeO_2	4.162
Sellaite MgF_2	3.148
Sepiolite $H_6Mg_8Si_{12}O_{30}(OH)_{10}{\cdot}6H_2O$	2.080
Serpentine $Mg_3(Si_2O_5)(OH)_4$	2.600
Siderite $FeCO_3$	3.944
Silicon Si	2.330
Silicon carbide SiC	3.217

Table 3 (continued)
GRAIN DENSITIES OF MINERALS

Name/description/formula	Grain density g/cc
Sillimanite Al_2SiO_5	3.247
Silver Ag	10.501
Silver oxide Ag_2O	7.140
Smaragdite $Ca_2(Mg,Fe)_5Si_8O_{22}(OH)_2$	3.400
Smithsonite $ZnCO_3$	4.435
Soda-niter $NaNO_3$	2.261
Sodalite $Na_4Al_3(SiO_4)_3Cl$	2.200
Sodium Na	.965
Sodium bromate $NaBrO_3$	3.339
Sodium carbonate Na_2CO_3	2.532
Sodium carbonate hydrogen $NaHCO_3$	2.159
Sodium hydroxide NaOH	2.130
Sodium monoxide Na_2O	2.395
Sodium perchlorate $NaClO_4$	2.020
Platinum Sperrylite $PtAs_2$	10.778
Spessartine $Mn_3Al_2Si_3O_{12}$	4.190
Sphaelerite ZnS	4.089
Spinel $MgAl_2O_4$	3.583
Spodumene $LiAlSi_2O_6$	3.188
Stannic sulfide SnS_2	4.500
Stannous tetrachloride $SnCl_4$	2.230
Stibnite Sb_2S_3	4.627
Stibnous bromide $SbBr_3$	4.148
Stibnous chloride $SbCl_3$	3.140
Stilbite $NaCa_2Al_5Si_{13}O_{36}{\cdot}14H_2O$	2.150
Stilleite ZnSe	5.420
Stishovite SiO_2	4.300
Stolzite $PbWO_4$	8.411
Strengite $FePO_4{\cdot}2H_2O$	2.740
Strontianite $SrCO_3$	3.784
Strontium Sr	2.583
Strontium bromide $SrBr_2$	4.210
Strontium nitrate $Sr(NO_3)_2$	2.986
Strontium oxide SrO	4.700
Strontium sulfide SrS	3.700
Strontium titanate $SrTiO_3$	5.110
Sulfur S	2.067
Sulfuryl chloride SO_2Cl_2	1.680
Sylvite KCl	1.987
Szomolnokite $FeSO_4{\cdot}H_2O$	2.970
Talc $Mg_3Si_4O_{10}(OH)_2$	2.784
Tantalite $(Fe,Mn)(Ta,Nb)_2O_6$	6.500
Tantalum Ta	16.676
Tantalum pentoxide Ta_2O_5	8.311
Tellurite TeO_2	5.751
Tellurium Te	6.232
Tenorite CuO	6.509
Tephroite Mn_2SiO_4	4.155
Terbium Tb	8.239
Tetradymite Bi_2Te_3 Tellurobismuthite	7.862
Thallium Tl	11.875
Thallous chloride TlCl	7.020
Thenardite Na_2SO_4	2.663
Thomsonite $NaCa_2[Al_5Si_5O_{20}]{\cdot}6H_2O$	2.300
Thorianite ThO_2	10.012
Thorium Th	11.726

Table 3 (continued)
GRAIN DENSITIES OF MINERALS

Name/description/formula	Grain density g/cc
Thulium Tm	9.320
Tiemannite HgSe	8.266
Tin Sn	7.287
Titanite $CaTiSiO_5$ Sphene	3.523
Titanium Ti	4.506
Titanium bromide $TiBr_4$	2.600
Titanium monoxide TiO	4.930
Titanium sesquioxide Ti_2O_3	4.574
Titanium trichloride $TiCl_3$	2.640
Titanomagnetite/ulvospinel Fe_2TiO_4	4.776
Topaz $Al_2(SiO_4)(F2)$	3.500
Topaz $Al_2(SiO_4)(OH)$	3.174
Tremolite $Ca_2Mg_5[Si_8O_{22}](OH)_2$	2.977
Trevorite $NiFe_2O_4$	5.370
Tridymite SiO_2	2.260
Troilite FeS	4.830
Trona $Na_2CO_3{\cdot}NaHCO_3{\cdot}2H_2O$	2.170
Tschermakite $Ca_4(Mg,Fe)6Al_8Si_{12}O_{44}(OH)_4$	3.200
Tungsten W	19.261
Tungsten dioxide WO_2	12.110
Tungsten trioxide WO_3	7.160
Tungstenite WS_2	7.500
Turquoise $CuAl_6(PO_4)_{4-}(OH)_8{\cdot}4H_2O$	2.700
Ulexite $NaCaB_5O_9{\cdot}8H_2O$	2.000
Uraninite UO_2	10.969
Uranium U	19.047
Uranium tetrachloride UCl_4	4.870
Uranium tetrafluoride UF_4	6.700
Uranium trichloride UCl_3	5.440
Uranium trioxide UO_3	7.290
Valentinite Sb_2O_3	5.829
Vanadinite $Pb_5Cl(VO_4)_3$	6.900
Vanadium V	6.101
Vanadium dichloride VCl_2	3.230
Vanadium monoxide VO	5.758
Vanadium pentoxide V_2O_5	3.357
Vanadium tetroxide V_2O_4	4.339
Vanadium trichloride VCl_3	3.000
Vaterite $CaCO_3$	2.715
Vermicullite $(Mg{\cdot}Ca)_x(Si_{8-x}Al_x)(Mg{\cdot}Fe)_6O_2O{\cdot}yH_2O$	2.300
Villiaumite NaF	2.790
Wairakite $Ca[AlSi_2O_6]_{2\,2}H_2O$	2.260
Water H_2O (liquid)	.997
Whitlockite $Ca_3(PO_4)_2$	3.140
Willemite Zn_2SiO_4	4.251
Witherite $BaCO_3$	4.308
Wollastonite $CaSiO_3$	2.909
Wulfenite $PbMoO_4$	6.817
Wurtzite ZnS	3.980
Wustite $Fe_{0.947}O$	5.722
Xenon Xe	.005
Ytterbium Yb	6.969
Ytterbium sesquioxide Yb_2O_3	9.170
Yttrium Y	5.912
Yttrium sesquioxide Y_2O_3	5.010

Table 3 (continued)
GRAIN DENSITIES OF MINERALS

Name/description/formula	Grain density g/cc
Zinc Zn	7.136
Zinc arsenide Zn_3As_2	5.578
Zinc stibnite ZnSb	6.383
Zinc telluride ZnTe	6.340
Zincite ZnO	5.676
Zinkosite $ZnSO_4$	4.330
Zircon $ZrSiO_4$	4.669
Zirconium Zr	6.508
Zoisite $Ca_2Al_3(SiO_4)_3(OH)$	3.328

Table 4
TYPICAL SOIL DENSITIES

Sample type	SGH	DBD[a]	DBD[b]	WBD[c]
Gravelly soil	2.68	1.66	1.77	1.42
Glacial soil	2.71	1.59	1.75	1.45
Sandy soil	2.67	1.44	1.56	1.43
Dune sand	2.59	1.61	1.76	1.47
Eolian sand	2.69	1.45	1.54	1.44
Glacial sand	2.66	1.44	1.58	1.44
Fire clay	2.66	1.46	1.61	1.42
Loess	2.66	0.99	1.09	1.41
Adobe	2.30	1.18	1.39	1.40
Brick clay	2.75	1.20	1.41	1.37
Sandy loam	2.66	1.42	1.62	1.43
Heavy blue loam	2.54	1.07	1.16	1.39
Silt loam	2.66	1.07	1.19	1.42
Fossiliferous soil	2.74	1.63	1.74	1.44
Greensand	2.93	1.35	1.52	1.49
Residual soil (horneblende schist)	2.85	1.07	1.20	1.43
Residual soil (siliceous oolite)	2.61	1.29	1.42	1.43
Peat	1.37	0.27	0.32	0.51
Peat moss	1.57	0.077	0.10	0.077
Muck	1.66	0.80	0.85	1.07

[a] Dry bulk density of "fluffed" sample.
[b] Dry bulk density of "tapped" sample.
[c] Wet bulk density of sample composed of equal weights specimen and water.

meaningful and usually significantly different from the mean in asymmetrical distributions as are common in Figures 4 through 22. More data is required for these rock types to properly calculate the mode.

The change of density with pressure is generally small in most rocks and minerals, except as pressure modifies the porosity of the rock. The change of density with temperature is also small, being caused mainly by thermal volumetric expansion, and roughly on the order of a few percent over 1000°C temperature change.[52] Changes in density of more than 10% are not common when changing from the solid to the melted state however (see Skinner[52] and Figure 23).

Table 5
DRY BULK DENSITY STATISTICS FOR VARIOUS ROCK TYPES

Rock type	Number of samples	Skewness	Mean	SD[a]	Mode	Median
All rocks	1647	0.29	2.73	0.26	2.65	2.86
Andesite	197	0.56	2.65	0.13	2.58	2.66
Basalt	323	−0.30	2.74	0.47	2.88	2.87
Diorite	68	−0.27	2.86	0.12	2.89	2.87
Dolerite-diabase	224	−0.49	2.89	0.13	2.96	2.90
Gabbro	98	−0.28	2.95	0.14	2.99	2.97
Granite	334	−0.02	2.66	0.06	2.66	2.66
Quartz porphyry	76	0.34	2.62	0.06	2.60	2.62
Rhyolite	94	−0.74	2.51	0.13	2.60	2.49
Syenite	93	0.27	2.70	0.10	2.67	2.68
Trachyte	71	−0.44	2.57	0.10	2.62	2.57
Sandstone	107	0.00	2.22	0.23	2.22	2.22

[a] Standard deviation.

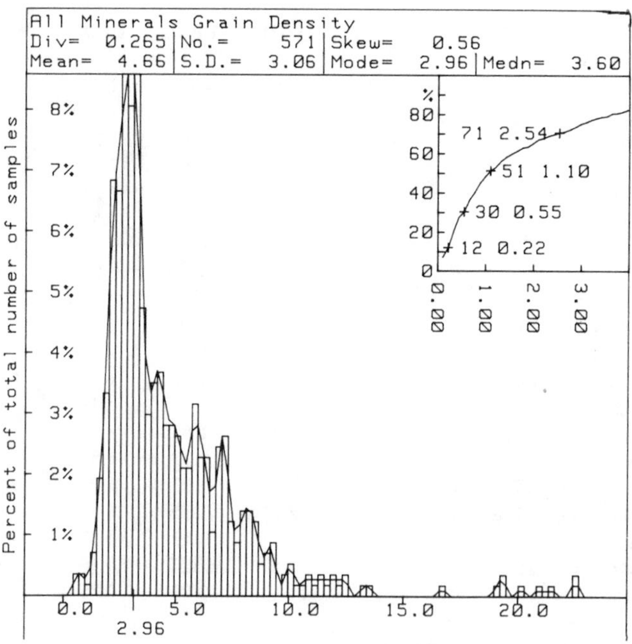

FIGURE 1. Histogram showing specific gravities of minerals from Table 1.

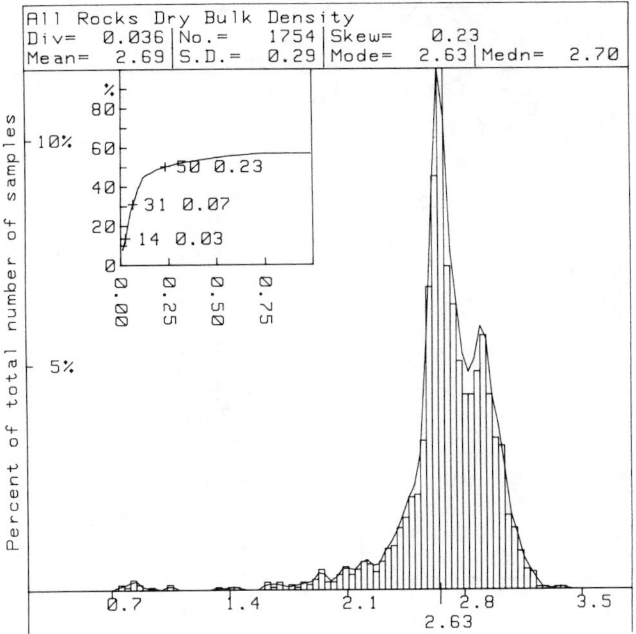

FIGURE 2. Histogram showing dry bulk densities of all rocks from Table 5.

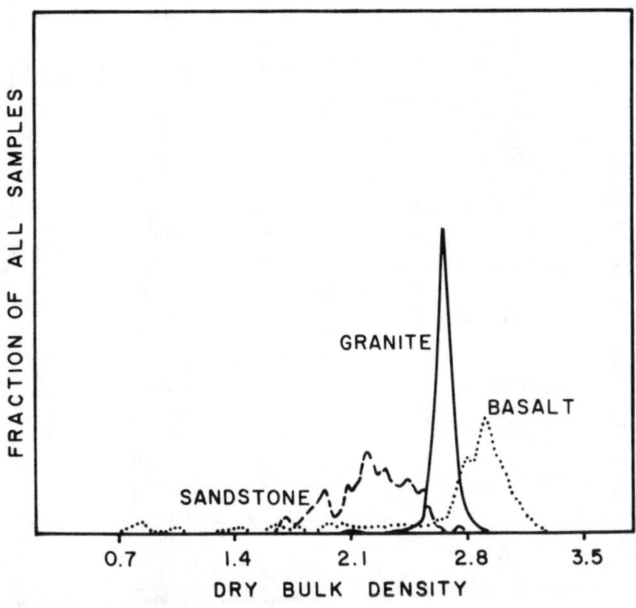

FIGURE 3. Comparison of frequency distributions of dry bulk densities of granites, basalts, and sandstone.

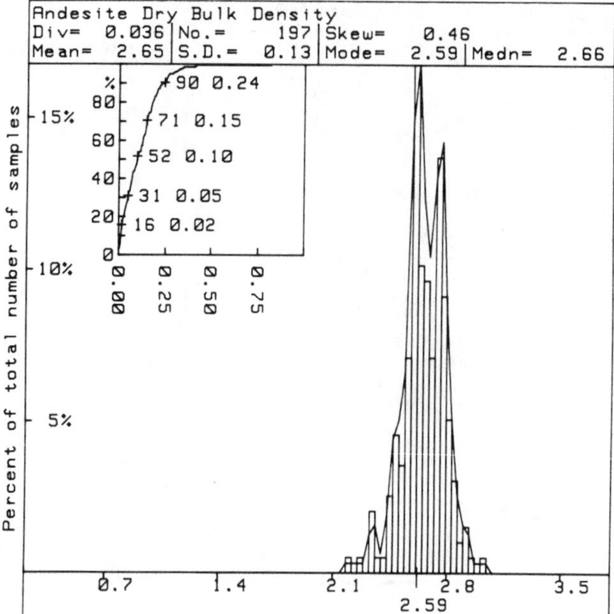

FIGURE 4. Histogram showing dry bulk densities of 197 samples of andesite.

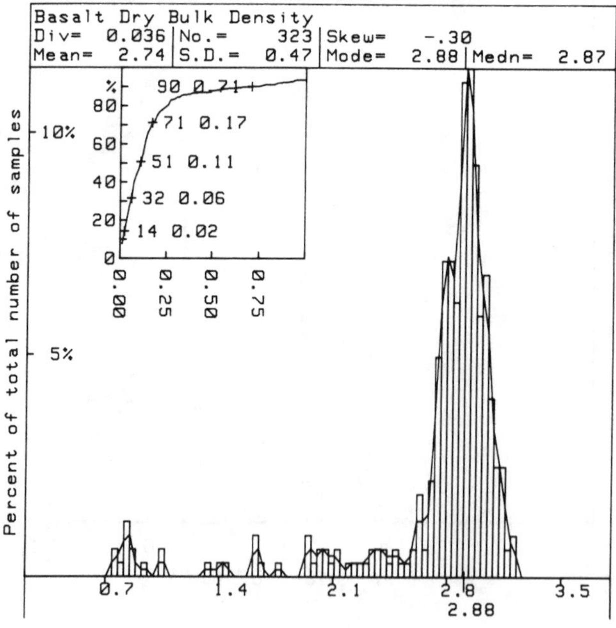

FIGURE 5. Histogram showing dry bulk densities of 323 samples of basalt.

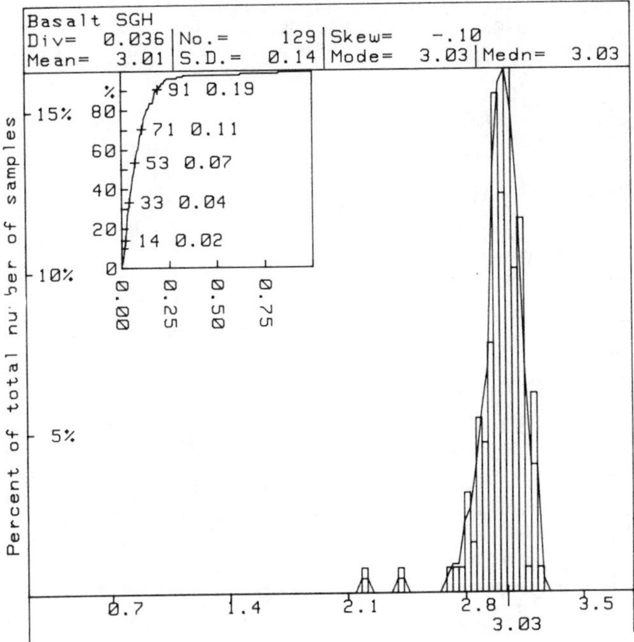

FIGURE 6.. Histogram showing grain densities of 129 samples of basalt.

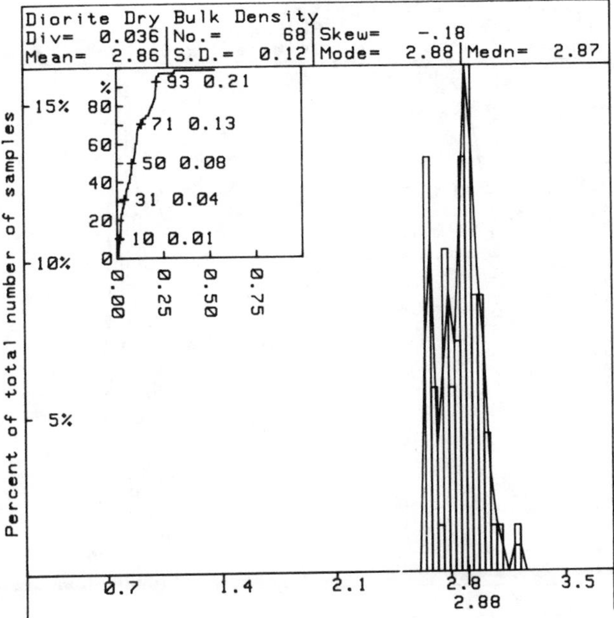

FIGURE 7. Histogram showing dry bulk densities of 68 samples of diorite.

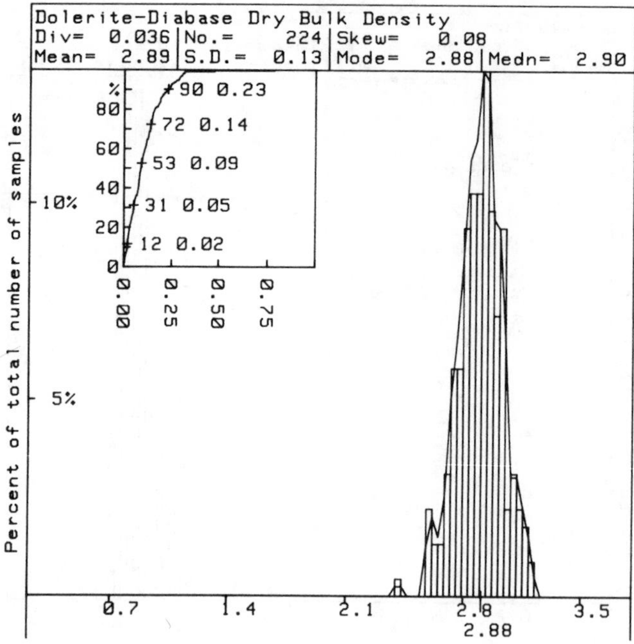

FIGURE 8. Histogram showing dry bulk densities 224 samples of dolerite-diabase.

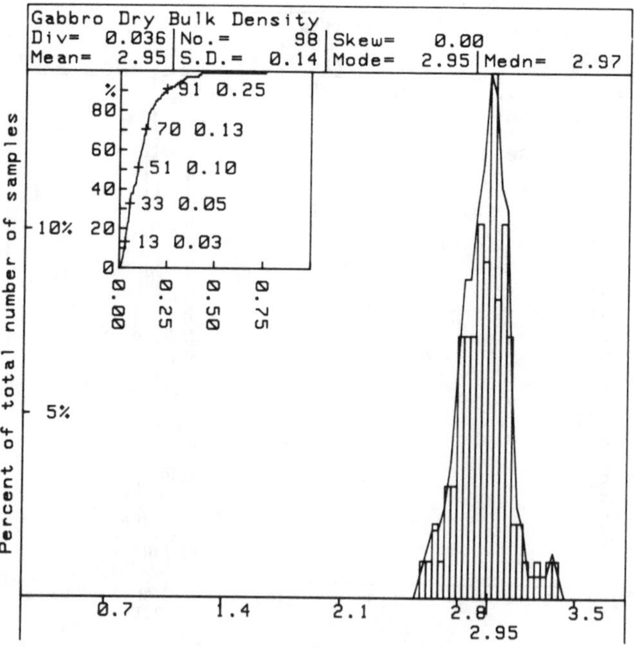

FIGURE 9. Histogram showing dry bulk densities of 98 samples of gabbro.

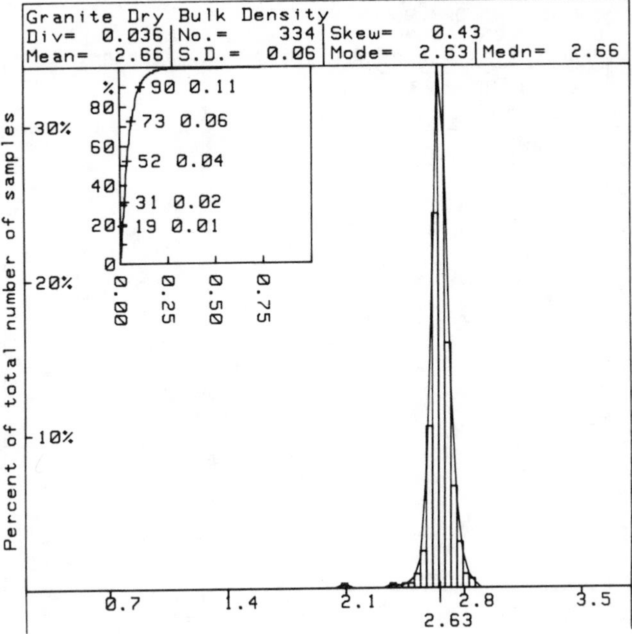

FIGURE 10. Histogram showing dry bulk densities of 334 samples of granite.

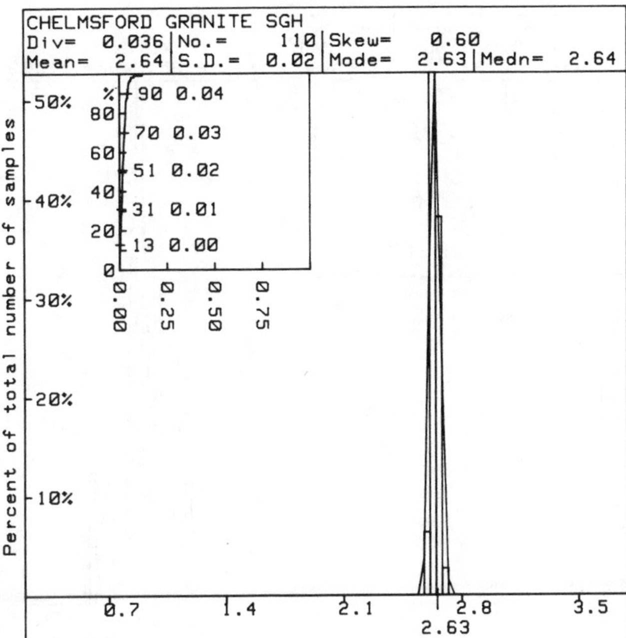

FIGURE 11. Histogram showing grain densities of 110 samples of Chelmsford granite.

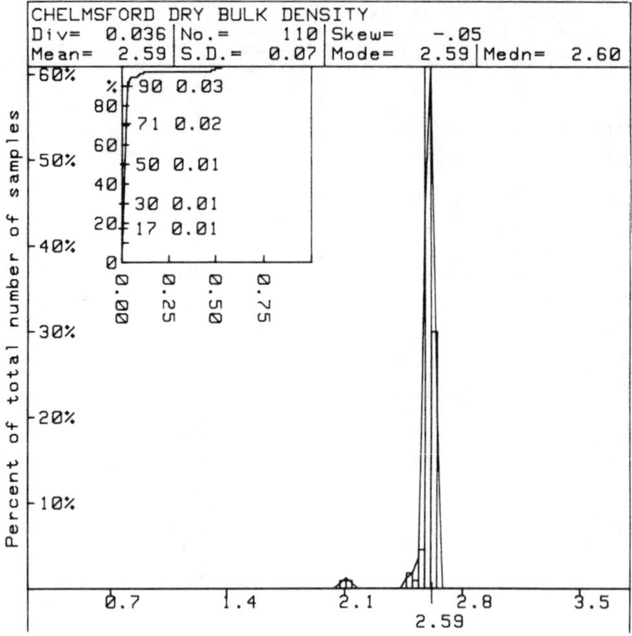

FIGURE 12. Histogram showing dry bulk densities of 110 samples of Chelmsford granite.

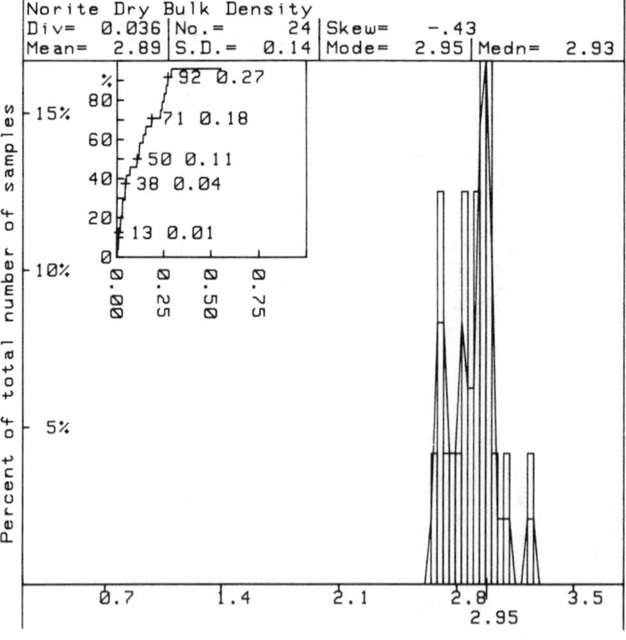

FIGURE 13. Histogram showing dry bulk densities of 24 samples of norite.

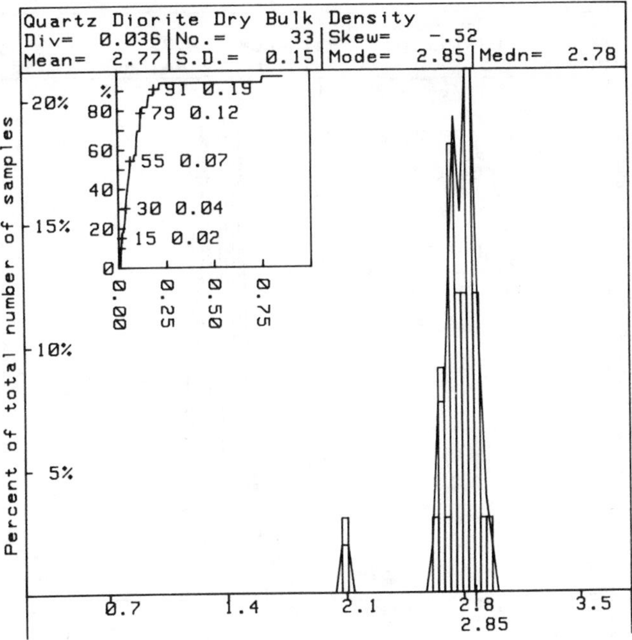

FIGURE 14. Histogram showing dry bulk densities of 33 samples of quartz diorite.

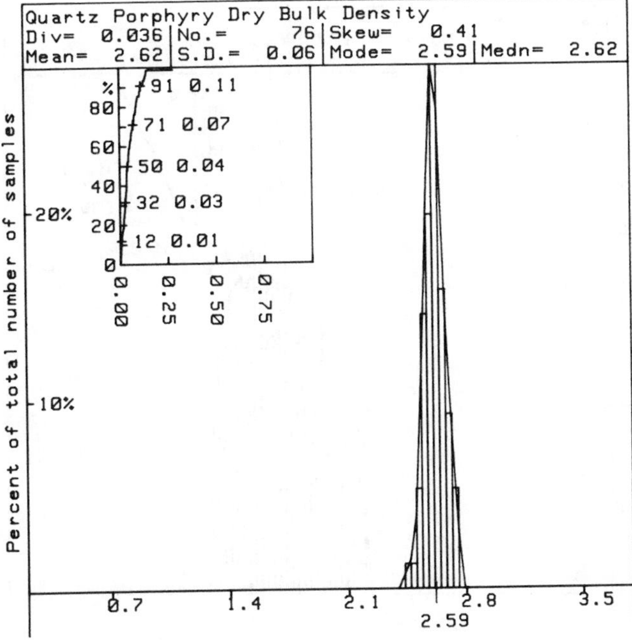

FIGURE 15. Histogram showing dry bulk densities of 76 samples of quartz porphyry.

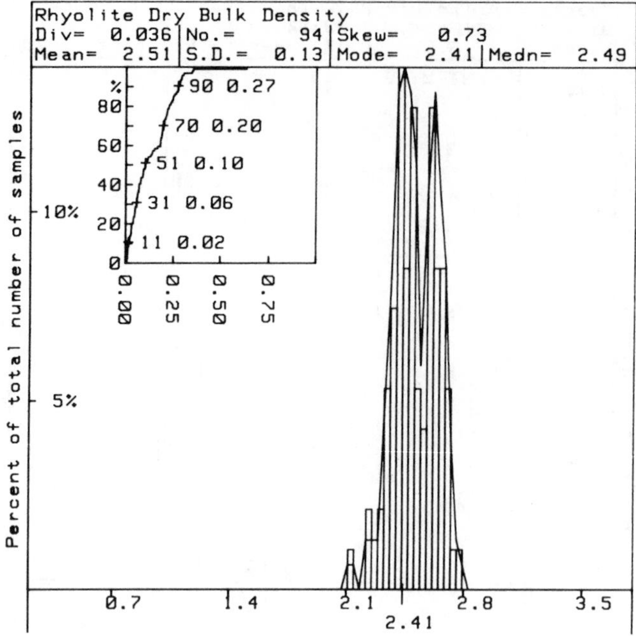

FIGURE 16. Histogram showing dry bulk densities of 94 samples of rhyolite.

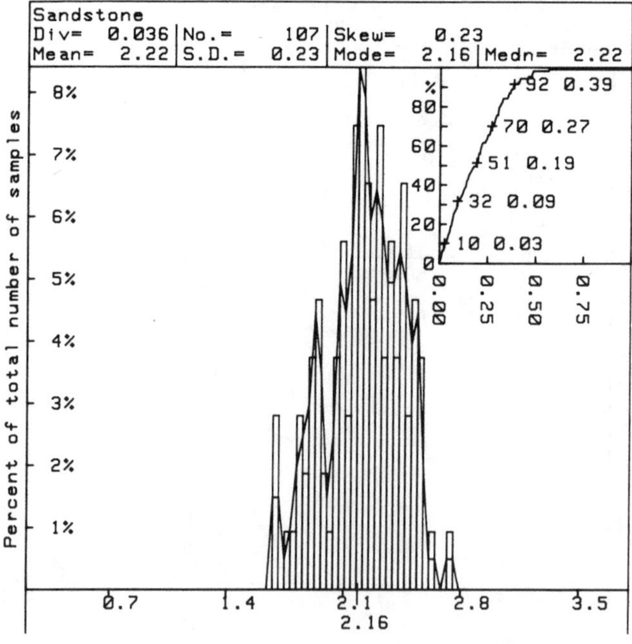

FIGURE 17. Histogram showing dry bulk densities of 107 samples of sandstone.

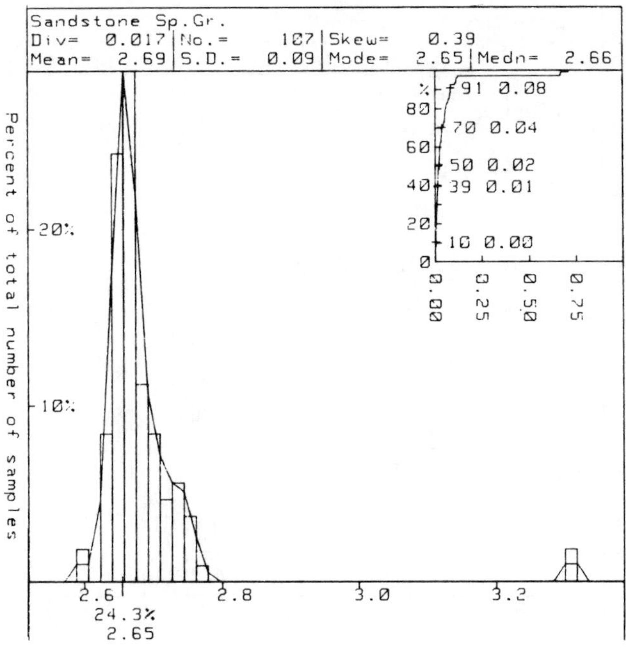

FIGURE 18. Histogram showing grain densities of 107 samples of sandstone.

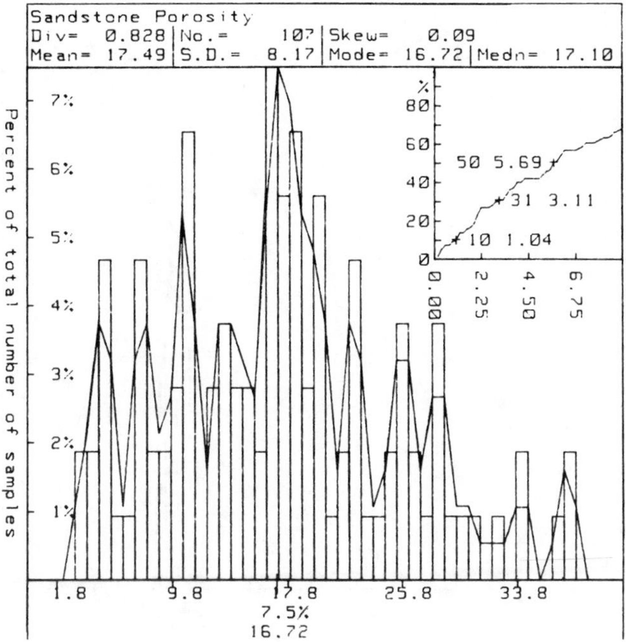

FIGURE 19. Histogram showing porosities of 107 samples of sandstone.

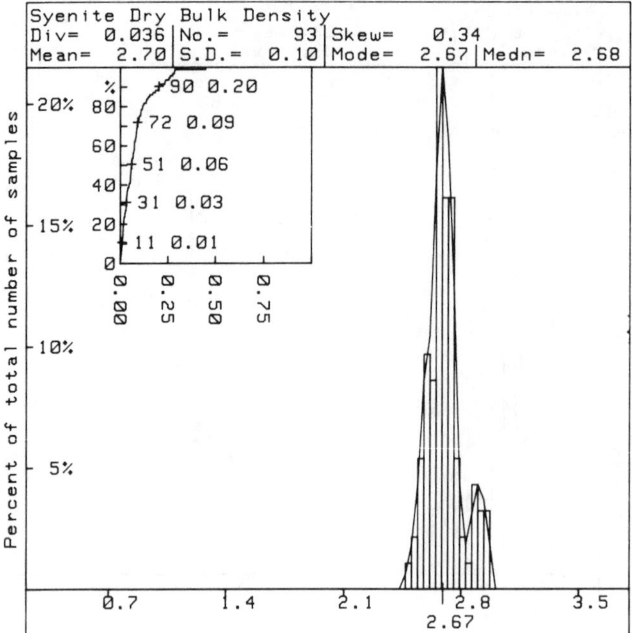

FIGURE 20. Histogram showing dry bulk densities of 93 samples of syenite.

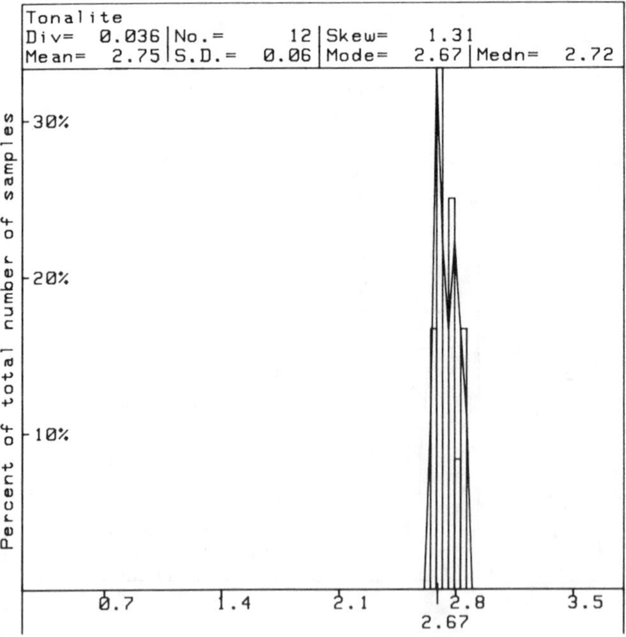

FIGURE 21. Histogram showing dry bulk densities of 12 samples of tonalite.

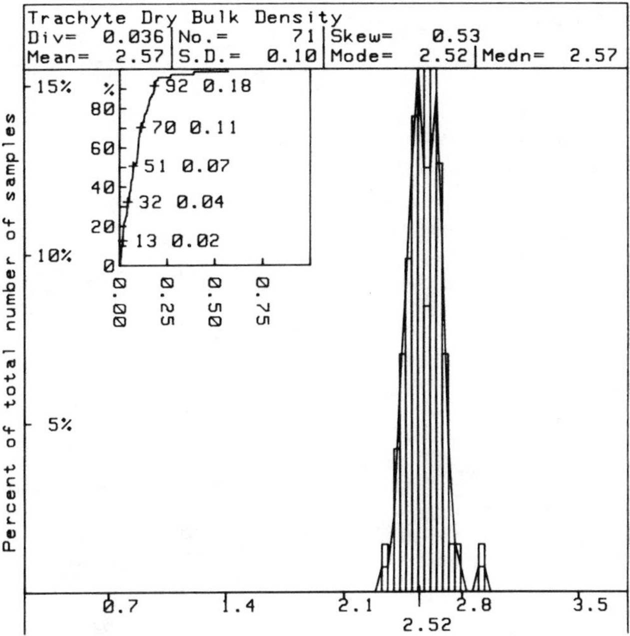

FIGURE 22. Histogram showing dry bulk densities of 71 samples of trachyte.

Table 6
RANGES OF BULK DENSITIES FOR ROCKS WITH INSUFFICIENT NUMBERS OF MEASUREMENTS TO COMPILE A HISTOGRAM

Rock type	Range of density	Rock type	Range of density
Albitite	2.61—2.77	Leucite tephrite glass	2.52—2.58
Amphibolite	2.79—3.14	Limestone	1.55—2.75
Andesite glass	2.400—2.573	Marble	2.67—2.75
Anhydrite	2.82—2.93	Marl	2.63
Anorthosite	2.64—2.92	Marlstone	2.26
Basalt glass	2.704—2.851	Norite	2.72—3.02
Bronzitite	3.26—3.29	Peridotite	3.152—3.276
Chalk	2.23	Pitchstone	2.321—2.370
Chlorite	2.79	Polyhalite	2.76
Diopside	3.24	Pyroxenite	3.24—3.31
Dolomite	2.72—2.84	Quartz diorite	2.798—2.906
Dunite	2.98—3.76	Quartz monzonite	2.64
Eclogite	3.32—3.45	Quartzite	2.647
Garnet	3.56—3.95	Rhyolite obsidian	2.330—2.413
Gneiss	2.59—2.84	Rocksalt	2.1—2.2
Granodiorite	2.668—2.785	Sand	1.44—2.40
Granulite	2.67—3.1	Schist	2.73—3.19
Graywacke	2.67—2.70	Serpentinite	2.44—2.80
Grossularite	3.49	Shale	2.06—2.67
Hornblendite	3.12—3.22	Slate	2.72—2.84
Jadeite	3.18—3.33	Trachyte obsidian	2.435—2.467

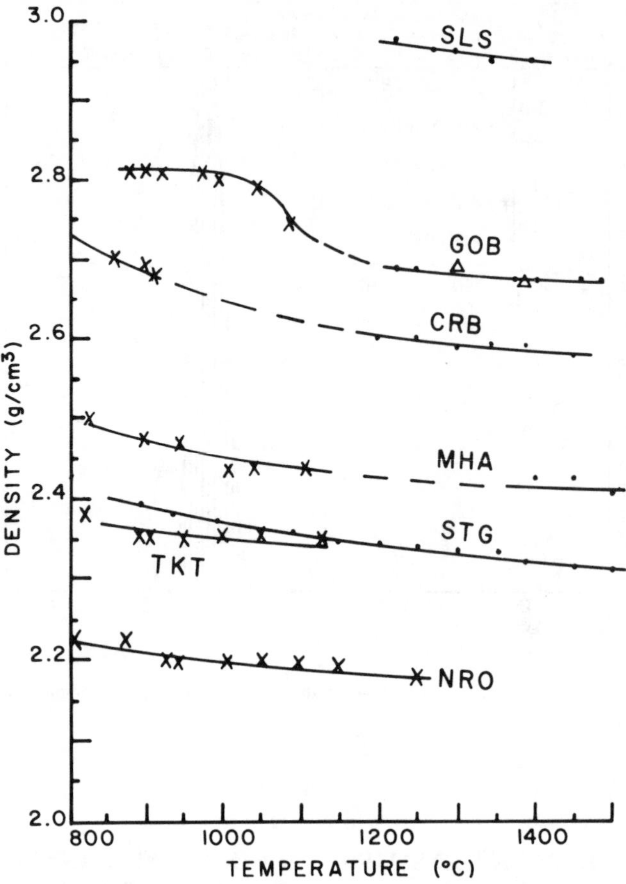

FIGURE 23. Density vs. temperature for Columbia River Basalt Group (CRB); Galapagos olivine basalt (GOB); Mt. Hood andesitic lavas (MHA); Newberry rhyolite obsidian flow (NRO); a synthetic lunar sample (SLS); and a National Bureau of Standards No. 710 standard glass (STG). (From Murase, T. and McBirney, A. R., *Geol. Soc. Am. Bull.*, 84, 3563, 1973. With permission.)

REFERENCES

1. **Mason, B.,** The determination of the density of solids, *Geol. Foeren. Stockholm Foerh.*, 66, 27, 1944.
2. International Society for Rock Mechanics, Commission on Standardization of Laboratory and Field Tests, Suggested Methods for Determining Water Content, Porosity, Density, Absorption and Related Properties and Swelling and Slake-Durability Index Properties, Committee on Laboratory Tests Document No. 2, Lisbon, Portugal, 1972.
3. ASTM, Density and specific gravity of solids, liquids, and gases. Definition of terms relating to. E-12-70, in *Book of ASTM Standards Including Tentatives,* American Society for Testing and Materials, Philadelphia, 1977, 41.
4. **Holmes, A.,** *Petrographic Methods and Calculations,* Thomas Murby & Co., London, 1930, chap. 2.
5. **Hidnert, P. and Peffer, E. L.,** Density of Solids and Liquids, *Natl. Bur. Stand. Circ.*, 487, 1950.
6. American Petroleum Institute, API Recommended Practice for Core-Analysis Procedure, API RP 40, 1st ed., Washington, D.C., 1960.
7. **Manger, G. E.,** Porosity and Bulk Density of Sedimentary Rocks, Geol. Surv. Bull. 1144-E, U.S. Geological Survey, Reston, Va., 1963.
8. **Muller, L. D.,** Density determination, in *Physical Methods in Determinative Mineralogy*, Zussman, J., Ed., Academic Press, London, 1977, chap. 13.
9. **Rall, C. G., Hamontre, H. C., and Taliaferro, D. E.,** Determination of porosity by a Bureau of Mines method; a list of porosities of oil sand, *Bur. Mines Rep. Invest.*, 5025, 1954.
10. **Manger, G. E.,** Method-Dependent Values of Bulk, Grain, and Pore Volume as Related to Observed Porosity, Geol. Surv. Bull. 1203-D, U.S. Geological Survey, Reston, Va., 1966.
11. **Hunt, G. R., Johnson, G. R., Olhoeft, G. R., Watson, D. E., and Watson, K.,** Initial Report of the Petrophysics Laboratory, Geol. Surv. Circ. 789, U.S. Geological Survey, Reston, Va., 1979, 71.
12. **Washburn, E. W. and Bunting, E. N.,** The determination of porosity by the method of gas expansion, *J. Am. Ceram. Soc.*, 5, 112, 1922.
13. **Smakula, A.,** High-precision density determinations of solids, in *Methods of Experimental Physics,* Vol. 6 (part A), Lark-Horovitz, K. and Johnson, V. A., Eds., Academic Press, New York, 1959, chap. 4.1.
14. **Athy, L. F.,** Density, porosity, and compaction of sedimentary rocks, *Bull. Am. Assoc. Pet. Geol.*, 14, 1, 1930.
15. **McIntyre, D. B., Welday, E. E., and Baird, A. K.,** Geologic application of the air pycnometer: a study of the precision of measurement, *Geol. Soc. Am. Bull.*, 76, 1055, 1965.
16. **Zenger, D. H.,** Determination of calcite and dolomite using the air comparison pycnometer, *J. Sediment. Pet.*, 38, 373, 1968.
17. **Reilly, J. and Rae, W. N.,** *Physico-Chemical Methods,* Vol. 1, 5th ed., Van Nostrand, New York, 1953, chap. 12.
18. **Midgeley, H. G.,** A quick method of determining the density of liquid mixtures, *Acta Crystallogr.*, 4, 565, 1951.
19. **Pelsmaekers, J. and Amelinck, S.,** Simple apparatus for comparative density measurements, *Rev. Sci. Instrum.*, 32, 828, 1961.
20. ASTM, Density of plastics by the density-gradient technique, test for D1505-68, in *Book of ASTM Standards Including Tentatives,* American Society for Testing and Materials, Philadelphia, 1975, 35.
21. **Zussman, J.,** X-ray diffraction, in *Physical Methods in Determinative Mineralogy,* Zussman, J., Ed., Academic Press, London, 1977, 426.
22. **Foote, F. and Jette, E. R.,** The fundamental relation between lattice constants and density, *Phys. Rev.*, 58, 81, 1940.
23. **Eliáš, M. and Uhmann, J.,** Densities of the rocks in Czechoslovakia, *Geol. Sur. Czech.*, 6, 1968.
24. **Tittman, J. and Wahl, J. S.,** The physical foundations of formation density logging (gamma-gamma), *Geophysics,* 30, 284, 1965.
25. **Baker, P. E.,** Density logging with gamma rays, *Trans. Am. Inst. Metall. Pet. Eng.*, 210, 289, 1957.
26. **Caldwell, R. L.,** Using nuclear methods in oil-well logging, *Nucleonics,* 16, 58, 1958.
27. **Campbell, J. L. P. and Wilson, J. C.,** Density logging in the Gulf coast area, *J. Pet. Technol.*, 10, 21, 1958.
28. **Pickell, J. J. and Heacock, J. G.,** Density logging, *Geophysics,* 25, 891, 1960.
29. **Wahl, J. S., Tittman, J., and Johnstone, C. W.,** The dual spacing formation density log, *J. Pet. Technol.*, 16, 1411, 1964.
30. **Scott, J. H.,** Borehole compensation algorithms for a small-diameter, dual-detector density well-logging probe, in Soc. Prof. Well Log Analysts 18th Annu. Logging Symp., 1977, chap. S.
31. **Davis, D. H.,** Estimating porosity of sedimentary rocks from bulk density, *J. Geol.*, 62, 102, 1954.
32. **Pirson, S. J.,** *Handbook of Well Log Analysis for Oil and Gas Formation Evaluation,* Prentice-Hall, Englewood Cliffs, N.J., 1963, chap. 17.

33. **Hilchie, D. W.,** *Applied Openhole Log Interpretation for Geologists and Petroleum Engineers,* Douglas W. Hilchie Inc., Golden, Colo., 1979, chap. 10.
34. **Airy, G. B.,** Account of pendulum-experiments undertaken in the Harton Collinery for the purpose of determining the mean density of the earth, *Philos. Trans. Soc. London,* 146, 297, 1856.
35. **Smith, N. J.,** The case for gravity data from boreholes, *Geophysics,* 15, 605, 1950.
36. **Hammer, S.,** Density determinations by undergound gravity measurements, *Geophysics,* 15, 637, 1950.
37. **Gilbert, R. L. G.,** Gravity observations in a borehole, *Nature (London),* 170, 424, 1952.
38. **Lukavchenko, P. I.,** Observations with gravimeters in boreholes and mineshafts, *Razved. Promysl. Geofiz.,* 43, 52, 1962.
39. **Goodell, R. R. and Fay, C. H.,** Borehole gravity meter and its application, *Geophysics,* 29, 774, 1964.
40. **Howell, L. G., Heintz, K. O., and Barry, A.,** The development and use of a high-precision downhole gravity meter, *Geophysics,* 31, 764, 1966.
41. **McCulloh, T. H., LaCoste, L. J. B., Schoellhamer, J. E., and Pampeyan, E. H.,** The U.S. Geological Survey — LaCoste and Romberg Precise Borehole Gravimeter System — Instrumentation and Support Equipment, Geol. Surv. Prof. Pap. 575-D, U.S. Geological Survey, Reston, Va., 1967.
42. **Schmoker, J. W.,** Accuracy of borehole gravity data, *Geophysics,* 43, 538, 1978.
43. **Robie, R. A., Hemingway, B. S., and Fisher, J. R.,** Thermodynamic Properties of Minerals and Related Substances at 298.15 K and 1 Bar (10^5 Pascals) Pressure and at Higher Temperatures, Geol. Surv. Bull. 1452, U.S. Geological Survey, Reston, Va., 1978.
44. **Rosenholtz, J. L. and Smith, D. T.,** Tables and Charts of Specific Gravity and Hardness for Use in the Determination of Minerals, Eng. Sci. Ser. No. 34, Rensselaer Polytechnic Institute, Troy, N.Y., 1931.
45. **Washington, H. S.,** Chemical Analysis of Igneous Rocks, Geol. Surv. Prof. Pap. 99, U.S. Geological Survey, Reston, Va., 1917.
46. **Piersol, R. J., Workman, L. E., and Watson, M. C.,** Porosity, total liquid saturation, and permeability of Illinois oil sands, *Ill. State Geol. Surv., Rep. Invest.,* 67, 1940.
47. **Daly, R. A., Manger, E., and Clark, S. P., Jr.,** Density of rocks, in *Handbook of Physical Constants,* Revised ed., Clark, S. P., Jr., Ed., Geological Society of America, Boulder, CO, 1966, chap. 4.
48. **Birch, F.,** Compressibility; elastic constants, in *Handbook of Physical Constants,* Revised ed., Clark, S. P., Jr., Ed., Geological Society of America, Boulder, CO, 1966, chap. 3.
49. **Press, F.,** Seismic velocities, in *Handbook of Physical Constants,* Revised ed., Clark, S. P., Jr., Ed., Geological Society of America, Boulder, CO, 1966, chap. 9.
50. **Clark, S. P., Jr.,** Thermal conductivity, in *Handbook of Physical Constants,* Revised ed., Clark, S. P., Jr., Ed., Geological Society of America, Boulder, CO, 1966, chap. 21.
51. **Schock, R. N., Bonner, B. P., and Louis, H.,** Collection of ultrasonic velocity data as a function of pressure for polycrystalline solids, Lawrence Livermore Laboratory TID-4500, UC-34, Physics-general, 1974.
52. **Skinner, B. J.,** Thermal expansion, in *Handbook of Physical Constantes,* Revised ed., Clark, S. P., Jr., Ed., Geological Society of America, Boulder, CO, 1966, chap. 6.
53. **Murase, T. and McBirney, A. R.,** Properties of Some Common Igneous Rocks and their Melts at High Temperatures, *Geol. Soc. Am. Bull.,* 84, 3563, 1973.
54. **Young, E. J. and Olhoeft, G. R.,** The Relation of Specific Gravity to Chemical Composition for Crystalline Rocks, Geol. Surv. Open-File Report, U.S. Geological Survey, Reston, Va., 1976, 14.
55. **van Krevelen, D. W.,** *Coal,* Elsevier, Amsterdam, 1961, 315.

Section III
Inelastic Properties of Rocks and Minerals: Strength and Rheology

By
Stephen H. Kirby and John W. McCormick

INTRODUCTION

Handin's[110] chapter on rock strength and ductility in the *Handbook of Physical Constants* was a major attempt to summarize data pertaining to rock strength. Written in the early 1960s, it is the most comprehensive collection of rock strength data published up to about 1964. Since it was written, a massive volume of rock strength data has appeared in the scientific literature, and periodic reviews and several excellent books on experimental rock mechanics have been written.[149,150,183-185,215,225,292] Given these facts and practical limitations of this chapter, it is not possible or even desirable to attempt to produce a summary with the scope of the Handin review. We rather view this work as a source on the science of rock mechanics and a representative summary of data pertaining to experimental rock strength which have appeared since 1964. No attempt has been made to summarize the engineering properties of rocks, such as the room temperature compressive strength, since comprehensive summaries of this kind already exist (see references cited in Handin,[110] Lama and Vukutari[183,184] Deere and Miller,[68] Hendron,[140] Ohnaka[218]).

Since the Handin review, there have been several major developments in the science of experimental rock mechanics:

1. The time-dependent aspects of rock strength have been extensively explored, not only in ductile flow regime, but also in brittle deformation and in rock friction.
2. The weakening effects of water on plastic flow, crack propagation, and related phenomena have led to the widespread belief that water plays a major, if not dominating role, in controlling deformation in the crust of the earth.
3. The phenomenon of brittle failure by faulting under compressive load has been illuminated by careful measurement of axial, radial, and volumetric strains and by detection of acoustical events associated with microfracturing. These developments have emphasized the important role of microfracturing prior to failure and have led to fundamental changes in prevailing ideas of the physics of brittle fracture. Recent work has emphasized the intrinsic propagation properties of cracks and the statistical interactions of three-dimensional arrays of microcracks.
4. The development of the solid-medium, piston-cylinder apparatus by David Griggs in the mid 1960s greatly expanded the pressure range for rock strength measurements, although the accuracies in stress and temperature measurement still do not rival those of fluid medium apparatus. It has been demonstrated that the pressure accommodated by most gas apparatus (up to about 8 kbar) is insufficient to suppress the weakening effects of microfracturing in most silicate rocks at temperatures below 1000°C. Thus the piston cylinder data are the only information available on the intrinsic plastic strength of most silicate rocks below 1000°C.
5. The conviction that rock friction measurements are relevant to the mechanics of large-scale faulting and earthquakes and to the engineering stability of rock structures prompted an explosion of data on rock friction. Recent work has focused on time-dependent effects and on the frictional behavior of artificial and natural fault gouge.

LABORATORY TESTING OF ROCK AND MINERAL STRENGTH

It is clearly desirable in experimental testing of rocks to maintain conditions in the parameters which control rock strength as uniformly as possible. In particular a uniform state of stress should be strived for, an objective which has probably never been fully realized in rock testing. Certain loading geometries approach this ideal most closely; specifically those experiments which have circular cylindrical geometry with a fluid pressure applied to the cylindrical surface and the specimen loaded differentially along the cylinder axis by compression or subject to a moment (Figure 1).* Cylindrical samples are usually enclosed in tubes of plastic or soft metal to deny the confining fluid accesses to the sample. Since the confining fluid cannot sustain significant shear stresses, the cylindrical surface bears no shear tractions and therefore, two of the principal stresses are normal to this surface and equal to the confining pressure, P, and the third is parallel to the cylinder axis. Following the usual convention, compressive stresses are considered positive and ranked $\sigma_1 > \sigma_2 > \sigma_3$. The principal sources of nonuniformity in the states of stress in cylindrical samples are nonuniformities in the elastic and inelastic properties of the sample material and nonuniformity due to end effects. The latter are thought to be due to mismatches in the elastic and thermal properties of the sample and pistons used to load the sample and due to the resulting frictional stresses at the sample-piston interfaces (see reviews on end effects by Jaeger and Cook,[150] Vukuturi et al.,[292] Paterson[225]). One of the ways of minimizing end effects is to increase the length ℓ to diameter d ratio. In triaxial compression tests ($\sigma_1 > \sigma_2 = \sigma_3 = P$, Figure 1A), ℓ/d is usually less than three to avoid buckling and greater than two to minimize end effects. Data on the buckling of rocks beams under confining pressure with $\ell/d \gg 3$ may be found in the papers on experimental folding by Handin et al.[114,115] In triaxial extension tests ($\sigma_1 = \sigma_2 = P > \sigma_3$, Figure 1B), end effects due to friction are lessened since the normal stress across the sample-piston interface is minimized, and evidentally the limitations of ℓ/d due to buckling problems do not apply. In triaxial torsion tests (Figure 1C and D), a circular cylinder (solid or hollow), is subject to a confining pressure, P, on the cylindrical surfaces and is torqued by a moment, M, and compressed along its axis by a stress σ_x.

In a simple torsion experiment on a solid cylinder where σ_x and P are zero, the maximum shearing stress $(\tau_{xy})_{max}$ on the cylindrical surface is:

$$(\tau_{xy})_{max} = 2M/(\pi r_o^3) \tag{1}$$

and τ_{xy} varies within the cylinder as the radial distance. This uncertainty due to the radial dependence is reduced if torque is applied to a thin-walled hollow cylinder and

$$(\tau_{xy})_{max} = 2M/[\pi r_o^3(1 - r_o^3/r_o^4)] \tag{2}$$

where again τ_{xy} within the cylinder wall varies as the radial distance. Both of these calculations rely on an elastic solution, but to a good approximation in a thin-walled hollow cylinder

$$\tau_{xy} = 3M/[2\pi(r_o^3 - r_i^3)] \tag{3}$$

and the radial dependence can be neglected. By combining torsion with axial compression or extension, and confining pressure P, the magnitude of σ_2 (always radially directed and equal to P) can be made to arbitrarily vary between the bounds of σ_1 and σ_3 (see Table 1). The relative values of σ_x, P, and M also determine the angle β between σ_1 and the x axis along the cylinder:

$$\tan \beta = 2\tau_{xy}/(\sigma_x - P) \tag{4}$$

Thus triaxial torsion-compression tests can be performed such that $\sigma_1 > \sigma_2 > \sigma_3$. The principal uncertainties are in the end effects, as in compression and extension tests. An

* All figures and tables follow the text.

empirical relationship between the results from solid and hollow cylinders has been put forward by Handin et al.,[117] which may permit solid cylinder testing in lieu of the more difficult tests on hollow cylinders. Data on the triaxial torsion of solid cylinders is summarized in Table 5.

The usual measure of rock strength in triaxial compression and extension tests is the sustained differential stress $\sigma = (\sigma_1 - \sigma_3) = |\sigma_x - P|$, which is directly related to the maximum shear stress $\tau_{max} = \sigma/2$ and the octahedral shear stress $\tau_{oct} = \sigma\sqrt{2/3}$. The latter is a measure of the average shear stress responsible for inelastic deformation.

Various attempts have been made to achieve a polyaxial state of stress ($\sigma_1 > \sigma_2 > \sigma_3$) in rectangular parallelopiped samples which are independently loaded normal to the three pairs of surfaces. Here, end effects are especially important and can pervade the whole sample. Typically, experiments where σ_2 is held equal to σ_3 compare poorly with triaxial compression tests under the same nominal conditions.[150] Other types of tests, such as three-point bending of beams, diametrical loading of circular discs, internal pressurization and axial compression of thick-wall cylinders, and indentation testing are much more difficult to interpret, because the stresses at failure or yield rely on an elastic solution of the stress distribution. In particular where failure occurs under local tensile stresses, the elastic properties of rocks are particularly difficult to estimate since the elastic moduli are strongly dependent on the elastic strain.[121,293,294] Data from experiments of these types are tabulated in Vukutari et al.[292] and in Jaeger and Cook.[150] Specialized loading geometries are used for friction measurements and are reviewed in a later section.

An elastic apparatus deflection accompanies the nonhydrostatic loading of a rock sample (Figure 2). This distortion includes the axial compression of the loading column and the elastic deflection in the force generating system (transmission driven by an electric motor or a hydraulic ram, etc.) Typically, the relationship between the differential axial load ΔF_x and the apparatus elastic distortion $\Delta\ell_d$ is strongly nonlinear at low load, but to a good approximation, the elastic deflection is given by $\Delta\ell_d = (k_a \, \Delta F_x) + \Delta\ell_t$ where k_a is the apparatus spring constant, and $\Delta\ell_t$ is the nonlinear apparatus take up at low load. The elastic deflection $\Delta\ell_d$ is usually subtracted from the total piston displacement $\Delta\ell_T$ and the relative shortening is defined as $\epsilon = (\Delta\ell_T - \Delta\ell_d)/\ell_o$ where ℓ_o is the original length.

The differential stress $\sigma = (\Delta F_x - \Delta F_c)/A_i$ where ΔF_x is the piston force above (or below in extension) that corresponding to the confining pressure P, ΔF_c is a correction for piston friction and for the force borne by the specimen jacket and A_i is the cross-sectional area of the specimen at the strain in question. In experiments on mineral single crystals, A_i depends on the specific geometry of plastic deformation. In rocks that are approximately isotropic, it is usually assumed that strain is homogeneous and constant volume, for which $A_i = A_0\,(1+\epsilon)$, where A_0 is the zero strain cross-sectional area, and compression is considered as positive strain. The uniform strain assumption may be judged by specimen shape. Paterson and Edmond[226] warn that the constant volume assumption may be substantially in error in experiments on specimens taken to large strain (>10%) where stable microfracturing and associated dilatency are important.

The stress-strain-time relations may be programed in a number of useful ways.

Nominally constant strain rate tests — Here, the force generator (synchronous motor, hydraulic pump, etc.) is set at constant speed such that the total shortening rate of the sample plus apparatus load column is constant. Since the apparatus distortion is nonlinear and the load changes with time, the axial strain rate, $\dot{\epsilon}$, is not fixed but varies with time and strain. In experiments where the stress drops due to plastic yielding or brittle fracture, the shape of the stress-strain curve after yield or failure can depend strongly on the apparatus distortion coefficient k_a[90,100] (see also review by Paterson[225]).

Creep (constant stress) tests — Here, the differential stress σ is fixed and the specimen shortening is monitored with time (Figures 14 and 16). Usually, the elastic strains upon loading are not included since time-dependent elastic strains are typically not significant. Constant load tests, in which changes in cross-sectional area with strain are not compensated by changes in axial load, are typically used in testing of mineral single crystals, since the compensation depends on the specific geometry of plastic strain and since it is more convenient to hold the axial load constant in dead load apparatuses. These type of experiments are used to explore the effects of time on rock strength.

Stress relaxation tests — These are performed by loading at constant strain rate and then turning off the force generating system. The load on the specimen is sustained by the elastic deflection of the apparatus and specimen $\Delta\ell_d + \Delta\ell_e$, and as the specimen shortens plastically by an amount $\Delta\ell_p$, the elastic distortion of the apparatus and sample decrease by a like amount, and the force on the specimen relaxes by $\Delta\ell_p/(k_a + k_s)$. A relationship between the differential stress σ and axial strain rate $\dot{\epsilon}$ can therefore be derived from the force-time record, the apparatus distortion coefficient, and the specimen dimensions. Relaxation tests are useful in exploring rheological relationships between stress and strain rate and may follow a conventional constant strain rate test. Although there are substantial problems of interpretation of such experiments, the technique has been applied to a wide range of rocks and minerals.[9,99,100,102,144,171,212,238,239,253,259,260]

Cyclic loading experiments — These tests, at low frequency ($\ll$1 Hz), have been used to determine the time-dependent (anelastic) elastic properties of rocks and to explore the cyclic fatigue failure of rocks under brittle conditions (see reviews by Paterson[225] and Brace[41]).

GENERAL MECHANICAL BEHAVIOR OF ROCKS IN SHORT-TERM TRIAXIAL TESTS

In the context of short-term constant strain rate (5×10^{-4} to 10^{-5} sec^{-1}) triaxial tests, "yield strength" is fundamentally defined as the differential stress σ_y above which a significant permanent change in length is sustained by the test specimen (Figure 3). This definition corresponds to the elastic limit or offset yield strength in tensile tests on metals and relates to a critical stress at onset of plastic deformation or brittle fracture. In practice, other definitions are often used. The "ultimate strength" is the maximum differential stress supported by a sample over a specified range of strain and can represent a relative maximum in strength or simply the strength at the end of a test. Strength can also be defined as the "stress at some specified strain". In constant strain rate experiments at high temperature and/or low strain rate, the stress-strain curve is flat over a wide range of strain; this stress is the "steady-state strength" (Figure 3).

Even in the earliest controlled triaxial experiments on rocks, it was recognized that rock strength increases markedly with increasing confining pressure (Figures 4, 5, and 8). Prominent changes in the mechanisms by which permanent changes in length are accommodated by test specimens also accompany increases in confining pressure (illustrated schematically in Figure 6). Broadly speaking, there are three major regimes of rock behavior.

Brittle regime — Displacements and strains are localized along discrete surfaces (fractures or faults). At the lowest pressures, extension fractures (axial splitting in compression tests) occur in orientations perpendicular to the least principal stress σ_3 and movement is primarily normal to such surfaces. Failure occurs apparently due to locally tensile stress to which rocks are weakly resistant. At yet higher pressures, shear fractures are developed at an angle θ to σ_1 (θ usually ranges from 10 to 35° depending on material and type of test). At elevated temperatures and pressures, loss of cohesion often does not accompany the localization of strain along shear surfaces; this process has been termed "faulting".[104] Collectively, the pressure interval over which extension and shear fracturing and faulting occur is the "brittle regime". In low porosity rocks, significant positive inelastic volumetric strains (dilatancy) occur prior to fracture (Figure 7). Stress-strain curves usually are characterized by stress drops or peaked curves corresponding to the development of fractures or faults and stresses in the post-failure segments of the curves typically decrease with increasing strain. The pressure sensitivity of strength is very high. The pressure sensitivity of fracture strength in triaxial tests is generally analyzed in terms of the Coulomb-Mohr failure criterion: $\tau_f = f(\sigma_n)$ where τ_f is the shear stress at failure on the shear fracture or fault and σ_n is the normal stress on the shear surface (at angle θ to σ_1) where $d\tau_f/d\sigma_n$ usually decreases with increasing σ_n. While useful as an empirical failure criterion relating τ, P, and θ at failure for a given type of test (see reviews by Handin,[110] Jaeger and Cook,[150] Paterson[225]), the experiments using general states of stress $\sigma_1>\sigma_2>\sigma_3$ clearly show that this criterion is inadequate to account for the systematic effects of σ_2.[111,117,206]

Semi-brittle regime — Macroscopic strains due to stable microfracturing (cataclasis) and to the mechanisms of crystal plasticity are distributed throughout the specimens. Large increases in volume typically are associated with the microfracturing in low porosity rocks and strains exceeding 20% can be sustained without fracture or faulting (Figure 8). Stress-strain curves are shifted upward in stress with increasing confining pressure. Strength increases nonlinearly with increasing confining pressure (Figures 4 and 5) and increasing confining pressure progressively increases the slope of the stress-strain curve in the post yield regime (Figure 8).

Ductile regime — At yet higher confining pressure, microfracturing can be essentially suppressed and the mechanisms of plastic glide (slip, twinning, and transformation glide) dominate at low to intermediate temperatures. The pressure sensitivity of strength (Figures 4 and 5) is usually very small ($d\sigma/dP < 0.1$) and the slope of the stress-strain curve (i.e., the work hardening rate $d\sigma/d\epsilon$) is insensitive to changes in confining pressure (Figure 8A). Inelastic volumetric strains are unimportant in low porosity rocks deformed in this regime (Figure 8B). Even though confining pressure is the principal factor influencing which deformation regime corresponds to the behavior of a rock in a given type of test, the state of stress has a marked effect on the magnitudes of the confining pressures at the transitions between the three regimes (Figures 5A and 5C). Paterson[225] has reviewed some of the recent literature on the brittle to semi-brittle transition. A review of the data on carbonate rocks by Kirby[165] suggests that transition data for various states of stress are consistent if it is assumed that the transitions occur at critical values of the least principal stress, independent of the state of stress (Figure 5B and 5D). We are not aware of data on crystalline rocks with which we can further test this hypothesis.

EFFECTS OF WATER AND OTHER FLUIDS

Five major weakening effects of water and other fluids have been identified.[99,225]

The effective stress phenomenon — The marked weakness of unjacketed specimens in triaxial tests on porous rocks compared to experiments on specimens jacketed to deny entry by the confining medium was recognized almost at the beginning of the modern stage of rock mechanics testing. The lack of such weakening effects in nonporous materials (crystals, metals) clearly established that the effect is related to pore pressure. Controlled pore pressure triaxial tests, where the pore pressure is independently controlled through a hollow piston, were first inspired by the application of Karl Terzaghi's effective stress law for soils to rocks by Hubbert and Rubey.[147] In porous rocks with sufficient permeability to equilibrate the pore pressure of an inert fluid during deformation, the effect of confining pressure P on rock strength is modified in such a way as if the effective confining pressure were $(P-p)$ or $(\sigma_3 - p)$ where p is the pore pressure (Figure 9). To a good approximation, this effective stress law accounts for the pore pressure effects on fracture strength (through the reduction of σ_n to $(\sigma_n - p)$), on frictional strength (also as an effect on σ_n (Figure 26), and on the ultimate strength of rocks in the semi-brittle regime (Figure 9). In rocks with low porosity, failure of the pore fluid pressure to communicate throughout a sample can lead to diminished weakening effect of fluids,[116] and concurrent increase in pore volume during deformation (dilatency) can locally reduce pore pressure and lead to a strengthening called dilatency hardening. Apparently, the strain rate (and thus the rate of increase of porosity) must be below some critical value relative to the viscosity of the fluid in order for the pore pressure to communicate throughout the pore space and render the pore pressure effective.[43,251] The effective stress phenomenon is purely mechanical and does not take into account any weakening due to chemical reaction of the pore fluid with a rock sample (see review by Paterson[225] on pore fluid chemical effects). In long-term creep tests (see next section), water often shows such chemical effects. Dramatic drops in strength upon heating hydrous minerals above their dehydration temperatures are observed in triaxial tests on sealed (undrained) samples (serpentine,[213,214,240] alabaster,[136,213] amphibole,[242] choritite and microdiorite[213,214]). Although some reduction in strength is observed in vented specimens[213,240] the major weakening effect appears to be associated with the generation of a pore water pressure approximately equal to the confining pressure (i.e., $\lambda = 1.0$) in the sealed tests, and the observed strengths are comparable with the uniaxial strengths of the same material (Figure 10), as in the controlled pore pressure triaxial tests. Also similar to controlled pore pressure tests, reductions of the effective confining pressure by dehydration in sealed capsules leads to an embrittlement[213,214,240] under conditions of confining pressure alone which would produce ductile or semi-brittle behavior. Provided that the pore fluid and sample do not react chemically, that the permeability is high enough for the pore fluid to equilibrate in the pore space, and that the rate of creation of porosity by dilatency during deformation is low relative to the permeability, the effective stress law provides a good approximation of the weakening effects of neutral pore fluids on porous rocks. Presumably, as the effective confining pressure $(P-p)$ increases and the porosity and permeability decrease, rocks in the "ductile field" as defined earlier will not exhibit such pore pressure effects, no matter what the imposed strain rate.

Hydrolytic weakening — Most single crystals and aggregates of high purity dry silicates are remarkably strong in triaxial tests at high confining pressures and temperatures below about half their melting temperatures. Typical strengths are greater than 10 kbars. In apparent contrast, crustal rocks subject to dynamic metamorphism deform plastically at tectonic stresses that are likely to be more than two orders of magnitude lower. Griggs and coworkers[27,29,99-102] have identified an intrinsic lowering of the plastic yield strength associated with water dissolved in the silicate structure. The source of the weakening is thought to be a chemical corrosion effect on the Si–O–Si links in silicates by a hydrolysis reaction to Si–OH:HO–Si. Slip would therefore only require breaking the weak hydrogen bond links. The magnitude of this hydrolytic weakening effect is most striking in single crystal quartz experiments, where "wet" crystals yield at stresses nearly two orders of magnitude lower than high purity "dry" crystals (Figure 11). Quartz crystals are rendered wet and weak by hydrothermal synthesis or by high-pressure hydrothermal treatment of initially "dry" quartz by water released from dehydration of a hydrous mineral used as a confining medium or by water sealed in capsule with sample.[99,101,102,156] Much recent work on the hydrolytic weakening process has centered on hydrothermally grown synthetic crystals with water incorporated during growth.[18,19,21,22,99,100,102,132,145,156,161,168,209,227] A systematic decrease in yield stress with increasing hydroxyl concentration C_{OH} has been identified in short-term triaxial tests on oriented single crystals of synthetic quartz (Figure 12). Time-dependent effects are clearly evident[18,25,161,168] and are reviewed in a later section. Application of these measurements are difficult without information on the equilibrium solubilities of water in the principal silicates as functions of hydrothermal pressure and temperature. The effects of the chemistry of aqueous fluids on the solubility also need to be established.

Stress corrosion of cracks — The importance of microfractures in the development of macroscopic fractures and faults and in semi-brittle behavior was noted earlier. Water has a specific and clear role in promoting crack growth in silicates by the corrosion of crack tips by a hydration reaction analgous to that in the hydrolytic weakening process. Here the replacement of Si–O–Si links across the crack tip by Si–OH: HO–Si links degrades the strength at the crack tip; only weak hydrogen bonds need be broken to sever the Si to Si links and extend the crack. Early experiments on rocks and minerals[60,270,271] established that time to failure of specimens subject to constant uniaxial load was systematically and strongly reduced by environmental water. In similar experiments on precracked quartz crystals,[197,198] the rate of extension of axial tensile cracks at a fixed uniaxial load increased systematically with the partial pressure of water vapor, and the temperature dependence matched that observed in the time to failure data in the static fatigue tests on quartz crystals of Scholz.[271] In controlled double torsion experiments in quartz, Atkinson[10] observed systematically higher velocities of stable crack growth in samples saturated with liquid water compared to those samples saturated with humid air (Figure 13). No significant stable cracking was observed below the critical stress intensity factor for catastrophic tensile failure in carefully vacuum-dried quartz samples. Similar effects were measured on novaculite[11] granite,[14] (Figure 13), and gabbro.[15] For many rock types, carefully dried samples have significantly higher fracture strengths than air-dired or saturated samples with atmospheric pore pressure (see references cited in Paterson,[225] p. 78). Indentation hardnesses of carefully dried minerals is significantly higher than samples exposed to air or liquid water.[302] Similar drying effects are observed in rock friction experiments.[72] As we note below, creep rates in carefully dried samples at room temperature are dramatically lower than saturated or air-exposed samples. The above phenomena may all be manifestations of stress corrosion effects on cracking associated with a hydration reaction at crack tips.

Solution transport — Silicates are generally very soluble in hydrothermal fluids, and the rates of mass transport of silicates in solution are far greater than the solid state diffusivities of the same species at moderate temperatures. In a saturated porous aggregate which is nonhydrostatically loaded, the stress distribution around grains is nonuniform, and considerable field evidence indicates that grain contacts at high normal stress are dissolved and reprecipitation occurs at grain contacts with water filled pores. Although theoretical work has been devoted to this problem (see reviews by Paterson,[223] Elliot,[83] and Robin[246]), and there is persuasive textural evidence that something like pressure solution is associated with reduction of porosity during compaction, convincing mechanical data are lacking which link nonhydrostatic stresses to distortional strains produced by this mechanism. This is a very difficult mechanism to explore experimentally since it is known to occur geologically at relatively low temperatures ($T < 400°C$). At similar conditions in the laboratory, strain rates associated with the process are exceedingly low. Increasing test temperature does not selectively aid in promoting the process, since it also helps to promote plastic deformation, stress corrosion cracking, and the reduction of porosity.

Other effects of water — At high temperature, water promotes recrystallization and deformation processes associated with bulk diffusion. Part of the weakening effect of water during high-temperature and high-pressure deformation may be related to the hydrolytic weakening process, since structurally incorporated water could aid diffusive mass transfer. Post[235,236] observed greater recrystallization and higher creep rates in creep tests on dunite exposed to water of dehydration compared to dry dunite. A similar contrast was noted by Tullis[285] and by Parrish et al.[221]

TIME-DEPENDENT STRENGTH EFFECTS

Engineers recognized almost at the inception of modern materials testing at the beginning of this century that the load-bearing capabilities of metal parts depend on the time duration of loading. The great range in load duration in geologic deformation and the great contrast in load duration between laboratory and geologic deformation compelled Griggs to initiate a program of creep tests on rocks in the 1930s[98,310] and this represents the beginning of modern work on the subject. The effects of the time scale of load duration generally is revealed by the effects of strain rate on strength in constant strain rate tests or by the variation of strain rate with time at constant differential stress in creep tests. Fundamentally, these time-dependent effects stem from the aid that thermal vibrations of atoms provide to deformation processes: the longer the duration of load, the greater the probability that a thermal vibration of sufficient amplitude to aid the deformation process will occur.

The subdivisions of rock behavior into brittle, semi-brittle, and ductile regimes based on differences in pressure sensitivity, strain distribution, and deformation mechanisms are also paralleled by differences in time-dependent effects, which we review below.

Brittle Creep

The time-dependent behavior of rocks in the brittle regime depend strongly on the initial porosity. The creep of high-porosity rocks is dominated by the elimination of porosity by consolidation processes. Brittle processes are undoubtedly involved in consolidation, but the strains realized from pore collapse are difficult to separate from the strains directly due to microcracking. The brittle creep of porous rocks is poorly understood for this reason.

Low-porosity crystalline rocks loaded to a constant differential stress near the short-term breaking strength generally show a time-delayed failure, or ''static fatigue''. Careful strain measurements indicate that a small amount of inelastic strain (generally less than 0.5% shortening) occurs prior to failure by fracture or faulting. That creep strains are due to extension of preexisting cracks and the production and growth of new cracks is clear from the direct measurement of cracks in crept rocks[179] and from a large number of indirect measurements, such as the measurement of time-dependent dilatent strains,[120,179,181,182,296,297,307] the detection and location of acoustical emissions associated with microfracturing,[120,192,193,307] and indirect measurements, such as the velocity of elastic waves and permeability.

A considerable literature on brittle creep has developed in the last 10 years, from which we can draw some general conclusions:

1. The form of the creep curves in experiments taken to near failure is sigmoidal or S-shaped, with an initial ''hardening stage'', where creep rates decrease with time followed by a ''stage of accelerating creep'', where creep rates increase continuously with time until failure (see Figure 14). This general form applies to axial, radial, and volumetric strains, and the creep rates at the inflection point are among the characteristic parameters of the creep curves. The reproducibility of the time to onset of tertiary creep and the time to failure typically are poor, often varying by as much as a factor of two. Nonetheless, Cruden[66] and Kranz[181] have analyzed the existing hardening stage creep curves using statistical fitting and criteria for goodness of fit and concluded that strain rates are proportional to power ''m'' of the time under load, where $m \simeq -1$. The transition between the hardening and accelerating creep stages occurs above some critical strain, which is a characteristic of the material and experimental conditions, especially pressure.[67,98,181,182,245] Kranz and Scholz[182] have emphasized that the fundamental strain parameter at the transition is the inelastic volumetric strain, since it is most closely related to the crack density or crack area per unit volume, which is extremely difficult to measure directly.[179]
2. Detection of acoustical emissions during creep is an indirect way of monitoring crack growth and development. The rate of emissions correspond to the changes in creep rates in the two stages of creep.[120,192,193,307] Lockner and Byerlee[193] show that the locations of microshocks in the hardening stage are randomly located throughout the

uniformly stressed sections of samples, and that they tend to localize on the future surface of failure only in the accelerating stage.

3. The principal factors controlling brittle creep rates are applied differential stress,[98,120,182,270,296,297] effective confining pressure,[181,244,297] and moisture content.[60,98,296,297,307] The powerful effects of applied stress and confining pressure on the inflection point axial strain rates of water-saturated Westerly granite[296,297] are shown in Figure 15. The strong effect of confining pressure[218] and weak effect of strain rate on fracture strength[42] in constant strain rate tests are consistent with the above generalizations.

The mechanism responsible for the time dependence of brittle creep is thought to be corrosion of crack tips by chemical hydration,[60,65,67,179,181,182,270,271] a process reviewed in an earlier section. The brittle creep of rocks under constant load is considered to be controlled by the rate of slow growth of tensile cracks. At low strains, tensile cracks extend from stress concentrators (cracks, pores, grain boundaries, etc.), and the rates of growth decrease as the number of concentrators is exhausted by cracking, and thus creep rates decrease with increasing time in the hardening stage. If the crack density (or crack area per unit volume) exceeds some critical value such that cracks are close enough to interact on a large scale, crack coalescence occurs to form a macroscopic fracture on which failure occurs. Although many of the microstructural observations and macroscopic mechanical data in short-term constant strain rate tests have their counterparts in samples under constant load, Kranz[179] has suggested that there may be fundamental differences in crack growth and development in the two types of tests.

We noted earlier that brittle creep is limited to total axial strain to about 0.5% in silicate rocks, and so the process is not capable of accommodating significant tectonic strains. Its importance lies in the insight that it provides into the process of brittle fracture and into the premonitory phenomena which are associated with it.

Semi-Brittle Creep

Carter and Kirby[59] have recently reviewed the meager data on the creep of rocks in the semi-brittle regime. This complex regime, in which both brittle and ductile processes operate, is very poorly understood, not only because of the limited amount of creep data available, but also due to lack of insight into how brittle and ductile processes interact. Kirby and Raleigh[171] and Carter and Kirby[59] have argued that most of the data on the high-temperature, room-pressure creep tests on fine-grained ceramics and crystalline rocks are dominated by microfracturing and by grain boundary sliding accommodated by void formation. The latter process is unlikely to be important at significant confining pressures within the Earth.

Griggs et al.[105] reported on an extensive series of short-term constant strain rate tests on crystalline rocks at a confining pressure of 5 kbars and at temperatures up to 800°C. The stress-strain curves from 500 to 800°C approached a steady-state stress by 5% strain. Stable microfracturing (cataclasis) and plastic deformation both occurred under these conditions, and subsequent work at higher pressure on these rock types indicates that true steady-state flow associated with ductile mechanisms could not have occurred under those conditions. Work hardening rates ($d\sigma/d\epsilon$) progressively increase with increasing confining pressure (Figure 8), with negative values at low pressure. At higher pressures, ($d\sigma/d\epsilon$) asymtotically approaches the intrinsic work hardening rates associated with crystal plasticity. This suggests that a progression of creep response should range from sigmoidal creep at low pressure, and to a quasi-steady-state creep at intermediate pressures, and to transient work hardening creep curves at higher pressures approaching the ductile regime.

Ductile Creep

The time-dependent behavior of rocks and minerals in the ductile regime has been reviewed extensively in the last decade,[55,95,112,171,215,286,301] so the topic will only be briefly outlined here.

In general terms, creep curves of rocks and minerals may be divided into two stages: an initial "transient stage" in which strain rates change continuously with time and a subsequent "steady-state stage" in which creep rates are constant (Figure 16). These creep stages have their counterparts in the stress-strain curves of constant strain rate tests; the variation of stress following the yield deflection from the elastic slope corresponds to the transient stage and the flat part of the stress-strain curve (such as in the bottom curve of Figure 3) corresponds to the steady-state stage. The transient stage in mineral single crystals can take on quite complex topologies, depending on the mineral, the orientation of the compression direction, the initial density of crystal defects, chemical composition, and details of the geometry of the plastic deformation mechanism.[19,22,25,38-40,103,145,168,290,299] The transient regime in rocks is generally of the decelerating type where creep rates decrease from a maximum just after loading and asymtotically approach steady-state creep rates with increasing strain and time (See Figure 16). In general, low temperature and high stresses (or high strain rates) promote transient flow relative to steady-state flow.

It is convenient and useful to consider the total creep strain under ductile conditions $\epsilon(t)$ as the sum of the "instantaneous" elastic and plastic strains upon loading ($\epsilon_e^o + \epsilon_p^o$), the transient creep strain (ϵ_t), and the steady-state strain ($\dot{\epsilon}_s t$):

$$\epsilon(t) = (\epsilon_e^o + \epsilon_p^o) + \epsilon_t(t) + \dot{\epsilon}_s t \qquad (5)$$

A major objective of rock rheology is to establish the effects of experimental conditions (temperature T, applied differential stress σ, confining pressure P, etc.) on the inelastic components of the total creep strain. Considerable progress has been made in recent years in understanding the effects of the environmental conditions on steady-state creep at high temperatures ($T > 0.5\ T_m$ where T_m is a melting temperature in °K). In this high-temperature range, creep is dominated by the steady-state component, and the steady-state creep rate generally follows a thermally activated power law rheology:

$$\dot{\epsilon}_s = A\, \sigma^n \exp[-(E^* + P\, V^*)/RT] \qquad (6)$$

where A, n, E*, and V* are material parameters and R is the gas constant (Figure 17). The pressure effect, through the PV* term, is generally small relative to E* and is very difficult to measure directly. The apparent activation energy for creep Q* at a given pressure is defined the sum $Q^* = E^* + PV^*$. The above material parameters for rocks and minerals are summarized in Tables 5 and 6.

The values of the material parameter A, n, E*, and V* are characteristic of the mechanisms which are responsible for creep. Broadly speaking, there are two classes of creep mechanisms:

1. Those due to the production and movement of defects within crystals. These defects include point defects (vancancies, interstitials impurities), dislocations (line defects that are defined by the localization of intracrystalline glide), and interfaces (twin boundaries, stacking faults, arrays of dislocations). These imperfections can strongly interact in crystals under stress at high temperature, and the nature of this interaction can lead to a rich variation of rheological behavior between materials of different composition and structure, and for a given material with variations of temperature and other experimental conditions.
2. Phenomena at grain boundaries can contribute to creep strains in two fundamental ways: the transport of mass along grain boundaries in polycrystalline media under

stress can change crystal shape by redistribution of crystal, a process called Coble creep. Grains can change relative positions by grains boundary sliding, a process largely responsible for superplastic behavior (a form of extreme ductility) in polyphase metal alloys. Fine grain size promotes Coble creep, grain boundary sliding, and Nabarro Herring creep (the bulk crystal counterpart of Coble creep).

The relative contributions to deformation mechanisms generally change with experimental conditions, and this is generally considered to be a result of competition of the various mechanisms, each with its own distinctive rheological laws.[272] For independent deformation mechanisms, the mechanism which contributes the largest creep rate at a given set of experimental conditions will dominate the rheological behavior and contribute the largest creep strain, and display the most microstructural evidence for its activity. The principal factors influencing this competition between mechanisms are the applied stress σ, the temperature T, and the grain size d. A useful technique for comparison between different materials is to predict the ranges of σ/μ and T/T_m over which a given mechanism will dominate (μ is the shear modulus and T_m the temperature of melting).[3,300] The creep rates associated with the operative mechanisms are contoured over the σ/μ–T/T_m field. This technique has been widely applied to rocks and minerals.[6,8,273,281,301] Deformation maps are generally constructed from limited experimental data and from theoretical predictions even for extensively explored crystalline materials. nonetheless, general conclusions that can be drawn from the deformation maps of rocks are that the processes involving grain size (Nabarro-Herring creep, Coble creep, and superplastic creep) are promoted largely by low stresses and fine grain size and that the transition from high-stress to low-stress mechanisms involves a marked lowering of the stress effect on steady-state creep rates (i.e., the n parameter decreases).

Although transient creep has been predicted for the grain-size sensitive mechanisms (Nabarro-Herring creep, Coble creep, and superplastic creep),[97,201] transient creep is generally lacking under conditions where these processes operate (see, for example, Schmid et al.[261]). Dislocation creep mechanisms produce significant transient creep strains in polycrystalline aggregates, and it is generally held that it represents the strain and time necessary to change the dislocation microstructure from that produced immediately upon loading to that appropriate to steady-state flow under the applied test conditions. Thus, the specific transient creep response depends on the dislocation microstructure of the starting material compared to the steady-state microstructure.[59] The limited ductile creep data on rocks taken to large creep strain indicate that the likely form of the transient creep equation is

$$\epsilon_t\,(t) = \epsilon_T\,(1 - \exp\,(-t/t_r)) \tag{7}$$

where ϵ_T is the total transient creep strain and t_r is a characteristic relaxation time which is the time necessary for the total transient strain to reach (l–l/e) of its final value.[59] If $t = 4t_r$, 98% of ϵ_T is achieved and this time can be taken as that necessary to reach steady, state creep $t_s = 4t_r$. Rearranging Equation 7,

$$\Delta\epsilon(t) \equiv \epsilon_T - \epsilon_t(t) = \epsilon_T \exp(-4t/t_s) \tag{8}$$

which is tested in Figure 18. The fit is satisfactory over all but the earliest part of transient creep.

The high-temperature rheology of Equation 6 generally does not apply at high stresses and low temperatures (note breakdown of this relationship (dashed lines) in the Yule marble data of Figure 17). This is generally interpreted to represent a change in the form of the steady-state rheological law,[163,212,236] but it is more likely to involve the greater importance of transient creep under high-stress and low-temperature conditions (note transient work hardening of low temperature stress-strain curves for Yule marble in Figure 3).

The important weakening effects of water on the ductile deformation of silicates was noted earlier. Time-dependent strength effects are associated with the hydrolytic weakening process[18,21,22,25,27,100,158,161,168] and flow laws of the form of Equation 6 have been applied to single crystal synthetic quartz creep test results (see Table 7). Recently, Kirby and Linker[167] have shown that creep rates of synthetic quartz crystals increase systematically with increasing grown-in hydroxyl concentration. It is generally recognized that the principal factors influencing the uptake of water by silicates are hydrothermal pressure PH_2O and temperature. Until these effects are measured, the hydrolytic weakening effect cannot be quantitatively applied to flow in the Earth.

ROCK FRICTION

Since the early work by Jaeger, the study of rock friction has enjoyed a renaissance. A remarkable number of papers have been published even in the last 5 years, and excellent recent reviews of the topic are available.[49,149,150,194,218,225] Especially noteworthy is the collection of papers of the Conference on Rock Friction, edited by Byerlee and Wys.[52]

The resistance to sliding of one material over another can be reduced to the mechanical elements of Figure 19. A mass M having nominal contact area A with a nominally flat substrate surface is subject to a normal force F_y and to a tangential force F_x through a loading column with spring constant k. The loading column is generally moved at a constant velocity V_x, and the tangential force F_x' necessary to produce an inelastic sliding displacement is a measure of the frictional resistance between mass and substrate. The coefficient of friction μ is the defined as

$$F_x' = \mu F_y \tag{9}$$

and to a good approximation μ is independent of the nominal area of contact A, hence

$$\tau = \mu\,\sigma_n \tag{10}$$

where $\tau = \tau_{xy}$ and $\sigma_n = \sigma_{yy}$.

Two general types of frictional behavior are observed. Following a small inelastic sliding marked by the deflection from the elastic behavior, sudden jerky sliding motion or stick slip can be observed. Byerlee[49] defines the frictional resistance at initial sliding as "initial friction" and the maximum resistance as "maximum friction" (**Figure 20**). Stable sliding occurs where sliding motion is smooth and initial and maximum friction can be defined in much the same way. Obviously, the frictional resistance to initial sliding (the static friction) is greater than the resistance to continued sliding (kinetic friction) and in general, the kinetic frictional resistance for a given velocity V_x is only clearly defined for the stable sliding case.

Some of the actual loading and sample configurations used are shown in **Figure 21**. Direct shear, double shear, and biaxial shear are generally limited to low normal stresses (a few hundred bars), and the triaxial and biaxial configurations suffer from the fact that τ and σ_n cannot be independently varied. All except the rotary shear technique are limited to relatively small displacements, but that technique is difficult to adapt to high-pressure testing. Jaeger and Cook[149,150] review the advantages and drawbacks of each technique.

The reproducibility of friction measurements, from sample to sample and between techniques, is generally no better than ±10%. By far, the largest effects on the frictional resistance of rocks are through the normal stress, the pore pressure, and the sliding displacement. All other effects are generally second order. Byerlee[49,50] and Jaeger and Cook[149,150] show that the frictional resistance of most rocks is rather independent of their mineralogical composition. In particular, Byerlee[49] shows that maximum friction for rocks follows the following friction laws at room temperature

$$\tau = 0.85\ \sigma_n \qquad \sigma_n < 2 \text{ kbar} \tag{11a}$$

$$\tau = 0.5 + 0.6\ \sigma_n \qquad 2 < \sigma_n < 17 \text{ kbar} \tag{11b}$$

The data on which these correlations are based are shown in Figures 22 and 23. The initial friction data show considerably more scatter, expecially in rocks which shown some ductility at room temperature. This suggests that plastic deformation is involved in the initial sliding of these rocks. Temperature has a very small effect on these frictional laws up to 300 or 400°C.[220,277-280] Decreasing displacement rate V_x or delaying loading between slip cycles can increase frictional resistance, but these effects are small and often masked by the significant effects of total displacement.[69-71,219,220,269,284]

The principal effect of water on rock friction is through the effective stress law where σ in Equation 10 is replaced by $(\sigma_n - p)$ where p is the pore pressure.[51,84,251,280,309] Careful drying of quartz and quartzite tends to remove the time-dependent aspects of friction in these materials, suggesting that chemical weakening effects may be important in friction experiments on nominally dry (air dried, etc.) specimens.[72]

ACKNOWLEDGMENTS

Hugh Heard, J. A. Tullis, Neville Carter, Hans Ave'Lallemant, Jim Blacic, John Ross, Barry Atkinson, John Handin, John Logan, Chris Williame, and John Christie generously provided tabulations of unpublished data. Barry Raleigh and Jim Byerlee furnished helpful reviews of the text. Permission to republish figures was universally granted by the authors and publishers.

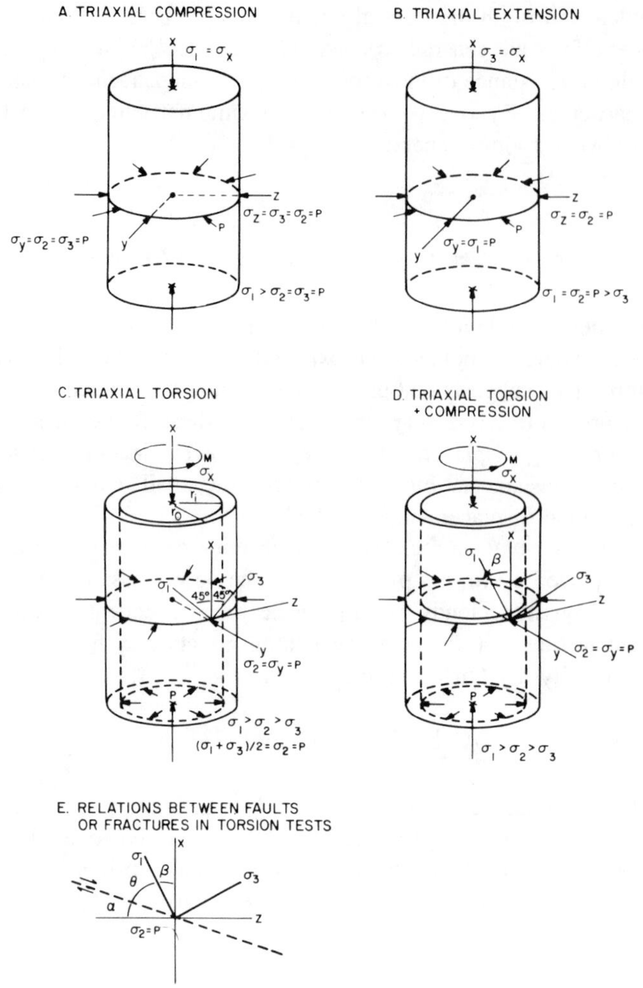

FIGURE 1. States of stress and angular relations in conventional tests on samples with cylindrical geometry. Principal stress magnitudes given in Table 1.

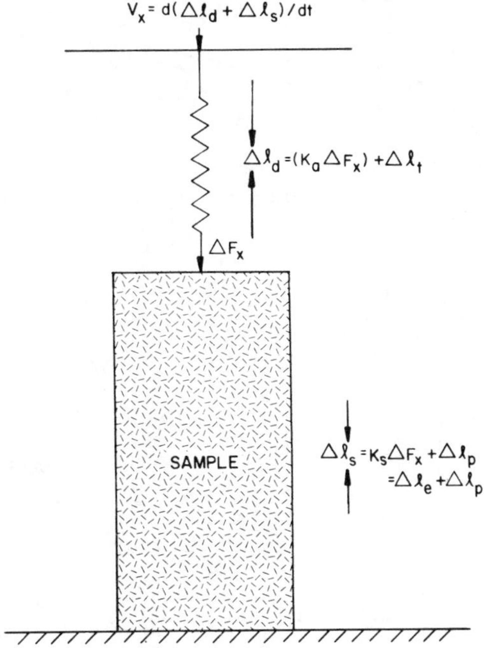

FIGURE 2. Schematic representation of relationships between axial differential force ΔF_x and sample shortening $\Delta \ell_s$. See text.

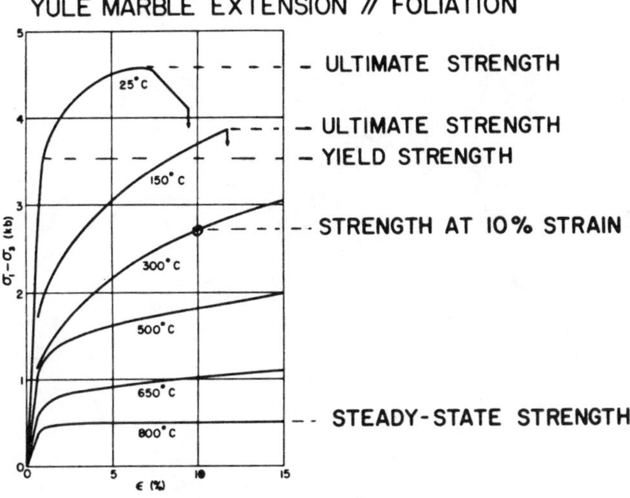

FIGURE 3. Definitions of strength parameters in constant strain rate triaxial tests, using as examples the stress-strain relations of Yule marble in extension at 5 kbar confining pressure. (From Griggs, D. T., Turner, F. J., and Heard, H. C., *Geol. Soc. Am. Mem.*, 79, 39, 1960. With permission.)

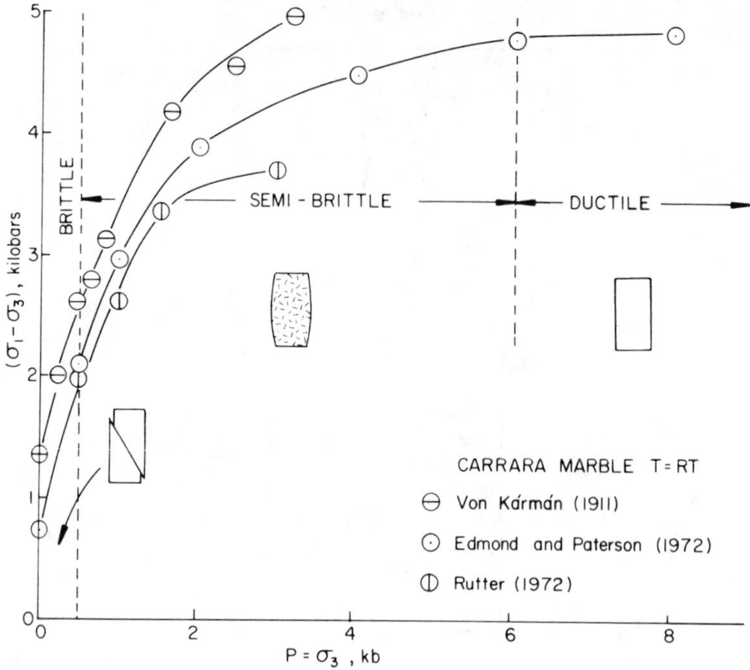

FIGURE 4. Influence of confining pressure on rock strength and regimes of rock deformation behavior. Carrara marble[155,82,251] at room temperature, strain rate = 10^{-4} to $10^{-5}s^{-1}$. Strength taken as ultimate strength or stress at 5% strain, whichever is greater.

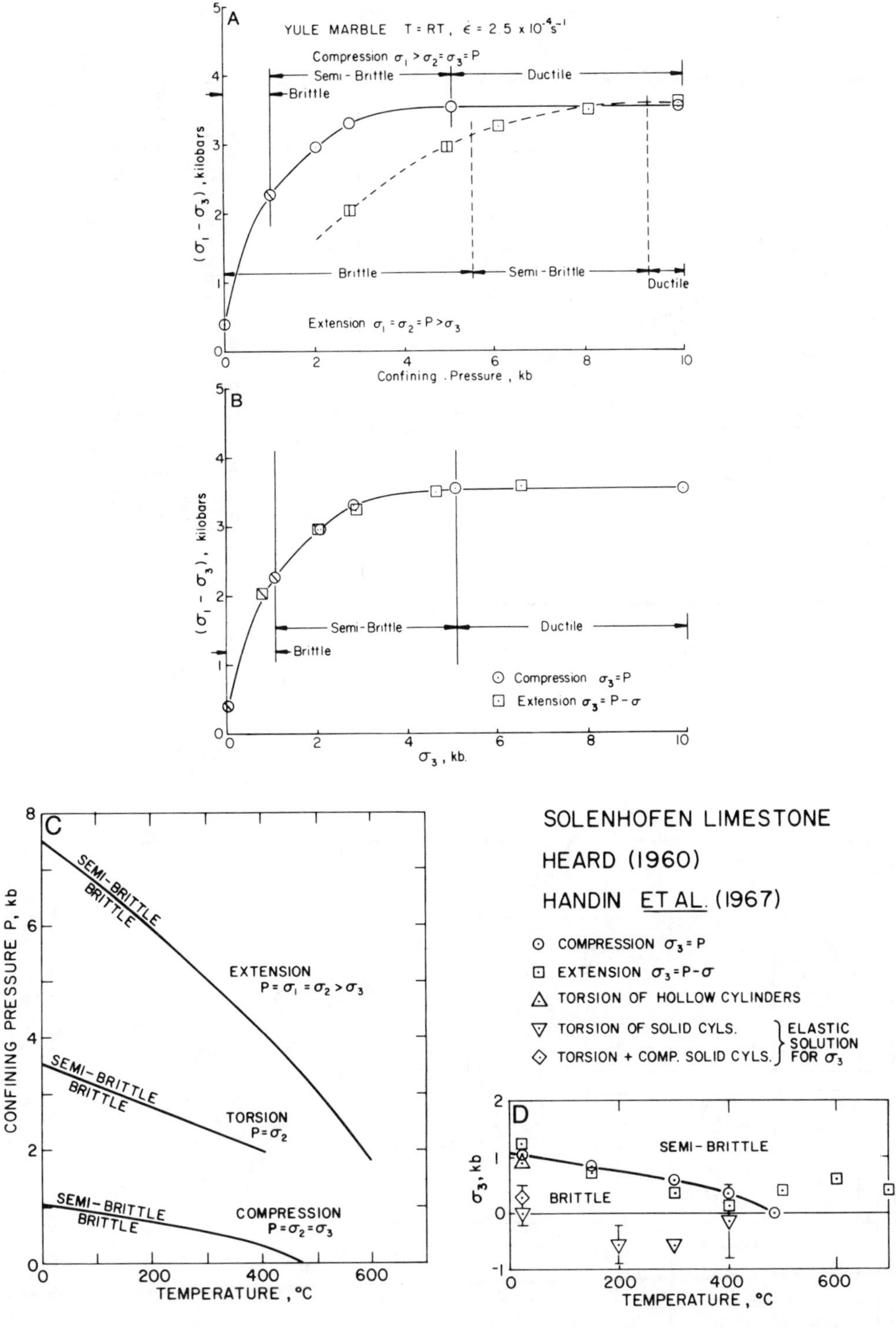

FIGURE 5. (A-B) Influence of confining pressure and state of stress on rock strength. Yule marble at room temperature in extension and compression, averaged for the two principal sample orientations. Same definitions of strength as in Figure 4. Slash mark through symbols indicates fracture or faulting. Data from Griggs and Miller[311] and Handin.[108,110] (A) Strength vs. confining pressure. (B) Strength vs. least principal compressive stress, (C-D) Influence of the state of stress on the brittle to semi-brittle transition in Solenhofen limestone at various temperatures. Diagram after Handin et al.[117] and Kirby.[165] (C) Variation of the critical confining pressure at the brittle to semi-brittle transition with state of stress and temperature. (D) Critical least principal stresses at the transition.

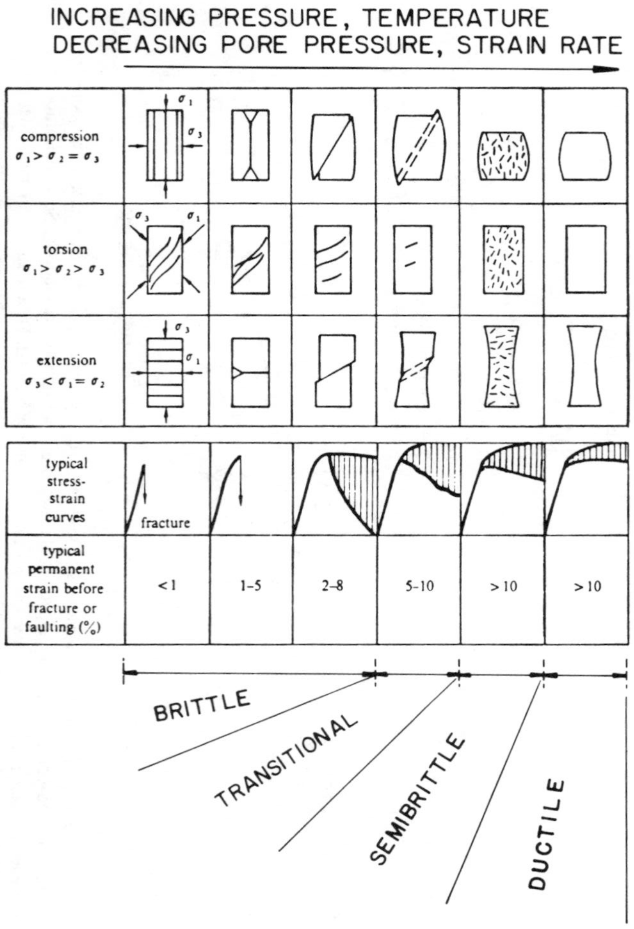

FIGURE 6. Schematic representation of the influences of environmental parameters on the macroscopic behavior, stress-strain relations, and ductility of rocks in triaxial tests. (From Heard, H. C., *Philos. Trans. R. Soc. London A*, 283, 173, 1976. With permission.)

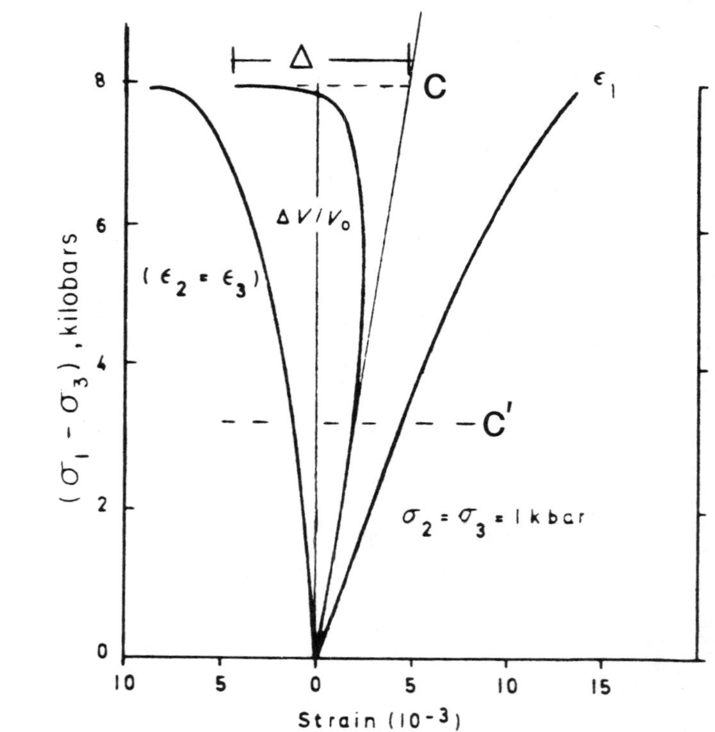

FIGURE 7. Complete stress-strain relations for Westerly granite in triaxial compression at room temperature. Strain homogeneity is assumed and compression is considered positive. An anomalous radial expansion $\epsilon_2 = \epsilon_3$ above a differential stress C′ leads to a departure of the volumetric strain from an elastic slope by a maximum of Δ at the fracture strength C. See Table 11 for tabulated values of Δ for various rock types. (From Brace, W. F. et al., *J. Geophys. Res.*, 71, 3939, 1966. With permission.)

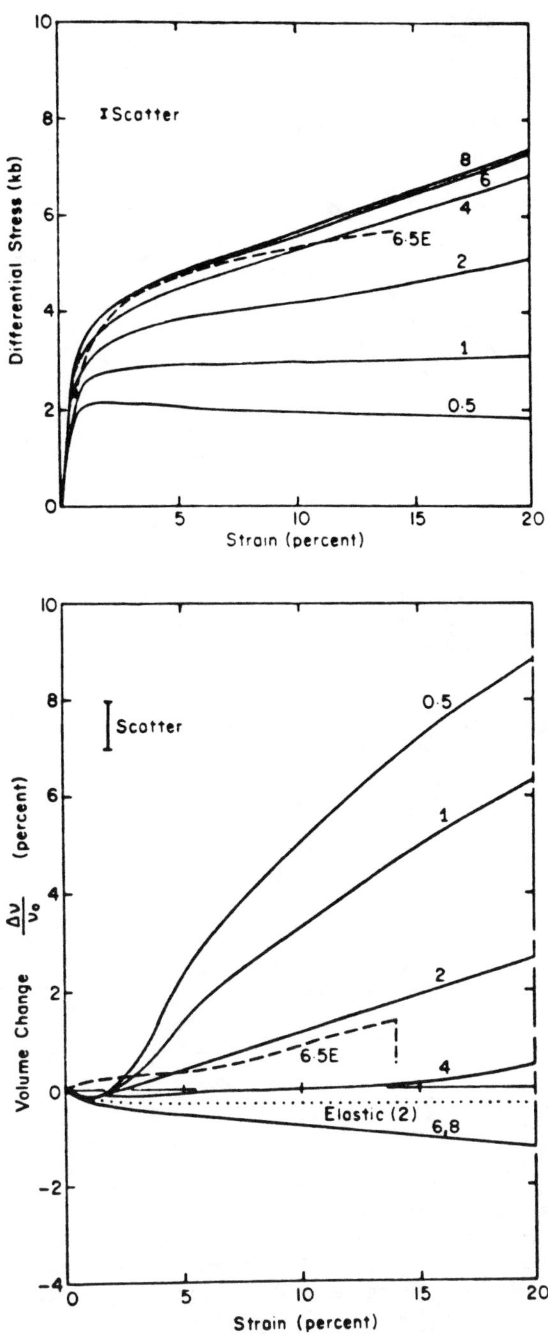

FIGURE 8. Differential stress vs. axial strain and volumetric strain vs. axial strain data for Carrara marble in triaxial compression at confining pressures between 0.5 and 8 kbar. Note large volumetric expansion at confining pressures below 4 kbar. (From Edmond, J. M. and Paterson, M. S., *Int. J. Rock Mech. Min. Sci.*, 9, 161, 1972. With permission.)

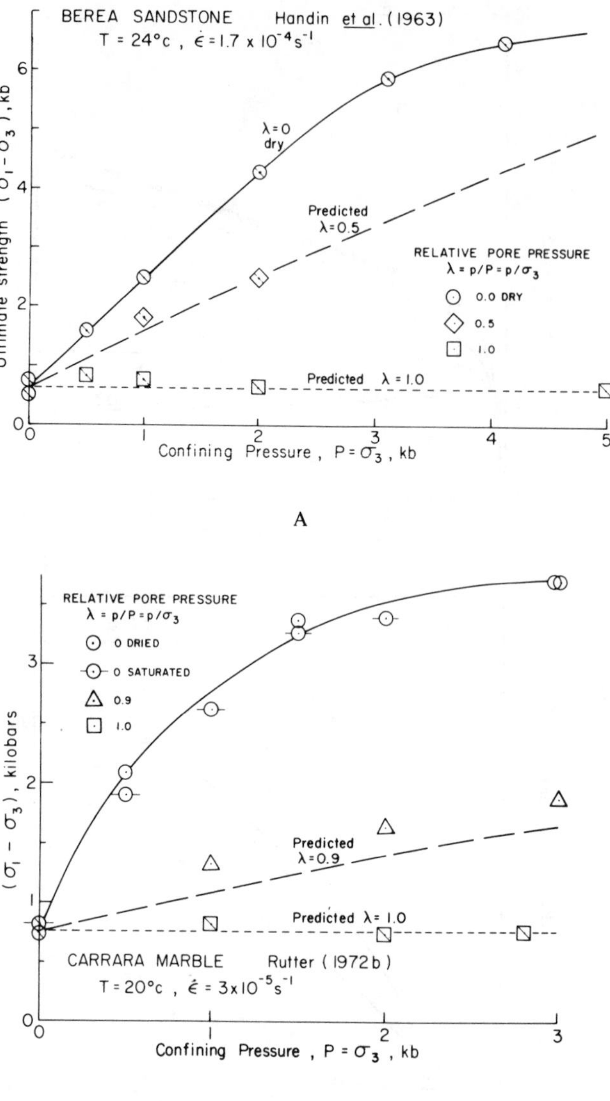

FIGURE 9. Pore pressure effects on the ultimate strengths of two rock types tested in triaxial compression: (A) Berea sandstone,[116] (B) Carrara marble.[251] Fluid medium: water. Slash mark through symbol indicates that specimen fractured or faulted. The pore pressure p is represented as a fraction λ of the confining pressure P, and the strengths for nonzero values of λ are predicted from the effective stress law (see text).

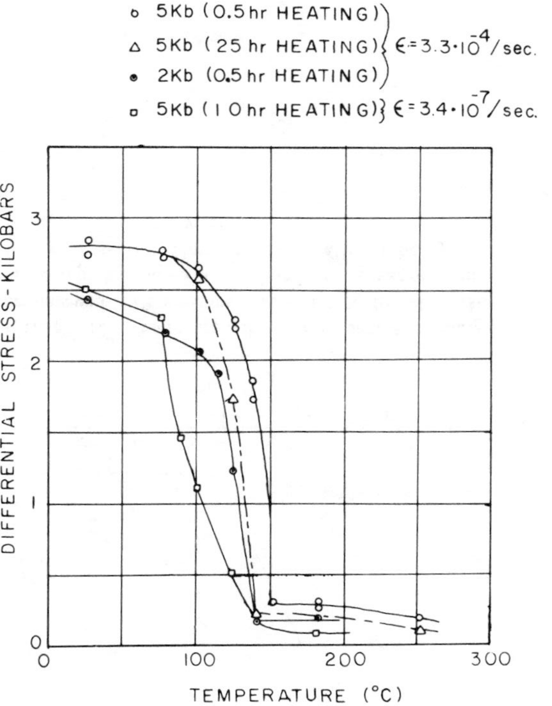

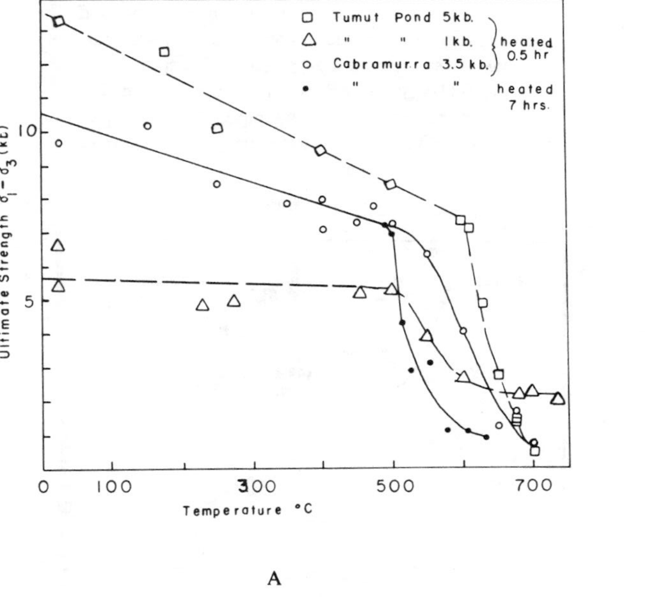

FIGURE 10. Effects of pore pressure generated by dehydration on the strengths of rocks containing hydrous minerals in sealed (undrained) triaxial compression tests. (A) Ultimate strength vs. temperature of two serpentinites at various confining pressures and heating times prior to testing. These serpentinites dehydrate above about 500°C. (From Raleigh, C. B. and Paterson, M. S., *J. Geophys. Res.*, 70, 3965, 1965. With permission. (B) Yield strength vs. temperature of alabaster at various confining pressures, strain rates, and heating times prior to testing. Gypsum dehydrates above about 80°C. (From Heard, H. C. and Rubey, W. W., *Geol. Soc. Am. Bull.*, 77, 741, 1966. With permission.)

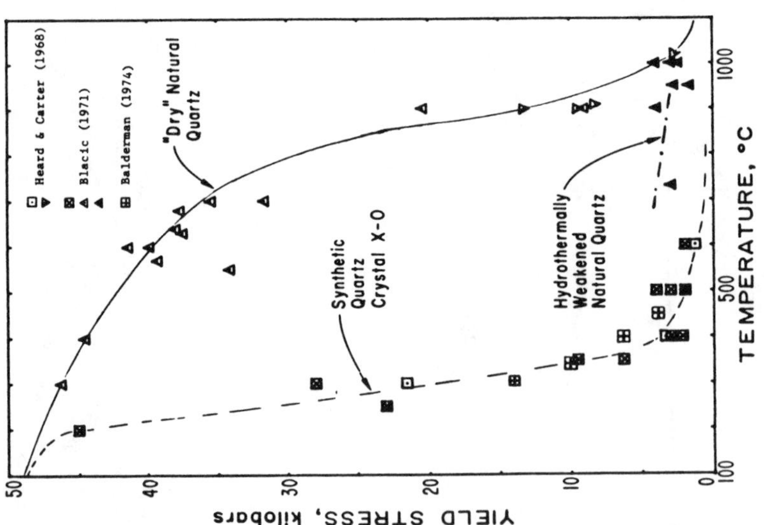

FIGURE 11. Comparative yield strengths of dry natural quartz, hydrothermally treated natural quartz, and a synthetic quartz crystal (with about 4000 ppm hydroxyl included during growth). Note dramatically lower yield strengths for "wet" quartz crystals. All crystals compressed at 45° to "a" and "c".

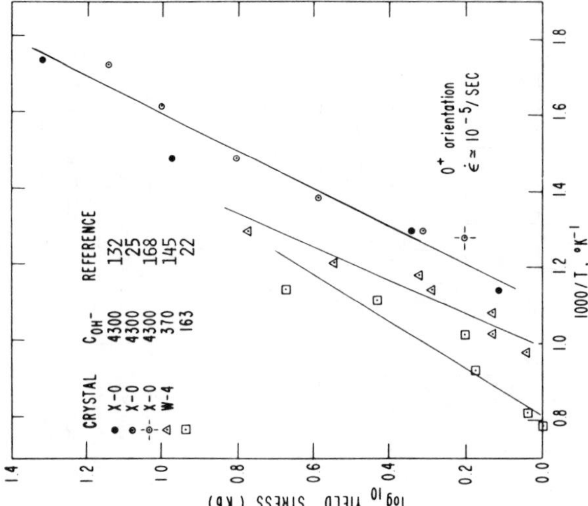

FIGURE 12. Yield strengths of three synthetic quartz crystals with different concentrations of hydroxyl. Crystals compressed at 45° to "a" and "c" at various confining pressures. Note systematically higher yield strengths for drier crystals.

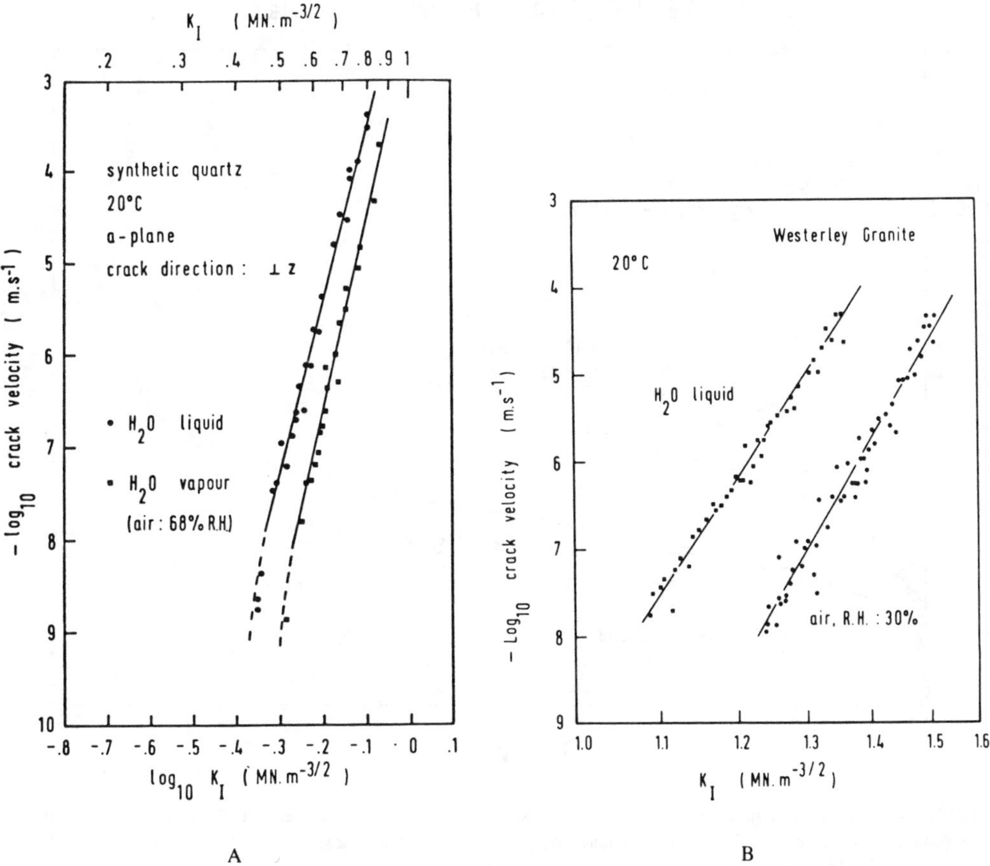

FIGURE 13. Crack velocity as a function of stress intensity factor K_I for (A) synthetic quartz and (B) Westerly granite in humid air and in liquid water.[10,13] K_I is a measure of the intensity of tensile stress at the crack tip (see Paterson[225]). Note systematically higher crack velocity in liquid water, which is interpreted to be a result of higher water concentration at crack tips.

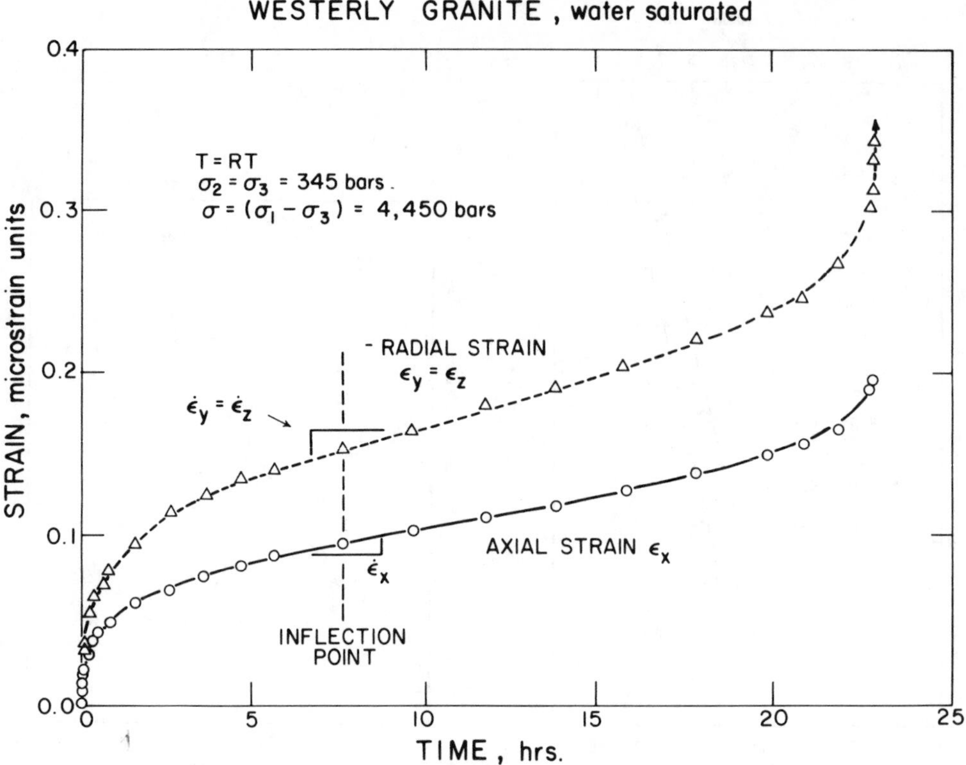

FIGURE 14. Complete creep curves for water-saturated Westerly granite at room temperature in triaxial compression. Note the characteristic brittle creep sigmoidal shape of the curves and large radial strain due to dilatancy (the radial strains should be half of the axial strains for volume to be conserved).[296,297]

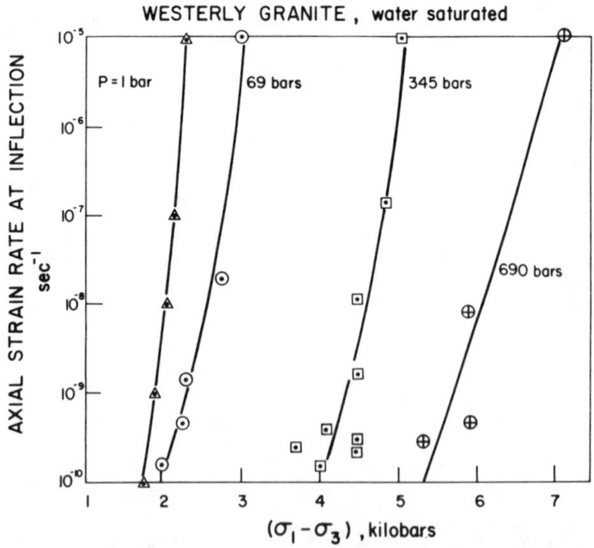

FIGURE 15. Influence of creep stress and confining pressure on the inflection point axial creep rates (defined in Figure 14) of water-saturated Westerly granite at room temperature[296,297]

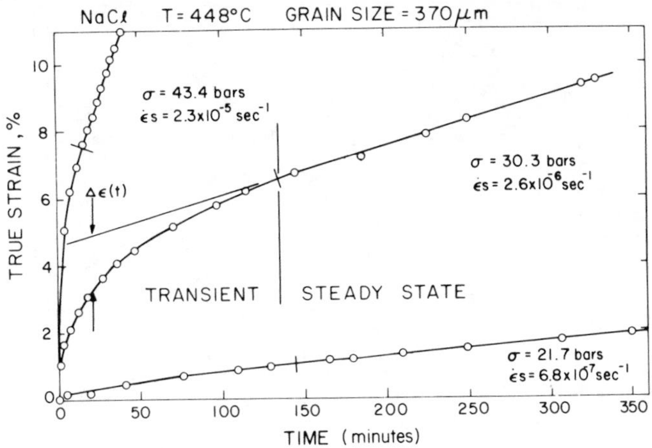

FIGURE 16. Ductile creep curves for polycrystalline halite. Note characteristic transient and steady-state creep stages and definition of the transient creep decay parameter $\Delta\epsilon(t)$ of Equation 8 in the text. (Courtesy of P.M. Burke, Department of Materials Science, Stanford University, 1968.)

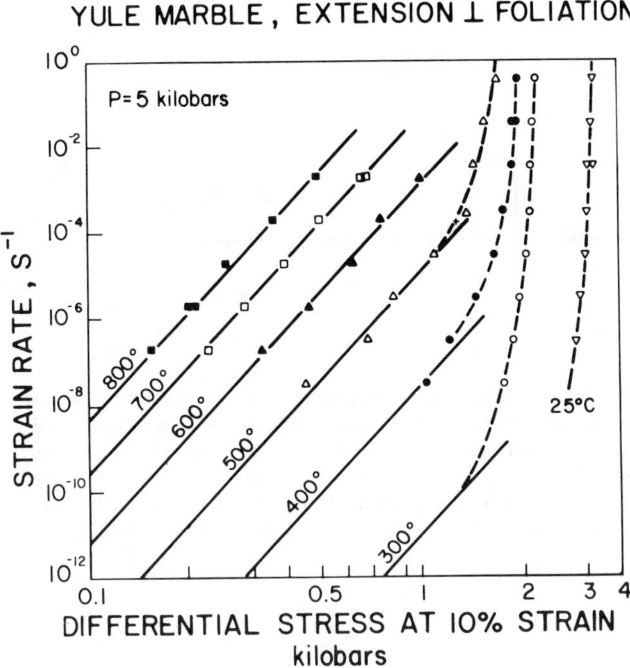

FIGURE 17. Ductile strength of Yule marble in triaxial extension as a function of temperature and strain rate. Note fit of the data to a steady-state flow law of Equation 6 (solid lines) at stresses below about 1 kbar. The departure from this law (dotted lines) probably represents a transition to another rate-controlling mechanism of flow. (From Heard, H. C. and Raleigh, C. B., *Geol. Soc. Am. Bull.*, 83, 935, 1972. With permission.)

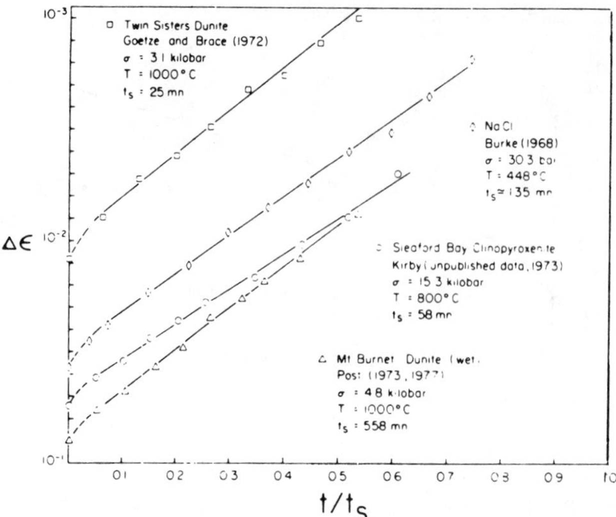

FIGURE 18. Fit of the large strain transient creep data for rocks in the ductile regime to an exponential decay law (Equation 7 in the text). (From Carter, N. L. and Kirby, S. H., *Pure Appl. Geophys.*, 116, 807, 1978. With permission.)

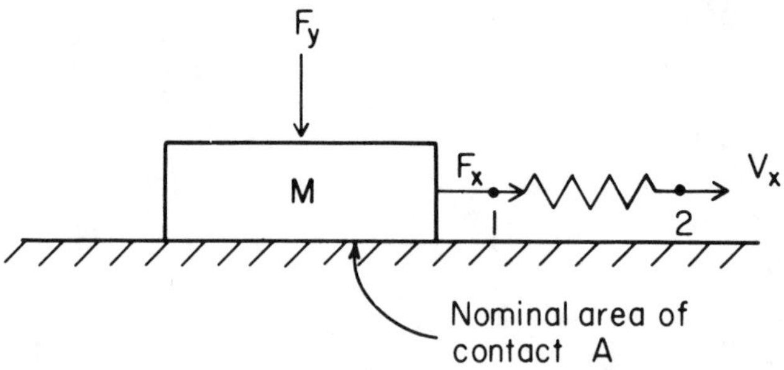

FIGURE 19. Schematic representation of the key physical elements in friction experiments. A slider with mass M is subject to a normal force F_y across a nominal area of contact A with a substrate and to a tangential force F_x. The tangential force necessary to produce an inelastic sliding displacement is a measure of the friction between the slider and substrate. V_x is the driving velocity on the tangential force generating column.

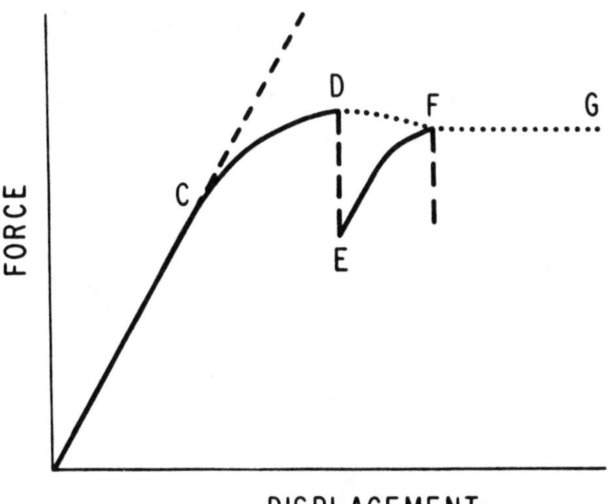

FIGURE 20. Tangential forces vs. tangential force column displacement (measured at 2 in Figure 19) in a hypothetical friction experiment. The curve deflects from an elastic segment at C (initial friction), reaches a maximum at D (maximum friction) and stable sliding may ensue (dotted line) or stick slip (solid line) may occur. (From Byerlee, J., *Pagoeph,* 116, 615, 1978. With permission.)

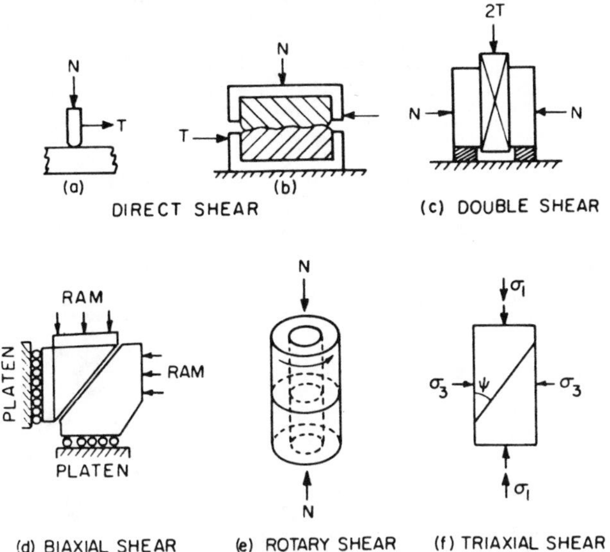

FIGURE 21. Types of loading arrangements in friction experiments. Techniques (a) through (d) are generally limited to a few hundred bars normal stress (N) since specimens are unconfined normal to N and to the tangential force (T). N and T cannot be independently varied in (d) and (f) and all techniques except (e) are limited to relatively small total displacements in a simple frictional loading cycle. Jaeger and Cook list additional complications in triaxial shear friction tests.[96,149,150] (From Jaeger, J. C. and Cook, N. G. W., *Fundamentals of Rock Mechanics,* first and 2nd ed., John Wiley & Sons, New York, 1969. With permission.)

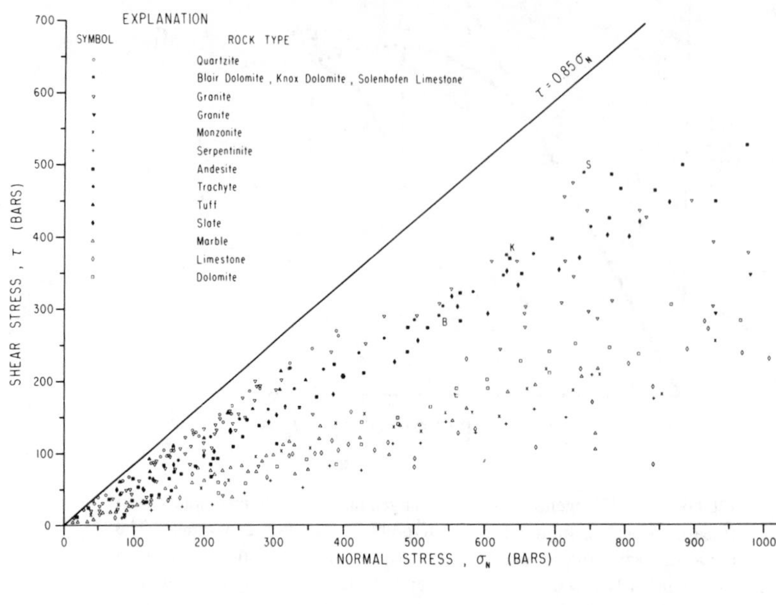

A

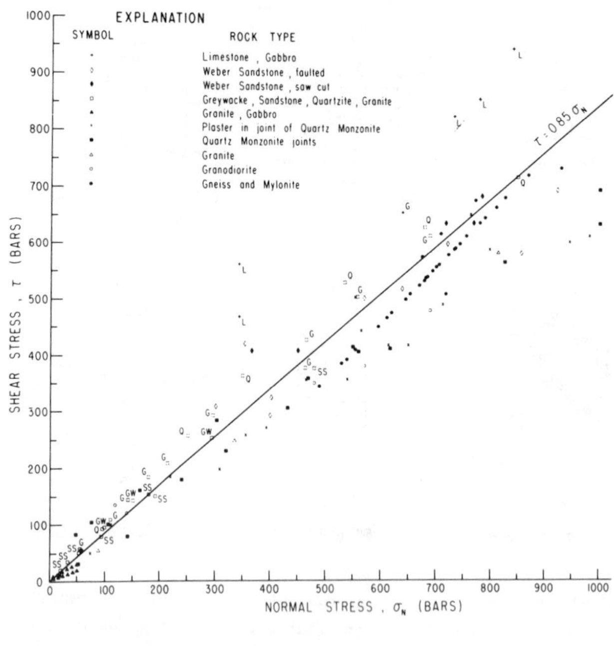

B

FIGURE 22. Shear stress at initial (A) and maximum (B) friction for rocks loaded to normal stresses up to 1000 bars (From Byerlee, J., *Pageoph*, 116, 615, 1978. With permission.)

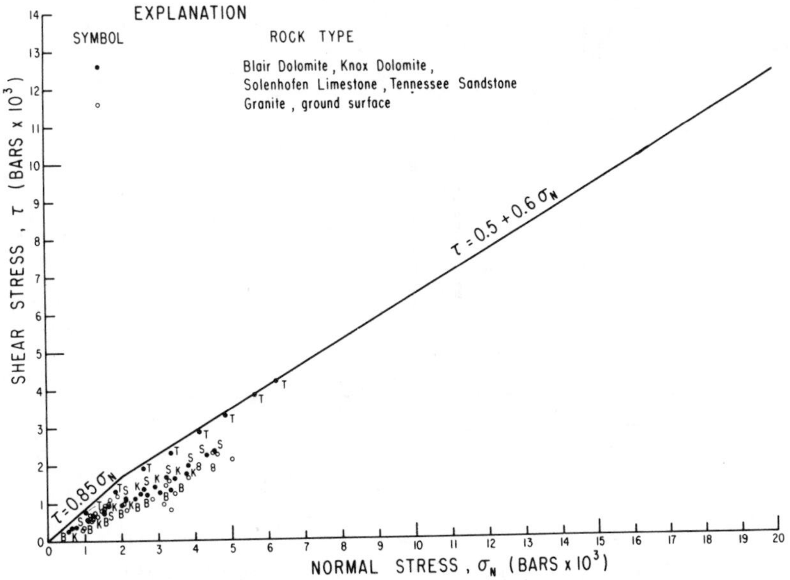

A

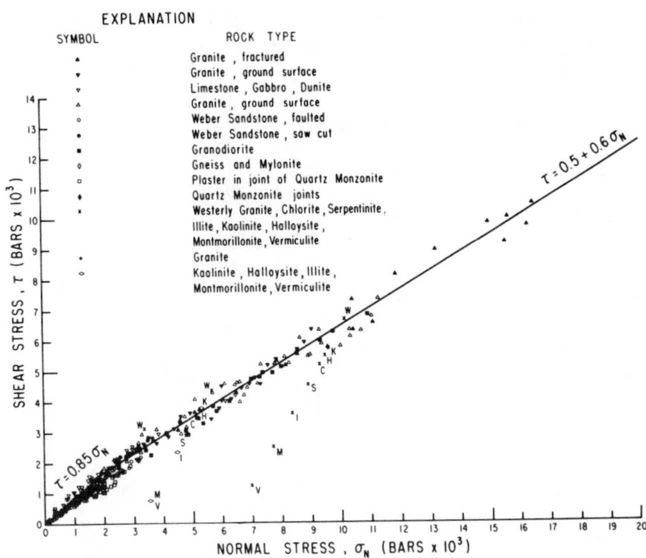

B

FIGURE 23. Shear stress at initial (A) and maximum (B) friction for rocks loaded to normal stresses up to 1000 bars (From Byerlee, J., *Pageoph*, 116, 615, 1978. With permission.)

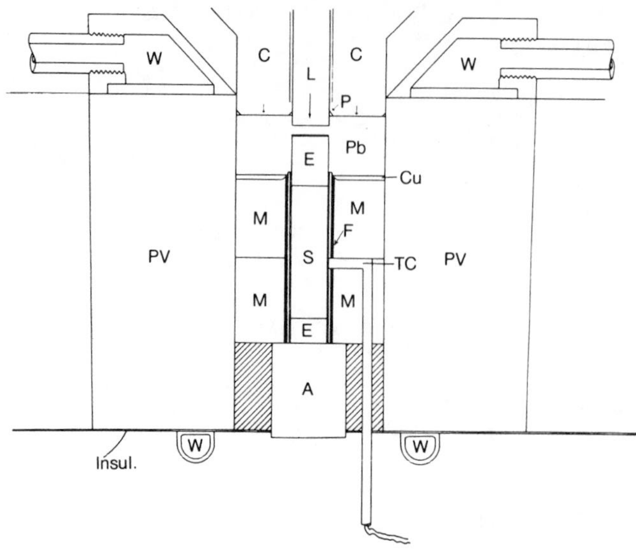

FIGURE 24. Piston cylinder apparatus schematic. Sample S loaded by axial piston L pushing through packing P and impinging on refractory end pieces E and base anvil A. Sample heated by graphite resistance furnace F and temperature measured by side thermocouple TC. Pressure generated by advance of annular piston C, compressing contents of pressure vessel (PV), including confining medium M. Pressure and axial force measured from external forces on pistons L and C, respectively. (Courtesy of J. D. Blacic, University of California, Los Angeles, 1971.)

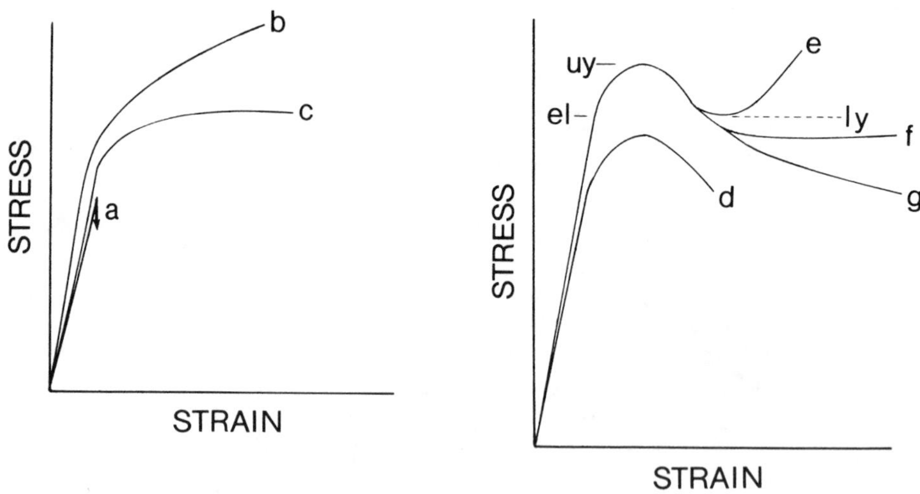

FIGURE 25. Key to stress-strain curve shapes (Table 2 and 3 and definitions of yield stress parameters tabulated in Table 3 for mineral single crystals. Yield stress symbols: el, elastic limit; uy, upper yield stress; ly, lower yield stress.

Table 1
PRINCIPAL STRESS COMPONENTS IN TESTS USING CYLINDRICAL SAMPLE GEOMETRY

Type of test	Principal stresses σ_1	σ_2	σ_3	Mean stress σ_m
Uniaxial compression $\sigma_y = \sigma_z = \tau_{xy} = 0$	σ_x	P	0	$\sigma_x/3$
Uniaxial tension $\sigma_y = \sigma_z = \tau_{xy} = 0$	0	0	$-\sigma_x$	$-\sigma_x/3$
Triaxial compression $\sigma_x \neq 0; \sigma_y = \sigma_z = P$	$\sigma_x = \Delta\sigma_x + P$	P	$P = \sigma_y = \sigma_z$	$P + \Delta\sigma_x/3$
Triaxial extension $\sigma_x \neq 0; \sigma_y = \sigma_z = P$	$P = \sigma_y = \sigma_z$	P	$\sigma_x = P - \Delta\sigma$	$P - \Delta\sigma_x/3$
Simple torsion $\tau_{xy} \neq 0$ $\sigma_x = \sigma_y = \sigma_z = 0$	τ_{xy}	0	$-\tau_{xy}$	0
Triaxial torsion $\tau_{xy} \neq 0$ $\sigma_x = \sigma_y = \sigma_z = P$	$P + \tau_{xy}$	P	$P - \tau_{xy}$	P
Triaxial compression and torsion $\sigma_x \neq 0$ $\tau_{xy} \neq 1$ $\sigma_y = \sigma_x = P$	[a]	P	[b]	$P + \Delta\sigma_x/3$

Note: Compressive stresses positive, coordinate system, and angular relations as shown in Figure 1. In the tests involving torsional moment M, τ_{xy} is calculated from Equations 1, 2, or 3 in text. P = confining pressure and σ_m is the mean of the three principal stresses. Compressive stresses considered positive in sign. $\Delta\sigma_x \equiv$ axial differential stress; $P \equiv$ confining pressure; τ_{xy} in torsion tests calculated from Equations 1, 2, or 3 in the text. Modified from Handin et al.[117]

[a] $\sigma_1 = (\sigma_x + \sigma_y)/2 + \{[(\sigma_x - \sigma_y)/2]^2 + \tau_{xy}^2\}^{1/2}$

[b] $\sigma_3 = (\sigma_x + \sigma_y)/2 - \{[(\sigma_x - \sigma_y)/2]^2 + \tau_{xy}^2\}^{1/2}$

Table 2
STRESS-STRAIN RELATIONS FROM CONSTANT STRAIN RATE TRIAXIAL TESTS ON ROCKS IN APPARATUS USING FLUID-CONFINING MEDIA

Material	Or	Temp (°C)	Pressure (kbar)	Strain rate (s^{-1})	Differential stress — kbar strain (%) 1	2	5	10	Ultimate strength (kbar)	Total strain (%)	Fault angle (°)	Comment	Ref.
Alabaster, Glebe Mine, Gotham, Nottinghamshire, England, >98% gypsum (gypsum dehydrated above about 110°C)	U	20	2.76	1.0×10^{-5}	1.25	1.74	2.24		2.51	7.3	NF	(b) Undrained test	213
	U	20	5.52	1.0×10^{-5}	1.36	2.27	3.39		3.58	6.7	NF	(b) Undrained test	
	U	100	5.52	1.0×10^{-5}	1.32	1.88	2.09		2.09	5.8	NF	(c) Undrained test	
	U	110	2.76	1.0×10^{-5}	1.07	1.35	1.60		2.21	5.4	NF	(b) Undrained test	
	U	120	2.76	1.0×10^{-5}	1.14	1.89	2.80		2.91	5.8	NF	(b)	
	U	120	2.76	1.0×10^{-5}	0.482	0.613	0.772		0.775	4.8	NF	(b) Undrained test, 5% data extrapolated	
	U	120	5.52	1.0×10^{-5}	0.87	1.32	1.26		1.53	6.2	NF	(d) Undrained test	
	U	145	5.52	1.0×10^{-5}	0.406	0.505	0.812		0.823	5.2	NF	(b) Undrained test	
	U	170	2.76	1.0×10^{-5}	1.00	1.66			1.93	4.4	NF	(c)	
	U	170	2.76	1.0×10^{-5}	0.346	0.471			0.567	3.6	F	(b) Undrained test	
	U	170	5.52	1.0×10^{-5}	0.318	0.362	0.450		0.450	5.2	NF	(b) Undrained test	
	U	270	5.52	1.0×10^{-5}	0.225	0.241			0.280	3.8	NF	(b) Undrained test	
	U	470	2.76	1.0×10^{-5}	0.74	1.26	2.13		2.18	5.3	NF	(b)	
	U	470	2.76	1.0×10^{-5}	0.182	0.323			0.431	3.5	F	(b) Undrained test	
	U	470	5.52	1.0×10^{-5}	0.077	0.115			0.154	3.3	F	(c) Undrained test	
Alabaster, Italy, preheated 0.5 hr before loading unless noted, undrained tests (gypsum dehydrated above about 77°C)	U	26	2.00	3.2×10^{-4}		2.22	2.40	2.38	2.43		NF	(f)	136
	U	26	5.00	3.1×10^{-4}		2.66	2.94	3.15		8.8	NF	(b) 10% Data extrapolated	
	U	26	5.00	3.1×10^{-4}		2.64	2.96	3.22		8.8	NF	(b) 10% Data extrapolated	
	U	26	5.00	3.4×10^{-7}		2.15	2.72	2.94			NF	(b)	
	U	77	2.00	3.2×10^{-4}		2.04	2.27	2.32			NF	(c)	
	U	77	5.00	3.1×10^{-4}		2.60	2.93	3.16		6.9	NF	(b) 10% Data extrapolated	
	U	77	5.00	3.1×10^{-4}		2.65	2.97	3.12		6.9	NF	(b) 10% Data extrapolated	
	U	77	5.00	3.5×10^{-7}		2.14	1.86	1.47	2.30			(d)	
	U	90	5.00	3.6×10^{-7}		1.43	0.92		1.45		NF	(d)	
	U	101	5.00	3.3×10^{-4}		2.34	2.71	2.58			NF	(c)	
	U	101	2.00	3.2×10^{-4}		1.93	2.16	2.22			NF	(c)	
	U	101	5.00	3.2×10^{-4}		2.38	2.78	2.80			NF	(c)	
	U	101	5.00	3.2×10^{-4}		2.18	2.77	2.80			NF	(c) Heated 25 hr before test	
	U	101	5.00	3.4×10^{-7}		1.06	5.80	5.00	1.11		NF	(f)	

	U	114	2.00	3.2×10^{-4}		1.86	1.96	1.84			NF	(d)	
	U	125	5.00	3.5×10^{-4}		1.88	2.02	1.43	2.22		NF	(d)	
	U	125	5.00	3.4×10^{-4}		2.10	2.08	1.68	2.26		NF	(d)	
	U	125	2.00	3.3×10^{-4}		1.13	1.18	0.89	1.23		NF	(d)	
	U	125	5.00	3.3×10^{-4}		1.50	1.73	1.47	1.74		NF	(d) Heated 25 hr before test	
	U	125	5.00	3.4×10^{-7}		0.360	0.220	0.080	0.500		NF	(d)	
	U	138	5.00	3.4×10^{-4}		1.62	1.17	0.64	1.72		NF	(g)	
	U	138	2.00	3.3×10^{-4}		0.170	0.140	0.100	0.180		NF	(d)	
	U	138	5.00	3.3×10^{-4}		1.85	1.29	0.89	1.85		NF	(g)	
	U	138	5.00	3.3×10^{-4}		0.230	0.200	0.130	0.240		NF	Heated 25 hr before test	
	U	152	5.00	3.2×10^{-4}		0.310	0.330	0.290			NF	(c)	
	U	152	5.00	3.2×10^{-4}		0.300	0.290	0.230			NF	(c)	
	U	177	2.00	3.3×10^{-4}		0.170	0.150	0.090	0.180		NF	(d)	
	U	177	5.00	3.3×10^{-4}		0.220	0.260	0.240	0.300		NF	(c)	
	U	177	5.00	3.3×10^{-4}		0.220	0.170	0.140	0.250		NF	(d)	
	U	177	5.00	3.3×10^{-7}		0.060	0.050		0.070		NF	(d)	
	U	250	5.00	3.3×10^{-4}		0.160	0.140	0.120	0.170		NF	(f)	
	U	250	5.00	3.3×10^{-4}		0.140	0.120	0.060	0.170		NF	(g)	
	U	250	5.00	3.3×10^{-4}		0.100	0.060	0.020	0.100		NF	(g) Heated 25 hr before test	
Amphibolite, "Hudson Highland", Putnam Co., New York	N	150	1.00	1.7×10^{-4}					6.54	2.9	21	(b)	36
	N	500	5.00	1.7×10^{-4}					11.37	20.0	35	(d)	
	P	150	1.00	1.7×10^{-4}					6.59	3.2	30	(a)	
	P	500	5.00	1.7×10^{-4}					13.50	3.6	35	(d)	
Andesite, "Mount Hood", Oregon	U	25	0.001	1.0×10^{-4}	0.84	1.37			1.50	2.2	F	(d)	113
	U	25	0.001	1.0×10^{-4}	0.740				1.00	1.8	F	(d)	
	U	25	0.001	1.0×10^{-4}	0.670				0.700	1.2	F	(b) Water saturated, undrained test	
	U	25	0.001	1.0×10^{-4}	0.580				0.700	1.1	F	(b) Water saturated, undrained test	
	U	25	0.500	1.0×10^{-4}	1.15	2.26			3.20	4.2	27	(d)	
	U	25	0.500	1.0×10^{-4}	1.15	2.03			2.90	4.5	26	(b)	
	U	25	0.500	1.0×10^{-4}	1.01	2.01	2.56		3.00	8.8	26	(g) Water saturated, undrained test	
	U	25	0.500	1.0×10^{-4}	1.09	1.89			2.00	4.0	27	(b) Water saturated, undrained test	
	U	25	0.001	1.0×10^{-7}					1.11		F		
	U	25	0.001	1.0×10^{-7}					0.910		F		
	U	25	0.001	1.0×10^{-7}					1.16		23		
	U	25	0.500	1.0×10^{-7}	1.01	1.92	2.20		2.20	6.7	34	(d)	
	U	25	0.500	1.0×10^{-7}	1.01	1.82	2.16		2.20	6.8	32	(d)	
	U	25	0.500	1.0×10^{-7}	0.670	0.770	0.670		0.770	5.0	35	(d) Water saturated, undrained test	

Table 2 (continued)
STRESS-STRAIN RELATIONS FROM CONSTANT STRAIN RATE TRIAXIAL TESTS ON ROCKS IN APPARATUS USING FLUID-CONFINING MEDIA

Material	Or	Temp (°C)	Pressure (kbar)	Strain rate (s^{-1})	Differential stress — kbar strain (%) 1	2	5	10	Ultimate strength (kbar)	Total strain (%)	Fault angle (°)	Comment	Ref.
	U	25	0.500	1.0×10^{-7}	0.87	1.24	1.16		1.24	6.3	F	(d) Water saturated, undrained test	
	U	400	0.001	1.0×10^{-4}	0.35	1.01			1.06	3.2	F	(d)	
	U	400	0.001	1.0×10^{-4}	0.54	1.36			1.40	2.6	F	(d)	
	U	400	0.001	1.0×10^{-4}	0.35	1.01			1.02	4.0	23	(d)	
	U	400	0.500	1.0×10^{-4}	0.93	1.79			2.30	4.6	32	(d)	
	U	400	0.500	1.0×10^{-4}	0.98	2.03			3.00	4.5	32	(b)	
	U	400	0.500	1.0×10^{-4}	0.370				0.400	2.9	42	(g) Water saturated, undrained test	
	U	400	0.500	1.0×10^{-4}	1.15	1.77	2.20	2.38	2.40	10.6	35	(b) Water saturated, undrained test	
	U	400	0.001	1.0×10^{-7}					1.13		F		
	U	400	0.001	1.0×10^{-7}					1.36		30		
	U	700	0.001	1.0×10^{-4}	0.760				1.12	1.5	25	(d)	
	U	700	0.001	1.0×10^{-4}	0.930				1.09	1.3	F	(d)	
	U	700	0.500	1.0×10^{-4}	1.56				2.30	1.9	F	(d)	
	U	700	0.500	1.0×10^{-4}	1.60	2.35			2.39	3.1	F	(d)	
	U	1000	0.001	1.0×10^{-4}	0.460	0.590			0.830	3.8	F	(d)	
	U	1000	0.001	1.0×10^{-4}	0.460	0.590			0.740	2.9	F	(d)	
Anhydrite, "Tafeljura", Riburg, Switzerland, 0.3 × 0.05 mm needles	N	300	1.50	8.2×10^{-5}	2.48	2.81	3.27	4.03	4.52	13.2	NF	(b)	212
	N	300	1.50	4.5×10^{-5}	2.56	2.88	3.39	4.08	4.46	13.6	NF	(b)	
	N	300	1.50	2.2×10^{-5}	2.42	2.72	3.16	3.75	4.35	16.0	NF	(b)	
	N	300	1.50	1.3×10^{-5}	2.32	2.57	3.06	3.69	3.78	11.2	NF	(b)	
	N	300	1.50	9.5×10^{-6}	2.18	2.36	2.73	3.32	3.85	14.6	NF	(b)	
	N	300	1.50	5.0×10^{-6}	1.92	2.22	2.63	3.20	3.32	11.1	NF	(b)	
	N	300	1.50	1.6×10^{-6}	1.98	2.18	2.56	3.09	3.16	10.7	NF	(b)	
	N	350	1.50	8.2×10^{-5}	2.51	2.74	3.18	3.72	3.76	1.0	NF	(b)	212
	N	350	1.50	4.5×10^{-5}	2.04	2.42	2.74	3.26	3.36	10.7	NF	(b)	
	N	350	1.50	2.2×10^{-5}	2.03	2.23	2.56	3.07	3.20	11.2	NF	(b)	
	N	350	1.50	9.5×10^{-6}	1.66	1.82	2.07	2.46	2.56	11.3	NF	(b)	
	N	350	1.50	3.6×10^{-6}	1.64	1.79	1.93	2.12	2.18	11.5	NF	(b)	

	N	350	1.50	1.6×10^{-6}	1.59	1.66	1.70	1.75	1.79	11.8	NF	(b)	
	N	350	1.50	9.3×10^{-7}	1.53	1.60	1.64	1.66	1.67	12.2	NF	(c)	
	N	400	1.50	8.2×10^{-5}	1.81	1.95	2.24	2.58	2.63	10.7	NF	(b)	
	N	400	1.50	4.5×10^{-5}	1.68	1.84	2.06	2.32	2.32	9.7	NF	(b) 10% Data extrapolated	
	N	400	1.50	2.2×10^{-5}	1.60	1.70	1.81	1.98	2.09	13.2	NF	(b)	
	N	400	1.50	1.3×10^{-5}	1.53	1.60	1.72	1.85	1.94	13.0	NF	(b)	
	N	400	1.50	5.0×10^{-6}	1.33	1.41	1.59	1.67	1.68	13.2	NF	(b)	
	N	400	1.50	3.6×10^{-6}	1.22	1.23	1.40	1.47	1.47	13.2	NF	(c)	
	N	400	1.50	2.6×10^{-6}	1.20	1.21	1.20	1.20	1.21	13.2	NF	(c)	
	N	400	1.50	1.6×10^{-6}	0.984	0.991	0.991	0.984	0.991	13.5	NF	(c)	
	N	400	1.50	9.3×10^{-7}	0.760	0.760	0.760	0.760	0.760	13.0	NF	(c)	
	N	450	1.50	8.2×10^{-5}	1.54	1.69	1.84	1.97	2.07	12.6	NF	(b)	
	N	450	1.50	3.5×10^{-5}	1.41	1.50	1.56	1.65	1.66	11.5	NF	(b)	
	N	450	1.50	5.0×10^{-6}	1.11	1.12	1.12	1.12	1.12	11.8	NF	(c)	
	N	450	1.50	3.6×10^{-6}	1.03	1.04	1.04	1.04	1.04	12.1	NF	(c)	
	N	450	1.50	1.6×10^{-6}	0.712	0.712	0.712	0.712	0.712	9.2	NF	(c) 10% Data extrapolated	
	N	450	1.50	1.3×10^{-6}	1.18	1.26	1.27	1.28	1.28	12.1	NF	(c)	
	N	450	1.50	9.3×10^{-7}	0.528	0.528	0.534	0.534	0.534	9.2	NF	(c) 10% Data extrapolated	
Anhydrite, ''Werra'', Osterode, Germany, Cores parallel to <001 > point maximum of original fabric		25	0.001	5.0×10^{-4}					0.760	0.9	F	(b)	211
		25	0.200	5.0×10^{-4}	1.94	1.83			1.21	3.5	NF	(c) Ultimate strength at 1.5%	
		25	0.300	5.0×10^{-4}	2.18	2.29	1.74		2.29	5.9	NF	(e) Ultimate strength at 2%	
		25	0.400	5.0×10^{-4}	2.32	2.55			2.55	4.6	NF	(e) Ultimate strength at 2%	
		25	0.500	5.0×10^{-4}	2.59	2.82	2.56		2.82	7.6	NF	(e) Ultimate strength at 2%	
		25	0.750	5.0×10^{-4}	2.73	3.18	3.30	3.03	3.45	14.2	NF	(e) Ultimate strength at 3.5%	
		25	1.00	5.0×10^{-4}	2.76	3.35	3.79	3.47	3.79	20.5	NF	(e) Ultimate strength at 5%	
		25	1.50	5.0×10^{-4}	3.36	3.94	4.39	4.41	4.41	20.0	NF	(c)	
		25	2.00	5.0×10^{-4}	3.41	3.97	4.71	4.97	5.14	23.5	NF	(b)	
		25	2.50	5.0×10^{-4}	3.42	4.09	4.92	5.50	5.91	18.5	NF	(b)	
		25	3.00	5.0×10^{-4}	3.44	4.12	5.29	6.03	6.67	23.4	NF	(b)	
		25	3.50	5.0×10^{-4}	3.45	4.39	5.45	6.24	6.97	21.5	NF	(b)	
		25	4.00	5.0×10^{-4}	3.48	4.67	5.85	6.85	8.33	27.2	NF	(b)	
		25	5.00	5.0×10^{-4}	3.64	4.85	6.23	7.29	9.41	27.7	NF	(b)	
		150	1.00	5.0×10^{-4}	1.73	2.28	3.25	3.28	3.46	20.0	NF	(f)	
		150	1.50	5.0×10^{-4}	2.52	3.11	3.91	4.48	4.58	20.5	NF	(c)	
		150	2.00	5.0×10^{-4}	2.18	3.22	4.48	5.26	5.75	20.2	NF	(b)	
		150	3.00	5.0×10^{-4}	2.79	3.56	4.85	6.16	7.35	16.2	NF	(b)	
		300	0.500	5.0×10^{-4}	1.05	1.64	2.80		3.10	8.2	NF	(b)	
		300	1.00	5.0×10^{-4}	1.82	2.42	3.44	4.16	4.37	14.0	NF	(b)	
		300	2.00	5.0×10^{-4}	2.11	2.78	4.02	5.07	6.43	26.0	NF	(b)	

Table 2 (continued)
STRESS-STRAIN RELATIONS FROM CONSTANT STRAIN RATE TRIAXIAL TESTS ON ROCKS IN APPARATUS USING FLUID-CONFINING MEDIA

Material	Or	Temp (°C)	Pressure (kbar)	Strain rate (s^{-1})	Differential stress — kbar strain (%) 1	2	5	10	Ultimate strength (kbar)	Total strain (%)	Fault angle (°)	Comment	Ref.
		300	3.00	5.0×10^{-4}	2.11	2.78	4.05	5.23	7.18	24.2	NF	(b)	
Anhydrite, "Werra", Osterode, Ger-		25	0.500	5.0×10^{-4}	1.76	2.48	2.98		2.98	6.4	NF	(d)	211
many, Cores parallel to <010>		25	1.00	5.0×10^{-4}	1.97	3.03	3.80	4.03	4.03	15.2	NF	(d)	
point maximum of original fabric		25	2.00	5.0×10^{-4}	2.33	3.47	4.27	5.00	5.15	12.3	NF	(b)	
		25	3.00	5.0×10^{-4}	2.33	3.55	4.77	5.62	6.85	28.7	NF	(b)	
		25	5.00	5.0×10^{-4}	2.33	3.77	5.26	6.53	9.21	26.7	NF	(b)	
		150	0.500	5.0×10^{-4}	1.09	2.17	2.67		2.67	5.4	NF	(d)	
		150	1.00	5.0×10^{-4}	1.09	2.19	3.19	3.55	3.61	14.2	NF	(c)	
		150	2.00	5.0×10^{-4}	1.09	2.19	3.81	4.77	5.68	23.0	NF	(b)	
		150	3.00	5.0×10^{-4}	1.09	2.19	4.05	5.31	7.59	21.7	NF	(b)	
		300	1.00	5.0×10^{-4}	0.59	1.21	2.61	3.64	4.25	15.2	NF	(b)	
		300	1.50	5.0×10^{-4}	0.76	1.44	2.78	3.78	5.02	20.0	NF	(b)	
		300	2.00	5.0×10^{-4}	1.09	1.90	3.19	4.34	6.22	24.4	NF	(b)	
		300	3.00	5.0×10^{-4}	1.09	1.90	3.34	4.61	6.54	21.5	NF	(b)	
Anorthosite, "Marcy", Essex Co.,	N	150	1.00	1.7×10^{-4}					5.94	2.6	29	(a)	36
New York	N	500	5.00	1.7×10^{-4}					9.39	32.1	28	(d)	
	U	23	0.001	5.0×10^{-5}					0.240		F		137
Basalt Breccia, Amchitka Island,	U	23	1.00	5.0×10^{-5}					0.420		F		
Alaska, 5410 ft below surface	U	23	2.00	5.0×10^{-5}					0.420				
	U	23	3.00	5.0×10^{-5}			0.600				NF		
Basalt, altered, Amchitka Island,	U	23	5.00	5.0×10^{-5}			1.00				NF		
Alaska, 5705 ft below surface	U	23	7.00	5.0×10^{-5}			0.980				NF		
	U	23	0.001	5.0×10^{-5}					0.340		F		137
	U	23	0.001	5.0×10^{-5}					0.860		F		
	U	23	0.001	5.0×10^{-5}					0.480		F		
	U	23	0.500	5.0×10^{-5}					1.40		F		
	U	23	1.00	5.0×10^{-5}					1.90		F		
	U	23	2.00	5.0×10^{-5}			2.32				NF		
	U	23	2.00	5.0×10^{-5}			1.92				NF		
	U	23	3.00	5.0×10^{-5}			2.14				NF		
	U	23	3.00	5.0×10^{-5}			3.34				NF		

	U	23	4.00	5.0×10^{-5}		4.90				NF		
	U	23	5.00	5.0×10^{-5}		4.50				NF		
	U	23	5.00	5.0×10^{-5}		3.74				NF		
	U	23	6.00	5.0×10^{-5}		4.90				NF		
	U	23	7.00	5.0×10^{-5}		3.60				NF		
	U	23	7.00	5.0×10^{-5}		2.76				NF		
Basalt, altered, Amchitka Island, Alaska, 5850 ft below surface	U	23	0.001	5.0×10^{-5}				0.580		F		137
	U	23	0.001	5.0×10^{-5}				0.340		F		
	U	23	1.00	5.0×10^{-5}				0.960		F		
	U	23	2.00	5.0×10^{-5}				0.680		F		
	U	23	3.00	5.0×10^{-5}				1.66				
	U	23	4.00	5.0×10^{-5}		1.64				NF		
	U	23	5.00	5.0×10^{-5}		1.60				NF		
	U	23	6.00	5.0×10^{-5}		1.96				NF		
	U	23	7.00	5.0×10^{-5}		2.86				NF		
Basalt, altered, Amchitka Island, Alaska, 5900 ft below surface	U	23	0.250	5.0×10^{-5}				1.24		F		137
	U	23	0.500	5.0×10^{-5}				1.30		F		
	U	23	1.00	5.0×10^{-5}				1.40		F		
	U	23	3.00	5.0×10^{-5}		1.76						
	U	23	5.00	5.0×10^{-5}		1.90				NF		
	U	23	7.00	5.0×10^{-5}		3.16				NF		
Breccia, "Banjo Point", Amchitka Island, Alaska, 898 ft below surface	U	23	0.001	5.0×10^{-5}				0.160		F		137
	U	23	1.00	5.0×10^{-5}				0.360		F		
	U	23	2.00	5.0×10^{-5}				0.360				
	U	23	3.00	5.0×10^{-5}		0.600						
	U	23	4.00	5.0×10^{-5}		0.440				NF		
	U	23	5.00	5.0×10^{-5}		0.980				NF		
	U	23	6.00	5.0×10^{-5}		0.740				NF		
	U	23	7.00	5.0×10^{-5}		0.860				NF		
Chalcopyrite ore, Creighton mine, Sudbury, Ontario, Canada, <10% impurities	U	24	1.00	7.2×10^{-5}	0.75	3.70		3.73	6.5	F	(b)	137
	U	24	2.00	7.2×10^{-5}				2.59	1.5	F	(b)	
	U	100	1.00	7.2×10^{-5}	2.65	3.72		4.01	8.5	F	(b)	
	U	100	1.00	7.2×10^{-5}	3.03	4.00	4.70		10.0	NF	(b)	
	U	200	0.500	7.2×10^{-5}	2.84	3.15		3.18	6.5	F	(b)	
	U	200	1.00	7.2×10^{-5}	2.68	2.96	3.18		23.0	NF	(b)	
	U	200	1.00	7.2×10^{-5}	2.51	2.98	3.23		22.5	NF	(b)	
	U	300	0.500	7.2×10^{-5}	1.26	1.64	2.05		11.7	NF	(b)	
	U	300	1.00	7.2×10^{-5}	1.93	2.23	2.67		9.5	NF	(b) 10% Data extrapolated	
	U	300	2.00	7.2×10^{-5}	1.35	1.48	1.66		23.5	NF	(b)	
	U	400	1.00	7.2×10^{-5}	0.900	0.940	0.940		5.4	NF	(b) 10% Data extrapolated	
	U	400	1.00	7.2×10^{-5}	1.97	2.17	2.51		13.3	NF	(b)	

Table 2 (continued)
STRESS-STRAIN RELATIONS FROM CONSTANT STRAIN RATE TRIAXIAL TESTS ON ROCKS IN APPARATUS USING FLUID-CONFINING MEDIA

Material	Or	Temp (°C)	Pressure (kbar)	Strain rate (s^{-1})	Differential stress — kbar strain (%) 1	2	5	10	Ultimate strength (kbar)	Total strain (%)	Fault angle (°)	Comment	Ref.
	U	400	2.00	7.2×10^{-5}		0.73	1.05	1.44		10.8	NF	(b)	
Chalcopyrite ore, Icon mine, Mistassini region, Quebec, Canada, <15% impurities	U	24	0.500	7.2×10^{-5}					4.70	0.8	F	(a)	137
	U	24	0.500	7.2×10^{-5}		3.92			3.92	2.0	F	(b)	
	U	24	1.00	7.2×10^{-5}					4.99	1.8	F	(b)	
	U	24	1.00	7.2×10^{-5}		4.01	4.03		4.03	5.7	F	(b)	
	U	24	1.00	7.2×10^{-5}		3.58			3.63	2.4	F	(b)	
	U	24	1.50	7.2×10^{-5}		4.00	4.27		4.27	6.3	F	(b)	
	U	24	1.50	7.2×10^{-5}		4.54	5.27		5.37	7.4	F	(b)	
	U	24	1.50	7.2×10^{-5}		3.84	4.77		5.03	7.0	F	(b)	
	U	24	2.00	7.2×10^{-5}		3.92	4.07	5.81		9.8	NF	(b) 10% Data extrapolated	
	U	24	2.00	7.2×10^{-5}		4.40	5.29	5.88		6.9	NF	(b) 10% Data extrapolated	
	U	100	1.00	7.2×10^{-5}		4.10	4.62		4.62	5.1	F	(b)	
	U	100	1.00	7.2×10^{-5}		4.55			4.59	2.5	F	(b)	
	U	200	0.500	7.2×10^{-5}		2.63	2.97		2.98	7.1	F	(b)	
	U	200	1.00	7.2×10^{-5}		3.33	3.67	4.06		10.5	NF	(b)	
	U	200	1.00	7.2×10^{-5}		3.86	4.17	4.42		7.8	NF	(b) 10% Data extrapolated	
	U	200	2.00	7.2×10^{-5}		3.70	4.10	4.87		7.1	NF	(b) 10% Data extrapolated	
	U	300	0.500	7.2×10^{-5}		2.39	2.80	2.98		10.6	NF	(b)	
	U	300	0.500	7.2×10^{-5}		2.03	2.28	2.55		11.2	NF	(b)	
	U	300	1.00	7.2×10^{-5}		1.77	2.11	2.45		10.3	NF	(b)	
	U	300	1.00	7.2×10^{-5}		2.53	2.91	3.41		8.7	NF	(b) 10% Data extrapolated	
	U	300	1.00	7.2×10^{-5}		2.41	2.80	3.28		8.2	NF	(b) 10% Data extrapolated	
	U	300	1.00	7.2×10^{-5}		1.90	2.26	2.49		5.9	NF	(b) 10% Data extrapolated	
	U	300	1.00	7.2×10^{-5}		2.12	2.60	3.01		8.2	NF	(b) 10% Data extrapolated	
	U	300	1.50	7.2×10^{-5}		1.92	2.34	2.77		8.3	NF	(b) 10% Data extrapolated	
	U	300	1.50	7.2×10^{-5}		1.78	2.15	2.49		10.2	NF	(b)	
	U	300	2.00	7.2×10^{-5}		1.62	1.22	2.27		3.9	NF	(b) 5% and 10% Data extrapolated	
	U	300	2.00	7.2×10^{-5}		2.21	2.58	3.08		10.1	NF	(b)	
	U	300	2.00	7.2×10^{-5}		1.78	2.13	2.47		8.7	NF	(b) 10% Data extrapolated	
	U	400	1.00	7.2×10^{-5}		1.39	1.55	1.57		5.6	NF	(c) 10% Data extrapolated	

	U	400	1.00	7.2×10^{-5}		2.02	2.42	2.83		8.5	NF	(b) 10% Data extrapolated	
	U	400	1.00	7.2×10^{-5}		1.06	1.16	1.16		5.3	NF	(c) 10% Data extrapolated	
	U	400	2.00	7.2×10^{-5}		1.39	1.55	1.71		9.5	NF	(b) 10% Data extrapolated	
	U	500	0.500	7.2×10^{-5}		1.07	1.09	1.11		3.1	NF	(b) 5% and 10% Data extrapolated	
	U	500	1.00	7.2×10^{-5}		0.670	0.680	0.680		5.0	NF	(c) 10% Data extrapolated	
	U	500	1.00	7.2×10^{-5}		0.850	0.960	0.960	0.990	10.2	NF	(b)	
Chalcopyrite ore, Mt. Isa, Australia, 80% chalcopyrite, 13% gangue, 3.5% pyrrhotite, 2% sphalerite	U	20	0.001	3.0×10^{-5}					0.880	0.4	F	(a)	4
	U	20	0.510	3.0×10^{-5}	3.72				3.93	1.7	F	(b)	
	U	20	0.790	3.0×10^{-5}	4.63	4.85	4.45		4.88	5.1	F	(d)	
	U	20	1.03	3.0×10^{-5}	4.42	4.85	5.24		5.24	6.5	F	(c)	
	U	20	1.47	3.0×10^{-5}	4.02	4.73	5.49	5.88	5.92	11.0	NF	(c)	
	U	20	2.92	3.0×10^{-5}	4.65	5.12	6.01	7.01	7.20	11.1	NF	(b)	
	U	200	0.001	3.0×10^{-5}	0.840				0.840	1.0	F	(a)	
	U	200	0.570	3.0×10^{-5}	2.39	2.67	2.90	3.00	3.00	11.2	NF	(c)	
	U	200	1.41	3.0×10^{-5}	2.29	2.68	3.20	3.81	3.92	11.1	NF	(b)	
	U	200	2.67	3.0×10^{-5}	2.29	2.70	3.34	4.02	4.15	11.1	NF	(b)	
	U	300	0.001	3.0×10^{-5}	0.760	0.630	0.380		0.760	7.8	F	(d)	
	U	300	0.540	3.0×10^{-5}	1.16	1.47	1.76	1.98	2.02	11.1	NF	(b)	
	U	300	1.45	3.0×10^{-5}	1.39	1.70	2.02	2.32	2.37	11.1	NF	(b)	
	U	300	2.75	3.0×10^{-5}	1.56	1.94	2.35	2.69	2.72	11.1	NF	(b)	
	U	400	0.001	3.0×10^{-5}	0.600	0.680	0.420		0.680	8.2	F	(d)	
	U	400	0.530	3.0×10^{-5}	0.73	0.95	1.21	1.38	1.40	11.1	NF	(b)	
	U	400	1.42	3.0×10^{-5}	0.82	1.04	1.27	1.44	1.46	11.1	NF	(b)	
	U	400	1.92	3.0×10^{-5}	0.69	1.01	1.43	1.60	1.62	11.1	NF	(b)	
	U	400	3.40	3.0×10^{-5}	0.95	1.21	1.59	1.86	1.91	11.1	NF	(b)	
Chalcopyrite ore, Termacami Lake, Ontario, Canada, 93% chalcopyrite, 6% calcite, 1% quartz and sphalerite, G.S. = 1mm	N	25	1.50	1.2×10^{-5}	3.22	3.71	4.50	5.05	5.06	12.1	NF	(d)	247
	N	25	1.50	8.3×10^{-6}	1.94	3.13	4.24	5.14	5.13	9.8	NF	(b) 10% Data extrapolated	
	N	100	1.50	2.3×10^{-4}	3.15	3.62	4.45	5.25	5.19	9.7	NF	(b) 10% Data extrapolated	
	N	100	1.50	7.4×10^{-5}	2.96	3.47	4.25	5.07	5.37	13.2	NF	(b)	
	N	100	1.50	7.1×10^{-6}	3.03	3.57	4.24	4.86	4.86	10.0	NF	(b)	
	N	150	1.50	2.3×10^{-5}	2.88	3.21	3.70	4.25	4.63	14.3	NF	(b)	
	N	150	1.50	7.4×10^{-6}	2.88	3.11	3.35	3.83	3.97	11.7	NF	(b)	
	N	200	1.50	5.6×10^{-3}	2.83	3.60	4.23	4.75	5.01	15.0	NF	(b)	
	N	200	1.50	4.5×10^{-4}	2.83	3.16	3.60	4.01	4.05	10.2	NF	(b)	
	N	200	1.50	4.0×10^{-5}	2.71	2.83	2.70	2.84	2.89	10.6	NF	(c)	
	N	200	1.50	7.1×10^{-6}	2.51	2.42	2.49	2.76	2.81	10.5	NF	(b)	
	N	200	1.50	1.0×10^{-6}		1.75				3.6		(f)	
	N	250	1.50	3.0×10^{-3}	2.84	2.97	3.10	3.46	3.47	10.6	NF	(b)	
	N	250	1.50	1.3×10^{-3}	2.62	2.70	2.85	3.21	3.46	12.7	NF	(b)	

Table 2 (continued)
STRESS-STRAIN RELATIONS FROM CONSTANT STRAIN RATE TRIAXIAL TESTS ON ROCKS IN APPARATUS USING FLUID-CONFINING MEDIA

Material	Or	Temp (°C)	Pressure (kbar)	Strain rate (s^{-1})	Differential stress — kbar strain (%)				Ultimate strength (kbar)	Total strain (%)	Fault angle (°)	Comment	Ref.
					1	2	5	10					
	N	250	1.50	1.4×10^{-4}	2.25	2.21	2.31	2.58	2.83	14.7	NF	(b)	247
	N	300	1.50	2.5×10^{-3}	1.46	1.49	1.84	2.20	2.31	14.3	NF	(b)	
	N	300	1.50	1.2×10^{-3}	1.67	1.55	1.97	2.57	2.93	14.8	NF	(b)	
	N	300	1.50	1.5×10^{-4}	0.99	1.01	1.18		1.19	6.0	NF	(b)	
	N	300	1.50	7.2×10^{-5}	1.18	1.27	1.67	2.06	2.20	11.6	NF	(b)	
	N	300	1.50	1.1×10^{-5}	0.80	0.84	1.00	1.18	1.30	12.3	NF	(b)	
	N	300	1.50	9.0×10^{-7}	0.797	0.835	0.604		0.848	4.1	NF	(f) 5% Data extrapolated	
Chalcopyrite ore, Timmins, Ontario, 70—80% chalcopyrite, 5—10% pyrite, 10—25% sphalerite, pyrrhotite, quartz, sericite, and chlorite	N	25	1.50	7.1×10^{-5}		4.56	5.50	6.40	6.54	11.6	NF	(b)	247
	N	25	1.50	4.5×10^{-5}		3.67	4.84	5.35	5.56	14.1	NF	(b)	
	N	25	1.50	3.3×10^{-5}		4.50	5.39	6.15	6.21	11.2	NF	(b)	
	N	25	1.50	9.0×10^{-6}		4.46	5.19	5.92	5.89	9.8	NF	(b) 10% Data extrapolated	
	N	100	1.50	7.9×10^{-4}	3.44	4.09	4.74	5.38	5.57	13.0	NF	(b)	
	N	100	1.50	6.8×10^{-5}	3.44	4.19	4.90	5.48	5.49	10.7	NF	(b)	
	N	100	1.50	7.1×10^{-6}	3.13	3.78	4.24		4.38	8.6	NF	(b)	
	N	200	1.50	7.9×10^{-4}	3.47	3.77	3.96	4.16	4.35	15.0	NF	(b)	
	N	200	1.50	1.1×10^{-4}	3.21	3.36	3.32	3.60	3.60	10.0	NF	(b)	
	N	200	1.50	4.4×10^{-5}	2.33	2.54	2.70	3.05	3.13	13.6	NF	(b)	
	N	200	1.50	7.1×10^{-6}	2.81	2.92	2.90	2.65	3.01	10.0	NF	(d)	
Chloritite, Piedmont Alps, Montgenevre, Italy, 85% chlorite, 6% sphene, 4% magnetite, 4% ilmenite, chlorite dehydrated at 618°C	U	20	1.38	1.0×10^{-5}	2.15	2.34	2.29		2.46	8.3	F	(e) Undrained test	213
	U	20	5.52	1.0×10^{-5}	2.42	4.42	4.83		4.95	6.7	F	(b) Undrained test	
	U	320	1.38	1.0×10^{-5}	1.19	2.03	1.87		1.88	5.2	F	(b) Undrained test	
	U	420	1.38	1.0×10^{-5}	0.96	1.27	1.62		1.62	5.0	F	(b) Undrained test	
	U	420	5.52	1.0×10^{-5}	1.69	2.59	3.38		3.29	4.6	NF	(b) Undrained test, 5% data extrapolated	
	U	580	5.52	1.0×10^{-5}	1.63	2.38	2.89		2.93	5.6	NF	(b) Undrained test	
	U	620	1.38	1.0×10^{-5}	0.269	0.297			0.600	4.6	F	(f) Undrained test	
	U	620	3.45	1.0×10^{-5}	1.55	2.42	2.83		2.89	5.6	NF	(b)	
	U	620	3.45	1.0×10^{-5}	1.42	2.04	2.52		2.56	5.7	NF	(b) Undrained test	
	U	620	5.52	1.0×10^{-5}	1.29	1.94	2.46		2.53	5.5	NF	(b) Undrained test	
	U	670	3.45	1.0×10^{-5}	1.39	1.78	2.00		2.09	6.0	NF	(b)	
	U	670	3.45	1.0×10^{-5}	0.850	0.850	0.730		1.19	5.2	F	(e) Undrained test	

	U	670	5.52	1.0×10^{-5}	0.81	0.78	1.04	1.05	5.1	NF	(e) Undrained test	
Coal, subbituminous, Kemmerer, Wyoming, partially water saturated, undrained tests	N	23	0.001	1.0×10^{-4}				0.180	2.3	F		34
	N	23	0.500	1.0×10^{-4}				0.530	3.5	F		
	N	25	1.00	1.0×10^{-4}				0.500	3.4	F		
	N	25	2.00	1.0×10^{-4}				0.590	4.0	F		
	N	25	3.00	1.0×10^{-4}				0.520	3.3	F		
	N	25	5.00	1.0×10^{-4}				0.720	5.4	F		
	N	25	5.00	1.0×10^{-4}				0.680	5.7	F		
	N	25	7.00	1.0×10^{-4}				0.600	5.7			
	P	25	1.00	1.0×10^{-4}				0.510	3.5	F		
	P	25	3.00	1.0×10^{-4}				0.500	3.3	F		
	P	25	5.00	1.0×10^{-4}				0.640	4.6	F		
Coal, subbituminous, Lincoln Co., Wyoming	N	20	0.001	5.0×10^{-5}				0.340		F		130
	N	20	0.001	5.0×10^{-5}				0.320		F		
	N	20	0.001	5.0×10^{-5}				0.430		F		
	N	20	0.500	5.0×10^{-5}				0.370		F		130
	N	20	1.00	5.0×10^{-5}				0.630		F		
	N	20	1.00	5.0×10^{-5}				0.630		F		
	N	20	2.00	5.0×10^{-5}				0.730		F		
	N	20	3.00	5.0×10^{-5}				0.840		F		
	N	20	4.00	5.0×10^{-5}				1.05		F		
	N	20	4.00	5.0×10^{-5}				0.750		F		
	N	20	5.00	5.0×10^{-5}				1.27		F		
	N	20	5.00	5.0×10^{-5}				0.930		F		
	N	20	5.00	5.0×10^{-5}				1.09		F		
	N	20	6.00	5.0×10^{-5}				1.29		F		
	N	20	6.00	5.0×10^{-5}				1.37		F		
	N	20	6.70	5.0×10^{-5}				1.56		F		
	N	20	7.00	5.0×10^{-5}				1.38		F		
	P	20	0.001	5.0×10^{-5}				0.230		F		
	P	20	0.001	5.0×10^{-5}				0.190		F		
	P	20	0.001	5.0×10^{-5}				0.240		F		
	P	20	0.001	5.0×10^{-5}				0.300		F		
	P	20	0.500	5.0×10^{-5}				0.440		F		
	P	20	1.00	5.0×10^{-5}				0.590		F		
	P	20	1.00	5.0×10^{-5}				0.490		F		
	P	20	2.00	5.0×10^{-5}				0.670		F		
	P	20	3.00	5.0×10^{-5}				0.750		F		
	P	20	4.00	5.0×10^{-5}				0.950		F		
	P	20	5.00	5.0×10^{-5}				0.850		F		
	P	20	5.00	5.0×10^{-5}				0.880		F		

Table 2 (continued)
STRESS-STRAIN RELATIONS FROM CONSTANT STRAIN RATE TRIAXIAL TESTS ON ROCKS IN APPARATUS USING FLUID-CONFINING MEDIA

Material	Or	Temp (°C)	Pressure (kbar)	Strain rate (s^{-1})	Differential stress — kbar strain (%) 1	2	5	10	Ultimate strength (kbar)	Total strain (%)	Fault angle (°)	Comment	Ref.
	P	20	5.00	5.0×10^{-5}					1.32		F		
	P	20	6.00	5.0×10^{-5}					1.40		F		
	P	20	6.00	5.0×10^{-5}					1.27		F		
	P	20	7.00	5.0×10^{-5}					1.31		F		
Diabase, "Tishomingo", Johnston	N	150	1.00	1.7×10^{-4}					5.10	1.9	28	(a)	36
Co., Oklahoma	N	500	5.00	1.7×10^{-4}					5.46	17.7	31	(e)	
Diorite, "Salem", Essex Co.,	N	150	1.00	1.7×10^{-4}					6.13	2.6	29	(a)	36
Massachusetts	N	500	5.00	1.7×10^{-4}					7.10	24.0	38	(d)	
Dolomite, "Blair"	2	23	2.00	1.0×10^{-4}					10.46		27		117
	2	23	2.00	1.0×10^{-4}					10.87		18		
	2	23	2.00	1.0×10^{-4}					11.45		30		
	2	23	3.00	1.0×10^{-4}					13.07		34		
	2	23	3.00	1.0×10^{-4}					11.98		32		
	2	23	4.00	1.0×10^{-4}					13.24		38		
	2	23	4.50	1.0×10^{-4}					13.66		30		
	2	23	4.50	1.0×10^{-4}					13.10		31		
	2	23	1.00	-1.0×10^{-4}					1.37		0		
	2	23	2.00	-1.0×10^{-4}					2.16		0		
	2	23	3.00	-1.0×10^{-4}					3.15		0		
	2	23	4.00	-1.0×10^{-4}					4.30		0		
	2	25	3.00	1.0×10^{-7}					8.01		21		
	2	25	3.50	1.0×10^{-7}					8.94		22		
	2	100	2.50	1.0×10^{-7}					9.52		34		117
	2	100	3.00	1.0×10^{-7}					10.48		31		
	2	100	3.50	1.0×10^{-7}					10.10		31		
	2	200	2.50	1.0×10^{-7}					9.24		30		
	2	200	3.00	1.0×10^{-7}					8.87		23		
	2	300	1.50	1.0×10^{-7}					7.92		27		
	2	300	2.00	1.0×10^{-7}					8.88		32		
	2	300	8.89	-1.0×10^{-7}					7.35		0		
	2	300	10.00	-1.0×10^{-7}					7.65		0		

	2	400	1.50	1.0×10^{-7}			10.02		
	2	400	2.00	1.0×10^{-7}			8.42	27	
	2	400	10.00	-1.0×10^{-7}			7.50	21	
	U	23	0.001	1.0×10^{-4}			4.82	F	
	U	23	0.001	1.0×10^{-4}			5.66	F	
	U	23	0.001	1.0×10^{-4}			4.38	F	
	U	23	0.001	1.0×10^{-4}			6.02	F	
	U	23	0.001	1.0×10^{-4}			1.81		134
	U	23	0.001	1.0×10^{-4}			2.80		
	U	23	0.500	1.0×10^{-4}			4.99	37	117
	U	23	0.750	1.0×10^{-4}			8.92	32	
	U	23	1.00	1.0×10^{-4}			9.10	26	
	U	23	1.00	1.0×10^{-4}			9.90	24	
	U	23	1.00	1.0×10^{-4}			7.04		134
	U	23	1.00	1.0×10^{-4}			6.76		
	U	23	1.50	1.0×10^{-4}			10.34	24	117
	U	23	2.00	1.0×10^{-4}			8.44		134
	U	23	3.00	1.0×10^{-4}			9.37		
	U	23	3.00	1.0×10^{-4}			9.86		
	U	23	4.00	1.0×10^{-4}			10.48		
	U	23	5.00	1.0×10^{-4}			11.67		
	U	23	5.00	1.0×10^{-4}			11.86		
	U	23	6.88	1.0×10^{-4}			14.00		
	U	23	6.29	-1.0×10^{-4}			6.48		
	U	23	8.60	-1.0×10^{-4}			8.39		
	U	23	9.80	-1.0×10^{-4}			8.75		
	U	23	10.30	-1.0×10^{-4}			9.24		
	U	23	11.41	-1.0×10^{-4}			9.72		
	U	23	12.69	-1.0×10^{-4}			9.87		
	U	23	15.05	-1.0×10^{-4}			11.70		
	U	23	16.97	-1.0×10^{-4}			12.72		
	U	23	21.20	-1.0×10^{-4}			15.12		
Dolomite, "Crevola Martle", Simplon, Switzerland (stresses given at 6% and 10% strain)	N	700	7.20	1.6×10^{-5}	6.27	7.51			124-126
	N	700	7.20	1.6×10^{-6}	5.60	6.68			
	N	700	7.20	1.5×10^{-6}	5.49	6.43			
	N	800	7.75	6.6×10^{-4}	5.93	6.48			
	N	800	7.75	1.7×10^{-4}	6.40	6.82			
	U	800	7.75	1.8×10^{-5}	5.55	5.85			
	N	800	7.75	1.7×10^{-5}	5.79	6.00			
	N	800	7.75	1.7×10^{-5}	5.90	6.03			

Table 2 (continued)
STRESS-STRAIN RELATIONS FROM CONSTANT STRAIN RATE TRIAXIAL TESTS ON ROCKS IN APPARATUS USING FLUID-CONFINING MEDIA

Material	Or	Temp (°C)	Pressure (kbar)	Strain rate (s^{-1})	Differential stress — kbar strain (%) 1	2	5	10	Ultimate strength (kbar)	Total strain (%)	Fault angle (°)	Comment	Ref.
	U	800	7.75	1.8×10^{-6}			4.62	4.62					
	N	800	7.75	1.8×10^{-7}			3.55	3.55					124-126
	N	900	9.00	7.0×10^{-4}			4.70	4.01					
	N	900	9.00	1.8×10^{-4}			4.62	4.65					
	N	900	9.00	1.7×10^{-4}			4.69	4.70					
	N	900	9.00	1.7×10^{-5}			3.89	3.90					
	N	900	9.00	1.8×10^{-6}			3.12	3.12					
Feldspar rock, various sources, plagioclase compositions	U	600	8.00	2.0×10^{-5}	3.44	6.48	8.40		8.55	7.5	NF	(d) An (44)	39
	U	600	10.00	2.0×10^{-5}	1.30	2.57	5.09	6.63	6.37	9.7	NF	(b) An (77) 10% data extrapolated	
	U	800	9.00	2.0×10^{-5}	2.78	4.75	6.57	6.67	6.65	9.0	NF	(b) An (44) 10% data extrapolated	
	U	800	9.00	2.0×10^{-5}	4.12	6.40	6.12		6.50	8.1	NF	(g) An (2) low	
	U	800	9.00	2.0×10^{-5}	2.74	4.60	6.48		6.54	8.5	NF	(c) An (44)	
	U	800	9.00	2.0×10^{-5}	1.18	2.30	3.59		3.72	7.6	NF	(c) An (95)	
	U	800	9.00	2.0×10^{-5}	0.63	1.31	2.93	3.82	3.86	11.7	NF	(b) An (77)	
	U	800	10.00	2.0×10^{-5}	0.66	1.30	3.02	3.94	4.04	11.5	NF	(b) An (77)	
	U	800	10.00	2.0×10^{-5}	4.34	6.58	6.21		6.58	8.0	NF	(g) An (2) low	
	U	800	10.00	2.0×10^{-5}	2.69	4.61	8.14		8.69	8.2	NF	(e) An (77) high	
	U	800	10.00	2.0×10^{-5}	1.68	3.31	7.43	9.88	9.97	11.5	NF	(b) An (2) high	
	U	800	10.00	2.0×10^{-5}	0.64	1.32	2.90	3.81	3.93	11.5	NF	(b) An (77) low	
Gabbro, "Elizabethtown", Essex Co., New York	N	150	1.00	1.7×10^{-4}					5.29	2.2	32	(a)	36
	N	500	5.00	1.7×10^{-4}					8.16	17.0	30	(d)	
	U	20	1.00	3.0×10^{-3}	1.15	1.58	2.18	2.62	2.79	18.9	NF	(b)	274
Galena ore, Bunker hill Mine, Kellog, Idaho	U	20	1.00	3.0×10^{-4}	0.99	1.37	1.95	2.33	2.60	18.2	NF	(b)	
	U	20	0.001	3.0×10^{-5}	0.620	0.510	0.290		0.650	5.1	F	(e)	
	U	20	0.100	3.0×10^{-5}	0.98	1.22	1.01		1.31	8.7	F	(e)	
	U	20	0.200	3.0×10^{-5}	0.98	1.23	1.43	1.32	1.45	13.0	F	(e)	
	U	20	0.250	3.0×10^{-5}	0.99	1.24	1.55	1.54	1.59	23.0	NF	(d)	
	U	20	0.300	3.0×10^{-5}	1.04	1.28	1.64	1.74	1.74	19.0	NF	(d)	

	U	20	0.350	3.0×10^{-5}	1.05	1.30	1.65	1.84	1.86	18.7	NF	(d)	
	U	20	0.500	3.0×10^{-5}	1.06	1.35	1.73	1.92	1.99	25.0	NF	(d)	
	U	20	1.00	3.0×10^{-5}	1.03	1.39	1.91	2.27	2.62	24.0	NF	(b)	
	U	20	1.00	3.0×10^{-5}	1.15	1.48	2.01	2.34	2.47	13.7	NF	(b)	
	U	20	1.00	3.0×10^{-5}	1.15	1.49	2.01	2.35	2.61	37.0	NF	(d)	
	U	20	1.00	3.0×10^{-5}	1.15	1.50	2.02	2.38	2.62	18.2	NF	(b)	
	U	20	1.00	3.0×10^{-5}	1.15	1.66	2.32	2.67	2.81	35.5	NF	(d)	
	U	20	3.00	3.0×10^{-5}	1.58	1.98	2.46	2.74	3.10	26.9	NF	(b)	
	U	20	1.00	3.0×10^{-6}	0.99	1.45	2.00	2.35	2.42	13.7	NF	(b)	
Galena ore, Federal Division Mine #12, St. Joseph Mineral Corp., Southeast Missouri, <10% impurities	U	24	0.500	7.2×10^{-5}		0.84	1.31	1.58		12.6	NF	(b)	254
	U	24	1.00	7.2×10^{-5}		0.86	1.37	1.75		11.3	NF	(b)	
	U	24	1.00	7.2×10^{-5}		0.86	1.23	1.55		26.5	NF	(b)	
	U	24	1.00	7.2×10^{-5}		0.89	1.25	1.56		20.5	NF	(b)	
	U	24	2.00	7.2×10^{-5}		0.73	1.20	1.68		13.2	NF	(b)	
	U	100	0.500	7.2×10^{-5}		0.61	0.90	1.17		12.5	NF	(b)	
	U	100	1.00	7.2×10^{-5}		0.77	1.22	1.52		14.7	NF	(b)	
	U	100	1.00	7.2×10^{-5}		0.85	1.33	1.62		25.7	NF	(b)	
	U	100	2.00	7.2×10^{-5}		0.56	0.87	1.18		13.0	NF	(b)	254
	U	100	2.00	7.2×10^{-5}		0.81	1.05	1.20		15.7	NF	(b)	
	U	200	0.500	7.2×10^{-5}		0.520	0.650	0.790		12.5	NF	(b)	
	U	200	1.00	7.2×10^{-5}		0.63	0.83	1.00		14.7	NF	(b)	
	U	200	1.00	7.2×10^{-5}		0.440	0.600	0.710		28.0	NF	(b)	
	U	200	2.00	7.2×10^{-5}		0.610	0.720	0.820		13.0	NF	(b)	
	U	200	2.00	7.2×10^{-5}		0.67	0.92	1.04		25.7	NF	(b)	
	U	300	0.500	7.2×10^{-5}		0.410	0.510	0.590		15.7	NF	(b)	
	U	300	1.00	7.2×10^{-5}		0.390	0.480	0.570		12.5	NF	(b)	
	U	300	1.00	7.2×10^{-5}		0.310	0.380	0.510		25.5	NF	(b)	
	U	300	2.00	7.2×10^{-5}		0.270	0.370	0.450		25.0	NF	(b)	
	U	300	2.00	7.2×10^{-5}		0.320	0.360	0.410		12.5	NF	(b)	
	U	400	0.500	7.2×10^{-5}		0.310	0.390	0.410		13.0	NF	(b)	
	U	400	1.00	7.2×10^{-5}		0.310	0.420	0.500		13.5	NF	(b)	
	U	400	2.00	7.2×10^{-5}		0.260	0.290	0.320		13.1	NF	(b)	
Galena ore, Mendota, Colo., only a few percent impurities	U	24	0.500	7.2×10^{-5}		1.11	1.68	1.96		16.4	NF	(b)	254
	U	24	1.00	7.2×10^{-5}		1.19	1.67	2.08		12.2	NF	(b)	
	U	100	1.00	7.2×10^{-5}		1.16	1.57	1.78		22.5	NF	(b)	
Galena ore, Minnie Moore Mine, Bellevue, Idaho, Specimens from three dissimilar blocks	U	20	0.001	3.0×10^{-5}	0.670	0.430	0.380		0.700	6.7	F	(g) Block A	274
	U	20	0.100	3.0×10^{-5}	0.830	0.910	0.700	0.720	0.910	11.8	F	(f) Block A	
	U	20	0.200	3.0×10^{-5}	0.92	1.06	1.18	1.08	1.18	12.2	F	(d) Block A	
	U	20	0.500	3.0×10^{-5}	0.99	1.16	1.44	1.66	1.81	19.5	NF	(b) Block A	
	U	20	0.500	3.0×10^{-5}	1.28	1.68	2.58	2.68	2.69	14.2	NF	(d) Block B	
	U	20	0.500	3.0×10^{-5}	1.21	1.66	2.09	2.33	2.38	17.9	NF	(d) Block C	

Table 2 (continued)
STRESS-STRAIN RELATIONS FROM CONSTANT STRAIN RATE TRIAXIAL TESTS ON ROCKS IN APPARATUS USING FLUID-CONFINING MEDIA

Material	Or	Temp (°C)	Pressure (kbar)	Strain rate (s^{-1})	Differential stress — kbar strain (%)				Ultimate strength (kbar)	Total strain (%)	Fault angle (°)	Comment	Ref.
					1	2	5	10					
	U	20	1.00	3.0×10^{-5}	0.84	1.17	1.66	2.03	2.22	20.0	NF	(d) Block A	
	U	20	1.00	3.0×10^{-5}	1.65	2.12	2.78	3.10	3.12	11.1	NF	(b) Block B	
	U	20	1.00	3.0×10^{-5}	1.44	1.88	2.50	2.84	3.03	18.0	NF	(b) Block C	
Galena ore, Mt. Isa, Australia, 78% galena, 10% sphalerite, 10% gangue, 1% pyrrhotite	U	20	0.500	5.0×10^{-4}	0.65	1.11	1.84	2.50			NF		
	U	20	1.00	5.0×10^{-4}	0.76	1.26	2.09	2.71			NF		196
	U	20	2.50	5.0×10^{-4}	0.92	1.51	2.41	3.06			NF		
	U	20	5.00	5.0×10^{-4}	1.30	1.96	2.83	3.60			NF		
	U	20	0.001	3.0×10^{-5}	0.890				0.890	1.0	F	(a)	4
	U	20	0.300	3.0×10^{-5}	1.71	2.36	1.44		2.36	5.4	F	(d)	
	U	20	0.500	3.0×10^{-5}	2.48	2.79	2.73	2.30	2.86	11.1	F	(d) No sharp stress drop	
	U	20	1.00	3.0×10^{-5}	2.38	2.73	3.25	3.62	3.71	11.2	NF	(b)	
	U	20	1.49	3.0×10^{-5}	2.49	2.94	3.52	3.89	3.95	11.2	NF	(b)	
	U	20	2.50	3.0×10^{-5}	3.02	3.62	4.24	4.52	4.59	11.2	NF	(b)	
	U	200	0.001	3.0×10^{-5}	0.530				0.640	1.5	F	(b)	
	U	200	0.360	3.0×10^{-5}	1.73	2.04	1.46	1.36	2.62		F	(d) >11.4% strain, no sharp stress drop	
	U	200	0.620	3.0×10^{-5}	2.18	2.43	2.60	2.47	2.61	11.2	NF	(d)	
	U	200	0.720	3.0×10^{-5}	2.06	2.33	2.60	2.64	2.65	11.2	NF	(c)	
	U	200	1.52	3.0×10^{-5}	1.97	2.27	2.54	2.70	2.70		NF	(c) >11.4% strain	
	U	200	2.59	3.0×10^{-5}	1.81	2.12	2.48	2.76	2.81	11.3	NF	(b)	
	U	300	0.001	3.0×10^{-5}	0.360	0.330	0.310		0.400	5.7	F	(d)	
	U	300	0.100	3.0×10^{-5}	0.540	0.610	0.480	0.400	0.610	11.2	F	(d) No sharp stress drop	
	U	300	0.370	3.0×10^{-5}	0.61	1.27	1.47	1.50	1.51	11.3	NF	(d)	
	U	300	1.45	3.0×10^{-5}	0.56	1.21	1.45	1.59	1.60	11.3	NF	(b)	4
	U	300	2.38	3.0×10^{-5}	0.47	1.07	1.38	1.66	1.72	11.3	NF	(b)	
	U	400	0.001	3.0×10^{-5}	0.190	0.210	0.070	0.040	0.210	10.7	F	(d) No sharp stress drop	
	U	400	0.150	3.0×10^{-5}	0.390	0.380	0.350	0.300	0.390	10.8	NF	(d)	
	U	400	0.500	3.0×10^{-5}	0.470	0.550	0.600	0.610	0.610	11.1	NF	(c)	
	U	400	1.44	3.0×10^{-5}	0.600	0.610	0.670	0.700	0.710	11.2	NF	(c)	
	U	400	2.40	3.0×10^{-5}	0.470	0.550	0.670	0.750	0.770	11.1	NF	(b)	

Galena ore, Mt. Isa, Australia, 87.6% galena, 6.4% quartz, 5.1% sphalerite, 0.9% pyrite, G.S. = 0.07 mm	U	20	1.50	2.5×10^{-4}	1.42	1.93	2.53	3.00	3.12	13.0	NF	(b)	6
	U	20	1.50	2.5×10^{-5}	1.54	1.93	2.39	2.82	2.93	13.0	NF	(b)	
	U	20	1.50	2.5×10^{-6}	1.21	1.59	2.16	2.63	2.75	12.8	NF	(b)	
	U	20	1.50	2.4×10^{-7}	1.21	1.59	2.12	2.39	2.45	13.0	NF	(b)	
	U	200	1.50	2.5×10^{-4}	0.99	1.24	1.48	1.68	1.76	13.0	NF	(b)	
	U	200	1.50	2.5×10^{-5}	0.88	1.09	1.40	1.59	1.66	12.8	NF	(b)	
	U	200	1.50	2.5×10^{-6}	0.77	1.01	1.25	1.38	1.42	13.0	NF	(b)	
	U	200	1.50	2.5×10^{-7}	0.84	0.98	1.14	1.25	1.27	13.0	NF	(b)	
	U	200	1.50	3.0×10^{-8}	0.59	0.78	0.97	1.00	1.02	12.7	NF	(b)	
	U	300	1.50	2.5×10^{-4}	0.86	1.01	1.13	1.24	1.29	12.2	NF	(b)	
	U	300	1.50	2.6×10^{-5}	0.68	0.78	0.91	1.06	1.09	12.2	NF	(b)	
	U	300	1.50	2.6×10^{-6}	0.525	0.611	0.728	0.833	0.863	12.2	NF	(b)	
	U	300	1.50	2.6×10^{-7}	0.349	0.450	0.566	0.641	0.656	12.2	NF	(b)	
	U	300	1.50	3.1×10^{-8}	0.225	0.300	0.379	0.450	0.458	12.2	NF	(b)	
	U	400	1.50	3.1×10^{-4}	0.59	0.72	0.90	1.05	1.12	12.5	NF	(b)	
	U	400	1.50	3.2×10^{-5}	0.698	0.735	0.784	0.825	0.829	12.8	NF	(b)	
	U	400	1.50	3.2×10^{-6}	0.326	0.450	0.551	0.596	0.600	12.5	NF	(b)	
	U	400	1.50	3.5×10^{-7}	0.203	0.281	0.371	0.413	0.416	12.2	NF	(b)	
	U	400	1.50	3.4×10^{-8}	0.105	0.169	0.251	0.311	0.319	12.3	NF	(b)	
Galena ore, No. 8 Mine, Bonne Terre, Mo., About 10% porosity	U	20	0.100	5.0×10^{-4}	0.430	0.570	0.760	0.750	0.770	13.6	F	(d)	274
	U	20	0.200	5.0×10^{-4}	0.450	0.610	0.860	0.940	0.960	21.0	F	(d)	
	U	20	0.300	5.0×10^{-4}	0.46	0.62	0.91	1.09	1.18	23.9	NF	(b)	
	U	20	0.400	5.0×10^{-4}	0.48	0.62	0.96	1.23	1.47	24.0	NF	(b)	
	U	20	0.500	5.0×10^{-4}	0.50	0.67	1.05	1.36	1.65	31.4	NF	(d)	
	U	20	0.750	5.0×10^{-4}	0.50	0.72	1.17	1.56	2.11	35.0	NF	(d)	
	U	20	1.00	5.0×10^{-4}	0.54	0.80	1.34	1.80	2.40	36.5	NF	(d)	
	U	20	1.50	5.0×10^{-4}	0.60	0.88	1.43	1.92	2.77	36.3	NF	(b)	
	U	20	2.00	5.0×10^{-4}	0.68	0.91	1.53	2.12	2.93	36.5	NF	(b)	
	U	20	2.50	5.0×10^{-4}	0.72	1.01	1.68	2.36	3.08	36.5	NF	(b)	
	U	20	3.00	5.0×10^{-4}	0.88	1.25	1.87	2.40	3.20	36.5	NF	(b)	
	U	20	3.50	5.0×10^{-4}	1.00	1.37	2.10	2.68	3.38	32.4	NF	(b)	
	U	20	4.00	5.0×10^{-4}	1.00	1.36	1.97	2.43	2.88	36.1	NF	(b)	
	U	20	5.00	5.0×10^{-4}	1.24	1.61	2.12	2.35	2.72	36.3	NF	(b)	
Gneiss, biotite, "Engel and Engel", St. Lawrence Co., New York	N	150	1.00	1.7×10^{-4}					5.89	2.5	31	(a)	36
	N	500	5.00	1.7×10^{-4}					11.29	14.8	35	(d)	
	P	150	1.00	1.7×10^{-4}					7.03	3.4	26	(a)	
	P	500	5.00	1.7×10^{-4}					10.75	8.6	35	(d)	
Gneiss, biotite, "Fordham", Manhattan, N.Y.	22	500	5.00	1.7×10^{-4}					4.32	10.2	NF	(b)	36
	45	500	5.00	1.7×10^{-4}					5.66	17.4	42	(d)	
	N	500	5.00	1.7×10^{-4}					7.65	11.2	36	(d)	

Table 2 (continued)
STRESS-STRAIN RELATIONS FROM CONSTANT STRAIN RATE TRIAXIAL TESTS ON ROCKS IN APPARATUS USING FLUID-CONFINING MEDIA

Material	Or	Temp (°C)	Pressure (kbar)	Strain rate (s^{-1})	Differential stress — kbar strain (%) 1	2	5	10	Ultimate strength (kbar)	Total strain (%)	Fault angle (°)	Comment	Ref.
	P	500	5.00	1.7×10^{-4}					8.79	13.5	34	(d)	36
Gneiss, granite, "Diana", St.	N	150	1.00	1.7×10^{-4}					6.32	2.9	31	(b)	36
Lawrence Co., New York	N	500	5.00	1.7×10^{-4}					11.50	10.8	33	(d)	
	P	150	1.00	1.7×10^{-4}					6.11	2.8	32	(a)	
	P	500	5.00	1.7×10^{-4}					12.89	7.9	32	(d)	
Granite, "Hoggar", Taburirt Tan	U	25	0.001	5.0×10^{-5}					2.09		F		265
Afella Massif, Algeria	U	25	0.001	5.0×10^{-5}					2.16		F		
	U	25	0.250	5.0×10^{-5}					5.44		F		
	U	25	0.510	5.0×10^{-5}					7.24		F		
	U	25	1.00	5.0×10^{-5}					9.27		F		
	U	25	1.00	5.0×10^{-5}					8.89		F		
	U	25	2.00	5.0×10^{-5}					13.03		F		
	U	25	2.01	5.0×10^{-5}					12.36		F		
	U	25	3.00	5.0×10^{-5}					13.89		F		
	U	25	3.01	5.0×10^{-5}					14.24		F		
	U	25	3.02	5.0×10^{-5}					14.46		F		
	U	25	3.67	5.0×10^{-5}					15.35		F		
	U	25	4.00	5.0×10^{-5}					15.94		F		
	U	25	5.05	5.0×10^{-5}					16.59		F		
	U	25	5.10	5.0×10^{-5}					16.07		F		
	U	25	6.06	5.0×10^{-5}					17.55		F		
	U	25	7.06	5.0×10^{-5}					18.70		F		
	U	25	7.06	5.0×10^{-5}					18.40		F		
	U	25	12.35	5.0×10^{-5}					21.50		F		
	U	25	15.36	5.0×10^{-5}					21.75		F		
	U	25	20.25	5.0×10^{-5}					22.53		F		
	U	25	21.93	5.0×10^{-5}					23.40		F		
Granite, "Silver Plume", Lyons,	N	150	1.00	1.7×10^{-4}					3.29	2.7	33	(d)	36
Colo.	N	500	5.0	1.7×10^{-4}					8.27	17.0	36	(d)	
Granite, "Westerly", Rhode Island	U	20	0.001	1.0×10^{0}					4.74	0.6	F	(b)	195
	U	20	1.00	1.0×10^{0}	8.79				11.23	1.4	F	(b)	

U	20	3.00	1.0×10^{0}	14.00	19.27	20.04	2.2	F	(b)	
U	20	5.00	1.0×10^{0}	14.80	22.13	23.45	2.4	F	(b)	
U	20	0.001	1.0×10^{-1}			4.21	.6	F	(b)	
U	20	1.00	1.0×10^{-1}	7.25		10.48	1.8	F	(b)	
U	20	3.00	1.0×10^{-1}	14.41		18.68	1.9	F	(b)	
U	20	5.00	1.0×10^{-1}	14.55	22.50	24.44	2.5	F	(b)	
U	20	0.001	1.0×10^{-2}			3.52	0.5	F	(a)	
U	20	1.00	1.0×10^{-2}	7.13		8.77	1.5	F	(b)	
U	20	2.00	1.0×10^{-2}	7.98	12.95	13.05	2.0	F	(b)	
U	20	5.00	1.0×10^{-2}	12.44	20.90	22.81	2.3	F	(b)	
U	23	0.001	1.0×10^{-4}			2.00			20% Water saturated, undrained test	76
U	23	0.001	1.0×10^{-4}			1.90			1.3% Water saturated, undrained test	
U	23	0.022	1.0×10^{-4}			2.36			0.8% Water saturated, undrained test	76
U	23	0.500	1.0×10^{-4}			6.00			0.8% Water saturated, undrained test	
U	23	0.500	1.0×10^{-4}			6.10			0.8% Water saturated, undrained test	
U	23	0.500	1.0×10^{-4}			5.90			0.8% Water saturated, undrained test	
U	23	1.00	1.0×10^{-4}			7.60				266
U	23	1.00	1.0×10^{-4}			2.10			Kerosene saturated, undrained test	76
U	23	1.00	1.0×10^{-4}			4.00			2.5% Water saturated, undrained test	
U	23	1.00	1.0×10^{-4}			7.20			1.3% Water saturated, undrained test	
U	23	1.00	1.0×10^{-4}			7.84			1.0% Water saturated, undrained test	
U	23	1.00	1.0×10^{-4}			8.10			0.8% Water saturated, undrained test	
U	23	2.00	1.0×10^{-4}			2.36			Kerosene saturated, undrained test	
U	23	2.00	1.0×10^{-4}			10.36			1.3% Water saturated, undrained test	
U	23	2.00	1.0×10^{-4}			10.53			1.0% Water saturated, undrained test	
U	23	3.00	1.0×10^{-4}			12.77				266
U	23	3.00	1.0×10^{-4}			13.50				

Table 2 (continued)
STRESS-STRAIN RELATIONS FROM CONSTANT STRAIN RATE TRIAXIAL TESTS ON ROCKS IN APPARATUS USING FLUID-CONFINING MEDIA

Material	Or	Temp (°C)	Pressure (kbar)	Strain rate (s^{-1})	Differential stress — kbar strain (%)				Ultimate strength (kbar)	Total strain (%)	Fault angle (°)	Comment	Ref.
					1	2	5	10					
	U	23	3.00	1.0×10^{-4}					1.72			20% Water saturated, undrained test	76
	U	23	3.00	1.0×10^{-4}					2.36			Kerosene saturated, undrained test	
	U	23	3.00	1.0×10^{-4}					4.40			2.5% Water saturated, undrained test	
	U	23	3.00	1.0×10^{-4}					12.50			1.0% Water saturated, undrained test	
	U	23	3.00	1.0×10^{-4}					15.39			0.8% Water saturated, undrained test	
	U	23	4.00	1.0×10^{-4}					4.20			Kerosene saturated, undrained test	
	U	23	4.00	1.0×10^{-4}					5.32			2.5% Water saturated, undrained test	
	U	23	4.00	1.0×10^{-4}					13.03			1.3% Water saturated, undrained test	
	U	23	4.00	1.0×10^{-4}					13.84			1.0% Water saturated, undrained test	
	U	23	5.00	1.0×10^{-4}					18.00				266
	U	23	5.00	1.0×10^{-4}					1.10			20% Water saturated, undrained test	76
	U	23	5.00	1.0×10^{-4}					5.20			Kerosene saturated, undrained test	
	U	23	5.00	1.0×10^{-4}					8.24			2.5% Water saturated, undrained test	
	U	23	5.00	1.0×10^{-4}					9.72			2.5% Water saturated, undrained test	
	U	23	5.00	1.0×10^{-4}					14.36			1.0% Water saturated, undrained test	
	U	23	6.00	1.0×10^{-4}					2.30			20% Water saturated, undrained test	

U	23	6.00	1.0×10^{-4}	8.60		Kerosene saturated, undrained test	
U	23	6.00	1.0×10^{-4}	11.44		2.5% Water saturated, undrained test	
U	23	6.00	1.0×10^{-4}	13.60		1.3% Water saturated, undrained test	
U	23	7.00	1.0×10^{-4}	21.09			266
U	23	9.29	1.0×10^{-4}	23.79			
U	23	11.89	1.0×10^{-4}	24.79			
U	23	12.57	1.0×10^{-4}	24.31			
U	23	14.58	1.0×10^{-4}	25.09			
U	23	17.91	1.0×10^{-4}	23.94			
U	23	19.02	1.0×10^{-4}	25.21			
U	23	20.14	1.0×10^{-4}	27.29			
U	23	0.001	5.0×10^{-5}	1.48	F	(b)	129
U	23	0.001	5.0×10^{-5}	2.08	F	(b)	
U	23	0.001	5.0×10^{-5}	1.68	F	(b)	
U	23	0.001	5.0×10^{-5}	1.38	F	(b)	
U	23	0.001	5.0×10^{-5}	1.66	F	(b)	
U	23	0.001	5.0×10^{-5}	1.40	F	(b)	
U	23	0.001	5.0×10^{-5}	1.82	F	(b)	
U	23	0.001	5.0×10^{-5}	1.72	F	(b)	
U	23	0.005	5.0×10^{-5}	2.38	F	(b)	
U	23	0.200	5.0×10^{-5}	4.36	F	(b)	
U	23	1.00	5.0×10^{-5}	8.08	F	(b)	
U	23	2.00	5.0×10^{-5}	11.25	F	(b)	
U	23	3.00	5.0×10^{-5}	13.20	F	(b)	
U	23	4.00	5.0×10^{-5}	15.53	F	(b)	
U	23	5.00	5.0×10^{-5}	16.98	F	(b)	
U	23	6.00	5.0×10^{-5}	18.50	F	(b)	
U	23	7.00	5.0×10^{-5}	20.00	F	(b)	
U	23	3.32	-1.0×10^{-4}	3.40			126
U	23	4.12	-1.0×10^{-4}	4.18			
U	23	6.81	-1.0×10^{-4}	6.88			
U	23	7.36	-1.0×10^{-4}	7.52			
U	23	7.81	-1.0×10^{-4}	7.92			
U	23	10.19	-1.0×10^{-4}	10.20			
U	23	12.22	-1.0×10^{-4}	12.39			
U	23	13.24	-1.0×10^{-4}	13.55			
U	23	14.71	-1.0×10^{-4}	13.74			
U	23	15.79	-1.0×10^{-4}	13.84			

Table 2 (continued)
STRESS-STRAIN RELATIONS FROM CONSTANT STRAIN RATE TRIAXIAL TESTS ON ROCKS IN APPARATUS USING FLUID-CONFINING MEDIA

Material	Or	Temp (°C)	Pressure (kbar)	Strain rate (s^{-1})	Differential stress — kbar strain (%) 1	2	5	10	Ultimate strength (kbar)	Total strain (%)	Fault angle (°)	Comment	Ref.
	U	23	17.75	-1.0×10^{-4}					15.48				126
	U	23	18.58	-1.0×10^{-4}					16.90				
	U	23	20.44	-1.0×10^{-4}					19.03				
Granodiorite, "Charcoal" or "St. Cloud Gray", Stevins County, Minnesota	N	150	1.00	1.7×10^{-4}					8.22	3.7	26	(b)	36
	N	500	5.00	1.7×10^{-4}					9.32	19.2	30	(d)	
	U	23	4.50	6.0×10^{0}					13.07	2.8	F		30
	U	23	2.50	3.7×10^{0}					12.70	2.0	F		
	U	23	3.47	3.6×10^{0}					8.39	1.1	F		
	U	23	3.47	3.5×10^{0}					8.39	1.3	F		
	U	23	4.50	3.3×10^{0}					12.70	2.7	F		
	U	23	4.50	2.9×10^{0}					12.50	2.7	F		
	U	23	0.500	2.9×10^{0}					6.70	1.3	F		
	U	23	0.001	2.7×10^{0}					4.09	0.4	F		
	U	23	0.500	2.6×10^{0}					5.60	1.1	F		
	U	23	0.001	2.3×10^{0}					3.48	0.5	F		
	U	23	0.001	2.2×10^{0}					3.68	0.3	F		
	U	23	0.001	2.0×10^{0}					3.17	0.5	F		
	U	23	4.50	1.5×10^{0}					13.50	2.8	F		
	U	23	2.50	1.5×10^{0}					10.00	2.0	F		
	U	23	0.500	9.5×10^{-1}					5.90	1.1	F		
	U	23	0.001	8.5×10^{-1}					2.96	0.6	F		
	U	23	2.50	3.5×10^{-1}					10.00	2.3	F		
	U	23	0.001	1.9×10^{-1}					2.96	0.5	F		
	U	23	4.50	8.1×10^{-2}					12.50	2.6	F		
	U	23	0.001	6.3×10^{-2}					2.86	0.5	F		
	U	23	0.500	5.9×10^{-2}					5.50	1.2	F		
	U	23	2.50	3.6×10^{-2}					10.20	1.0	F		
	U	23	0.500	2.2×10^{-2}					5.60	1.2	F		
	U	23	4.50	1.7×10^{-2}	6.19	10.94			13.05	2.8	F	(b)	
	U	23	0.001	1.7×10^{-2}					2.86	0.5	F		
	U	23	4.50	1.6×10^{-2}					12.70	2.5	F		

	U	23	0.001	1.4 × 10^{-2}				2.76	0.5	F	(a)	
	U	23	0.500	1.3 × 10^{-2}				5.60	1.1	F		
	U	23	0.500	1.3 × 10^{-2}	5.62			5.90	1.1	F	(b)	
	U	23	2.50	8.7 × 10^{-3}	6.15	10.36		10.78	2.2	F	(b)	
	U	23	2.50	7.6 × 10^{-3}				10.20	2.1	F		
	U	25	0.001	1.0 × 10^{-4}	1.39	2.83		3.60	2.5	F	(a)	113
	U	25	0.001	1.0 × 10^{-4}	1.53	3.10		3.40	2.2	F	(a)	
	U	25	0.001	1.0 × 10^{-4}	1.44	2.20		2.30	2.3	F	(b) Water saturated, undrained test	
	U	25	0.001	1.0 × 10^{-4}	1.51	3.11		3.20	2.2	F	(a) Water saturated, undrained test	
	U	25	5.00	1.0 × 10^{-4}	1.44	2.87		6.29	4.4	29	(a)	
	U	25	5.00	1.0 × 10^{-4}	1.84	3.67		6.20	2.9	F	(a)	
	U	25	5.00	1.0 × 10^{-4}	1.78	3.59		6.30	3.0	35	(a)	
	U	250	5.00	1.0 × 10^{-4}				4.90		F		
	U	400	0.001	1.0 × 10^{-4}	0.43	1.49		1.68	2.2	F	(d)	
	U	400	0.001	1.0 × 10^{-4}	0.57	1.66		1.66	3.4	F	(d)	
	U	400	5.00	1.0 × 10^{-4}	1.34	2.61		4.60	3.6	24	(a)	
	U	400	5.00	1.0 × 10^{-4}	1.25	2.45		5.10	3.8	F	(a)	
	U	400	5.00	1.0 × 10^{-4}	1.61	3.18		4.40	2.8	F	(a)	113
	U	400	5.00	1.0 × 10^{-7}	1.81	3.58		5.27	3.0	F	(a)	
	U	400	5.00	1.0 × 10^{-4}	1.34	2.81		4.40	3.4	32	(a) Water saturated, undrained test	
	U	400	5.00	1.0 × 10^{-4}	1.59	3.22		5.20	3.9	26	(a) Water saturated, undrained test	
	U	400	0.001	1.0 × 10^{-4}				1.63		F		
	U	400	0.001	1.0 × 10^{-7}				1.64		F		
	U	700	0.001	1.0 × 10^{-4}	0.380			0.470	1.8	F	(d)	
	U	700	0.001	1.0 × 10^{-4}	0.540	0.170		0.690	2.4	F	(d)	
	U	700	5.00	1.0 × 10^{-4}	1.46	2.93		3.42	3.5	29	(a)	
	U	700	5.00	1.0 × 10^{-4}	1.42	2.79		3.30	4.3	29	(a)	
	U	1000	0.001	f381.0 × 10^{-4}	0.100	0.320		0.410	3.4	30		
	U	1000	0.001	1.0 × 10^{-4}	0.130	0.300		0.300	3.5	30		
Granodiorite, "Climax Stock", Nevada	U	20	0.001	1.0 × 10^{-4}	0.58	1.26		2.19	3.3	F	(b)	268
	U	20	0.200	1.0 × 10^{-4}		1.42		2.99	4.1	F	(b)	
	U	20	0.500	1.0 × 10^{-4}		1.53		3.15	4.3	F	(b)	
Granodiorite, "microgranodiorite", Penmaemawr, Caernarvonshire, England, chlorite dehydrated between 506 and 870°C	U	20	4.48	1.0 × 10^{-5}	1.86	4.16	9.76	10.64	8.0	F	(d) Undrained test	214
	U	20	4.48	1.0 × 10^{-5}	0.93	2.33	3.44	3.49	5.6	F	(d) Water saturated, undrained test	
	U	20	4.48	1.0 × 10^{-5}	1.33	2.28	3.49	3.58	5.5	F	(b) Water saturated, undrained test	

Table 2 (continued)
STRESS-STRAIN RELATIONS FROM CONSTANT STRAIN RATE TRIAXIAL TESTS ON ROCKS IN APPARATUS USING FLUID-CONFINING MEDIA

Material	Or	Temp (°C)	Pressure (kbar)	Strain rate (s^{-1})	Differential stress — kbar strain (%) 1	2	5	10	Ultimate strength (kbar)	Total strain (%)	Fault angle (°)	Comment	Ref.
	U	130	4.48	1.0×10^{-5}	1.07	1.33			1.49	3.8	F	(c) Water saturated, undrained test	
	U	320	4.48	1.0×10^{-5}	0.67	1.53	2.79		2.79	5.4	F	(b) Water saturated, undrained test	
	U	320	4.48	1.0×10^{-5}	1.88	4.14	8.84		9.12	6.5	F	(d)	
	U	320	4.48	1.0×10^{-5}	1.14	2.35			2.67	4.9	F	(e) Undrained test	
	U	520	4.48	1.0×10^{-5}	0.58	1.00			1.09	4.7	NF	(b) Water saturated, undrained test	
	U	520	4.48	1.0×10^{-5}	0.91	1.47			1.86	4.9	F	(e) Undrained test	
	U	670	4.48	1.0×10^{-5}	0.349	0.488	0.256		0.535	5.0	NF	(d) Undrained test	
	U	720	4.48	1.0×10^{-5}	0.163	0.302	0.558		0.605	7.8	NF	(b) Water saturated, undrained test	
Graphite, strong preferred orientation	P	20	1.00	4.0×10^{-4}	0.27	0.51	1.01	1.34	1.58	20.0	NF	(b)	226
	P	20	2.00	4.0×10^{-4}	0.35	0.62	1.14	1.59	2.06	20.0	NF	(b)	
	P	20	4.00	4.0×10^{-4}	0.43	0.76	1.48	2.10	2.79	20.0	NF	(b)	
	P	20	6.00	4.0×10^{-4}	0.49	0.87	1.60	2.28	3.09	20.0	NF	(b)	
	P	20	8.00	4.0×10^{-4}	0.60	1.02	1.76	2.52	3.38	20.0	NF	(b)	
Halite, annealed isotropic aggregates with 0.5—1.0 mm average crystal size	U	23	0.001	1.0×10^{-4}					0.600		F		128
	U	23	0.001	1.0×10^{-4}					0.560		F		
	U	23	0.001	1.0×10^{-4}					0.670		F		
	U	23	0.001	1.0×10^{-4}					0.640		F		
	U	23	0.001	1.0×10^{-4}					0.590		F		
	U	23	0.001	1.0×10^{-4}					0.530		F		
	U	23	0.001	1.0×10^{-4}					0.690		F		
	U	23	0.005	1.0×10^{-4}					.620		F		
	U	23	0.015	1.0×10^{-4}					.770		F		
	U	23	0.100	1.0×10^{-4}		0.810	0.840				NF	Stresses at 3% and 5% strain	
	U	23	0.200	1.0×10^{-4}		0.750	0.780				NF	Stresses at 3% and 5% strain	
	U	23	0.250	1.0×10^{-4}		0.820	0.820	0.870			NF	Stresses at 3%, 5%, and 10% strain	

C1	C2	C3	C4	C5	C6	C7	C8	C9	C10	C11	C12	C13	C14
	U	23	0.500	1.0×10^{-4}		0.780	0.820	0.890			NF	Stresses at 3%, 5%, and 10% strain	
	U	23	0.750	1.0×10^{-4}		0.790	0.830	0.900			NF	Stresses at 3%, 5%, and 10% strain	
	U	23	1.00	1.0×10^{-4}		0.790	0.830	0.880			NF	Stresses at 3%, 5%, and 10% strain	
	U	23	1.00	1.0×10^{-4}		0.710	0.750	0.810			NF	Stresses at 3%, 5%, and 10% strain	
	U	23	1.00	1.0×10^{-4}		0.750	0.790	0.840			NF	Stresses at 3%, 5%, and 10% strain	
	U	23	1.00	1.0×10^{-4}		0.740	0.770	0.830			NF	Stresses at 3%, 5%, and 10% strain	
	U	23	1.00	1.0×10^{-4}		0.850	0.900				NF	Stresses at 3% and 5% strain	
	U	23	1.00	1.0×10^{-4}		0.850	0.890				NF	Stresses at 3% and 5% strain	
	U	23	1.00	1.0×10^{-4}		0.810	0.840				NF	Stresses at 3% and 5% strain	
	U	23	1.00	1.0×10^{-4}		0.780	0.830				NF	Stresses at 3% and 5% strain	
	U	23	2.00	1.0×10^{-4}		0.780	0.820	0.910			NF	Stresses at 3%, 5%, and 10% strain	
	U	23	3.00	1.0×10^{-4}		0.820	0.870	0.910			NF	Stresses at 3%, 5%, and 10% strain	
	U	23	4.00	1.0×10^{-4}		0.850	0.880				NF	Stresses at 3% and 5% strain	
	U	23	4.00	1.0×10^{-4}		0.860	0.870	0.870			NF	Stresses at 3%, 5%, and 10% strain	
Halite, annealed isotropic aggregates with 2—3 mm average crystal size	U	23	2.00	-1.5×10^{-8}		0.280		0.310			NF	(b) 10% Data extrapolated	123
	U	23	2.00	-1.5×10^{-7}	0.200	0.280	0.320	0.340	0.340	11.1	NF	(b)	
	U	23	2.00	-1.5×10^{-5}	0.280	0.330	0.410	0.470	0.480	10.5	NF	(b)	
	U	100	2.00	-1.5×10^{-8}	0.100	0.110	0.110	0.120	0.120	11.2	NF	(b)	
	U	100	2.00	-1.5×10^{-7}	0.120	0.140	0.160	0.180	0.190	11.7	NF	(b)	
	U	100	2.00	-1.5×10^{-6}	0.130	0.160	0.190	0.220	0.230	12.2	NF	(b)	
	U	100	2.00	-1.5×10^{-5}	0.160	0.200	0.240	0.270	0.280	12.5	NF	(b)	
	U	100	2.00	-1.5×10^{-4}	0.140	0.190	0.260	0.310	0.320	12.6	NF	(b)	
	U	100	2.00	-1.5×10^{-3}	0.220	0.270	0.340	0.380	0.390	12.3	NF	(b)	
	U	200	2.00	-1.2×10^{-8}	0.032				0.032	1.2	NF	(c)	
	U	200	2.00	-1.2×10^{-7}	0.046	0.050	0.054	0.054	0.054	11.3	NF	(c)	
	U	200	2.00	-1.2×10^{-6}	0.069	0.073	0.082	0.090	0.090	11.5	NF	(b)	
	U	200	2.00	-1.2×10^{-5}	0.082	0.100	0.120	0.140	0.140	11.1	NF	(b)	
	U	200	2.00	-1.2×10^{-4}	0.100	0.120	0.150	0.170	0.180	11.3	NF	(b)	
	U	200	2.00	-1.5×10^{-4}	0.110	0.140	0.160	0.180	0.180	11.3	NF	(b)	
	U	200	2.00	-1.5×10^{-3}	0.160	0.170	0.200	0.220	0.220	11.2	NF	(b)	
	U	200	2.00	-1.8×10^{-2}	0.170	0.200	0.240	0.280	0.290	11.7	NF	(b)	
	U	248	2.00	-1.1×10^{-7}	0.036	0.037	0.040	0.041	0.041	11.1	NF	(c)	

Table 2 (continued)
STRESS-STRAIN RELATIONS FROM CONSTANT STRAIN RATE TRIAXIAL TESTS ON ROCKS IN APPARATUS USING FLUID-CONFINING MEDIA

Material	Or	Temp (°C)	Pressure (kbar)	Strain rate (s^{-1})	Differential stress — kbar strain (%)				Ultimate strength (kbar)	Total strain (%)	Fault angle (°)	Comment	Ref.
					1	2	5	10					
	U	248	2.00	-1.1×10^{-6}	0.044	0.049	0.057	0.059	0.059	10.6	NF (c)		
	U	248	2.00	-1.1×10^{-5}	0.066	0.071	0.081	0.090	0.090	11.2	NF (b)		
	U	248	2.00	-1.1×10^{-4}	0.089	0.102	0.122	0.134	0.136	10.8	NF (b)		
	U	248	2.00	-1.5×10^{-3}	0.126	0.140	0.162	0.182	0.186	11.3	NF (b)		126
	U	248	2.00	-1.8×10^{-2}	0.134	0.161	0.192	0.220	0.227	11.7	NF (b)		
	U	248	2.00	-1.8×10^{-1}		0.157	0.205	0.246	0.249	10.7	NF (b)		
	U	300	2.00	-1.1×10^{-8}	0.015	0.016			0.016	3.9	NF (c)		
	U	300	2.00	-1.1×10^{-7}	0.020	0.021	0.021	0.022	0.022	10.8	NF (c)		
	U	300	2.00	-1.2×10^{-6}	0.023	0.024	0.027	0.031	0.031	11.2	NF (b)		
	U	300	2.00	-1.2×10^{-5}	0.027	0.031	0.041	0.049	0.049	11.0	NF (b)		
	U	300	2.00	-1.5×10^{-4}	0.032	0.043	0.061	0.074	0.075	11.0	NF (b)		
	U	300	2.00	-1.1×10^{-3}	0.057	0.069	0.091	0.110	0.112	11.2	NF (b)		
	U	300	2.00	-1.8×10^{-2}	0.072	0.083	0.110	0.142	0.150	12.0	NF (b)		
	U	300	2.00	-1.8×10^{-1}	0.082	0.110	0.150	0.176	0.181	11.7	NF (b)		
	U	400	2.00	-1.1×10^{-7}	0.015	0.016	0.016	0.016	0.016	10.3	NF (c)		
	U	400	2.00	-1.2×10^{-6}	0.019	0.020	0.021	0.022	0.022	11.3	NF (c)		
	U	400	2.00	-1.2×10^{-5}	0.023	0.027	0.033	0.035	0.035	11.1	NF (b)		
	U	400	2.00	-1.2×10^{-4}	0.035	0.039	0.047	0.052	0.052	11.2	NF (b)		
	U	400	2.00	-1.1×10^{-3}	0.047	0.054	0.067	0.075	0.077	11.6	NF (b)		
	U	400	2.00	-1.4×10^{-2}	0.058	0.071	0.092	0.110	0.110	11.2	NF (b)		
	U	400	2.00	-1.9×10^{-1}	0.054	0.073	0.100	0.130	0.130	10.6	NF (b)		
Limestone, "Annona Chalk", Caddo Parish, Louisiana (stresses at 3% and 5% strain)	N	23	0.001	1.0×10^{-4}		0.280					F		
	N	23	0.001	1.0×10^{-4}		0.270					F		298
	N	23	0.100	1.0×10^{-4}		0.470					F		
	N	23	0.250	1.0×10^{-4}		0.660	0.680				NF		
	N	23	0.500	1.0×10^{-4}		0.840	0.970				NF		
	N	23	1.00	1.0×10^{-4}		1.02	1.30				NF		
	N	23	1.00	1.0×10^{-4}		0.700	0.700				NF		
	N	23	2.00	1.0×10^{-4}		1.24	1.32				NF		
	N	23	3.00	1.0×10^{-4}		1.34	1.42				NF		
	N	23	3.00	1.0×10^{-4}		2.02	2.28				NF		

	N	23	3.00	1.0×10^{-4}		0.980	0.960				NF		
	N	23	5.00	1.0×10^{-4}		1.39	1.31				NF		
Limestone, "Carrara Marble", Italy	U	20	0.500	4.0×10^{-4}	2.04	2.15	2.09	1.96	2.16		NF	(d)	82
	U	20	1.00	4.0×10^{-4}	2.53	2.76	2.98	3.00	3.10		NF	(c)	
	U	20	2.00	4.0×10^{-4}	2.86	3.36	3.90	4.26			NF	(b)	
	U	20	4.00	4.0×10^{-4}	3.18	3.73	4.51	5.33			NF	(b)	
	U	20	6.00	4.0×10^{-4}	3.40	4.00	4.78	5.61			NF	(b)	
	U	20	8.00	4.0×10^{-4}	3.41	4.00	4.86	5.73			NF	(b)	
	U	20	0.001	3.0×10^{-5}					0.730	0.9	F	(d)	251
	U	20	0.500	3.0×10^{-5}	1.87	1.92	1.98		1.98	7.8	NF	(c)	
	U	20	0.500	3.0×10^{-5}	1.87	1.89	1.89		1.89	7.5	NF	(c) wet	
	U	20	1.00	3.0×10^{-5}	2.04	2.26	2.62	2.81	2.87	12.1	NF	(b)	
	U	20	1.00	3.0×10^{-5}	1.38	1.38	1.32		1.38	5.0	NF	(d) Wet, lamda = 0.9	
	U	20	1.50	3.0×10^{-5}	2.39	2.76	3.36	3.76	3.88	12.0	NF	(b)	
	U	20	1.50	3.0×10^{-5}	2.36	2.73	3.26	3.63	3.67	12.0	NF	(b) Wet	
	U	20	2.00	3.0×10^{-5}	2.43	2.83	3.38	4.00	4.07	11.3	NF	(b) Wet	
	U	20	2.00	3.0×10^{-5}	1.64	1.69	1.63		1.69	7.0	NF	(d) Wet, lamda = 0.9	
	U	20	3.00	3.0×10^{-5}	2.56	3.01	3.71	4.41	4.74	12.6	NF	(b)	
	U	20	3.00	3.0×10^{-5}	2.56	3.01	3.71	4.41	4.74	12.7	NF	(b) Wet	
	U	20	3.00	3.0×10^{-5}	1.80	1.88			1.88	4.3	NF	(a) Wet, lamda = 0.9	
	U	500	0.001	3.0×10^{-5}	0.420	0.570	0.640		0.640	5.5	F	(b)	251
	U	500	0.250	3.0×10^{-5}	0.68	0.96	1.25	1.35	1.36	10.7	NF	(b)	
	U	500	0.500	3.0×10^{-5}	1.10	1.28	1.71		1.99	9.7	NF	(b)	
	U	500	1.00	3.0×10^{-5}	1.23	1.42	1.75	1.96	1.96	11.0	NF	(b)	
	U	500	1.50	3.0×10^{-5}	1.23	1.40	1.66		1.91	9.7	NF	(b)	
Limestone, "Hauptrogenstein", oolitic	P	500	3.00	1.0×10^{-4}	1.36	2.01	2.18	2.32	2.54	20.0	NF	(b)	261
	P	600	3.00	1.0×10^{-3}	1.36	1.52	1.68	1.76	1.84	20.0	NF	(b)	
	P	700	3.00	1.0×10^{-3}	0.409	0.627	0.737	0.818	0.968	20.0	NF	(b)	
	P	800	3.00	1.0×10^{-3}	0.372	0.429	0.472	0.503	0.544	20.0	NF	(b)	
	P	800	3.00	1.0×10^{-5}	0.060	0.068	0.082	0.101	0.134	20.0	NF	(b)	
	P	900	3.00	1.0×10^{-5}	0.016	0.018	0.027	0.035	0.055	20.0	NF	(b)	
Limestone, "Indiana"	U	23	0.001	1.0×10^{-4}					0.620		F		127
	U	23	0.001	1.0×10^{-4}					0.640		F		
	U	23	0.250	1.0×10^{-4}					0.990		F		
	U	23	0.500	1.0×10^{-4}			1.23				NF		
	U	23	1.00	1.0×10^{-4}			1.71				NF		
	U	23	2.00	1.0×10^{-4}			2.58				NF		
	U	23	3.00	1.0×10^{-4}			3.12				NF		
	U	23	4.00	1.0×10^{-4}			3.62				NF		
	U	23	5.00	1.0×10^{-4}			4.30				NF		
	U	23	6.00	1.0×10^{-4}			4.93				NF		

Table 2 (continued)
STRESS-STRAIN RELATIONS FROM CONSTANT STRAIN RATE TRIAXIAL TESTS ON ROCKS IN APPARATUS USING FLUID-CONFINING MEDIA

Material	Or	Temp (°C)	Pressure (kbar)	Strain rate (s^{-1})	Differential stress — kbar strain (%) 1	2	5	10	Ultimate strength (kbar)	Total strain (%)	Fault angle (°)	Comment	Ref.
	U	23	7.00	1.0×10^{-4}			5.23				NF		
Limestone, "Indiana", 50% saturated	U	23	0.001	1.0×10^{-4}					0.400		F		127
	U	23	0.001	1.0×10^{-4}					0.420		F		
	U	23	0.250	1.0×10^{-4}					0.880		F		
	U	23	0.500	1.0×10^{-4}			1.18				NF		
	U	23	1.00	1.0×10^{-4}			1.72				NF		
	U	23	2.00	1.0×10^{-4}			2.24				NF		
	U	23	3.00	1.0×10^{-4}			2.92				NF		
	U	23	4.00	1.0×10^{-4}			2.92				NF		
	U	23	5.00	1.0×10^{-4}			2.96				NF		
	U	23	6.00	1.0×10^{-4}			3.20				NF		
	U	23	7.00	1.0×10^{-4}			3.26				NF		
Limestone, "Indiana", 100% saturated	U	23	0.001	1.0×10^{-4}					0.380		F		127
	U	23	0.001	1.0×10^{-4}					0.420		F		
	U	23	0.250	1.0×10^{-4}					0.480		F		
	U	23	0.500	1.0×10^{-4}			0.870				NF		
	U	23	1.00	1.0×10^{-4}			0.870				NF		
	U	23	1.00	1.0×10^{-4}			1.03				NF		
	U	23	2.00	1.0×10^{-4}			0.960				NF		
	U	23	3.00	1.0×10^{-4}			1.13				NF		
	U	23	4.00	1.0×10^{-4}			1.08				NF		
	U	23	5.00	1.0×10^{-4}			1.17				NF		
	U	23	6.00	1.0×10^{-4}			1.38				NF		
	U	23	7.00	1.0×10^{-4}			1.53				NF		
Limestone, "Indiana", medium grained bioclastic	U	23	0.500	3.2×10^{0}	2.01						NF		
	U	23	2.50	3.2×10^{0}	2.07						NF		30
	U	23	0.200	3.1×10^{0}					1.66	0.6	F		
	U	23	2.50	1.9×10^{0}	1.36						NF		
	U	23	0.500	1.9×10^{0}	1.36						NF		
	U	23	0.001	1.7×10^{0}					0.890	0.3	F		
	U	23	0.001	1.7×10^{0}					1.30	0.2	F		

	U	23	0.001	1.5×10^{0}					0.830	0.3			
	U	23	0.001	1.5×10^{0}					1.01	0.3	F		
	U	23	0.001	1.4×10^{0}					0.710	0.3	F		
	U	23	0.001	1.1×10^{0}					1.42	0.3	F		
	U	23	0.001	7.8×10^{-1}					0.590	0.4	F		
	U	23	0.500	5.7×10^{-1}	1.01						NF		
	U	23	0.001	5.6×10^{-1}					0.830	0.5	F		
	U	23	0.001	5.5×10^{-1}					0.590	0.4	F		
	U	23	2.50	5.2×10^{-1}	1.24								
	U	23	0.200	4.1×10^{-1}					0.950	0.6	F		
	U	23	2.50	1.5×10^{-1}	1.07						NF		
	U	23	0.500	1.5×10^{-1}	0.950						NF		
	U	23	0.001	1.2×10^{-1}					0.530	0.4	F		
	U	23	0.001	1.1×10^{-1}					0.590	0.4	F		
	U	23	2.50	5.7×10^{-2}	1.05	1.49			1.76	2.8	NF	(b)	
	U	23	0.050	5.4×10^{-2}					0.650	0.5	F		
	U	23	2.50	5.1×10^{-2}	1.07						NF		
	U	23	0.500	3.8×10^{-2}	0.99	1.12			1.15	2.9	NF	(b)	
	U	23	0.500	3.8×10^{-2}	1.07						NF		
	U	23	0.001	2.9×10^{-2}					0.530	0.3	F		
	U	23	0.100	2.9×10^{-2}	0.580	0.480			0.640	3.0	F	(f)	
	U	23	0.050	2.7×10^{-2}					0.710	0.5	F		
	U	23	0.001	2.6×10^{-2}					0.590	0.3	F	(a)	
Limestone, ''Solenhofen'', Bavaria, Germany	N	600	3.00	6.6×10^{-3}					2.42		NF	(c)	259
	N	600	3.00	3.4×10^{-3}					2.47		NF	(c)	
	N	600	3.00	1.0×10^{-3}	1.23	1.82	2.05	2.08	2.09	13.2	NF	(c)	260
	N	600	3.00	9.9×10^{-4}					2.19		NF	(c)	259
	N	600	3.00	9.8×10^{-4}					2.14		NF	(c)	
	N	600	3.00	9.5×10^{-4}					2.04		NF	(c)	
	N	600	3.00	3.1×10^{-4}					1.89		NF	(c)	
	N	600	3.00	1.5×10^{-4}					1.87		NF	(c)	
	N	600	3.00	8.6×10^{-5}					1.66		NF	(c)	
	N	600	3.00	3.5×10^{-5}					1.31		NF	(c)	
	N	600	3.00	1.0×10^{-5}	0.65	0.89	0.97	1.01	1.02	12.1	NF	(c)	260
	N	600	3.00	1.0×10^{-5}					1.07		NF	(c)	259
	N	600	3.00	1.0×10^{-5}					0.990		NF	(c)	
	N	600	3.00	5.4×10^{-6}					0.640		NF	(c)	
	N	700	3.00	6.3×10^{-3}					1.69		NF	(c)	
	N	700	3.00	2.0×10^{-3}					1.23		NF	(c)	
	N	700	3.00	1.0×10^{-3}	1.02	1.17	1.17	1.17	1.17	12.5	NF	(c)	260
	N	700	3.00	1.0×10^{-3}					1.02		NF	(c)	

Table 2 (continued)
STRESS-STRAIN RELATIONS FROM CONSTANT STRAIN RATE TRIAXIAL TESTS ON ROCKS IN APPARATUS USING FLUID-CONFINING MEDIA

Material	Or	Temp (°C)	Pressure (kbar)	Strain rate (s^{-1})	Differential stress — kbar strain (%) 1	2	5	10	Ultimate strength (kbar)	Total strain (%)	Fault angle (°)	Comment	Ref.
	N	700	3.00	1.0×10^{-3}					1.15		NF (c)		259
	N	700	3.00	9.9×10^{-4}					1.14		NF (c)		
	N	700	3.00	3.5×10^{-4}					0.930		NF (c)		259
	N	700	3.00	1.5×10^{-4}					0.680		NF (c)		
	N	700	3.00	1.1×10^{-4}					0.560		NF (c)		
	N	700	3.00	2.1×10^{-5}					0.350		NF (c)		
	N	700	3.00	1.4×10^{-5}					0.230		NF (c)		
	N	700	3.00	1.0×10^{-5}	0.090	0.110	0.140	0.200	0.220	11.6	NF (c)		260
	N	700	3.00	1.0×10^{-5}					0.210		NF (c)		259
	N	700	3.00	5.2×10^{-6}					0.150		NF (c)		
	N	800	3.00	6.4×10^{-3}					0.860		NF (c)		
	N	800	3.00	6.3×10^{-3}					0.770		NF (c)		
	N	800	3.00	2.0×10^{-3}					0.650		NF (c)		
	N	800	3.00	1.0×10^{-3}	0.460	0.610	0.620	0.630	0.630	12.3	NF (c)		260
	N	800	3.00	1.0×10^{-3}					0.620		NF (c)		259
	N	800	3.00	7.6×10^{-4}					0.620		NF (c)		
	N	800	3.00	6.1×10^{-4}					0.450		NF (c)		
	N	800	3.00	3.5×10^{-4}					0.400		NF (c)		
	N	800	3.00	1.4×10^{-4}					0.180		NF (c)		
	N	800	3.00	1.3×10^{-4}					0.310		NF (c)		
	N	800	3.00	1.1×10^{-4}					0.220		NF (c)		260
	N	800	3.00	1.1×10^{-4}					0.150		NF (c)		
	N	800	3.00	1.0×10^{-4}					0.145		NF (c)		
	N	800	3.00	8.5×10^{-5}					0.240		NF (c)		259
	N	800	3.00	7.4×10^{-5}					0.120		NF (c)		260
	N	800	3.00	7.1×10^{-5}					0.120		NF (c)		
	N	800	3.00	7.1×10^{-5}					0.120		NF (c)		259
	N	800	3.00	3.3×10^{-5}					0.090		NF (c)		260
	N	800	3.00	3.2×10^{-5}					0.090		NF (c)		
	N	800	3.00	3.1×10^{-5}					0.070		NF (c)		
	N	800	3.00	1.1×10^{-5}					0.040		NF (c)		

N	800	3.00	1.0×10^{-5}	0.020	0.030	0.030	0.040	0.040	11.5	NF	(c)	
N	800	3.00	9.4×10^{-6}					0.090		NF	(c)	259
N	800	3.00	9.0×10^{-6}					0.090		NF	(c)	
N	900	3.00	7.1×10^{-3}					0.370		NF	(c)	
N	900	3.00	1.0×10^{-3}	0.230	0.270	0.280	0.280	0.280	12.6	NF	(c)	260
N	900	3.00	6.7×10^{-4}					0.280		NF	(c)	259
N	900	3.00	3.8×10^{-4}					0.130		NF	(c)	
N	900	3.00	2.5×10^{-4}					0.060		NF	(c)	260
N	900	3.00	1.2×10^{-4}					0.050		NF	(c)	
N	900	3.00	9.9×10^{-5}					0.030		NF	(c)	
N	900	3.00	2.1×10^{-5}					0.020		NF	(c)	
N	900	3.00	1.1×10^{-5}					0.010		NF	(c)	
U	20	0.001	1.0×10^{-1}					4.09	0.9	F	(b)	195
U	20	1.00	1.0×10^{-1}	4.77				5.66	1.8	F	(b)	
U	20	1.00	1.0×10^{-1}	4.77	5.91	5.83		6.00	5.5	NF	(d)	
U	20	2.00	1.0×10^{-1}	4.78	6.79	7.55		7.65	5.8	NF	(b)	
U	20	3.00	1.0×10^{-1}	4.52	5.57	7.56		8.13	9.0	NF	(b)	
U	20	1.00	4.0×10^{-4}	3.24	3.48	3.41	3.38	3.49	12.3		(c) Wet	250
U	20	0.300	3.0×10^{-4}	1.45				2.89	1.8		(f) Wet	
U	20	0.600	3.0×10^{-4}	2.89	3.03	2.73	2.26	3.05	9.1		(f) Wet, 10% data extrapolated	
U	20	0.600	3.0×10^{-4}	2.46	2.87	2.92	2.56	2.97	12.5		(g) Wet, lamda = 0.5	
U	20	0.600	1.6×10^{-4}	2.81	3.09	2.96	2.22	3.11	12.2		(g) Wet, lamda = 0.5	250
U	20	0.600	4.0×10^{-5}	3.13	3.14	2.85	2.14	3.16	12.1		(g) Wet	
U	20	0.300	3.0×10^{-5}	2.50	1.90			2.65	4.3		(g) Wet	
U	20	0.600	3.0×10^{-5}	2.82	3.12	2.56	1.25	3.12	11.2		(g) Wet, lamda = 0.5	
U	20	3.00	3.0×10^{-5}	3.32	3.92	4.72	5.72	5.91	11.6	NF	(b)	251
U	20	3.00	3.0×10^{-5}	3.32	3.80	4.37	5.02	5.18	11.7	NF	(b) Wet	
U	20	3.00	3.0×10^{-5}	2.82	3.19	3.46	3.82	3.91	11.7	NF	(b) Wet, lamda = 0.5	
U	20	3.00	3.0×10^{-5}	2.57	2.68	2.67		2.68	7.7	NF	(c) Wet, lamda = 0.9	
U	20	3.00	3.0×10^{-5}	2.04	1.40			2.43	2.3	F	(d) Wet, lamda = 1.0	
U	20	1.00	4.5×10^{-6}	3.03	3.26	3.36	3.46	3.46	12.3		(b) Wet	250
U	20	0.300	4.2×10^{-6}	2.61	2.55	1.36		2.64	8.3		(f) Wet	
U	20	0.600	3.0×10^{-6}	2.75	3.13	3.00	1.45	3.13	9.7		(g) Wet, 10% data extrapolated	
U	20	0.600	1.5×10^{-6}	2.63	2.71	2.00		2.71	8.0		(g) Wet, lamda = 0.5	
U	20	0.600	6.0×10^{-7}	2.61	2.66	1.30		2.72	7.6		(g) Wet, lamda = 0.5	
U	20	1.00	3.3×10^{-7}	2.88	3.15			3.26	3.8		(b) Wet	
U	20	0.300	3.0×10^{-7}	2.55	2.64	1.54		2.69	8.1		(f) Wet	
U	20	0.600	3.0×10^{-7}	2.49	2.88	2.85	1.96	2.89	9.7		(g) Wet, 10% data extrapolated	

Table 2 (continued)
STRESS-STRAIN RELATIONS FROM CONSTANT STRAIN RATE TRIAXIAL TESTS ON ROCKS IN APPARATUS USING FLUID-CONFINING MEDIA

Material	Or	Temp (°C)	Pressure (kbar)	Strain rate (s^{-1})	Differential stress — kbar strain (%)				Ultimate strength (kbar)	Total strain (%)	Fault angle (°)	Comment	Ref.
					1	2	5	10					
	U	20	1.00	6.0×10^{-8}	2.58	3.03	3.33	3.54	3.54	12.5		(b) Wet	
	U	20	0.300	4.5×10^{-8}	2.23	2.60	1.89		2.63	7.3		(g) Wet	
	U	20	0.600	4.2×10^{-8}	2.53	2.75	2.74	1.89	2.76	9.7		(g) Wet, 10% data extrapolated	
	U	23	0.001	1.0×10^{-4}					3.47				117
	U	23	0.001	1.0×10^{-4}					3.98		F		
	U	23	0.001	1.0×10^{-4}					4.09		F		
	U	23	0.001	1.0×10^{-4}					3.98		F		
	U	23	0.001	1.0×10^{-4}					2.75		F		
	U	23	0.001	1.0×10^{-4}					2.72		F		
	U	23	0.200	1.0×10^{-4}					4.73		F		
	U	23	0.300	1.0×10^{-4}					3.53		25		
	U	23	0.350	1.0×10^{-4}					3.68		24		
	U	23	0.350	1.0×10^{-4}					4.30		16		
	U	23	0.350	1.0×10^{-4}					3.71		25		
	U	23	0.400	1.0×10^{-4}					4.93		20		
	U	23	0.600	1.0×10^{-4}					4.85		22		
	U	23	0.690	1.0×10^{-4}					4.44		26		
	U	23	0.690	1.0×10^{-4}					4.40				
	U	23	0.690	1.0×10^{-4}					3.93		26		
	U	23	0.760	1.0×10^{-4}					4.75		25		
	U	23	0.800	1.0×10^{-4}					5.14		NF		
	U	23	0.980	1.0×10^{-4}					3.92		NF		
	U	23	1.00	1.0×10^{-4}					4.65				
	U	23	1.00	1.0×10^{-4}					5.35		NF		
	U	23	1.03	1.0×10^{-4}					4.90		F		
	U	23	1.03	1.0×10^{-4}					4.60		17		
	U	23	1.03	1.0×10^{-4}					4.85		31		
	U	23	1.03	1.0×10^{-4}					4.63		30		
	U	23	1.04	1.0×10^{-4}					4.27		F		
	U	23	1.27	1.0×10^{-4}					5.03		30		

U	23	1.38	1.0×10^{-4}					5.25		26		
U	23	1.38	1.0×10^{-4}					4.59		NF		
U	23	1.53	1.0×10^{-4}					5.07				
U	23	1.96	1.0×10^{-4}					5.98		NF		117
U	23	2.94	1.0×10^{-4}					5.98		NF		
U	23	3.00	1.0×10^{-4}					4.76		NF		
U	23	3.92	1.0×10^{-4}					7.94		NF		
U	23	5.00	1.0×10^{-4}					7.64		NF		
U	23	1.00	-1.0×10^{-4}					1.11		0		
U	23	1.00	-1.0×10^{-4}					1.15		0		
U	23	2.00	-1.0×10^{-4}					2.12		0		
U	23	2.00	-1.0×10^{-4}					2.11		0		
U	23	3.00	-1.0×10^{-4}					3.05		0		
U	23	3.00	-1.0×10^{-4}					3.14		0		
U	23	4.06	-1.0×10^{-4}					4.02		10		
U	23	5.08	-1.0×10^{-4}					4.66		25		
U	23	7.10	-1.0×10^{-4}					5.84		F		
U	23	7.62	-1.0×10^{-4}					6.30		22		
U	25	0.750	1.0×10^{-7}					4.85		31		
U	25	1.00	1.0×10^{-7}					4.80		27		
U	25	5.90	-1.0×10^{-7}					4.95		19		
U	25	6.40	-1.0×10^{-7}					5.45		22		
U	150	0.300	1.0×10^{-7}					3.80		26		
U	150	0.400	1.0×10^{-7}					4.35		26		
U	150	0.500	1.0×10^{-7}					4.30		NF		
U	150	0.750	1.0×10^{-7}					4.85		NF		
U	200	1.50	3.0×10^{-5}	2.85	3.20	3.61	4.11	4.30	11.7	NF	(b)	251
U	200	1.50	3.0×10^{-5}	1.94	2.29	2.49	2.76	2.80	11.3	NF	(b) Wet, lamda = 0.1	
U	200	1.50	3.0×10^{-5}	1.90	2.09	2.30	2.48	2.49	11.2	NF	(b) Wet, lamda = 0.5	
U	200	1.50	3.0×10^{-5}	2.17	2.29	2.27	2.17	2.29	10.8	NF	(d) Wet, lamda = 0.9	
U	200	1.50	3.0×10^{-5}	1.57	0.89			1.74	2.2	F	(d) Wet, lamda = 1.0	
U	225	0.001	1.0×10^{-7}					4.35		30		117
U	300	2.20	3.0×10^{-5}	2.67	3.04	3.54	4.07	4.14	11.2	NF	(b)	251
U	300	2.20	3.0×10^{-5}	1.58	2.03	2.44	2.81	2.86	11.3	NF	(b) Wet, lamda = 0.1	
U	300	2.20	3.0×10^{-5}	1.65	2.18	2.67	2.77	2.78	10.8	NF	(b) Wet, lamda = 0.5	
U	300	2.20	3.0×10^{-5}	1.95	2.12	2.21	2.22	2.22	11.3	NF	(c) Wet, lamda = 0.9	
U	300	2.20	3.0×10^{-5}	1.84	0.57			1.84	2.0	F	(d) Wet, lamda = 1.0	
U	300	0.001	1.0×10^{-7}					3.85		NF		117
U	300	0.300	1.0×10^{-7}					4.50		NF		
U	300	0.400	1.0×10^{-7}					4.00		NF		
U	300	0.500	1.0×10^{-7}					4.05		NF		

Table 2 (continued)
STRESS-STRAIN RELATIONS FROM CONSTANT STRAIN RATE TRIAXIAL TESTS ON ROCKS IN APPARATUS USING FLUID-CONFINING MEDIA

Material	Or	Temp (°C)	Pressure (kbar)	Strain rate (s^{-1})	Differential stress — kbar strain (%) 1	2	5	10	Ultimate strength (kbar)	Total strain (%)	Fault angle (°)	Comment	Ref.
	U	300	4.00	-1.0×10^{-7}					3.92		0		
	U	300	4.00	-1.0×10^{-7}					4.04		0		
	U	400	3.00	3.0×10^{-5}	2.17	2.60	3.32	3.96	4.09	12.5	NF	(b)	251
	U	400	3.00	3.0×10^{-5}	2.17	2.48	2.67	2.67	2.67	11.5	NF	(c) Wet, lamda = 0.1	
	U	400	3.00	3.0×10^{-5}	1.99	2.35	2.71	2.94	2.99	11.8	NF	(b) Wet, lamda = 0.5	
	U	400	3.00	3.0×10^{-5}	1.97	2.16	2.30	2.37	2.37	10.2	NF	(b) Wet, lamda = 0.9	
	U	400	3.00	3.0×10^{-5}	1.89				1.93	1.9	F	(d) Wet, lamda = 1.0	
	U	400	2.50	-1.0×10^{-7}					2.06		0		117
	U	400	3.00	-1.0×10^{-7}					2.83		0		
	U	400	3.00	-1.0×10^{-7}					2.67		0		
	U	400	3.50	-1.0×10^{-7}					3.26		0		
	U	450	2.00	-1.0×10^{-7}					1.26		0		
	U	500	0.001	3.0×10^{-5}	1.35	2.22	2.54	2.75	2.79	10.7	NF	(b)	251
	U	500	0.250	3.0×10^{-5}	2.21	2.56	2.87	3.11	3.12	11.2	NF	(b)	251
	U	500	0.500	3.0×10^{-5}	2.21	2.61	2.97	3.23	3.24	11.0	NF	(b)	
	U	500	1.50	3.0×10^{-5}	2.21	2.51	2.81	3.11	3.18	11.3	NF	(b)	
	U	500	2.00	-1.0×10^{-7}					0.480		NF		117
Limestone, "Yule Marble", "Leadville Limestone", Colorado	N	25	5.00	-3.3×10^{-7}				2.94					135
	N	25	5.00	-3.3×10^{-6}				3.02					
	N	25	5.00	-3.3×10^{-5}				3.13					
	N	25	5.00	-3.3×10^{-4}				3.15					
	N	25	5.00	-4.0×10^{-3}				3.14					
	N	25	5.00	-4.0×10^{-3}				3.27					
	N	25	5.00	-4.0×10^{-2}				3.18					
	N	25	5.00	-4.0×10^{-1}				3.28		8.0		10% Data extrapolated	
	N	300	5.00	-3.3×10^{-8}				1.76					
	N	300	5.00	-3.3×10^{-7}				1.87					
	N	300	5.00	-3.3×10^{-6}				1.96					
	N	300	5.00	-3.3×10^{-5}				2.06		8.0		10% Data extrapolated	
	N	300	5.00	-3.3×10^{-4}				2.11					
	N	300	5.00	-4.0×10^{-3}				2.13					

N	300	5.00	-4.0×10^{-2}			2.15			
N	300	5.00	-4.0×10^{-1}			2.21			
N	400	5.00	-3.3×10^{-8}			1.04			
N	400	5.00	-3.3×10^{-7}			1.21			
N	400	5.00	-3.3×10^{-6}			1.46			
N	400	5.00	-3.3×10^{-5}			1.66			
N	400	5.00	-3.3×10^{-4}			1.77			
N	400	5.00	-4.0×10^{-3}			1.87			
N	400	5.00	-4.0×10^{-2}			1.93			
N	400	5.00	-4.0×10^{-2}			1.88			
N	400	5.00	-4.0×10^{-1}			1.94			
N	500	5.00	-3.3×10^{-8}			0.450			
N	500	5.00	-3.3×10^{-7}			0.690			
N	500	5.00	-3.3×10^{-6}			0.820			
N	500	5.00	-3.3×10^{-5}			1.10			
N	500	5.00	-1.8×10^{-4}	0.96	1.13	1.28		5% Data interpolated	
N	500	5.00	-1.8×10^{-4}	0.92	1.11	1.28		5% Data interpolated	
N	500	5.00	-3.3×10^{-4}			1.34			
N	500	5.00	-3.3×10^{-4}			1.32			
N	500	5.00	-4.0×10^{-3}			1.45	6.0	10% Data extrapolated	
N	500	5.00	-4.0×10^{-3}			1.42			
N	500	5.00	-4.0×10^{-2}			1.55			
N	500	5.00	-4.0×10^{-1}			1.69			
N	600	5.00	-1.9×10^{-7}	0.320	0.330	0.330		5% Data interpolated	
N	600	5.00	-1.9×10^{-6}	0.410	0.450	0.460		5% Data interpolated	
N	600	5.00	-1.9×10^{-5}	0.550	0.590	0.620		5% Data interpolated	
N	600	5.00	-1.9×10^{-4}	0.600	0.680	0.750		5% Data interpolated	
N	600	5.00	-1.9×10^{-3}	0.750	0.880	0.980		5% Data interpolated	
N	700	5.00	-1.9×10^{-7}	0.230	0.230	0.230		5% Data interpolated	
N	700	5.00	-1.9×10^{-6}	0.280	0.280	0.280		5% Data interpolated	
N	700	5.00	-1.9×10^{-5}	0.350	0.380	0.390		5% Data interpolated	
N	700	5.00	-1.9×10^{-4}	0.440	0.470	0.490		5% Data interpolated	
N	700	5.00	-1.9×10^{-3}	0.580	0.630	0.680		5% Data interpolated	135
N	700	5.00	-1.9×10^{-3}	0.560	0.610	0.660		5% Data interpolated	
N	800	5.00	-1.9×10^{-7}	0.140	0.150	0.150		5% Data interpolated	
N	800	5.00	-1.9×10^{-6}	0.190	0.210	0.210		5% Data interpolated	
N	800	5.00	-1.9×10^{-6}	0.170	0.190	0.200		5% Data interpolated	
N	800	5.00	-1.9×10^{-5}	0.240	0.250	0.260		5% Data interpolated	
N	800	5.00	-1.9×10^{-4}	0.320	0.340	0.360		5% Data interpolated	
N	800	5.00	-1.9×10^{-3}	0.420	0.460	0.490		5% Data interpolated	
P	300	5.00	-3.3×10^{-8}			2.24			

Table 2 (continued)
STRESS-STRAIN RELATIONS FROM CONSTANT STRAIN RATE TRIAXIAL TESTS ON ROCKS IN APPARATUS USING FLUID-CONFINING MEDIA

Material	Or	Temp (°C)	Pressure (kbar)	Strain rate (s^{-1})	Differential stress — kbar strain (%) 1	2	5	10	Ultimate strength (kbar)	Total strain (%)	Fault angle (°)	Comment	Ref.
	P	300	5.00	-3.3×10^{-7}				2.48					
	P	300	5.00	-3.3×10^{-6}				2.65		6.5		10% Data extrapolated	
	P	300	5.00	-3.3×10^{-6}				2.71		6.0		10% Data extrapolated	
	P	300	5.00	-3.3×10^{-6}				2.70		9.2		10% Data extrapolated	
	P	300	5.00	-3.3×10^{-5}				2.79					
	P	300	5.00	-3.3×10^{-4}				2.93					
	P	350	5.00	-3.3×10^{-7}				2.12					
	P	350	5.00	-3.3×10^{-7}				2.01					
	P	400	5.00	-3.3×10^{-8}				1.19					
	P	400	5.00	-3.3×10^{-7}				1.50					
	P	400	5.00	-3.3×10^{-7}				1.53					
	P	400	5.00	-3.3×10^{-6}				1.86					
	P	400	5.00	-3.3×10^{-5}				2.22					
	P	400	5.00	-3.3×10^{-4}				2.53					
	P	450	5.00	-3.3×10^{-7}				1.07					
	P	500	5.00	-3.3×10^{-8}				0.540		9.5		10% Data extrapolated	
	P	500	5.00	-3.3×10^{-7}				0.770					
	P	500	5.00	-3.3×10^{-7}				0.830					
	P	500	5.00	-3.3×10^{-6}				1.08					
	P	500	5.00	-3.3×10^{-5}				1.43					
	P	500	5.00	-3.3×10^{-4}				1.68					
	P	500	5.00	-4.0×10^{-3}				1.97					
	P	500	5.00	-4.0×10^{-2}				2.14					
	P	500	5.00	-4.0×10^{-1}				2.27					
	P	600	5.00	-1.9×10^{-6}		0.460	0.490	0.510				5% Data interpolated	
	P	600	5.00	-1.9×10^{-5}		0.620	0.650	0.670				5% Data interpolated	
	P	600	5.00	-1.9×10^{-4}		0.880	0.950	0.980				5% Data interpolated	
	P	600	5.00	-1.9×10^{-3}		1.02	1.16	1.22				5% Data interpolated	
	P	700	5.00	-1.9×10^{-6}		0.310	0.330	0.340				5% Data interpolated	
	P	700	5.00	-1.9×10^{-5}		0.420	0.440	0.450				5% Data interpolated	
	P	700	5.00	-1.9×10^{-4}		0.540	0.570	0.580				5% Data interpolated	

	P	700	5.00	-1.9×10^{-3}		0.710	0.770	0.800				5% Data interpolated	
	P	800	5.00	-1.9×10^{-6}		0.210	0.210	0.220				5% Data interpolated	
	P	800	5.00	-1.9×10^{-5}		0.280	0.290	0.290				5% Data interpolated	
	P	800	5.00	-1.9×10^{-4}		0.370	0.390	0.400				5% Data interpolated	
	P	800	5.00	-1.9×10^{-3}		0.490	0.530	0.550				5% Data interpolated	
Limestone, Lithographic (presumed "Solenhofen")	U	20	1.00	4.0×10^{-4}	3.94	4.24	4.28	4.16	4.31		NF	(d)	82
	U	20	2.00	4.0×10^{-4}	3.95	4.60	5.00	5.40			NF	(b)	
	U	20	4.00	4.0×10^{-4}	3.73	4.26	5.16	6.21			NF	(b)	
	U	20	6.00	4.0×10^{-4}	2.90	3.96	5.30	6.61			NF	(b)	
	U	20	8.00	4.0×10^{-4}	2.55	3.54	5.31	6.86			NF	(b)	
Magnetite ore, Kiruna-type, Scandinavia, 3% (series 4) to 35% (series 3) impurities	U	20	0.001	5.0×10^{-4}					1.04	0.9	F	(a) Series 3	210
	U	20	0.001	5.0×10^{-4}					0.760	0.9	F	(a) Series 4	
	U	20	0.500	5.0×10^{-4}	5.40	6.60			6.73	2.0	F	(d) Series 3	
	U	20	0.500	5.0×10^{-4}	3.78	5.02			5.14	2.1	F	(d) Series 4	
	U	20	1.00	5.0×10^{-4}	5.02	6.53			6.70	2.7	22	(d) Series 4	
	U	20	1.00	5.0×10^{-4}	7.49	8.88			9.05	2.5	F	(d) Series 3	
	U	20	1.50	5.0×10^{-4}	5.32	6.97			7.78	3.5	F	(d) Series 4	
	U	20	1.50	5.0×10^{-4}	7.68	9.83			9.87	2.2	F	(b) Series 3	
	U	20	2.00	5.0×10^{-4}	6.87	8.89	9.83		9.87	7.4	F	(d) Series 4	
	U	20	2.00	5.0×10^{-4}		10.25			11.10	2.9	F	(b) Series 3	
	U	20	2.50	5.0×10^{-4}		11.14			12.04	3.5	F	(d) Series 3	
	U	20	3.00	5.0×10^{-4}	6.17	8.87	11.05		12.17	16.5	F	(d) Series 4	
	U	20	3.00	5.0×10^{-4}	9.14	12.32			12.50	2.8	F	(d) Series 3	
	U	20	3.00	5.0×10^{-4}		6.81	8.54	9.17	9.19	12.8	NF	(d) Series 1	
	U	20	3.00	5.0×10^{-4}		7.07	9.73	10.23	10.29	11.1	NF	(d) Series 6	
	U	20	3.00	5.0×10^{-4}	5.39	8.29	10.85		11.24	8.8	NF	(d) Series 2	
	U	20	3.00	5.0×10^{-4}	7.57	11.11			11.55	3.2	F	(d) Series 8	
	U	20	4.00	5.0×10^{-4}		10.01	12.30	14.12	14.87	16.2	NF	(b) Series 4	
	U	20	4.00	5.0×10^{-4}	10.79	13.20	13.83	14.00	14.02	11.1	F	(d) Series 3	
	U	20	5.00	5.0×10^{-4}		10.02	14.49	15.72	18.88	19.5	NF	(b) Series 4	
	U	20	5.00	5.0×10^{-4}		12.34	13.98	15.72	16.08	18.2	NF	(d) Series 3	
	U	20	5.00	5.0×10^{-4}	5.40	8.13	9.94	11.55	12.89	19.2	NF	(d) Series 1	
	U	20	5.00	5.0×10^{-4}	4.08	7.35	11.32	13.23	15.50	19.2	NF	(b) Series 6	
	U	20	5.00	5.0×10^{-4}	7.57	10.32	12.59	14.24	15.71	15.2	NF	(b) Series 2	
	U	20	5.00	5.0×10^{-4}	6.70	9.42	12.36	14.70	16.98	17.0	NF	(b) Series 8	
	U	25	2.00	5.0×10^{-4}	6.80	8.69	9.79		9.86	7.5	NF	(d) Series 4	
	U	25	2.00	5.0×10^{-4}	7.57	10.59			11.42	3.4	F	(d) Series 3	
	U	25	3.00	5.0×10^{-4}	6.14	8.05	10.42	11.10	11.20	18.9	NF	(d) Series 7	
	U	25	3.00	5.0×10^{-4}	7.57	10.82			11.13	2.8	F	(b) Series 8	
	U	25	3.00	5.0×10^{-4}	10.59	11.69			12.50	2.5	F	(d) Series 3	
	U	300	2.00	5.0×10^{-4}	1.25	1.53	6.42	7.66	7.71	11.7	NF	(d) Series 4	

Table 2 (continued)
STRESS-STRAIN RELATIONS FROM CONSTANT STRAIN RATE TRIAXIAL TESTS ON ROCKS IN APPARATUS USING FLUID-CONFINING MEDIA

Material	Or	Temp (°C)	Pressure (kbar)	Strain rate (s^{-1})	Differential stress — kbar strain (%) 1	2	5	10	Ultimate strength (kbar)	Total strain (%)	Fault angle (°)	Comment	Ref.
	U	300	2.00	5.0×10^{-4}	1.34	3.66			7.49	3.9	F	(d) Series 3	
	U	300	3.00	5.0×10^{-4}	4.40	6.14	7.54	8.92	10.00	15.7	NF	(b) Series 7	
	U	300	3.00	5.0×10^{-4}	1.83	4.36	9.63	9.83	10.08	9.8	NF	(d) Series 3, 10% data extrapolated	
	U	300	3.00	5.0×10^{-4}	4.54	6.55	8.83	10.28	10.61	11.2	NF	(b) Series 8	
MgO, sintered synthetic, 0.3—0.5 mm grains, 1% porosity	U	23	0.001	1.0×10^{-3}					2.36	0.3	F	(a) Thick rubber jacket	228
	U	23	1.00	1.0×10^{-3}	4.19	4.78			4.78	2.1	F	(b) Thick rubber jacket	
	U	23	1.50	1.0×10^{-3}	6.18	6.46			6.46	4.2	F	(d) Thick rubber jacket	
	U	23	2.00	1.0×10^{-3}	6.97				7.13	1.4	F	(b) Thick rubber jacket	
	U	23	2.00	1.0×10^{-3}	4.49	4.78	5.17		5.17	5.0	NF	(b) Latex jacket	
	U	23	3.00	1.0×10^{-3}	7.33	7.87	8.42	8.71	8.71	13.0	NF	(d) Thick rubber jacket	
	U	23	5.00	1.0×10^{-3}	8.02	8.67	9.51	10.45	10.78	12.3	NF	(b) Thick rubber jacket	
	U	23	5.00	1.0×10^{-3}	8.39	8.71	9.24		9.65	8.0	NF	(b) Latex jacket	
	U	23	8.00	1.0×10^{-3}	10.78	11.49	12.38	13.75	14.61	13.5	NF	(b) Thick rubber jacket	
	U	23	10.00	1.0×10^{-3}	11.17	11.90	13.67	16.00	16.38	11.0	NF	(b) Thick rubber jacket	
	U	23	10.00	1.0×10^{-3}	8.92	9.44	10.45	11.71	12.08	11.8	NF	(b) Later jacket	
	U	100	2.00	1.0×10^{-3}	6.52	6.83	7.30	7.30	7.33	10.8	NF	(c)	
	U	100	5.00	1.0×10^{-3}	8.48	8.67	9.24	9.88	10.13	11.7	NF	(b)	228
	U	300	2.00	1.0×10^{-3}	7.22	7.81	8.39	8.14	8.42	9.5	NF	(d) 10% Data extrapolated	
	U	300	5.00	1.0×10^{-3}	9.77	10.64	12.11	12.91	12.98	10.8	NF	(b)	
	U	400	2.00	1.0×10^{-3}	7.81	8.25	8.67		8.67	4.5	NF	(b) 5% Data extrapolated	
	U	400	5.00	1.0×10^{-3}	7.89	8.87	10.66	12.08	12.15	10.7	NF	(b)	
	U	500	2.00	1.0×10^{-3}	6.12	6.91	7.89	7.78	8.02	9.5	NF	(d) 10% Data extrapolated	
	U	500	5.00	1.0×10^{-3}	7.02	7.92	9.79	10.92	10.96	10.0	NF	(b)	
	U	600	2.00	1.0×10^{-3}	5.11	5.87	7.05	7.98	8.09	10.8	NF	(b)	
	U	600	5.00	1.0×10^{-3}	5.98	6.69	7.95	8.78	8.89	10.5	NF	(b)	
	U	750	2.00	1.0×10^{-3}	4.24	4.94	6.26	7.13	7.08	9.2	NF	(b) 10% Data extrapolated	
	U	750	5.00	1.0×10^{-3}	4.02	4.75	6.15	7.13	7.22	10.6	NF	(b)	
MgO, sintered synthetic, 10—15-μm grains, nearly theoretical density	U	23	0.001	1.0×10^{-3}					2.53	0.3	F	(a) Latex jacket	228
	U	23	0.500	1.0×10^{-3}					8.78	0.6	F	(a) Thick rubber jacket	
	U	23	0.500	1.0×10^{-3}					6.94	0.6	F	(a) Latex jacket	

	U	23	1.00	1.0×10^{-3}					9.79	0.6	F	(a) Thick rubber jacket	
	U	23	1.00	1.0×10^{-3}					8.14	0.6	F	(a) Latex jacket	
	U	23	2.00	1.0×10^{-3}					12.36	0.6	F	(a) Thick rubber jacket	
	U	23	2.00	1.0×10^{-3}					9.49	0.7	F	(a) Latex jacket	
	U	23	3.00	1.0×10^{-3}			11.21	11.28	13.02	10.0	NF	(e) Thick rubber jacket U.S.	
												at upper yield	
	U	23	3.00	1.0×10^{-3}					10.62	0.7	F	(a) Latex jacket	
	U	23	4.00	1.0×10^{-3}		11.65	12.27	12.88	13.02	10.0	NF	(e) Thick rubber jacket U.S.	
												at upper yield	
	U	23	4.00	1.0×10^{-3}	12.36	12.11	11.87		12.41	7.2	F	(g) Latex jacket	
	U	23	5.00	1.0×10^{-3}		13.30	14.03	14.91	14.94	11.6	NF	(e) Thick rubber jacket U.S.	
												at 11.6% strain	
	U	23	5.00	1.0×10^{-3}	13.16	13.25	14.03	14.88	15.87	12.6	F	(e) Latex jacket	
	U	23	8.00	1.0×10^{-3}	17.19	17.44	17.42	18.30	18.30	10.0	NF	(e) Thick rubber jacket U.S.	
												at 10% strain	
	U	23	8.00	1.0×10^{-3}	15.02	15.00	16.84	17.89	18.48	12.7	F	(e) Latex jacket	
MgO, sintered synthetic, 30-μm	U	23	0.001	1.0×10^{-3}					4.21	0.3	F	(a) Latex jacket	228
grains, 97% of theoretical density	U	23	1.00	1.0×10^{-3}	7.42	7.58			7.58	3.4	F	(b) Latex jacket	
	U	23	2.50	1.0×10^{-3}	8.23	8.85	9.51		9.54	8.0	F	(d) Latex jacket	
	U	23	5.00	1.0×10^{-3}	11.46	12.19	12.95	13.25	13.28	11.3	NF	(b) Latex jacket	
	U	23	8.00	1.0×10^{-3}	14.91	15.50	16.00		16.28	8.7	NF	(b) Latex jacket	
	U	23	10.00	1.0×10^{-3}	16.79	17.16	17.64		17.70	7.5	NF	(b) Latex jacket	
Migmatite, "Engel and Engel", St.	N	150	1.00	1.7×10^{-4}					7.25	2.9	25	(a)	36
Lawrence Co., New York	N	500	5.00	1.7×10^{-4}					9.86	16.0	35	(d)	
	P	150	1.00	1.7×10^{-4}					7.58	2.2	29	(a)	
	P	500	5.00	1.7×10^{-4}					6.70	11.3	27	(b)	
Monzonite, San Juan Co., Colorado	N	150	1.00	1.7×10^{-4}					6.40	7.5	26	(d)	36
	N	500	5.00	1.7×10^{-4}					9.64	14.0	35	(d)	
Peridotite, "Fidalgo" Fidalgo Is-	U	25	3.50	7.0×10^{-4}					7.90		NF		240
land, partly serpentinized	U	200	3.50	7.0×10^{-4}					6.90		NF	Preheated 1.0 hr	
	U	300	3.50	7.0×10^{-4}					6.40		NF	Preheated 1.0 hr	
	U	355	3.50	7.0×10^{-4}					4.90			Preheated 1.0 hr	
	U	405	3.50	7.0×10^{-4}					2.80		F	Preheated 1.0 hr	
Peridotite, olivine and pyroxene,	N	20	3.10	1.0×10^{-5}	2.74	5.79	8.00	7.79	8.41	10.0	F	(f) Undrained test	213
60% altered to serpentine, amphi-	N	20	5.52	1.0×10^{-5}	2.93	6.33	12.47	13.28	13.26	9.5	NF	(b) Undrained test, 10% data	
bole, and brucite, (serpentine dehy-												extrapolated	
drated above 702°C	N	170	5.52	1.0×10^{-5}	2.71	5.64	9.36		9.58	6.9	NF	(b) Undrained test	
	N	320	3.10	1.0×10^{-5}	2.45	4.50	5.79		5.86	6.0	NF	(b) Undrained test	
	N	420	3.10	1.0×10^{-5}	1.80	2.69	3.38		3.41	5.1	NF	(b) Undrained test	
	N	420	5.52	1.0×10^{-5}	2.24	3.58	5.07		5.11	6.2	NF	(b) Undrained test	
	N	520	5.52	1.0×10^{-5}	1.13	1.76			2.38	3.7	NF	(e) Undrained test	

Table 2 (continued)
STRESS-STRAIN RELATIONS FROM CONSTANT STRAIN RATE TRIAXIAL TESTS ON ROCKS IN APPARATUS USING FLUID-CONFINING MEDIA

Material	Or	Temp (°C)	Pressure (kbar)	Strain rate (s^{-1})	Differential stress — kbar strain (%) 1	2	5	10	Ultimate strength (kbar)	Total strain (%)	Fault angle (°)	Comment	Ref.
	N	670	3.10	1.0×10^{-5}	0.178	0.378			0.623	2.7	F	(e) Undrained test	
	N	670	3.10	1.0×10^{-5}	1.62	2.24			2.71	4.8	F	(b)	
	N	780	5.52	1.0×10^{-5}	0.222	0.267			0.644	2.0	F	(f) Undrained test	
Peridotite, serpentinized, Lowell, Vt	N	150	1.00	1.7×10^{-4}					5.08	2.4	29	(b)	36
	N	500	5.00	1.7×10^{-4}					3.97	13.2	29	(d)	
	P	150	1.00	1.7×10^{-4}					4.48	1.7	32	(a)	
	P	500	5.00	1.7×10^{-4}					3.64	22.5	34	(d)	
Peridotite, xenolith from basalt in Snowy Mountains, Australia, olivine, enstatite, and diopside, G.S. = 0.5 mm	U	800	5.00	5.0×10^{-5}	6.64	8.50			9.16	3.9	NF	(b)	237
	U	1000	5.00	2.0×10^{-4}	3.57	4.73			6.32	4.8	NF	(b)	
Pyrite, 89.2% pyrite, 4% sphalerite, 6.8% gangue, 1% porosity, G.S. = 0.04 mm	U	20	0.001	3.0×10^{-5}					2.89		F	(a)	5
	U	20	0.500	3.0×10^{-5}					6.89		F	(a)	
	U	20	1.00	3.0×10^{-5}					8.41		F	(a)	
	U	20	1.95	3.0×10^{-5}					9.97		F	(a)	
Pyrite, 91.7% pyrite, 4.7% sphalerite, 3.6% gangue, 8% porosity, G.S. = 0.04 mm	U	20	0.001	3.0×10^{-5}	1.16				1.38	1.2	F	(a)	5
	U	20	0.620	3.0×10^{-5}	4.17				4.39	1.3	F	(b)	
	U	20	0.790	3.0×10^{-5}	4.94				4.99	1.7	F	(d)	
	U	20	1.46	3.0×10^{-5}	6.15	6.84	6.32	5.52	6.90	10.7	F	(d)	
	U	20	1.93	3.0×10^{-5}	7.09	7.37	6.07	5.79	7.50	10.8	F	(e)	
	U	20	2.43	3.0×10^{-5}	8.44	8.36	8.00	7.42	8.50	11.0	NF	(d)	
	U	20	2.91	3.0×10^{-5}	6.84	7.81	8.79	8.44	8.83	11.1	NF	(d)	
	U	200	0.001	3.0×10^{-5}					1.32	0.7	F	(a)	
	U	200	0.530	3.0×10^{-5}					4.04	0.2	F	(a)	
	U	200	1.38	3.0×10^{-5}	4.95	6.10	4.67	3.85	6.16	11.5	NF	(d)	
	U	200	2.65	3.0×10^{-5}	4.87	5.77	6.96	7.51	7.56	11.3	NF	(b)	
	U	300	0.001	3.0×10^{-5}					1.05	0.5	F	(a)	
	U	300	0.520	3.0×10^{-5}	3.72	3.53	2.54		3.89	6.1	F	(d)	
	U	300	1.46	3.0×10^{-5}	6.01	5.99	4.74	4.03	6.15	11.5	NF	(d)	
	U	300	2.48	3.0×10^{-5}	6.32	6.98	7.14	6.73	7.23	11.3	NF	(f)	
	U	400	0.001	3.0×10^{-5}					0.910	0.7	F	(a)	

	U	400	0.500	3.0×10^{-5}	3.28	2.48			3.56	2.0	F	(d)	5
	U	400	1.43	3.0×10^{-5}	5.77	6.34			6.37	3.2	F	(d)	
	U	400	1.99	3.0×10^{-5}	6.34	6.90	6.07		7.01	6.3	F	(d)	
	U	400	2.43	3.0×10^{-5}	5.21	6.12	7.23	7.48	7.48	11.2	NF	(d)	
	U	400	2.88	3.0×10^{-5}	4.91	5.93	7.50	7.92	7.92	11.3	NF	(b)	
Pyroxenite, Webster, N.C.	N	150	1.00	1.7×10^{-4}					5.31	3.1	28	(b)	36
	N	500	5.00	1.7×10^{-4}					6.40	14.3	36	(d)	
Pyrrhotite ore, Araca Mine, Bolivia, only a few percent impurities	U	100	1.00	7.2×10^{-5}		3.70			4.21	4.9	F		63
	U	200	1.00	7.2×10^{-5}		2.24	2.17	2.38			NF		
	U	200	1.00	7.2×10^{-5}		2.20	2.60	2.85			NF		
	U	300	1.00	7.2×10^{-5}		1.00	1.08	1.24			NF		
	U	400	1.00	7.2×10^{-5}		0.370	0.460	0.610			NF	The 10% data is extrapolated	
Pyrrhotite ore, Strathcona Mine, Sudbury, Ontario, Canada, <10% impurities	U	24	0.500	7.2×10^{-5}		3.21			4.32	3.1	F	(b)	63
	U	24	1.00	7.2×10^{-5}		3.24			5.89	4.9	F	(b)	
	U	24	1.50	7.2×10^{-5}		3.68	6.47	6.89			NF	(d)	
	U	24	2.00	7.2×10^{-5}		2.80	5.73	6.72			NF	(d)	
	U	100	1.00	7.2×10^{-5}		2.96			4.32	4.0	F		
	U	200	0.500	7.2×10^{-5}		2.73			3.69	4.7	F		
	U	200	1.00	7.2×10^{-5}		2.44	3.45		3.45	8.2	F	(d)	
	U	200	1.50	7.2×10^{-5}		2.70	3.45	3.90			NF		
	U	200	2.00	7.2×10^{-5}		2.96	3.86	3.82	3.86		NF		
	U	250	0.500	7.2×10^{-5}		1.86	1.88	1.93			NF		
	U	250	1.00	7.2×10^{-5}		2.33	2.40	2.28	2.43		NF		
	U	250	2.00	7.2×10^{-5}		2.00	2.20	2.42			NF		
	U	300	0.500	7.2×10^{-5}		1.04	1.04	1.01	1.05		NF	(c)	
	U	300	1.00	7.2×10^{-5}		0.570	0.650	0.870			NF	(b)	
	U	300	1.50	7.2×10^{-5}		0.74	0.96	1.26			NF	(b)	
	U	300	2.00	7.2×10^{-5}		1.15	1.31	1.43			NF	(b)	
	U	400	0.500	7.2×10^{-5}		0.630	0.650	0.820			NF		
	U	400	1.00	7.2×10^{-5}		0.520	0.580	0.600			NF	(c)	
	U	500	1.00	7.2×10^{-5}		0.200	0.270	0.280			NF	(c)	
Pyrrhotite ore, Sudbury, Ontario, Canada, 80% pyrrhotite, 8% pentlandite, 8% gangue, 1% pyrite, 1% sphalerite, 1% chalcopyrite	U	20	1.50	3.1×10^{-4}	4.31	4.97	5.55	5.63	5.63	13.1	NF	(c)	7
	U	20	1.50	1.5×10^{-5}	3.52	4.19	4.82	5.01	5.09	12.7	NF	(b)	
	U	20	1.50	3.0×10^{-6}	3.29	3.95	4.65	4.89	4.94	12.6	NF	(b)	
	U	20	1.50	3.0×10^{-7}	3.11	3.77	4.50	4.77	4.86	13.2	NF	(b)	
	U	200	1.50	3.1×10^{-4}	2.53	2.98	3.35	3.41	3.42	13.1	NF	(c)	
	U	200	1.50	3.2×10^{-5}	1.86	2.35	2.80	2.89	2.92	12.7	NF	(c)	
	U	200	1.50	3.3×10^{-6}	1.44	1.94	2.41	2.53	2.53	13.0	NF	(c)	
	U	200	1.50	2.7×10^{-7}	1.12	1.53	1.98	2.07	2.08	13.2	NF	(c)	
	U	300	1.50	3.0×10^{-4}	0.80	1.04	1.33	1.48	1.56	12.8	NF	(b)	
	U	300	1.50	3.1×10^{-5}	0.67	0.86	1.06	1.24	1.33	12.7	NF	(b)	

Table 2 (continued)
STRESS-STRAIN RELATIONS FROM CONSTANT STRAIN RATE TRIAXIAL TESTS ON ROCKS IN APPARATUS USING FLUID-CONFINING MEDIA

Material	Or	Temp (°C)	Pressure (kbar)	Strain rate (s^{-1})	Differential stress — kbar strain (%) 1	2	5	10	Ultimate strength (kbar)	Total strain (%)	Fault angle (°)	Comment	Ref.
	U	300	1.50	3.2×10^{-6}	0.62	0.81	1.02	1.10	1.11	12.5	NF	(b)	
	U	300	1.50	3.3×10^{-7}	0.659	0.763	0.852	0.881	0.889	11.6	NF	(c)	
	U	300	1.50	6.1×10^{-8}	0.511	0.637	0.733	0.741	0.748	11.6	NF	(c)	
	U	400	1.50	3.2×10^{-4}	0.469	0.615	0.755	0.857	0.901	12.5	NF	(b)	
	U	400	1.50	3.6×10^{-5}	0.557	0.637	0.733	0.806	0.828	12.5	NF	(b)	7
	U	400	1.50	3.2×10^{-6}	0.344	0.476	0.630	0.681	0.696	12.5	NF	(c)	
	U	400	1.50	3.5×10^{-7}	0.256	0.350	0.484	0.579	0.601	12.3	NF	(b)	
	U	400	1.50	3.5×10^{-8}	0.271	0.352	0.469	0.505	0.520	12.5	NF	(b)	
Pyrrhotite ore, Sudbury, Ontario,	U	20	0.001	3.0×10^{-5}					0.610	.7	F	(a)	4
Canada, 80% pyrrhotite, 8% pen-	U	20	0.180	3.0×10^{-5}	2.04	2.16	1.31	1.10	2.16	10.7	F	(d) No sharp stress drop	
tlandite, 8% gangue, 1% pyrite, 1%	U	20	0.310	3.0×10^{-5}	2.13	2.41	1.89	1.59	2.44	11.3	F	(d) No sharp stress drop	
sphalerite, 1% chalcopyrite	U	20	0.710	3.0×10^{-5}	2.47	2.91	3.02	2.62	3.14	11.3	F	(d) No sharp stress drop	
	U	20	0.950	3.0×10^{-5}	2.77	3.32	3.72	3.48	3.72	11.3	NF	(d)	
	U	20	1.46	3.0×10^{-5}	3.11	3.48	3.96	4.12	4.15	11.3	NF	(d)	
	U	20	2.17	3.0×10^{-5}	3.35	3.90	4.73	5.24	5.27	11.2	NF	(c)	
	U	20	2.91	3.0×10^{-5}	3.99	4.66	5.61	6.25	6.31	11.1	NF	(b)	
	U	200	0.001	3.0×10^{-5}	0.390				0.470	1.8	F	(b)	
	U	200	0.530	3.0×10^{-5}	1.32	1.57	1.73	1.60	1.74	11.3	NF	(d)	
	U	200	1.48	3.0×10^{-5}	1.70	2.03	2.44	2.55	2.56	11.3	NF	(c)	
	U	200	2.55	3.0×10^{-5}	1.81	2.19	2.64	2.78	3.03	11.3	NF	(b)	
	U	300	0.001	3.0×10^{-5}	0.180	0.250	0.250	0.140	0.260	11.2	F	(d) No sharp stress drop	
	U	300	0.530	3.0×10^{-5}	0.690	0.760	0.800	0.800	0.800	11.0	NF	(c)	
	U	300	1.51	3.0×10^{-5}	0.57	0.74	0.93	1.10	1.12	11.1	NF	(b)	
	U	300	2.61	3.0×10^{-5}	0.88	1.11	1.35	1.46	1.47	11.0	NF	(b)	
	U	400	0.001	3.0×10^{-5}	0.080	0.120	0.120	0.050	0.130	11.1	F	(d) No sharp stress drop	
	U	400	0.520	3.0×10^{-5}	0.330	0.410	0.500	0.550	0.560	11.1	NF	(b)	
	U	400	1.49	3.0×10^{-5}	0.470	0.550	0.650	0.710	0.710	11.2	NF	(b)	
	U	400	2.41	3.0×10^{-5}	0.520	0.600	0.700	0.770	0.780	11.0	NF	(b)	
	U	400	2.88	3.0×10^{-5}	0.570	0.650	0.760	0.820	0.830	11.2	NF	(b)	
Rhyolite, "Tishomingo", Johnston	N	150	1.00	1.7×10^{-4}					8.00	2.6	24	(a)	36
Co., Oklahoma	N	500	5.00	1.7×10^{-4}					10.47	30.5	33	(a)	

Sandstone, "Berea", medium grained	U	23	0.001	2.9×10^{0}					1.18	0.3	F		30
	U	23	0.001	2.3×10^{0}					0.070	0.4	F		
	U	23	0.001	2.3×10^{0}					0.120	0.4	F		
	U	23	0.001	2.0×10^{0}					0.830	0.4	F		
	U	23	0.001	1.9×10^{0}					0.770	0.3	F		
	U	23	0.001	6.5×10^{-1}					0.650	0.4	F		
	U	23	0.001	1.7×10^{-1}					0.530	0.5	F		
	U	23	2.50	4.7×10^{-2}	1.75	2.43			2.64	3.0	NF	(b)	
	U	23	0.001	3.7×10^{-2}					0.470	0.5	F		
	U	23	0.001	3.5×10^{-2}					0.470	0.5	F	(a)	
	U	23	0.500	2.4×10^{-2}	1.71	1.79			1.89	3.0	NF	(d)	
Sandstone, "Fort Union", Rio Blanco County, Colorado, medium to fine-grained graywacke, 50% water saturated, undrained tests	U	25	0.001	1.0×10^{-4}					0.680		F		267
	U	25	0.001	1.0×10^{-4}					0.840		F		
	U	25	0.001	1.0×10^{-4}					0.920		F		
	U	25	1.00	1.0×10^{-4}					4.72		F		
	U	25	2.00	1.0×10^{-4}					6.78		F		
	U	25	3.00	1.0×10^{-4}			8.55				NF		
	U	25	4.00	1.0×10^{-4}			10.08				NF		
	U	25	5.00	1.0×10^{-4}			12.00				NF		
	U	25	6.00	1.0×10^{-4}			13.89				NF		
Sandstone, "Gosford", New South Wales, Australia	U	20	1.00	4.0×10^{-4}	1.79	2.28	2.22	2.17	2.29		NF	(d)	82
	U	20	2.00	4.0×10^{-4}	1.41	2.08	2.97	3.24	3.50		NF	(c)	
	U	20	4.00	4.0×10^{-4}	1.79	2.78	4.77	6.05	6.23		NF	(c)	
	U	20	6.00	4.0×10^{-4}	2.06	3.50	6.58	8.51	8.61		NF	(c)	
Sandstone, "Kayenta", Mesa County, Colorado	U	23	0.001	1.0×10^{-4}					0.320		F		75
	U	23	0.001	1.0×10^{-4}					0.320		F		
	U	23	0.250	1.0×10^{-4}					1.24		F		
	U	23	0.500	1.0×10^{-4}					1.74		F		
	U	23	1.00	1.0×10^{-4}					2.10		F		
	U	23	1.00	1.0×10^{-4}			2.46				NF	Prepressurized to 7 kb	
	U	23	2.00	1.0×10^{-4}			3.88				NF	Prepressurized to 7 kb	
	U	23	3.00	1.0×10^{-4}			3.00				NF		
	U	23	3.00	1.0×10^{-4}			6.08				NF	Prepressurized to 7 kb	
	U	23	3.00	1.0×10^{-4}			5.06				NF	Prepressurized to 7 kb	
	U	23	4.00	1.0×10^{-4}			3.94				NF		
	U	23	4.00	1.0×10^{-4}			6.30				NF	Prepressurized to 7 kb	
	U	23	4.50	1.0×10^{-4}			5.90				NF		
	U	23	4.50	1.0×10^{-4}			6.98				NF	Prepressurized to 7 kb	
	U	23	5.00	1.0×10^{-4}			8.29				NF		
	U	23	5.00	1.0×10^{-4}			8.34				NF		
	U	23	6.00	1.0×10^{-4}			9.10				NF		

Table 2 (continued)
STRESS-STRAIN RELATIONS FROM CONSTANT STRAIN RATE TRIAXIAL TESTS ON ROCKS IN APPARATUS USING FLUID-CONFINING MEDIA

Material	Or	Temp (°C)	Pressure (kbar)	Strain rate (s^{-1})	Differential stress — kbar strain (%) 1	2	5	10	Ultimate strength (kbar)	Total strain (%)	Fault angle (°)	Comment	Ref.
	U	23	6.00	1.0×10^{-4}			8.60				NF		
	U	23	7.00	1.0×10^{-4}			10.34				NF		
Sandstone, "Lance", Wyoming	N	20	0.001	1.0×10^{-4}					1.39	0.5	F	(a)	264
	N	20	1.00	1.0×10^{-4}	3.65				4.48	1.6	F	(b)	
	N	20	3.00	1.0×10^{-4}	3.65	5.36	6.40		6.40	5.4	F	(c)	
	N	20	5.00	1.0×10^{-4}	3.65	5.84			8.79	4.5	NF	(b)	
	N	20	7.00	1.0×10^{-4}	3.65	5.84	10.32		13.03	7.3	NF	(b)	
	N	20	9.00	1.0×10^{-4}	2.97	5.88	13.37	22.91	24.66	11.5	NF	(b)	
Sandstone, "Laurencekirk"	U	23	0.001	2.5×10^{-5}					0.710	0.6		(a)	255
	U	23	0.172	2.5×10^{-5}					1.13	0.8		(b)	
	U	23	0.372	2.5×10^{-5}	1.51				1.53	1.1		(b)	
	U	23	0.690	2.5×10^{-5}	1.66				1.92	1.8		(b)	
	U	23	1.03	2.5×10^{-5}	1.66	2.12						(b)	
	U	23	1.38	2.5×10^{-5}	1.68	2.28						(b)	
Sandstone, "Mesa Verde", Rio Blanco County, Colorado, medium to fine-grained graywacke, 50% water saturated, undrained tests	U	25	0.001	1.0×10^{-4}					0.700		F		267
	U	25	0.001	1.0×10^{-4}					0.780		F		
	U	25	0.001	1.0×10^{-4}					0.680		F		
	U	25	0.001	1.0×10^{-4}					0.720		F		
	U	25	0.001	1.0×10^{-4}					0.920		F		
	U	25	1.00	1.0×10^{-4}					3.82		F		
	U	25	2.00	1.0×10^{-4}					5.44				
	U	25	3.00	1.0×10^{-4}			7.06				NF		
	U	25	4.00	1.0×10^{-4}			9.37				NF		
	U	25	5.00	1.0×10^{-4}			11.15				NF		
	U	25	6.00	1.0×10^{-4}			11.87				NF		
Sandstone, "Nugget", Utah, orthoquartzite	U	25	0.001	5.0×10^{-5}					2.30		F		264
	U	25	0.001	5.0×10^{-5}					2.60		F		
	U	25	1.00	5.0×10^{-5}					8.60		F		
	U	25	2.03	5.0×10^{-5}					12.10		F		
	U	25	3.00	5.0×10^{-5}					13.96		F		
	U	25	3.00	5.0×10^{-5}					13.77		F		

	U	25	3.00	5.0×10^{-5}					14.50		F		
	U	25	3.00	5.0×10^{-5}					14.24		F		
	U	25	3.01	5.0×10^{-5}					14.51		F		
	U	25	4.02	5.0×10^{-5}					16.01		F		
	U	25	5.02	5.0×10^{-5}					18.75		F		
	U	25	6.07	5.0×10^{-5}					20.85		F		
	U	25	7.01	5.0×10^{-5}					22.50		F		
Sandstone, "Simpson Orthoquartz-ite", Oklahoma	U	500	6.09	6.0×10^{-6}					21.40		NF		132
	U	500	8.00	6.0×10^{-6}	6.02	11.04	19.14		21.79	9.2	F	(f)	
	U	500	10.00	6.0×10^{-6}					24.90				
	U	500	8.00	7.2×10^{-7}	4.97	9.37	19.25	21.64	22.17	11.7	NF	(d)	
	U	600	8.00	6.4×10^{-5}			15.91	18.67	18.67		NF	(c)	
	U	600	10.00	6.7×10^{-6}					18.29				
	U	600	8.00	6.5×10^{-6}			14.74	16.60	16.65	11.2	NF	(d)	
	U	600	6.00	6.4×10^{-6}					14.70				
	U	600	6.00	6.3×10^{-6}					14.00				
	U	600	8.00	7.0×10^{-7}	4.38	8.28	17.49	15.27	18.42	10.3	F		
	U	700	8.00	6.4×10^{-5}	3.32	6.23	13.75	16.30	16.69	13.1	NF	(d)	
	U	700	6.00	7.1×10^{-6}					13.50				
	U	700	10.00	6.6×10^{-6}					13.10				
	U	700	8.00	6.4×10^{-6}			12.86	13.78	13.95	9.2	NF	(d) 10% Data extrapolated	
	U	700	8.00	3.3×10^{-6}			11.78	13.25	13.12	8.7	NF	(b) 10% Data extrapolated	
	U	700	8.00	5.1×10^{-7}			9.80		10.86	6.3	F		
	U	800	8.00	6.6×10^{-4}	3.67	7.43	14.87	16.48	16.49	14.5	NF	(d)	
	U	800	8.00	3.7×10^{-5}			11.25		12.25	8.2	NF	(b)	
	U	800	8.00	3.0×10^{-5}		6.23	12.30		13.01	6.1	NF	(b)	
	U	800	8.00	3.0×10^{-5}			12.44		13.47	6.9	NF	(b)	
	U	800	10.00	3.9×10^{-6}					7.80				
	U	800	6.00	3.7×10^{-6}					8.10				
	U	800	8.00	3.4×10^{-6}		6.63	8.89		9.51	7.0	F	(b)	
	U	800	8.00	2.5×10^{-6}					8.39				
	U	800	8.00	5.9×10^{-7}		5.50	5.90		6.78	5.3	F		
	U	900	8.00	6.7×10^{-4}			9.74	10.82	11.17	11.1	NF	(d)	
	U	900	8.00	3.7×10^{-4}	2.51	4.94	8.53	9.41	9.49	11.5	NF	(d)	
	U	900	8.00	3.7×10^{-5}		4.29	6.95		7.16	8.0	NF	(d)	
	U	900	8.00	3.7×10^{-5}					8.79				
	U	900	8.00	3.7×10^{-5}					8.50				
	U	900	8.00	3.6×10^{-5}					9.20				
	U	900	8.00	3.9×10^{-6}	2.16	3.80	5.32		5.35	7.5	NF	(d)	
	U	900	10.00	3.9×10^{-6}					5.20				
	U	900	6.00	3.5×10^{-6}					11.10				

Table 2 (continued)
STRESS-STRAIN RELATIONS FROM CONSTANT STRAIN RATE TRIAXIAL TESTS ON ROCKS IN APPARATUS USING FLUID-CONFINING MEDIA

Material	Or	Temp (°C)	Pressure (kbar)	Strain rate (s^{-1})	Differential stress — kbar strain (%) 1	2	5	10	Ultimate strength (kbar)	Total strain (%)	Fault angle (°)	Comment	Ref.
	U	900	10.00	3.5×10^{-6}					5.00				
	U	900	8.00	2.5×10^{-6}					5.60				
	U	900	8.00	4.5×10^{-7}	2.28	3.38				2.9	F	Strain rate approximate	
	U	950	8.00	4.0×10^{-4}	3.58	7.15	10.51	11.30	11.35	10.7	NF	(d)	
	U	950	8.00	2.5×10^{-5}	3.20	4.92	6.60		6.60	5.0	NF	(b)	132
	U	1000	8.00	3.5×10^{-4}	2.84	5.85	9.57	10.40	10.40	14.7	NF	(d) Strain rate approximate	
	U	1000	8.00	3.0×10^{-5}	2.50	3.69	4.71		4.73	6.5	NF	(c) Strain rate approximate	
	U	1000	6.00	3.0×10^{-6}					2.40			Strain rate approximate	
	U	1000	10.00	3.0×10^{-6}					2.10			Strain rate approximate	
Sandstone, "Sulfur Creek", Rio Blanco County, Colorado, graywacke	U	25	0.001	1.0×10^{-4}					0.560		F		
	U	25	1.00	1.0×10^{-4}					1.86		F		267
	U	25	2.00	1.0×10^{-4}			2.76						
	U	25	3.00	1.0×10^{-4}			3.58				NF		
	U	25	4.00	1.0×10^{-4}			4.64				NF		
	U	25	5.00	1.0×10^{-4}			5.18				NF		
	U	25	5.00	1.0×10^{-4}			5.32				NF		
	U	25	6.00	1.0×10^{-4}			6.70				NF		
	U	25	7.00	1.0×10^{-4}			7.78				NF		
Schist, mica, Keystone, S.D.	N	150	1.00	1.7×10^{-4}					2.24	8.2	32	(f)	36
	N	500	5.00	1.7×10^{-4}					6.10	22.2	NF	(d)	
	N	500	5.00	-1.7×10^{-4}					2.69	31.4	F	(d)	
	P	150	1.00	1.7×10^{-4}					1.76	7.9	33	(c)	
	P	500	5.00	1.7×10^{-4}					5.31	19.2	NF	(d)	
Serpentinite, "Cabramurra", New South Wales, antigorite-chrysotile, undrained tests unless noted (dehydration above about 500°C)	U	25	1.00	7.0×10^{-4}					6.00		F		240
	U	25	2.00	7.0×10^{-4}					8.20				
	U	25	3.50	7.0×10^{-4}	5.94	10.58	10.30		11.10	4.5	NF	(d) 5% Data extrapolated	
	U	25	3.50	7.0×10^{-4}					9.70				
	U	25	5.00	7.0×10^{-4}					11.10		NF		
	U	150	3.50	7.0×10^{-4}	7.81	10.13	9.20		10.20	5.3	F	(d) Preheated 0.5 hr	
	U	250	3.50	7.0×10^{-4}					8.50			preheated 0.5 hr	
	U	350	3.50	7.0×10^{-4}	5.11	7.63	7.23		7.90		NF	(d) Preheated 0.5 hr	

	U	400	3.50	7.0×10^{-4}				8.00			Preheated 0.5 hr	
	U	400	3.50	7.0×10^{-4}				7.00			Preheated 0.5 hr	
	U	450	3.50	7.0×10^{-4}				7.20			Preheated 0.5 hr	
	U	475	3.50	7.0×10^{-4}				7.70			Preheated 0.5 hr	
	U	500	3.50	7.0×10^{-4}	4.31	7.06	6.53	7.20	6.3	F	(d) Preheated 0.5 hr	
	U	500	3.50	7.0×10^{-4}				6.90			Preheated 7.0 hr	
	U	513	3.50	7.0×10^{-4}				4.30		F	Preheated 7.0 hr	
	U	527	3.50	7.0×10^{-4}				2.90		F	Preheated 7.0 hr	
	U	550	3.50	7.0×10^{-4}				6.30		F	Preheated 0.5 hr	
	U	550	3.50	7.0×10^{-4}				3.10		F	Preheated 7.0 hr	
	U	575	3.50	7.0×10^{-4}				1.10		F	Preheated 7.0 hr	
	U	600	3.50	7.0×10^{-4}	3.31	3.09		4.00	3.0	F	(d) Preheated 0.5 hr	
	U	600	3.50	7.0×10^{-4}				1.10		F	Preheated 7.0 hr	
	U	630	3.50	7.0×10^{-4}				0.900		F	Preheated 7.0 hr	
	U	650	3.50	7.0×10^{-4}				1.20	.5	F	(a) Preheated 0.5 hr	
	U	675	3.50	7.0×10^{-4}				1.70		F	Preheated 0.5 hr	
	U	700	3.50	7.0×10^{-4}				0.700		F	Preheated 0.5 hr	
	U	700	3.50	7.0×10^{-4}				3.60		NF	Preheated 0.5 hr, vented specimen	
Serpentinite, "Fidalgo", Fidalgo Island, undrained tests unless noted (dehydration above about 300°C)	U	25	3.50	7.0×10^{-4}				8.00		NF		240
	U	200	3.50	7.0×10^{-4}				6.30		NF	Preheated 1.0 hr	
	U	340	3.50	7.0×10^{-4}				5.70		NF	Preheated 0.5 hr	
	U	340	3.50	7.0×10^{-4}				2.40		F	Preheated 7.0 hr	
	U	365	3.50	7.0×10^{-4}				2.10		F	Preheated 7.0 hr	
	U	380	3.50	7.0×10^{-4}				2.80		F	Preheated 0.5 hr	
	U	390	3.50	7.0×10^{-4}				1.50		F	Preheated 1.0 hr	
	U	425	3.50	7.0×10^{-4}				3.20		F	Preheated 0.5 hr	
	U	445	3.50	7.0×10^{-4}				2.50		F	Preheated 0.5 hr	
	U	565	3.50	7.0×10^{-4}				0.800		F	Preheated 0.5 hr	
	U	605	3.50	7.0×10^{-4}				0.600		F	Preheated 2.0 hr	
Serpentinite, "Tumut Pond", New South Wales, antigorite-chrysotile, undrained tests unless noted (dehydrated above about 500°C)	U	25	0.200	7.0×10^{-4}				3.90		F		240
	U	25	0.500	7.0×10^{-4}				4.70		F		
	U	25	1.00	7.0×10^{-4}				5.50		F		
	U	25	1.00	7.0×10^{-4}				6.60		F		
	U	25	2.00	7.0×10^{-4}				7.80		F		
	U	25	3.50	7.0×10^{-4}				11.20				
	U	25	5.00	7.0×10^{-4}	5.13	10.30	14.22	14.05	4.1	NF	(b) 5% Data extrapolated	
	U	175	5.00	7.0×10^{-4}				12.39		NF	Preheated 0.5 hr	
	U	230	1.00	7.0×10^{-4}				4.90		F	Preheated 0.5 hr	
	U	250	5.00	7.0×10^{-4}	5.13	9.08	9.69	10.10	5.7	NF	(d) Preheated 0.5 hr	
	U	275	1.00	7.0×10^{-4}				4.90		F	Preheated 0.5 hr	

Table 2 (continued)
STRESS-STRAIN RELATIONS FROM CONSTANT STRAIN RATE TRIAXIAL TESTS ON ROCKS IN APPARATUS USING FLUID-CONFINING MEDIA

Material	Or	Temp (°C)	Pressure (kbar)	Strain rate (s^{-1})	Differential stress — kbar strain (%) 1	2	5	10	Ultimate strength (kbar)	Total strain (%)	Fault angle 1.9(°)	Comment	Ref.
	U	400	5.00	7.0×10^{-4}					9.50		NF	Preheated 0.5 hr	
	U	454	1.00	7.0×10^{-4}					5.30		F	Preheated 0.5 hr	
	U	500	1.00	7.0×10^{-4}					5.30		F	Preheated 0.5 hr	
	U	500	5.00	7.0×10^{-4}	4.23	7.63	8.59		8.89	5.3	NF	(d) Preheated 0.5 hr	
	U	550	1.00	7.0×10^{-4}					4.00		F	Preheated 0.5 hr	
	U	600	1.00	7.0×10^{-4}					2.70		F	Preheated 0.5 hr	
	U	600	5.00	7.0×10^{-4}	3.39	5.98	6.59		7.30	7.0		(d) Preheated 0.5 hr	
	U	605	5.00	7.0×10^{-4}					7.10			Preheated 0.5 hr	
	U	625	5.00	7.0×10^{-4}	3.53	4.81	3.89		4.90	4.6	F	(d) 5% Data extrapolated preheated 0.5 hr	
	U	650	1.00	7.0×10^{-4}					4.20		F	Preheated 0.5 hr, vented specimen	
	U	650	5.00	7.0×10^{-4}	2.69	1.88	1.23		2.80	4.3	F	(d) 5% Data extrapolated preheated 0.5 hr	
	U	675	5.00	7.0×10^{-4}					1.50		F	Preheated 0.5 hr	
	U	675	5.00	7.0×10^{-4}					1.20		F	Preheated 0.5 hr	
	U	680	1.00	7.0×10^{-4}					2.20		F	Preheated 0.5 hr	
	U	695	1.00	7.0×10^{-4}					2.20		F	Preheated 0.5 hr	
	U	700	5.00	7.0×10^{-4}					0.500	0.6	F	(a)	
	U	735	1.00	7.0×10^{-4}					2.10		F	Preheated 0.5 hr	
Serpentinite, Mayaquez, Puerto Rico, 49-m core, undrained tests unless noted	N	25	1.25	1.7×10^{-4}					3.58	1.7	26	(a)	109
	N	25	1.25	1.7×10^{-4}					3.40	1.6	30	(a)	
	P	25	0.750	1.7×10^{-4}					3.12	1.5	22	(a)	
	P	25	0.750	1.7×10^{-4}					3.15	1.3	22	(a)	
	P	25	1.25	1.7×10^{-4}					3.38	1.7	26	(a)	
	P	25	1.75	1.7×10^{-4}	3.64				3.64	1.0	30	(a)	
	P	25	1.75	1.7×10^{-4}					3.45	1.3	27	(a)	
	P	100	1.75	1.7×10^{-4}					2.63	1.9	31	(a)	
	P	200	1.75	1.7×10^{-4}		2.46			2.46	2.0	23	(a) Vented specimen	
	P	200	1.75	1.7×10^{-4}					1.55	7.3	24		

Serpentinite, Mayaquez, Puerto Rico, 180-m core, undrained tests unless noted	N	25	1.25	1.7×10^{-4}				5.30	1.2	33	(a)	109
	N	25	1.25	1.7×10^{-4}				5.11	1.1	27	(a)	
	P	25	0.750	1.7×10^{-4}				2.40	2.4	21	(a)	
	P	25	0.750	1.7×10^{-4}	3.50			4.35	1.2	25	(a)	
	P	25	1.25	1.7×10^{-4}	4.12			6.15	1.5	27	(a)	
	P	25	1.25	1.7×10^{-4}				5.98	1.1	27	(a)	
	P	25	1.75	1.7×10^{-4}	6.20			6.72	1.1	35	(a)	
	P	100	1.75	1.7×10^{-4}				4.65	0.8	26	(a)	
	P	200	1.75	1.7×10^{-4}	4.60			4.60	1.0	35	(a) Vented specimen	
	P	200	1.75	1.7×10^{-4}	2.76	2.50	2.50	2.76	6.1	22	(f)	
Serpentinite, Mayaquez, Puerto Rico, 270-m core, undrained tests unless noted	N	25	1.25	1.7×10^{-4}				3.30	1.8	32	(a)	109
	N	25	1.25	1.7×10^{-4}				3.38	1.2	28	(a)	
	P	25	0.750	1.7×10^{-4}				4.45	1.3	24	(a)	
	P	25	0.750	1.7×10^{-4}				4.14	1.8	F	(a)	
	P	25	1.25	1.7×10^{-4}				5.10	1.7	22	(a)	
	P	25	1.25	1.7×10^{-4}				4.92	1.3	33	(a)	
	P	25	1.75	1.7×10^{-4}				5.90	0.8	30	(a)	
	P	100	1.75	1.7×10^{-4}	3.69			3.69	1.0	35	(a)	
	P	200	1.75	1.7×10^{-4}				3.54	6.6	30	Vented specimen	
	P	200	1.75	1.7×10^{-4}				2.33	5.2	32		
Serpentinite, Piedmont Alps, Mont-genevre, Italy, 90% serpentine (lizardite), 8% magnetite, 2% olivine and pyroxene	U	20	0.690	1.0×10^{-5}	1.60	3.68	2.81	4.21	7.9	F	(f) Undrained test	213
	U	20	3.31	1.0×10^{-5}	1.98	4.11	6.56	6.56	8.3	NF	(c) Undrained test	
	U	20	4.83	1.0×10^{-5}	3.12	4.38	9.48	9.50	8.6	NF	(c) Undrained test	
	U	120	0.690	1.0×10^{-5}	1.60	3.34	2.77	3.45	7.2	F	(g) Undrained test	
	U	220	0.690	1.0×10^{-5}	1.60	3.45		3.45	4.2	F	(g) Undrained test	
	U	320	0.690	1.0×10^{-5}	2.14	1.69		2.96	2.4	F	(c) Undrained test	
	U	320	3.31	1.0×10^{-5}	1.98	4.11	6.22	6.22	6.5	NF	(c) Undrained test	
	U	320	4.83	1.0×10^{-5}	2.39	4.55	7.12	7.25	6.5	NF	(c) Undrained test	
	U	420	0.690	1.0×10^{-5}	2.14	1.74		2.81	3.3	F	(g) Undrained test	
	U	420	3.31	1.0×10^{-5}	1.98	4.11	5.87	6.00	5.9	NF	(c) Undrained test	
	U	420	4.83	1.0×10^{-5}	2.39	4.38	6.90	7.45	6.5	F	(b) Undrained test	
	U	520	0.690	1.0×10^{-5}	1.60	1.83		2.55	2.9	F	(g) Undrained test	
	U	520	3.31	1.0×10^{-5}	1.87	3.31	4.60	4.60	5.4	NF	(c) Undrained test	
	U	520	4.83	1.0×10^{-5}	1.94	2.36	2.67	2.67	5.0	F	(b) Undrained test	
	U	520	4.83	1.0×10^{-5}	1.66	3.20	2.96	3.70	5.9	F	(f)	
	U	620	0.690	1.0×10^{-5}	0.577	0.526		0.798	3.3	F	(e) Undrained test	
	U	620	3.31	1.0×10^{-5}	1.29	1.22		1.38	2.9	F	(e) Undrained test	
	U	620	4.83	1.0×10^{-5}	1.52	1.39		1.97	4.0	F	(f) Undrained test	
	U	620	4.83	1.0×10^{-5}	0.95	1.57	2.65	2.86	7.0	NF	(b)	
	U	720	0.690	1.0×10^{-5}	0.441	0.554	0.311	0.673	4.8	F	(g) Undrained test, 5% data extrapolated	

Table 2 (continued)
STRESS-STRAIN RELATIONS FROM CONSTANT STRAIN RATE TRIAXIAL TESTS ON ROCKS IN APPARATUS USING FLUID-CONFINING MEDIA

Material	Or	Temp (°C)	Pressure (kbar)	Strain rate (s^{-1})	Differential stress — kbar strain (%)				Ultimate strength (kbar)	Total strain (%)	Fault angle (°)	Comment	Ref.
					1	2	5	10					
	U	720	3.31	1.0×10^{-5}	0.578	0.311			0.689	3.6	F	(e) Undrained test	
	U	720	4.83	1.0×10^{-5}	0.420	0.243			0.420	2.6	F	(f) Undrained test	
Shale, "Middle Gust", Colorado	N	23	0.001	1.0×10^{-4}					0.017		F		35
	N	23	0.001	1.0×10^{-4}					0.018		F		
	N	23	0.020	1.0×10^{-4}					0.046		F		
	N	23	0.500	1.0×10^{-4}					0.003		F		
	N	23	1.00	1.0×10^{-4}					0.003		F		
	N	23	1.00	1.0×10^{-4}					0.012		F		
	N	23	1.00	1.0×10^{-4}					0.017		F		
	N	23	1.00	1.0×10^{-4}					0.075		F		
	N	23	1.00	1.0×10^{-4}					0.026		F		
	N	23	2.00	1.0×10^{-4}					0.004		F		
	N	23	2.00	1.0×10^{-4}					0.008		F		
	N	23	2.00	1.0×10^{-4}					0.011		F		
	N	23	2.00	1.0×10^{-4}					0.091		F		
	N	23	3.00	1.0×10^{-4}					0.044		F		
	N	23	5.00	1.0×10^{-4}			0.102				NF		
	N	23	6.00	1.0×10^{-4}			0.142				NF		
Shale, "Pierre", Colorado	N	23	0.020	1.0×10^{-4}			0.018				NF		1
	N	23	0.250	1.0×10^{-4}			0.006				NF		
	N	23	0.250	1.0×10^{-4}			0.022				NF		
	N	23	0.250	1.0×10^{-4}			0.028				NF		
	N	23	0.500	1.0×10^{-4}			0.008				NF		
	N	23	0.500	1.0×10^{-4}			0.028				NF		
	N	23	0.500	1.0×10^{-4}			0.028				NF		
	N	23	0.500	1.0×10^{-4}			0.028				NF		
	N	23	0.500	1.0×10^{-4}			0.016				NF		
	N	23	1.00	1.0×10^{-4}			0.014				NF		
	N	23	1.00	1.0×10^{-4}			0.010				NF		
	N	23	1.00	1.0×10^{-4}			0.034				NF		
	N	23	1.00	1.0×10^{-4}			0.028				NF		

	N	23	2.00	1.0×10^{-4}		0.016				NF		
	N	23	2.00	1.0×10^{-4}		0.050				NF		
	N	23	2.00	1.0×10^{-4}		0.066				NF		
	N	23	3.00	1.0×10^{-4}		0.050				NF		
	N	23	3.00	1.0×10^{-4}		0.094				NF		
	N	23	3.00	1.0×10^{-4}		0.104				NF		
	N	23	3.50	1.0×10^{-4}		0.082				NF		
	N	23	5.00	1.0×10^{-4}		0.164				NF		
	N	23	5.00	1.0×10^{-4}		0.218				NF		
	N	23	5.00	1.0×10^{-4}		0.186				NF		
	N	23	5.00	1.0×10^{-4}		0.038				NF		
	N	23	7.00	1.0×10^{-4}		0.420				NF		
	N	23	7.00	1.0×10^{-4}		0.410				NF		
	N	23	7.00	1.0×10^{-4}		0.440				NF		
	N	23	7.00	1.0×10^{-4}		0.280				NF		
Slate, "Mettawee", Granville, N.Y.	45	500	5.00	1.7×10^{-4}				3.50	24.2	45	(d)	36
	N	150	1.00	1.7×10^{-4}				5.04	2.7	29	(a)	
	N	500	5.00	1.7×10^{-4}				6.43	21.0	30	(d)	
	P	150	1.00	1.7×10^{-4}				5.19	2.9	22	(a)	
	P	500	5.00	1.7×10^{-4}				5.82	27.0	NF	(d)	
	P	500	5.00	1.7×10^{-4}				6.32	23.4	NF	(d)	36
	P	500	5.00	1.7×10^{-4}				5.90	11.8	NF	(d)	
	P	500	5.00	1.7×10^{-4}				4.82	5.4	NF	(d)	
Sphalerite ore, Central Tennessee-Knox District, <1% iron, <20% impurities	U	24	0.250	7.2×10^{-5}	2.38			2.38	2.1	F		63
	U	24	0.500	7.2×10^{-5}	3.04	3.31		3.37	6.0	F		
	U	24	1.00	7.2×10^{-5}	3.10	3.76	4.19			NF		
	U	24	1.50	7.2×10^{-5}	2.97	3.66	4.46			NF		
	U	24	2.00	7.2×10^{-5}	3.05	4.02	5.02			NF		
	U	100	1.00	7.2×10^{-5}	2.18	2.90	3.64			NF		
	U	200	0.250	7.2×10^{-5}	1.88			2.01	2.9	F		
	U	200	1.00	7.2×10^{-5}	2.31	2.68	3.37			NF		
	U	300	0.500	7.2×10^{-5}	1.79	2.18	2.60			NF		
	U	300	1.00	7.2×10^{-5}	1.74	2.57	3.33			NF		
	U	300	2.00	7.2×10^{-5}	1.67	2.09	2.69			NF		
	U	400	1.00	7.2×10^{-5}	1.85	2.11	2.49			NF		
	U	500	0.500	7.2×10^{-5}	1.47	1.88	2.24			NF		
	U	500	1.00	7.2×10^{-5}	1.60	1.83	2.16			NF		
Sphalerite ore, East Tennessee, <1% iron	U	24	1.50	7.2×10^{-5}	3.02	5.47		5.81	8.6	F		63
	U	24	2.00	7.2×10^{-5}	2.86	4.52	5.21			NF		
	U	100	0.500	7.2×10^{-5}	3.26			3.74	4.2	F		
	U	100	1.00	7.2×10^{-5}	2.66	4.68		5.04	9.0	F		

Table 2 (continued)
STRESS-STRAIN RELATIONS FROM CONSTANT STRAIN RATE TRIAXIAL TESTS ON ROCKS IN APPARATUS USING FLUID-CONFINING MEDIA

Material	Or	Temp (°C)	Pressure (kbar)	Strain rate (s^{-1})	Differential stress — kbar strain (%) 1	2	5	10	Ultimate strength (kbar)	Total strain (%)	Fault angle (°)	Comment	Ref.
	U	100	1.50	7.2×10^{-5}		2.20	4.03	4.43			NF	<10% Strain 10% data extrapolated	
	U	100	2.00	7.2×10^{-5}		2.71	4.18	5.20			NF		
	U	200	1.00	7.2×10^{-5}		2.11	2.87		2.87	8.2	F		
	U	200	1.50	7.2×10^{-5}		2.81	4.06	4.76			NF		
	U	200	2.00	7.2×10^{-5}		1.83	2.75	3.18			NF		
	U	300	0.500	7.2×10^{-5}		2.38	2.96		3.04	7.2	F		
	U	300	1.00	7.2×10^{-5}		2.52	3.10	3.50			NF		
	U	350	2.00	7.2×10^{-5}		1.47	1.92	2.32			NF		
	U	400	1.00	7.2×10^{-5}		2.86	2.91		2.91	5.6	F		
Sphalerite ore, Hurnigskopf mine, Hurnig/Ahr, Germany, very fine grained	U	20	4.00	3.0×10^{-4}	1.54	5.57	7.83		8.83	8.0	NF	(b)	258
	U	20	5.00	3.0×10^{-4}	3.48	6.17	8.02	9.50	10.45	16.0	NF	(b)	
	U	20	5.00	3.0×10^{-4}	3.48	6.51	8.37	9.99	11.04	19.7	NF	(d)	
Sphalerite ore, Nikolaus-Phonix mine, Markelsbach/Siegkreis, Germany	U	20	0.001	3.0×10^{-4}	0.450				0.460	1.6	0	(d)	258
	U	20	0.001	3.0×10^{-4}	0.720				0.740	1.7	0	(d)	
	U	20	0.050	3.0×10^{-4}							15		
	U	20	0.100	3.0×10^{-4}	2.77	1.89			2.80	2.3	20	(d)	
	U	20	0.100	3.0×10^{-4}	2.82	2.45			3.02	2.4	25	(d)	
	U	20	0.250	3.0×10^{-4}	3.95	3.62			4.17	2.5	26	(d)	
	U	20	0.500	3.0×10^{-4}	3.64	4.48			4.48	4.0	29	(d)	
	U	20	0.750	3.0×10^{-4}	3.68	4.44	4.47		4.64	6.2	27	(d)	
	U	20	1.00	3.0×10^{-4}	4.05	4.53	4.98	3.75	5.04	10.5	34	(d)	
	U	20	1.50	3.0×10^{-4}							34		
	U	20	1.50	3.0×10^{-4}		4.80	5.85		6.05	7.5	NF	(b)	258
	U	20	2.00	3.0×10^{-4}		4.88	6.25	6.92	6.93	13.5	32	(d)	
	U	20	2.00	3.0×10^{-4}	3.16	4.32	5.57	6.18	6.18	9.8	NF	(b) 10% Data extrapolated	
	U	20	3.00	3.0×10^{-4}	3.89	4.99	6.43	7.37	7.38	10.3	NF	(d)	
	U	20	3.00	3.0×10^{-4}	4.01	5.27	6.83	7.70	7.76	13.5	NF	(d)	
	U	20	4.00	3.0×10^{-4}	3.95	5.51	7.30	8.29	8.95	16.9	NF	(d)	
	U	20	5.00	3.0×10^{-4}		4.80	6.32	7.72	8.89	18.4	34	(d)	
	U	20	5.00	3.0×10^{-4}	4.00	5.14	6.52	8.08	9.37	21.5	NF	(d)	

	U	20	5.00	3.0×10^{-4}	4.00	5.14	6.52	8.15	9.33	20.7	NF	(d)	
	U	20	5.00	3.0×10^{-4}	4.08	5.27	6.95	8.49	10.16	20.9	NF	(d)	
Sphalerite ore, Sullivan Mine, British Columbia, ''iron rich''	U	100	1.00	7.2×10^{-5}		2.56	3.02	3.63			NF		63
	U	200	1.00	7.2×10^{-5}		1.13	1.67	2.28			NF		
	U	300	1.00	7.2×10^{-5}		1.22	1.62	2.24			NF		
	U	400	1.00	7.2×10^{-5}		1.19	1.47	1.86			NF		
Syenite, Victor, Colo.	N	150	1.00	1.7×10^{-4}					5.32	2.4	25	(a)	36
	N	500	5.00	1.7×10^{-4}					4.19	17.4	37	(d)	
Talc, ''Three Springs'', West Australia	U	20	2.00	4.0×10^{-4}	1.20	1.28	1.20	1.10	1.29		NF	(f)	82
	U	20	4.00	4.0×10^{-4}	1.26	1.60	1.66	1.66			NF	(c)	
	U	20	6.00	4.0×10^{-4}	1.44	1.85	1.96	2.03			NF	(b)	
	U	20	8.00	4.0×10^{-4}	1.44	1.90	2.05	2.21			NF	(b)	
Tuff, ''Indian Trail'', Nevada, partially water saturated, undrained tests unless noted	U	23	0.001	5.0×10^{-5}					0.090		F		276
	U	23	0.001	5.0×10^{-5}					0.060		F		
	U	23	0.001	5.0×10^{-5}					0.080		F		
	U	23	0.001	5.0×10^{-5}					0.060		F		
	U	23	0.500	5.0×10^{-5}			0.380				NF		
	U	23	0.500	5.0×10^{-5}			0.680				NF		
	U	23	0.500	5.0×10^{-5}			0.520				NF		
	U	23	0.500	5.0×10^{-5}			0.200				NF		
	U	23	0.500	5.0×10^{-5}			0.200				NF		
	U	23	0.500	5.0×10^{-5}			0.200				NF		
	U	23	1.00	5.0×10^{-5}			0.660				NF		
	U	23	1.00	5.0×10^{-5}			0.900				NF		
	U	23	1.00	5.0×10^{-5}			0.440				NF		
	U	23	1.00	5.0×10^{-5}			0.120				NF		
	U	23	1.00	5.0×10^{-5}			0.160				NF		
	U	23	1.50	5.0×10^{-5}			0.380				NF		
	U	23	1.50	5.0×10^{-5}			0.900				NF		
	U	23	1.50	5.0×10^{-5}			0.280				NF		
	U	23	1.50	5.0×10^{-5}			0.200				NF		
	U	23	2.00	5.0×10^{-5}			0.700				NF		
	U	23	2.00	5.0×10^{-5}			0.840				NF		
	U	23	2.00	5.0×10^{-5}			0.860				NF		
	U	23	2.00	5.0×10^{-5}			0.220				NF		
	U	23	2.50	5.0×10^{-5}			0.560				NF		
	U	23	2.50	5.0×10^{-5}			0.800				NF		
	U	23	2.50	5.0×10^{-5}			0.500				NF		
	U	23	2.50	5.0×10^{-5}			0.240				NF		276
	U	23	2.50	5.0×10^{-5}			0.200				NF		
	U	23	2.70	5.0×10^{-5}			0.280				NF		

Table 2 (continued)
STRESS-STRAIN RELATIONS FROM CONSTANT STRAIN RATE TRIAXIAL TESTS ON ROCKS IN APPARATUS USING FLUID-CONFINING MEDIA

Material	Or	Temp (°C)	Pressure (kbar)	Strain rate (s^{-1})	Differential stress — kbar strain (%)				Ultimate strength (kbar)	Total strain (%)	Fault angle (°)	Comment	Ref.
					1	2	5	10					
	U	23	3.00	5.0×10^{-5}			0.460				NF		
	U	23	3.00	5.0×10^{-5}			0.280				NF		
	U	23	3.20	5.0×10^{-5}			1.04				NF		
	U	23	3.50	5.0×10^{-5}			1.24				NF		
	U	23	3.50	5.0×10^{-5}			0.540				NF		
	U	23	3.50	5.0×10^{-5}			0.360				NF		
	U	23	3.50	5.0×10^{-5}			0.480				NF		
Tuff, "Mt. Helen", Nevada, 2%	U	25	0.001	1.0×10^{-4}					0.460		F		131
water saturation, dry density 1.51	U	25	0.001	1.0×10^{-4}					0.400		F		
g/cc, undrained tests unless noted	U	25	0.001	1.0×10^{-4}					0.400		F		
	U	25	0.250	1.0×10^{-4}			0.580				NF		
	U	25	0.490	1.0×10^{-4}			0.980				NF		
	U	25	0.500	1.0×10^{-4}			1.00				NF		
	U	25	1.00	1.0×10^{-4}			0.840				NF		
	U	25	1.50	1.0×10^{-4}			1.42				NF		
	U	25	2.00	1.0×10^{-4}			1.50				NF		
	U	25	2.50	1.0×10^{-4}			1.94				NF		
	U	25	2.50	1.0×10^{-4}			2.62				NF		
	U	25	3.00	1.0×10^{-4}			2.78				NF		
	U	25	4.00	1.0×10^{-4}			3.58				NF		
	U	25	5.00	1.0×10^{-4}			4.88				NF	Dry	
	U	25	5.00	1.0×10^{-4}			4.78				NF	Dry	
	U	25	5.00	1.0×10^{-4}			3.94				NF		
	U	25	6.00	1.0×10^{-4}			4.44				NF		
	U	25	7.00	1.0×10^{-4}			4.90				NF		
Tuff, "Mt. Helen", Nevada, 90%	U	25	0.500	1.0×10^{-4}			0.440						131
water saturation, dry density 1.51	U	25	1.00	1.0×10^{-4}			0.800				NF		
g/cc, undrained tests unless noted	U	25	3.00	1.0×10^{-4}			1.26				NF		
	U	25	3.00	1.0×10^{-4}			1.46				NF		
	U	25	5.00	1.0×10^{-4}			1.82				NF		
	U	25	7.00	1.0×10^{-4}			2.00				NF		

Tuff, "Mt. Helen", Nevada, 100%	U	25	0.001	1.0×10^{-4}		0.200	F	131
water saturation, dry density 1.51	U	25	0.001	1.0×10^{-4}		0.260	F	
g/cc, undrained tests unless noted	U	25	0.001	1.0×10^{-4}		0.220	F	
	U	25	0.500	1.0×10^{-4}		0.220	F	
	U	25	0.500	1.0×10^{-4}		0.260	F	
	U	25	1.00	1.0×10^{-4}	0.260			
	U	25	3.00	1.0×10^{-4}	0.400		NF	
	U	25	3.00	1.0×10^{-4}	0.360		NF	
	U	25	5.00	1.0×10^{-4}	0.580		NF	
	U	25	7.00	1.0×10^{-4}	0.760		NF	
	U	25	7.00	1.0×10^{-4}	0.580		NF	
	U	25	7.00	1.0×10^{-4}	0.680		NF	
Tuff, Nevada, 100% water saturated,	U	25	0.001	1.0×10^{-4}		0.120	F	74
dry density 1.54 g/cc, undrained	U	25	0.250	1.0×10^{-4}		0.120	F	
tests unless noted	U	25	0.500	1.0×10^{-4}		0.140	F	
	U	25	0.750	1.0×10^{-4}		0.140	F	
	U	25	1.00	1.0×10^{-4}		0.200	F	
	U	25	2.00	1.0×10^{-4}		0.220	F	
	U	25	3.00	1.0×10^{-4}	0.240		NF	

Note: This table follows the general format of Handin's[110] Table 11-3. We also include the test strain rate and an indication of the stress-strain curve shape. The tabulated data cover the period 1966 to 1979. The rock material description includes rock type, common name, localty, and (when available) mineral modes, porosity, and grain size. The orientations of the test cylinders are usually referenced to bedding or foliation and the following symbols apply: P, cylinder parallel to bedding or foliation; N, cylinder normal to bedding or foliation; U, cylinders not uniformly oriented or no orientation given. The nominal strain rate is usually not corrected for apparatus elastic distortion and is indicated as a positive quantity for triaxial compression and negative for triaxial extension. The axial force-displacement data are generally corrected for apparatus friction and elastic distortion of the loading column, and for perminent strains greater than 5%, the differential stress is corrected for changes in specimen cross-sectional area. The ultimate strength is the maximum strength over the indicated total strain. Fault or fracture characteristics are noted as follows under the column "fault angle": NF, no through-going fault or fracture observed; F, fault or fracture observed but angle with specimen axes not recorded; X, through-going fault or fracture observed at an angle X to maximum principal compressive stress. A curve-shape index letter is provided when original stress-grain curves are available (Figure 27). In controlled pore pressure tests, the pore pressure is indicated as a fraction λ of the confining pressure. In tests at elevated temperature, rocks with water included as pore water or in hydrous minerals, the specimens are typically vented to the atmosphere (the so-called drained test), unless otherwise noted. Experimental accuracies in the tabulated data vary in the data tabulated here, but the following values may be used as a guide: temperature ± 1% of the temperature in °C. Confining and pore pressure ± 1/2 to 1%. Strain rate ± 10% excluding elastic distortion correction. Differential stress ± 1 to 2% ± 20 bars. Strain ± 3 to 5% of the total strain. Fault angle ± 5°. Reproducibility of stress-strain curves varies with rock type and specific apparatus but often is no better than ± 5% in differential stress at a given strain.

Table 3
STRESS-STRAIN RELATIONS IN CONSTANT STRAIN RATE TRIAXIAL COMPRESSION TESTS ON ROCKS AND MINERALS: PISTON-CYLINDER SOLID MEDIUM APPARATUS

					Experimental set-up			Differential stress — Kb strain (%)						
Material	Or	Temp (°C)	Conf. pressure (Kb)	Strain rate (s^{-1})	Conf. media	Pist. Mtrl.	Wet dry	5	10	15	Flow stress (Kb)	F/NF	Comment	Ref.
Amphibolite, Hope, B.C., Canada, magnesiahastingsite	P	700	9.79	1.1×10^{-5}	Talc	WC	W:TD	20.09	25.90					249
	P	700	10.20	1.1×10^{-6}	Talc	WC	W:TD	22.00						
	P	800	10.10	1.1×10^{-4}	Talc	WC	W:TD	15.50	15.50	14.60				
	P	800	10.00	1.1×10^{-5}	Talc	WC	W:TD	15.20	16.00					
	P	800	10.10	1.1×10^{-6}	Talc	WC	W:TD	21.00	21.59					
	P	825	10.00	1.1×10^{-4}	Talc	WC	W:TD	15.20	16.70	18.50				
	P	850	10.10	1.1×10^{-5}	Talc	WC	W:TD	9.89	12.20					
	P	850	10.00	1.1×10^{-6}	Talc	WC	W:TD	7.10	7.20	6.60				
	P	900	10.20	1.1×10^{-4}	Talc	WC	W:TD	6.60	7.30					
	P	950	10.10	1.1×10^{-5}	Talc	WC	W:TD	3.20	5.10					
	P	950	10.20	1.1×10^{-6}	Talc	WC	W:TD	5.50	5.30					
	P	1000	10.10	1.1×10^{-4}	Talc	WC	W:TD	6.50	6.20					
Amphibolite, Mt. Stuart Batholith, Washington, >70% hornblende, <30% plagioclase, G.S. = 0.4mm	U	600	10.00	1.0×10^{-5}	Talc	WO	D:A	13.70				NF	"5%" Data taken at 3% strain	73
	U	650	10.00	1.0×10^{-5}	Talc	WO	D:A	13.00				NF	5% Data taken at 3% strain	
	U	750	10.00	1.0×10^{-5}	Talc	WO	D:A	6.60				NF	5% Data taken at 3% strain	
	U	750	10.00	1.0×10^{-5}	Talc	WO	D:A	9.20				NF	5% Data taken at 3% strain	
	U	750	10.00	1.0×10^{-5}	Talc	WO	D:A	8.10				NFs		
	U	800	10.00	1.0×10^{-5}	Talc	WO	D:A	3.60				NFs		
	U	800	10.00	1.0×10^{-5}	Talc	WO	D:A	3.20				NFs		
	U	850	10.00	1.0×10^{-5}		WO	D:A	.900				NFs		
Anhydrite	N	500	5.10	1.2×10^{-5}	ASM	WC	D:O	11.60	12.79			NF		249
	N	600	4.80	1.2×10^{-5}	ASM	WC	D:O	7.40	7.30			NF		
	N	650	5.10	1.2×10^{-5}	ASM	AlO	D:O	7.40	7.60			NF		
	N	700	5.00	1.2×10^{-5}	ASM	WC	D:O	6.30	6.10			NF		
	N	750	5.00	1.2×10^{-5}	ASM	AlO	D:O	11.00	12.70	12.39		NF		

	N	750	5.10	1.2×10^{-6}	ASM	WC	D:O	11.20	12.89	12.60		NF		
	N	800	4.90	1.2×10^{-6}	ASM	WC	D:O	5.40	5.50			NF		
	N	850	4.90	1.2×10^{-5}	ASM	WC	D:O	4.60	4.80			NF		
	N	900	5.00	1.2×10^{-6}	ASM	WC	D:O	4.10	4.30			NF		
	N	950	4.90	1.2×10^{-6}	ASM	WC	D:O	3.80	3.60			NF		
	N	1000	4.90	1.2×10^{-5}	ASM	WC	D:O	4.70	4.50			NF		
Aplite, veins cutting quarry granite, Enfield, Vermont, 36% quartz, 36% microcline, 26% plagioclase, 2% muscovite, G.S. = 0.2mm	U	25	5.00	3.0×10^{-6}	NaCl	AlO	D:O	21.20	24.00			F		289,287
	U	300	5.00	3.0×10^{-6}	NaCl	AlO	D:O	15.20	14.50	14.39		F		
	U	600	7.50	3.0×10^{-6}	NaCl	AlO	D:O	17.70	18.09	17.59		NF		
	U	600	10.00	3.0×10^{-6}	NaCl	AlO	D:O	16.59	17.79	16.79		NF		
	U	600	15.00	3.0×10^{-6}	NaCl	AlO	D:O	16.79	18.09	18.09		NF		
	U	700	5.00	3.0×10^{-6}	NaCl	AlO	D:O	7.20	8.39	8.79		F		
	U	900	5.00	3.0×10^{-6}	NaCl	AlO	D:O	9.39				F		
Clinopyroxene crystal, Finland, Cr-diopside, oriented 45° to [001] & 45° to a* in acute angle between [100] & [001]		500	10.00	7.0×10^{-8}	NaCl	AlO	D:O				3.80		Twinned on (100)[001]	176—178
		600	10.00	5.0×10^{-7}	KCl	AlO	D:O				2.00		Twinned	
		800	10.00	9.0×10^{-6}	NaCl	AlO	D:O				1.80		Twinned	
		800	10.00	5.0×10^{-7}	KCl	AlO	D:O				1.60		Twinned	
		800	10.00	7.0×10^{-8}	NaCl	AlO	D:O				3.60		Twinned	
		800	10.00	9.0×10^{-9}	NaCl	AlO	D:O				1.40		Twinned	
		1000	10.00	5.0×10^{-7}	KCl	AlO	D:O				1.40		Twinning and slip	
		1200	10.00	5.0×10^{-7}	KCl	AlO	D:O				.300		Twinning and slip	
Clinopyroxene crystal, Finland, Cr-diopside, oriented 45° to [001] & 45° to a* in obtuse angle between [100] & [001]		500	10.00	7.0×10^{-6}	KCl	AlO	D:O				8.89		Slip on (100)[001]	176—178
Clinopyroxene crystal, hedenbergite, oriented at 45° to [001] & 45° to a* in acute angle between [100] & [001]		400	10.00	9.0×10^{-8}	KCl	AlO	D:O				2.75		Twinned	176—178
		700	10.00	9.0×10^{-8}	NaCl	AlO	D:O				3.30		Twinned on (100)[001]	
		700	10.00	8.0×10^{-8}	NaCl	AlO	D:O				2.70		Twinned on (100)[001]	
		700	10.00	9.0×10^{-9}	NaCl	AlO	D:O				3.20		Twinned on (100)[001]	
		800	10.00	8.0×10^{-8}	NaCl	AlO	D:O				2.70		Twinned on (100)[001]	
		900	10.00	8.0×10^{-8}	NaCl	AlO	D:O				2.50		Twinned on (100)[001]	
		1000	10.00	8.0×10^{-8}	NaCl	AlO	D:O				1.50		Twinning and slip	
Clinopyroxene, hedenbergite, oriented at 45° to [001] & 45° to a* in obtuse angle between [100] & [001]		600	10.00	8.0×10^{-7}	NaCl	AlO	D:O				12.10		Slip on (100)[001]	176—178
		700	10.00	9.0×10^{-7}	NaCl	AlO	D:O				10.20		Slip on (100)[001]	
		700	10.00	8.0×10^{-7}	NaCl	AlO	D:O				5.10		Slip on (100)[001]	
		800	10.00	9.0×10^{-7}	NaCl	AlO	D:O				7.10		Slip on (100)[001]	

Table 3 (continued)
STRESS-STRAIN RELATIONS IN CONSTANT STRAIN RATE TRIAXIAL COMPRESSION TESTS ON ROCKS AND MINERALS: PISTON-CYLINDER SOLID MEDIUM APPARATUS

					Experimental set-up			Differential stress — Kb strain (%)						
Material	Or	Temp (°C)	Conf. pressure (Kb)	Strain rate (s^{-1})	Conf. media	Pist. Mtrl.	Wet dry	5	10	15	Flow stress (Kb)	F/NF	Comment	Ref.
Diabase, Frederick,	U	600	5.00	3.0×10^{-6}	NaCl	AlO	D:O	10.00	11.50			F		287
Maryland, 36%	U	600	10.00	3.0×10^{-6}	NaCl	AlO	D:O	12.89	10.00	8.79		F		
plagioclase (An70),	U	600	15.00	3.0×10^{-6}	NaCl	AlO	D:O	20.09	23.20			NF		
58% clinopyroxene, 3%	U	600	15.00	3.0×10^{-6}	NaCl	AlO	D:O	16.29	19.29	20.00		NF		
chlorite, 3% opaque	U	800	5.00	3.0×10^{-6}	NaCl	AlO	D:O	10.00	8.29	8.39		F		
minerals	U	800	10.00	3.0×10^{-6}	NaCl	AlO	D:O	7.00				NF		
	U	800	15.00	3.0×10^{-6}	NaCl	AlO	D:O	10.00	12.20	12.60		NF		
	U	800	15.00	3.0×10^{-6}	NaCl	AlO	D:O	6.40	7.30	7.30		NF		
	U	900	15.00	3.0×10^{-5}	NaCl	AlO	D:O	7.20	7.90	8.10		NF		
	U	900	5.00	3.0×10^{-6}	NaCl	AlO	D:O	8.50	8.39	7.20		NF		
	U	900	15.00	3.0×10^{-6}	NaCl	AlO	D:O	5.00	5.30	4.90		NF		
Dunite, Mt. Burnet,	U	975	15.00	2.2×10^{-5}	Talc	WC	W:TD		5.12			NF		57, 55
Alaska, 99% olivine	U	975	15.00	1.9×10^{-5}	Talc	WC	W:TD		5.98			NF		
(Fo 92), 1% chromite,	U	975	15.00	1.8×10^{-5}	Talc	WC	W:TD		6.10			NF		
grain size = 1mm	U	975	15.00	2.6×10^{-6}	Talc	WC	W:TD	7.33				NF		
	U	975	15.00	2.1×10^{-6}	Talc	WC	W:TD	2.06				NF		
	U	975	15.00	1.8×10^{-6}	Talc	WC	W:TD	2.01				NF		
	U	975	15.00	1.8×10^{-7}	Talc	WC	W:TD			1.53		NF		
	U	1100	15.00	3.4×10^{-4}	Talc	WC	W:TD		4.00			NF		
	U	1100	15.00	3.2×10^{-4}	Talc	WC	W:TD	4.77				NF		
	U	1100	15.00	2.6×10^{-4}	ASM	WC	D:A	7.76				NF		
	U	1100	15.00	2.1×10^{-4}	Talc	WC	W:TD		7.97			NF		
	U	1100	15.00	4.5×10^{-5}	Talc	WC	W:TD	2.83				NF		
	U	1100	15.00	4.1×10^{-5}	Talc	WC	W:TD	2.11				NF		
	U	1100	15.00	4.0×10^{-5}	ASM	WC	D:A	1.20				NF		
	U	1100	15.00	3.9×10^{-5}	ASM	WC	D:A	7.59				NF		
	U	1100	15.00	3.4×10^{-5}	Talc	WC	W:TD		2.13			NF		
	U	1100	15.00	3.2×10^{-5}	Talc	WC	W:TD		3.56			NF		

	U	1100	15.00	2.7×10^{-5}	ASM	WC	D:A	5.62			NF		
	U	1100	15.00	2.6×10^{-5}	Talc	WC	W:TD		3.76		NF		
	U	1100	15.00	2.5×10^{-5}	ASM	WC	D:A		6.17		NF		
	U	1100	15.00	1.7×10^{-5}	ASM	WC	D:A			5.89	NF		
	U	1100	15.00	1.4×10^{-5}	ASM	WC	D:A		6.03		NF		
	U	1100	15.00	3.5×10^{-6}	Talc	WC	W:TD		.690		NF		
	U	1100	15.00	2.5×10^{-6}	ASM	WC	D:A		4.37		NF		
	U	1100	15.00	1.9×10^{-6}	ASM	WC	D:A			4.17	NF		
	U	1100	15.00	1.5×10^{-6}	ASM	WC	D:A		4.57		o		
	U	1100	15.00	2.5×10^{-7}	ASM	WC	D:A		2.69		NF		
	U	1100	15.00	1.9×10^{-7}	ASM	WC	D:A			2.09	NF		
	U	1225	15.00	4.5×10^{-4}	ASM	WC	D:A	3.72			NF		
	U	1225	15.00	3.0×10^{-4}	ASM	WC	D:A		4.27		NF		
	U	1225	15.00	2.4×10^{-4}	Talc	WC	W:TD		2.53		NF		
	U	1225	15.00	2.2×10^{-4}	Talc	WC	W:TD			3.90	NF		
	U	1225	15.00	3.7×10^{-5}	Talc	WC	W:TD	.850			NF		
	U	1225	15.00	3.2×10^{-5}	Talc	WC	W:TD		1.14		NF		
	U	1225	15.00	2.8×10^{-5}	ASM	WC	D:A	4.47			NF		
	U	1225	15.00	2.5×10^{-5}	Talc	WC	W:TD			.730	NF		
	U	1225	15.00	2.2×10^{-5}	ASM	WC	D:A	3.72			NF		
	U	1225	15.00	2.1×10^{-5}	ASM	WC	D:A		3.47		NF		
	U	1225	15.00	2.8×10^{-6}	ASM	WC	D:A		2.57		NF		
	U	1225	15.00	2.2×10^{-6}	ASM	WC	D:A		2.09		NF		
	U	1225	15.00	2.1×10^{-6}	ASM	WC	D:A		2.09		NF		
	U	1225	15.00	2.8×10^{-7}	ASM	WC	D:A		1.78		NF		
	U	1225	15.00	2.2×10^{-7}	ASM	WC	D:A			.930	NF		
	U	1225	15.00	2.2×10^{-7}	ASM	WC	D:A		.910		NF		
	U	1350	15.00	2.3×10^{-4}	Talc	WC	W:TD		1.21		NF		
	U	1350	15.00	2.2×10^{-4}	Talc	WC	W:TD			1.84	NF		
	U	1350	15.00	1.9×10^{-4}	ASM	WC	D:A	2.24			NF		
	U	1350	15.00	2.0×10^{-5}	ASM	WC	D:A		1.55		NF		
	U	1350	15.00	2.0×10^{-6}	ASM	WC	D:A		.810		NF		
Feldspar crystal, Eifel mountains, West Germany, Sanidine, oriented 41° to [100], 45° to [010], & 71° to [001]		700	15.00	1.7×10^{-6}	NaCl	AlO	D:O	6.80			NF		304—306
		800	15.00	1.7×10^{-6}	NaCl	AlO	D:O	4.80			NF		
Feldspar crystal, Eifel mountains, West Germany, Sanidine, oriented 64° to [100], 50° to [010], & 68° to [001]		700	15.00	1.7×10^{-6}	NaCl	AlO	D:O	9.39			NF	5% Data extrapolated	304—306
		900	15.00	1.5×10^{-6}	NaCl	AlO	D:O	2.80			NF	5% Data extrapolated	

Table 3 (continued)
STRESS-STRAIN RELATIONS IN CONSTANT STRAIN RATE TRIAXIAL COMPRESSION TESTS ON ROCKS AND MINERALS: PISTON-CYLINDER SOLID MEDIUM APPARATUS

Material	Or	Temp (°C)	Conf. pressure (Kb)	Strain rate (s^{-1})	Experimental set-up: Conf. media	Experimental set-up: Pist. Mtrl.	Experimental set-up: Wet dry	Differential stress — Kb, strain (%): 5	10	15	Flow stress (Kb)	F/NF	Comment	Ref.
Feldspar rock,	U	600	15.00	3.0×10^{-6}	NaCl	AlO	D:O	10.60	11.39	12.50		NF		287
Adirondack Mountains,	U	800	15.00	3.0×10^{-4}	NaCl	AlO	D:O	9.70	10.10			NF		
New York, anorthite	U	800	15.00	3.0×10^{-6}	NaCl	AlO	D:O	3.70	4.30	4.40		NF		
composition	U	900	15.00	3.0×10^{-6}	NaCl	AlO	D:O	2.60	3.00	3.10		NF		
Feldspar rock, Hale	U	500	5.00	3.0×10^{-6}	NaCl	AlO	D:O	13.00	12.60	12.89		F		287
pegmatite quarry,	U	600	15.00	3.0×10^{-6}	NaCl	AlO	D:O	18.50	19.70	19.70		NF		
Middletown, Conn.	U	700	5.00	3.0×10^{-6}	NaCl	AlO	D:O	9.39	8.10	7.50		F		
(Ab96, An3, Or1), G.S.	U	700	15.00	3.0×10^{-6}	NaCl	AlO	D:O	16.50	17.00	16.70		NF		
= 0.2mm	U	700	15.00	3.0×10^{-6}	NaCl	AlO	D:O	2.00	14.79	14.39		NF		
	U	800	10.00	3.0×10^{-6}	NaCl	AlO	D:O	10.39	9.79	8.89		NF		
	U	800	15.00	3.0×10^{-6}	NaCl	AlO	D:O	13.29	12.39	12.00		NF		
	U	900	7.50	3.0×10^{-6}	NaCl	AlO	D:O	4.70				NF		
	U	900	10.00	3.0×10^{-6}	NaCl	AlO	D:O	4.30	4.30	4.10		NF		
	U	900	15.00	3.0×10^{-6}	NaCl	AlO	D:O	4.40	4.20	4.40		NF		
	U	900	15.00	3.0×10^{-6}	NaCl	AlO	D:O	3.60	3.80	3.80		NF		
Granite, Bonner Quarry,	45	25	5.00	3.0×10^{-6}	NaCl	AlO	D:O	15.29	16.00	16.40		F		289
Westerly, Rhode	45	200	5.00	3.0×10^{-6}	NaCl	AlO	D:O	13.00	13.79	14.00		F		
Island, 30% quartz,	45	200	10.00	3.0×10^{-6}	NaCl	AlO	D:O	17.09	19.59	20.70		F		
30% oligoclase, 30%	45	300	5.00	3.0×10^{-6}	NaCl	AlO	D:O	12.50	13.00	13.20		F		
microcline, 5-10%	45	300	10.00	3.0×10^{-6}	NaCl	AlO	D:O	15.79	18.00	18.79		NF		
biotite, G.S. = 0.75 mm	45	400	5.00	3.0×10^{-6}	NaCl	AlO	D:O	12.10	12.29	12.29		F		
	45	500	5.00	3.0×10^{-6}	NaCl	AlO	D:O	10.39	10.79	11.20		F		
	45	500	7.50	3.0×10^{-6}	NaCl	AlO	D:O	13.39	14.29	14.29		NF		
	45	500	10.00	3.0×10^{-6}	NaCl	AlO	D:O	14.60	16.29	16.00		NF		
	45	600	5.00	3.0×10^{-6}	NaCl	AlO	D:O	9.89	10.60	11.50		F		
	45	600	7.50	3.0×10^{-6}	NaCl	AlO	D:O	11.89	11.39	10.10		F		
	45	600	10.00	3.0×10^{-6}	NaCl	AlO	D:O	12.60	12.70	12.39		NF		
	45	600	15.00	3.0×10^{-6}	NaCl	AlO	D:O	15.00	16.70	16.20		NF		289
	45	600	15.00	3.0×10^{-6}	NaCl	AlO	D:O	14.00	15.20	15.10		NF		

	45	600	5.00	3.0×10^{-7}	NaCl	AlO	D:O	8.00	7.00	7.80		F	
	45	700	5.00	3.0×10^{-6}	NaCl	AlO	D:O	8.10	8.00	8.10		F	
	45	700	5.00	3.0×10^{-6}	NaCl	AlO	D:O	7.40	7.10	6.90		F	
	45	700	5.00	3.0×10^{-6}	NaCl	AlO	D:O	7.40	5.80	4.60		F	
	45	700	10.00	3.0×10^{-6}	NaCl	AlO	D:O	8.70	8.70	8.20		NF	
	45	700	15.00	3.0×10^{-6}	NaCl	AlO	D:O	11.79	12.00	11.39		NF	
	45	900	5.00	3.0×10^{-6}	NaCl	AlO	D:O	5.40	4.60			F	
	45	900	7.50	3.0×10^{-6}	NaCl	AlO	D:O	2.40	2.30			NF	
	45	900	10.00	3.0×10^{-6}	NaCl	AlO	D:O	4.00	3.60	2.90		NF	
	45	900	15.00	3.0×10^{-6}	NaCl	AlO	D:O	4.50	4.50	4.10		NF	
	45	900	15.00	3.0×10^{-6}	NaCl	AlO	D:O	4.80	4.00	3.10		NF	
	N	900	15.00	3.0×10^{-6}	NaCl	AlO	D:O	1.80	2.30	2.30		NF	
Olivine crystal, gem quality, phenocrysts in basalt, St. John's Island, Red Sea, Fo 92, oriented 45° to [010] and [001]		800	10.00	5.0×10^{-5}	Talc*	AlO	D:A	6.50					232
Olivine crystal, gem quality, phenocrysts in basalt, St. John's Island, Red Sea, Fo 92, oriented 45° to [100] and [001]		600	10.00	5.0×10^{-5}	Talc	AlO	D:A	4.20					232
		600	10.00	5.0×10^{-5}	Talc	AlO	D:A	4.20					
		800	10.00	5.0×10^{-5}	Talc	AlO	D:A	4.40					
		800	10.00	5.0×10^{-5}	Talc	AlO	D:A	5.90					
		1000	10.00	5.0×10^{-5}	Talc	AlO	D:A	4.90					
		1000	10.00	5.0×10^{-5}	Talc	AlO	D:A	3.80					
Olivine crystals, gem quality, phenocrysts in basalt, St. John's Island, Red Sea, Fo 92, oriented 45° to [100] & [010]		800	10.00	5.0×10^{-5}	Talc	AlO	D:A	10.89					232
		1000	10.00	5.0×10^{-5}	Talc	AlO	D:A	4.30					
Orthopyroxene single crystal, Bamble, Norway, oriented about 45° to [100] and [001]		900	15.00	8.0×10^{-6}	NaCl	AlO	D:O				3.10		176—178
Pyroxene single crystal (diopside), Canyon Mountain Ophiolite Complex, Oregon, 20° to [100], 80° to [010], 45° to [001]		200	5.50	7.8×10^{-5}	Talc	WC	D:A	6.50	8.50	9.70		NF	16
		500	6.00	7.8×10^{-5}	Talc	WC	D:A	5.00	6.40	7.10		NF	
		600	17.00	7.8×10^{-5}	Talc	WC	D:A	7.90	10.70	12.39		NF	
		800	8.00	7.8×10^{-5}	Talc	WC	W:TD	5.10	5.90	6.40		NF	
		800	18.00	7.8×10^{-5}	Talc	WC	D:A	3.80	5.00	5.70		NF	
		950	8.00	7.8×10^{-5}	Talc	WC	W:TD	3.70	4.60	5.20		NF	
		1000	7.50	7.8×10^{-5}	Talc	WC	W:TD	3.30	3.80	4.30		NF	
		1000	17.00	7.8×10^{-5}	Talc	WC	W:TD	2.80	3.20	3.50		NF	
		1050	6.50	7.8×10^{-5}	Talc	WC	W:TD	2.10	3.20	3.50		NF	

Table 3 (continued)
STRESS-STRAIN RELATIONS IN CONSTANT STRAIN RATE TRIAXIAL COMPRESSION TESTS ON ROCKS AND MINERALS: PISTON-CYLINDER SOLID MEDIUM APPARATUS

					Experimental set-up			Differential stress — Kb strain (%)						
Material	Or	Temp (°C)	Conf. pressure (Kb)	Strain rate (s^{-1})	Conf. media	Pist. Mtrl.	Wet dry	5	10	15	Flow stress (Kb)	F/NF	Comment	Ref.
Pyroxene single crystal		400	6.00	7.8×10^{-5}	Talc	WC	D:A	8.79	12.70	15.20		F		16
(diopside), Canyon		700	7.00	7.8×10^{-5}	Talc	WC	D:A	11.00	13.00	13.60		F		
Mountain Ophiolite		1000	8.00	7.8×10^{-5}	Talc	WC	W:TD	1.90	3.00	3.40		NF		
Complex, Oregon, 61° to														
[100], 90° to [010], 45°														
to [001]														
Pyroxenite, Sleaford	U	400	14.39	1.1×10^{-3}	NaCl	AlO	D:VO	18.70	19.59	20.40		NF		166
Bay, South Australia,	U	400	14.89	1.1×10^{-4}	NaCl	AlO	D:VO	17.09	18.20	19.00		NF		
>99% clinopyroxene	U	400	14.89	1.1×10^{-5}	NaCl	AlO	D:VO	16.79	18.00	19.29		NF		
(salite), G.S. = 0.6mm	U	400	14.70	1.1×10^{-6}	NaCl	AlO	D:VO	16.79	17.70	18.29		NF		
	U	400	14.39	1.1×10^{-7}	NaCl	AlO	D:VO	14.60	16.00	17.00		NF		
	U	600	14.39	1.1×10^{-3}	NaCl	AlO	D:VO		17.50			NF		
	U	600	14.20	1.1×10^{-4}	NaCl	AlO	D:VO	16.29	17.00	17.20		NF		
	U	600	1.70	1.1×10^{-5}	NaCl	AlO	D:VO	4.50	5.00	4.90		F		
	U	600	3.80	1.1×10^{-5}	NaCl	AlO	D:VO	8.79	9.29	9.60		F		
	U	600	5.00	1.1×10^{-5}	NaCl	AlO	D:VO	10.20	10.70	10.89		NF		
	U	600	7.30	1.1×10^{-5}	NaCl	AlO	D:VO	11.89	12.60	13.10		NF		
	U	600	8.60	1.1×10^{-5}	NaCl	AlO	D:VO	11.60	12.79	13.50		NF		
	U	600	10.00	1.1×10^{-5}	NaCl	AlO	D:VO	13.60	14.60	15.29		NF		
	U	600	13.89	1.1×10^{-5}	NaCl	AlO	D:VO	14.89	16.20	16.59		NF		
	U	600	14.79	1.1×10^{-5}	NaCl	AlO	D:VO	16.09	16.70	17.50		NF		
	U	600	14.79	1.1×10^{-5}	NaCl	AlO	D:VO		16.70			NF		
	U	600	20.09	1.1×10^{-5}	NaCl	AlO	D:VO	16.00	17.20	18.29		NF		
	U	600	14.60	1.1×10^{-6}	NaCl	AlO	D:VO	15.00	16.20	17.00		NF		
	U	800	14.79	1.1×10^{-3}	NaCl	AlO	D:VO	15.20	16.20	16.50				
	U	800	16.09	1.1×10^{-4}	NaCl	AlO	D:VO	15.29	15.60	15.79				
	U	800	15.20	1.1×10^{-5}	NaCl	AlO	D:VO	14.79	15.20	14.70				
	U	800	14.60	1.1×10^{-6}	NaCl	AlO	D:VO	10.89	11.20	10.50				
	U	800	15.10	1.1×10^{-7}	NaCl	AlO	D:VO	8.70	7.60					

	U	900	14.70	1.1×10^{-3}	NaCl	AlO	D:VO	15.60	16.29	17.09			
	U	900	14.29	1.1×10^{-4}	NaCl	AlO	D:VO	13.10	13.10	13.89			
	U	900	15.39	1.1×10^{-5}	NaCl	AlO	D:VO	13.39	13.00	11.10			
	U	900	15.00	3.0×10^{-6}	NaCl	AlO	D:O	6.30	5.70	5.20	NF		287
	U	900	14.70	1.1×10^{-6}	NaCl	AlO	D:VO	8.79	8.10	7.40			166
	U	900	14.89	1.1×10^{-6}	NaCl	AlO	D:VO	8.79	6.80	5.50			
	U	900	15.39	1.1×10^{-6}	NaCl	AlO	D:VO	10.10	8.60	6.90			
	U	900	14.60	1.1×10^{-7}	NaCl	AlO	D:VO	5.70	4.80	4.00			
	U	1000	15.50	1.1×10^{-3}	NaF	AlO	D:VO	11.79	13.10	13.29			
	U	1000	15.70	1.0×10^{-3}	NaF	AlO	D:VO	15.50	16.29	16.59			
	U	1000	14.60	1.1×10^{-4}	NaF	AlO	D:VO	8.39	8.79	9.10			
	U	1000	14.89	1.1×10^{-5}	NaF	AlO	D:VO	6.60	7.00	6.90			
	U	1000	15.89	1.1×10^{-5}	NaF	AlO	D:VO	5.70	4.70	3.90			
	U	1000	12.60	1.1×10^{-6}	NaF	AlO	D:VO	4.00	3.50	3.30			
	U	1000	13.60	1.1×10^{-6}	NaF	AlO	D:VO	4.50	4.80	3.20			
	U	1100	15.29	1.1×10^{-3}	NaF	AlO	D:VO	12.10	14.00	16.50			
	U	1100	15.70	1.1×10^{-5}	NaF	AlO	D:VO	5.40	5.40	5.10			
Pyroxenite, Stillwater	U	600	15.00	1.1×10^{-4}	NaCl	AlO	D:VO	12.70	13.50	14.00	NF		173
Complex, Montana, 90%	U	600	1.50	1.1×10^{-5}	NaCl	AlO	D:VO	5.00	5.20	5.30	F	Stress drop at 3.5% strain and 5.7 Kb stress	
orthopyroxene													
(bronzite), 10%	U	600	5.00	1.1×10^{-5}	NaCl	AlO	D:VO	9.20	9.60	9.89	NF		
clinopyroxene, G.S. =	U	600	10.00	1.1×10^{-5}	NaCl	AlO	D:VO	11.89	12.20	12.20	NF		
1—2mm	U	600	10.00	1.1×10^{-5}	NaCl	AlO	D:VO	12.89	13.70	14.29	NF		
	U	600	15.00	1.1×10^{-6}	NaCl	AlO	D:VO	11.70	12.70	13.39	NF		
	U	900	15.00	1.1×10^{-5}	NaCl	AlO	D:VO	11.39	12.29	12.29	NF		
Pyroxenite, Webster,	N	1000	10.10	1.1×10^{-5}	Talc	WC	W:TD	8.50	9.10		NF		249
N.C., orthopyroxene,	N	1000	9.79	1.1×10^{-6}	Talc	WC	W:TD	4.20	4.40	4.30	NF		
G.S. = 1mm	N	1050	10.29	1.1×10^{-5}	Talc	WC	W:TD	5.90	6.50	6.50	NF		
	N	1050	10.00	1.1×10^{-6}	Talc	WC	W:TD	2.60	2.70	2.70	NF		
	N	1100	9.89	1.1×10^{-4}	Talc	WC	W:TD	9.50	10.39		NF		
	N	1100	9.70	1.1×10^{-5}	Talc	WC	W:TD	4.30	4.70	4.50	NF		
	N	1150	10.10	1.1×10^{-6}	Talc	WC	W:TD	2.00	2.20	2.10	NF		
	N	1150	10.10	1.1×10^{-6}	Talc	WC	W:TD	1.30	1.50	1.50	NF		
	N	1200	9.89	1.1×10^{-4}	Talc	WC	W:TD	6.30	6.40		NF		
	N	1200	10.20	1.1×10^{-5}	Talc	WC	W:TD	2.60	2.80	2.70	NF		
	N	1250	10.10	1.1×10^{-4}	Talc	WC	W:TD	4.40	4.70		NF		
	N	1250	10.29	1.1×10^{-5}	Talc	WC	W:TD	1.90	2.10	2.00	NF		
	N	1280	10.00	1.1×10^{-4}	Talc	WC	W:TD	3.60	4.00		NF		
	N	1300	10.20	1.1×10^{-4}	Talc	WC	W:TD	2.20	2.20		NF		
	N	1300	10.10	1.1×10^{-5}	Talc	WC	W:TD	1.50	1.70	1.70	NF		

Table 3 (continued)
STRESS-STRAIN RELATIONS IN CONSTANT STRAIN RATE TRIAXIAL COMPRESSION TESTS ON ROCKS AND MINERALS: PISTON-CYLINDER SOLID MEDIUM APPARATUS

					Experimental set-up			Differential stress — Kb strain (%)						
Material	Or	Temp (°C)	Conf. pressure (Kb)	Strain rate (s^{-1})	Conf. media	Pist. Mtrl.	Wet dry	5	10	15	Flow stress (Kb)	F/NF	Comment	Ref.
Quartz crystal, Brazil,	0+	300	14.50	7.8×10^{-6}	Talc	WC	D:O				46.30			27, 28
dry (<50ppm OH),	0+	300	15.00	7.7×10^{-8}	Talc	WC	D:O				41.09			
compression at 45°	0+	400	15.50	7.7×10^{-5}	Talc	WC	D:O				46.69			
to [a] and	0+	400	15.00	8.2×10^{-6}	Talc	WC	D:O				44.59			
[c], designated 0^+	0+	450	15.00	7.8×10^{-7}	Talc	WC	D:O				38.40			
	0+	500	15.50	1.0×10^{-5}	KCl	AlO	D:O				40.00			
	0+	600	14.50	7.7×10^{-6}	Talc	WC	D:O				39.90			
	0+	600	14.50	7.7×10^{-7}	Talc	WC	D:O				31.29			
	0+	600	16.00	7.7×10^{-8}	Talc	WC	D:O				30.59			
	0+	700	14.79	1.1×10^{-5}	NaCl	AlO	D:O				38.00			
	0+	700	9.00	7.7×10^{-6}	Talc	WC	D:O				35.69			
	0+	800	15.50	7.7×10^{-5}	Talc	WC	D:O				41.69			
	0+	800	14.00	1.1×10^{-5}	Talc	WC	W:SC				1.00			
	0+	1000	13.50	1.1×10^{-5}	Talc	WC	W:TD				2.30			
	0+	1000	14.50	1.0×10^{-5}	NaCl	AlO	D:O				33.00			
	0+	1100	14.00	1.1×10^{-5}	NaCl	AlO	D:O				24.00			
Quartz crystal, Brazil,	⊥r	300	14.00	7.7×10^{-6}	Talc	WC	D:O				50.30			27, 28
dry (<50ppm OH),	⊥r	400	14.50	7.7×10^{-6}	Talc	WC	D:O				45.30			
compression normal to	⊥r	500	15.00	7.8×10^{-6}	Talc	WC	D:O				45.50			
first order	⊥r	600	14.00	7.8×10^{-6}	Talc	WC	D:O				44.00			
rhombohedron $(10\bar{1}1)$ =	⊥r	740	15.00	7.7×10^{-6}	Talc	WC	D:O				35.09			
⊥r	⊥r	850	15.50	7.7×10^{-6}	Talc	WC	W:TD				19.40			
	⊥r	870	15.50	7.7×10^{-6}	Talc	WC	W:TD				13.89			
	⊥r	900	17.50	7.9×10^{-6}	ASM	WC	D:O				28.59			27, 28
	⊥r	900	15.50	7.7×10^{-6}	Talc	WC	W:TD				25.20			
	⊥r	1000	14.50	7.8×10^{-6}	ASM	WC	D:O				15.10			
Quartzite, "Black Hills", Northern Black Hills, South Dakota, G.S. = 0.075mm	U	600	15.00	1.0×10^{-5}	NaCl	WC	D:A		20.00					285
Quartzite, "Canyon	U	960	10.10	5.1×10^{-5}	Talc	WC	W:TD	3.47	3.45	3.46				248
Creek", Big Hole	U	960	10.10	2.1×10^{-5}	Talc	WC	W:TD	2.45	2.44	2.44				

Canyon, Montana	U	1060	10.10	5.1×10^{-6}	NaCl/CaF	WC	W:SC	.890	.880					
	U	1060	10.10	5.1×10^{-6}	Talc	WC	W:TD	.780	.790	.780				
	U	1060	10.10	5.1×10^{-6}	Talc	WC	W:TD	.720	.710	.710				
	U	1060	9.89	2.1×10^{-6}	Talc	WC	W:TD	.580	.560	.560				
	U	1060	10.10	2.1×10^{-6}	NaCl/CaF	WC	W:SC	.480	.460	.460				
	U	1160	9.89	2.1×10^{-4}	NaCl/CaF	WC	W:SC	1.91	1.90	1.91				
	U	1160	9.79	2.1×10^{-5}	NaCl/CaF	WC	W:SC	1.55	1.54	1.54				
	U	1160	10.00	5.1×10^{-6}	Talc	WC	W:TD	1.48	1.46	1.46				
	U	1160	10.10	5.1×10^{-6}	Talc	WC	W:TD	.560	.550	.550				
	U	1160	10.10	5.1×10^{-7}	NaCl/CaF	WC	W:SC	.370	3.80	.380				
Quartzite, "Heavitree",	U	600	15.00	3.0×10^{-6}	NaCl	AlO	D:O	15.50	21.00			NF		287
Australia, grain size	U	900	15.00	3.0×10^{-5}	NaCl	AlO	D:O	5.00	6.00	6.40		NF		
= 0.2mm	U	900	15.00	3.0×10^{-6}	NaCl	AlO	D:O	2.00	2.10			NF		
Quartzite, "Quadrant"	U	400	14.50	1.0×10^{-5}	Talc	WC	D:A		40.00			F		285
or "Canyon Creek",	U	400	14.00	1.0×10^{-6}	Talc	WC	D:A		36.69			NF		
Beaverhead County,	U	400	14.50	1.0×10^{-6}	Talc	WC	D:A		30.09			NF		
Montana, G.S. = 0.15mm	U	400	15.00	1.0×10^{-7}	Talc	WC	D:A		32.80			NF		
	U	500	14.50	1.0×10^{-5}	Talc	WC	D:A		34.50			NF		
	U	500	15.00	1.0×10^{-7}	Talc	WC	D:A		23.59			NF		
	U	600	14.50	1.0×10^{-5}	Talc	WC	D:A		26.29			NF		
	U	600	13.60	1.0×10^{-6}	NaCl	WC	D:A		20.29			NF		
	U	600	15.00	1.0×10^{-6}	Talc	WC	D:A		19.59			NF		
	U	600	15.00	1.0×10^{-7}	Talc	WC	D:A		15.00			NF		
	U	700	14.50	1.0×10^{-5}	Talc	WC	D:A		22.90			NF		
	U	700	14.50	1.0×10^{-6}	Talc	WC	D:A		23.90			NF		
	U	700	15.00	1.0×10^{-7}	Talc	WC	D:A		9.20			NF		
	U	800	14.50	1.0×10^{-5}	Talc	WC	D:A		15.50			NF		
	U	800	15.00	1.0×10^{-6}	Talc	WC	D:A		7.00			NF		
	U	800	15.00	1.0×10^{-7}	Talc	WC	D:A		4.60			NF		
	U	850	14.50	1.0×10^{-5}	Talc	WC	D:A		7.40			NF		
	U	850	15.50	1.0×10^{-6}	Talc	WC	W:TD		4.00			NF		
	U	900	14.50	1.0×10^{-5}	Talc	WC	W:TD		6.80			NF		
	U	900	15.00	1.0×10^{-6}	Talc	WC	W:TD		2.70			NF		
	U	1000	14.50	1.0×10^{-5}	Talc	WC	W:TD		4.10			NF		
Quartzite, "Simpson",	U	700	15.00	1.0×10^{-5}	Talc	WC	D:A		21.29					285
12,956 ft level in	U	850	15.00	1.0×10^{-5}	Talc	WC	W:TD		7.70					
Sunray Cullen well,	U	1000	15.00	1.0×10^{-5}	Talc	WC	W:TD		3.10					
Oklahoma, G.S. = 0.2mm														
Spinel single crystal, synthetic	<001>	400	14.00	2.0×10^{-5}	NaCl	AlO	D:O				19.40	NF	Activated {110}<$1\bar{1}0$> slip	172
(Union Carbide), MgO: 1.1(Al_2O_3)	<011>	400	14.00	2.0×10^{-5}	NaCl	AlO	D:O				43.00	NF	Activated {111}<$1\bar{1}0$> slip	

Table 3 (continued)
STRESS-STRAIN RELATIONS IN CONSTANT STRAIN RATE TRIAXIAL COMPRESSION TESTS ON ROCKS AND MINERALS: PISTON-CYLINDER SOLID MEDIUM APPARATUS

Material	Or	Temp (°C)	Conf. pressure (Kb)	Strain rate (s^{-1})	Experimental set-up: Conf. media	Pist. Mtrl.	Wet dry	Differential stress — Kb strain (%): 5	10	15	Flow stress (Kb)	F/NF	Comment	Ref.
	<111>	400	14.00	2.0×10^{-5}	NaCl	AlO	D:O				37.19	NF	Activated {100}<1$\bar{1}$0> slip	
Websterite, Webster	P	200	5.80	1.0×10^{-4}	Talc	WC	D:A	17.79	20.09	21.50		NF		16
Co., North Carolina,	P	200	4.80	1.0×10^{-5}	Talc	WC	D:A	14.79	16.40	17.59		NF		
68% clinopyroxene	P	200	5.50	1.0×10^{-6}	Talc	WC	D:A	16.20	17.59	18.40		NF		
(diopside, 0.4—1.7mm),	P	300	5.30	1.0×10^{-4}	Talc	WC	D:A	16.09	17.20	18.40		NF		
32% orthopyroxene	P	400	5.20	1.0×10^{-4}	Talc	WC	D:A	13.89	15.89	17.20		NF		
(bronzite, 1.7mm)	P	500	5.50	1.0×10^{-4}	Talc	WC	D:A	13.50	15.10			NF		
	P	700	5.70	1.0×10^{-4}	Talc	WC	D:A	12.39	14.00	15.10		NF		
	P	800	6.30	1.0×10^{-4}	Talc	WC	D:A	11.39	12.89	14.29		NF		
	P	800	6.50	1.0×10^{-5}	Talc	WC	D:A	12.00	13.20	13.79		NF		
	P	800	5.50	1.0×10^{-6}	Talc	WC	D:A	9.89	11.00	11.60		NF		
	P	900	6.30	1.0×10^{-4}	Talc	WC	W:TD	9.10	9.79	10.00		NF	SS = 10.2 Kb	
	P	900	6.10	1.0×10^{-6}	Talc	WC	W:TD	5.40	6.60	6.80		NF	SS = 6.8 Kb	
	P	950	7.00	1.0×10^{-4}	Talc	WC	W:TD	6.60	8.00	9.00		NF	SS = 9.3 Kb	
	P	1000	6.60	1.0×10^{-4}	Talc	WC	W:TD	5.80	6.70	7.10		NF	SS = 7.2 Kb	
	P	1000	6.50	1.0×10^{-5}	Talc	WC	W:TD	3.70	5.00	5.40		NF	SS = 6.1 Kb	
	P	1000	6.80	1.0×10^{-5}	Talc	WC	W:TD	2.90	4.30	4.80		NF	SS = 5.4 Kb	
	P	1000	7.00	1.0×10^{-5}	Talc	WC	W:TD	2.80	4.00	4.90		NF	SS = 5.6 Kb	
	P	1000	7.00	1.0×10^{-5}	Talc	WC	W:TD	2.90	4.30	4.90		NF	SS = 4.9 Kb	
	P	1000	10.20	1.0×10^{-5}	ASM	WC	D:O	7.10	8.70	9.29		NF	SS = 10.4 Kb	
	P	1000	10.00	1.0×10^{-6}	ASM	WC	D:O	8.10	9.60	10.20		NF		
	P	1000	10.00	1.0×10^{-6}	ASM	WC	D:O	6.80	8.00	8.70		NF	SS = 9.1 Kb	
	P	1000	10.50	1.0×10^{-6}	ASM	WC	D:O	6.20	7.50	8.39		NF	SS = 9.4 Kb	
	P	1000	10.60	1.0×10^{-6}	ASM	WC	D:O	5.90	7.20	8.00		NF	SS = 8.6 Kb	
	P	1000	10.00	1.0×10^{-7}	ASM	WC	D:O	3.40	5.20	6.30		NF	SS = 6.6 Kb	
	P	1050	6.20	1.0×10^{-4}	Talc	WC	W:TD	2.90	4.30	5.00		NF	SS = 5.6 Kb	
	P	1050	6.50	1.0×10^{-5}	Talc	WC	W:TD	2.40	3.50	3.90		NF	SS = 4.3 Kb	
	P	1050	10.50	1.0×10^{-6}	ASM	WC	D:O	2.30	4.60	7.20		NF	SS = 7.6 Kb	
	P	1100	10.00	1.0×10^{-4}	ASM	WC	D:O	8.29	9.39	9.89		NF	SS = 11.1 Kb	

P	1100	10.50	1.0×10^{-4}	ASM	WC	D:O	7.30	8.79	9.60	NF	SS = 11.1 Kb	
P	1100	10.39	1.0×10^{-5}	ASM	WC	D:O	3.80	5.80	6.80	NF	SS = 7.9 Kb	
P	1100	10.00	1.0×10^{-6}	ASM	WC	D:O	1.60	3.40	4.70	NF	SS = 7.0 Kb	
P	1100	10.20	1.0×10^{-6}	ASM	WC	D:O	4.20	5.80	6.60	NF	SS = 8.1 Kb	
P	1100	10.00	1.0×10^{-7}	ASM	WC	D:O	2.10	3.40	4.20	NF	SS = 4.9 Kb	
P	1200	10.00	1.0×10^{-5}	ASM	WC	D:O	1.50	2.80	4.40	NF	SS = 7.0 Kb	
P	1200	11.29	1.0×10^{-6}	ASM	WC	D:O	2.50	4.30	4.90	NF	SS = 5.4 Kb	
U	1000	10.00	1.0×10^{-6}	ASM	C	D:O	8.10	9.60	10.20	NF	0	16
U	1200	10.00	1.0×10^{-5}	ASM	C	D:O	1.50	2.80	4.40	NF	SS = 7.0	

Note: A host of practical problems generally limit the maximum test temperature and pressure in rock deformation apparatus utilizing fluids as pressure media. Experimental data at confining pressures in excess of 8 kbar and temperatures above 800°C are rare, and only recently has gas apparatus design permitted triaxial tests at temperatures above 1000°C even at pressures below 5 kbar. These conditions are adequate to promote flow well into the regime of ductile flow for certain ''soft'' rocks and minerals (carbonates, alkali, and alkaline earth halides, sulfates, sulfides, and ''wet'' synthetic quartz). For most silicates, these limitations in pressure and temperature are inadequate to suppress the contribution of brittle processes at practical laboratory strain rates and creep stresses. At confining pressures which are convenient and practical for these types of apparatuses, some rock types (such as the feldspathic rocks), are brittle right up to the temperature of first melting.[88] In short, the ductile strengths of most silicates cannot currently be characterized in apparatuses of this kind. In the early 1960s Griggs[99] set out to design a triaxial apparatus based on the solid-medium piston cylinder hydrostatic apparatus developed by the Geophysical Laboratory. The cylindrical sample is sheathed in a nominally weak solid medium which is compressed and pressurized by the advance of a large confining pressure piston which closely fits the bore of the pressure vessel (Figure 27). The specimen is axially loaded by a smaller axial piston (L) acting through carbide or ceramic spacers (E) and bearing upon a carbide anvil A. The specimen is heated by a cylindrical graphite resistance furnace coaxial with the specimen and temperature is monitored by a ceramic sheathed thermocouple (TC) which touches the side of the sample. The differential stress on the sample is subject to the following uncertainties: (1) Since the axial force on the sample is measured external to the pressure vessel, the piston friction through the pressure packing (P) contributes to the measured total axial force. Uncertainties in the piston friction correction lead to uncertainties in the differential stress on the specimen. (2) Viscous and frictional drag on the upper spacer (E) and the sample support part of the measured force. (3) The radial expansion of the sample must be accommodated by flow in the confining medium and significant strength of the medium increases the total force necessary to achieve a given specimen strain. (4) Since the advance of the load piston (L) decreases the volume within the vessel, the confining pressure increases with increasing piston displacement. The uncertainty in piston friction can be minimized by operating in the constant strain rate mode so that the piston velocity is constant throughout the test and by determining the force at zero differential stress directly from the deflection of the force curve. This procedure introduces an uncertainty of about ±0.4 kbar in the differential stress measurement. Drag effects (2) and the resistance to radial expansion (3) are minimized by the selection of a weak confining medium. Blacic[27] has estimated the sum of these errors for the confining media that are typically used and these estimates, extended by us, are summarized as follows:

Medium	T°C	P kbar
Talc	<800	3
	>800	1
Alsimag 222	<900	Very large
	900—1100	5
	1200	<2
NaCl	<500	<1
	>500	<0.5
NaF	<900	1
	>900	0.5

Table 3 (continued)
STRESS-STRAIN RELATIONS IN CONSTANT STRAIN RATE TRIAXIAL COMPRESSION TESTS ON ROCKS AND MINERALS: PISTON-CYLINDER SOLID MEDIUM APPARATUS

Obviously these errors decrease with increasing temperatures. The increase in confining pressure with increasing piston displacement depends upon the bulk modulus of the pressure cell contents and the "compliance" of the pressure vessel. The use of alkali halides as confining media is attractive, not only because of their low strength, but also due to their high compressibilities. Pressure increases due to piston displacement in alkali halide cells can be held to less than 0.5 kbar at 20% specimen strain and can be corrected for. The absolute confining pressure based on the cell diameter and total force exerted on the confining pressure piston (C) exceed the actual pressure in the cell due to: (1) friction between the vessel bore and the piston, the piston seal, and the surface of the sample assembly; (2) pressure gradients due to the strength of the confining medium. Mirwald et al.[203] show that the use of alkali halide confining media and friction mitigation measures can reduce the uncertainty in pressure in piston-cylinder apparatuses to about (±1% + 0.5 kbar). In practice, the uncertainties in pressure (based on the calibrations in the petrological piston cylinder apparatus) are likely to be (±0.5 kbar ±2%) for alkali halide cells with nominal temperatures above 300°C and (±0.5 kbar ±2% — (5 to 10%)) for other confining media. Uncertainties in the specimen temperature are primarily due to uncertainties in the pressure correction of the thermocouple emf (~10°C at 15 kbar), to axial temperature gradients caused by heat conduction out the axial spacers (E), and to gradients introduced by the perturbation of the temperature field by the thermocouple. Spatial temperature variations depend upon the specific assembly design but the following may be used as a guide:

Radial temperature variation	±5 to 10% of nominal T, °C
Axial temperature variation	
Tungsten carbide pistons	−25 to −50% of nominal T, °C
Aluminum oxide pistons	
No metal jacket	−10 to −20% of nominal T, °C
With metal jacket	−3 to −10% of nominal T, °C

Recent developments in furnace design may reduce the temperature uncertainties to less than 1%. Table explanation and symbol key — The columns pertaining to the sample characterization and experimental conditions are the same as Table 2 except as follows: all experiments are triaxial compression. Confining medium (self explanatory), piston material (WC, tungsten carbide; AlO, aluminum oxide (corundum)); wet/dry (D:A, air dried, D:O, oven dried, W:TD, wet, water from talc dehydration, W:SC, wet, water in sealed capsule). Flow stress is defined as the stress at which a strong deflection from linear (quasi-elastic) stress-strain behavior. SS is the steady-state stress, where the stress becomes independent of strain.

Table 4
STRENGTH DATA FROM TORSION EXPERIMENTS ON SOLID CYLINDERS OF ROCKS AND MINERALS

Material	Orientation	Temp (°C)	Press (Kbar)	Axial stress (Kbar)	Twist rate (rad/min)	Torque (Dyne-cm × 10^{-8})	Maximum shear stress (Kbar)	Angular relations α	β	θ	F/NF	Ref.
Calcite, single	⊥a	20	2.800	0	0.10	1.34	0.94				NF	37
crystals, yield torque	⊥a	20	1.000	1	0.10	1.30	0.81				NF	
indicated as maximum	⊥a	20	2.700	3	0.10	3.04	1.86				NF	
torque (solid	⊥c	20	2.800	0	0.10	2.21	1.54				NF	
cylinders)	⊥c	20	1.000	1	0.10	1.37	0.85				F	
	⊥c	20	2.700	3	0.10	4.10	2.54				NF	
	⊥e	20	2.800	0	0.10	1.79	1.25				NF	
	⊥e	20	1.000	1	0.10	1.22	0.76				F	
	⊥e	20	2.700	3	0.10	2.70	1.67				F	
	⊥r	20	2.800	0	0.10	0.58	0.41				NF	
	⊥r	20	2.800	0	0.10	0.64	0.45				NF	
	⊥r	20	1.000	1	0.10	1.35	0.84				NF	
	⊥r	20	2.700	3	0.10	2.70	2.00				NF	
Limestone, ''Carrara	U	20	0.001	0	0.10	9.29	0.18				F	32
Marble (solid	U	20	0.300	0	0.10	35.59	0.67				F	
cylinders)	U	20	0.470	0	0.10	39.80	0.75				F	
	U	20	1.020	1	0.10	76.69	1.44				NF	
	U	20	1.280	1	0.10	69.00	1.30				F	
	U	20	1.640	2	0.10	9.54	1.81				NF	
	U	20	1.940	2	0.10	9.79	1.85				NF	
	U	20	0.570	2	0.10	7.30	1.38				F	
	U	20	2.380	2	0.10	10.20	1.92				NF	
	U	20	0.650	3	0.10	8.48	1.60				F	
	U	20	0.880	3	0.10	11.50	2.17				F	
Limestone, ''Solenhofen'',	U	20	0.001	0	0.10	0.23	0.17		45		F	117
Bavaria, Germany (solid	U	20	0.001	0	0.10	0.25	0.18	45	45		F	
cylinders)	U	20	0.001	0	0.10	0.50	0.35	48	45	−3	F	
	U	20	0.500	0	0.10	1.90	1.36	48	45	−3	F	

Table 4 (continued)
STRENGTH DATA FROM TORSION EXPERIMENTS ON SOLID CYLINDERS OF ROCKS AND MINERALS

Material	Orientation	Temp (°C)	Press (Kbar)	Axial stress (Kbar)	Twist rate (rad/min)	Torque (Dyne-cm × 10^{-8})	Maximum shear stress (Kbar)	Angular relations α	β	θ	F/NF	Ref.
	U	20	1.000	0	0.10	2.40	1.72	24	45	21	F	
	U	20	2.000	0	0.10	2.40	2.74	21	45	24	F	
	U	20	1.000	0	0.10	2.50	1.78	21	45	24	F	
	U	20	3.500	0	0.10	2.65	5.24	15	45	30	F	
	U	20	2.000	0	0.10	3.35	2.40	20	45	25	F	
	U	20	3.000	0	0.10	3.37	3.86	20	45	25	F	
	U	20	3.000	0	0.10	4.27	3.05	15	45	30	F	
	U	20	3.000	0	0.10	5.56	3.97	19	45	26	F	
	U	20	4.500	0	0.10	6.12	4.38		45		NF	
	U	20	4.000	0	0.10	6.16	4.40	16	45	29	F	
	U	20	0.001	0	0.10	0.65	0.46	55	33	2	F	
	U	20	0.001	0	0.10	0.24	0.17	45	19	26	F	
	U	20	0.001	1	0.10	1.06	0.76	50	30	10	F	
	U	20	1.000	1	0.10	2.25	1.60	37	38	15	F	
	U	20	1.000	1	0.10	2.38	1.70	28	38	24	F	
	U	20	1.000	1	0.10	2.99	2.14	27	39	24	F	
	U	20	1.000	1	0.10	2.74	1.95	28	38	24	F	
	U	20	2.000	1	0.10	4.04	2.88	41	33	−19	F	
	U	20	1.500	1	0.10	4.07	2.91	32	41	17	F	
	U	20	2.000	1	0.10	4.60	3.29	24	41	25	F	
	U	20	1		0.10	5.37	3.83	16	42	32	F	
	U	20	3.000	1	0.10	5.58	3.98	15	42	33	F	
	U	20	0.001	1	0.10	0.65	0.47	62	23	5	F	
	U	20	0.500	1	0.10	2.03	1.45	27	36	27	F	
	U	20	4.000	1	0.10	6.52	4.66	64	42	−16	F	
	U	20	0.001	1	0.10	1.27	0.91		29		F	
	U	20	2.000	2	0.10	4.60	3.29	25	38	27	F	
	U	20	2.000	2	0.10	5.08	3.63	34	38	18	F	
	U	20	1.000	2	0.10	2.80	2.07	24	34	32	F	
	U	20	2.000	2	0.10	4.33	3.09	24	37	29	F	

	U	20	3.000	2	0.10	5.91	4.21	27	39	24	F	
	U	20	4.000	2	0.10	3.37	3.86	61	39	10	F	
	U	20	0.500	2	0.10	2.15	1.54	54	30	6	F	
	U	20	2.000	2	0.10	5.93	4.23	21	38	31	F	
	U	20	1.000	2	0.10	3.71	2.65	40	34	16	F	
	U	20	1.500	2	0.10	4.00	2.86	51	34	5	F	
	U	20	3.000	3	0.10	5.10	3.64		35		F	
	U	20	2.000	3	0.10	5.40	3.85	32	35	23	F	
	U	20	0.500	3	0.10	2.18	1.56	50	24	16	F	
	U	20	1.000	3	0.10	3.29	2.35	33	29	28	F	
	U	20	1.000	3	0.10	3.48	2.48	35	30	25	F	
	U	20	2.000	3	0.10	6.15	4.38	75	36	−21	F	
	U	20	0.001	3	0.10	1.90	1.35		21		F	
	U	20	1.000	4	0.10	2.90	2.07		24		NF	
	U	20	1.000	4	0.10	3.43	2.45	50	27	13	F	
	U	20	2.000	4	0.10	5.91	4.20	11	33	46	F	
	U	20	3.000	4	0.10	5.08	3.63		31		NF	
	U	20	2.500	4	0.10	4.90	3.50		31		NF	
	U	20	3.000	4	0.10	6.26	4.47		33		NF	
Limestone, "Yule Marble", (solid cylinders)	N	20	1.000	1	0.10	1.22	0.72				F	118
	N	20	2.000	2	0.10	3.40	2.16				F	
	P	20	1.000	1	0.10	1.85	1.07				F	
	P	20	1.000	1	0.10	2.30	1.41				F	
	P	20	2.000	2	0.10	2.60	2.29				F	
	P	20	2.000	3	0.10	2.49	1.48				F	
	P	20	2.700	3	0.10	5.20	3.39				F	
	P	20	2.000	3	0.10	2.20	1.32				F	
	P	20	2.700	4	0.10	7.95	4.68				NF	
	P	150	2.000	2	0.10	2.35	2.07				F	
	P	300	2.000	2	0.10	2.24	1.83				NF	

Note: The formats for material identification, orientation, temperature, pressure, and fault characteristics are as indicated in the Table 2. Typically, the torque vs. twist and axial differential stress vs. axial strain data are not presented in the literature on these types of tests, so we tabulate here the torque at yield or fracture for a given fixed axial stress σ_x and compute the maximum shear stress $(\tau_{xy})_{max}$ from Equation 1 in the text. The parameters describing the angular relations between cylinder axis x direction and fault or fracture planes are defined in Figure 1E and presented in the table in degrees. In the experiments with nonzero axial stress σ_x, the stress is applied at a strain rate of about 10^{-4} s^{-1} and then fixed at a value below the corresponding triaxial strength of the material. The torque is then applied to failure or yield. Additional data on rock strength in torsion on hollow cylinders may be found in Handin et al.[117] and Durand[77] for rocks and in Steinemann,[312] Kamb,[154] Byers,[53] and Duval[81] for ice.

Table 5
STEADY-STATE FLOW LAW PARAMETERS FOR THE HIGH-TEMPERATURE FLOW OF ROCKS TESTED IN FLUID MEDIUM APPARATUS

Material	$\log_{10}A$ ($kbar^{-n}s^{-1}$)	n	Δσ (kbar)	Q* (kcal/mol)	ΔT (°C)	P (kbar)	Comments[b]	Ref.
Anhydrite, Riberg, Switzerland, >90% pure, 0.3 × 0.05 mm G.S.	5.48	2	0.2—0.6[a]	36.4 ± 2.2	350—450	1.5	CSR, SR	212
Chalcopyrite, Ontario, Canada, 70—95% pure, G.S. = 1 mm	5.2 ± 1.1	8.6 ± 1.0	1.0—4.9	30.0 ± 2.7	150—250	1.5	CSR, SRD	247
Diabase, Frederick, Md., 36% plag(An70), 58% cpx, 3% chlorite, 3% opaque minerals, G.S. = 0.2 mm		3.0 ± 0.3	<2.8[a]				CS, CSR, T = 990°C	54
			>2.8				CS, CSR, T = 990°C	54
		13						
Dolomite, "Crevola marble", Simplon, Switzerland, nominally pure	5.3 ± 2.6	9.1 ± 0.9	3.1—7.5	83.2 ± 6.0	700—900	7.2—9.0	CSR	125
Galena, Mt. Isa, Australia, 88% galena, 5% sphalerite, 6% quartz, 1% pyrite, G.S. = 0.07 mm	3.5 ± 0.6	7.3 ± 0.6	0.3—1.2	22.5 ± 3.7	200—400	1.5	CSR	6
Halite, synthetic, nonporous, <10 at. ppm divalent impurities, G.S. = 0.2—3 mm	14.3	5.5 ± 0.4	0.014—0.138	38 ± 6	365—513	0.001	CS, dry nitogen atm	47
	14.0	3.6 ± 0.4	0.003—0.009	48 ± 6	603—742	0.001	CS	47
Halite, synthetic, <1% porosity, <10 at. ppm divalent impurities, G.S. = 2—3 mm	10.9 ± 0.8	5.5 × 0.4	0.02—0.14	23.5 ± 1.9	100—400	2.0	CSR, extension	123
Ice Ih, synthetic, nominally pure, grown from boiled, deionized distilled water, G.S. = 1 to 2 mm	26.3	3.2	0.001—0.010	32	−13—0	0.001	CS	91
	14.1	3.1 ± 0.1	0.004—0.038[a]	17.8	−8 — −45	0.001	CS	26
	23.6	3.2	0.001— 0.038[a]	29.0	−2 — −8	0.001	CS	26
		2.6	0.001—0.010			0.001	CS, combined torsion and compression of hollow cylinders, T = −10—0	53

Limestone, lithographic, "Solenhofen", Germany, weak preferred orientation, G.S. = 4 μm	12.8	4.7	0.7—2.0[a]	71.1	600—900	3.0	CSR, SR, CS	260
	7.8	1.7	0.01—0.07[a]	50.9	600—900	3.0	CSR, SR, CS	260
Marble, "Yule", Colorado, >98% calcite, G.S. = 0.3 mm, strong preferred orientation: c-axis maximum ⊥ foliation								
Cylinders ‖ foliation	11.8 ± 0.9	7.7 ± 0.5	0.22—1.44[a]	60.7 ± 2.9	400—800	5.0	CSR, extension	135
Cylinders ⊥ foliation	12.7 ± 0.7	8.3 ± 0.4	0.16—1.12[a]	62.0 ± 2.9	400—800	5.0	CSR, extension	135
Quartzite, "Simpson", Oklahoma, >95% quartz, random fabric, G.S. = 0.2 mm	1.3 ± 1.9	5.7 ± 0.6	5.0—16.0	58.1 ± 5.0	600—900	6—10	CSR	125, 132

Note: The material parameters A, n, and the apparent activation energy Q* = (E* + PV*) in the steady-state ductile flow law of Equation 6 in the text are tabulated here, together with the ranges of stress and temperature and the confining pressure to which the parameters apply. In all cases, material parameters were checked so that the original steady-state strain rates were reproduced at the test temperature and stress. Differences between our values of these constants and those in the original sources are due to our corrections made to reproduce the original data. We regard the data from simple creep and constant strain rate tests on rocks to be superior to stress relaxation and strain rate change tests, since the latter two types of runs can incorporate unknown primary creep effects. More comprehensive tabulations of data on ice may be found in Weertman,[299] Glen,[92] and in a recent volume on ice properties (Symposium on the Physics and Chemistry of Ice, J. Glaciology, 1978). High-temperature creep data on polycrystalline ceramics, which are closely related to rock creep, are summarized in Breteau et al.[46] The equilibrium point defect concentrations in oxide ceramics are influenced by the ambient oxygen fugacity f_{O2} as well as by temperature. The point defect concentrations in turn, can influence creep rates, especially at high temperatures when point defects are mobile. The oxygen fugacity was not controlled in the above experiments, except indirectly by possible buffering by the surrounding metal jackets and pistons. Thus an unknown effect of f_{O2} may be present in the data tabulated, possibly in the A parameter. The systematic effects of structurally bound water on the ductile flow of rocks was reviewed in the text. All of the tests except those on Ice Ih, were carried out on nominally dry samples. It is not certain however, that the drying procedures used were sufficient to eliminate all significant mechanical effects of structural water. We have specifically excluded rheological data on silicate rocks tested at low confining pressure, since direct evidence (cited in the text) indicates that brittle and semi-brittle processes dominate in polycrystalline silicates at low pressure, even up to the temperature of first melting.

[a] Transition to greater stress sensitivity at stresses above this range.
[b] CSR, constant strain rate; CS, constant stress, creep; SR, stress relaxation.
[c] All experiments in triaxial compression, unless otherwise noted.

Table 6
STEADY-STATE RHEOLOGICAL CONSTANTS FOR ROCKS AND MINERALS TESTED IN SOLID MEDIUM PISTON-CYLINDER APPARATUS

Material	$\log_{10}A$ ($kbar^{-n}s^{-1}$)	n	$\Delta\sigma$ (kbar)	Q* ($kcal\ mol^{-1}$)	ΔT (°C)	P (kbar)	Comment[b]	Ref.
Dunite, Mt. Burnet, Alaska, 99% olivine $(Mg_{0.92}Fe_{0.08})_2SiO_4$) minor serpentine 1% chromite, grain size 0.3—4mm, 1mm av., nominally anhydrous	10.1	4.8 ± 0.4	1—9	120 ± 17	1100—1300	15	CSR, Alsimag 222 medium, drying probably not sufficient to remove all water	
	9.7	3.3	1—9	111	1100—1300	15	Ditto, corrected for nonuniform strain (bulging)	55
	8.0	3	1—7	100 ± 15	1100—1200	15	CSR, SR, SDC, pyrex glass medium	238, 171
	8.9	3.6	1—8	126 ± 15	1100—1400	10—15	CS, SDC, TDC, CSR, copper and nickel-confining media	235—236
	8.6	3	1—3[a]	126 ± 15	1100—1400	10—15	CS, SDC, TDC, CSR, copper and nickel-confining media	235—236
Dunite, Mt. Burnet, Alaska, 99% olivine $(Mg_{0.92}Fe_{0.08})_2SiO_4$), 1% chromite, minor serpentine, grain size 0.3 to 4mm av. 1 mm, samples deformed under hydrous conditions (water released from dehydration of talc confining medium)	6.8	2.4 ± 0.2	1—7	80 ± 8	975—1350	10—15	CSR, talc-confining medium	57
	3.1	2.1	1—7	54	975—1350	10—15	CSR, talc-confining medium corrected for nonuniform strain (bulging)	55
	7.7	5.1 ± 0.3	1—8	94 ± 3	800—1150	5—15	CS, SR, TDC, SDC, CSR, talc-confining medium	235—236

	8.6	3	1—3[a]	94 ± 3	800—1150	5—15	CS, SR, TDC, SDC, CSR, talc-confining medium	235—236
Pyroxenite, Sleaford Bay, South Australia, >99% clinopyroxene (salite), grain size 0.6 mm	8.0	6.4 ± 0.2	4.5—13[a]	106 ± 7	900—1200	15	CSR, NaCl, and NaF confining media, stresses picked at 10% strain (steady-state)	166
	−94	83.7 ± 13.2	13—20	50 ± 8	400—900	15	CSR, NaCl, and NaF-confining media, stresses picked at 10% straim (steady-state) work hardening observed, stresses picked at 10% strain	
Pyroxenite, Webster, N.C. Orthopyroxene ($Ca_{0.03}(Mg_{0.89}Fe_{0.8})Si_2O_7$, grain size 1mm, samples deformed under hydrous conditions (water released from talc dehydration)	3.4	2.8 ± 0.2	1.5—11	65 ± 2	960—1300	10	CSR, talc-confining medium	249
	−8.5	5.4 ± 0.3	3—13	7.3 ± 1.8	800—1000	10	CSR, talc-confining medium	249
Pyroxene single crystal, Canyon Mt. Ophiolite Complex, Oregon, clinopyroxene (diopside), compressed at 61° to [100], 90° to [010], and 45° to [001], samples deformed under hydrous conditions (water released from talc dehydration)	4.5 ± 4.1	4.3 ± 0.4	1.5—4.5	68 ± 14	1000—1050	5—15	CSR, SRD, talc-confining medium	16
Pyroxene single crystal, source unknown, clinopyroxene (hedenbergite), compressed at 45° to [001], 90° to [010] in acute angle between [100] and [001]							CSR, NaCl, and KCl-confining media, yield stress = 3.0 ± 0.3 kbar, independent of strain rate and temperature (10^{-6}—10^{-8} s^{-1},	176—178

Table 6 (continued)
STEADY-STATE RHEOLOGICAL CONSTANTS FOR ROCKS AND MINERALS TESTED IN SOLID MEDIUM PISTON-CYLINDER APPARATUS

Material	$\log_{10}A$ ($kbar^{-n}s^{-1}$	n	$\Delta\sigma$ (kbar)	Q* ($kcal\ mol^{-1}$)	ΔT (°C)	P (kbar)	Comment[b]	Ref.
							400 to 800°C), mechanical twins on (100) produced	
Pyroxene single crystal, source unknown, clinopyroxene (hedenbergite), compressed at 45° to [001], 90° to [010] in obtuse angle between [100] and [001]	5.5	4.0 ± 0.6	3—10	59 ± 11	700—900	10	CSR, NaCl, and KCl-confining media	178
Quartzite, "Simpson", 12,956 ft level in Sunray Cullen well, Oklahoma, grain size, 0.2 mm, nominally pure and anhydrous	0.9 ± 0.3	2.86 ± 0.18	2—30	36.0 ± 4.0	800—900	10—12	CSR, copper-confining medium, α-quartz stability field	61—62
Quartzite, "Simpson", samples deformed under hydrous conditions (water released from dehydration of talc jacket)		2.9	2—20			10	CSR, copper-confining medium, T = 850°C, α- quartz stability field	61—62
Quartzite, "Canyon Creek", Big Hole Canyon, Montana, nominally pure, grain size 0.1 mm, samples deformed under hydrous conditions (water released from dehydration of talc-confining medium)	3.9 ± 0.7	2.6 ± 0.4	1—4	55 ± 7	860—1160	10	CSR, SRD, SR, talc-confining medium (dehydrated), tests largely in α-quartz stability field	221
Websterite, Webster N.C., 68% clinopyroxene, 32% orthopyroxene, grain size 1—2 mm, nominally anhydrous	2.2 ± 1.6 2.0 ± 1.8	4.3 ± 0.6 5.8 ± 0.4	3—7[a] 3—13	80 ± 4 84 ± 8	1000—1200 1000—1200	10 10	CSR, SRD, SR, Alsimag-confining medium	16
Websterite, Webster, N.C., samples deformed under hydrous conditions (water released from dehydration of talc-confining medium)	10.1 ± 1.2 5.3 ± 0.4	3.3 ± 0.1 5.9 ± 2.1	1.6—4[a] 4—10	111 ± 4 91 ± 6	1000—1050 950—1050	5 5	CSR, SRD, SR, talc-confining medium (dehydrated)	16

Note: The general format and explanations of Table 6 apply to this table. Results from simple constant strain rate and creep tests should be preferred over those from differential tests (where one parameter is changed stepwise) and over stress relaxation tests because of the unknown contributions of transient creep. The apparent activation energy for creep Q^* is subject to the largest potential error, since increases in temperature can systematically reduce the amount of cell friction and viscous resistance and thus bias the temperature effects. The reliability of these rheological constants should be considered in the order of experimental accuracy for the various confining media listed in the explanation of Table 4. Additional experimental details on these suites of experiments may also be found in Table 4.

[a] Transition to greater stress sensitivity above this stress range.

[b] CSR, Constant strain rate; CS, constant differential stress, creep test; SRD, strain rate differential tests (strain rate changed stepwise); TDC, temperature differential creep test (temperature changed stepwise at constant differential stress); SDC, stress differential creep test; SR, stress relaxation test.

Table 7
FRACTURE TOUGHNESS OF ROCKS AND MINERALS

Material	Testing method[a]	K_{Ic}(MN/m$^{3/2}$)	Comments	Ref.
Tennessee sandstone	2	0.454 ± 0.002		12
Sandstone, porous	4	0.57—1.46		64
Gabbro, black	2	2.884 ± 0.049		14
Chelmsford granite	3	0.592—0.636		229
Arkansas novaculite	2	1.335 ± 0.075		11
Synthetic quartz crystal cracks on "a" {2110} plane	2	1.002 ± 0.048	Direction ⊥ r {0110}	10
	2	0.852 ± 0.045	Crack direction ⊥ r {0111}	14
Silica glass	3	0.0753		303
Indiana limestone	1	0.929		263
Indiana limestone	3	0.990		262
Saint-Pons marble	3	0.746—1.33		141
Carrara marble	2	0.664 ± 0.021		12
Ice, synthetic, columnar crystals in aggregate grown from distilled water	1	0.092—0.100, T = −4°C		191
		0.111—1.163, T = −46°C		191

Note: The fracture strength of rocks depends on the initial density and configuration of preexisting cracks and hence is not an intrinsic property of rock type. Modern work on rock fracture emphasizes the progapation properties of individual cracks in the presence of an applied nonhydrostatic stress and how microcracks interact in rocks. The intrinsic propagation properties of cracks in given a rock or mineral are generally analyzed using linear elastic fracture mechanics (see Paterson[225]). In this approach, the measure of the stress intensity at a crack tip is the stress intensity factor K, which is generally related to the applied stress σ_a by

$$K = \beta \, \sigma_a \sqrt{c}$$

where 2c is the crack length and β is a factor which depends on loading geometry and crack geometry. For cracks subject to tensile stresses tending to open the crack (the so-called mode I loading at crack tip), unstable crack propagation to rupture occurs above some critical value of tensile stress and thus above some critical stess intensity factor K_{Ic}, also called the fracture toughness. Below K_{Ic}, slow stable crack propagation can also occur in the presence of water (Figure 13). Compiled from the table in Atkinson[12] and subsequent data. Data obtained at room temperature and ambient humidity, unless otherwise noted.

[a] 1, Tensile specimen with single edge notch; 2, double torsion; 3, three or four point bending; 4, prenotched internally pressurized thick-walled hollow cylinders.

Table 8
DILATANCY OF ROCKS AT FAILURE

Rock	Confining pressure p (kbar)	Porosity at test pressure η_p	Dilatant strain at failure Δ	$\Delta\eta_p$
Waldhams anorthosite	1.5	0.002	0.006	3.0
Spruce pine dunite	1.5	0.002	0.014	7.0
Cape granodiorite	1.5	0.003	0.006	2.0
Rutland quartzite	1.5	0.004	0.004	1.0
	0.001	0.005	0.003	0.6
Witwatersrand quartzite	0.03	0.005	0.013	3.0
	0.1	0.005	0.014	3.0
	0.3	0.005	0.019	4.0
Westerly granite	1.6	0.007	0.002	0.3
	3.0	0.007	0.003	0.35
	5.0	0.007	0.004	0.6
Climax granodiorite	0.2	0.007	0.002	0.3
Barre granite	0.001	0.014	0.001	0.1
Tension	—	0.014	0.0002	0.1
Blair dolomite	1.0	0.009	0.012	1.3
	3.0	0.009	0.013	1.4
Nugget sandstone	0.001	0.03	0.009	0.3
	1.0	0.02	0.004	0.2
Pottsville sandstone	1.0	0.025	0.007	0.3
Lance sandstone	1.0	0.08	0.008	0.1
Gosford sandstone	2.0	0.11	0.02	0.2
	6.0	0.08	0.02	0.3
Kayenta sandstone	1.9	0.23	0.06	0.25
NTS tuff	2.0	0.26	0.07	0.35

Note: The inelastic volumetric strain Δ at failure has been tabulated for a number of rocks tested in compression at room temperature by Brace,[41] which is reproduced here in edited form. Brace notes that Δ values are much larger for porous rocks. The data suggest a systematic increase in Δ with increasing confining pressure for low porosity rocks and the reverse seems to be true for porous rocks. See Brace[41] for references to original sources.

REFERENCES

1. **Abey, A. E., Bonner, B. P., Heard, H. C., and Schock, R. N.,** Mechanical Properties of a Shale from Site U-2, Rep. UCID-16023, Lawrence Livermore Laboratory, 1972, 20.
2. **Anderson, O. L. and Grew, P. C.,** Stress corrosion theory of crack propagation with applications to geophysics, *Rev. Geophys. Space Phys.*, 15, 77, 1977.
3. **Ashby, M. F.,** A first report on deformation-mechanism maps in *Acta Metall.*, 20, 887, 1972.
4. **Atkinson, B. K.,** Experimental deformation of polycrystalline galena, chalcopyrite, and pyrrhotite, *Trans. Section B Inst. Min. Metall.*, 83, B19, 1974.
5. **Atkinson, B. K.,** Experimental deformation of polycrystalline pyrite: effects of temperature, confining pressure, strain rate, and porosity, *Econ. Geol.*, 70, 473, 1975.
6. **Atkinson, B. K.,** The temperature- and strain rate-dependent mechanical behavior of a polycrystalline galena ore in *Econ. Geol.*, 71, 513, 1976.

7. **Atkinson, B. K.,** A preliminary study of the influence of temperature and strain rate on the rheology of a polycrystalline pyrrhotite, *N. Jb. Miner. Mh.*, 11, 483, 1976.
8. **Atkinson, B. K.,** The kinetics of ore deformation: its illustration and analysis by means of deformation-mechanism maps, *Geol. Foeren. Stockholm Foerh.*, 99, 186, 1977.
9. **Atkinson, B. K.,** High-temperature stress relaxation of synthetic, polycrystalline galena, *Phys. Chem. Miner.*, 2, 305, 1978.
10. **Atkinson, B. K.,** A fracture mechanics study of subcritical tensile cracking of quartz in wet environments, *Pure Appl. Geophys.*, 117, 1011, 1979.
11. **Atkinson, B. K.,** Stress corrosion and the rate-dependent tensile failure of a fine-grained quartz rock, *Tectonophysics*, in press, 1979.
12. **Atkinson, B. K.,** Fracture toughness of Tennessee sandstone and Carrara marble using the double torsion testing method, *Int. J. Rock Mech. Min. Sci. Geomech. Abstr.*, 16, 49, 1979.
13. **Atkinson, B. K.,** unpublished data, 1979.
14. **Atkinson, B. K.,** Acoustic Emission During Sucritical and Fast Tensile Cracking of Westerly Granite and a Gabbro, unpublished manuscript, 1979.
15. **Atkinson, B. K. and Rawlings, R. D.,** Acoustical emission during stress corrosion cracking in rocks, in Earthquake Prediction: An International Rewiew, Maurice Ewing Series, 4, GOS, Am. Geophys. Union, 1981.
16. **Ave'Lallemant, H. G.,** Experimental deformation of diopside and websterite, *Tectonophysics*, 48, 1, 1978.
17. **Ave'Lallemant, H. G. and Carter, N. L.,** Syntectonic recrystallization of olivine and modes of flow in the upper mantle, *Geol. Soc. Am. Bull.*, 81, 2203, 1970.
18. **Ayensu, A. and Ashbee, K. H. G.,** The creep of quartz single crystals, with special reference to the mechanism by which water accommodates dislocation glide, *Philos. Mag.*, 36, 713, 1977.
19. **Baëta, R. D. and Ashbee, K. H. G.,** Plastic deformation and fracture of quartz at atmospheric pressure, *Philos. Mag.*, 15, 931, 1967.
20. **Baëta, R. D. and Ashbee, K. H. G.,** Slip systems in quartz. I. Experiments, *Am. Min.*, 54, 1551, 1969.
21. **Baëta, R. D. and Ashbee, K. H. G.,** Mechanical deformation of quartz. II. Stress relaxtion and thermal activation parameters, *Philos. Mag.*, 22, 624, 1970.
22. **Baëta, R. D. and Ashbee, K. H. G.,** Mechanical deformation of quartz. I. Constant strain-rate compression experiments, *Philos. Mag.*, 22, 601, 1970.
23. **Baker, R. W.,** The influence of ice-crystal size on creep, *J. Glaciol.*, 85, 485, 1978.
24. **Balderman, M. A.,** Relationship of Yield Stress and Strain-Rate in Hydrolytically Weakened Synthetic Quartz, M.Sc. thesis, University of California, Los Angeles, 1972, 119.
25. **Balderman, M. A.,** The effect of strain rate and temperature on the yield point of hydrolytically weakened synthetic quartz, *J. Geophys. Res.*, 79, 1647, 1974.
26. **Barnes, P., Tabor, D., and Walker, J. C. F.,** The friction and creep of polycrystalline ice in *Proc. R. Soc. London A*, 324, 127, 1971.
27. **Blacic, J. D.,** Hydrolytic weakening of Quartz and Olivine, Ph.D. thesis, University of California, Los Angeles, 1971, 205.
28. **Blacic, J. D.,** unpublished data, 1978.
29. **Blacic, J. D.,** Plastic deformation mechanisms in quartz: the effect of water, *Tectonophysics*, 27, 271, 1975.
30. **Blanton, T. L.,** Effect of Strain Rates from 0.01 to 10/sec in Triaxial Compression Tests on Three Rocks, Ph.D. thesis, Texas A&M University, 1976, 67.
31. **Blum, W. and Ilschner, B.,** Uber das Kriechverhalten von NaCl-Einkristallen, *Phys. Stat. Sol.*, 20, 629, 1967.
32. **Böker, R. von,** Die Mechanik der bleibenden Formanderung in kristallinisch aufgebauten Korpern, *Ver. Dtsch. Ing. Mitt. Forsch.*, 175, 1, 1915.
33. **Boland, J. N., Hobbs, B. E., and McLaren, A. C.,** The defect structure in natural and experimentally deformed kyanite, *Phys. Stat. Sol. A*, 39, 631, 1977.
34. **Bonner, B. P. and Abey, A. E.,** High-pressure deformation of coal form Powder River Basin, Wyoming, *Fuel*, 54, 165, 1975.
35. **Bonner, B. P., Abey, A. E., Heard, H. C., and Schock, R. N.,** High Pressure Mechanical Properties of Shales and Regolith from the Middle Gust Site, Rep. UCID-16103, Lawrence Livermore Laboratory, 1972, 16.
36. **Borg, I. Y. and Handin, J.,** Experimental deformation of crystalline rocks, *Tectonophysics*, 3, 249, 1966.
37. **Borg, I. Y. and Handin, J.,** Torsion of calcite single crystals, *J. Geophys. Res.*, 72, 641, 1967.
38. **Borg, I. Y. and Heard, H. C.,** Mechanical twinning and slip in experimentally deformed plagioclases, *Contr. Mineral. Pet.*, 23, 128, 1969.
39. **Borg, I. Y. and Heard, H. C.,** Expermental deformation of plagioclases, in *Experimental and Natural Rock Deformation*, Paulitsch, P., Ed., Springer-Verlag, Berlin, 1970, 375.

40. **Borg, I. Y. and Heard, H. C.,** Mechanical twinning in Sphene at 8 Kbar, 25 to 500°C, *Geol. Soc. Am. Mem.*, 132, 585, 1972.
41. **Brace, W. F.,** Volume changes during fracture and frictional sliding: a review, *Pure Appl. Geophys.*, 116, 627, 1978.
42. **Brace, W. F. and Jones, A. H.,** Comparison of uniaxial deformation in shock and static loading of three rocks, *J. Geophys. Res.*, 76, 4913, 1971.
43. **Brace, W. F. and Martin, R. J.,** A test for the law of effective stress for crystalline rocks of low porosity, *Int. J. Rock Mech. Min. Sci.*, 5, 415, 1968.
44. **Brace, W. F., Paulding, B. W., and Scholtz, C., Dilatency in the fracture of crystalline rocks,** ***J. Geophys. Res.*, 71, 3939, 1966.**
45. **Brace, W. F. and Walsh, J. B.,** Some direct measurements of the surface energies of quartz and orthoclase, *Am. Miner.*, 47, 1111, 1962.
46. **Bretheau, T., Castaing, J., Veyessière, P., and Rabier, J.,** Mouvement des dislocations et plasticite a haute temperature des oxydes binaire et ternaire, *Adv. Phys.*, 28, 835, 1979.
47. **Burke, P. M.,** High Temperature Creep of Polycrystalline Sodium Chloride, Ph.D. thesis, Department of Materials Science, Stanford University, 1968, 122.
48. **Butkovich, T. R. and Landauer, J. K.,** The flow law for ice, in Int. Union of Geodesy and Geophysics, Symp. Chamonix, Phys. Movement of the Ice, Publ. No. 47, International Association of Scientific Hydrology, 1958, 318.
49. **Byerlee, J.,** Friction of rocks, *Pageoph*, 116, 615, 1978.
50. **Byerlee, J. D.,** Brittle-ductile transition in rocks, *J. Geophys. Res.*, 73, 4741, 1968.
51. **Byerlee, J. D., The fracture strength and frictional strength of Weber sandstone,** ***Int. J. Rock Mech. Min. Sci.*, 12, 1, 1975.**
52. **Byerlee, J. D. and Wys, M., Eds.,** Rock friction and earthquake prediction, *Pure Appl. Geophys.*, 116, 583, 1978.
53. **Byers, B. A.,** Secondary Creep of Polycrystalline Ice under Biaxial Stress, Ph.D. thesis, University of Washington, Seattle, 1973, 136.
54. **Caristan, Y. and Goetze, C.,** High temperature plasticity of Maryland diabase (abstract), *Trans. Am. Geophys. Union*, 59, 375, 1978.
55. **Carter, N. L.,** Steady state flow of rocks, *Rev. Geophys. Space Phys.*, 14, 301, 1976.
56. **Carter, N. L.,** unpublished data, 1977.
57. **Carter, N. L. and Ave'Lallemant, H. G.,** High temperature flow of dunite and peridotite, *Geol. Soc. Am. Bull.*, 81, 2181, 1970.
58. **Carter, N. L. and Heard, H. C.,** Temperature and rate dependent deformation of halite, *Am. J. Sci.*, 269, 193, 1970.
59. **Carter, N. L. and Kirby, S. H.,** Transient creep and semibrittle behavior of crystalline rocks, *Pure Appl. Geophys.*, 116, 807, 1978.
60. **Charles, R. J.,** The strength of silicate glasses and some crystalline oxides, in *Fracture, Proc. Int. Conf. Atomic Mechanisms of Fracture, Swampscott, April 1959*, Averback, B. L., Felbeck, D. K., Hahn, G. T., Thomas, D. A., Eds., John Wiley & Sons, New York, 1959, 225.
61. **Christie, J. M., Koch, P. S., and George, R. P.,** Flow law of quartzite in the alpha-quartz field, *Trans. Am. Geophys. Union*, 60, 948, 1979.
62. **Christie, J. M., Koch, P. S., and George, R. P.,** Flow law of quartzite in the alpha-quartz field, unpublished manuscript, 1980.
63. **Clark, B. R. and Kelly, W. C.,** Sulfide deformation studies. I. Experimental deformation of pyrrhotite and sphalerite to 2,000 bars and 500 degrees C, *Econ. Geol.*, 68, 332, 1973.
64. **Clifton, R. J., Simonsen, E. R., Jones, A. H., and Green, S. J.,** Determination of the critical stress intensity factor from internally pressurized thick walled vessels in *Exp. Mech.*, 16, 233, 1976.
65. **Cruden, D. M.,** A theory of brittle creep in rocks under uniaxial compression, *J. Geophys. Res.*, 75, 3431, 1970.
66. **Cruden, D. M.,** The form of the creep law for rock under uniaxial compression, *Int. J. Rock Mech. Min. Sci.*, 8, 105, 1971.
67. **Cruden, D. M.,** The static fatigue of brittle rock under uniaxial compression, *Int. J. Rock Mech. Min. Sci.*, 11, 67, 1974.
68. **Deere, D. U. and Miller, R. P.,** Engineering Classification and Index Properties for Intact Rock, Tech. Rep. AF-TR-65-116, Air Force Weapons Laboratory, Kirtland Air Force Base, New Mexico, 1966, 300.
69. **Dieterich, J. H.,** Time-dependent friction in rocks, *J. Geophys. Res.*, 77, 3690, 1972.
70. **Dieterich, J. H.,** Time-dependent friction and the mechanics of stick-slip, *Pure Appl. Geophys.*, 116, 790, 1978.
71. **Dieterich, J. H.,** Modeling of rock friction. I. Experimental results and constitutive equations, *J. Geophys. Res.*, 84, 2161, 1979.

72. **Dieterich, J. H. and Conrad, G.,** Effect of humidity and adsorbed water on time and velocity-dependent friction in Rocks, *J. Geophys. Res.*,in press, 1983.
73. **Dollinger, G. and Blacic, J. D.,** Deformations mechanisms in experimentally and naturally deformed amphiboles, *Earth Planet. Sci. Lett.*, 26, 409, 1975.
74. **Duba, A., Abey, A. E., and Heard, H. C.,** High Pressure Mechanical Properties of an Area 12, Nevada Test Site Tuff, Rep. UCID-16377, Lawrence Livermore Laboratory, 1973, 20.
75. **Duba, A. G., Abey, A. E., Bonner, B. P., Heard, H. C., and Schock, R. N.,** High-Pressure Mechanical Properties of Kayenta Sandstone, Rep. UCRL-51526, Lawrence Livermore Laboratory, 1974, 22.
76. **Duba, A. G., Heard, H. C., and Santor, M. L.,** Effect of Fluid Content on the Mechanical Properties of Westerly Granite, Lawrence Livermore Laboratory, Rep. UCRL-51626, 1974.
77. **Durand, E.,** L'essai de torsion et la resistance au cisallement des roches, *Rock Mech.*, 7, 199, 1975.
78. **Durham, W. B. and Froidevaux, C.,** Transient and steady-state creep of pure forsterite at low stress, *Phys. Earth Planetary Inter.*, 19, 263, 1979.
79. **Durham, W. B. and Goetze, C.,** A comparison of the creep properties of pure forsterite and iron-bearing olivine, *Tectonophysics*, 40, T15, 1977.
80. **Durham, W. B. and Goetze, C.,** Plastic flow of oriented single crystals of olivine. I. Mechanical data, *J. Geophys. Res.*, 82, 5737, 1977.
81. **Duval, P.,** Creep and recrystallization of polycrystalline ice, *Bull. Mineral. Soc. Fr. Miner. Cryst.*, 102, 80, 1979.
82. **Edmond, J. M. and Paterson, M. S.,** Volume changes during the deformation of rocks at high pressures, *Int. J. Rock Mech. Min. Sci.*, 9, 161, 1972.
83. **Elliott, D.,** Diffusion flow laws in metamorphic rocks, *Geol. Soc. Am. Bull.*, 84, 2645, 1973.
84. **Engelder, J. T., Logan, J. M., and Handin, J.,** The sliding characteristics of quartz fault gouge, *Pure Appl. Geophys.*, 113, 69, 1975.
85. **Etheridge, M. A., Hobbs, B. E., and Paterson, M. S.,** Experimental deformation of single crystals of biotite, *Contr. Mineral. Petrol.*, 38, 21, 1973.
86. **Friedman, M., Handin, J., and Alani, G.,** Fracture-surface energy of rocks, *Int. J. Rock Mech. Min. Sci.*, 9, 757, 1972.
87. **Friedman, M., Handin, J., and Alani, G.,** Fracture-surface energy of rocks, *Int. J. Rock Mech. Min. Sci.*, 9, 757, 1972.
88. **Friedman, M., Handin, J., Higgs, N. G., and Lantz, J. R.,** Strength and ductility of four dry igneous rocks at low pressures and temperatures to partial melting, in 20th U.S. Symp. Rock Mechanics, Austin, Texas, Gray, K., Ed., 1979, 35.
89. **Gilman, J. J.,** Direct measurement of surface energies of crystals, *J. Appl. Phys.*, 31, 2208, 1960.
90. **Gilman, J. J.,** *Micromechanics of Flow in Solids*, McGraw-Hill, New York, 1969, 294.
91. **Glen, J. W.,** The creep of polycrystalline ice, *Proc. R. Soc. London A*, 228, 519, 1955.
92. **Glen, J. W.,** The Mechanics of Ice, Cold Regions Science and Engineering Monograph II-C2b, 1975, 41.
93. **Glen, J. W. and Jones, S. J.,** The deformation of ice single crystals at low temperatures, in Physics of Snow and Ice, Oura, H., Ed., The Institute of Low Temperature Science, Hokkaido University, 1967, 267.
94. **Goetze, C.,** High temperature rheology of westerly granite, *J. Geophys. Res.*, 76, 1223, 1971.
95. **Goetze, C. and Brace, W. F.,** Laboratory observations of high temperature rheology of rocks, *Tectonophysics*, 13, 583, 1972.
96. **Goodman, R. E. and Sundaram, P. N.,** Fault and system stiffness and stick-slip phenomena, *Pure Appl. Geophys.*, 116, 873, 1978.
97. **Green, H. W.,** Diffusional flow in polycrystalline materials, *J. Appl. Phys.*, 41, 3899, 1970.
98. **Griggs, D. T.,** Experimental flow of rocks under conditions favoring recrystallization, *Geol. Soc. Am. Bull.*, 51, 1001, 1940.
99. **Griggs, D. T.,** Hydrolytic weakening of quartz and other silicates, *Geophys. J. R. Abstr. Soc.*, 14, 19, 1967.
100. **Griggs, D. T.,** A model of hydrolytic weakening in quartz, *J. Geophys. Res.*, 79, 1655, 1974.
101. **Griggs, D. T. and Blacic, J. D.,** The strength of quartz in the ductile regime, *Trans. Am. Geophys. Union*, 45(Abstr.), 102, 1964.
102. **Griggs, D. T. and Blacic, J. D.,** Quartz: anomalous weakness of synthetic crystals, *Science*, 147, 292, 1965.
103. **Griggs, D. T. and Coles, N. E.,** Creep of single crystals of ice in *Snow, Ice Permafrost Establishment (SIPRE)*, 11, 1, 1954.
104. **Griggs, D. T. and Handin, J.,** Observations of fracture and a hypothesis of earthquakes, *Geol. Soc. Am. Mem.*, 79, 347, 1960.
105. **Griggs, D. T., Turner, F. J., and Heard, H. C.,** Deformation of rocks at 500° to 800°C, *Geol. Soc. Am. Mem.*, 79, 39, 1960.
106. **Guillope, M. and Poirier, J-P.,** Dynamic recrystallization during creep of single crystal halite, an experimental study, *J. Geophys. Res.*, 84, 5557, 1979.

107. **Haimson, B. C.,** Mechanical behavior of rock under cyclic loading, in *Advances in Rock Mechanics,* Vol. 2 (Part 1) Ed., National Academy Sciences, Washington, D.C., 1974.
108. **Handin, J.,** An application of high pressure in geophysics: experimental rock deformation, *Trans. Am. Soc. Mech. Eng.,* 75, 315, 1953.
109. **Handin, J.,** Strength at high confining pressure and temperature of serpentinite from Mayaguez, Puerto Rico, in *A Study of Serpentinite,* Publ. 1188, National Academy of Sciences, Washington, D.C., 1964, 126.
110. **Handin, J.,** Strength and ductility, *Geol. Soc. Am. Mem.,* 97, 223, 1966.
111. **Handin, J.,** On the Coulomb-Mohr failure criterion, *J. Geophys. Res.,* 74, 5343, 1969.
112. **Handin, J. and Carter, N. L.,** The rheology of rocks at high temperatures, *Proc. Fourth Int. Congr. on Rock Mechanics, Int. Soc. for Rock Mechanics,* 3, 97, 1980.
113. **Handin, J. and Friedman, M.,** Mechanical Properties of Rocks at High Temperature and Pressure, 3rd Annu. Prog. Rep. Contract No. 82-9794, Sandia Laboraries, Albuquerque, N.M., 1977, 62.
114. **Handin, J., Friedman, M., Logan, J. M., Pattison, L. J., and Swolfs, H. S.,** Experimental folding of rocks under confining pressure: buckling of single-layer rock beams, in Flow and Fracture of Rocks, Geophysical Monograph 16, The Griggs Volume, Heard, H. C., Borg, I. Y., Carter, N. L., and Raleigh, C. B., Eds., American Geophysical Union, Washington, D.C., 1972, 1.
115. **Handin, J., Friedman, M., Min, K. D., and Pattison, L. J.,** Experimental folding of rocks under confining pressure. II. Buckling and multilayered rock beams, *Geol. Soc. Am. Bull.,* 87, 1035, 1976.
116. **Handin, J., Hager, R. V., Friedman, M., and Feather, J. N.,** Experimental deformation of sedimentary rocks under confining pressure: pore pressure effects, *Bull. Am. Assoc. Pet. Geol.,* 47, 717, 1963.
117. **Handin, J., Heard, H. C., and Magouirk, J. N.,** Effects of the intermediate principal stress on the failure of limestone, dolomite, and glass at different temperatures and strain rates, *J. Geophys. Res.,* 72, 611, 1967.
118. **Handin, J., Higgs, D. V., and O'Brien, J. K.,** Torsion of Yule marble under confining pressure, *Geol. Soc. Am. Mem.,* 79, 245, 1960.
119. **Hardy, H. R. and Chugh, Y. P.,** Failure of geologic materials under low-cycle fatigue, in Proc. 6th Can. Rock Mech. Symp., Montreal, Department of Mineral Engineering, Pennsylvania State University, University Park, 1970, 33.
120. **Hardy, H. R., Kim, R. Y., Stefanko, R., and Wang, Y. J.,** Creep and microseismic activity in geological materials, in Rock Mechanics-Theory and Practice, Proc. 11th Symp. on Rock Mechanics, Berkeley, Calif., Somerton, W. H., Ed., AIME, New York, N.Y., 1970, 377.
121. **Hawkes, I., Mellor, M., and Gariepy, S.,** Deformation of rocks under uniaxial tension, *Int. J. Rock Mech. Min. Sci.,* 10, 493, 1973.
122. **Heard, H. C.,** Transition from brittle fracture to ductile flow in Solenhofen limestone, *Geol. Soc. Am. Mem.,* 79, 193, 1960.
123. **Heard, H. C.,** Steady-state flow in polycrystalline halite at pressure of 2 kilobars, in Flow and Fracture of Rocks, Geophysical Monograph 16, The Griggs Volume, Heard, H. C., Borg, I. Y., Carter, N. L., and Raleigh, C. B., Eds., American Geophysical Union, Washington, D.C., 1972, 191.
124. **Heard, H. C.,** unpublished data, 1975.
125. **Heard, H. C.,** Comparison of the flow properties of rocks at crustal conditions, *Philos. Trans. R. Soc. London A,* 283, 173, 1976.
126. **Heard, H. C.,** unpublished data, 1976.
127. **Heard, H. C., Abey, A. E., and Bonner, B. P.,** High Pressure Mechanical Properties of Indiana Limestone, Rep. UCID-16501, Lawrence Livermore Laboratory, 1974, 17.
128. **Heard, H. C., Abey, A. E., Bonner, B. P., and Duba, A.,** Stress-Strain Behavior of Polycrystalline NaCl to 3.2 GPa, Rep. UCRL-51743, Lawrence Livermore Laboratory, 1975, 16.
129. **Heard, H. C., Abey, A. E., Bonner, B. P., and Schock, R. N.,** Mechanical Behavior of Dry Westerly Granite at High Pressure, Rep. 51642, Lawrence Livermore Laboratory, 1974, 14.
130. **Heard, H. C., Bonner, B. P., Costantino, M. S., Schock, R. N., and Weed, H. C.,** Mechanical Response of Saturated Kemmerer Coal to 4 GPa, Rep. UCRL-52063, Lawrence Livermore Laboratory, 1976, 28.
131. **Heard, H. C., Bonner, B. P., Duba, A. G., Schock, R. N., and Stephens, D. R.,** High Pressure Mechanical Properties of Mt. Helen, Nevada, Tuff, Rep. UCID-16261, Lawrence Livermore Laboratory, 1973, 39.
132. **Heard, H. C. and Carter, N. L.,** Experimentally induced "natural" intragranular flow in quartz and quartzite, *Am. J. Sci.,* 266, 1, 1968.
133. **Heard, H. C. and Duba, A.,** Capabilities for Measuring Physicochemical Properties at High Pressure, Rep. UCRL-52420, Lawrence Livermore Laboratory, 1978, 44.
134. **Heard, H. C., Duba, A., Abey, A. E., and Schock, R.,** Mechanical Properties of Blair Dolomite, Rep. UCRL 51465, University of California Lawrence Livermore Laboratory, 1973.

135. **Heard, H. C. and Raleigh, C. B.,** Steady-state flow of marble at 500 to 800 degrees C, *Geol. Soc. Am. Bull.*, 83, 935, 1972.
136. **Heard, H. C. and Rubey, W. W.,** Tectonic implications of gypsum dehydration, *Geol. Soc. Am. Bull.*, 77, 741, 1966.
137. **Heard, H. C., Stephens, D. R., and Schock, R. N.,** High-Pressure Equation-of State Measurements for Altered Basalts and Bressias from Amchitka Island, Alaska, Rep. UCID-16165, Lawrence Livermore Laboratory, 1972, 17.
138. **Heard, H. C., Turner, F. J., and Weiss, L. E.,** Studies of heterogeneous strain in experimentally deformed calcite, marble, and phyllite, *Univ. Calif. Publ. Geol. Sci.*, 46, 81, 1965.
139. **Barber, D. J., Heard, H. C., and Wenk, H. R.,** Deformation of dolomite single crystals from 20 to 800°C, *Phys. Chem. Minerals*, 7, 271, 1981.
140. **Hendron, A. J., Jr.,** Mechanical properties of rock, *Rock Mechanics in Engineering Practice*, Stagg, K. G. and Zienkiewicz, O. C., Eds., John Wiley & Sons, New York, 1968, 21.
141. **Henry, J.-P. and Paquet, J.,** Mechanique de la Rupture de Roches calcitique in *Bull. Soc. Geol. Fr.*, 18, 1573, 1976.
142. **Higashi, A.,** Mechanisms of plastic deformation in ice single crystals, in Physics of Snow and Ice: Int. Conf., Sopporo, Japan, Vol. 1, Oura, H., Ed., Institute of Low Temperature Science, Hokkaido University, 1967, 277.
143. **Hoagland, R. G., Halm, G. T., and Rosenfield, A. R.,** Influence of microstructure on fracture propagation in rocks, in Semiannual Report, Batelle Columbus Laboratories, 1971.
144. **Hobbs, B. E.,** Recrystallization of single crystals of quartz, *Tectonophysics*, 6, 353, 1968.
145. **Hobbs, B. E., McLaren, A. C., and Paterson, M. S.,** Plasticity of single crystals of synthetic quartz, in *Flow and Fracture of Rocks, Geophysical Monograph 16, The Griggs Volume*, Heard, H. C., Borg, I. Y., Carter, N. L., and Raleigh, C. B., Eds., American Geophysical Union, Washington, D.C., 1972, 29.
146. **Homer, D. R. and Glen, J. W.,** The creep activation energies of ice, *J. Glaciol.*, 85, 429, 1978.
147. **Hubbert, M. K. and Rubey, W. W.,** Role of fluid pressure in the mechanics of overthrust faulting, *Geol. Soc. Am. Bull.*, 70, 115, 1959.
148. **Iida, K. and Kumazawa, M.,** Viscoelastic properties of rocks, *J. Earth Sci. Nagoya Univ.*, 5, 68, 1957.
149. **Jaeger, J. C. and Cook, N. G. W.,** *Fundamentals of Rock Mechanics*, 1st ed., John Wiley & Sons, New York, 1969.
150. **Jaeger, J. C. and Cook, N. G. W.,** *Fundamentals of Rock Mechanics*, 2nd ed., John Wiley & Sons, New York, 1976, 515.
151. **Jones, S. J. and Brunet, J.-G.,** Deformation of ice single crystals close to the melting point, *J. Glaciol.*, 85, 445, 1978.
152. **Jones, S. J. and Glen, J. W.,** The mechanical properties of single crystals of ice at low temperatures, in Reports and Discussions, Commission of Snow and Ice, General Assembly of Bern, Int. Union Geodesy and Geophysics, Publ. No. 79, International Association of Scientific Hydrology, Reading, England, 1968, 326.
153. **Jones, S. J. and Glen, J. W.,** The mechanical properties of single crystals of pure ice, *J. Glaciol.*, 8, 463, 1969.
154. **Kamb, B.,** Experimental recrystallization of ice under stress in *Flow and Fracture of Rocks, the Griggs Volume*, Heard, H. C., Borg, I., Carter, N. L., and Raleigh, C. B., Eds., American Geophysical Union, Washington, D.C., 1972, 211.
155. **Kármán, T. von,** Festigkeitsversuche unter allseitigem Druck, *Ver. Deut. Ingr.*, 55, 1749, 1911.
156. **Kekulawala, K. R. S. S., Paterson, M. S., and Boland, J. N.,** Hydrolytic weakening in quartz, *Tectonophysics*, 46, T1, 1978.
157. **Kelly, W. C. and Clark, B. R.,** Sulfide Deformation Studies. III. Experimental deformation of chalcopyrite to 2,000 bars and 500 degrees C, *Econ. Geol.*, 70, 431, 1975.
158. **Kirby, S. H.,** Creep of Synthetic Alpha Quartz, Ph.D. thesis, University of California, Los Angeles, 1975, 193.
159. **Kirby, S. H.,** Creep of synthetic quartz, *Trans. Am. Geophys. Union*, 56(Abstr.), 1062, 1975.
160. **Kirby, S. H.,** The alpha/beta inversion in quartz: effects of temperature on creep rates, *Trans. Am. Geophys. Union*, 57(Abstr.), 1001, 1976.
161. **Kirby, S. H.,** The effects of the alpha-beta phase transformation on the creep properties of hydrolytically-weakened synthetic quartz, *Geophys. Res. Lett.*, 4, 97, 1977.
162. **Kirby, S. H.,** Micromechanical interpretation of the incubation stage of the creep of hydrolytic weakened quartz crystals, *Trans. Am. Geophys Union*, 58(Abstr.), 1239, 1977.
163. **Kirby, S. H.,** State of stress in the lithosphere: inferences from the flow laws of olivine, *Pure Appl. Geophys.*, 115, 245, 1977.
164. **Kirby, S. H.,** Rheology of olivine: a critical review, *Trans. Am. Geophys. Union*, 59, 374, 1978.

165. **Kirby, S. H.,** Tectonic stresses in the lithosphere: constraints provided by the experimental deformation of rocks, *J. Geophys. Res.*, 85, 6353, 1980.
166. **Kirby, S. H. and Kronenberg, A. K.,** Ductile strength of clinopyroxenite: evidence for a transition in flow mechanisms, *Trans. Am. Geophys. Union*, 59(Abstr.), 376, 1978; *J. Geophys. Res.*, in press, 1983.
167. **Kirby, S. H. and Linker, M. F.,** Creep of hydrolytically-weakened synthetic quartz crystals at atmospheric pressure: effects of hydroxyl concentration, *Trans. Am. Geophys. Union*, 60, 949, 1979.
168. **Kirby, S. H. and McCormick, J. W.,** Creep of hydrolytically weakened synthetic quartz crystals oriented to promote $\{2\bar{1}\bar{1}0\}$<0001> slip: a brief summary of work to date, *Bull. Mineral.*, 102, 124, 1979.
169. **Kirby, S. H. and McCormick, J. W.,** Experimental creep and dislocation micromechanics of synthetic quartz, unpublished manuscript, 1980.
170. **Kirby, S. H., McCormick, J. W., and Linker, M.,** The effect of water concentration on creep rates of hydrolytically-weakened synthetic quartz single crystals, *Trans. Am. Geophys. Union*, 58(Abstr.), 1239, 1977.
171. **Kirby, S. H. and Raleigh, C. B.,** Mechanisms of high-temperature, solid-state flow in minerals and ceramics and their bearing on creep behavior of the mantle, *Tectonophysics*, 19, 1965, 1973.
172. **Kirby, S. H. and Veyssière, P.,** Plastic deformation of MgO 1.1(Al_2O_3)1.1 spinel at 0.28 Tm: preliminary results, *Philos. Mag.*, 41, 129, 1979.
173. **Kirby, S. H. and Zateslo, T.,** unpublished data, 1979.
174. **Kohlstedt, D., Goetze, C., and Durham, W. B.,** Experimental deformation of single crystal olivine with application to flow in the mantle, in *The Physics and Chemistry of Minerals and Rocks*, Strens, R. G. J., Ed., John Wiley & Sons, London, 1976, 35.
175. **Kohlstedt, D. L. and Goetze, C.,** Low-stress high-temperature creep in olivine single crystals, *J. Geophys. Res.*, 79, 2045, 1974.
176. **Kollé, J. J., and Blacic, J.D.,** Deformation of single crystal clinopyroxenes, *J. Geophys. Res* 87, 4019, 1982.
177. **Kollé, J. J. and Blacic, J. D.,** Preliminary deformation characteristics of single crystal clinopyroxene, *Trans. Am. Geophys. Union*, 58(Abstr.), 513, 1977.
178. **Kollé, J. J. and Blacic, J. D.,** Mechanical deformation of a single crystal hypersthene, *Trans. Am. Geophys. Union*, 59(abstr.), 1185, 1978.
179. **Kranz, R. L.,** Crack growth and development during creep of Barre granite, *Int. J. Rock Mech. Min. Sci.*, 16, 23, 1979.
180. **Kranz, R. L.,** Crack-crack and crack-pore interactions in stressed granite, *Int. J. Rock Mech. Min. Sci.*, 16, 37, 1979.
181. **Kranz, R. L.,** The Static Fatigue and Hydraulic Properties of Barre Granite, Ph.D. thesis, Columbia University, 1979, 192.
182. **Kranz, R. L. and Scholz, C.,** Critical dilatant volume of rocks at the onset of tertiary creep, *J. Geophys. Res.*, 82, 4893, 1977.
183. **Lama, R. D. and Vutukuri, V. S.,** *Handbook on Mechanical Properties of Rocks*, Vol. 2, Trans Tech Publications, Rockport, MA, 1978, 481.
184. **Lama, R. D. and Vutukuri, V. S.,** *Handbook on Mechanical Properties of Rocks*, Vol. 3, Trans Tech Publications, Rockport, MA, 1978, 406.
185. **Lama, R. D. and Vutukuri, V. S.,** *Handbook on Mechanical Properties of Rocks*, Vol. 4, Trans Tech Publications, Rockport, MA, 1978, 515.
186. **Langdon, T. G. and Pask, J. A.,** Mechanical behavior of single-crystal and polycrystalline MgO, in *High Temperature Oxides*, Vol. 3, Academic Press, New York, 1970, 53.
187. **Lawn, B. R. and Wilshaw, T. R.,** *Fracture of Brittle Solids*, Cambridge University Press, 1975, 204.
188. **Lile, R. C.,** The effect of anisotropy on the creep of polycrystalline ice, *J. Glaciol.*, 85, 475, 1978.
189. **Linker, M. F.,** Experimental Creep of Hydrolytically Weakened Synthetic Quartz Crystals Oriented to Promote <a> and <c> Slip, B.A. thesis, Earth Sciences Board, University of California, Santa Cruz, 1979, 108.
190. **Linker, M. F. and Kirby, S. H.,** Creep of hydrolytically-weakened synthetic quartz: experiments with samples oriented to promote duplex {1010} <a> slip, Trans. Am. Geophys. Union, 59(Abstr.), 1185, 1978; Geophys. Monogr. 24, Am. Geophys. Union, 29, 1981.
191. **Liu, H. W. and Miller, K. J.,** Fracture toughness of fresh-water ice, *J. Glaciol.*, 86, 135, 1979.
192. **Lockner, D. and Byerlee, J.,** Acoustical emission and creep in rocks at high confining pressure and differential stress, *Seismol. Soc. Am. Bull.*, 67, 243, 1977.
193. **Lockner, D. and Byerlee, J.,** Development of fracture planes during creep in granite in *Proc. 2nd Conf. Acoustical Emission/Microseismic Activity in Geologic Structures and Materials*, Hardy, H. R., Jr. and Leighton, F. W., Eds., Trans Tech Publications, Rockport, MA, 1979.
194. **Logan, J. M.,** Friction in rocks, *Rev. Geophys. Space Phys.*, 13, 358, 1975.
195. **Logan, J. M. and Handin, J.,** Triaxial compression testing at intermediate strain rates, in Dynamic Rock Mechanics, 12th Symp. Rock Mechanics, AIME, New York, N.Y., 1971, 167.

196. **Lyall, K. D. and Paterson, M. S.,** Plastic deformation of galena (lead sulphide), *Acta Metall.*, 14, 371, 1966.
197. **Martin, R.,** Time-dependent crack growth in quartz and its application to the creep of rocks, *J. Geophys. Res.*, 77, 1406, 1972.
198. **Martin, R. J. and Durham, W. B.,** Mechanisms of crack growth in quartz, *J. Geophys. Res.*, 80, 4837, 1975.
199. **McClay, K. R. and Atkinson, B. K.,** Experimentally induced kinking and annealing of single crystals of galena, *Tectonophysics*, 39, 175, 1977.
200. **McCormick, J. W.,** Transmission Electron Microscopy of Experimentally Deformed Synthetic Quartz, Ph.D. thesis, University of California, Los Angeles, 1977, 171.
201. **McKenzie, D. P.,** The geophysical importance of high temperature creep, in *The History of the Earth's Crust*, Phinney, R. A., Ed., Princeton University Press, 1968, 28.
202. **Mellor, M. and Testa, R.,** Effect of temperature on the creep of ice, *J. Glaciol.*, 8, 131, 1969.
203. **Mirwald, P. W., Getting, I. C., and Kennedy, G. C.,** Low friction cell for piston-cylinder high-pressure apparatus, *J. Geophys. Res.*, 80, 1519, 1975.
204. **Moavenzadeh, F., Williamson, R. B., and Wissa, A. E. Z.,** Rock Fracture Research, Department of Civil Engineering, Research Report, MIT Press, Cambridge, 1966, 757.
205. **Mogi, K.,** Pressure dependence of rock strength and transition from brittle fracture to ductile flow, *Bull. Earthquake Res. Inst. Tokyo Univ.*, 44, 215, 1966.
206. **Mogi, K.,** Effect of the intermediate principal stress on rock failure, *J. Geophys. Res.*, 72, 5117, 1967.
207. **Mogi, K.,** Fracture and flow of rocks under high triaxial compression, *J. Geophys. Res.*, 76, 1255, 1971.
208. **Mogi, K.,** Fracture and flow or rocks, *Tectonophysics*, 13, 541, 1972.
209. **Morrison-Smith, D. J., Paterson, M. S., and Hobbs, B. E.,** An electron microscope study of plastic deformation in single crystals of synthetic quartz, *Tectonophysics*, 33, 43, 1976.
210. **Müller, P. and Siemes, H.,** Zur festigkeit und gefugeregelung von experimentell verformten magnetiterzen, *N. Jb. Miner. Abh.*, 117, 39, 1972.
211. **Müller, P. and Siemes, H.,** Festigkeit, verformbarkeit und gefugeregelung von anhydrit — experimentelle stauchverformung unter manteldrucken bis 5 kbar bei temperaturen bis 300° C, *Tectonophysics*, 23, 105, 1974.
212. **Müller, W. H. and Briegel, U.,** The rheological behavior of polycrystalline anhydrite, *Ecol. Geol. Helv.*, 71, 397, 1978.
213. **Murrell, S. A. F. and Ismail, I. A. H.,** The effect of decomposition of hydrous minerals on the mechanical properties of rocks at high pressures and temperatures, *Tectonophysics*, 31, 207, 1976.
214. **Murrell, S. A. F. and Ismail, I. A. H.,** The effect of temperature on the strength at high confining pressure of granodiorite containing free and chemically-bound water, *Contrib. Mineral. Pet.*, 55, 317, 1976b.
215. **Nicolas, A. and Poirier, J.-P.,** *Crystalline Plasticity and Solid State Flow in Metamorphic Rocks*, John Wiley & Sons, London, 1976, 444.
216. **Nye, J. F.,** *Physical Properties of Crystals*, Oxford at Clarendon Press, 1957, 322.
217. **Obreimoff, J. W.,** The splitting strength of mice, *Proc. R. Soc. London*, A127, 290, 1930.
218. **Ohnaka, M.,** The quantitative effect of hydrostatic confining pressure on the compressive strength of crystalline rocks, *J. Phys. Earth*, 21, 125, 1973a.
219. **Ohnaka, M.,** Frictional characteristics of typical rocks, *J. Phys. Earth*, 23, 87, 1975.
220. **Olsson, W. A.,** Effects of temperature, pressure, and displacement rate on the frictional characteristics of a limestone, *Int. J. Rock Mech. Min. Sci. Geomech. Abstr.*, 11, 267, 1974.
221. **Parrish, D. K., Krivz, A., and Carter, N. L.,** Finite element folds of similar geometry, *Tectonophysics*, 32, 183, 1976.
222. **Paterson, M. S.,** Effect of pressure on stress-strain properties of materials, *Geophys. J. R. Abstr. Soc.*, 14, 13, 1967.
223. **Paterson, M. S.,** Nonhydrostatic thermodynamics and its geologic applications, *Rev. Geophys. Space Phys.*, 11, 355, 1973.
224. **Paterson, M. S.,** Some current aspects of experimental rock deformation, *Philos. Trans. R. Soc. London A*, 283, 163, 1976.
225. **Paterson, M. S.,** *Experimental Rock Deformation — The Brittle Field*, Springer-Verlag, New York, 1978, 254.
226. **Paterson, M. S. and Edmond, J. M.,** Deformation of graphite at high pressures, *Carbon*, 10, 29, 1972.
227. **Paterson, M. S. and Kekulawala, K. R. S. S.,** The role of water in quartz deformation, *Bull. Mineral.*, 102, 92, 1979.
228. **Paterson, M. S. and Weaver, C. W.,** Deformation of polycrystalline MgO under pressure, *J. Am. Ceram. Soc.*, 53, 463, 1970.
229. **Peng, S. and Johnson, A. M.,** Crack growth and faulting in cylindrical specimens of Chelmsford granite, *Int. J. Rock Mech. Min. Sci.*, 9, 37, 1972.

230. **Perkins, T. K. and Bartlett, L. E.,** Surface energies of rocks measured during cleavage, *Soc. Pet. Eng. J.*, 3, 307, 1963.
231. **Perkins, T. K. and Krech, W. W.,** Effect of cleavage rate and stress level on the apparent surface energies of rocks in *Soc. Pet. Eng.*, 6, 308, 1966.
232. **Phakey, P., Dollinger, G., and Christie, J. M.,** Transmission electron microscopy of experimentally deformed olivine crystals, in *Flow and Fracture of Rocks, Geophysical Monograph 16, The Griggs Volume*, Heard, H. C., Borg, N. L., Carter, N. L., and Raleigh, C. B., Eds., American Geophysical Union, Washington, D.C., 1972, 117.
233. **Poirier, J.-P.,** High temperature creep in single crystalline sodium chloride. I. Creep controlling mechanism, *Philos. Mag.*, 26, 701, 1972.
234. **Poirier, J.-P. and Guillope, M.,** Deformation induced recrystallization of minerals, *Bull. Mineral.*, 102, 67, 1979.
235. **Post, R. L.,** The Flow Laws of Mt. Burnett Dunite, Ph.D. thesis, University of California, Los Angeles, 1973, 272.
236. **Post, R. L.,** High temperature creep of Mt. Burnet dunite, *Tectonophysics*, 42, 75, 1977.
237. **Raleigh, C. B.,** Mechanisms of plastic deformation of olivine, *J. Geophys. Res.*, 73, 5391, 1968.
238. **Raleigh, C. B. and Kirby, S. H.,** Creep in the upper mantle, *Miner. Soc. Am. Spec. Pap.*, 3, 113, 1970.
239. **Raleigh, C. B., Kirby, S. H., Carter, N. L., and Ave'Lallemant, H. G.,** Slip and the clinoenstatite transformation as competing rate processes in enstatite, *J. Geophys. Res.*, 76, 4011, 1971.
240. **Raleigh, C. B. and Paterson, M. S.,** Experimental deformation of serpentinite and its tectonic implications, *J. Geophys. Res.*, 70, 3965, 1965.
241. **Ramsier, R. O.,** Growth and Mechanical Properties of River and Lake Ice, Ph.D. thesis, Universite Laval, Quebec, Canada, 1971.
242. **Riecker, R. E. and Rooney, T. P.,** Water-induced weakening of hornblende and amphibolite, *Nature (London)*, 224, 1299, 1969.
243. **Riley, N. W., Noll, G., and Glen, J. W.,** The creep of NaCl-doped ice monocrystals, *J. Glaciol.*, 85, 501, 1978.
244. **Robertson, E. C.,** Creep of Solenhofen limestone under moderate hydrostatic pressure in Rock Deformation, Griggs, D. T., Handin, J., Ed., Geological Society of America, Boulder, CO, 1960, 227.
245. **Robertson, E. C.,** Viscoelasticity of rocks, in *State of Stress in the Earth's Crust*, Judd, W. R., Ed., Elsevier, New York, 1964, 181.
246. **Robin, Y. E.,** Pressure solution at grain-to-grain contacts, *Geochim. Cosmochim. Acta*, 42, 1383, 1978.
247. **Roscoe, W. E.,** Experimental deformation of natural chalcopyrite at temperatures up to 300°C over the strain rate range 10^{-2} to 10^{-6} sec^{-1}, *Econ. Geol.*, 70, 454, 1975.
248. **Ross, J. V.,** unpublished data, 1979.
249. **Ross, J. V. and Nielsen, K. C.,** High temperature flow of wet polycrystalline enstatite, *Tectonophysics*, 44, 233, 1978.
250. **Rutter, E. H.,** The effects of strain-rate changes on the strength and ductility of Solenhofen limestone at low temperatures and confining pressures, *Int. J. Rock Mech. Min. Sci.*, 9, 183, 1972.
251. **Rutter, E. H.,** The influence of interstitial water on the rheological behaviour of calcite rocks, *Tectonophysics*, 14, 13, 1972.
252. **Rutter, E. H.,** On the creep testing of rocks at constant stress and constant force, *Int. J. Rock Mech. Min. Sci.*, 9, 191, 1972.
253. **Rutter, E. H., Atkinson, B. K., and Mainprice, D. H.,** On the use of the stress relaxation testing method in studies on the mechanical behavior of geological materials, *Geophys. J. R. Abstr. Soc.*, 55, 155, 1978.
254. **Salmon, B. C., Clark, B. R., and Kelly, W. C.,** Sulfide deformation studies. II. Experimental deformation of Galena to 2,000 bars and 400 degrees C, *Econ. Geol.*, 69, 1, 1974.
255. **Sangha, C. M. and Dhir, R. K.,** Strength and deformation of rock subject to multiaxial compressive stresses, *Int. J. Rock Mech. Min. Sci. Geomech. Abstr.*, 12, 277, 1975.
256. **Santhanan, A. T. and Gupta, Y. P.,** Cleavage surface energy of calcite, *Int. J. Rock Mech. Min. Sci.*, 5, 253, 1968.
257. **Sawbridge, P. T. and Sykes, E. C.,** Dislocation glide in UO2 single crystals at 1600 K, *Philos. Mag.*, 24, 33, 1971.
258. **Saynisch, H. J.,** Festigkeits- und Gefugeuntersuchungen an Experimentell und Naturlich Verformten Zinkblendeerzen, in *Experimental and Natural Rock Deformation*, Paulitsch, P., Ed., Springer-Verlag, Berlin, 1970, 209.
259. **Schmid, S. M.,** Rheological evidence for changes in the deformation mechanism of Solenhofen limestone towards low stress, *Tectonophysics*, 31, T21, 1976.
260. **Schmid, S. M., Boland, J. N., and Paterson, M. S.,** Superplastic flow in finegrained limestone, *Tectonophysics*, 43, 257, 1977.

261. **Schmid, S. M. and Paterson, M. S.,** Strain analysis in an experimentally deformed oolitic limestone, in *Energetics of Geological Processes*, Saxena, S. K. and Bhattacharji, S., Eds., Springer-Verlag, New York, 1977, 67.
262. **Schmidt, R. A.,** Fracture toughness testing of limestone, *Exp. Mech.*, 16, 161, 1976.
263. **Schmidt, R. A. and Huddle, C. W.,** Effect of confining pressure on fracture toughness of Indiana limestone, *Int. J. Rock. Mech. Min. Sci.*, 14, 289, 1977.
264. **Schock, R. N., Abey, A. E., Bonner, B. P., Duba, A., and Heard, H. C.,** Mechanical Properties of Nugget Sandstone, Rep. UCRL-51447, Lawrence Livermore Laboratory, 1973, 19.
265. **Schock, R. N., Abey, A. E., Heard, H. C., and Louis, H.,** Mechanical Properties of Granite from the Taourirt Tan Afella Massif, Algeria, Rep. UCRL-51296, Lawrence Livermore Laboratory, 1972, 21.
266. **Schock, R. N. and Heard, H. C.,** Static mechanical properties and shock loading response of granite, *J. Geophys. Res.*, 79, 1662, 1974.
267. **Schock, R. N., Heard, H. C., and Stephens, D. R.,** Mechanical Properties of Rocks from the Site of the Rio Blanco Gas Stimulation Experiment, Rep. UCRL-51260, Lawrence Livermore Laboratory, 1972, 22.
268. **Schock, R. N., Heard, H. C., and Stephens, D. R.,** Stress-strain behavior of a granodiorite and two graywackes on compression to 20 kilobars, *J. Geophys. Res.*, 78, 5922, 1973.
269. **Scholtz, C., Molnar, P., and Johnson, T.,** Detailed studies of frictional sliding of granite and implications for the earthquake mechanism, *J. Geophys. Res.*, 77, 6392, 1972.
270. **Scholz, C. H.,** Mechanism of creep in brittle rock, *J. Geophys. Res.*, 73, 3295, 1968.
271. **Scholz, C. H.,** Static fatigue of quartz, *J. Geophys. Res.*, 77, 2104, 1972.
272. **Sherby, O. D. and Burke, P. M.,** Mechanical behavior of crystalline solids at elevated temperatures, *Prog. Met. Sci.*, 13, 325, 1968.
273. **Shoji, H. and Higashi, A.,** A deformation mechanism map of ice, *J. Glaciol.*, 85, 419, 1978.
274. **Siemes, H.,** Experimental Deformation of Galena Ores, in *Experimental and Natural Rock Deformation*, Paulitsch, P., Ed., Springer-Verlag, Berlin, 1970, 165.
275. **Siemes, H., Saynisch, H. J., and Borges, B.,** Experimentelle verformung von zinkblendeeinkristallen bei raumtemperatur und 5000 bar manteldruck, *N. Jb. Miner. Abh.*, 119, 65, 1973.
276. **Stephens, D. R., Heard, H. C., and Schock, R. N.,** High-Pressure Mechanical Properties of Tuff from the Diamond Dust Site, Rep. UCRL-50858, Lawrence Livermore Laboratory, 1970, 14.
277. **Stesky, R. M.,** Mechanisms of high temperature frictional sliding in Westerly granite, *Can. J. Earth Sci.*, 15, 361, 1978.
278. **Stesky, R. M.,** Rock friction — effect of confining pressure, temperature, and pore pressure, *Pure Appl. Geophys.*, 116, 690, 1978.
279. **Stesky, R. M. and Brace, W. F.,** Estimation of frictional stress on the San Andreas fault from laboratory measurements, in *Proc. Conf. Tectonic Problems of the San Andreas Fault System*, Vol. 13, Kovach, R. L. and Nur, A., Eds., Stanford University Publications on Geological Sciences, 1973, 206.
280. **Stesky, R. M., Brace, W. F., Riley, D. K., and Robin, P.-Y. F.,** Friction in faulted rock at high temperature and pressure, *Tectonophysics*, 23, 177, 1974.
281. **Stocker, R. L. and Ashby, M. F.,** On the rheology of the upper mantle, *Rev. Geophys. Space Phys.*, 11, 391, 1973.
282. **Swain, M. V. and Atkinson, B. K.,** Fracture surface energy of olivine, *Pure Appl. Geophys.*, 116, 866, 1978.
283. **Tapponnier, P. and Brace, W. F.,** Development of stress-induced microcracks in Westerly granite, *Int. J. Rock Mech. Min. Sci.*, 13, 103, 1976.
284. **Teufel, L. W. and Logan, J. M.,** Effect of displacement rate on the real area of contact and temperature generated during frictional sliding of Tennessee sandstone, *Pure Appl. Geophys.*, 116, 840, 1978.
285. **Tullis, J. A.,** Preferred Orientations in Experimentally Deformed Quartzites, Ph.D. thesis, Department of Geology, University of California, Los Angeles, 1971, 344.
286. **Tullis, J. A.,** High temperature deformation of rocks and minerals, *Rev. Geophys. Space Phys.*, 17, 1137, 1979.
287. **Tullis, J. A.,** unpublished data, 1979.
288. **Tullis, J. A., Shelton, G. L., and Yund, R. A.,** Pressure dependence of rock strength: implications for hydrolytic weakening, *Bull. Mineral.*, 102, 110, 1979.
289. **Tullis, J. and Yund, R. A.,** Experimental deformation of dry westerly granite, *J. Geophys. Res.*, 82, 5705, 1977.
290. **Turner, F. J., Griggs, D. T., and Heard, H. C.,** Experimental deformation of calcite crystals, *Geol. Soc. Am. Bull.*, 65, 883, 1954.
291. **Turner, F. J. and Heard, H. C.,** Deformation of calcite single crystals at different strain rates, *Univ. Calif. Publ. Geol. Sci.*, 46, 103, 1965.
292. **Vutukuri, V. S., Lama, R. D., and Saluja, S. S.,** *Handbook on Mechanical Properties of Rocks*, Vol. 1, Trans Tech Publications, Rockport, MA, 1974, 280.

293. **Walsh, J. B.,** The effects of cracks on the compressibility of rock, *J. Geophys. Res.*, 70, 381, 1965.
294. **Walsh, J. B.,** The effect of cracks on the uniaxial elastic compression of rocks, *J. Geophys. Res.*, 70, 399, 1965.
295. **Wawersik, W. R.,** Time-dependent rock behaviour in uniaxial compression, in *New Horizons in Rock Mechanics, Proc. 14th Symp. Rock Mech. Penn. State Univ., 1972,* Hardy, H. R. and Stefanko, R., Eds., American Society of Civil Engineers, New York, 1973, 85.
296. **Wawersik, W. R.,** Time-dependent behavior of rock in compression, in *Advances in Rock Mechanics, Proc. 3rd Congr. Int. Soc. Rock Mech.*, Vol. 2, (Part A), National Academy Science, Washington, D.C., 1974, 357.
297. **Wawersik, W. R. and Brown, W. S.,** Creep Fracture of Rock, Advance Research Projects Agency, Arpa Order No. 1579, Program Code 2F10, Department of Defense, Washington, D.C., 1973, 72.
298. **Weed, H. C. and Heard, H. C.,** Mechanical Properties of Annona Chalk to 3.8 GPa, Rep UCID-16675, Lawrence Livermore Laboratory, 1975, 17.
299. **Weertman, J.,** Creep of ice, in Physics and Chemistry of Ice, Whalley, E., Jones, S. J., and Gold, L. W., Eds., Royal Society of Canada, Ottawa, 1973, 320.
300. **Weertman, J. and Weertman, J. R.,** Mechanical properties, strongly temperature-dependent, in *Physical Metallurgy*, Cahn, R. W., Ed., Elsevier, 1970, 983.
301. **Weertman, J. and Weertman, J. R.,** High temperature creep of rock and mantle viscosity, *Ann. Rev. Earth Planet. Sci.*, 3, 293, 1975.
302. **Westbrook, J. H. and Jorgensen, P. J.,** Effects of water desorption on indentation microhardness anisotropy in minerals, *Am. Miner.*, 53, 1899, 1968.
303. **Wiederhorn, S. M., Evans, A. G., and Roberts, D. E.,** A fracture mechanics study of the Skylab windows, in *Fracture Mechanics of Ceramics*, Bradt, R. C., Hasselman, D. P. H., and Lang, F. F., Eds., Plenum Press, New York, 1974, 829.
304. **Willaime, C.,** unpublished data, 1978.
305. **Willaime, C., Christie, J. M., and Kovacs, M. P.,** Experimental deformation of K-feldspar single crystals, *Bull. Miner. Soc. Fr. Miner. Cryst.*, 102, 168, 1979.
306. **Willaime, C. and Gandais, M.,** Electron microscope study of plastic defects in experimentally deformed alkali feldspars in *Bull. Soc. Fr. Mineral. Cristallogr.*, 100, 263, 1977.
307. **Wu, F. T. and Thomsen, L.,** Microfracturing and deformation of Westerly granite under creep conditions, *Int. J. Rock Mech. Min. Sci.*, 12, 167, 1975.
308. **Zoback, M. D. and Byerlee, J. D.,** The effect of cyclic differential stress on dilatancy in Westerly granite under uniaxial and triaxial conditions in *J. Geophys. Res.*, 80, 1526, 1975.
309. **Byerlee, J. D.,** Frictional characteristics of granite under high confining pressure, *J. Geophys. Res.*, 72, 3639, 1967.
310. **Griggs, D. T.,** Creep of rocks, *J. Geol.*, 47, 225, 1939.
311. **Griggs, D. T. and Miller, W. B.,** Deformation of Yule marble, *Geol. Soc. Am. Bull.*, 62, 722, 1951.
312. **Steinemann, S.,** Experimentelle Untersuchung zur Plastizitat von Eis in Beitrage zur Geologie der Schweis, Geotechnische Serie, Hydrologie, No. 10, 1958, 72.

Section IV
Magnetic Properties of Minerals and Rocks

By
Robert S. Carmichael

INTRODUCTION

Since rocks consist of mineral grains and crystals, their magnetization arises from those mineral constituents present which are measurably magnetic. The fraction of the total rock which has magnetic minerals may be only a few percent. Since it is this relatively small proportion and its chemical and physical state which determines the bulk magnetic properties and magnetization, two consequences result:

1. Magnetic properties can be quite variable within a given rock body or structure, depending on chemical inhomogeneity, depositional or crystallization conditions, and postformational geologic history.
2. The magnetic characteristics are not necessarily closely predictable by the lithology (rock type, and name). This is because the geological name or classification of the rock is generally given on the basis of the gross mineralogy, dominantly silicate minerals. However, it is the minor fraction of, say, iron oxides which controls the magnetization.

Despite the variability of magnetic properties of typical rocks, it is still true that the properties bear general relationships to rock type and overall composition. The prospects for properly predicting and interpreting magnetic properties are enhanced if there is an understanding of:

1. Basic phenomenology of magnetization of crystalline materials
2. Magnetic characteristics of magnetic minerals, and how the properties vary with chemical composition, grain size, mechanical condition, temperature and pressure, and other factors
3. Properties of typical rocks, and their variation with geological condition

Applications of rock magnetism include:

1. "Mapping" the subsurface by magnetic prospecting, and interpretation of the depth, size, magnetic mineralogy, proportion of magnetic material, and inferred lithology (rock type) of buried rocks. Such magnetic prospecting, whether conducted at the ground surface, by aircraft sensing, or on shipboard, is a good first tool in the exploration for and interpretation of buried geological structure, lithology, and economic mineral deposits. Magnetic exploration for resources can either locate rock bodies (e.g., iron ore) directly, or by inference. An example of the latter is determining the structural habitat favorable for oil and gas from topography of the buried "basement" crystalline rock underlying a thick sedimentary section.
2. Understanding the origin and character of the magnetism of rocks for paleomagnetic work in which the objective is interpreting the remanent magnetization. This can indicate the character of the Earth's magnetic field at the time the rock was formed, and has use for stratigraphic correlation, age-dating, and reconstruction of past movements of the Earth's crust. The latter includes structural deformation and seafloor spreading and continental drift — "plate tectonics". It is the magnetic properties and magnetization of rocks, and pattern of magnetization in rock sequences, which gave the initial body of quantitative evidence to demonstrate convincingly that the sections of the Earth's crust had undergone large lateral displacements over geologic time. To indicate this contribution,

"Magnetism could be the key to reconstructing the history of the ocean floor and movements of the continents."*
"The study of paleomagnetism has produced a revolution in the earth sciences over the past decade."**

3. Use in materials science, as in creating materials of desired magnetic properties such as ferrite memory cores, magnetic tapes, or permanent magnets.

Notation and Units

Magnetism of rocks and natural materials involves the study of (1) geomagnetism — the Earth's magnetic field; origin and source, character and configuration, and change with time, (2) rock magnetism — the magnetic properties and behavior of magnetic minerals and rocks, and (3) paleomagnetism — the study of the remanent magnetization retained in rocks, as a means of deducing the history and nature of the Earth's field through geologic time. Of interest are the direction, intensity, polarity, and configuration of the geomagnetic field.

Such work has a long historical legacy. A few thousand years ago the Chinese were using magnetic rocks for tricks and navigation. In 1269 Petro Peregrinus studied the polarity characteristics of a sphere of rock and wrote what was perhaps the first significant scientific treatise, *Epistola de Magnete* although it was not published until 1558. In 1600, in another of the earliest scientific books, William Gilbert published *De Magnete* in England, based on his experimental observations of the magnetic force field around a sphere of lodestone (magnetite), and deducing that the Earth was like a magnet.

The units of historical and conventional use have been in the cgs-emu system. Virtually all literature values are in these cgs units. Full transition to work in SI units is yet to come, because of a lack of general agreement on how the units are to be converted. This is because of the special problems associated with definition of basic quantities in magnetism. For example, whether they should be based on magnetic poles or electric currents, and the choice between rationalized or unrationalized units for magnetics.

Table 1 shows the cgs and mks units and their conversion. Because of the existing body of data and the general preference of workers in rock magnetism, the cgs units have been used in the tables and figures to follow.

MAGNETIZATION: TYPES, AND PARAMETERS

Types of Intrinsic Magnetization

The basic types of magnetization in a lattice are

diamagnetism
paramagnetism
ferromagnetism
antiferromagnetism
spin-canted (anti)ferromagnetism
ferrimagnetism
superparamagnetism

* Heirtzler, J. R., Seafloor spreading, *Sci. Am.*, December, 1968.
** McElhinny, M. W., *Paleomagnetism and Plate Tectonics*, Cambridge University Press, Cambridge, 1973.

Table 1
MAGNETIC NOTATION AND UNITS

Symbol	Name	CGS unit (cgs-emu)	MKS unit (SI)
		(with equivalence of magnitude of units)[a]	
ϕ	Magnetic flux	10^8 maxwell	= weber
B	Flux density (induction)[b]	10^4 gauss, or maxwell/cm^2 $1\gamma = 10^{-5}$ gauss $B = H + 4\pi J$	= tesla, or weber/m^2 = nanotesla $B = \mu_o (H + J)$
H	Magnetic field strength (force field)	$4\pi \times 10^{-3}$ oersted or line/cm^2 1 oe = 79.577 amp-turn/m	= amp-turn/m
μ	Permeability	$(10^7/4\pi)$ gauss/oersted $\mu_o = 1$ $B = \mu H$	= weber/amp-m, or henry/m $\mu_o = 4\pi \times 10^{-7}$ $B = \mu_r \mu_o H$ $\mu_r = 1 + k$
k	Susceptibility[d]	4π[c]	
χ	Specific susceptibility, $\chi = k/\varrho$	$J = kH$	$J = kH$
m	Magnetic pole strength	10 emu 10^8 unit poles	= amp-m
m.d	Magnetic (dipole) moment	10^3 "emu", or (gauss/cm^3) 10^{10} pole-cm	= amp-m^2
J	Magnetization (intensity), dipole moment per unit volume	10^{-3} "emu" (gauss)	= amp-m^2/m^3, or amp/m
I	Magnetization, dipole moment per unit mass	gauss-cm^3/gm	
	Coulomb's law	$F = (m_1 m_2)/\mu r^2$	$F = (m_1 m_2)/(4\pi \mu_r \mu_o r^2)$

[a] The maxwell is smaller than a weber, by a factor of 10^8.

[b] The Earth's surface flux density is about 0.6 gauss, or 60,000 gammas (0.00006 tesla, in SI units). This results from a magnetic field strength (force) in SI units of:

$$H = \frac{B}{\mu_o} = \frac{0.6 \times 10^{-4}}{4\pi \times 10^{-7}} = 47.75 \text{ amp-turns/m}$$

In cgs units this H is $47.75 \times 4\pi \times 10^{-3} = 0.6$ oersted.

[c] In rationalized system, the rationalized unit is $(1/4\pi) \times$ unrationalized unit.

[d] Susceptibility is dimensionless, but is frequently given units to indicate whether it is based on J in emu/cm^3 (i.e., for k) or in emu/gm (i.e., for χ).

Diamagnetic — Material in which the magnetic induction is slightly less than the applied field; that is, the atomic moments act to oppose the external field. Diamagnetism is due to electrons orbiting around the nucleus. The susceptibility, $k = J/H$, is negative and is of the order of -10^{-6} emu/cm^3. The effect is present in all materials, is weak, and exists only in the presence of an applied field.

Paramagnetic — Material in which the magnetic induction is slightly greater than the applied (external) magnetic field. There is a partial alignment of the atomic dipole moments to augment the net magnetization in the direction of the applied field. Paramagnetism is due to electron spin of unpaired electrons. The magnetic susceptibility is positive and small, being of the order of $+10^{-4}$ to 10^{-6} emu/cm^3. The effect is weak and exists only in the presence of an applied field.

Ferromagnetic* — Material in which the atomic magnetic moments tend strongly to align parallel to one another because of exchange interaction energy, including when no external field is applied. The moments are oriented in the same sense. There is thus a spontaneous, or intrinsic, magnetization and a remanent magnetization can be retained. Complete ordering is achieved only at absolute zero temperature where the magnetization would be exactly the sum of the moments of all magnetic atoms. Above absolute zero, thermal energy begins to disorder the magnetic moments. At a temperature characteristic of the material — the Curie temperature, T_c — the long range interaction ordering of the moments is lost. Below T_c, a ferromagnetic material in a magnetic field has the magnetization aligned so that the induction is much greater than for the field alone. The susceptibility is positive and large, being of the order of $10 - 10^4$ emu/cm^3. Above the Curie temperature the material is paramagnetic. The effect is present in a few materials — Fe, Co, Ni especially — and is strong and residual.

Antiferromagnetic — Material in which the atomic magnetic moments tend to assume an ordered antiparallel arrangement in the absence of an applied field such that there is no net magnetization for the sample. An example of this type of material would be one in which there are two sublattices of magnetic atoms with equal but oppositely directed moments. This could be brought about by equal numbers of atoms with the same moment in each sublattice, or unequal numbers with moments such that the oppositely directed moments balance. The susceptibility is comparable to that of paramagnetic materials. Above a temperature called the Néel temperature, T_N, the magnetic interaction is completely disordered by thermal energy and the material becomes paramagnetic. *Spin-canted (anti)ferromagnetism* is a condition when antiparallel magnetic moments are deflected from the antiferromagnetic plane, resulting in a weak or "parasitic" magnetism. The magnetization disappears at the material Curie temperature. Hematite (αFe_2O_3) is an example.

Ferrimagnetic or ferrite — Material in which the atomic magnetic moments are antiparallel but an appreciable net magnetization results. The magnetic sublattices do not have balancing moments, either because of unequal numbers of spins or unequal dipole moments. There is a remanent magnetization. Above the Curie temperature, the material is paramagnetic. Examples are magnetite (Fe_3O_4), maghemite (γFe_2O_3), and several minerals with spinel structure.

Superparamagnetic — Material with magnetic grains or regions so small (less than the order of 0.01 μm) that a cooperative alignment of atomic dipole moments is overcome by the disordering effect of thermal energy. At a given temperature, there is a "blocking" size below which the relaxation time for magnetic alignment is very short.

Types of Remanent Magnetization

Rocks and minerals may retain a variety of forms of remanent magnetization, depending on their magnetic properties and geologic origin and history. The following kinds are distinguished: ARM, CRM, DRM, IRM, NRM, PRM, TRM, VRM.

ARM — (anhysteretic remanent magnetization) is that produced in a sample by applying a constant external magnetic field during application of a decaying alternating field. The latter is usually produced by putting an alternating current through a solenoid or coil. The directed field serves to bias the final orientation of magnetic moments, while the alternating field is being reduced to zero. It is used for lab study and characterization of samples.

CRM — (chemical, or crystallization, remanent magnetization) is that acquired by a magnetic phase as it undergoes some physicochemical change after deposition or

* In general usage, particularly in engineering, the term "ferromagnetic" refers to any material which is appreciably magnetic.

crystallization. The change might be an oxidation or reduction, phase change, dehydration, precipitation of cement, exsolution, recrystallization, or grain growth. The process usually occurs in the ambient, or Earth's, magnetic field, and at constant temperature. It can be important in some (red) sediments, and metamorphic rocks.

DRM — (depositional, or detrital, remanent magnetization) is that formed in clastic sediments (and thus clastic sedimentary rock) by the deposition of fine particles on the floor of a body of water. In the simplest case, it is due to the falling of grains through still water onto a flat-lying plane, probably accompanied by some postdepositional rotation and adjustment. The grains are preferentially oriented by the Earth's magnetic field, giving a net moment in that direction. It can be important in marine sediments, lake sediments, and varved clays.

IRM — (isothermal remanent magnetization) is the moment remaining in a sample after application of a magnetic field. This is done at a constant temperature, usually room temperature. It may be similar to lightning-induced remanence. It is useful to analyze magnetic characteristics of a rock in the laboratory.

NRM — (natural remanent magnetization) is the magnetization found in a sample in its natural, or in situ, condition as collected. The term is a general one, and represents one or a combination of the other types of magnetization described here. The example, the NRM might be a DRM of grains with a TRM originally, plus a small VRM picked up from postformational residence in the Earth's field.

PRM — (pressure remanent magnetization, or piezoremanent magnetization) is that acquired through the process of mechanical deformation while the sample is in a magnetic field. It could thus be termed deformation, or strain, remanent magnetization. The applied stress may be in the elastic or plastic range, and may be directed tectonic stress, hydrostatic (confining) pressure, or shock impact. The most pronounced effects are associated with irreversible structural changes in the magnetic minerals.

TRM — (thermoremanent magnetization) is acquired by a sample when it is cooled to normal temperature from above its Curie temperature, in the presence of a magnetic field. TRM is the most important remanence, in general, because of its known mode of origin, stability, widespread occurrence in igneous rocks and sedimentary rocks derived from them, and reliability for paleomagnetic applications. *PTRM* (partial thermoremanent magnetization) is a fraction of total TRM developed by cooling the sample while the field is applied, over only a specified interval of temperature, i.e., $T_2 - T_1$, where both are below T_c. *ITRM* (inverse thermoremanent magnetization) is produced by heating a sample from low temperature to normal temperature, in a magnetic field. It occurs in those magnetic crystals (materials) that undergo certain structural or magnetic changes at temperatures lower than T_c.

VRM — (viscous remanent magnetization) is that acquired gradually with time in a sample, while it is in a small external field. In the usual case, this occurs at low temperature and in the Earth's field. The irreversible increase in moment is generally a logarithmic function of time. The process, one of developing a preferred orientation of moments, is due to thermal agitation. This remanence is relatively weak and unstable, but is present in most rocks because of their having been in the Earth's field since their original magnetization was acquired.

Total magnetization is the sum of remanent magnetization, J_r, and induced magnetization. For small fields, the latter is proportional to the field: $J_{total} = J_r + J_{induced} = J_r + kH$.

Terms and Parameters

In classifying, describing, and comparing the magnetic properties of rocks, the following concepts and terms are used:

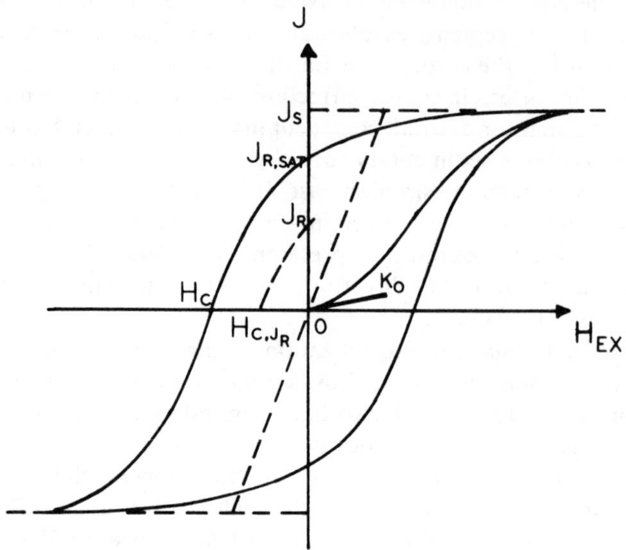

FIGURE 1. General magnetic hysteresis curve, with parameters.

Magnetic Hysteresis

This is the variation of magnetization, J, with applied field, H_{ex}, or of flux density (magnetic induction) B with H_{ex}. Figure 1 shows an idealized hysteresis curve. It has parameters:

1. Saturation (or "spontaneous") magnetization, J_s. The bulk magnetic moment of a sample when all atomic moments are aligned in their maximum ordered configuration. This can be achieved by spontaneous alignment, as in a single-domain-size grain, or by application of a sufficiently large magnetic field. J_s is temperature-dependent, the magnetization decreasing to zero at the Curie temperature.
2. Saturation isothermal remanence, $J_{r,sat}$. The remanent magnetization left in the sample from a saturating field, when the applied field is reduced to zero.
3. Remanent magnetization, J_r. The residual magnetization left from some process of magnetization acquisition other than IRM saturation, e.g., TRM, NRM.
4. Coercive force, or field, H_c. The required field to reduce a $J_{r,sat}$ to zero. This is sometimes termed remanent coercivity, H_{cr}. The remanence is thereby not permanently reduced to zero; when the field is removed, the magnetization curve will follow a path upward to a residual value on the J ordinate. Other parameters of use in classifying and identifying rock magnetic characteristics are the reversed field which while applied will reduce the remanent magnetization (e.g., $J_{r,NRM}$) to zero — see $H_{c,Jr}$ on Figure 1 — or the reversed field, H_{rc}, required to reduce the residual remanence to zero ($H_{rc} > H_c > H_{c,Jr}$). (See also figure in footnotes, Table 9.)
5. Magnetic susceptibility, k. This represents the ease of magnetization of the material in an external field. In general it is $k = J/H$, that is, the slope of the J versus H curve. However, this varies depending on the location of H (abscissa) value. It will typically be low at small fields, increase as the field increases, and then decrease as J_s is approached. Susceptibility (or permeability, μ, for the B vs H curve) can be measured anywhere on the curve, and experimental values can thus vary widely. For work in rock magnetism and paleomagnetism, an appropriate susceptibility is one measured in zero field, or for practical use one not

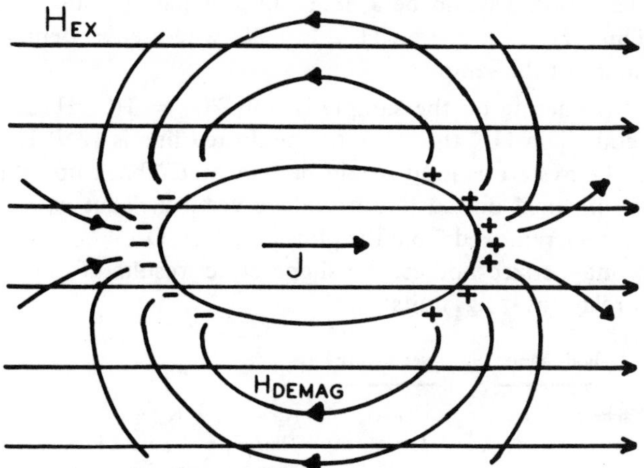

FIGURE 2. Effect of demagnetizing field.

exceeding the Earth's field — H ∼ 0.5 oersted. The value of susceptibility determines the induced magnetization in a rock so that it is useful for diagnostic studies and interpretation of magnetic maps in geological exploration for buried structure and rock or ore bodies.

Standard parameters would be

initial susceptibility

$$k_o = \left.\frac{dJ}{dH}\right]_{H=O} \qquad \text{i.e., for } J = 0$$

reversible susceptibility

$$k_{rev} = \left.\frac{\Delta J}{\Delta H}\right]_{H \pm \delta H}$$

where H = 0 or some H less than 0.5 oe

remanence susceptibility

at J_r, i.e., for $J = J_r$

$$k_r = \left.\frac{\Delta J_r}{\Delta H}\right]_{\pm \delta H}$$

Demagnetizing, or Shape, Effect

The hysteresis curve for a general sample slopes to the upper right (see Figure 1), having a trend shown by the dotted line. This is due to the "shape effect", illustrated in Figure 2. In the absence of domination of the magnetic energy state by other factors, e.g., magnetocrystalline anisotropy, magnetization tends to lie in the direction of greatest dimension in the sample (grain, crystallite). This minimizes the magnetostatic energy associated with the creation of magnetic free poles. The magnetic field arising from the latter is a demagnetizing field, H_{demag}, and this is proportional to the value of J. This magnetization J could be J_s (e.g., in a single-domain-size grain) or some remanence J_r. Thus, $H_{demag} \alpha J$, i.e., $H_{demag} = NJ$ where N = demagnetizing factor; depends on the shape of the sample.

The effective field acting on the sample is thus $H_{eff} = H - H_{demag} = H - NJ$. In Figure 1, for the abscissa H_{ex}, the slope of the dotted line is 1/N. If H_{eff} was plotted for the abscissa, the hysteresis loop would be "sheared" back upright (counterclockwise), so that the inclined dotted line would be vertical. That is, the demagnetizing effect would have been removed from the plot, i.e., N = O.

For uniformly magnetized bodies, the shape effect results in the following demagnetization factor relations in cgs units:

body shape	see Figure	
sphere	3(A) a = b = c	$N_a + N_b + N_c = 4\pi$ Thus $N_a = 4/3\pi$ and in direction with direction cosines α_1, α_2, α_3 $N = N_a\alpha^2_1 + N_b\alpha^2_2 + N_c\alpha^2_3$
infinite cylinder (or long needle)	3(B)	$N_a = N_b = 2\pi$ $N_c = 0$
plate (or disc)	3(C)	$N_c = 4\pi$ $N_a = N_b = 0$

Table 2 lists values of the demagnetizing factor, in units of $N/4\pi$, for ellipsoids, cylinders, and rectangular prisms.

The shape of a grain has an important influence, particularly for elongated grains or magnetic regions in crystals, in affecting magnetic properties such as k and H_c.

Magnetic Domains

To minimize the magnetic energy of a grain or crystal with magnetization, the region will tend to reorder or subdivide into "domains" of differently-oriented magnetization. Each domain retains the saturation magnetization J_s. The domains are separated by domain walls, planar regions in which atomic moments are progressively deflected. The bulk magnetic moment, being the vector sum of the domains' magnetizations, is thus less than J_s. With complete rearrangement of domains, if energetically permitted, the sample can be "demagnetized" to $J_r = 0$.

The creation, size, shape, and orientation and stability of domains depends on such physical factors as grain size, shape, and mechanical condition (e.g., presence of microcracks, dislocations, grain boundaries, nonmagnetic chemical inclusions, internal stress state), and temperature. They also depend on crystallization and magnetic factors such as exchange energy, magnetocrystalline anisotropy energy, and domain wall energy.

There are two types of domain walls: Bloch walls, in which the vector of atomic moment rotates out of planes containing the vectors in the adjoining domains, and Néel walls in which the vector rotates in that plane (e.g., in thin films). The magnetic vector changes direction, in a Bloch wall, by 90°, 180°, 70.5°, or 109.5°, depending on the anisotropy of the crystal.

Figure 4 shows schematic cross-sections of grains of varying sizes, ranging from single-domain to multidomain.

Table 3 gives typical domain wall thicknesses for the mineral magnetite (Fe_3O_4) and the ferromagnetic metals.

Very small magnetic grains cannot retain a coherent alignment of atomic moments; they are superparamagnetic. Larger grains can accommodate such alignment, having one domain with magnetization J_s; they are "single domain" grains, with distinct mag-

Table 2
DEMAGNETIZING FACTORS, AS $N_a/4\pi$, FOR ELLIPSOIDS, CYLINDERS, AND RECTANGULAR PRISMS[4,54]

	$N_a/4\pi$		
Dimension ratio a/b[a]	Ellipsoid[b]	Cylinder[c]	Prism[c]
0.01	0.9845	0.9650	0.9660
0.1	0.8608	0.7967	0.8051
0.2	0.7505	0.6802	0.6942
0.4	0.5882	0.5281	0.5482
0.8	0.3944	0.3619	0.3843
1[d]	0.3333	0.3116	0.3333
2	0.1736	0.1819	0.1983
3	0.1087	0.1278	0.1404
4	0.0754	0.0984	0.1085
6	0.0432	0.0673	0.0745
8	0.0284	0.0511	0.0567
10	0.0203	0.0412	0.0457
100	0.00043	0.00423	0.00472

[a] Dimension ratio is a/b for ellipsoid (a/b < 1 for oblate ellipsoid, a = b = c for sphere, a/b > 1 for prolate ellipsoid), h/dia. for cylinder of length h, and h/w for rectangular prism of length h and width w.

[b] From Stoner, E. C., *Philos. Mag.*, 36, 803, 1945; and Osborn, J. A., *Phys. Rev.*, 67, 351, 1945. Ellipsoid can have uniform magnetization, and thus uniform demagnetizing field.

[c] From Joseph, R. I., *Geophysics*, 41, 1052, 1976; calculated using "magnetometric" method, averaging the spatially varying demagnetizing factor over the volume of the sample. Sample assumed to be uniformly magnetized.

[d] Sphere (see ellipsoid column), $N_a = 4\pi/3 = 0.3333 \cdot 4\pi$.

Table 3
MAGNETIC DOMAIN WALL THICKNESSES, FOR MAGNETITE (Fe_3O_4) AND FERROMAGNETICS

Material	Wall type	Notes	Wall thickness (μm)	Ref.
Magnetite	180°	Typical	~0.15	
		Calculated	0.046	77
		Calculated	0.138	78
		Observed	0.50	79
Iron	180°		0.10	80
			0.07—0.141	4
	90°		0.05	4
		Calculated, for multidomain grain size (μm)		
		0.02 μm	0.012	14
		0.04	0.016	
		0.06	0.019	
		0.10	0.023	
Nickel	180°	Typical	~0.015	
			0.1	80
			0.206	41
	70.5°/109.5°		0.09—0.11	4
Cobalt	180°		0.016	4

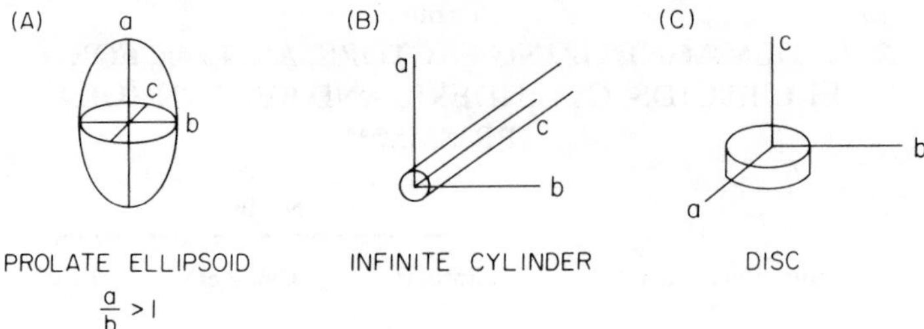

FIGURE 3. Shape anisotropy, for calculating N.

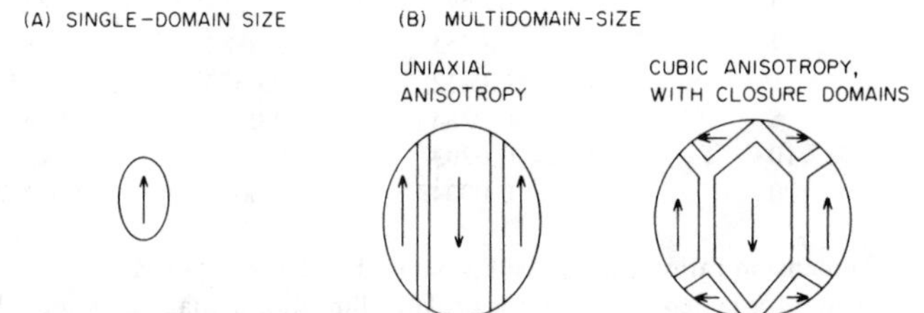

FIGURE 4. Grain sizes with representative domain configurations.

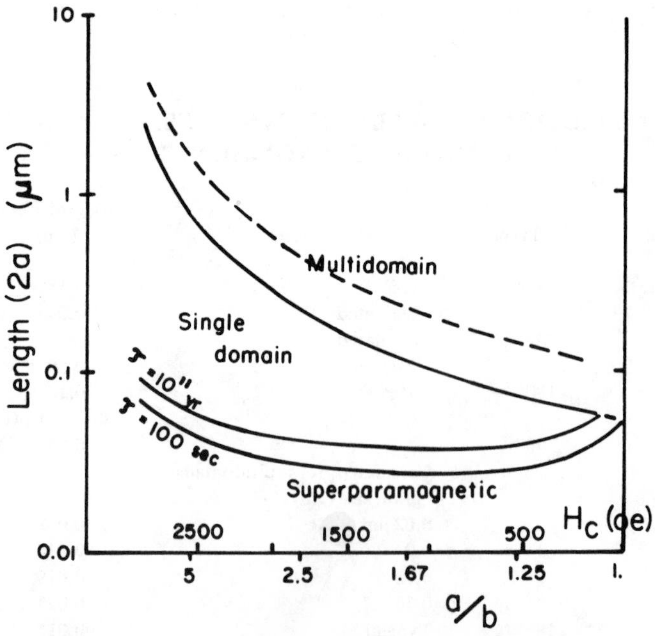

FIGURE 5. Calculated domain-size range for spheroidal magnetite (Fe_3O_4) grains of varying size and shape. Dimensions (a/b) as in Figure 3 (A). Coercive force is calculated from shape anisotropy. (From Reference 8, p. 54. Data from Evans and McElhinny, 1969.)

netic properties and behavior. Their magnetization changes by a rotation or flipping of the J_s vector, under the influence of, say, an applied magnetic field, large stress, or elevated temperature. Grains larger than a "critical single-domain size" have more than one domain. They may exhibit "multidomain" properties and behavior, or have a "pseudo-single-domain" transition size in which both single-domain and multidomain behavior and characteristics are present. In multidomain behavior, magnetization changes as in an applied field by growth of some domains at the expense of others. This is accomplished by lateral motion of the domain walls. At sufficiently high fields, the magnetization within domains themselves may rotate.

Effect of Grain Size

The size of a magnetic grain or region in a crystal is thus very important in determining the nature of intrinsic magnetization (e.g., superparamagnetism or ferromagnetism) and its behavior, e.g., multidomain or single domain. This affects the acquisition, retention, and stability of remanent magnetization, and such properties as k, H_c, and intensity of remanence. Table 4 shows the effect of grain size on magnetic state — whether superparamagnetic (SPM), single-domain (SD), pseudo-single-domain (PSD), or multidomain (MD).

Figure 5 shows the SPM-SD-MD size ranges, calculated as a function of grain size (ordinate) and shape (abscissa) for magnetite grains. Shape is given as a/b ratio (see Figure 3(A); $a/b > 1$ and $c = b$ for prolate spheroid), for a/b varying from 1 to 10. The single-domain to multidomain transition could extend up to the dotted line in the figure.

Figure 6 also shows the SPM-SD-MD ranges, for prolate ellipsoids with a/b ranging from 2 to 20. The range transitions are given for three different temperatures.

The effect of grain size on coercive force, H_c, is illustrated by the measurements on magnetite powder, Table 5, and dispersed magnetite powder, Figure 7. At large diameters (over 70 μm), coercivity is less dependent on grain size because the multidomain arrangement is not determined primarily by grain size then. The general relationship of k and H_c to grain size is shown in Figure 8. In general, the two parameters are inversely related to one another.

Magnetocrystalline Anisotropy

Magnetocrystalline anisotropy is the tendency for magnetization (magnetic moments) to lie in certain crystallographic directions preferentially. This is because of crystal symmetry. A magnetic crystal has a lattice with a regular array of magnetic atoms. There are interactions between these atomic moments, depending on their spatial orientation and interatomic distance. There are certain directions which have associated with them a lower energy for alignment of magnetization. Because of this anisotropy, the magnetic moment will tend to lie in the direction(s) in which the magnetocrystalline anisotropy energy, E_K, is minimum. These directions are the "easy" directions or axes. Work is required to turn the magnetization from these directions. For a crystal with uniaxial anisotropy, $E_K = K_o + K_1\sin^2\phi + K_2\sin^4\phi + \ldots$, where K_o = constant (an "isotropic" term), K_1, K_2 = magnetocrystalline anisotropy constants, and ϕ = angle between magnetization and the preferred (easy) axis. This is a series expansion, with the condition of symmetry around $\phi = 0$ eliminating the odd terms in $\sin\phi$. The easy directions are for $\phi = 0$ and 180°, and both are equally favorable energetically. In a crystal with cubic symmetry,

$$E_K = K_o + K_1(\alpha_1^2\alpha_2^2 + \alpha_2^2\alpha_3^2 + \alpha_3^2\alpha_1^2) + K_2\alpha_1^2\alpha_2^2\alpha_3^2 + \cdots$$

where α's = direction cosines of magnetization vector with respect to the cubic axes.

Table 4
EFFECT OF GRAIN SIZE ON MAGNETIC DOMAIN BEHAVIOR, AT ROOM TEMPERATURE (GRAIN SIZES IN μm)

Material	Notes	Superpara-magnetic	→	Single-domain	→	"Pseudo-single-domain"	→	Multidomain	Ref.
Magnetite					0.03—0.1				
		———	0.03	———	0.05	———	15—17	———	68
	Theoretical; spherical grains				0.03				8
	L/d∼10[a]				3				8
Hematite		———	0.03	———	———	———	10	———	68
			0.03			100			
	Theoretical; spherical grains						1500		8
Iron		———	.013	———	———	0.018	———	———	14,68
Nickel						0.5			
	Theoretical; spherical grains					0.05			4
	L/d∼10[a]					0.2			

[a] L/d is length/diameter ratio.

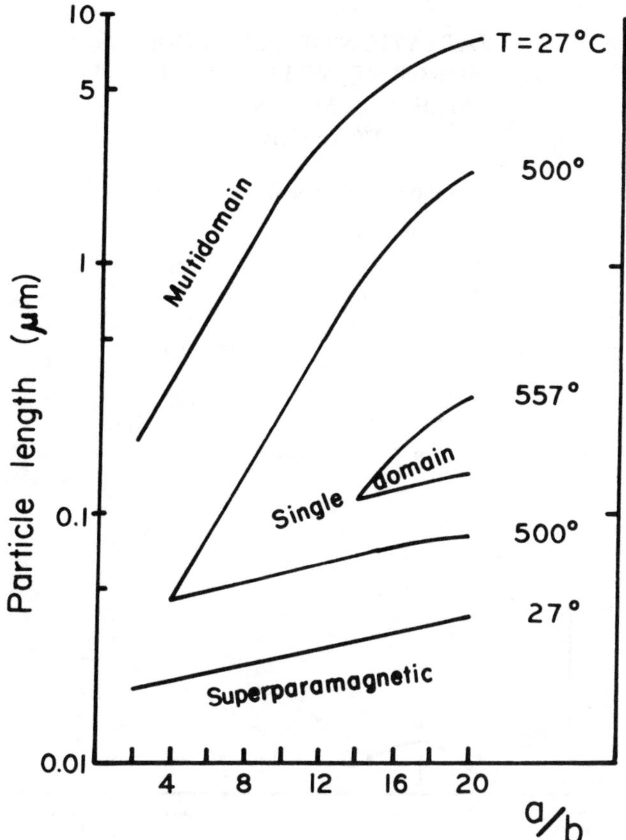

FIGURE 6. Domain-size ranges for spheroidal grains of varying size and shape. Dimensions (a/b) as in Figure 3 (A). (From Reference 63, p. 141. Data from Strangway, Larson, and Goldstein, *J. Geophys. Res.*, 73, 3787, 1968.)

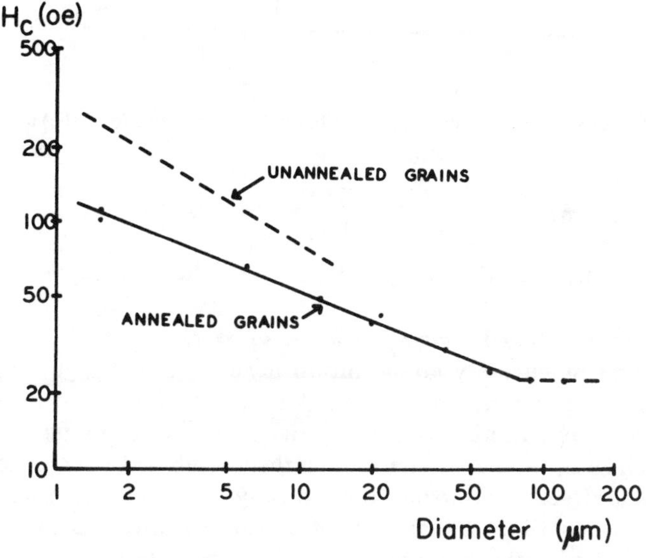

FIGURE 7. Variation of coercive force with grain size, for dispersed magnetite powder.[11, 51]

Table 5
VARIATION OF COERCIVE FORCE (H_c) WITH GRAIN SIZE, FOR MAGNETITE POWDER

Mean grain diameter (μm)	H_c (oe)
80	10
40	20
12	50
4	100
2	200
∼0.08	250
∼0.05	420

Data from T. Nagata and K. Kobayashi.

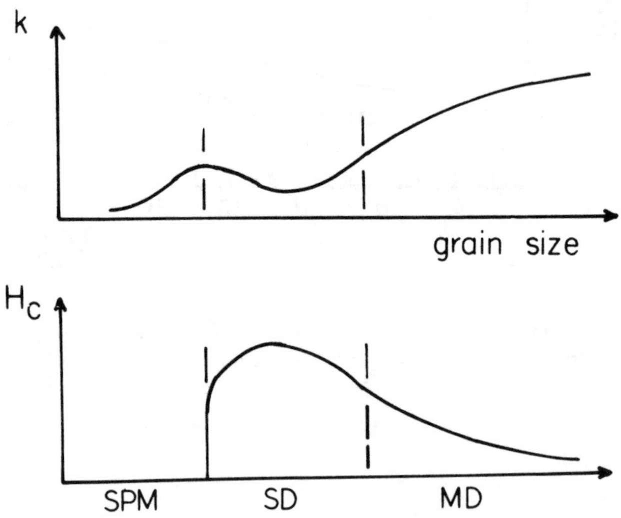

FIGURE 8. General relationship of susceptibility (k) and coercive force (H_c) to grain size.

For orthorhombic symmetry,

$$E_K = K_0 + K_1\alpha_1^2 + K_2\alpha_2^2 + K_{11}\alpha_1^4 + K_{12}\alpha_1^2\alpha_2^2 + K_{22}\alpha_2^4 + \cdots$$

where α_3^2 has been eliminated using $\alpha_2^1 + \alpha_2^2 + \alpha_2^3 = 1$.

The values of the magnetocrystalline anisotropy constants change with temperature. The energy difference between easy and hard directions vanishes at the Curie temperature. At a lower temperature in some materials, the anisotropy terms may mutually cancel, approach zero, or change sign. When their combined effect is to reduce E_K to zero, the crystal is effectively isotropic magnetically. The preferred orientation of magnetization may switch from one axis or set of axes to others as this temperature is passed. This is termed a magnetocrystalline anisotropy transition, occurs at temperature T_k, and is found in a variety of materials.

Magnetostriction Anisotropy

Magnetostriction is the phenomenon of the dependence of change of sample dimension and mechanical deformation on changes in the magnitude and direction of its magnetization. The alignment of atomic dipole moments causes elastic strains in a crystal. The changes in dimension, i.e., strains, depend on the extent of magnetic coherence; that is, the intensity of magnetization J. The elastic deformation produces a magnetoelastic anisotropy energy, in addition to changes in the magnetocrystalline anisotropy energy. The magnetoelastic energy depends on the stress (σ) magnitude, the orientation of stress with respect to magnetization vector, and symmetry properties of the crystal as reflected in magnetostriction constants.

For a crystal with cubic symmetry, the magnetostriction anistropy energy is

$$E_{\lambda\sigma} = \frac{3}{2} \lambda_{100}\sigma(\alpha_1^2\beta_1^2 + \alpha_2^2\beta_2^2 + \alpha_3^2\beta_3^2) + 3\lambda_{111}\sigma(\alpha_1\alpha_2\beta_1\beta_2 + \alpha_2\alpha_3\beta_2\beta_3 + \alpha_3\alpha_1\beta_3\beta_1)$$

where σ = stress (compressive stress with positive sign); α = direction cosines of magnetization vector, J; β = direction cosines of stress, σ; λ_{110} = magnetostriction constant for [100] axis, and λ_{111} = magnetostriction constant for [111] axis.

The λ's are constants for saturation, that is, the strain in the respective directions produced by a saturation field (and J_s) in that direction. For a polycrystalline sample with random orientation of crystallites, the bulk (average) saturation magnetostriction constant is denoted by λ_s.

If a crystal is deformed elastically by an applied stress, and magnetization and its anisotropy are affected, this process is termed inverse magnetostriction, or piezomagnetism.

Table 6 lists values of the magnetocrystalline anisotropy constants, K, and magnetostriction anisotropy constants, λ, for some magnetic materials.

Table 6
MAGNETOCRYSTALLINE AND MAGNETOSTRICTION ANISOTROPY CONSTANTS

Titanomagnetite: ulvospinel-magnetite
$xFe_2TiO_4 . (1\text{-}x)Fe_3O_4$
(values at 17°C)

		Magnetocrystalline		Magnetostriction				
x	J_s (emu/gm)	K_1	K_2	λ_{100}	λ_{111}	λ_{110}	λ_s	Ref.
		(in 10^{-5} ergs/cm^3)		(in 10^{-6} cm/cm)				
0	93	−1.36	−0.44	−20	78	60	39	12, 64
0.04	90	−1.94	−0.18	−6	87		50	
0.10	82	−2.50	+0.48	4	96		59	
0.18	73	−1.92		47	109		84	
0.31	59	−1.81		67	104		89	
0.56	29	−0.70		170	92		139	
0.68	15	+0.18						

Iron: (values at room temperature)

K_1	K_2	λ_{100}	λ_{111}	λ_s	Ref.
(in 10^{-5} ergs/cm^3)		(in 10^{-6} cm/cm)			
4.2	1.5			−7	38
		20	−20		1

Ilmenohematite: ilmenite-hematite
$yFeTiO_3 . (1-y)Fe_2O_3$
(values at room temperature)

y	J_s (emu/gm)	$\lambda_{\bar{1}\bar{1}2}$[a]	λ_{111}[b]	λ_s	Ref.
		(in 10^{-6} cm/cm)			
0	0.5	8	1.3		85
				8	81

Nickel: (values at room temperature)

	K_1	K_2	λ_{100}	λ_{111}	λ_s	Ref.
	(in 10^{-5} ergs/cm^3)		(in 10^{-6} cm/cm)			
typical	−0.48	−0.38				
	−0.51					4
	−0.5	−0.42				55
	−0.46	−0.36				72
			−46	−24		40
			−46	−25		1
					−34	38

[a] In basal plane (0001).
[b] Normal to (0001).

MAGNETIC MINERALOGY, CRYSTALLINE AND MAGNETIC PROPERTIES

The magnetization of rocks is retained in certain magnetic minerals. These minerals and their physicochemical state control the intensity and stability over time of the remanent and induced magnetization. Proper geological interpretation, as for magnetic prospecting or paleomagnetism, of the magnetic character of rock samples depends on knowledge of the mineralogic, structural, magnetic, and mechanical properties of the minerals present.

Representative values for important properties are given here. The values are "typical". There is sometimes considerable discrepancy between published values; this is often due to the differing conditions and composition of the "natural" minerals and rocks studied.

Important Minerals in Rock Magnetism

The major minerals having magnetization, or being of interest in studies of magnetic minerology are

Iron oxides
- titanomagnetite series: ulvospinel-magnetite $xFe_2TiO_4.(1-x)Fe_3O_4$
- ilmenohematite series: ilmenite-hematite $yFeTiO_3.(1-y)Fe_2O_3$
- maghemite γFe_2O_3
- martite αFe_2O_3
- geothite $\alpha FeOOH$ (the most common of the natural hydrous ferric oxides, i.e., "limonite", $Fe_2O_3.H_2O$)
- lepidocrocite $\gamma FeOOH$
- akaganeite $\beta FeOOH$

Sulfides
- pyrrhotite series: troilite-pyrrhotite $yFeS.(1-y)Fe_{1-x}S$
- pyrite FeS_2
- marcasite FeS_2
- mackinawite FeS

Carbonates
- siderite $FeCO_3$
- magnesite $MgCO_3$

Iron Oxides

The most important and common rock-forming magnetic minerals for the study of rock magnetism and paleomagnetism are the iron oxides. They can be represented in the ternary diagram FeO-TiO_2-Fe_2O_3 shown in Figure 9. There are three main solid solution series:.

1. Titanomagnetite series — cubic structure (inverse spinel)

ulvospinel - magnetite

$$xFe_2^{2+}Ti^{4+}O_4.(1-x)Fe^{3+}(Fe^{2+}Fe^{3+})O_4$$

2. Ilmenohematite series — hexagonal/rhombohedral structure

ilmenite - hematite

$$yFe^{2+}Ti^{4+}O_3.(1-y)Fe_2^{3+}O_3$$

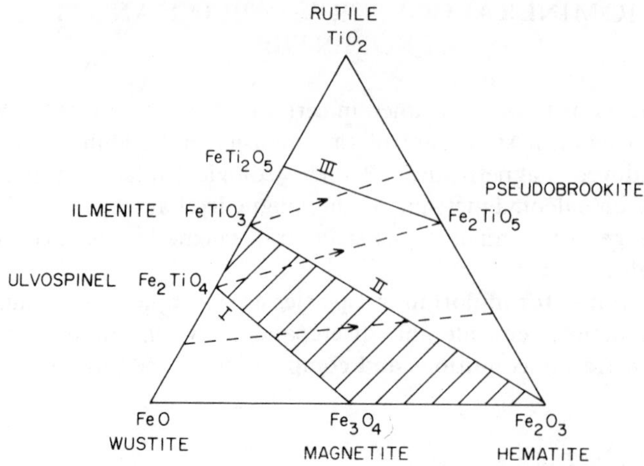

FIGURE 9. Compositions in ternary diagram FeO-TiO_2-Fe_2O_3. Solid-solution series are I — titanomagnetite, II — ilmenohematite, III — pseudobrookite. Dotted lines represent direction of oxidation.

3. Pseudobrookite series — orthorhombic structure

$$zFe_2^{3+}Ti^{4+}O_5.(1\text{-}z)Fe^{2+}Ti_2^{4+}O_5$$

Possible impurity phases are MnO, MgO, Al_2O_3, and V_2O_3. The solid solution region of greatest interest in nature is shaded in Figure 9. The dotted lines are lines of constant Fe:Ti ratio and represent trend of oxidation/reduction. The oxidation increases to the right.

TiO_2 has three polymorphs. It is called rutile with a tetragonal structure (c axis = 0.6 Å) or anatase if tetragonal (c axis = 1.8 Å), and brookite if orthorhombic.

The diagram is drawn on the basis of molecular ratios. Thus,

$$FeTiO_3 = FeO \cdot TiO_2$$

$$Fe_2TiO_4 = \frac{2}{3} FeO \cdot \frac{1}{3} TiO_2$$

$$Fe_3O_4 = FeO \cdot Fe_2O_3$$

Figure 10 shows crystallographic structure and type of magnetism in the ternary diagram. In the ilmenohematite series, the magnetism varies with the composition as

$y \approx 1$	antiferromagnetic	(Fe and Ti both occupying all cation layers equally)
$0.45 \leq y \leq 0.95$	Ferrimagnetic	(ordered state of Fe and Ti, with Ti ions occupying every second cation layer perpendicular to the "c" axis)
$0 \leq y \leq 0.45$	Antiferromagnetic	(parasitic (spin-canted) antiferromagnetism; Fe and Ti occupying all cation layers, i.e., disordered state)

For $0.45 < y < 0.60$, synthetic specimens show self-reversal of magnetization.

Summarized in Figure 11 are some magnetic properties of the minerals. The Curie temperatures for the end-members of the series are shown in degrees centigrade. They vary uniformly from one end-member to the other. The increase in lattice parameter "a" is shown by arrows, likewise for saturation magnetization J_s, and the constants K for magnetocrystalline anisotropy and λ for magnetostriction. An auxiliary series is magnetite — hausmannite, $(1\text{-}x)Fe_3O_4.xMn_3O_4$. For $0 \leq x < 0.6$, it is cubic structure. For $0.6 < x \leq 1$ it is tetragonal.

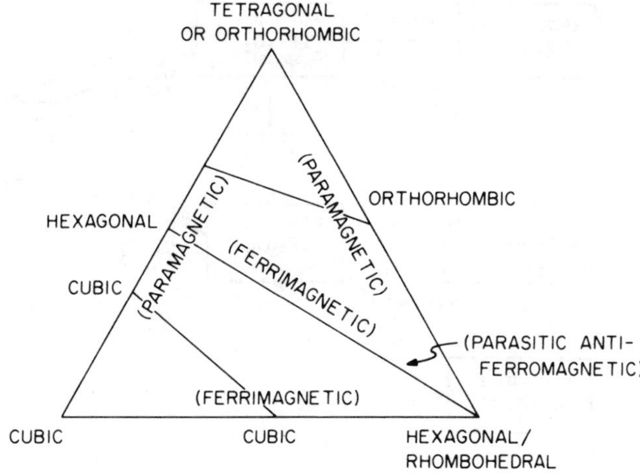

FIGURE 10. Structure and magnetism (at room temperature) in ternary diagram.

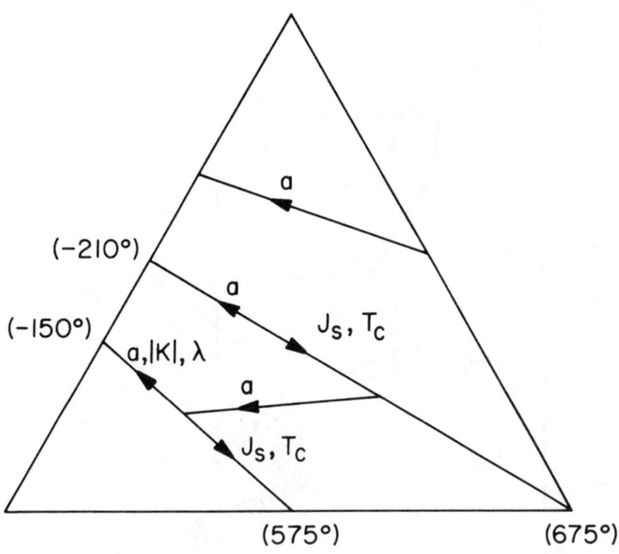

FIGURE 11. Magnetic properties in the ternary diagram. Curie temperatures in °C, "a" is cell dimension. Arrows indicate direction of increasing parameters.

The relationships between the main iron oxides are shown in Figure 12. The temperatures given are approximate. Values vary widely in the literature, depending on the sample condition and experimental conditions such as ambient atmosphere (air, vacuum, etc.). The structural conversion from maghemite to hematite is pressure-dependent. The goethite to hematite conversion could occur during consolidation of sediments, as $2FeOOh \rightarrow Fe_2O_3 + H_2O$. The magnetite to hematite oxidation could occur during initial cooling of igneous rock, at temperatures of about 600 to 1000°C. The magnetite to maghemite conversion could occur as low-temperature oxidation, as in late cooling or weathering.

The conditions for precipitation of different iron minerals are shown in Figure 13. The stability fields are outlined by the parameters Eh (volts) representing oxidizing

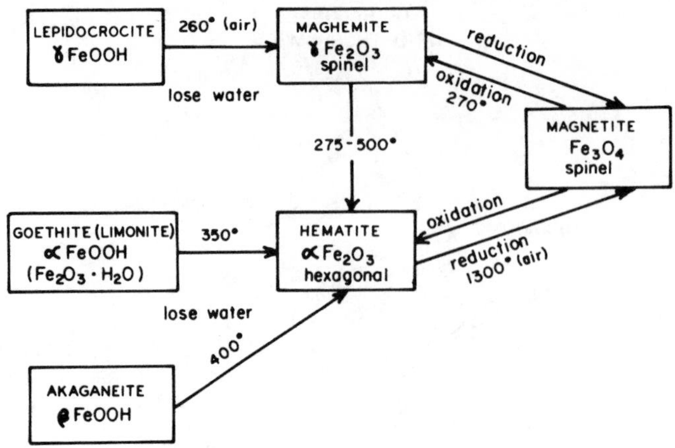

FIGURE 12. Conversion of iron-oxide magnetic minerals. Temperatures are approximate.

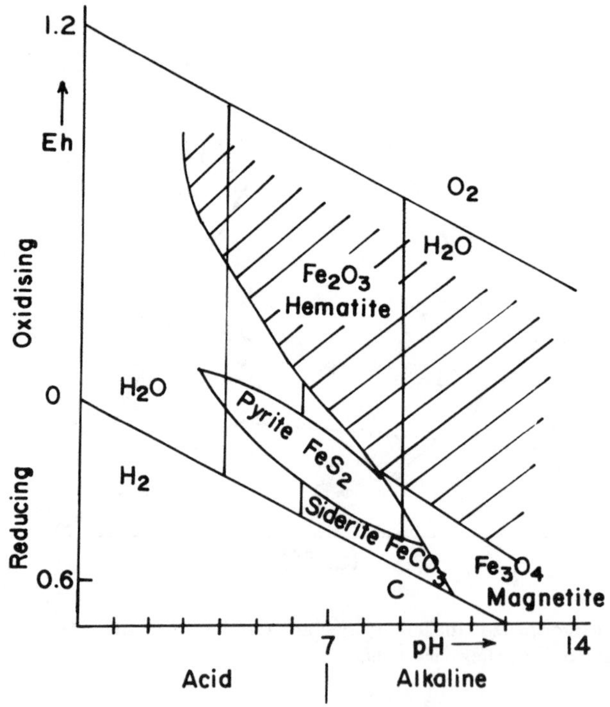

FIGURE 13. Conditions for precipitation of different iron minerals. (From Gass, I., et al., *Understanding the Earth*, MIT Press, Cambridge, MA, 1971, 179. Article by Watson, E. K.)

potential, and pH (concentration of H ions) representing acidity/alkalinity. The stability fields also depend on the concentration of components (e.g., Fe) in solution. ABCD is the range of naturally occurring solutions. The upper hatched area is the stability field of αFe_2O_3.

The main minerals will now be described briefly. These oxides and oxyhydrates can be described in terms of stacked "almost-close-packed" layers of oxygen atoms with iron atoms behaving as interstitials. These irons have an atomic radius such that they will not quite fit into the available sites of the oxygen lattice without distorting it. The distances between the oxygen planes in all the structures are about the same, ranging from 2.3 to 2.5 Å. The stacking sequences are characteristic of close-packed lattices, i.e., either the -ab-ab-ab- (hexagonal close packed) type, or the -abc-abc- (face centered cubic) type.

In the following, Z = number of formula units per unit cell, T_k = temperature of magnetocrystalline transition, T_c = Curie temperature, T_N = Neel temperature, E = Young's modulus of elasticity, and G = shear modulus of elasticity (modulus of rigidity).

Values are at room temperature unless otherwise noted.

Magnetite

1. Chemical formula: Fe_3O_4; Z = 8; percent by weight: Fe^{2+}(24.1), Fe^{3+}(48.3), O^{2-}(27.6).

2. Structure: cubic, spinel of inverse type (see Figure 14). Stacking of hexagonal close packed oxygen planes with an -abc-abc- sequence along the ⟨111⟩ direction. The general spinel structure is $A^{+2}B_2^{+3}O_4^{-2}$, where A occupies a tetrahedral site and B octahedral sites. In the unit cell, there are 32 oxygens, 64 avalable A sites, and 32 available B sites. The "normal" spinel has 8 A's in tetrahedral sites and 16 B's in octahedral sites. The "inverse" spinel has 8 B's in tetrahedral sites, and 8 A's and 8 B's in octahedral sties. Cations occupy layers between two adjacent oxygen planes; successive layers have cations either in the octahedral (B) sites between individual oxygens, or one third in the octahedral and two thirds in the tetrahedral sites. The inverse spinel arrangement for magnetite is $Fe^{3+}(Fe^{2+}Fe^{3+})\ O_4^{2-}$. The occupied tetrahedral sites form a diamond-type lattice. The octahedral sites have sixfold coordination, and the tetrahedral fourfold. The oxygen atoms form a face-centered-cubic lattice. The structure converts from inverse spinel to orthorhombic below about -155°C. The transition is abrupt for synthetic magnetite crystals, but can be spread over about 10° for natural crystals. The presence of impurities lowers the transition temperature. At high pressures (about 225 to 240 kbar), Fe_3O_4 undergoes a phase transition to a monoclinic phase with density about 6.4 gm/cm³. Cell dimensions (cubic phase): a = 8.394 Å; distance between {111} oxygen planes ~ 2.9 Å; distance between Fe and O, octahedral site — 2.06 Å, tetrahedral site — 1.87 Å; angle between O-Fe-O, octahedral site — 88.1°, 90°, 91.9°, tetrahedral site — 109.5°.

3. Mineral characteristics: crystals are generally octahedral form {111}, but may occur as cubic {100} or dodecahedral {110}; twinning and parting on {111}; hardness about 6 on Moh's scale; density = 5.18 gm/cm³.

4. Magnetic properties: ferrimagnetic at room temperature; saturation magnetization J_s ~ 98 emu/gm (at 0°K); ~ 92 emu/gm (at room temperature) = 480 emu/cm³; critical single-domain size about 0.1-1 μm; magnetic anisotropy — magnetocrystalline transition at T_k = −140°C;

$T_k < T < T_c$	--	easy axis <111> hard axis <100>	K_1 negative
$-155° < T < T_k$	--	easy axis <100> hard axis <111>	K_1 positive
$T < -155°$	--	easy axis orthorhombic "c" axis	

$K_1 \sim -1.35 \times 10^5$ ergs/cm³
$K_2 \sim -0.48 \times 10^5$ ergs/cm³

T_k transition suppressed if grain size too small (less than about 0.1 μm); $\lambda_{100} \sim -20 \times 10^{-6}$ cm/cm; $\lambda_{110} \sim 60 \times 10^{-6}$; $\lambda_{111} \sim 78 \times 10^{-6}$; $\lambda s \sim 40 \times 10^{-6}$; T_c = 575°C; domain (Bloch) walls thickness about 500 to 1500 Å and energy about 1 erg/cm³.

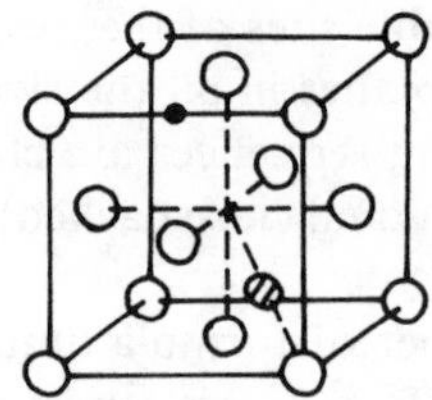

● OCTAHEDRAL (B) SITE
⊘ TETRAHEDRAL (A) SITE
○ OXYGEN IONS

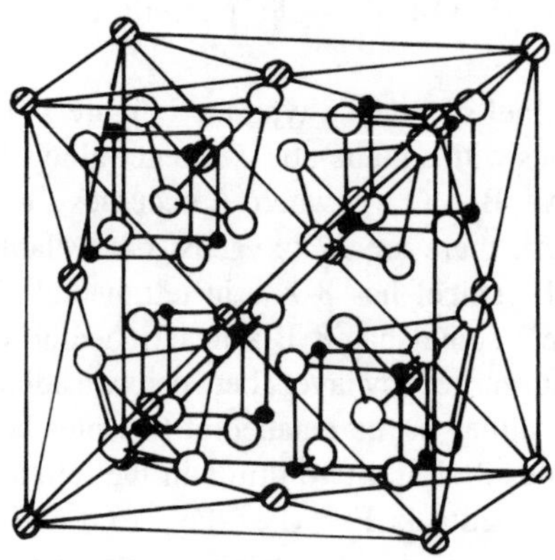

FIGURE 14. Spinel structure of magnetite. Top cube is 1/8 of unit cell.

5. Mechanical and other properties: E ∼ 1.6×10^6 kg/cm²; G ∼ 0.5×10^6 kg/cm²; Poisson's ratio σ ∼ 0.3; slip plane {111}, slip direction <110> for dislocations; ultimate compressive strength (crystal) ∼ 1000 kg/cm²; electrical resistivity ϱ ∼ 10^{-3}–10^2 ohm-meter (places it in the range of semiconductors, 10^{-5} to 10^8 ohm-meters. Other ferrites are generally much less conductive.)

Ulvospinel

1. Chemical formula: Fe_2TiO_4;
   percent by weight: Fe^{2+}(50), Ti^{4+}(21.4), O^{2-}(28.6)
2. Structure: cubic, spinel of inverse type (see Figure 14). Fe^{2+} and Ti^{4+} replace $2Fe^{3+}$ in magnetite. Cell dimension: a ∼ 8.5 Å
3. Mineral characteristics: occurs as intergrowth on {100} faces of magnetite; density about 4.8 g/cm³.
4. Magnetic properties: paramagnetic at room temperature: T_c ∼ −150°C; weakly ferrimagnetic below T_c (should be antiferromagnetic, in theory).

Maghemite

1. Chemical formula: γFe_2O_3
2. Structure: a polymorph of hematite (αFe_2O_3); cubic, defect spinel of inverse type (see Figure 14). Isostructural with magnetite except 1/9 of cations missing. These vacancies are distributed either randomly throughout octahedral and tetrahedral sites, or in octahedral sites alone, as in $4\ Fe_2^{3+}O_3 \rightarrow 3[Fe^{3+}O.Fe_{5/3}^{3+}.\square_{1/3}O_3]$ where □ represents a vacancy. Cell dimension: a∼8.35 Å; transition to αFe_2O_3 at T∼250 to 500°C; distance between Fe and O, octahedral site — 2.05 Å, tetrahedral site — 1.86 Å

3. Mineral characteristics: may form as oxidation product of magnetite ($3Fe^{2+} \rightarrow 2Fe^{3+}$), or converted from lepidocrocite (see Figure 12). The tetrahedral sites are less available for oxidation than the octahedral sites, since the former are covalently bonded and the latter ionically bonded; Density = 4.88 g/cm^3.
4. Magnetic properties: ferrimagnetic; J_s = 83.5 emu/g (407 emu/cm^3); $T_c \sim$ 675 to 750°C. (depends on structural state).

Wustite

1. Chemical formula: FeO
2. Structure: cubic, NaCl type; stacking of hexagonal close-packed oxygen planes with -abc-abc- sequence along the ⟨111⟩ direction. Lattice is like two interpenetrating f.c.c. lattices, one of oxygen and the other of irons. Defect structure, deficient in Fe, i.e., $Fe_{0.83\text{-}.95^\circ}$; cell dimension: a$\sim$4.30 A; distance between Fe and O — 2.15 A; angle between O-Fe-O — 90°.

Hematite

1. Chemical formula: αFe_2O_3; percent by weight: Fe^{2+}(70), O^{2-}(30); Z = 6 for hexagonal unit cell; Z = 2 for rhombohedral unit cell.
2. Structure: hexagonal system, trigonal subsystem, rhombohedral class (see Figure 15). Stacking of slightly distorted hexagonal-close-packed oxygen planes with -ab-ab-ab- sequence, along the [0001] direction, corundum-type structure. Groups of three oxygen atoms form a common face of two neighboring octahedra with a face parallel to {0001}. The stoichiometric number of iron atoms occupy octahedral interstices (sixfold coordination) between oxygen planes. They form Fe_2O_3 groups in the form of a trigonal dipyramid. Two thirds of the available cation positions are filled, with each Fe being surrounded by six oxygen in a near-octahedron. Cell dimensions (refer to Figure 15): hexagonal unit cell — a = 5.035 Å; b = 13.75 Å; rhombohedral unit cell — a_{rh} = 5.426 Å; α = 55°16′; distance between oxygen planes about 2.16 Å; distance between Fe and O — $r_1 = r_2 = r_3$ = 2.08 Å, $d_1 = d_2 = d_3$ = 1.95 Å; angle between O-Fe-O (angle defined by bonds listed) — $r_1r_2 = r_2r_3 = r_1r_3$ = 77°; $r_1d_2 = r_2d_3 = r_3d_1$ = 86.1°; $r_3d_2 = r_2d_1 = r_1d_3$ = 90.8°; $d_1d_2 = d_2d_3 = d_1d_3$ = 102.5°.
3. Mineral characteristics: crystals have form of positive rhombohedron $\{10\bar{1}1\}$, negative rhombohedron $\{01\bar{1}2\}$, or pinacoid {0001}; parting on {0001} or rhombohedral plane $\{01\bar{1}2\}$ due to twinning. Twinning on {0001} as penetration twins and on $\{01\bar{1}2\}$ usually lamellar. Hardness 5-6; density = 5.27 gm/cm^3.
4. Magnetic properties: each sheet of Fe is ferromagnetic, but sheets are coupled antiferromagnetically, i.e., two antiferromagnetic sublattices. There is a systematic deviation from oppositely-directed spin configuration, resulting in weak spin-canted (anti)ferromagnetism. Magnetic anisotropy: magnetocrystalline transition T_k at about −23°C (−13°C for synthetic material); $-23° < T < T_c$ — atomic moments in (0001) plane. Weak ferromagnetism. Fe atoms on trigonal axis of rhombohedral unit cell have spins parallel in groups of two, pointing in direction of "a" axis of hexagonal unit cell. Easy axis is $<10\bar{1}0>$, hard axis is [0001]. $T < -23°$ — atomic moments point along trigonal axis. Perfectly antiferromagnetic (no net magnetization). Successive spins directed oppositely. Easy axis is [0001]. Transition suppressed for grain size sufficiently small. $T_c < T < 725°C(T_N)$ — antiferromagnetic. Has a hard isotropic stable magnetization which endures until T_N (Néel temperature). Probably associated with unbalanced spins due to lattice defects or impurities, and is found in natural, not synthetic, material. $J_s \sim$ 0.45 emu/gm (2.4 emu/cm^3); decreases with grain size; T_c (weak ferromagnetism disappears) = 675°C; T_N (antiferromagnetism disappears) = 725°C. Criti-

cal single domain size, about 10 to 100 μm; $\lambda_{111} \sim 1.3 \times 10^{-6}$ cm/cm (i.e., normal to (0001)); $\lambda_{112} \sim 8 \times 10^{-6}$ cm/cm (i.e., in (0001) plane). Both λ's become negative below region of magnetocrystalline transition.

5. Mechanical properties: E $\sim 2 \times 10^6$ kg/cm^2; G $\sim 0.8 \times 10^6$ kg/cm^2; Poisson's ratio $\sigma \sim 0.27$.

Ilmenite

1. Chemical formula: $FeTiO_3$; percent by weight: Fe^{2+}(36.8), Ti^{4+}(31.6), O^{2-}(31.6); Z = 6.
2. Structure: hexagonal system, trigonal subsystem, rhombohedral class. Layers of Ti separated by Fe layers of alternating magnetic polarity. Cell dimensions: hexagonal unit cell — a = 5.08 Å; c = 14.13 Å; rhombohedral unit cell — a_{rh} = 5.52 Å; α = 54°51′.
3. Mineral characteristics: twinning on {0001}, parting on {0001},{01$\bar{1}$2}; hardness 5-6; density = 4.78 gm/cm^3; may occur as lamellae on octahedral {111} of Fe_3O_4.
4. Magnetic properties: paramagnetic at room temperature; $T_N \sim -210$°C; $T < T_N$, antiferromagnetic.

Martite

1. Chemical formula: αFe_2O_3
2. Structure: polymorphous after magnetite. Has cubic spinel outer form but rhombohedral internal structure.

Iron Oxyhydroxides

(Minerals of the family of hydrous ferric oxides, generally designated as Limonite)

Goethite

1. Chemical formula: αFeOOH, or $2HFeO_2$; Z = 4.
2. Structure: orthorhombic, corresponds to hematite. Stacking of hexagonal close-packed oxygen planes with -ab-ab-ab- sequence along [001]. Iron atoms occupy only the octahedral positions. Converts to hematite (see Figure 15). Cell dimensions: a = 4.60 Å; b = 9.95 Å; c = 3.02 Å.
4. Magnetic properties: antiferromagnetic at room temperature, but may have some stable remanence due to spin-canting or unbalanced spins. Fe atoms alternate direction of spin along axis of magnetization, [001]. $T_N \sim 120$°C.

Lepidocrocite

1. Chemical formula: γFeOOH, or 2(FeO.OH); Z = 4.
2. Structure: orthorhombic. Stacking of oxygen-hydrogen planes with an -abc-abc-abc- sequence along <051> direction of the orthorhombic unit cell (corresponds to <111> direction of a distorted cubic arrangement). Hexagonal packing of oxygen atoms in a sheet is not regular. Iron atoms occupy only octahedral sites. Each H atom is associated with an O atom, forming a discrete hydroxyl group. Cell dimensions: a = 3.06 Å; b = 12.4 Å; c = 3.87 Å.
3. Mineral characteristics: {010} cleavage.

Akaganeite

1. Chemical formula: βFeOOH; Z = 8.
2. Structure: tetragonal.
4. Magnetic properties: T_N from −160°C to 20°C.

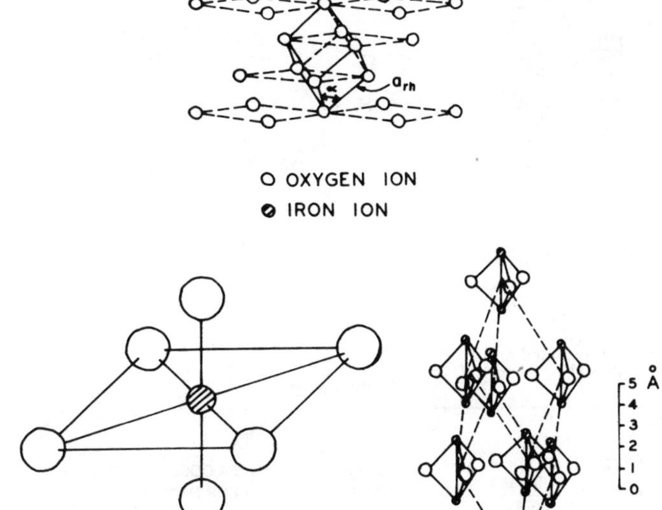

FIGURE 15. Hematite structure. Top — rhombohedron in hexagonal system; Left — schematic of Fe-O configuration; Right — rhombohedron of Fe_2O_3 units.

Sulfides

The main magnetic sulfides of interest to rock magnetism are in the troilite-pyrrhotite series, troilite-pyrrhotite, $yFeS.(1 - y)Fe_{1-x}S$. Troilite is hexagonal in structure, with Fe and S atom layers alternating. It has a niccolite (NiAs) structure (see Figure 16). It is antiferromagnetic, with a Neel temperature of 320°C. The tetragonal form of FeS is termed mackinawite. For x = ½, or pyrite (FeS_2), the structure is cubic and the crystal is paramagnetic at room temperature.

Pyrrhotite

1. Chemical formula: $Fe_{1-x}S$, where $0<x<0.125$.
   $Z = 2$
2. Structure: hexagonal system, rhombohedral class. In $Fe_{1-x}S$, x = 0 - hexagonal; $0<x<0.07$ - mixed structure; $0.07<x<0.1$ - hexagonal; $0.1<x<.125$ - monoclinic. When hexagonal, it has a defect NiAs structure (see Figure 16). It is related to FeS in that Fe^{3+} replaced Fe^{2+}, leaving some cation vacancies. Cell dimensions (for x = 0.115); a = 3.446 Å; c = 5.848 Å.
3. Mineral characteristics: crystals usually tabular on {0001}; hardness 3.5 to 4.5; density ~ 4.6 gm/cm³.
4. Magnetic properties: $0< x <0.09$ - antiferromagnetic; $0.09< x <0.14$ - ferrimagnetic; $J_s \sim 18$ emu/gm (83 emu/cm³) for monoclinic phase; this maximum J_s occurs for Fe_7S_8, or $Fe_{0.875}S$ (i.e., x = 0.125); $T_c \sim 320$°C; hard axis of magnetization [0001].

Carbonates

Siderite

1. Chemical formula: $FeCO_3$;
   $Z = 6$.

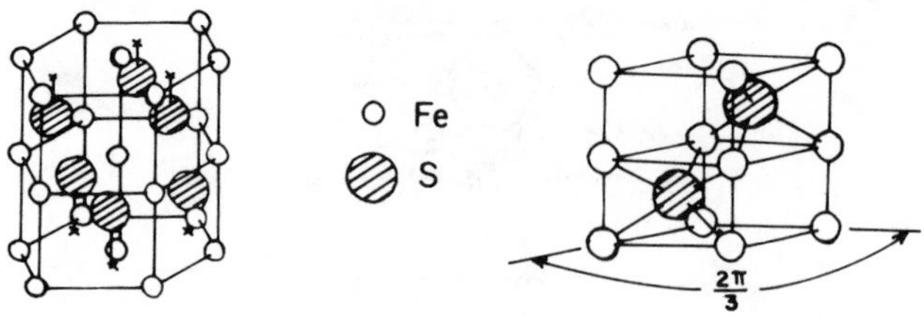

FIGURE 16. Hexagonal (NiAs, niccolite) structure for pyrrhotite.

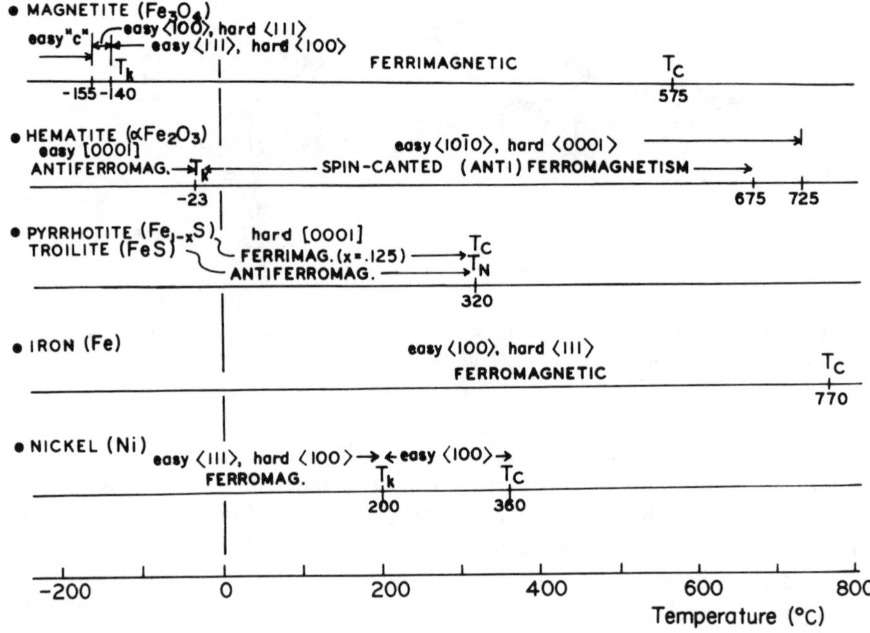

FIGURE 17. Variation of magnetism with temperature, for magnetic minerals and ferromagnetics.

2. Structure: hexagonal system, trigonal subsystem, rhombohedral class. Fe atoms are at points of a face-centered rhombohedron, CO_3 groups midway between Fe's. CO_3 groups are trigonal, planar parallel to [0001]. Cell dimensions: a = 4.72 Å; c = 15.46 Å.
3. Magnetic properties: antiferromagnetic; $T_N \sim -230°C$; magnetization along [0001].
4. Mineral characteristics: crystals commonly rhombohedral $\{10\bar{1}1\}$, sometimes $\{01\bar{1}2\}$, $\{02\bar{2}1\}$, $\{40\bar{4}1\}$, or tabular on {0001}; cleavage $\{10\bar{1}1\}$; hardness 4; density = 3.96 gm/cm^3.

Figure 17 shows the variation of magnetic properties with temperature for magnetic minerals and ferromagnetics.

MAGNETIC PROPERTIES

Saturation Magnetization, Curie/Néel Temperatures, and Cell Dimensions

Table 7
J_s AND T_c/T_N OF MAGNETIC MINERALS, ROCKS, AND METALS

Material	Composition	Notes a	Density (gm/cm^3) b	T_c, T_N (°C) c	J_s (emu/cm^3)	J_s (emu/gm)	Ref.
Minerals							
Magnetite	Fe_3O_4	Ferrimagnetic					
			(pure) 5.18				
			(natural) 5–5.4				
				578	480[d]	92.3(24°C)	10
					510[d]	98.2(0°K)	10
					471		10
				580			63
				575	480		
		0.6% Ti			410–430		16
		0.06% Ti			480–500		16
	Titanomagnetite, x = 0.55			210			37
Ulvospinel	Fe_2TiO_4	antiferro-magnetic	4.8	−153			63
Maghemite	γFe_2O_3	Ferrimagnetic	4.88	675[e]	407[d]	83.5(24°C)	10
					417		10
					545–675[e]		63
					750		73
Hematite	αFe_2O_3	Spin-canted (anti) ferromagnetic or ferrimagnetic					
			5.26	675	2.6	0.5(24°C)	10
		(Synthetic)		670	2.1	0.39(24°C)	10
				680	2.6	0.5	63
				680	2.2	0.42	8
		"Defect" parasitic antiferromagnetic[f]		725			
	Ilmenohematite, y= 0.75	max. ferrimagnetism in series		~10		19.5	34
Ilmenite	$FeTiO_3$	Antiferro-magnetic	4.72	−205		~0	74
				−216			63
Wustite	FeO	Antiferro-magnetic		−87			74
				−83			10
Goethite	$\alpha FeOOH$	Antiferro-magnetic	4.27	120			63
Akaganeite	$\beta FeOOH$	Antiferro-magnetic		−196 to 23			63
Magnesio-ferrite	$MgFe_2O_4$	Ferrimagnetic[g]	4.18			24.5	63
				310	110		10
					110	23	74
				440	110	25	4
					140	29(0°K)	4

Table 7 (continued)
J_s AND T_c/T_N OF MAGNETIC MINERALS, ROCKS, AND METALS

Material	Composition	Notes a	Density (gm/cm³) b	T_c, T_N (°C) c	J_s (emu/cm³)	J_s (emu/gm)	Ref.
Jacobsite	$MnFe_2O_4$	Ferrimagnetic[g]	4.95	300	416	84	63
					400	81	4
					408		10
					560	112(0°K)	4
Chromite	$FeCr_2O_4$	Ferrimagnetic[g]	4.5–5.1	–185			10, 63
Trevorite	$NiFe_2O_4$	Ferrimagnetic[g]	5.35	585	270	51	63, 74
					300	56(0°K)	4
					267		10
Franklinite	$ZnFe_2O_4$	Ferrimagnetic[g]	5.1–5.3	–258			74
				–264			10
Fayalite	Fe_2SiO_4	Antiferro-magnetic		–147			63
Pyroxene	$FeSiO_3$	Antiferro-magnetic		–233			63
Pyrrhotite	$Fe_{1-x}S$	Ferrimagnetic	4.6	300–325	62	13.5[d]	10
				320	90[d]	19.5 max. at Fe_7S_8	63
				300			8
Troilite	FeS	Antiferro-magnetic	4.83	320			74
				340			10
Pyrolusite	MnO_2	Antiferro-magnetic	5.06	–189			10, 63
Chromium dioxide	CrO_2	Ferrimagnetic		119	515		74
Siderite	$FeCO_3$	Antiferro-magnetic	3.96	–233			63
Rhodochrosite	$MnCo_3$	Antiferro-magnetic	3.70	–243			63
Metals							
Iron	Fe	Ferromagnetic	7.87	770	1714	218	2
					1760	224(0°K)	2
Nickel	Ni	Ferromagnetic	8.9	358			38
					485	54	2, 38
					510	57(0°K)	2, 38
				360			4
				370			82
Cobalt	Co	Ferromagnetic	8.85	1120	1420	160	2
					1445	163(0°K)	2
High permeability metals:							
4% Si-Fe				690			74
45 Permalloy				440			74
Mumetal				400			74
Supermalloy				400			74

Table 7 (continued)
J_s AND T_c/T_N OF MAGNETIC MINERALS, ROCKS, AND METALS

Material	Composition	Notes a	Density (gm/cm³) b	T_c, T_N (°C) c	J_s (emu/cm³)	J_s (emu/gm)	Ref.
2V Permendure				980			74
Rocks							
Sandstones	Red; Britain; 12 sites, Triassic and Devonian				(0.2–1.2) × 10^{-2}		75, 20
Basalt	Titanomagnetite x = 0.6			200–400			37
Magnetite ore	Lodestone (Fe_3O_4)					50–80	66
Lunar rocks							
Soils		8 samples				0.9–1.5	28
Breccia		16 samples				0.05–2	28
		Anorthosite breccia		765		0.145	44
Anorthosite		2 samples				0.7–2.6	28
Basalt				760		0.2–2.2	44
		10 samples				0.1–2	28
		3 samples				0.1–3(0°K)	28
Gabbro		2 samples				0.2–0.7	28
Igneous rocks				760–790			28
Breccia and fines				745–790			28
Meteorites							
4 Iron meteorites	(ie., Ni–Fe)			600–780			32

[a] Magnetic state is at temperatures below T_c or T_N; would be paramagnetic above.
[b] Used to convert emu/cm³ to emu/gm or vice versa; density values from *Handbook of Materials Science*, Vol. I, Lynch, C. T., Ed., CRC Press, Boca Raton, Fla., 1974, p. 235, and Vol. III, p. 184.
[c] T_c is Curie temperature, for ferrimagnetic or ferromagnetic materials; T_N is Néel temperature for antiferromagnetics.
[d] Calculated, from other J_s units.
[e] May have been converted to hematite.
[f] Present in natural specimens, due to effect of lattice imperfections.
[g] Spinel structure, i.e., $A^{+2}B_2^{+3}O_4$

Variation of J_s, T_c, and Cell Dimension With Chemical Composition

The titanomagnetite series is ulvospinel-magnetite, $xFe_2TiO_4.(1-x)Fe_3O_4$. Figure 18 shows the variation of Curie temperature (T_c) and cell dimension (cubic lattice parameter, i.e., unit cell size, "a") with "x" in the above series.

The relationships for titanomagnetite can be represented by

cell parameter, a = 8.395 + 0.135x Å
i.e., a = 8.395 Å for magnetite (x = 0)
a = 8.53 for ulvospinel (x = 1)
Curie temperature, T_c = 575 − 725x + 600x (½ − x)(1 − x) °C.
i.e., T_c = 575 °C for magnetite
= −150 for ulvospinel

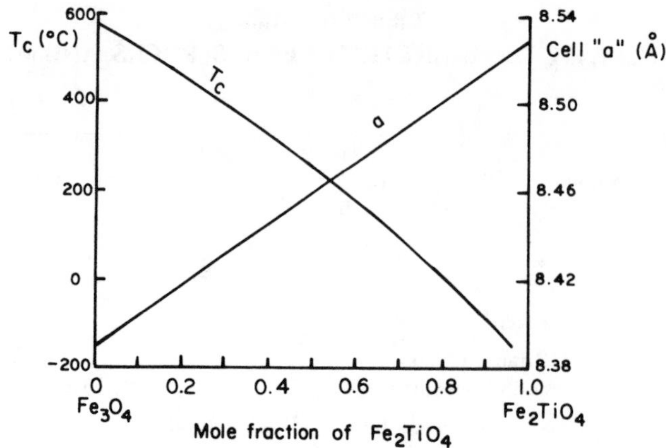

FIGURE 18. Variation of T_c and cell parameter "a" in titanomagnetite series.[8]

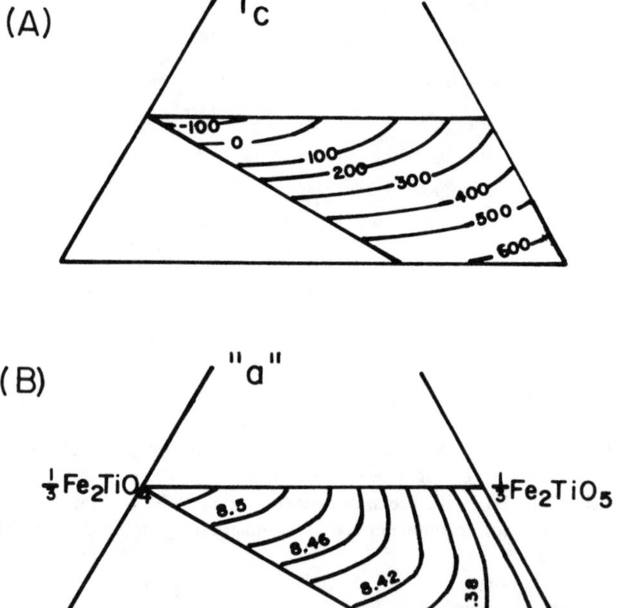

FIGURE 19. Variation of Curie temperature (T_c) and cell parameter "a" in FeO-TiO_2-Fe_2O_3 ternary diagram. (A) for T_c, in °C,(B) for "a", in Å. (Data from Reference 49.)

Figure 19 shows the variation of T_c and "a" in the compositional range of interest in the FeO-TiO_2-Fe_2O_3 ternary diagram.

The variation of saturation magnetization, J_s, for the titanomagnetite series at room temperature is given by $J_s = 92(1 - x) - 42x(1 - x)$ emu/gm, i.e., $J_s = 92$ emu/gm for magnetite, $J_s = 0$ for ulvospinel.

Table 8 shows a typical experimental determination of the variation of J_s with composition in the titanomagnetite series.

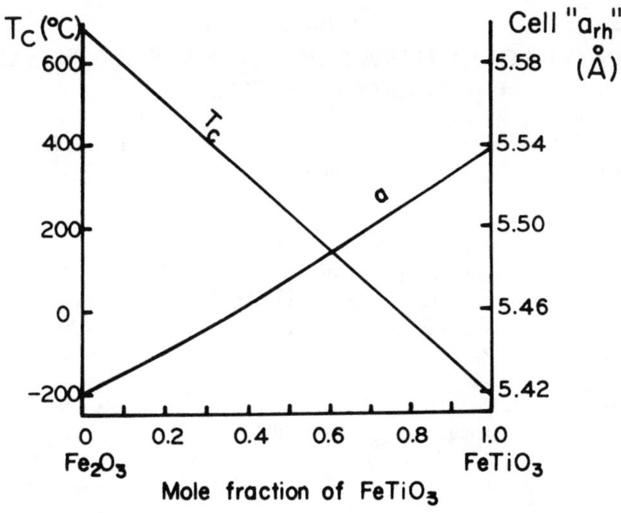

FIGURE 20. Variation of T_c and rhombohedral cell parameter "a_{rh}" in ilmenohematite series.[8]

Table 8
VARIATION OF J_s WITH COMPOSITION IN THE TITANOMAGNETITE AND ILMENOHEMATITE SERIES[12,34]

Titanomagnetite $xFe_2TiO_4.(1-x)Fe_3O_4$		Ilmenohematite $yFeTiO_3.(1-y)Fe_2O_3$	
x	J_s (emu/gm)	y	J_s (emu/gm)
0	93	0	0.5[a]
0.04	90	0.25	0.6[a]
0.10	82	0.45	0.6[a]
0.18	73	0.5	2.9[b]
0.31	59	0.6	9.2[b]
0.56	29	0.7	17.4[b]
0.68	15	0.75	19.5[b]
		0.8	18.2[b,d]
		0.9	5.3[b,d]
		0.95	1.6[b,d]
		1.0	0[c,d]

[a] Spin-canted (anti)ferromagnetic.
[b] Ferrimagnetic.
[c] Antiferromagnetic.
[d] For $x \geqslant 0.75$, T_c is below room temperature, so the material is paramagnetic at room temperature.

The ilmenohematite series is ilmenite-hematite, $yFeTiO_3.(1-y)Fe_2O_3$.

Figure 20 shows the variation of T_c and cell parameter a_{rh} with "y" in the series.

Relationships can be represented by Curie temperature $T_c = 680 - 890y$ °C, i.e., $T_c = 680$ °C for hematite ($y = 0$), $T_c = -210$ for ilmenite ($y = 1$); volume of unit cell, $V = 100.4 + 3.8y$ Å^3, i.e., $V = 100.4$ Å^3 for hematite, $V = 104.2$ for ilmenite. The variation of J_s with composition "y" is shown in Table 8.

Table 9
COERCIVE FORCE DATA FOR MINERALS, ROCKS, AND FERROMAGNETICS

Material	Notes	H_c (oe) for $J_{r,sat}$[a]	Ref.
Minerals			
Magnetite	Pure, J_s = 92 emu/gm (24°C)	20	10
	Pure, single crystals	10—25	16
Hematite	Pure, J_s = 0.5 emu/gm (24°C)	7600	10
	Synthetic, J_s = 0.39 emu/gm (24°C)	430	10
Pyrrhotite	Pure	15—20	10
Rocks			
Magnetite ore	Lodestone, < 1% TiO_2, J_s~60–70 emu/gm; $H_{c,Jn}$ = 100—300 oe	300—960	66
Sandstones	Red, Britain, 12 sites	2500—6000	75
	Typical	3000—7000	
Basalt		70—300	43
	NSW Australia, 7 sites	160—225	6
Lunar Rocks			
Soils	8 samples	20—35	28
Breccia	16 samples	15—120	28
Gabbro		10	28
	2 samples (4°K)	30—87	28
Anorthosite	2 samples	8—15	28
Basalt		8—20	44
	6 samples	8—45	28
	4 samples (4°K)	21—100	
Meteorites			
Irons (i.e., Ni-Fe)		.5—1	32
Stony-irons		.7—10	32
Metals			
Iron	Annealed	.1—2	2
Nickel	Annealed	.5—3	17
	Cold worked	15—25	17
	Cold worked	36	69
Cobalt	Annealed	5—20	17
	Cold worked	30—90	17

[a] Field required to reduce saturation IRM ($J_{r,sat}$) to zero; on occasion termed remanent coercivity H_{cr}. Distinguish from H_{rc}, field required to reduce residual remanence in zero.

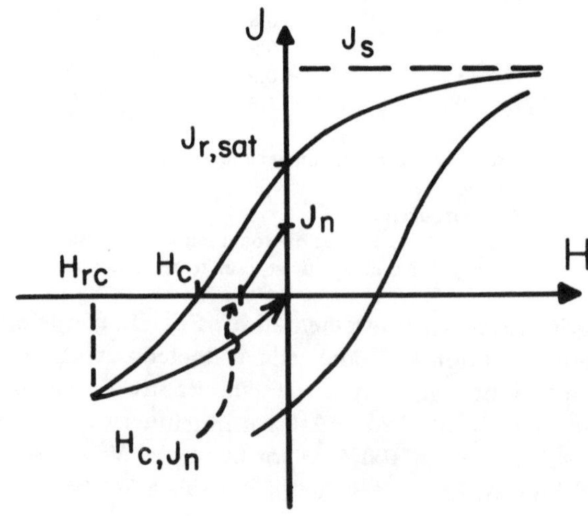

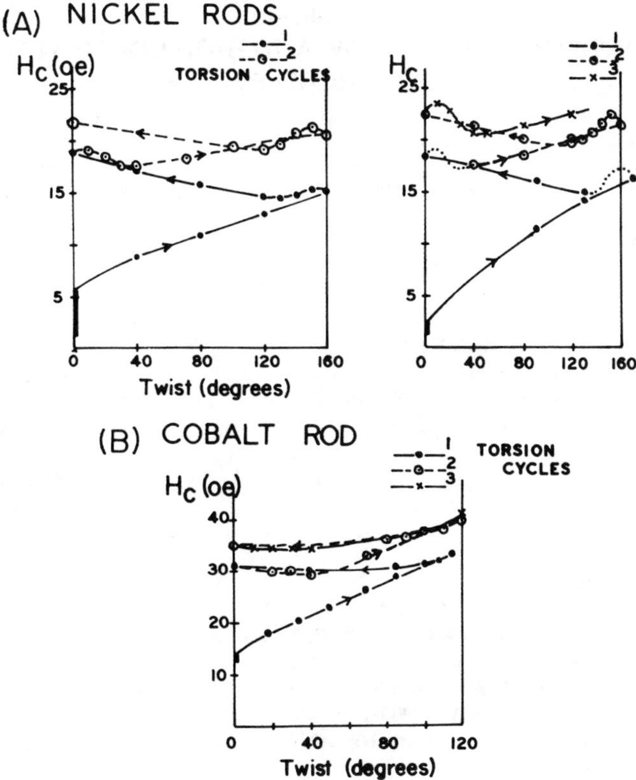

FIGURE 21. Variation of coercive force, H_c, with deformation of (A) nickel and (B) cobalt. Rods cyclically deformed. Magnetic H_c reflects mechanical hardness and internal stress condition and changes in that condition.[18]

Coercive Force

Table 9 lists values for coercive force, for magnetic minerals, rocks, lunar material, and ferromagnetic metals. The values are representative, but coercive force is a parameter which can vary widely depending on grain size and shape, magnetocrystalline anisotropy, and mechanical condition of the sample. Coercive force generally reflects mechanical hardness; that is, it is low for annealed or recrystallized material, and increases as degree of deformation (e.g., cold working, plastic deformation) increases. This can be illustrated with ferromagnetic metals which are deformable at room temperature and pressure. The variation of coercive force with cyclic (fatigue) deformation is shown in Figure 21 for nickel and cobalt. The change in mechanical hardness (e.g., yield point stress) is faithfully mirrored by changes in coercive force. The metal rods were in an initially annealed state. The decrease in H_c (and hardness) just after reversing the deformation reflects the Bauschinger effect (fatigue softening), which is then followed by renewed work hardening as deformation is continued. Analogous behavior has been found for rocks which are plastically deformed and fatigue (reversely) deformed.

Table 10 lists values of coercive force, saturation flux density, and maximum permeability for some high-permeability and high-coercivity (permanent magnet) materials.

Magnetic Susceptibility

Susceptibility is a parameter of considerable diagnostic and interpretational use when studying rocks. This is true whether in the laboratory or indirectly in studying magnetic fields to deduce the structure and lithologic character of buried rock bodies.

Table 10
MAGNETIC PROPERTIES OF LOW AND HIGH COERCIVITY MAGNETIC METALS[86]

Material	Notes	H_c (oe)	Saturation B (gauss)	Maximum permeability (gauss/oe)
High-permeability Materials				
Steel	Cold rolled, 98.5% Fe	1.8	21,000	2000
Iron	99.91% Fe	1.	21,500	5000
4% Si-Fe	96% Fe	0.5	19,700	7000
45 Permalloy	54.7% Fe, 45% Ni, 0.3% Mn	0.07	16,000	50,000
Mumetal	18% Fe, 75% Ni, 2% Cr, 5% Cu	0.05	6,500	100,000
Supermalloy	15.7% Fe, 79% Ni, 5% Mo, 0.3% Mn	0.002	8,000	800,000
2V Permendure	49% Fe, 49% Co, 2% V	2.	24,000	4500
		H_c (oe)	Remanent induction, B_r (gauss)	
Permanent magnet materials				
Carbon steel	98.1% Fe, 1% Mn, 0.9% C	50	10,000	
Cobalt steel	71.75% Fe, 17% Co, 0.75% C, 2.5% Cr, 8% W	150	9,500	
Alnico I	63% Fe, 12% Al, 20% Ni, 5% Co	440	7,200	
Alnico IV	55% Fe, 12% Al, 28% Ni, 5% Co	700	5,500	
Alnico VI	49% Fe, 8% Al, 15% Ni, 24% Co, 3% Cu, 1% Ti	750	10,000	
Vectolite	30% Fe_2O_3, 44% Fe_3O_4, 26% Cr_2O_3	1000	1,600	
Platinum-cobalt	77% Pt, 23% Co	3600	5,900	

However, susceptibility of a sample can vary widely, depending on such factors as magnetic mineralogy, grain size and shape, magnitude of external field, and relative magnitude of remanent magnetization present. In general, k increases with grain size (in the multidomain range) and magnitude of saturation magnetization J_s, and is inversely related to remanent magnetization (J_r), temperature, and coercive force (H_c). The measuring technique can also introduce a large range of values (see section on Magnetization: Types and Parameters.)depending on whether initial k_o (at $H = 0$, $J_r = 0$), or remanence k_r (at $H = 0$, $J = J_r$) or some susceptibility at $H = H_E \sim 0.5$ oersted, is measured. Further, the measured k could depend on the frequency used, if dynamic (alternating-field) method is employed. It is desirable to use a frequency low enough so that there is negligible electrical conductivity response of the sample. For magnetite, this means less than about 2000 Hz.[62]

Some susceptibilities of importance are

1. k_m — true susceptibility of the magnetic material
2. k' — effective susceptibility, dependent on the shape of the sample or of its constituent grains (i.e., depends on the demagnetizing factor N)
3. k_{app} — apparent susceptibility, measured when the sample has a remanent magnetization as well as the induced magnetization

The magnetization induced by a small applied field H_{ex}, such as the Earth's field H_E, is proportional to the field, as $J_{induced} \propto H_{ex} = kH_{ex}$, i.e., $J_{induced} = kH_{ex}$.

Table 11
COMPARISON OF CALCULATED VALUES OF EFFECTIVE SUSCEPTIBILITY WITH TRUE (MINERAL) SUSCEPTIBILITY, FOR DISSEMINATED SPHERICAL[a] PARTICLES[31,46]

Mineral susceptibility k_m (emu/cm³)	Effective susceptibility k' (emu/cm³)
0.01	0.0096
0.1	0.0705
1.0	0.193
5	0.228
10	0.233
100	0.238
∞	0.239

[a] $N = 4\pi/3$ for spherical particles.

Considering the shape (demagnetizing) effect, for a magnetic material,

$$J_{induced} = \frac{k_m H_{ex}}{1 + Nk_m}$$

where k_m = susceptibility of the magnetic mineral (i.e., "true" susceptibility of the magnetic material present).
If the rock or material has a volume fraction "V" of magnetic material, $J_{induced,\ rock} = V \cdot J_{induced}$ where $0 \leqslant V \leqslant 1.0$.
Thus the measured susceptibility would be

$$k = \frac{J}{H_{ex}} = \frac{V \cdot J_{induced}}{H_{ex}}$$

$$= \frac{V \cdot k_m}{1 + NK_m}$$

$$= V \cdot k'$$

For a uniformly-magnetized spherical grain, $N = 4\pi/3$ (in cgs units), and thus the effective susceptibility is

$$k' = \frac{k_m}{1 + Nk_m}$$

$$\sim \frac{k_m}{1 + 4.2k_m}$$

Thus the effective susceptibility is reduced from the "true" susceptibility (of the magnetic mineral alone) by a factor of $1/(1 + Nk_m)$.

Table 11 gives calculated values of effective k′ for disseminated spherical grains, compared to the true mineral susceptibility k_m.

For more elongated grains, N is smaller and thus the effective susceptibility is not reduced as much from the true mineral k_m. The effective susceptibility k′ for disseminated magnetite grains is about 0.25 (about 0.2 for spherical grains), compared to k_m for magnetite of $\sim$ 1 to 2.5 emu/cm³.

Thus for magnetite, the dominant magnetic constituent of most rocks, effective k′ $\sim$ (0.2 to 0.25) emu/cm³ = 200,000–250,000 × 10^{-6} emu/cm³.
Measured susceptibility would be, for 1% magnetite, k = V·k′ = 0.01 × 0.25 = 0.0025 emu/cm³ = 2500 × 10^{-6} emu/cm³; for 10% magnetite, k = 0.1 × 0.25 = 25,000 × 10^{-6} emu/cm³ and the induced magnetization would use this susceptibility, $J_{induced} = kH\cdot_{ex}$.

Table 12a lists some measured effective susceptibilities for rocks.

Table 12b also lists susceptibilities, for a collection of specimens from various volcanic units. For them, the magnetite content was about 2 to 3 % by volume, with average mineral grains having N $\sim$ 3.6. The table also shows the maximum susceptibility anisotrophy, i.e.,

$$\frac{k_{max} - k_{min}}{k} \times 100\%$$

These specimens had some susceptibility anisotropy; amother group of specimens from the area had little or no measurable anisotropy.

Apparent susceptibility, k_{app}, is measured or used for modelling interpretation of rocks when the total magnetization includes both induced and remanent magnetization. The remanence might be a TRM, CRM, or simply denoted NRM for natural remanent magnetization. Thus, $J_{total} = J_{induced} + J_{remanent}$.
Consider the case for the rock in an external field of the Earth, and define a parameter — the Koenigsberger ratio,

$$Q_n = \frac{J_{remanent}}{kH_E}$$

Thus, $J_{total} = kH_E + Q_n \cdot kH_E = k(1 + Q_n)H_E = k_{app}H_E$.
The Koenigsberger ratio could be

$$Q_n = \frac{J_{NRM}}{kH_E}$$

in the general case, or

$$Q_T = \frac{J_{TRM}}{kH_E}$$

if TRM is the known remanence. Rocks, Q is generally inversely related to magnetic grain size, since smaller grains (particularly if single-domain) have larger remanence (J_r) and smaller susceptibility, and thus a larger Q.

Table 12a
MEASURED EFFECTIVE SUSCEPTIBILITIES (K′) FOR DISSEMINATED MAGNETITE, IN TYPICAL MAGNETITE-BEARING ROCKS

k′ (emu/cm³)	Ref.
0.2—.25	46
0.28	59
0.16	31
0.12—0.38; ave. 0.29	42
0.25—0.30	52

Table 12b
MEASURED SUSCEPTIBILITIES AND ANISOTROPY FOR VOLCANIC ROCKS: FROM LITHOLOGIC UNITS IN UPPER CRETACEOUS ELKHORN MOUNTAINS (MONTANA)[89]

Rock type	k (10^{-4} emu/cm³)	k anisotropy (%)
Andesite flow	24.5	1.4
	7.6	1.5
	7.6	.1
	4.0	5.7
	34.	.4
Vesicular flow	22.8	4.9
Welded tuff	29.9	2.2
	47.9	1.2
Vitric tuff	6.4	2.2
	67.7	1.1
Vesicular tuff	23.6	4.2
Basalt flow	34.	1.9
	101.6	1.8
	59.1	.8
	28.	.01
	5.1	3.4
Ash flow	34.3	1.7
	30.	.6
Crystal tuff	16.1	.5
	51.9	2.4
	31.9	5.9
	14.9	1.2
Hypabyssal intrusive	61.9	5.1
	98.9	4.9
Metamorphosed flow	513.9	2.9
	419.8	1.3
Metamorphosed intrusive	505.3	1.5

Table 13 gives typical values of apparent susceptibility for rock types. Table 14 lists values of susceptibility for minerals. Values are given in both emu/cm³ and emu/gm, with conversion from the given datum using the density. Compilation is by this author.

Table 13
APPROXIMATE APPARENT SUSCEPTIBILITY FOR ROCK TYPES[10]

Material	k_{app}
Iron ores	>0.1
Basalt	10^{-2}
Andesite	10^{-3}
Dacite; metamorphic rocks	10^{-4}
Sedimentary rocks	10^{-5}

Table 15 is another compilation of susceptibility values for minerals and other materials. The values are generally considerably higher than published elsewhere. Table 16 gives magnetic properties of rocks, including sedimentary, igneous, metamorphic, ores, and lunar. The properties are intensity of natural remanent magnetization (J_n), susceptibility, and typical Koenigsberger ratio. Also included for comparison with igneous rocks are some data for magnetite and titanomagnetite powder dispersions.

For comparison with the preceding Tables 12 to 16, the following are data for calculated and measured susceptibilities.

Table 17 has calculated susceptibilities for various rock types, based on their typical content of magnetite and ilmenite, and using $k_{magnetite} = 0.30$ emu/cm^3, $k_{ilmenite} = 0.137$ emu/cm^3.

Figure 22 shows susceptibilities, as measured in the laboratory on samples from the surface and from subsurface cores. The range of susceptibility, and averages for rock types, are given.

Table 18 gives the distribution of measured susceptibilities for major rock types. The data reflect the general trend that, for k values, basic extrusive > basic intrusive > acidic igneous > sedimentary.

There have been a number of attempts to empirically relate apparent susceptibility of rocks to their magnetite content. Magnetite is taken as a reference, since it dominates the magnetization of rocks, particularly for crystalline (igneous, metamorphic) rocks and ores.

For example, for rocks with magnetite grains as the principal magnetic constituent[67],

$$k_{app} = \frac{V \cdot k_m}{1 + N \cdot k_m(1 - V^{1/6})}$$

where V = volume fraction of magnetic material, e.g., V = 0.03 is 3% magnetite; N = demagnetizing factor ($4\pi/3$ for spherical grains); k_m = true (mineral) susceptibility.

Another empirical formula, for magnetiferous igneous rock, is $k_{app} = 0.289.V^{1.01}$ for $.002 < V < 0.04$, i.e., 0.2 to 4% magnetite.The data for this are shown in Figure 23.

For metamorphic rocks from the Adirondacks, USA, an empirical relation is $k_{app} = 0.26 \cdot V^{1.11}$ for $0.002 < V < 0.1$. This is shown in Figure 24.

Figure 25 combines the data from Figure 24 with earlier data on rocks and ores to extend the range of observed susceptibilities. The vertical extent of data lines shows the effect of susceptibility anisotropy of individual specimens.

Figure 26 gives another compilation of susceptibility versus magnetite content for iron formations in Minnesota. The line fits data in addition to the points shown, and is $k_{app} = 0.116 \cdot V^{1.39}$.

Table 14
MAGNETIC SUSCEPTIBILITY OF MINERALS
(AT ROOM TEMPERATURE, UNLESS OTHERWISE NOTED)

Material	Composition	Notes	Density (gm/cm^3) a	k in 10^{-6} emu/cm^3 c	k in 10^{-6} emu/gm	Ref.
Iron minerals						
Magnetite	Fe_3O_4		(pure) 5.18 (natural) 5–5.4	100,000–1,600,000; average		65
				500,000	96,500[b]	65
				200,000–1,300,000		53
				300,000–4,000,000		26
				300,000–800,000	58,000–154,000[b]	25
	Powder, 2% magnetite by wgt dispersion in epoxy;					
	grain size: 1–2 μm			1140		37
	4–8 μm			1390		37
	37–75 μm			1470		37
	75–150 μm			2200		37
	Powder, 1.7% by wgt dispersion in epoxy; grain size 2–150 μm			580		47
	Powder, 30% magnetite by wgt dispersion in epoxy; grain size 20 × 20 μm			80,000		47
Titanomagnetite						
	x = 0.55 in series, 2% by wgt dispersion in epoxy;					
	grain size 1–2 μm			340		37
	5–15 μm			320		37
	75–150 μm			980		37
Hematite	αFe_2O_3		5.26	40–3000		65
				ave. 550	105[b]	
		Specular		430–3200	80–600[b]	53
		Amorphous		40–500		53
Ilmenite	$FeTiO_3$		4.72	135,000	28,600[b]	25
		46.4% Fe		400[b]	80–90	23
				135,000–252,000		53
				25,000–300,000 ave. 150,000		65
Limonite	$Fe_2O_3 \cdot H_2O$			220		65
Pyrrhotite	$Fe_{1-x}S$		4.6	125,000	27,200[b]	25, 53
				100–500,000 ave. 125,000		65
Pyrite	FeS_2			4–420 ave. 130		65

Table 14 (continued)
MAGNETIC SUSCEPTIBILITY OF MINERALS (AT ROOM TEMPERATURE, UNLESS OTHERWISE NOTED)

Material	Composition	Notes	Density (gm/cm^3) a	k in 10^{-6} emu/cm^3 c	k in 10^{-6} emu/gm	Ref.
Iron minerals (continued)						
Chalcopyrite	$CuFeS_2$			32		65
Siderite	$FeCO_3$		3.96	400[b]	100	23
				388[b]	98	63
				100–310		65
Chromite	$FeCr_2O_4$		4.5–5.1	240–9400		65
				ave. 600		
Jacobsite	$MnFe_2O_4$		4.95	2000		53
Franklinite	$ZnFe_2O_4$		5.1–5.3	36,000		65
Pyroxene	$FeSiO_3$			285[b]	73	63
	orthopyroxene: 24% FeO, 1% Fe_2O_3			140	40	23
Biotites	19.2% FeO, 7.9% Fe_2O_3			200[b]	63	23
			2.7–3.1	190[b]	58–78	63
Fayalite	Fe_2SiO_4		4.39	439[b]	100	63
Other minerals						
Quartz	SiO_2		2.65	−1.0		65
				−1.1 to −1.2		53
				−1.32[b]	−0.50	63
				−1.2		58
Rock salt	NaCl		2.16	−1.12[b]	−0.52	63
				−0.8		58
				−1.		65
				−0.82 to −1.3		53
Calcite	$CaCO_3$		2.71	−1.03[b]	−0.38	63
				−0.6 to −1.		65
Sphalerite	ZnS			60		65
Galena	PbS		7.58	−2.58[b]	−0.34	63
Cuprite	Cu_2O		6.14	−0.86[b]	−0.14	63
Hausmannite	Mn_3O_4		4.84	261[b]	54	63
Rhodochrosite	$MnCO_3$		3.7	370[b]	100	63
Anhydrite	$MgSO_4$		2.96	−1.0		65
				−1.1 to −1.2		53
Rutile	TiO_2		4.3	0.3[b]	0.07	63
Illite	a clay: 1.4% FeO, 4.7% Fe_2O_3			34[b]	12	23
Montmorillonite	a clay: 2.8% FeO,3% Fe_2O_3			26[b]	11	23
Nontronite	Fe-rich montmorill-onite: 0.2% FeO, 28% Fe_2O_3			140[b]	52	23
Metals, etc.						
Gold	Au		19.3	−2.7[b]	−0.14	63
Silver	Ag		10.5	−1.9[b]	−0.18	63

Table 14 (continued)
MAGNETIC SUSCEPTIBILITY OF MINERALS
(AT ROOM TEMPERATURE, UNLESS OTHERWISE NOTED)

Material	Composition	Notes	Density (gm/cm$_3$) a	k in 10^{-6} emu/cm^3 c	k in 10^{-6} emu/gm c	Ref.
Metals, etc. (continued)						
Sulfur	S		2.07	-1.0[b]	-0.48	63
Water	H_2O; liquid		1.0	-0.72[b]	-0.72	63
	Solid (ice)		0.92	-0.64[b]	-0.70	63
	Solid (ice), 0°C		0.92	-0.7	-0.76[b]	58

a Used to convert emu/gm to emu/cm³ or vice versa; density values from *Handbook of Materials Science*, Vol. I, Lynch, C. I., Ed., CRC Press, 1974, 235, with data from R. Kretz in *Handbook of Chemistry and Physics*, 55th ed., Weast, R. C., Ed., CRC Press, 1974, B-192, or from *Handbook of Physical Constants*, revised ed., Clark, S. P., Ed., Memoir 97 of Geological Society of America, 1966, 60.

b Calculated, using density value.

c Susceptibility is dimensionless, J/H, but can be determined either for J in emu/cm³ or in emu/gm.

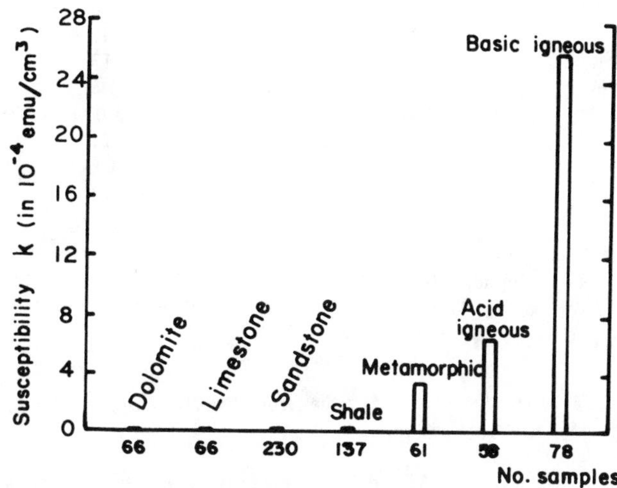

FIGURE 22. Measured susceptibilities for different rock types. (Reference 25, from earlier work by J. W. Peters.)

Table 15
SUSCEPTIBILITY OF MINERALS AND OTHER MATERIALS (AT ROOM TEMPERATURE, UNLESS OTHERWISE NOTED)

Material	Composition	Notes	Density (gm/cm^3) a	k in 10^{-6} cgs/cm^3 b	k in 10^{-6} cgs/gm	Ref. c
Minerals						
Hematite	αFe_2O_3	760°C	5.26	18,862	3,586	
Wustite	FeO	20°C	5.75	41,400	7,200	
Troilite	FeS	20°C	4.83	5,187	1,074	
Siderite	$FeCO_3$	20°C	3.96	44,750	11,300	
Corundum	Al_2O_3		4.02	−149	−37	
Cassiterite	SnO_2		6.99	−287	−41	
Rutile	TiO_2		4.25	25	5.9	
Quartz	SiO_2		2.65	−78	−29.6	
Calcite	$CaCO_3$		2.71	−103	−38	
Magnesite	$MgCO_3$		3-3.4	~−100	−32	
Anhydrite	$MgSO_4$	21°C	2.96	−148	−50	
Galena	PbS		7.58	−637	−84	
Millerite	NiS	20°C	5.5	1045	190	
Rock salt	NaCl		2.16	−65	−30	
Metals, etc.						
Gold	Au	23°C	19.3	−540	−28	
Copper	Cu	23°C	8.95	−49	−5.5	
Tin	Sn	gray; 7°C	5.75	−213	−37	
Sodium	Na		0.97	15.5	16	
Gadolinium	Gd	27°C	7.90	5,965,000	755,000	
Diamond	C		3.51	−21	−5.9	
				−1.8		53
Graphite	C		2.1−2.2	−13	−6	
				−6 to −16		58
Water	H_2O	liquid, 100°C			−13.1	
		liquid, 0°C	1.0	−12.9	−12.9	
		solid (ice), 0°C	0.92	−11.7	−12.7	
Oxygen	O_2	gas, 20°C			3449	

[a] Used to convert cgs/gm to cgs/cm^3; density values from *Handbook of Materials Science*, Vol. I, 1974 or *Handbook of Physical Constants*, 1966.

[b] Calculated from cgs/gm.

[c] Table's values for k (except for Ref. 53 and 58) are taken from *Handbook of Chemistry and Physics*, Weast, R. C., Ed., 55th edition, CRC Press, 1974, E-121; and quoted in *Handbook of Materials Science*, Vol. I, Lynch, C. T., Ed., CRC Press, Boca Raton, Fla., 1974, 214.

Table 16
MAGNETIC PROPERTIES (J_n, K, Q) OF ROCKS

Material	Notes	J_n (in 10^{-6} emu/cm^3) a	k (in 10^{-6} emu/cm^3)	Q_n b	Ref.
Sedimentary rocks					
Soils	Typical		1–100		29
	Texas Coast		5		
Marine sediments					
	Sandy clay, Texas continental slope	10–150	15–35	~5	
Silty shale	Ventura basin, Calif.	5–40	20–120	~5	
Siltstone	Precambrian, Britain; 4 samples			0.02–2	6
Clays			20		65
Shale	137 samples		5–1478 ave. 52		25
Sandstone	230 samples		0–1665 ave. 20–30		25
	Redbeds, Precam-			1.6–6	6
	brian, 9 samples		100		26
	Redbeds, U.S., 82 sites	2–20	0.4–40	1–3	45
			3–76		87
	Redbeds, Wyoming, Triassic	4–29	2–13	ave. 4.4	21, 22
	Britain, 12 sites	0.2–3.3	10–28		75, 20
	Redbeds, typical	0.5–50			21, 22
Limestone	66 samples		2–280 ave. 23		25
			0–5		26
Dolomite	66 samples		8		25
Coal			2		65
Typical sedimentary rocks, average		1–100	3–300	0.02–10	
Igneous rocks					
Typical igneous rocks, average		100–40,000	50–5000	1–40	
Granite	Pluton, Yosemite Calif.	100–800	1000–4000	0.3–1	58, 62, 24
			10		26
	97 samples, Okla.	1000–180,000		28	62, 24
	41 samples		280–2000		
			30–2700		25
	Minnesota, 31 samples <1.4% Fe_3O_4		0–4000 ave. 470		42
	Without Fe_3O_4		1–5		58
	Intrusives, Japan			0.1–0.5	58
Acidic intrusives					
	58 samples		3–6527 ave. 647		
			30–60		76

Table 16 (continued)
MAGNETIC PROPERTIES (J_n, K, Q) OF ROCKS

Material	Notes	J_n (in 10^{-6} emu/cm³) a	k (in 10^{-6} emu/cm³)	Q_n b	Ref.
	Minnesota, 17 samples		350		42
Granodiorite	Nevada			0.1–0.2	13, 58
Diorite			200		26
Dolerite (diabase/dikes)					
	sills, England, 5 samples			2–3.5	6
	dikes, India, 28 samples		55–1100 ave. 337		53
Diabase	Typical	1900–4000	1500–2300	2–3.5	
			80–1000		25
	Minnesota, 19 samples <3.4% Fe_3O_4		800–12000 ave. 2600		42
	dikes, Precambrian		100–20000	0.2–4	62
Gabbro	Minnesota, 37 samples, <0.9% Fe_3O_4		80–6100 ave. 1000		42
	Minnesota			1–8	
			2000		26
	Sweden			9.5	58
			70–2400		25
Intrusives	Sudbury basin, Ontario	1000–60,000	20–5000	0.1–20	62
	Precambrian, basic		2000–9000	1–2	76
Basic	78 samples		44–9711 ave. 2596		
	Precambrian, India, 5 samples		3675–4300		53
Basalt/diabase	Minn., 64 samples		2500		42
Volcanics	Montana, Eocene, 455 samples	11,000	700	~30	62
	rapidly-cooled			30–50	
Basalt	Australia, Cenozoic, 127 samples	2100	900	~5	58, 62
	Minn., 37 samples <2.5% Fe_3O_4		20–8400 ave. 2950		42
			3000–8000		
			40–9600		58
	Iceland, Tertiary, 70 samples			6	6
	NSW Australia, Tertiary, 7 sites	2000–30,000			6
	India, Deccan traps, 60 samples		1000–6000 ave. 2300		53
	W. Greenland, Tertiary			1–39	58
			2000		26
	Seamounts, N. Pacific			8–57	71
	Seafloor, EM-7 Mo-hole, NE Pacific			15–105 ave. 40	71
	Seafloor, mid-Atlantic ridge		24–2900	1–160 ave. 48	71

Table 16 (continued)
MAGNETIC PROPERTIES (J_n, K, Q) OF ROCKS

Material	Notes	J_n (in 10^{-6} emu/cm^3) a	k (in 10^{-6} emu/cm^3)	Q_n b	Ref.
	Seafloor, depth 1–6 meters	5000–8000	300–600	25–45	
(and for comparison:)				Q_T[c]	
	Titanomagnetite grains,				
	x = 0.6, size 1–2 μm			35	37
	10 μm			17	
	20 μm			10	
	Magnetite powder, 2% by wt dispersion in epoxy, grain size 1–2 μm			20	37
	4–8 μm			5.4	
	37–75 μm			1.5	
	75–150 μm			1.3	
	Titanomagnetite powder, x = 0.55, 2% by wt dispersion in epoxy, grain size				
	1–2 μm			39	37
	5–15 μm			23	
	75–150 μm			1	
	Magnetite powder dispersed, grain size:	J_T[d] (emu/cm^3 of magnetite)	(per cm^3 of magnetite)	Q_T	
	1.5 μm	0.55	0.19	7.2	11, 51
	6	0.15	0.19	2.0	
	19	0.041	0.19	0.54	
	21	0.032	0.21	0.38	
	58	0.046	0.22	0.52	
	88	0.041	0.21	0.49	
	120	0.044	0.24	0.46	
Metamorphic rocks		J_n		Q_n	
Metasediments	Precambrian		20–200		76
Granite/gneiss	Precambrian, India, 12 samples		30–100 ave. 59		53
Gneiss			0–240		25
Slate	Minn., 26 samples, <.2% Fe_3O_4		0–100 ave. 50		42
Greenstone	Precambrian, 8 samples		10–60		76
	Minn., 15 samples, <.2% Fe_3O_4		40–880 ave. 100		42
Basic metaigneous					
	Precambrian		200–4000	0.5–2	76
Serpentinite			250–6000		87
	61 samples		0–5824 ave. 349		25
Peridotite			12,500		25
			5000		26

Table 16 (continued)
MAGNETIC PROPERTIES (J_n, K, Q) OF ROCKS

Material	Notes	J_n (in 10^{-6} emu/cm³) [a]	k (in 10^{-6} emu/cm³)	Q_n [b]	Ref.
Ores					
Magnetite ore	Sweden			1–10	70
	Sweden, 31–63% Fe_3O_4		240,000–490,000		67
	Sweden, 86–95% Fe_3O_4		1,000,000–1,120,000		67
	76% by wt Fe_3O_4		350,000		47
	90% Fe_3O_4, grain size 0.2 x .2 mm		400,000		47
	Lodestone iron ore, 7 samples	5,000,000–35,000,000			66
	Lodestone ore, Arkansas	13,100,000	323,000	94	10
	Lodestone ore, Japan	11,100,000	311,000	80	10
	Biwabik & Soudan, U.S.; 15–26% Fe_3O_4		50,000–120,000		42
	6 samples	200,000–800,000			66
Hematite ore	Sweden		330–800		10
	Precambrian		60–750		76
Chromite ore	$FeCr_2O_4$; 27–58% Fe		600–100,000		50
Hausmannite ore	Mn_3O_4; Sweden		130		67
Pyrite ore	Sweden		420		50
	Sweden		8–400		50
Pyrrhotite ore	Sweden		60		50
Lunar rocks		(in 10^{-6}emu/gm)	(in 10^{-6}emu/gm)		
	typical	0.1–1000	500–2000	0.001–1[e]	
Soils	7 samples		1100–3500		28
Breccia	15 samples		50–3300		28
Anorthosite	2 samples		200–900		28
	Breccia		400 x 10^{-6}emu/cm³		44
Gabbro			50 x 10^{-6}emu/gm		28
Basalt	8 samples		50–700 x 10^{-6} emu/gm		28
			100–300 x 10^{-6}emu/cm³		44
Meteorites		(in cgs)	(in cgs)		
Meteorites	60 samples (48 irons, i.e., Ni-Fe; 12 stony-irons)	0.05–.3	0.2–4		32

[a] J_n is natural remanent magnetization (NRM).
[b] Q_n is Koenigsberger ratio for NRM, i.e., $Q_n = J_{NRM}/KH_E$ where H_E is earth's field (about 0.5 oe).
[c] Q_T is Koenigsberger ratio for TRM, i.e., $Q_T = J_{TRM}/KH_E$ where J_{TRM} is TRM acquired in Earth's field.
[d] Lab TRM, in H = 0.4 oe.
[e] In H = 0.5 oe.

Table 17
SUSCEPTIBILITIES OF ROCK TYPES, CALCULATED FROM THEIR MAGNETITE AND ILMENITE CONTENT[25,59]

	Magnetite Content and Susceptibility,[a] cgs units							
	Minimum		Maximum		Average		Ilmenite, average	
Material	%	k × 10⁶	%	k × 10⁶	%	k × 10⁶	%	k × 10⁶
Quartz porphyries	0.0	0	1.4	4,200	0.82	2,500	0.3	410
Rhyolites	0.2	600	1.9	5,700	1.00	3,000	0.45	610
Granites	0.2	600	1.9	5,700	0.90	2,700	0.7	1000
Trachyte-syenites	0.0	0	4.6	14,000	2.04	6,100	0.7	1000
Eruptive nephelites	0.0	0	4.9	15,000	1.51	4,530	1.24	1700
Abyssal nephelites	0.0	0	6.6	20,000	2.71	8,100	0.85	1100
Pyroxenites	0.9	3000	8.4	25,000	3.51	10,500	0.40	5400
Gabbros	0.9	3000	3.9	12,000	2.40	7,200	1.76	2400
Monzonite-latites	1.4	4200	5.6	17,000	3.58	10,700	1.60	2200
Leucite rocks	0.0	0	7.4	22,000	3.27	9,800	1.94	2600
Dacite-quartz-diorite	1.6	4800	8.0	24,000	3.48	10,400	1.94	2600
Andesites	2.6	7800	5.8	17,000	4.50	13,500	1.16	1600
Diorites	1.2	3600	7.4	22,000	3.45	10,400	2.44	4200
Peridotites	1.6	4800	7.2	22,000	4.60	13,800	1.31	1800
Basalts	2.3	6900	8.6	26,000	4.76	14,300	1.91	2600
Diabases	2.3	6900	6.3	19,000	4.35	13,100	2.70	3600

[a] Using $k_{magnetite} = 0.30$ emu/cm³; $k_{ilmenite} = 0.137$ emu/cm³.

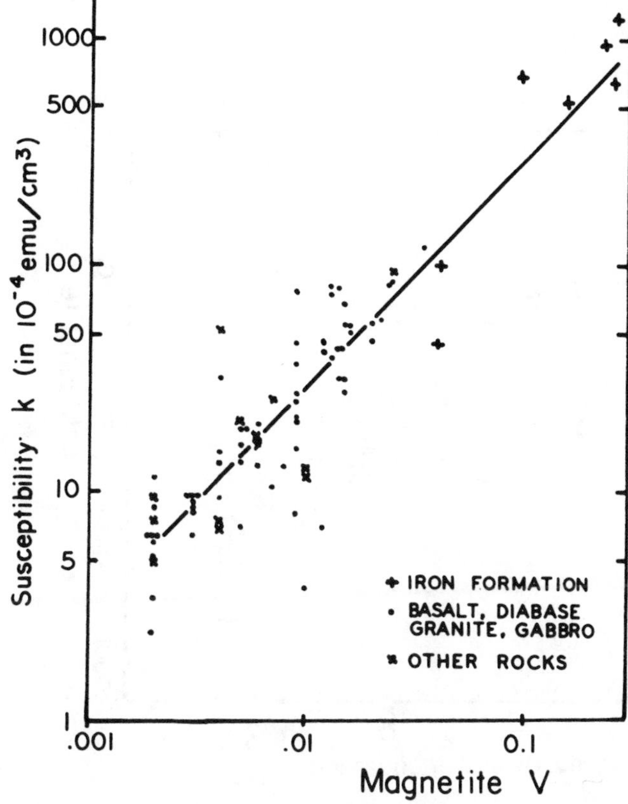

FIGURE 23. Relation of susceptibility to magnetite content for some igneous and ore rocks (Precambrian, of Minnesota).[42]

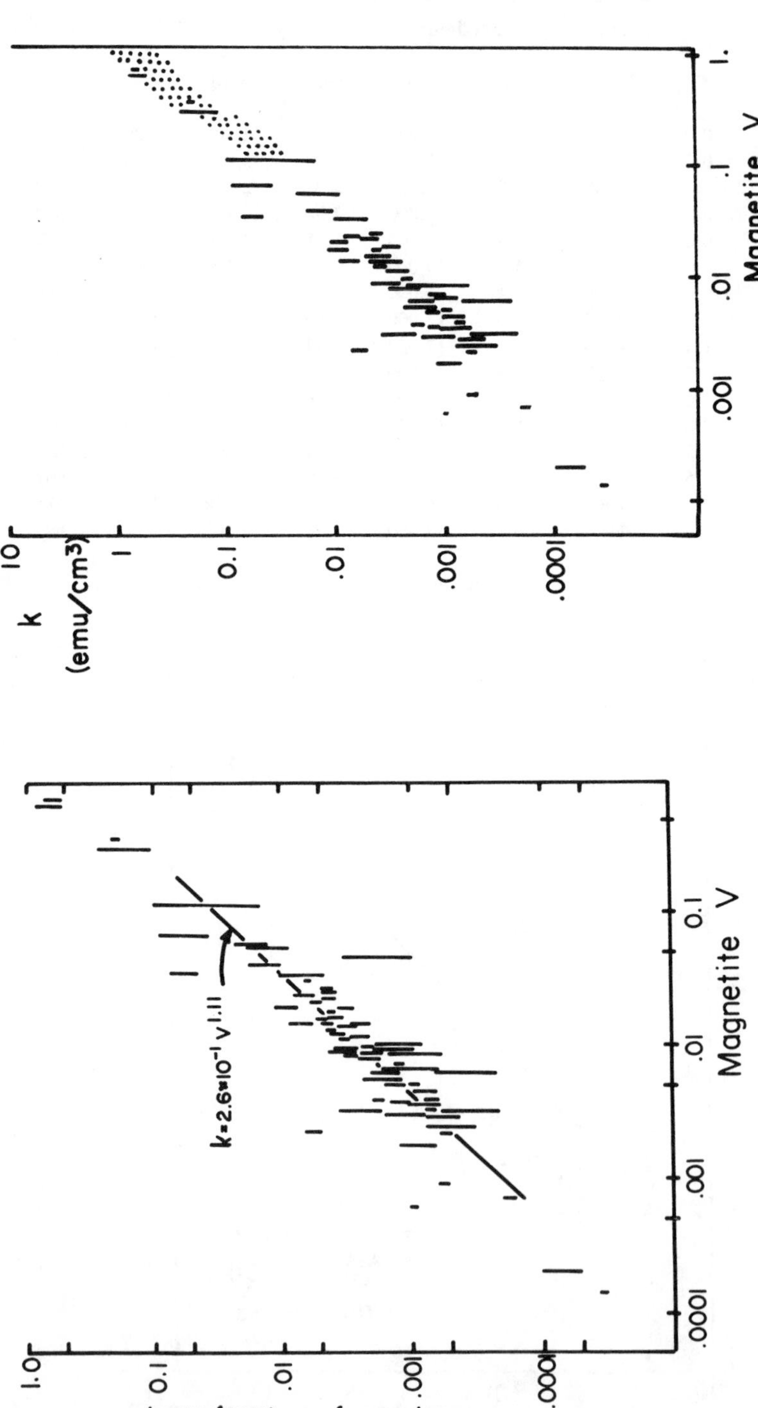

FIGURE 24. Relation of susceptibility to magnetite content for some metamorphic rocks (Adirondacks.) (Reference 29, from data of Reference 15).

FIGURE 25. Susceptibility and magnetite content of rocks and ores. (Reference 7; data from Reference 15 for lines, Reference 67 for stippled area.)

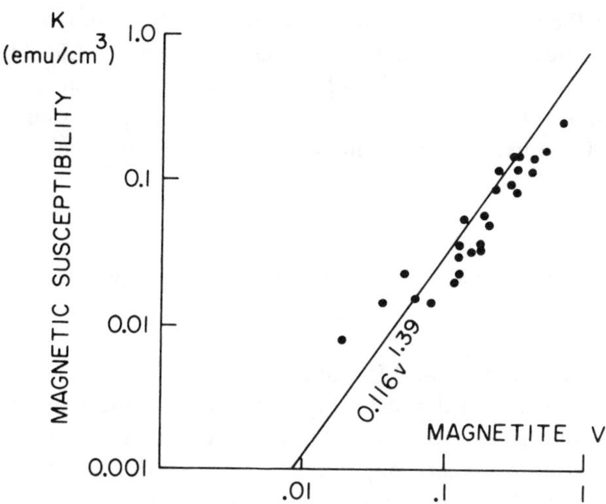

FIGURE 26. Susceptibility and magnetite content for iron formation ore (Reference 50, from data of Reference 36).

Table 18
DISTRIBUTION OF MEASURED SUSCEPTIBILITIES FOR MAJOR ROCK TYPES[7,60]

Rock type	Number of samples	Percent of samples with k (in 10^{-6} emu/cm³)			
		<100	100—1000	1000—4000	>4000
Basic extrusive	97	5	29	47	19
Basic intrusive	53	24	27	28	21
Granite etc. (i.e., acidic igneous)	74	60	23	16	1
Metamorphic (gneiss, schist, slate)	45	71	22	7	0
Sedimentary	48	73	19	4	4

Note: Where the terms indicate: basic = mafic; high in iron/magnesium silicates, i.e., the denser and darker-colored ferromagnesian minerals; acidic = siliceous; high in quartz; extrusive = formed by cooling after extruding onto the land surface, or seafloor; typically finer-grained; intrusive = plutonic; formed by cooling at depth; typically coarser-grained.

Figure 27 shows the variation of bulk susceptibility with magnetite content, for magnetite-rich ore samples from Sweden. V varies from 0.1 to 0.8. The axis scales are linear, not logarithmic as in the previous figures. Lines for various formulas are drawn, using $k_m = 1.5$ emu/cm^3 for magnetite. The Werner formula is empirical, with $N = 4\pi/3$ for spherical grains. The Puzicha formula[52] is theoretical, for noninteracting grains.

Permeability

Table 19 gives values of relative permeability μ_r, with respect to free space, for some minerals and rocks.

Effects of Depth, Temperature, and Pressure

Temperature increases with depth in the Earth, and spontaneous magnetization decreases and finally diminishes to zero at the Curie temperature. There is thus a depth below which there is no remanent or induced magnetization in the rocks. This depth, and the thickness of the layer of "magnetic" rock above it, are important for modelling of long-wavelength magnetic-field anomalies arising from deep magnetic source bodies and structure. This is particularly true now that satellite magnetometry is being used to map such large-scale magnetic features. Deep and laterally extensive geologic features are useful in interpreting tectonically-active spreading and subduction zones in the crust, and areas of volcanic and geothermal activity.

The depth of the Curie-temperature isotherm, or base of the magnetic layer, will depend on the local geothermal gradient, the titanomagnetite composition (proportion of the mineral series, and "x" composition in it), any change in T_c with hydrostatic pressure with depth, and the rate of decrease of J_s and the remanent magnetization with depth and thus temperature, up to the Curie temperature. The geothermal gradient is known generally in various geologic provinces, as illustrated in Figure 28. It is fairly low (about 10°C/km) in ancient shield areas, and higher (about 17°C/km) in typical ocean areas. These gradients are averages for the upper 40-km layer of the lithosphere; the gradient at the Earth's surface itself is about 25 to 30°C/km on a global average. The geothermal gradient can be higher in local areas, and the Curie-temperature isotherm will be correspondingly shallower. The Curie temperatures for different titanomagnetite compositions ("x"), as affected by increasing hydrostatic pressure, are plotted as lines A, B, and C in Figure 28.

Calculated depths of the T_c isotherm are given in Table 20. These would be maximum depths for the layer of appreciable magnetization, because the spontaneous magnetization would decrease as T_c is approached with depth. These calculations presume the magnetic carrier is magnetite as listed. Estimates of the "effective" bottom of deep magnetic sources, obtained from magnetic modelling, are listed in Table 20.

The effect of pressure on Curie temperature is shown in Table 21, for titanomagnetite and ilmenohematite. The effect of pressure on magnetocrystalline (K) and magnetostriction (λ) anisotropy coefficients of titanomagnetites is given in Table 22, with the variation in K_1 shown in Figure 29.

The variation of saturation (spontaneous) magnetization J_s with temperature from 0°K up to the Curie temperature, is shown in Figure 30 for a "typical ferromagnetic" material, and for nickel.

The variation of magnetostriction anisotropy coefficients with temperature for titanomagnetite is shown in Figure 31. Data are compiled for λ_{111} and λ_{100}, for $x = 0, 0.1$, and 0.3.

The variation of magnetocrystalline anisotropy coefficients K_1 and K_2 for titanomagnetite is shown in Figure 32.

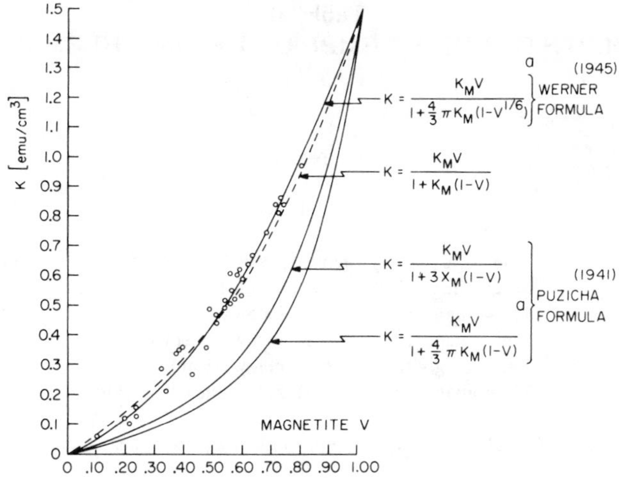

FIGURE 27. Variation of susceptibility with magnetite content for magnetite-rich ores (Sweden) with various formulas. Using $K_m = 1.5$ emu/cm^3.[62,67] a—with $N = 4\pi/3$.

Table 19
RELATIVE PERMEABILITY OF MINERALS AND ROCKS[88]

Material		μ_r (relative permeability)
Minerals		
Quartz (diamagnetic)		0.999985
Calcite (diamagnetic)		0.999987
Rutile (paramagnetic)		1.0000035
Pyrite		1.0015
Hematite		1.053
Ilmenite		1.55
Pyrrhotite		2.55
Magnetite (ferrimagnetic)		5.0
Rocks	% magnetite	
	0	~1.0
Granites	0.2	1.006
	0.5	1.017
	1.0	1.04
Basalts	2.0	1.08
	3.0	1.12
	5.0	1.18
Iron ore	10.0	1.34
	20.0	1.56

Table 20
DEPTH OF CURIE-TEMPERATURE ISOTHERM IN EARTH

Depth (km)	Notes	Ref.
	Theoretical	
33	Magnetite (x = 0), below oceanic crust	19
60	Magnetite (x = 0), below ancient shields	19
29	Titanomagnetite (x = 0.1), below oceanic crust	19
51	Titanomagnetite (x = 0.1), below ancient shields	19
23	Titanomagnetite (x = 0.3), below oceanic crust	19
40	Titanomagnetite (x = 0.3), below ancient shields	19
	Calculated, from magnetic modeling on continents	
20—48	5 studies, North America	30
15—35	3 studies, USSR	30
15—26	1 study, Britain	30

Table 21
PRESSURE DEPENDENCE OF CURIE TEMPERATURES OF TITANOMAGNETITE AND ILMENOHEMATITE

Material	Notes	dT_c/dp (°C/kbar)	Ref.
Magnetite	Natural, Sweden	1.8—2.3 ± 10%	57
	Natural	2.05 ± 5	57
	Natural, Austria	1.9 ± 10	57
	Synthetic titanomagnetite		
	x = 0	1.9 ± 5	
	= 0.1	1.75	
	= 0.2	1.6	
	= 0.3	1.5	
	= 0.4	1.45	
	= 0.6	1.3	
	= 0.8	0.8	
Ilmenohematite		0.8	83

Table 22
PRESSURE DEPENDENCE OF MAGNETOCRYSTALLINE AND MAGNETOSTRICTION ANISOTROPY COEFFICIENTS FOR TITANOMAGNETITE[19,56]

	dK_1/dp (%/Kb)	dK_2/dp (%/Kb)	$d\lambda_{111}/dp$ (%/Kb)	$d\lambda_{100}/dp$ (%/Kb)
Magnetite (x = 0)	−2	−0.2 to 10 kb	+14	+14
Titanomagnetite (x = 0.1)	−1	−0.1 to 10 kb	+14	+14

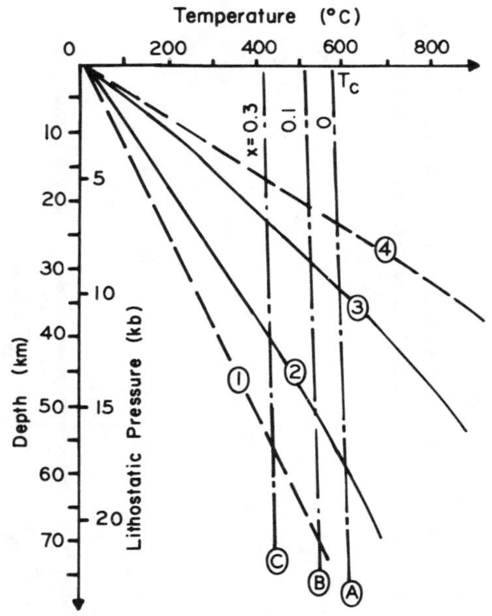

FIGURE 28. Geothermal gradients in geologic areas, as a basis for calculating depth of Curie-temperature isotherms. (1) Low gradient, e.g., Sierra Nevada, USA; (2) Typical ancient shield; (3) Typical oceanic crust; (4) High gradient, e.g., Basin and Range, USA. A, B, and C are Curie-temperature gradients for titanomagnetite of composition x = 0, 0.1, 0.3, respectively. (From Carmichael, R., *Earth Planet. Sci. Lett.*, 36, 309, 1977; based on data from Blackwell, D., *AGU Geophys. Monogr.*, 14, 169, 1971; Wyllie, P., *The Dynamic Earth*, John Wiley & Sons, New York, 1971; and Roy, R., et al., *Earth Planet. Sci. Lett.*, 5, 1, 1968.)

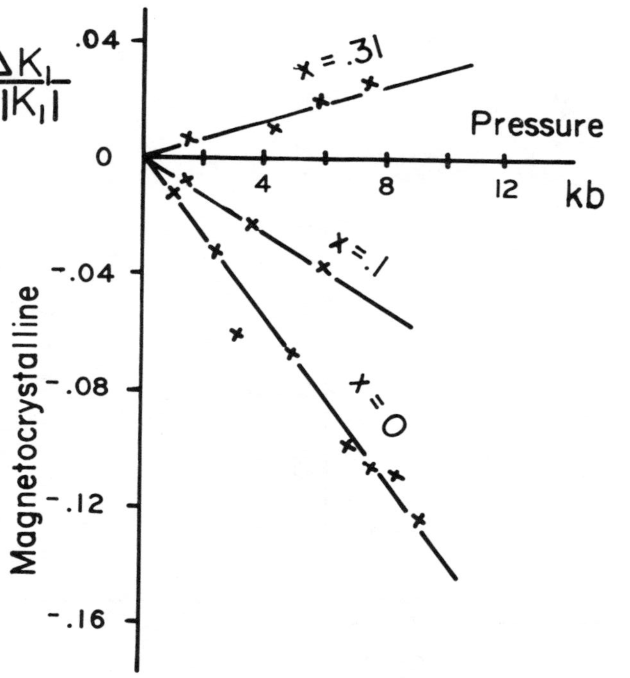

FIGURE 29. Pressure dependence of magnetocrystalline anisotropy coefficient, K_1, of titanomagnetite with x = 0,0.1,0.31.[56]

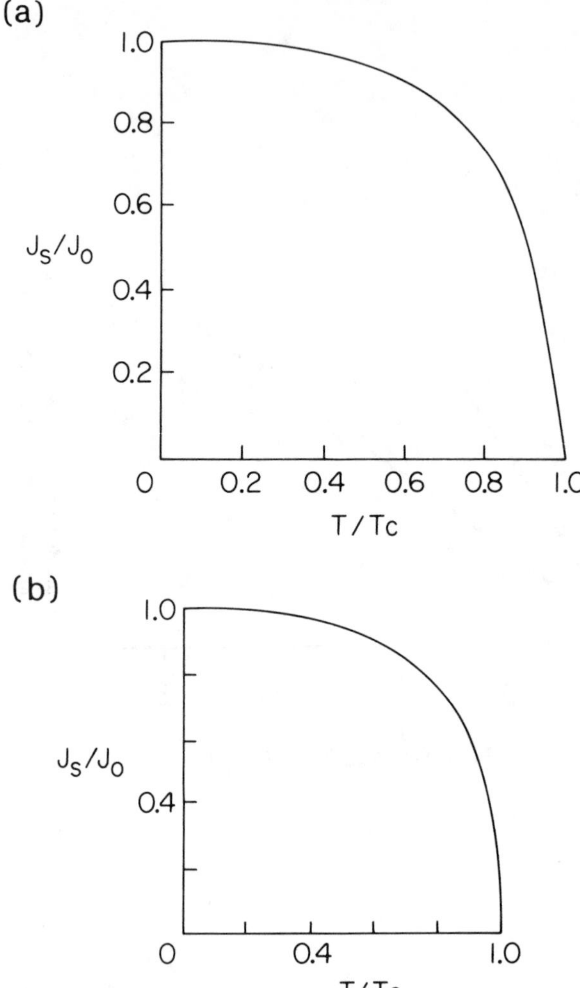

FIGURE 30. Variation of spontaneous magnetization J, with temperature up to the Curie temperature, for (A) typical ferromagnetic, and (B) nickel. Ordinate is J_s, normalized to its value at 0°K. Abscissa is absolute temperature, normalized to T_c.[2]

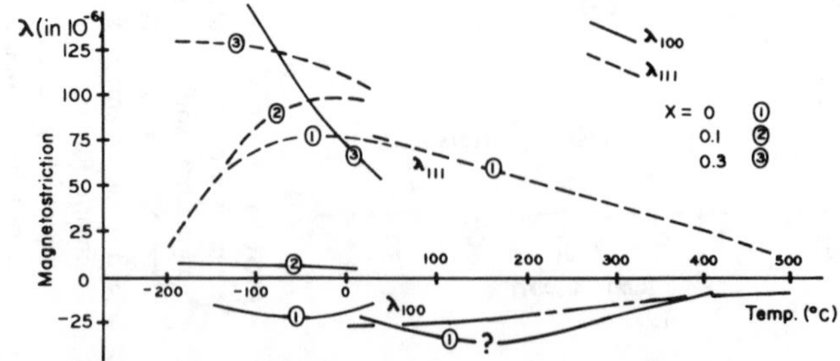

FIGURE 31. Variation of magnetostriction constants (λ) with temperature for titanomagnetite. Line 1 for x = 0, 2 for x = 0.1, and 3 for x = 0.3. (From Reference 19; data below 30°C from Reference 64, above 30°C from Reference 39.)

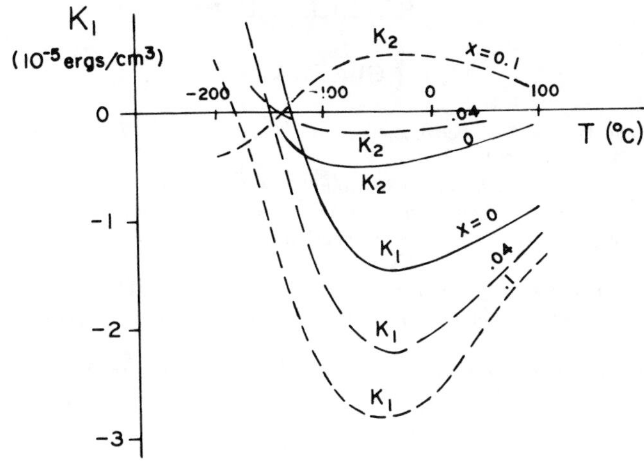

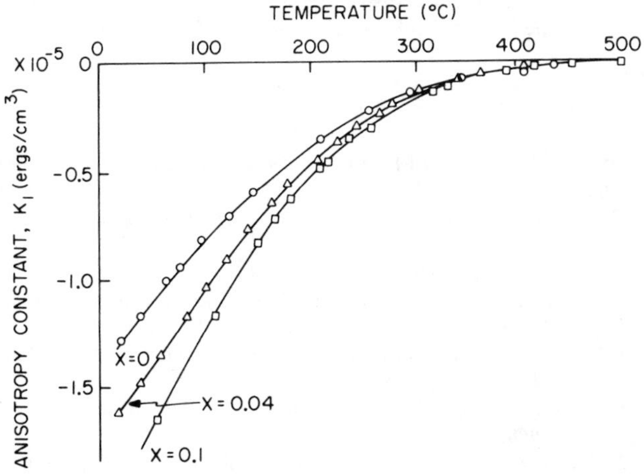

FIGURE 32. Variation of magnetocrystalline anisotropy constants (K) with temperature, for titanomagnetite. (A) K_1 and K_2, for -180° < T < 90°C. (B) K_1, for 30° < T < 500°C. (A from Reference 64 and B from Reference 27.)

REFERENCES

GENERAL

1. **Becker, R. and Doring, W.,** *Ferromagnetismus*, Springer-Verlag, Berlin, 1939; reproduced by Edwards Bros., Ann Arbor, MI, 1943.
2. **Bozorth, R. M.,** *Ferromagnetism*, Van Nostrand, New York, 1951.
3. **Chikazumi, S.,** *Physics of Magnetism*, John Wiley & Sons, New York, 1964.
4. **Craik, D. J. and Tebble, R. S.,** *Ferromagnetism and Ferromagnetic Domains*, Interscience, New York, 1965.
5. **Haalck, H.,** *Der Gesteinsmagnetismus*, Becker and Erler, Leipzig, 1945.
6. **Irving, E.,** *Paleomagnetism and its Application to Geological and Geophysical Problems*, John Wiley & Sons, 1964.
7. **Lindsley, D. H., Andreasen, G. E., and Balsley, J. R.,** Magnetic properties of rocks and minerals, in *Handbook of Physical Constants*, Clark, S. P., Ed., Memoir 97, Geological Society of America, Boulder, Colo., 1966.
8. **McElhinny, M. W.,** *Paleomagnetism and Plate Tectonics*, Cambridge University Press, New York, 1973.
9. *Mining Geophysics — Theory*, Vol. 2, Society of Exploration Geophysicists, Tulsa, 1967.
10. **Nagata, T., Ed.,** *Rock Magnetism*, revised ed., Maruzen Company, Tokyo, 1961.
11. **Stacey, F. D. and Banerjee, S. K.,** *Physical Principles of Rock Magnetism*, Elsevier, Amsterdam, 1974.

OTHER

12. **Akimoto, S.,** Magnetic properties of $FeO-Fe_2O_3-TiO_2$ system as a basis of rock magnetism, *J. Phys. Soc. Jpn.*, 17, 706, 1962.
13. **Allingham, J. W. and Zietz, I.,** Geophysical data on the Climax stock Nevada Test Site, *Geophysics*, 27, 599, 1962.
14. **Amar, L. F.,** *J. Appl. Phys.*, 29, 989, 1958.
15. **Balsley, J. R. and Buddington, A. F.,** Iron-titanium oxide minerals rocks and aeromagnetic anomalies of the Adirondack area, New York, *Econ. Geol.*, 53, 777, 1958.
16. **Carmichael, R. S.,** Remanent and transitory effects of elastic deformation of magnetic crystals, *Philos. Mag.*, 17, 911, 1968.
17. **Carmichael, R. S.,** Stable strain-induced magnetic remanence in nickel cobalt and magnetite, *Jpn. J. Appl. Phys.*, 7, 1247, 1968.
18. **Carmichael, R. S.,** Magnetomechanical behavior of nickel and cobalt particularly during fatigue deformation, *Acta Metall.*, 17, 261, 1969.
19. **Carmichael, R. S.,** Depth calculation of piezomagnetic effect for earthquake prediction, *Earth Planet. Sci. Lett.*, 36, 309, 1977.
20. **Clegg, J. A., Almond, M., and Stubbs, P.,** The remanent magnetism of some sedimentary rocks in Britain, *Philos. Mag.*, 45, 583, 1954.
21. **Collinson, D. W.,** Origin of remanent magnetization in certain red sediments, *Geophys. J. R. Astron. Soc.*, 9, 203, 1965.
22. **Collinson, D. W.,** The remanent magnetization and magnetic properties of red sediments, *Geophys. J.*, 10, 105, 1965.
23. **Collinson, D. W.,** An estimate of the hematite content of sediments by magnetic analysis, *Earth Planet. Sci. Lett.*, 4, 417, 1968.
24. **Currie, R. G., Gromme, C. S., and Verhoogen, J.,** Remanent magnetization of some upper Cretaceous granitic plutons in the Sierra Nevada California, *J. Geophys. Res.*, 68, 2263, 1963.
25. **Dobrin, M. B.,** *Introduction to Geophysical Prospecting*, 3rd ed., McGraw-Hill, New York, 1976.
26. **Dohr, G.,** *Applied Geophysics*, Halsted/John Wiley & Sons, New York, 1974.
27. **Fletcher, E. J. and O'Reilly, W.,** Contribution of Fe^{2+} ions to the magnetocrystalline anisotropy constant K_1 of $Fe_{3-x}Ti_xO_4$ $(0< x< 0.1)$, *J. Phys. Chem.*, 7, 171, 1974.
28. **Fuller, M.,** Lunar magnetism, *Rev. Geophys. Space Phys.*, 12, 23, 1974.
29. **Grant, F. S. and West, G. F.,** *Interpretation Theory in Applied Geophysics*, McGraw-Hill, New York, 1965.
30. **Green, A. G.,** Interpretation of Project MAGNET aeromagnetic profiles across Africa, *Geophys. J. R. Astron. Soc.*, 44, 203, 1976.
31. **Grenet, G.,** Sur les proprietes magnetiques des roches, *Ann. Geophys.*, 13, 263, 1930.
32. **Guskova, Y. G.,** Study of natural remanent magnetization of iron and stony-iron meteorites, *Geomagn. Aeron. USSR*, 5, 91, 1965.

33. Clark, S. P., Ed., *Handbook of Physical Constants,* revised ed., Memoir 97, Geological Society of America, Boulder, Colo., 1966.
34. **Ishikawa, Y. and Akimoto, S.**, Magnetic property and crystal chemistry of ilmenite and hematite system. II. Magnetic property, *J. Phys. Soc. Jpn.,* 13, 1298, 1958.
35. **Ishikawa, Y. and Syono, Y.**, Order-disorder transformation and reverse thermoremanent magnetism in the $FeTiO_3$-Fe_2O_3 system, *Phys. Chem. Solids,* 24, 517, 1963.
36. **Jahren, C. E.**, Magnetic susceptibility of bedded iron-formation, *Geophysics,* 28, 756, 1963.
37. **Kean, W. F., Day, R., Fuller, M., and Schmidt, V.**, The effect of uniaxial compression on the initial susceptibility of rocks as a function of grain size and composition of their constituent titanomagnetites, *J. Geophys. Res.,* 81, 861, 1976.
38. **Kittel, C.**, Physical theory of ferromagnetic domains, *Rev. Modern Phys.,* 21, 541, 1949.
39. **Klapel, G. and Shive, P.**, High-temperature magnetostriction of magnetite, *J. Geophys. Res.,* 79, 2629, 1974.
40. **Lee, E. W.**, *Rep. Prog. Phys.,* 18, 184, 1955.
41. **Lilley, B. A.**, *Philos. Mag.,* 41, 792, 1950.
42. **Mooney, H. M. and Bleifuss, R.**, Magnetic susceptibility measurements in Minnesota. II. Analysis of field results, *Geophysics,* 18, 383, 1953.
43. **Nagata, T.**, Anisotropic magnetic susceptibility of rocks, *Pure Appl. Geophys.,* 78, 110, 1970.
44. **Nagata, T.**, Piezoremanent magnetization of lunar rocks, *Pure Appl. Geophys.,* 110, 2022, 1973.
45. **Nesbitt, J. D.**, Variation of the ratio intensity to susceptibility in red sandstones, *Nature (London),* 210, 618, 1966.
46. **Nettleton, L. L.**, *Gravity and Magnetics in Oil Prospecting,* McGraw-Hill, New York, 1976.
47. **Nulman, A., Shapiro, V., Maksimovskikh, S., Ivanov, N., Kim, J., and Carmichael, R. S.**, Magnetic susceptibility of magnetite under hydrostatic pressure and implications for tectonomagnetism, *J. Geomag. Geoelectr.,* 30, 585, 1978.
48. **Osborn, J. A.**, Demagnetizing factors of the general ellipsoid, *Phys. Rev.,* 67, 351, 1945.
49. **O'Reilly, W. and Readman, P.**, The preparation and unmixing of cation-deficient titanomagnetites, *Z. Geophys.,* 37, 321, 1971.
50. **Parasnis, D. S.**, *Mining Geophysics,* revised ed., Elsevier, Amsterdam, 1973.
51. **Parry, L. G.**, Magnetic properties of dispersed magnetite powder, *Philos. Mag.,* 11, 303, 1965.
52. **Puzicha, K.**, Der magnetismus der gesteine als funktion ihrer magnetithaltes, *Beitr. Angew. Geophys.,* 9, 158, 1941.
53. **Ramachandra Rao, M. B.**, *Outlines of Geophysical Prospecting,* Wesley Press, Mysore, India, 1975.
54. **Rhodes, P. and Rowlands, G.**, *Proc. Leeds Phil. Lit. Soc.,* 191, 1954.
55. **Rodbell, D. S.**, Magnetic resonance of high quality ferromagnetic metal single crystals, *Physics,* 1, 279, 1965.
56. **Sawaoka, A. and Kawai, N.**, The effect of hydrostatic pressure on the magnetic anisotropy of ferrous and ferric ions in ferrites with spinel structure, *J. Phys. Soc. Jpn.,* 25, 133, 1968.
57. **Schult, A.**, Effect of pressure on the Curie temperature of titanomagnetites, *Earth Planet. Sci. Lett.,* 10, 81, 1970.
58. **Sharma, P. V.**, *Geophysical Methods in Geology,* Elsevier, Amsterdam, 1976.
59. **Slichter, L. B.**, Certain aspects of magnetic surveying, *Trans. Am. Inst. Min. Metall. Pet. Eng.,* 81, 238, 1929.
60. **Slichter, L. B.**, Magnetic properties of rocks, in *Handbook of Physical Constants,* Birch, F., Ed., Geological Society of America, Boulder Colo., 1942.
61. **Stoner, E. C.**, Demagnetizing factors for ellipsoids, *Philos. Mag.,* 36, 803, 1945.
62. **Strangway, D. W.**, Mineral magnetism, and Magnetic characteristics of rocks, in *Mining Geophysics — Theory,* Vol. 2, Society of Exploration Geophysicists, Tulsa, 1967.
63. **Strangway, D. W.**, *History of the Earth's Magnetic Field,* McGraw-Hill, New York, 1970.
64. **Syono, Y.**, Magnetocrystalline anisotropy and magnetostriction of Fe_3O_4-Fe_2TiO_4 series with special application to rock magnetism, *Jpn. J. Geophys.,* 4, 71, 1965.
65. **Telford, W., Geldart, L., Sheriff, R., and Keys, D.**, *Applied Geophysics,* Cambridge University Press, New York, 1976.
66. **Wasilewski, P. J.**, Magnetic and microstructural properties of some lodestones, *Phys. Earth Planet. Inter.,* 15, 349, 1977.
67. **Werner, S.**, Determinations of the magnetic susceptibility of ores and rocks from Swedish iron ore deposits, *Swed. Geol. Surv.,* 39, 1, 1945.
68. **Dunlop, D. J.**, Proceedings of Nagata Conference, University of Pittsburgh, June 1974.
69. **Kersten, M.**, *Zeit. Angew. Phys.,* 80, 496, 1936.
70. **Parasnis, D. S.**, *Principles of Applied Geophysics,* 2nd ed., Chapman & Hall, London, 1972.
71. **Vacquier, V.**, *Geomagnetism in Marine Geology,* Elsevier, Amsterdam, 1972.
72. **Reich, K. H.**, *Phys. Rev.,* 101, 1647, 1956.

73. **Fuller, M.**, Geophysical aspects of paleomagnetism, *Crit. Rev. Solid State Sci.*, 1, 137, 1970.
74. **Lynch, C. T., Ed.**, *Handbook of Materials Science*, Vol. 3, CRC Press, Cleveland, 1974, 183.
75. **Blackett, P. M. S.**, *Lectures on Rock Magnetism*, Weizmann Science Press, Jerusalem, 1956.
76. **Meshref, W. and Hinze, W.**, Report on investigations 12, *Mich. Geol. Surv.*, 1970.
77. **Soffel, H.**, Ph.D. thesis, Ludwig-Maximilians University, Munich, 1964.
78. **Morrish, A. and Yu, S.**, Dependence of the coercive force on the density of some iron oxide powders, *J. Appl. Phys.*, 26, 1049, 1955.
79. **Hanss, R.**, Thermochemical etching reveals domain structure in magnetite, *Science*, 146, 398, 1964.
80. **Vicena, F.**, On the influence of dislocations on the coercive force of ferromagnetites, *Czech. J. Phys.*, 5, 480, 1955.
81. **Eaton, J. and Morrish, A.**, Magnetic domains in hematite at and above the Morin transition, *J. Appl. Phys.*, 40, 3180, 1969.
82. **Owen, E. and Yates, E.**, X-ray measurements of the thermal expansion of nickel, *Philos. Mag.*, 21, 809, 1936.
83. **Zvegintsev, A., Grankin, P., and Bugayev, M.**, Self-reversal of thermoremanent magnetization of synthesized solid solutions of hemoilmenites under pressure, Izvestiya Acad. Sci. USSR Earth Physics, 8, 47, 1974.
84. **Dunlop, D. J.**, Magnetic Fields — Past and Present, in Proc. Takesi Nagata Conf., Goddard Space-Flight Center, Maryland, 1975.
85. Urquhart and Goldman, 1956.
86. **Weast, R. C., Ed.**, *Handbook of Chemistry and Physics*, 47th ed., CRC Press, Boca Raton, Fla., 1966, E-102.
87. Parasnis, D. S., 1971.
88. **Keller, G. V. and Frischknecht, F. C.**, *Electrical Methods in Geophysical Prospecting*, Pergamon Press, Elmsford, N. Y., 1966, 57.
89. **Hanna, W. F.**, Weak-field magnetic susceptibility anisotropy and its dynamic measurement, *U. S. Geol. Survey Bull.*, 1418, 68, 1977.

Section V
Electrical Properties

By
George V. Keller

INTRODUCTION

A number of compilations of the electrical properties of rocks and minerals have been published previously. Among the earliest was the *Handbook of Physical Constants* (Special Paper #36, published by the Geological Society of America in 1942) where the editor sought to compile for the first time a wide variety of physical constants needed for geological and geophysical calculations. As time passed, and many more scientific investigations into the properties of earth materials were carried out, this pioneer compilation was followed by the second *Handbook of Physical Constants* (1966) compiled by S. P. Clark, Jr., and containing a section on the electrical properties of rocks and minerals by Keller.[55] Other recent compilations include a book by Parkhomenko[84] covering the electrical properties of rocks, and a second volume by the same author[85] on the electrification phenomena which can take place in rocks. Keller,[56,57] Keller and Frischknecht,[61] and Olhoeft[80] have also prepared recent summaries of data on the electrical characteristics of rocks in the earth's crust. Shuey[100] has compiled a monographic study of the properties of ore minerals. Hutton[44] has summarized information on electrical conductivity as a function of depth in the earth.

Not all compilations have appeared in the English language; two extensive compilations have appeared in Russian, one by Kobranova,[72] and one by Dortman et al.[22] It should also be noted that the two volumes by Parkhomenko[84,85] were first published in Russian.

The electrical properties of a material include the primary properties of conductivity and dielectric constant, and many secondary properties such as the coefficients which determine the rate of change of the primary properties with such parameters as frequency, temperature, etc. Also, there are a number of secondary electrical properties which represent the ease with which other forms of energy are transformed to electrical energy; among these are included the Hall coefficient, the thermoelectric coefficient, the piezoelectric coefficient, and the pyroelectric coefficient. Relatively specific values can be given for the properties of some minerals which occur with a specific chemical composition, but more commonly for minerals with variable composition, and in particular for rocks with no exact compositional formula, the electrical properties of a specifically named rock or mineral cover a very wide range, with the variation depending on chemical and mineral composition. The fact that many rocks and minerals are not characterized by a specific value for the various electrical properties even under standard conditions makes it difficult to tabulate a meaningful set of numbers for these properties. As a consequence, the tabulation must represent the variability of the properties under discussion, and the parameters on which the properties depend. This leads to a dual method of classification for data on electrical properties of rock and minerals. One classification might be termed "parametric", with the relationships between properties and the other parameters which cause the properties to vary being specified. The second mode of presentation might be termed "existentialist", where compilations of properties of rocks and minerals as they exist in the earth are given, with no reference being made to the parameters that control them. Both modes of classification will be used in this compilation; the parametric data will be presented first, and the existentialist data will follow.

ELECTRICAL CONDUCTIVITY AND DIELECTRIC CONSTANT OF MINERALS AND DRY ROCKS

In accordance with modern usage, the MKS unit for conductivity, which is the S/m is used in these tables. Frequently, in geophysical practice, the reciprocal of conductivity, called the resistivity and measured in ohm-meters, is used. The MKS unit for dielectric constants should properly be the farad per meter, but by convention, dielectric constants are compiled in terms of the ratio:

$$\text{dielectric constant} = \frac{\text{specific capacity of material}}{\text{specific capacity of vacuum}} \qquad (1)$$

To determine the specific capacity of a material from the data listed in these tables, it is necessary to multiply the listed values by the specific capacity of a vacuum, which is 8.85×10^{-12}F/m.

Materials are classified generally as conductors, semiconductors, and insulators. If the conductivity of a material is greater than 10^5 S/m, it is classed as a conductor; materials where conductivities less than 10^{-8} S/m are classed as insulators; materials in the intermediate range are semiconductors.

Metallic Conduction

Native metals occur infrequently in nature, but such occurrences can be of considerable economic importance. Two of the more important native metals are copper and gold. Platinum, iridium, osmium, and iron occur in elemental form, but are extremely rare. Carbon occurs commonly in the form of graphite, which may be a metallic conductor.

The resistivities of several refined metals and of naturally occurring metallic minerals are listed in Table 1.

Electronic Semiconduction

Conduction in semiconductors is by movement of electrons through the material, but the conductivity is much lower than that observed in true metals. The lower conductivity is explained by a smaller number of electrons contributing to conduction. Semiconductors differ from metals in that an appreciable energy is required to move an electron from one atom to the next in the crystal structure. This energy is most commonly provided by heat, so that in semiconductors, the number of conduction electrons is found to increase with temperature according to the law:

$$n_e = ae^{\frac{-E}{2KT}} \qquad (2)$$

where n_e is the number of conduction electrons per unit volume, a is a constant of proportionality, E is the activation energy required to transfer an electron from one atom to the next, K is Boltzmann's constant, and T is the absolute temperature.

The reported values of resistivity for semiconducting minerals generally show very wide ranges of values for a single mineral. Published studies of conductivity of these minerals have been reviewed by Parkhomenko[84] and Shuey.[100] The variability of resistivity values is to be expected in semiconductors inasmuch as the electrical properties are determined primarily by small amounts of impurities and by imperfections, rather than by the bulk composition of the material. Harvey[38] also points out that in many cases, erroneously high values of resistivity can be measured on samples which have hairline cracks. In this case, the lower resistivities reported for a mineral are probably more correct than the higher values.

Reported values for the resistivities of semiconductors are listed in Table 2. Some minerals are characterized by a relatively narrow range of values reported for resistivity. These narrow ranges should be viewed with suspicion inasmuch as the lack of scatter is more probably due to a limited number of measurements than to a uniformity of character for the mineral. Parkhomenko[84] and Shuey[100] have compiled probability distribution curves (histograms) for the values of resistivity reported for several minerals for which numerous measurements are available (Figures 1 and 2).

The third of the general types of conduction which may take place in the solid state is electrolytic conduction.[76] Most rock-forming minerals act as solid electrolytes, with the transfer of electric current being by motion of ions through a crystal lattice. Electrolytic conduction takes place most readily in ionic-bonded crystals. When an electric field is applied to an ionic-bonded structure, the force exerted on each ion by the field is small compared to the strength of the coulomb binding forces, so that in an ideal crystal one would not expect any electrolysis, or conduction by ion movement, to take place. However, there are several types of defect in a crystal lattice which may provide ions of conduction.

The conductivity of an ionic material depends on the number of charge carriers (ions) and the mobility, where apparent velocity with which they move under an implied field of 1 V/m. The mobilities of ions will depend on the relative sizes of the moving ions and the interstices of the crystal lattice. Approximate ionic radii are listed in Table 3.

Conductivity in ionic crystals can usually be represented by the equation:

$$\sigma = A_1 e^{\frac{-E_1}{KT}} + A_2 e^{\frac{-E_2}{KT}} \qquad (3)$$

where the constants A_1 and A_2 are functions of the number of ions available for conduction and their mobility through the lattice, and E_1 and E_2 are the activation energies required to liberate these ions. In this equation, the term with the smaller activation energy dominates at low temperature while the other term dominates at high temperature. The low-temperature conductivity term is called extrinsic or structure sensitive conductivity, and is associated with the presence of weakly bonded impurities or defects in a crystal. The high-temperature conductivity term is intrinsic conductivity, contributed by ions displaced from the regular lattice by thermal vibrations.

Extensive studies of the conductivity of rocks and minerals as a function of temperature have been carried out in recent years. Most of the work has involved the minerals present in ultrabasic rocks that occur deep in the earth where high temperatures are expected to occur. Hughes[42] was among the first to carry out such studies, using peridote as the mineral in his study. Olivine has been much studied.[23,25,26,95] The properties of enstatite have been studied by Dvorak and Schoessin.[28] Zablocki[119] made measurements on serpentinite. Dvorak[27] made measurements on olivinite, peridotite, and dunite. Basalts have been studied by Hansen et al.[36] and by Khitarov et al.[68] Basalt-like rocks have been studied extensively as part of the lunar program with samples returned from the moon.[2,13,14,17,79,81-83,105,106] Pyroxene has been the subject of a study by Duba et al.[24] A few measurements have been made of rocks other than ultrabasic rocks by Keller[54,57] and by Housely and Oliver.[41]

These studies generally have shown that the conductivity in a dry rock at elevated temperature is primarily a function of temperature and to a lesser degree a function of composition. At relatively low temperatures where extrinsic conduction mechanisms are important, the more acidic rocks tend to have a higher resistivity than the more basic rocks. At high temperatures where intrinsic conduction dominates, there is relatively little difference in conductivity that can be explained by compositional changes. However, it has been found that conduction is sensitive to the partial pressure of volatile gases that are in contact with the minerals; several studies have been made of the effect to the partial pressure of oxygen and of water on the conductivity of rocks.[90,115] Some studies have been carried to temperatures high enough for melting to take place but few if any reliable data are available on the electrical properties of rocks in the molten state at present.[98]

A typical behavior of resistivity as a function of frequency and temperature is shown for a single sample of granodiorite in Figures 3 and 4. Similar data for dunite, published by Dvorak[27] are shown in Figures 5, 6, and 7. In these figures, conductivity has been measured as a function of temperature and pressure, at a very low frequency, taken to be DC. Plots of resistivity as a function of frequency and temperature for four samples of serpentinite from Puerto Rico are shown in Figure 8.[119]

At very low frequencies ($\ll$1 Hz at normal temperatures) resistivity measured on rock samples tends to approach a constant value which depends on temperature but is independent of further reduction in frequency. A summary of resistivity at very low frequencies (as a function of temperature) is shown for igneous rocks in Figure 9.[57-59] Table 4 is a listing of the parameters defined in Equation 3 for the various major rock types.

The total conduction in a dry rock may be written as the sum of two terms, an AC conductivity and a DC conductivity:

$$\sigma_{total} = \sigma_{AC} + \sigma_{DC} \tag{4}$$

The DC conductivity term is the same as the intrinsic conductivity term in Equation 3, while the AC term is the impurity conduction term. Little is known about the AC component of conduction in dry rocks, but some data suggest that it varies with frequency according to the law:

$$\sigma_{AC} = a(T)\omega^{n(T)} \tag{5}$$

where ω is frequency and a(T) and n(T) are functions of temperature determined empirically. The multiplying parameter, a(T), determines the amount of AC conduction, while the exponent, n(T), determines the rate with which AC conduction changes with temperature and frequency (Figure 10). The amplitude parameter, a(T), exhibits the usual characteristics of a thermally activated conductivity. When it is plotted as a function of temperature, as in Figure 10, it can be described over a limited temperature range by the equation:

$$A(T) = a_0 e^{-T_c/T} \tag{6}$$

where $T_c = W/k$ represents an excitation energy and a_o is a base level for the function a(T). These curves also consist of two segments, with transitions from low-temperature to high-temperature behavior taking place at about 100 to 150°C. This is apparently the transition from a region where one type of impurity, such as absorbed water, is important to a region where a second type of impurity takes over. Values for T_c and a_o are listed in Table 5 for granite, metabasalt, and metarhyolite, the only three rocks for which data are available.

The value for the exponent n(T) in Equation 5 also shows a systematic variation with temperature and rock type. The value is closer to unity at low temperature and for granitic rocks than at high temperatures or for basaltic rocks. The behavior of the exponent is shown graphically in Figure 11.

Dielectric Constants of Dry Rocks

Dielectric polarization is a subject treated in a number of books, including Böttcher,[11] Frolich,[34] and Von Hippel.[113,114] Mechanisms involved in dielectric polarization include:

1. Electron polarization: an applied electric field will tend to shift the center of mass of the electron swarm about an atomic nucleus with respect to the center of the nucleus. The shift of the electrons comprises a displacement current at or above optical frequencies for most minerals and rocks. Below optical frequencies, the contribution of electron polarization to the dielectric constant can be expressed in terms of the optical refractive index.

$$\epsilon/\epsilon_o = n^2 \tag{7}$$

   where ε is the dielectric constant in F/m, ε_o is 8.85×10^{-12} F/m, and n is the index of refraction at optical frequencies. Values for dielectric constants due to electron polarization are listed in Table 6.
2. Atomic polarization: if the period of oscillation of an applied electric field is sufficiently long, the charged nucleus in an atom also has time to polarize. Normally, in earth materials, atomic polarization becomes significant at frequencies above normal radio frequencies, but below optical frequencies. Values for dielectric constants of some rock-forming minerals and common rocks are listed in Table 7, and it may be noted that they are higher than the values measured at optical frequencies by ratios of two or four. Because the number of charges per unit volume available for displacement is roughly proportional to density, dielectric constant correlates to some degree with mass density (Figure 12).[82,84]
3. Molecular polarization: some molecules exhibit a permanent electric dipole configuration, resulting from the center of positive charge distribution not being coincident with the center of negative charge distribution. When an external electric field is applied to such a material, the polar molecules will attempt to rotate to align their axes with the externally applied field. This motion constitutes a displacement current. Retarding forces in the liquid state are much weaker than in the solid state, and molecular polarization takes place at a much higher frequency in a liquid than in a solid. The relatively high dielectric constant of water at radio and lower frequencies is caused by molecular polarization. For solid water (ice), molecular polarization can take place only at very low frequencies as illustrated in Figure 13, the dielectric constant of water and ice as a function of frequency. Water is the only common rock-forming substance that has a marked molecular polarizability.

Determinations of dielectric constant as a function of frequency and temperature often yield relationships in which the dielectric constant is very large at low frequencies. Generally, the high dielectric constants observed at low frequencies are also associated with a substantial amount of conduction, and it is likely that the high dielectric constants result from interfacial polarization when conduction is interrupted at grain boundaries or between domains in a mineral. Examples of this behavior of dielectric constant are shown in Figures 14 to 18. Figure 14 shows the behavior for dielectric constant measured on an olivine as a function of temperature and frequency.[49] Figure 15 shows measurements made on granite.[41] Figure 16 shows dielectric constants for serpentinite, a mineral and rock type in which a significant amount of water is present in the crystal structure.[119] Figure 17 shows the variation of dielectric constant in a group of sandstone samples containing minor amounts of water.[51,52] Figure 18 shows dielectric constant values measured on various types of basaltic material, including one sample from the lunar surface. The sample from the lunar surface shows none of the dispersion in dielectric constant observed for all the Earth materials. The lunar sample may be the only sample which is completely devoid of moisture.

Resistivity of Water-Bearing Rocks

Water carrying various amounts of salt in solution is the rock component which is most important in determining the conductivity of rocks within the first few kilometers of the earth's surface. The resistivity of a water-bearing rock will depend on the amount of water present, the salinity of this water, the temperature, and the way in which the water is distributed through the rock.

When salt is dissolved in water, the constituent ions in the salt separate and are free to move in the solution. When an electric field is applied, cations in the solution will be accelerated towards the negative pole of the field and anions to the positive pole. This acceleration is opposed by viscous drag which limits the maximum velocity to which the ions can be accelerated. The terminal velocity obtained with a unit electric field (1 V/m) is defined as the mobility of an ion.

Mobility is a function of both temperature and concentration of salt in solution. Increasing the temperature decreases the viscosity of water, permitting higher terminal velocities for the same voltage gradient. If a solution contains a high concentration of ions, the motion of one ion may be inhibited by the motion of other ions close to it, reducing the terminal velocity. For these reasons, the temperature and concentration for which a value of ion mobility applies must be specified. The mobilities of common ions at 25°C in dilute solution are given in Table 8. Many groundwater analyses have been published, and Chebotarev[16] has given a summary of groundwater salinities (Table 9). The resistivity of an electrolyte consisting of NaCl in solution is given in Table 10 as a function of concentration, for a temperature of 18°C. Electrolyte conductivity increases very nearly linearly with concentration except in highly concentrated solutions. The graphical relationship between electrolyte resistivity and concentration is shown in Figure 19 for several temperatures ranging from 20 to 100°C. A comparison of the resistivity-concentration curves for solutions of various salts, all at 20°C, is shown in Figure 20. Quist and Marshall[91,92] have published tables giving the resistivity of saline solutions at high temperatures and pressures. Data on NaCl solutions are given in Table 11, while data on $KHSO_4$ solutions are given in Table 12.

A great deal of research has been done in correlating resistivity of a rock with the water content. Such studies have led to the widespread use of an empirical formula relating resistivity and porosity, known as Archie's law:[6]

$$\rho_t = a\rho_w\phi^{-m} \tag{8}$$

where ϱ_t is the bulk resistivity of the rock, ϱ_w is the resistivity of the water contained in the pore structure, ϕ is the porosity expressed as a fraction for unit volume of rock, and a and m are parameters whose values are assigned arbitrarily to make the equation fit a particular group of measurements.

Figure 21, a collection of 793 measurements reported by Carothers,[15] shows the scatter typical of such correlation plots of resistivity and water content. The scatter is not the consequence of measurement error, but indicates a variability in texture between small rock samples. As more and more measurements are made, the average resistivity measured on samples with a given porosity approaches closer and closer to a relationship such as Archie's law. If instead of plotting each individual value of resistivity against each individual value of porosity, as was done in Figure 21, the measurements are separated into groups on the basis of porosity, and only the average resistivity of each group is plotted, a correlation such as that shown in Figure 22 is obtained (sandstone samples, reported by Keller).[51,52] At porosities above 10%, the linear relationship predicted by Archie's law holds very closely for the average values.

Studies of the correlation between water content and resistivity have also been carried out in some nonsedimentary rocks. Figure 22 shows the results of measurements made on samples of volcanic tuff from the Oak Springs Formation in southern Nevada.[53] Figure 23 shows measurements made on basalt taken from a drill hole at Kilauea Volcano in Hawaii as reported by Keller et al.,[63] Zablocki et al.,[120] and Keller et al.[65] Recently, Kirkpatrick[69] has published such a correlation for samples of basalt recovered from deep sea drill holes (Figure 24).

Resistivity-porosity correlations have also been carried out for samples of unconsolidated marine deposits recovered from the seafloor.[43,48,66,108] The data reported by Kermabom are shown in Figure 25 with an interpretation of the data in terms of Archie's law being shown in Figure 26.

The Archie's law relationship is a particularly simple description of the correlation between resistivity and water content, but has not been substantiated by any physical models that would lead to its explanation. There has been extensive research attempting to relate water content and resistivity on the basis of mathematical models of pore structures.[32,73,77,99] No simple geometrical description of the pore structure has been found to synthesize the Archie's law of behavior although many have been proposed. Kobranova[72] has listed expressions for the relationship between resistivity and water content for geometrically deterministic pore structures (Table 13). Models with a nondeterministic or statistical description of pore geometries show more promise, as has been indicated by the results of Kwon[73] and Madden.[77]

In Archie's law, the parameters *a* and *m* often correlate with the lithology and texture of a rock. The value of the parameter *a* varies from slightly less than one in clastic detrital rocks to values as large as 3.5 in vesicular to tuffaceous rock. The parameter *m* varies from values as small as 1.3 in packed sand to values as large as 2.3 in tightly cemented clastic rocks. Expressions in the form of Archie's law for various types of rocks are listed in Table 14.

Induced Polarization

The measurement of induced electric polarization has been an accepted exploration tool in mining geophysics for several decades. The method has been described in text books on geophysical exploration, including an early book edited by Wait,[116] one by Keller and Frischknecht,[61] and two recent monographs by Sumner[107] and by Bertin and Loeb.[9]

The source of induced polarization lies in chemical reactions which must take place when electrical current flows between an electrolytic solution in the pore spaces of a rock and an electronically semiconducting sulfide mineral blocking these pores. To transfer charge from the electrolyte to the semiconductor it is necessary that oxidation or reduction take place at the interface. As with all other chemical reactions, oxidation and reduction are rate-controlled by the concentrations of reactions that buildup as current flows through the interface. Prolonged current flow will lead to the development of concentrations of ions near the interface. This accelerates the chemical reaction which permits charge to be transferred across the interface. It also represents a storage of charge at the interface, or "polarization".

A variety of quantitative measures of induced polarization has come into use in the practice of geophysics. When induced polarization is detected by observing the transient decay of voltage after the earth has been energized with a current step, the effect is measured in terms of "polarizability" or by "chargeability". Polarizability is defined as

$$P_t^T = \frac{\Delta V_{IP}}{\Delta V} \tag{9}$$

where the superscript T indicates the length of excitation of the earth by a current step, the subscript t indicates the length of time after excitation has ceased that the transient voltage, ΔV_{IP}, is measured, and ΔV is the voltage across the receiving electrodes during excitation. Polarizability is measured in % or mv/v.

Chargeability is defined as the integral of polarizability over a time interval from t_1 to t_2 following the cessation of current flow:

$$M_{t_1, t_2}^T = \frac{1}{\Delta V} \int_{t_1}^{t_2} \Delta V_{IP} dt. \tag{10}$$

This quantity is normally expressed in milliseconds.

Another way to measure the effects of induced polarization is by determining the voltage drop across the sample using two like currents at different frequencies. Because of the effect of induced polarization, the voltage drop measured at the higher frequency is not as great as that measured at the lower frequency. This behavior manifests itself as an apparent change in the electrical resistivity of the sample with frequency. This dispersion of resistivity with frequency is measured as a percentage per decade charge in the frequency (PFE).

Evjen[31] reported that induced polarization effects existed even in rocks with no electronic mineral content. In making measurements in normal soil, Evjen found changes in apparent resistivity with frequency that indicated that the relative dielectric constant of the earth should be of the order of 10^8 to 10^{11} at frequencies below 1 Hz. The phenomenon of nonelectronic induced polarization effect, or "background" effect as it has become to be known, was studied further by Keller and Licastro.[62,64] Their measurements indicated that in rocks with a relatively high water content, the electrical conductivity varied with frequency as did the dielectric constants. Some of the values for relative dielectric constant reported were in the range from 10^8 to 10^9 at frequencies below 100 Hz.

An interesting feature of the observed low-frequency low polarization in rocks is that the loss tangent, defined as $\sigma/\omega\varepsilon$ where ω is frequency, remains very nearly constant for frequencies between 0.001 and 1000 Hz. In view of the constancy of the loss tangent, it is a convenient measure of the electrical characteristics of a rock at low frequencies. Typical values for low frequency loss tangent are listed in Table 15.

The relative dielectric constants associated with these loss tangents range from 10^4 to 10^7, and it is important to note that despite the fact that the dielectric constant is so large, conduction in these rocks is essentially ohmic. As the frequency is raised above 1000 Hz, the ratio $\omega\varepsilon/\sigma$ begins to increase, reflecting the fact that the more conventional polarization mechanisms become dominant. At higher frequencies, particularly at low radio frequencies and above, the dielectric constant becomes more nearly constant. Information on the parametric behavior of induced polarization is available in a number of articles published in recent years.[1,5,18,33,51,52,86,97,101,111] Several reports[4,97] indicate that in mineral bearing rocks, induced polarization depends on the size of the grains of the polarizing mineral. The induced polarization effect may exhibit a maximum for intermediate grain sizes as indicated by the data in Figure 27 or it may increase monotonically with decreasing grain size, as shown in Figure 28.

A number of investigators[4,50,70,71] have found that the induced polarization effect is nonlinear; that is, its intensity varies with the strength of the current used to excite it. Examples of the nonlinear relationship between applied voltage and current flow in sulfide samples reported by Collet and Katsube[18] are shown in Figure 29.

A tendency for the rate of decay of transient current in a rock as the induced polarization effect increases has long been noted. Pelton et al.[86] have reported on extensive measurements correlating the size of induced polarization and the time constant observed in a variety of rocks; their results are shown in Figures 30 to 34.

"EXISTENTIALIST" SUMMARIES OF ELECTRICAL PROPERTIES

Sedimentary Rock Sequences

All of the information concerning electrical properties contained in the preceding paragraphs is parametric in nature; it represents measurements made in a laboratory environment after the rock had been removed from its natural environment. Considering the strong dependence of many of the electrical properties on moisture content, temperature, and vapor pressure, it is difficult to argue that an electrical property measurement made on a sample in the laboratory is the value for that property for the sample in its original state. To complete a tabulation of electrical properties, it is therefore necessary to review data on electrical properties of rocks as they exist in place in the earth. Two sets of data reflecting the existentialist properties of rocks are available; some of these data are from electrical surveys run in boreholes drilled in mining and oil recovery operations, while other summaries are available from the interpretation of surface-based electrical surveying methods.

The average electrical properties of a sequence of layered rock can be described with a set of five parameters:

1. Average resistivity along the bedding planes, ϱ_l
2. Average resistivity transverse to the bedding planes ϱ_{tr}
3. Integrated conductance along the bedding planes, S
4. Integrated resistance transverse to the bedding planes, T
5. Coefficient of anisotropy, λ

These parameters are defined in terms of a column of rock cut perpendicular to the layering, the column having dimensions of 1 m on a side and a height equal to the thickness of the unit under consideration. This column consists of M beds, each with its own characteristic resistivity ϱ_i and thickness h_i. The total thickness is H. The total resistance to current flow vertically in a column is:

$$T = \sum_{i=1}^{M} \rho_i h_i \tag{11}$$

The average resistivity to current flowing across the bedding is:

$$\rho_{tr} = \frac{T}{H} \tag{12}$$

The conductance for current flowing horizontally through the column of the rock is

$$S = \sum_{i=1}^{M} \sigma_i h_i \tag{13}$$

The average conductivity for horizontal current flow is

$$\sigma_\ell = \frac{S}{H} \tag{14}$$

The longitudinal resistivity $1/\varrho\ell$ is always less than the transverse resistivity, ϱ_{tr}, unless the column is uniform, in which case, the two resistivity values are equal. The dependence of resistivity on direction of current flow constitutes anisotropy. A layered medium appears to be anisotropic, even though each layer in the medium is individually isotropic. The coefficient of anisotropy is defined as

$$\lambda = (ST/H^2)^{1/2} \tag{15}$$

Sedimentary Rocks in the Mississippi Embayment

Using the definitions of electrical parameters given in the equations above, Keller[56] has given tabulations of properties derived from electric logs in various sedimentary basins in the U.S. Figure 35 shows locations of wells in the Mississippi Embayment Area of the southeastern U.S. which comprised one such compilation. The section penetrated by the wells from which logs were taken consists chiefly of sand, clay, marl, limestone, and chalk, all poorly consolidated. Calcareous sediments tend to be more abundant in the lower part of the column than in the upper part, and increase in importance towards the Gulf Coast. All logs exhibited a similar overall character; surficial resistivities are high, decreasing gradually to a minimum resistivity at a depth of several thousand feet, and then increasing again slowly at greater depths. The logs were subdivided into four geoelectric units designated with the letters U, M, or L, indicating the upper part of the section, the middle part, and the lower part. In some cases the section is further subdivided and the letter designator was subscripted to indicate the sequence from top down. Thus, if the section above the interval with minimum resistivity were broken into two parts, the upper part was designated U_1 and the lower part U_2. Lithologically, the upper sequence, U, consists of formations of Miocene in later age, while the mid part of the section consists generally of formations of Eocene and Oligocene age, including the Wilcox, Claiborne, Jackson, and Vicksburg formations. The lower part of the sequence, L, includes beds of Cretaceous age, though none of the wells penetrated the entire Cretaceous section.

All resistivities compiled in Table 16 were taken from 6FF40 induction logs (see Pirson, 1963, for a discussion of induction log characteristics).

Paleozoic to Cenozoic Section of the Colorado Plateau

Jackson[46,47] tabulated average resistivities derived from borehole logs from wells in the Colorado Plateau area of the states of Colorado, Utah, New Mexico, and Arizona (Figure 36). The contours on this map are the elevation of the Precambrian basement. Jackson[46] divided the sedimentary sequence into four geoelectric units, although all four were not necessarily present in all the wells considered. The units considered were

1. Sedimentary rocks, predominantly shale and sandstone, of upper Cretaceous and lower Cenozoic age
2. Sedimentary rocks, primarily sandstones, of Permian to lower Cretaceous age
3. In places, the Paradox formation of Pennsylvanian age is sufficiently thick to be considered a separate electrical unit
4. Sedimentary rocks, mainly limestones and dolomites, of Cambrian to Permian age

Designators for these four units are M, L_1, L_2, and L_3, defined analogously to those used for the Gulf Coast area. A summary of formation names and lithology is given in Table 17. Compiled resistivities for wells from the Colorado Plateau's province are listed in Table 18.

Sedimentary Sequence in Pennsylvania

Anderson[3] has compiled average resistivities from borehole geophysical logs run in eight deep wells in Pennsylvania at the locations indicated on the map in Figure 37. The sections log were mainly of Paleozoic age, consisting of Cambrian to Pennsylvanian shale and limestone bed. The compiled electrical properties for these eight wells are listed in Table 19.

Coal

In recent years, the study of the electrical properties of coal has become important because radar probing is being used to locate inhomogeneities in coal seams that are being opened with automatic mining machinery. The resistivity of coal depends on both water content, usually associated with the presence of clay minerals, and free carbon, which is relatively more abundant in higher rank coals. Parkhomenko[84] and Grechukhin[35] have presented data on the correlation between resistivity of coal, and grade, expressed in terms of ash content Figure 38). Cook[19] has given the results of measurements of both resistivity and dielectric constant of coals at frequencies between 1 and 100 MHz (Figure 40a, 40b). Cook[20] has combined these measurements to provide a "radar probing distance", defined as the distance a radar signal can travel through coal before being attenuated by 100 to 150 dB (Figure 40).

Permafrost

The electrical properties of frozen rocks and soil are of importance because approximately one seventh of the earth's surface lies within areas of permafrost occurrence. Extensive studies of the electrical properties of permafrost have been reported by Yakupov[117] and by Borovinskiy.[10] Freezing of water in the pore structure of a rock increases the resistivity of that rock by a factor which may be only slightly greater than unity, to factors of 1000 or more (Figure 41). The effect of freezing is least marked in fine-grained rocks, most marked in coarse-grained rocks. A histogram of observed resistivities in frozen rock is shown in Figure 42.

Resistivity Profile in the Deep Interior of the Earth

Over the years, many investigators have attempted to determine the variation of resistivity with depth within the earth to depths of tens and hundreds of kilometers.

These investigations have been carried out using a variety of surface-based geophysical methods, including geomagnetic deep sounding,[7,8,21,29,74,78,88,102,109,118] magnetotelluric sounding,[39] and direct current resistivity surveys.[60,104,112] All of these studies have shown a consistent pattern for the variation of resistivity with depth; surficial rocks are generally moderately to highly conductive, underlying crustal rocks commonly have very high resistivities and at greater depths, rocks again become conductive. Presumably, rocks at depths of many tens of kilometers become conductive because of thermally excited conduction mechanisms, rather than because of the necessary presence of water. The depth to conductive rocks has been found to vary greatly from one geologic problem to another. In some areas, such rocks are found at depths as shallow as 10 or 20 km, while in other areas, the depth is measured in hundreds of kilometers. It has been speculated that the difference depends on heat flow from the mantle to the crust. In the areas of high heat flow, as in areas of current volcanic activity, the conductive rocks appear to be found at very shallow depths. In areas of low heat flow from the mantle, particularly where the surface has been stable for long periods in the geologic sense, the depth to the conductive zone may be 500 km or more.

Several summaries of the resistivity profile in the Earth are shown in Figures 43 to 46, and in the Moon, in Figure 47.

Table 1
RESISTIVITIES OF METALS AND METALLIC MINERALS (ZERO FREQUENCY)

Refined metals at 0°C	ϱ (ohm-m)
Lithium	8.5×10^{-8}
Beryllium	5.5×10^{-8}
Sodium	4.3×10^{-8}
Magnesium	4.0×10^{-8}
Aluminum	2.5×10^{-8}
Potassium	6.3×10^{-8}
Calcium	4.2×10^{-8}
Titanium	83×10^{-8}
Chromium	15.3×10^{-8}
Iron	9.0×10^{-8}
Cobalt	6.3×10^{-8}
Nickel	6.3×10^{-8}
Copper	1.6×10^{-8}
Zinc	5.5×10^{-8}
Gallium	41×10^{-8}
Arsenic	35×10^{-8}
Rubidium	11.6×10^{-8}
Strontium	33×10^{-8}
Zirconium	42×10^{-8}
Molybdenum	4.3×10^{-8}
Ruthenium	11.7×10^{-8}
Rhenium	4.5×10^{-8}
Palladium	10.0×10^{-8}
Silver	1.5×10^{-8}
Cadmium	6.7×10^{-8}
Indium	8.5×10^{-8}
Tin	10.0×10^{-8}
Antimony	36×10^{-8}
Cesium	18×10^{-8}
Barium	59×10^{-8}
Lanthanum	59×10^{-8}
Cerium	71×10^{-8}
Praesodymium	62×10^{-8}
Hafnium	29×10^{-8}
Tantalum	14×10^{-8}
Tungsten	5.0×10^{-8}
Osmium	9.1×10^{-8}
Iridium	5.0×10^{-8}
Platinum	9.8×10^{-8}
Gold	2.0×10^{-8}
Tellurium	14×10^{-8}
Lead	19×10^{-8}
Bismuth	100×10^{-8}
Metallic minerals	
Native copper	1.2 to 30×10^{-8} ohm-m
Graphite (carbon)	36 to 100×10^{-8} (current flow parallel to cleavage)
	28 to 9900×10^{-6} (current flow across cleavage)
Ulmanite, NiSbS	9.0 to 120×10^{-8}
Breithauptite, NiSb	3.0 to 50×10^{-8}

Table 2
RESISTIVITIES OF SEMICONDUCTING MINERALS (ZERO FREQUENCY)

Native elements	ϱ (ohm-m)
Diamond (C)	2.7
Sulfides	
Argentite, Ag_2S	1.5 to 2.0×10^{-3}
Bismuthinite, Bi_2S_3	3 to 570
Bornite, $Fe_2S_3 \cdot nCu_2S$	1.6 to 6000×10^{-6}
Chalcocite, Cu_2S	80 to 100×10^{-6}
Chalcopyrite, $Fe_2S_3 \cdot Cu_2S$	150 to 9000×10^{-6}
Covellite, CuS	0.30 to 83×10^{-6}
Galena, PbS	6.8×10^{-6} to 9.0×10^{-2}
Haverite, MnS_2	10 to 20
Marcasite, FeS_2	1 to 150×10^{-3}
Metacinnabarite, 4HgS	2×10^{-6} to 1×10^{-3}
Millerite, NiS	2 to 4×10^{-7}
Molybdenite, MoS_2	0.12 to 7.5
Pentlandite, $(Fe, Ni)_9S_8$	1 to 11×10^{-6}
Pyrrhotite, Fe_7S_8	2 to 160×10^{-6}
Pyrite, FeS_2	1.2 to 600×10^{-3}
Sphalerite, ZnS	2.7×10^{-3} to 1.2×10^{4}
Antimony-sulfur compounds	
Berthierite, $FeSb_2S_4$	0.0083 to 2.0
Boulangerite, $Pb_5Sb_4S_{11}$	2×10^{3} to 4×10^{4}
Cylindrite, $Pb_3Sn_4Sb_2S_{14}$	2.5 to 60
Franckeite, $Pb_5Sn_3Sb_2S_{14}$	1.2 to 4
Hauchecornite, $Ni_9(Bi, Sb)_2S_8$	1 to 83×10^{-6}
Jamesonite, $Pb_4FeSb_6S_{14}$	0.020 to 0.15
Tetrahedrite, Cu_3SbS_3	0.30 to 30,000
Arsenic-sulfur compounds	
Arsenopyrite, FeAsS	20 to 300×10^{-6}
Cobaltite, CoAsS	6.5 to 130×10^{-3}
Enargite, Cu_3AsS_4	0.2 to 40×10^{-3}
Gersdorffite, NiAsS	1 to 160×10^{-6}
Glaucodote, (Co, Fe)AsS	5 to 100×10^{-6}
Antimonide	
Dyscrasite, Ag_3Sb	0.12 to 1.2×10^{-6}
Arsenides	
Allemonite, $SbAs_3$	70 to 60,000
Lollingite, $FeAs_2$	2 to 270×10^{-6}
Nicollite, NiAs	0.1 to 2×10^{-6}
Skutterudite, $CoAs_3$	1 to 400×10^{-6}
Smaltite, $CoAs_2$	1 to 12×10^{-6}
Tellurides	
Altaite, PbTe	20 to 200×10^{-6}
Calavarite, $AuTe_2$	6 to 12×10^{-6}
Coloradoite, HgTe	4 to 100×10^{-6}
Hessite, Ag_2Te	4 to 100×10^{-6}
Nagyagite, $Pb_6Au(S, Te)_{14}$	20 to 80×10^{-6}
Sylvanite, $AgAuTe_4$	4 to 20×10^{-6}
Oxides	
Braunite, Mn_2O_3	0.16 to 1.0
Cassiterite, SnO_2	4.5×10^{-4} to 10,000
Cuprite, Cu_2O	10 to 50
Hollandite, $(Ba, Na, K)Mn_8O_{16}$	2 to 100×10^{-3}
Ilmenite, $FeTiO_3$	0.001 to 4
Magnetite, Fe_3O_4	52×10^{-6}
Manganite, $MnO \cdot OH$	0.018 to 0.5
Melaconite, CuO	6000
Psilomelane, $KMnO \cdot MnO_2 \cdot nH_2O$	0.04 to 6000
Pyrolusite, MnO_2	0.007 to 30
Rutile, TiO_2	29 to 910
Uraninite, UO	1.5 to 200

Table 3
APPROXIMATE RADII OF ATOMS OR IONS (IN ANGSTROMS)

$O^{=}$	1.40	K^{+}	1.33
Cl^{-}	1.81	Ca^{++}	0.99
H^{-}	2.08	Fe^{++}	0.76
Na^{+}	0.95	Fe^{+++}	0.64
Mg^{++}	0.65	Mn^{++}	0.66
Al^{+++}	0.50	Mn^{++++}	0.54
Si^{++++}	0.41		

Table 4
PARAMETERS DEFINING THE TEMPERATURE DEPENDENCE OF RESISTIVITY IN SOLID ELECTROLYTES

Rock	A_1	A_2	U_1	U_2
Granite	5×10^{-4} mho/cm	10^{5} mho/cm	0.62 eV	2.5 eV
Gabbro	7×10^{-3}	10^{5}	0.70	2.2
Basalt	7×10^{-3}	10^{5}	0.57	2.0
Peridotite	4×10^{-2}	10^{5}	0.81	2.3
Andesite	6×10^{-3}		0.7	1.6

Table 5
PARAMETERS DESCRIBING AC CONDUCTIVITY IN DRY, IONIC ROCKS

Rock type	a_o (mho/m)	T_c (°K)
Metabasalt	3.7×10^{-4} to 1.35×10^{-3}	5140° to 5240°K
Granite	2.95×10^{-5} to 1.5×10^{-4}	6400° to 6960°K
Metarhyolite	0.61	13,500°K

Table 6
DIELECTRIC CONSTANTS OF MINERALS, ROCKS, AND SOILS AT HIGH FREQUENCIES

		Dielectric constant	
Mineral or rock	Source and orientation of sample	Radio frequencies	Optical frequencies
	Sulfide Minerals		
Galena, PbS		17.9	
Sphalerite, ZnS	Titibu, Japan	7.90	5.61 to 6.10
	Nakatatu, Japan	12.1	
	Joplin, Missouri	69.9—7.90	
	Harz, Germany	69.7—7.88	
	Oxide Minerals		
Corundum, Al_2O_3		11.0—13.2	
	Along optic axis	—	3.10
	Across optic axis	—	3.14
Cassiterite, SnO_2		23.4—24.0	
	Along optic axis	—	3.98
	Across optic axis	—	4.36
Diaspore, AlO(OH)	Mituisi, Japan		
	Along a axis	7.70	2.90
	Along b axis	8.38	2.96
	Along c axis	7.27	3.05
Hematite, Fe_2O_3		25.0	
	Along optic axis	—	8.65
	Across optic axis	—	10.33
Rutile, TiO_2		31.0—170	
	Along optic axis	—	6.82
	Across optic axis	—	8.42
Anatase, TiO_2		425	

Table 6 (continued)
DIELECTRIC CONSTANTS OF MINERALS, ROCKS, AND SOILS AT HIGH FREQUENCIES

Mineral or rock	Source and orientation of sample	Dielectric constant: Radio frequencies	Dielectric constant: Optical frequencies
	Halide Minerals		
Halite, NaCl		5.70—6.20	2.39
Fluorite, CaF_2		6.79	2.06
	Akenobe, Japan	6.26	
	Saxony	6.61	
	Okuno, Japan	6.27	
	Switzerland	6.25	
	Durham, England	6.30	
	Freiburg, Saxony	6.60	
Sylvite, KCl		4.39—6.20	2.20
	Carbonate Minerals		
Aragonite, $CaCO_3$	Bohemia		
	Along a axis	6.46	2.34
	Along b axis	9.72	2.82
	Along c axis	7.55	2.84
Calcite, $CaCO_3$		7.80—8.50	
	Across optic axis	—	2.21
	Along optic axis	—	2.75
Dolomite, $CaMg(CO_3)_2$		6.80—8.00	
	Across optic axis	7.53	2.28
	Along optic axis	6.11	2.85
	Phosphate Minerals		
Apatite, $Ca_5(F, Cl)(PO_4)_3$		7.40—10.47	
	Asio, Japan		
	Across optic axis	7.60	2.69
	Along optic axis	10.0	2.71
	Kamioka, Japan		
	Across optic axis	7.43	
	Along optic axis	6.07	
Vivianite, $Fe_3(PO_4)_2 \cdot 8H_2O$		6.07	2.49—2.67
	Sulfate Minerals		
Anglesite, $PbSO_4$		74.0—500	3.52—3.59
Anhydrite, $CaSO_4$		5.70—6.30	
	Along a axis	—	2.48
	Along b axis	—	2.49
	Along c axis	—	2.61
Barite, $BaSO_4$		6.99—12.2	
	Along a axis	7.10	2.68
	Along b axis	8.85—10.0	2.38—2.69
	Along c axis	6.72—7.60	2.40—2.71
	Trintington, England (nine samples)		
	Along a axis	7.85 ± (0.09)	
	Along b axis	12.31 ± (0.05)	
	Along c axis	7.88 ± (0.05)	

Table 6 (continued)
DIELECTRIC CONSTANTS OF MINERALS, ROCKS, AND SOILS AT HIGH FREQUENCIES

Mineral or rock	Source and orientation of sample	Dielectric constant: Radio frequencies	Dielectric constant: Optical frequencies
Celestite, $SrSO_4$	Sicily		
	Along a axis	7.60	2.62
	Along b axis	—	2.64
	Along c axis	8.26	2.66
Gypsum, $CaSO_4 12H_2O$		5.00—11.5	
	Along a axis	11.2	2.31
	Along b axis	12.0	2.32
	Along c axis	5.40	2.34
	Silicate Minerals		
Analcime, $NaAlSi_2O_6 \cdot H_2O$	Tyrol	5.88	2.21
Augite, $Ca(Mg, Fe, Al)(Al, Si)_2O_6$		6.90—10.27	
	Along a axis	8.60	2.92
	Along b axis	6.90	2.95
	Along c axis	7.10	3.01
Beryl, $Be_3Al_2Si_6O_{18}$		5.48—7.80	
	Across optic axis	6.59	2.53
	Along optic axis	6.16	2.56
	Urals		
	Along optic axis	6.18	
	Across optic axis	5.67	
Biotite, $K(Mg, Fe)_3AlSi_3O_{10}(OH)_2$		6.19—9.30	
	Along a axis	—	2.50
	Along b axis	—	2.68
	Along c axis	—	2.68
Epidote (average of 14 samples)	Tyrol		
$Ca_2(Al, Fe)_3(SiO_4)_3OH$	Along a axis	7.60 ± .13	3.01
	Along b axis	9.99 ± .14	3.11
	Along c axis	15.36 ± .04	3.17
Leucite, $KAlSi_2O_6$	Italy	7.13	
	Across optic axis	—	2.27
	Along optic axis	—	2.27
Muscovite, $KAl_3Si_3O_{10}(OH)_2$		6.19—8.00	
	Along a axis	—	2.46
	Along b axis	—	2.55
	Along c axis	—	2.60
Opal, $SiO_2 \cdot nH_2O$	Bodai, Japan	7.15	2.10
	Takarasaka, Japan	7.43	
Opal, var. Hyalite	Tateyama, Japan	4.21	
Orthoclase feldspar var. adularia, $KAlSi_3O_8$	Along a axis	5.55	2.30
	Along b axis	5.80	2.33
	Along c axis	4.50	2.34
Phlogopite, $KMg_2Al_2Si_3O_{10}(OH)_2$		5.90—6.50	
	Along a axis	—	2.44
	Along b axis	—	2.58
	Along c axis	—	2.58
Plagioclase feldspar			
var. albite $Ab_{97}An_3$	Basi-bergwerk, Japan	5.58	2.33
var. albite $Ab_{99}An_1$	Switzerland	5.45	2.34
var. albite $Ab_{95}An_5$	Tyrol	5.57	2.36

Table 6 (continued)
DIELECTRIC CONSTANTS OF MINERALS, ROCKS, AND SOILS AT HIGH FREQUENCIES

Mineral or rock	Source and orientation of sample	Dielectric constant	
		Radio frequencies	Optical frequencies
var. albite $Ab_{96}An_4$	Switzerland	5.52	2.36
var. albite $Ab_{98}An_2$	Italy	5.55	2.34
var. albite $Ab_{94}An_6$	Norway	5.63	2.36
var. albite $Ab_{99}An_1$	Urals	5.55	2.33
var. albite $Ab_{98}An_2$	Bavaria	5.39	2.34
var. oligoclase $Ab_{76}An_{24}$	North Carolina	6.03	2.39
var. oligoclase $Ab_{77}An_{23}$	Norway	6.06	2.39
var. andesine $Ab_{65}An_{35}$	Norway	6.20	2.41
var. andesine $Ab_{61}An_{39}$	Korea	6.47	2.41
var. andesine $Ab_{52}An_{48}$	Nakasiota, Japan	6.30	2.43
var. labradorite $Ab_{43}An_{57}$	Labradore, Italy	6.61	2.45
var. labradorite $Ab_{45}An_{55}$	North America	6.51	2.45
var. anorthite Ab_4An_{96}	Otaru, Japan	7.24	2.51
var. anorthite Ab_2An_{98}	Tsushima, Japan	7.14	2.51
var. anorthite Ab_7An_{93}	Hokkaido, Japan	7.05	2.49
var. anorthite Ab_4An_{96}	Miyakeshima, Japan	7.15	2.49
Quartz SiO_2		4.19—5.00	
	Across optic axis	4.96	2.36
	Along optic axis	5.05	2.41
	Naegi, Japan		
	Normal to optic axis	4.11	
	Parallel to optic axis	4.27	
	Kinbuzan, Japan		
	Normal to optic axis	4.13	
	Parallel to optic axis	4.27	
Quartz, bipyramidal	Cumberland, Maryland		
	Normal to optic axis	4.09	
	Parallel to optic axis	4.27	
Sericite		19.55—25.35	
Sillimanite		4.80	
Al_2SiO_5	Along a axis	—	2.78
	Along b axis	—	2.79
	Along c axis	—	2.84
Topaz, $Al_2SiO_4(F, OH)_2$		6.30—7.60	
	Along a axis	6.65	2.66
	Along b axis	6.70	2.66
	Along c axis	6.30	2.68
	Naegi, Japan		
	Along a axis	6.31	
	Along b axis	6.43	
	Along c axis	6.27	
Tourmaline		5.60—7.10	
	Ceylon		
	Normal to optic axis	6.75	2.76
	Parallel to optic axis	5.52	2.89
	Cumberland, Maryland		
	Normal to optic axis	6.76	
	Parallel to optic axis	5.45	
Zircon, $ZrSiO_4$		8.59—12.0	~3.84

Table 7
DIELECTRIC CONSTANTS OF MINERALS (ELECTRON AND ION POLARIZATION ONLY)

Mineral	Dielectric constant (pF/m)[a]
Galena, Pbs	158
Sphalerite, ZnS	69.7
Corundum, Al_2O_3	97.2 to 117
Cassiterite, SnO_2	207 to 212
Hematite, Fe_2O_3	221
Rutile, TiO_2	274 to 1500
Water, H_2O	721
Halite, NaCl	50.4 to 54.8
Fluorite, CaF_2	55.4 to 60.0
Sylvite, KCl	38.8 to 54.8
Aragonite, $CaCO_3$	57.1 to 86.0
Calcite, $CaCO_3$	69.0 to 75.2
Dolomite, $CaMg(CO_3)_2$	60.1 to 70.7
Apatite, $Ca_5(F, Cl)(PO_4)_3$	65.5 to 92.8
Anglesite, $PbSO_4$	644 to 4400
Anhydrite, $CaSO_4$	50.5 to 55.7
Barite, $BaSO_4$	69.5 to 109
Celestite, $SrSO_4$	67.2
Gypsum, $CaSO_4 \cdot 2H_2O$	47.7 to 106
Analcime, $NaAlSi_2O_6 \cdot H_2O$	52.0
Augite, $Ca(Mg, Fe, Al)(Al, Si)_2O_6$	61.0 to 76.0
Beryl, $Be_3Al_2Si_6O_{18}$	50.2 to 58.3
Biotite, $K(Mg, Fe)_3AlSi_3O_{10}(OH)_2$	54.8 to 82.3
Epidote, $Ca(Al, Fe)_3(SiO_4)_3OH$	67.3 to 136
Leucite, $KAlSi_2O_6$	63.0
Muscovite, $KAl_3Si_3O_{10}(OH)_2$	54.8 to 70.7
Orthoclase, $KAlSi_3O_8$	39.8 to 51.3
Plagioclase	
var. albite	48.2 to 49.1
var. oligoclase	53.3 to 53.6
var. andesine	54.8 to 57.2
var. labradorite	57.6 to 58.5
var. anorthite	62.3 to 64.0
Quartz, SiO_2	36.4 to 37.8
Sericite	173 to 224
Topaz, $Al_2SiO_4(F, OH)_2$	55.7 to 59.2
Zircon, $ZrSiO_4$	76.0 to 106

[a] pF, picofarad or 10^{-12} farad.

Table 8
ION MOBILITIES MEASURED[a] IN LOW CONCENTRATIONS AT 25°C

Ion	Mobility
H^+	36.2×10^{-8}
OH^-	20.5×10^{-8}
$SO_4^{=}$	8.3×10^{-8}
Na^+	5.2×10^{-8}
Cl^-	7.9×10^{-8}
K^+	7.6×10^{-8}
NO_3^-	7.4×10^{-8}
Li^+	4.0×10^{-8}
HCO_3^-	4.6×10^{-8}

[a] Meters per second/volts per meter.

Table 9
RESISTIVITIES OF NATURAL WATERS

Average values on a regional basis

Source of water samples	Number of samples	Resistivity[a] at 20°C	
		Median	Range
Igneous rocks, Europe	314	7.6	3.0—40
Igneous rocks, South Africa	175	11.0	0.50—80
Metamorphic rocks, South Africa	88	7.6	0.86—80
Metamorphic rocks, Precambrian of Australia	31	3.6	1.5—8.6
Recent and Pleistocene continental sediments, Europe	610	3.9	1.0—27
Recent and Pleistocene sediments, Australia	323	3.2	0.38—80
Tertiary sediments, Europe	993	1.40	0.70—3.5
Tertiary (Miocene and Oligocene) sedimentary rocks, Australia	240	3.2	1.35—10
Mesozoic sedimentary rocks, Europe	105	2.5	0.31—47
Paleozoic sedimentary rocks, Europe	161	0.93	0.29—7.1
Chloride waters from oil fields	967	0.16	0.049—.95
Sulfate waters from oil fields	256	1.20	0.43—5.0
Bicarbonate waters from oil fields	630	0.98	0.24—10

[a] Ohm-meters.

Table 10
RESISTIVITY IN OHM-METERS OF SODIUM CHLORIDE ELECTROLYTES

Temperature (°C)	NaCl (g/l)					
	58.45	29.23	5.845	2.933	0.5845	0.2923
0	0.211	0.386	1.73	3.36	15.82	31.2
2	0.200	0.368	1.65	3.19	15.1	29.6
4	0.190	0.352	1.57	3.02	14.3	28.1
6	0.182	0.336	1.49	2.86	13.7	26.7
8	0.174	0.320	1.42	2.73	12.9	25.0
10	0.165	0.304	1.35	2.57	12.3	23.9
12	0.157	0.288	1.28	2.43	11.7	22.6
14	0.149	0.274	1.21	2.31	11.1	21.4
16	0.142	0.260	1.15	2.19	10.5	20.3
18	0.135	0.248	1.09	2.09	9.8	19.3
20	0.129	0.238	1.04	2.00	9.5	18.4
22	0.123	0.228	1.00	1.92	9.0	17.6
24	0.117	0.219	0.96	1.34	8.6	16.8
26		0.210	0.93	6	8.2	16.2
28		0.200	0.87	18	7.9	15.6
30		0.191	0.84	1.61	7.5	14.9
32		0.183	0.80	1.55	7.2	14.3
34		0.176	0.77	1.49	6.9	13.7

Table 11
EQUIVALENT CONDUCTANCES ($cm^2\ ohm^{-1}\ equiv^{-1}$) OF 0.1000 *m* NaCl SOLUTIONS AT INTEGRAL TEMPERATURES AND DENSITIES (g/cm^3)

Temp. °C	0.35	0.40	0.45	0.50	0.55	0.60	0.65	0.70	0.75	0.80	0.85	0.90	0.95	1.00
100													(300)	290
150													405	380
200											(535)	520	490	450
250											580	570	535	500
300								(695)	675	635	615	600	575	535
350								680	675	645	635	620	590	
400	260	350	410	490	540	610	650	670	670	655	645	630	600	
450	190	270	360	450	520	590	630	655	660	655	645			
500	140	220	320	410	490	565	610	640	650	655	640			
550	105	185	290	380	470	540	585	625	640	655				
600	85	160	260	360	450	520	565	605	625	645				
650	70	140	235	330	420	495	545	585	610					
700	60	125	215	305	400	470	525	570	600					
750	50	110	195	285	380	450	510							
800	50	100	180	265	350	430	490							

Note: Equivalent conductances is defined as $\Lambda = 1000\ \sigma/c$ where σ is the conductivity and c is the number of equivalents of electrolyte per 100 cm^3 of solution.

Table 12
EQUIVALENT CONDUCTANCES OF $KHSO_4$ SOLUTIONS

The Molar Conductances ($cm^2\ ohm^{-1}\ mole^{-1}$) of 0.000817 *m* $KHSO_4$ Solutions at Integral Temperatures and Densities (g/cm^3)

Temp. °C	0.40	0.45	0.50	0.55	0.60	0.65	0.70	0.75	0.80	0.85	0.90	0.95
0												
25												
100												(900)
150												
200										(905)	945	1000
250										870	900	980
300							(900)	880	860	850	870	940
350							915	880	840	820	840	
400	1030	1080	1070	1030	1000	950	910	880	830	810	800	
450	990	1060	1060	1030	1000	950	905	870	825	800		
500	950	1040	1050	1020	995	950	905	870	820	795		
550	920	1010	1040	1015	990	945	900	860	820			
600	880	990	1030	1010	990	945	900	860	810			
650	840	970	1020	1010	985	940	895	860				
700	800	950	1010	1005	980	940	890	850				

The Molar Conductances ($cm^2\ ohm^{-1}\ mole^{-1}$) of 0.00240 *m* $KHSO_4$ Solutions at Integral Temperatures and Densities (g/cm^3)

Table 12 (continued)
EQUIVALENT CONDUCTANCES OF $KHSO_4$ SOLUTIONS

Temp. °C	0.40	0.45	0.50	0.55	0.60	0.65	0.70	0.75	0.80	0.85	0.90	0.95
0												
25												
100												(750)
150												
200										(765)	760	810
250										775	765	810
300							(870)	840	805	780	765	795
350							880	840	805	780	765	775
400	830	905	950	965	950	920	875	840	805	775	760	750
450	790	880	935	955	940	910	870	835	805	770	750	
500	750	855	920	940	930	905	865	830	795	760	740	
550	710	835	905	930	920	895	860	820	790			
600	680	810	890	920	910	885	850	815	780			
650	640	785	875	910	905	875	845	810				
700	600	765	860	895	895	870	835	805				

The Molar Conductances ($cm^2\ ohm^{-1}\ mole^{-1}$) of 0.00505 m $KHSO_4$ Solutions at Integral Temperatures and Densities (g/cm^3)

Temp. °C	0.40	0.45	0.50	0.55	0.60	0.65	0.70	0.75	0.80	0.85	0.90	0.95	1.00
0													290
25													455
100												(605)	700
150													765
200										(720)	700	725	790
250										735	710	730	785
300							(850)	820	780	740	720	725	755
350							850	820	780	745	720	715	
400	680	790	850	875	880	870	840	810	775	740	720	700	
450	630	750	820	850	860	850	830	800	765	735	710		
500	590	715	790	830	845	835	820	790	755	725	705		
550	540	680	760	810	825	820	810	780	740				
600	490	640	730	785	810	800	790	770	730				
650	450	610	700	765	790	785	770	750					
700	400	570	670	745	775	770	750	730					

From Quist, A. S. and Marshall, W. L., *J. Phys. Chem.*, 72, 684, 1968. With permission.

Table 13
FORMULAS FOR COMPUTING RESISTIVITY AND FORMATION FACTOR ($\rho_{1,2}$) FOR VARIOUS MODELS OF TWO-PHASE ROCKS AS A FUNCTION OF THE RELATIVE ABUNDANCE OF THE TWO PHASES AND THE RATIO OF THEIR RESISTIVITIES

Description of model	Formulas for computing: Resistivity	Formulas for computing: Formation factor
Spheres of uniform size with a resistivity ρ_2 stacked in cubic, hexagonal, and rhombohedral packings, and with the interstices filled with a medium having a resistivity ρ_1. The porosity, ω, ranges from 0.260–0.474 for the various packing schemes. The ratio ρ_2/ρ_1 is finite.	$\rho_{1,2(x)} = \rho_{1,2(y)} = \rho_{1,2(z)}$ $= \rho_{1,2} = \dfrac{\rho_1\omega + (3-\omega)\rho_2}{\rho_1(3-2\omega) - 2\omega\rho_2}\rho_1$ $= \dfrac{\omega + (3-\omega)\dfrac{\rho_2}{\rho_1}}{(3-2\omega) + 2\omega\dfrac{\rho_2}{\rho_1}}\rho_1$	$P_{1,2} = \dfrac{\omega + (3-\omega)\rho_2/\rho_1}{(3-2\omega) + 2\omega\rho_2/\rho_1}$
The same, but $\rho_2/\rho_1 \to \infty$	$\rho_{1,2(x)} = \rho_{1,2(y)} = \rho_{1,2(z)} = \rho_{1,2}$ $\approx \dfrac{3-\omega}{2\omega}\rho_1$	$P_{1,2} \approx \dfrac{3-\omega}{2\omega}$
The same, but $\rho_2/\rho_1 \to 0$	$\rho_{1,2(x)} = \rho_{1,2(y)} = \rho_{1,2(z)}$ $\approx \dfrac{\omega}{3-2\omega}\rho_1$	$P_{1,2} \approx \dfrac{\omega}{3-2\omega}$
η-series of spherical inclusions, each of which is present in the same proportion ω_0 in a two-phase rock. The resistivity of the sphere is ρ_2 and that of the surrounding materials $\rho_1 \cdot \rho_2/\rho_1 \to \infty$	$\bar{\rho}_{1,2} = \left(\dfrac{3-\omega_0}{2\omega_0}\right)^{\frac{\lg\omega}{\lg\omega_0}}\rho_1$	$P_{1,2} = \left(\dfrac{3-\omega_0}{2\omega_0}\right)^{\frac{\lg\omega}{\lg\omega_0}}$
The same, but $\rho_2/\rho_1 = 0$	$\bar{\rho}_{1,2} = \left(\dfrac{\omega_0}{3-2\omega_0}\right)^{\frac{\lg\omega}{\lg\omega_0}}\rho_1$	$P_{1,2} = \left(\dfrac{\omega_0}{3-2\omega_0}\right)^{\frac{\lg\omega}{\lg\omega_0}}$
Elongate ellipsoids of revolution of uniform size and various axial ratios and resistivities, ρ_2 packed in a cubic packing and with the interstices having a variable fractional volume ω, and filled with a material having a resistivity, ρ_1; ρ_2/ρ_1 is finite.	The resistivity in a direction parallel of the axes of revolution of the ellipsoids. $\rho_{1,2(z)} = \dfrac{\omega\rho_1 - \rho_2[k-(1-\omega)]}{[1-(1-\omega)k]\rho_1 - \omega k\rho_2}\rho_1$ $= \dfrac{\omega - [k-(1-\omega)]\rho_2\rho_1}{[1-(1-\omega)k] - \omega k\rho_2/\rho_1}\rho_1$	$P_{1,2(z)} = \dfrac{\omega - [k-(1-\omega)]\rho_2/\rho_1}{[1-(1-\omega)k] - \omega k\rho_2/\rho_1}$
	The resistivity in a direction perpendicular to the axes of revolution of the ellipsoids.	$P_{1,2(x)} = P_{1,2(y)}$ $= \dfrac{\omega - [l-(1-\omega)]\rho_2/\rho_1}{[1-(1-\omega)l] - \omega l\rho_2/\rho_1}$

(continued next page)

Table 13 (continued)
FORMULAS FOR COMPUTING RESISTIVITY AND FORMATION FACTOR ($\rho_{1,2}$) FOR VARIOUS MODELS OF TWO-PHASE ROCKS AS A FUNCTION OF THE RELATIVE ABUNDANCE OF THE TWO PHASES AND THE RATIO OF THEIR RESISTIVITIES

Description of model	Formulas for computing: Resistivity	Formulas for computing: Formation factor
	$\rho_{1,2(x)} = \rho_{1,2(y)}$ $= \dfrac{\omega\rho_1 - [l - (1-\omega)]\rho_2}{[1 - (1-\omega)l]\rho_1 - \omega l\rho_2}\rho_1$ $= \dfrac{\omega - [l - (1-\omega)]\rho_2\rho_1}{[1 - (1-\omega)l] - \omega l\rho_2/\rho_1}\rho_1$ k and l are coefficients which vary as follows: k→0 as $\rho\to-\infty$ to k→$-\infty$ as $\rho\to-1$	
The same, but $\rho_2/\rho_1\to\infty$	$\rho_{1,2(z)} = \dfrac{[k - (1-\omega)]\rho_2}{(1-\omega)k\rho_1 - \omega k\rho_2}\rho_1$ $= \dfrac{[k - (1-\omega)]\rho_2/\rho_1}{(1-\omega)k - \omega k\rho_2/\rho_1}\rho_1$	$\mathrm{P}_{1,2(z)} = \dfrac{[k - (1-\omega)]\rho_2/\rho_1}{(1-\omega)k - \omega k\rho_2/\rho_1}$
	$\rho_{1,2(x)} = \rho_{1,2(y)}$ $\rho_{1,2} = \dfrac{[l - (1-\omega)]\rho_2}{(1-\omega)l\rho_1 - \omega l\rho_2}\rho_1$ $= \dfrac{[l - (1-\omega)\rho_2/\rho_1}{(1-\omega)l - \omega l\rho_2/\rho_1}\rho_1$	$\mathrm{P}_{1,2(x)} = \mathrm{P}_{1,2(y)}$ $= \dfrac{[l - (1-\omega)]\rho_2/\rho_1}{[(1-\omega)l] - \omega l\rho_2/\rho_1}$
The same, but $\rho_2/\rho_1\to 0$	$\rho_{1,2(z)} = \dfrac{\omega\rho_1 - k\rho_2}{[1 - (1-\omega)k]\rho_1 - \omega k\rho_2}\rho_1$ $= \dfrac{\omega - k\rho_2/\rho_1}{[1 - k(1-\omega)] - \omega k\rho_2/\rho_1}\rho_1$	$\mathrm{P}_{1,2(z)} = \dfrac{\omega - k\rho_2/\rho_1}{[1-k(1-\omega)] - \omega k\rho_2/\rho_1}$
	$\rho_{1,2(x)} = \rho_{1,2(y)}$ $= \dfrac{\omega\rho_1 - l\rho_2}{[1 - l(1-\omega)]\rho_1 - \omega l\rho_2}\rho_1$ $= \dfrac{\omega - l\rho_2/\rho_1}{[1 - l(1-\omega)] - \omega l\rho_2/\rho_1}\rho_1$	$\mathrm{P}_{1,2(x)} = \mathrm{P}_{1,2(y)}$ $= \dfrac{\omega\rho_1 - l\rho_2/\rho_1}{[1 - l(1-\omega)] - \omega l\rho_2/\rho_1}$

Table 13 (continued)

FORMULAS FOR COMPUTING RESISTIVITY AND FORMATION FACTOR ($\rho_{1,2}$) FOR VARIOUS MODELS OF TWO-PHASE ROCKS AS A FUNCTION OF THE RELATIVE ABUNDANCE OF THE TWO PHASES AND THE RATIO OF THEIR RESISTIVITIES

Description of model	Formulas for computing: Resistivity	Formulas for computing: Formation factor
Markedly elongate uniform-size ellipsoids of revolution in a cubic packing, but lying at an angle with respect to their axes— $c/a \to \infty$, $k = -\infty$, $\rho = -1$, and a resistivity ρ_2. The interstices have a volume ω (as a fraction) and are filled with a medium having a resistivity, $\rho_1 \cdot \rho_2/\rho_1$ is finite.	$\rho_{1,2(z)} = \dfrac{\rho_1\rho_2}{(1-\omega)\,\rho_1 + \omega\,\rho_2} = \dfrac{\rho_2/\rho_1}{(1-\omega) + \omega\,\rho_2/\rho_1}\,\rho_1$ $\rho_{1,2(x)} = \rho_{1,2(y)} = \dfrac{\omega\,\rho_1 + (2-\omega)\,\rho_2}{(2-\omega)\,\rho_1 + \rho_2\,\omega}\,\rho_1 = \dfrac{\omega + (2-\omega)\,\rho_2/\rho_1}{(2-\omega) + \omega\,\rho_2/\rho_1}\,\rho_1$	$P_{1,2(z)} = \dfrac{\rho_2/\rho_1}{(1-\omega) + \omega\,\rho_2/\rho_1}$ $P_{1,2(x)} = P_{1,2(y)} = \dfrac{\omega + (2-\omega)\,\rho_2/\rho_1}{(2-\omega) + \omega\,\rho_2/\rho_1}$
Disc-like ellipsoids of revolution of uniform size with a resistivity ρ_2, in a cubic packing with the interstices filled with a material having resistivity ρ_1 and fractional volume ω. $c/a \to 0$, $k \to 0$, and $\ell \to -\infty$, ρ_2/ρ is finite.	$\rho_{1,2(z)} = \omega\,\rho_1 + (1-\omega)\,\rho_2 = [\omega + (1-\omega)\,\rho_2/\rho_1]\,\rho_1$ $\rho_{1,2(x)} = \rho_{1,2(y)} = \dfrac{\rho_1\rho_2}{(1-\omega)\,\rho_1 + \rho_2} = \dfrac{\rho_2/\rho_1}{(1-\omega) + \omega\,\rho_2\rho_1}\,\rho_1$	$P_{1,2(z)} = \omega + (1-\omega)\,\rho_2/\rho_1$ $P_{1,2(x)} = P_{1,2(y)} = \dfrac{\rho_2/\rho_1}{(1-\omega) + \omega\,\rho_2/\rho_1}$
Cubes of uniform dimensions, in a cubic pattern, with the space between cubes comprising a volume fraction ω, and having a resistivity $\rho_1; \rho_2 >> \rho_1$	$\rho_{1,2(B)} = \dfrac{(1+\gamma)^2\,(\rho_2+\gamma\rho_1)}{\gamma\,(\gamma+2)\,\rho_2 + (1+\gamma+2\gamma^2+\gamma^3)\,\rho_1}\,\rho_1$, where $\gamma = \sqrt[3]{\dfrac{1}{1-\omega}} - 1$; $\rho_{1,2(B)}$ is the upper limit for the rock resistivity $\gamma = \delta/a$, where δ is the thickness of the layer between the $\rho_{1,2(H)} = \rho_1 \times \dfrac{\gamma\,\rho_1 + (1+\gamma)^2\,\rho_2 + \gamma^2\,(\gamma+2)\,\rho_2}{(1+\gamma)\,[\rho_1+\gamma\,(\gamma+2)\,\rho_2]}$ $\rho_{1,2(H)}$ is the lower limit for the rock resistivity.	$P_{1,2(B)} = \dfrac{(1-\gamma)^2\,(\rho_2-\gamma\rho_1)}{\gamma\,(\gamma+2)\,\rho_2 + (1+\gamma+2\gamma^2+\gamma^3)\,\rho_1}$ $P_{1,2(H)} = \dfrac{\gamma\rho_1 + (1+\gamma)^2\,\rho_2 + \gamma^2\,(\gamma+2)\,\rho_2}{(1+\gamma)\,[\rho_1 + \gamma\,(\gamma+2)\,\rho_2]}$

Table 13 (continued)
FORMULAS FOR COMPUTING RESISTIVITY AND FORMATION FACTOR ($\rho_{1,2}$) FOR VARIOUS MODELS OF TWO-PHASE ROCKS AS A FUNCTION OF THE RELATIVE ABUNDANCE OF THE TWO PHASES AND THE RATIO OF THEIR RESISTIVITIES

Description of model	Formulas for computing: Resistivity	Formulas for computing: Formation factor
Cubes of uniform size, and high resistivity, distributed in a cubic pattern, with the space between the cubes having a material having finite resistivity, ρ_1.	$\rho_{1,2} = \dfrac{\rho_1}{1 - \sqrt[3]{(1-\omega)^2}}$	$P_{1,2} = \dfrac{1}{1 - \sqrt[3]{(1-\omega)^2}}$
Cubes of uniform size with high resistivity and distributed in a staggered cubic pattern. The space between the cubes has a volume fraction ω and is filled with a material having resistivity ρ_1.	$\rho_{1,2} = \dfrac{1 + 0.5\sqrt[3]{1-\omega}}{1 - \sqrt[3]{(1-\omega)^2}}\,\rho_1$	$P_{1,2} = \dfrac{1 + 0.5\sqrt[3]{1-\omega}}{1 - \sqrt[3]{(1-\omega)^2}}$
Equant octahedrons of uniform size with a resistivity $\rho_1 \rightarrow 0$, distributed in a cubic pattern. The space between octahedrons forms a volume fraction ω and is filled with a material having resistivity $\rho_2 \rightarrow \infty$.	$\rho_{1,2} = 0.5\,\rho_1\gamma(\sqrt{2} + \gamma)$ $(1 - \omega) = \dfrac{4(\sqrt{2} + \gamma)^3 + 3(2\sqrt{2} + \gamma)^2\gamma}{4(\sqrt{2} + \gamma)^3}$	$P_{1,2} = 0.5\,\gamma(\sqrt{2} + \gamma)$[a]
Three sets of mutually perpendicular circular cylindrical tubes of material with a resistivity ρ_1, embedded in a material with a resistivity $\rho_2 \rightarrow \infty$.	$\rho_{1,2} = \dfrac{3}{\omega}\,\rho_1$	$P_{1,2} = \dfrac{3}{\omega}$[b]

Table 13 (continued)
FORMULAS FOR COMPUTING RESISTIVITY AND FORMATION FACTOR ($\rho_{1,2}$) FOR VARIOUS MODELS OF TWO-PHASE ROCKS AS A FUNCTION OF THE RELATIVE ABUNDANCE OF THE TWO PHASES AND THE RATIO OF THEIR RESISTIVITIES

Description of model	Formulas for computing: Resistivity	Formulas for computing: Formation factor
The same, but the lengths of the tubes are $1/\sin 45° = 1.42$ times the edge diameter of a cube.	$\rho_{1,2} = \frac{4.25}{\omega} \rho_1$	$P_{1,2} = \frac{4.25}{\omega}$ [b]

[a] This expression is valid for small distances between grains.
[b] This expression is approximate – tubes intersect one another.

Table 14
FORMS OF ARCHIE'S LAW WHICH MAY BE USED WHEN LITHOLOGY OF A ROCK IS KNOWN

Description of rock	a	m
1. Weakly-cemented detrital rocks, such as sand, sandstone and some limestones, with a porosity range from 25 to 45%, usually Tertiary in age	0.88	1.37
2. Moderately well cemented sedimentary rocks, including sandstones and limestones, with a porosity range from 18 to 35%, usually Mesozoic in age	0.62	1.72
3. Well-cemented sedimentary rocks with a porosity range from 5% to 25%, usually Paleozoic in age	0.62	1.95
4. Highly porous volcanic rocks, such as tuff, aa and pahoehoe, with porosity in the range 20% to 80%	3.5	1.44
5. Rocks with less than 4% porosity, including dense igneous rocks and metamorphosed sedimentary rocks	1.4	1.58

Table 15
INVERSE LOSS TANGENTS FOR ROCKS OVER THE FREQUENCY RANGE FROM 0.001 to 1000 Hz

Rock type	Inverse loss tangent
Light-colored igneous rocks with little clay or magnetite mineral content (granite, granodiorite)	0.001—0.005
Dark colored igneous rocks with no pyrite or graphite (labradorite, eclogite, gabbro)	0.003—0.01
Volcanic rocks (tuff, rhyolite, basalt)	0.01—0.10
Marble, clay-free limestone and dolomite	0.0001—0.001
Shale	0.0005—0.002
Siltsone	0.01—0.03
Sandstone	0.005—0.02
Mineralized rock, ore deposits	0.03—1

Table 16
ELECTRICAL RESISTIVITIES SUMMARIZED FROM INDUCTION LOGS OF WELLS PENETRATING MESOZOIC-CENOZOIC ROCKS OF THE EAST GULF COAST, USA

Index no. of well	Location	Depth interval (ft)	Designator	Transverse resistivity (Ω-m)	Longitudinal resistivity (Ω-m)	Coefficient of anisotropy
1.	Walthall County,	2,130—3,150	U	3.19	1.70	1.37
	Mississippi	3,160—6,800	M	0.71	0.66	1.04
	Sec. 18, 2N, 12E	6,810—7,650	L_1	1.25	1.21	1.02
		7,660—11,000	L_2	2.28	1.50	1.23
2.	Smith County,	2,580—5,350	U	2.40	2.00	1.10
	Mississippi	5,350—6,020	M	0.89	0.86	1.02
	Sec. 6, 10N, 16W	6,020—11,080	L_1	2.28	1.73	1.15
		11,090—14,000	L_2	7.44	6.35	1.08
3.	Forrest County,	2,500—5,330	M_1	0.66	0.64	1.01
	Mississippi	5,330—5,830	M_2	0.93	0.80	1.08
	Sec. 17, 1S, 13W	5,830—8,380	L	1.62	1.51	1.04
4.	Marion County,	100—1,740	U_2	10.1	3.59	1.68
	Mississippi	1,750—2,470	U_2	3.48	2.16	1.27
	Sec. 3, 4N, 19W	2,480—6,350	M_1	0.85	0.77	1.05
		6,360—7,300	M_2	0.98	0.93	1.02
5.	Stone County,	1,550—2,500	U	4.56	2.86	1.26
	Mississippi	2,510—5,770	M	0.76	0.69	1.05
	Sec. 1, 3S, 12W	5,700—8,590	L	1.66	1.42	1.08
6.	Wayne County,	1,720—3,650	U	2.25	2.08	1.04
	Mississippi	3,650—4,140	M_1	0.61	0.60	1.01
	Sec. 16, 7N, 9W	4,140—8,650	L_1	1.26	0.93	1.17
		8,650—12,010	L_2	2.06	1.36	1.23
7.	Green County	9,870—13,050	L_2	3.58	3.22	1.05
	Mississippi					
	Sec. 10, 4N, 5W					
8.	George County,	830—2,330	U	1.76	1.42	1.11
	Mississippi	2,340—5,510	M_1	0.59	0.57	1.02
	Sec. 15, 3S, 6W	5,510—7,950	M_2	1.04	0.97	1.04
		7,960—8,570	M_3	0.66	0.59	1.06
9.	Perry County,	1,980—2,670	M_1	0.77	0.76	1.01
	Mississippi	2,680—5,760	M_2	0.67	0.65	1.02
	Sec. 18, 3N, 9W	5,770—8,650	L_1	1.41	1.04	1.17
		8,660—12,350	L_2	2.43	1.68	1.21
10.	Rankin County,	5,040—8,790	L_1	1.33	0.96	1.18
	Mississippi	8,800—12,500	L_2	2.67	1.76	1.25
	Sec. 14, 5N, 5E					
11.	Lamar County,	1,760—2,480	U	3.84	2.24	1.31
	Mississippi	2,490—5,900	M_1	0.66	0.63	1.02
	Sec. 19, 2N, 16W	5,910—6,650	M_2	0.83	0.81	1.01
		6,650—10,820	L_1	2.75	1.91	1.20
12.	Madison County	1,450—3,830	U	5.02	3.45	1.21
	Mississippi	3,830—4,600	M	0.86	0.84	1.01
	Sec. 1, 8N, 2E	4,600—7,530	L_1	1.67	1.08	1.24
		7,540—9,740	L_2	3.11	1.36	1.51
		9,740—11,000	L_3	2.86	2.23	1.13
13.	Scott County,	660—2,260	U_1	4.25	2.75	1.24
	Mississippi	2,260—4,030	U_2	3.34	2.31	1.20
	Sec. 15, 5N, 6E	4,030—4,560	M	0.63	0.61	1.02
		4,660—7,500	L_1	1.45	1.19	1.10
14.	Smith County,	2,500—5,340	U	2.25	1.08	1.44
	Mississippi	5,350—6,020	M	0.57	0.55	1.02
	Sec. 6, 10N, 16W	6,020—11.080	L_1	1.55	0.94	1.28
		11,090—14,000	L_2	5.08	2.82	1.34

Table 16 (continued)
ELECTRICAL RESISTIVITIES SUMMARIZED FROM INDUCTION LOGS OF WELLS PENETRATING MESOZOIC-CENOZOIC ROCKS OF THE EAST GULF COAST, USA

Index no. of well	Location	Depth interval (ft)	Designator	Transverse resistivity (Ω-m)	Longitudinal resistivity (Ω-m)	Coefficient of anisotropy
15.	Pearl River County,	1,250—2,070	U_1	7.58	6.12	1.11
	Mississippi	2,070—2,850	U_2	3.20	2.61	1.12
	Sec. 11, 2S, 15W	2,850—5,480	M_1	0.82	0.66	1.11
		5,480—6,340	M_2	1.05	1.03	1.01
		6,340—8,710	L_1	1.84	1.63	1.06
		8,710—10,200	L_2	1.78	1.02	1.32
16.	Hinds County,	3,210—5,900	U	1.20	0.88	1.17
	Mississippi	5,900—6,820	M	0.98	0.96	1.01
	Sec. 30, 6N, 4W	6,830—9,720	L_1	1.89	1.33	1.19
		9,730—11,100	L_2	4.03	2.83	1.19
17.	Pike County,	1,500—3,160	U	4.27	2.21	1.39
	Mississippi	3,160—7,350	M	0.80	0.70	1.07
	Sec. 1, 4N, 7E	7,360—8,500	L_1	1.31	1.23	1.03
		8,500—11,100	L_2	3.15	2.87	1.05
18.	Jones County,	2,390—5,020	U	1.49	0.85	1.32
	Mississippi	5,030—5,660	M	0.68	0.66	1.01
	Sec. 16, 7N, 10W	5,670—10,020	L_1	1.36	1.04	1.14
		10,030—14,860	L_2	3.84	2.84	1.16
19.	Covington County,	3,050—5,750	U	1.04	0.76	1.17
	Mississippi	5,760—6,740	M	0.82	0.80	1.01
	Sec. 4, 7N, 17W	6,750—10,650	L_1	2.20	1.68	1.14
		10,660—13,460	L_2	3.59	3.01	1.09
20.	Lincoln County,	1,700—3,400	U	3.13	1.54	1.42
	Mississippi	3,410—7,700	M	0.82	0.73	1.06
	Sec. 32, 5N, 6E	7,710—8,870	L_1	1.33	1.30	1.01
		8,880—11,250	L_2	3.05	2.78	1.05
21.	Copiah County,	2,110—3,130	U	4.04	3.74	1.04
	Mississippi	3,140—7,110	M_1	0.76	0.73	1.02
	Sec. 9, 9N, 7E	7,110—7,970	M_2	0.80	0.79	1.01
		7,980—10,900	L_1	2.03	1.49	1.16
		10,910—12,340	L_2	3.68	3.55	1.02
22.	Jasper County,	2,520—2,980	U	2.81	2.50	1.06
	Mississippi	2,980—3,570	M	0.79	0.77	1.03
	Sec. 12, 3N, 11E	3,580—8,850	L_1	1.04	0.79	1.15
		8,860—14,330	L_2	4.51	1.39	1.80
23.	Simpson County,	2,410—6,080	U	7.23	6.39	1.06
	Mississippi	6,080—6,780	M	0.63	0.62	1.01
	Sec. 10, 10N, 19W	6,790—9,990	L_1	1.60	1.13	1.18
		9,990—12,980	L_2	3.80	2.94	1.14
24.	Green County,	1,990—5,070	M	0.89	0.81	1.05
	Mississippi	5,070—8,270	L_1	1.33	1.10	1.10
	Sec. 10, 4N, 5W	8,270—9,870	L_2	1.36	0.89	1.23
25.	Franklin County,	240—1,200	U	3.82	2.13	1.34
	Florida	1,210—4,340	M_1	1.05	0.74	1.18
		4,350—9,620	M_2	1.12	0.53	1.44
		9,630—10,560	L_1	4.32	2.95	1.19
26.	Baldwin County,	1,400—5,200	U	1.02	0.83	1.11
	Alabama	5,210—6,720	M	0.71	0.42	1.30
	Sec. 16, 2N, 4E					
27.	Amite County,	1,760—2,430	U	8.52	6.03	1.19
	Mississippi	2,440—6,000	M	1.01	0.59	1.29
	Sec. 28, 1N, 3E	6,010—9,510	L_1	1.40	1.11	1.12
		9,520—10,620	L_2	2.64	2.49	1.03

Table 16 (continued)
ELECTRICAL RESISTIVITIES SUMMARIZED FROM INDUCTION LOGS OF WELLS PENETRATING MESOZOIC-CENOZOIC ROCKS OF THE EAST GULF COAST, USA

Index no. of well	Location	Depth interval (ft)	Designator	Transverse resistivity (Ω-m)	Longitudinal resistivity (Ω-m)	Coefficient of anisotropy
28.	Clarke County,	520—2,380	U_1	9.75	6.17	1.25
	Mississippi	2,390—4,120	U_2	1.57	1.39	1.06
	Sec. 6, 2N, 15E	4,130—6,120	M	1.18	0.46	1.59
29.	Adams County,	1,880—4,100	M	0.54	0.38	1.18
	Mississippi	4,160—6,990	L_1	1.65	0.86	1.40
	Sec. 36, 7N, 1W	6,990—13,000	L_2	2.30	1.51	1.23
30.	Adams County,	1,140—4,300	U	1.35	0.74	1.46
	Mississippi	4,310—7,820	M_1	0.89	0.70	1.13
	Sec. 4, 6N, 2W	7,830—9,410	L_1	1.49	1.18	1.13
		9,410—10,810	L_2	4.46	3.17	1.19
31.	Clarke County,	460—3,000	U	5.72	4.07	1.18
	Alabama	3,000—3,930	M	0.92	0.62	1.21
	Sec. 11, 10N, 2W					
32.	Claiborne County,	2,220—3,760	U	1.23	0.76	1.27
	Mississippi	3,770—6,970	M	0.89	0.52	1.31
	Sec. 27, 13N, 2E	6,980—8,050	L_1	0.98	0.96	1.01
		8,060—10,430	L_2	3.15	2.06	1.23
33.	Decatur County,	1,200—6,150	L	2.65	1.45	1.35
	Georgia					
34.	Sumpter County,	80—3,950	L_3	33.6	14.4	1.52
	Alabama	3,960—10,020	L_4	190.	158.	1.10
	Sec. 9, 23N, 3W					
35.	Mobile County,	1,840—5,300	M	1.05	0.90	1.05
	Alabama	5,310—8,720	L_1	2.09	1.54	1.16
	Sec. 16, 3S, 2W	8,730—13,030	L_2	10.8	5.35	1.42
36.	Clay County,	2,050—8,800	L_3	80.0	43.3	1.35
	Mississippi					
	Sec. 24, 16S, 5E					
37.	Escambia County,	2,350—4,580	M	0.94	0.88	1.03
	Florida	4,580—7,590	L_1	1.57	1.47	1.03
	Sec. 31, 2S, 31W	7,600—10,420	L_2	3.89	2.52	1.24
		10,430—12,500	L_3	7.35	4.59	1.26

Note: All resistivities are compiled from the 6FF40 induction log.

Table 17
FORMATION NAMES FROM THE COLORADO PLATEAU

Electrical designator	Formation names	Age	Thickness	Lithology
M	Mancos, Mesa Verde, Lewis Shale, Fruitland, Kirtland, Animas	Cretaceous and younger	to 6,000′	Shale, sandstone
L_1	Cutler, Dolores, La Plata, McElmo, Dakota, Moenkopi, Shinarump, Chinle, Glen Canyon, San Rafael, Morrison	Permian to Cretaceous	to 4,000′	Sandstone, Shale, limestone
L_2	Paradox, Cutler, Kaibab, Coconino	Pennsylvanian to Permian	to 2,000′	Limestone, evaporites
L_3	Hermosa, Cutler, Ignacio, Elbert, Ouray	Cambrian to Permian	to 2,000′	Limestones

Table 18
AVERAGE RESISTIVITIES COMPILED FROM ELECTRIC LOGS RUN IN WELLS ON THE COLORADO PLATEAUS

Index no. of well	Location	Depth interval (ft)	Designator	Transverse resistivity (Ω-m)	Longitudinal resistivity (Ω-m)	Coefficient of anisotropy
1.	Grand County,	310—2.398	M	17.8	14.7	1.10
	Utah	2.400—4.123	L_1	37.9	24.5	1.24
	23, 20S, 21E	4.125—4.745	L_2	1,050	582	1.35
2.	Grand County,	310—1.930	M	22.0	13.2	1.29
	Utah	1,932—3,370	L_1	47.1	35.0	1.16
	20, 21S, 23E	3,372—3,800	L_2	1,530	691	1.49
3.	McKinley County,	1,200—2,148	M	14.7	9.4	1.25
	N.M.	2,150—2,230	—	1,550	570	1.65
	14, 14N, 8W	2,736—6,206	L_1	52.5	21.2	1.47
4.	Montrose County,	1,700—5,400	M	19.7	14.6	1.16
	Colo.	5,430—11,250	L_1	142	81.3	1.32
	21, 47N, 19W					
5.	La Plata County,	980—1,800	M	21.8	15.2	1.20
	Colo.	6,920—10,025	L_1	446	83.2	2.32
	17, 34N, 11W					
6.	San Juan County,	255—535		15.2	13.2	1.08
	Utah	532—805		51.3	35.2	1.21
	7, 40S, 26E	807—3,437	M	25.6	17.9	1.20
		3,427—5,576		25.7	17.9	1.20
		5,528—7,890	L_2	800	49.9	4.11
7.	San Juan County,	420—743		9.7	8.4	1.07
	N.M.	745—1,330	M	39.8	22.0	1.34
	30, 26N, 19W	1,332—3,770		16.2	9.4	1.31
		3,776—7,004	L_1	135	41.8	1.80
8.	Garfield County,	1,400—4,505	M	62.3	38.0	1.28
	Utah	4,507—6,080		156	137	1.07
	18, 36S, 10E	6,082—8,360	L_1	620	179	1.86
9.	Navajo County,	140—590		25.2	23.6	1.03
	Ariz.	592—1.980	M	35.1	13.9	1.59
	6, 19N, 23E	3,022—3,340		39.6	28.4	1.18
10.	Coconino County,	605—2,370	L_1	350	297	1.09
	Ariz.	2,372—3,090	L_2	1,100	640	1.32
	35, 28N, 1W	3,092—3,522	L_3	67	63	1.03
11.	McKinley County,	450—1,160		41.3	30.4	1.16
	N.M.	1,162—2,115	M	21.8	14.2	1.24
	14, 19N, 3W	2,117—7,776		10.6	8.0	1.15
		7,777—8,600	L_1	43.9	21.2	1.44
		8,602—9,626	L_2	335	38.7	2.95
12.	Emery County,	0—1,020	L_1	145	120	1.10
	Utah	1,022—3,675	L_2	1,280	200	2.52
	6, 22S, 12E	3,677—4,180	L_3	89	68	1.14
13.	Garfield County,	175—700		6.3	5.7	1.04
	Utah	700—1,290		28.1	18.6	1.23
	12, 36S, 1E	1,290—2,350		11.6	9.6	1.10
		2,350—4,710	M	38.3	34.3	1.06
		4,710—5,150		8.1	6.7	1.10
		5,150—5,970		28.6	25.8	1.06
		5,970—6,438		340	310	1.04
		6,438—6,828	L_1	2,850	2,125	1.16
		6,828—8,040		95	68	1.18
		8,040—8,767		193	82	1.55
		8,767—10.098	L_2	750	53	3.76

Table18 (Continued)
AVERAGE RESISTIVITIES COMPILED FROM ELECTRIC LOGS RUN IN WELLS ON THE COLORADO PLATEAUS

Index	No. of well	Location	Depth interval	Designator	Transverse resistivity	Coefficient of anisotropy
14.	Delta County,	100—970		10.0	8.8	1.07
	Colo.	970—1,202	M	32.0	26.4	1.10
	16, 15S, 95W	1,202—2,025		10.8	9.9	1.12
		2.025—7,841	L_1	57.3	45.6	1.12
15.	San Juan County,	218—740		50.6	45.4	1.06
	Utah	740—1,745	M	414	382	1.04
	33, 37S, 15E	1,745—2,195		90	66	1.17
		2,195—5,023	L_1	560	76	2.71
16.	San Juan County,	70—835	U	128	83	1.24
	Utah	835—5,050	M	36.0	18.2	1.41
	8, 31S, 22E	5,050—7,800	L_2	1,920	97	4.50
17.	Montezuma County,	84—1,582		75.8	53.7	1.19
	Colo.	1,582—4,064	M	54.0	35.2	1.24
	19, 39N, 14W	4,064—5,034		178	28.0	2.50
		5,034—6,506	L_1	580	39.5	3.75
		6,506—8,700	L_2	3,250	171	4.29
18.	San Juan County	500—2,875	M	21.6	15.7	1.17
	Utah	2,875—4,160	L_1	232	35.2	2.56
	3, 27S, 19E	4,160—7,000	L_2	545	85.5	2.52
		7,000—8,000	L_3	177	43.7	2.02
19.	Garfield County,	10—670		78.8	69.6	1.06
	Utah	670—1,200	M	245	199	1.11
	33, 32S, 15E	1,200—1,770		57.1	49.5	1.07
		1,770—4,955	L_1	340	143	1.54
20.	La Plata County,	300—1,720		15.2	12.8	1.09
	Colo.	1,722—2,920	M	49.7	29.2	1.30
	15, 33N, 7W	3,000—6,400		19.2	15.6	1.11
		6,402—13,050	L_1	163	31.8	2.27
21.	Coconino County,	300—2,935	M	31.4	15.6	1.42
	Ariz.	2,940—5,305	L_1	98	64	1.24
	28, 37N, 14E	5,310—7,210	L_2	458	93	2.22
22.	Navajo County,	450—1,310	M	44.2	41.6	1.03
	Ariz.	1,313—2,550	L_1	200	81.8	1.57
	12, 41N, 18E	2,553—4,530	L_2	950	210	2.13
23.	San Juan County,	70—675		6.6	5.2	1.13
	N.M.	680—930	M	50.1	20.2	1.57
	19, 29N, 16W	935—3,930		8.4	6.4	1.15
		3,935—5,456	L_1	14.0	11.7	1.09
		5,460—7,456	L_2	250	41.6	2.45
24.	San Juan County,	750—3,150	M	14.2	9.8	1.20
	Utah	3,170—6,146	L_1	56.2	30.6	1.36
	36, 41S, 20E	6,150—7,560	L_3	106	70.3	1.23

Table 19
RESISTIVITIES COMPILED FROM LOGS RUN IN DEEP WELLS IN PENNSYLVANIA

Location	Depth interval	Transverse resistivity	Longitudinal resistivity	Coefficient of anistropy
1. Sullivan County	3,070—8,050	190	151	1.12
	8,100—8,400	4.41	2.83	1.25
	8,470—12,310	205	72.3	1.68
2. Indiana County	1,380—6,460	39.5	32.7	1.12
	6,500—8,190	72.0	56.8	1.12
3. Indiana County	1,400—6,570	39.1	31.1	1.12
	6,600—8,760	44.8	38.5	1.11
4. Indiana County	2,020—6,620	41.2	31.4	1.14
5. Cameron County	910—4,620	65.3	50.2	1.14
	4,630—5,850	204	154	1.15
6. Mercer County	4,880—6,510	170	139	1.10
	6,570—7,360	1,800	1,490	1.11
	7,370—8,200	606	303	1.42
7. Fayette County	1,800—3,570	28.1	22.6	1.11
	3,600—9,060	73.7	63.6	1.07
8. Fayette County	1,300—3,090	44.2	37.6	1.09
	3,100—8,000	79.1	65.7	1.09

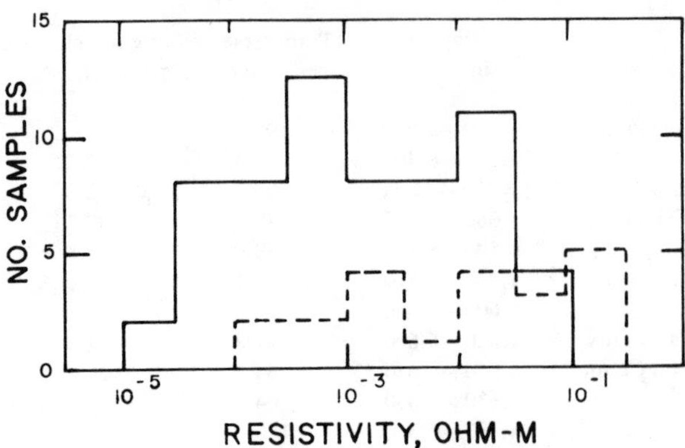

FIGURE 2B. Resistivity distribution for galena. Solid line is for n-type, dashed line is for p-type, and samples of mixed type are omitted.

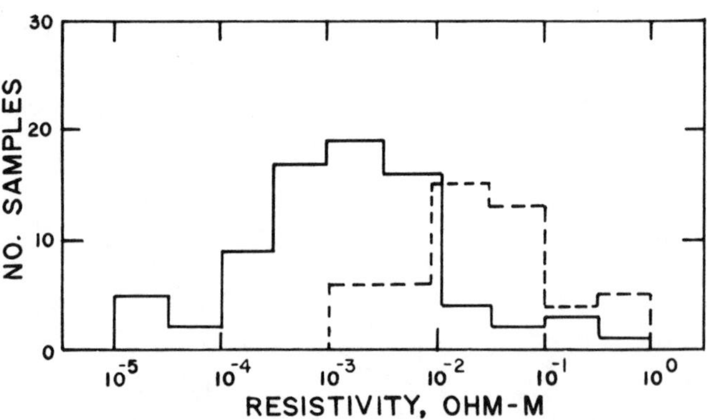

FIGURE 2C. Resistivity distribution for pyrite. Solid line is for n-type, dashed line is for p-type, and samples of mixed type are omitted.[100]

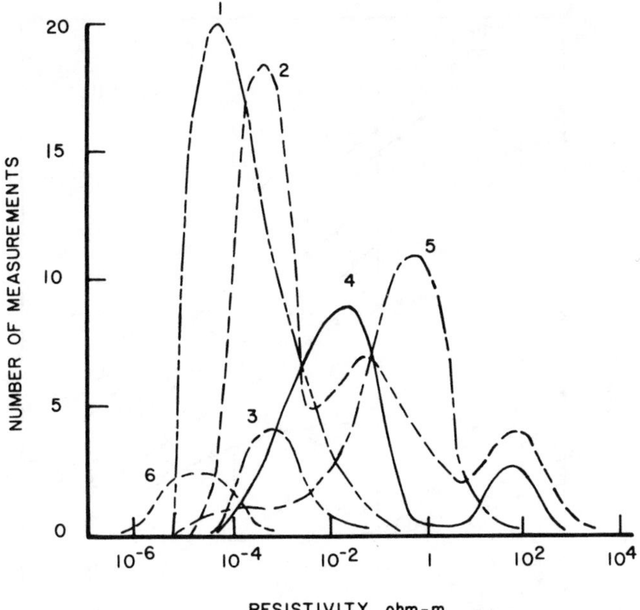

FIGURE 1. Smoothed histograms for the reported values of resistivity for the common minerals with high conductivity.[84] 1. Bornite, 2. Magnetite, 3. Pyrrhotite, 4. Arsenopyrite, 5. Galena, 6. Covellite.

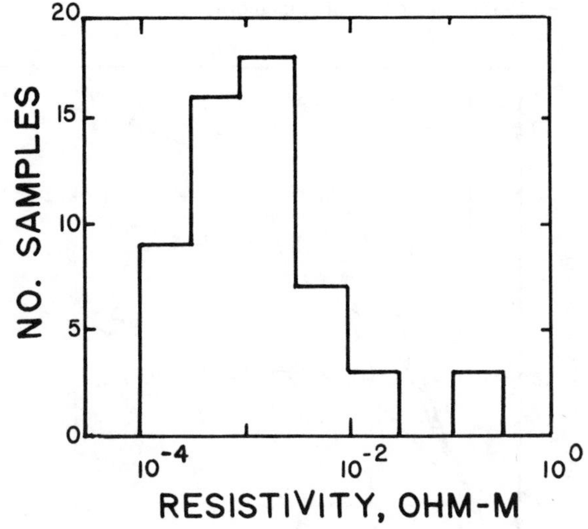

FIGURE 2A. Resistivity histogram for natural chalcopyrite.

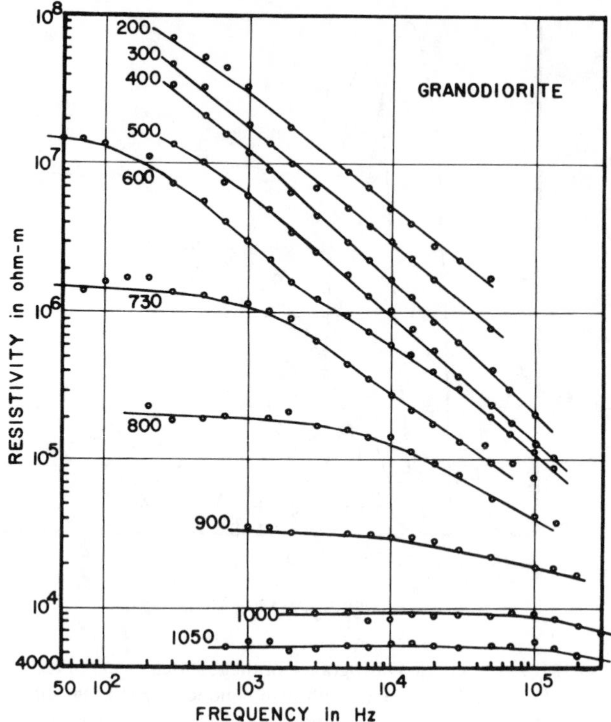

FIGURE 3. Resistivity as a function of frequency and temperature (°C) for Sample 1100 of granodiorite.[54]

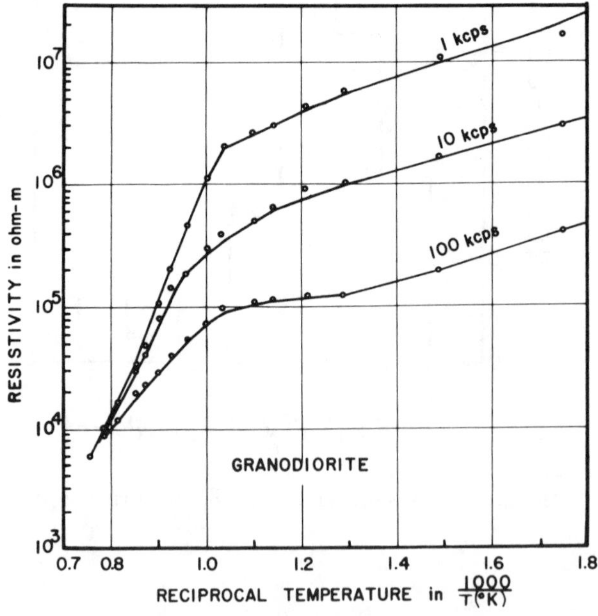

FIGURE 4. Resistivity as a function of absolute temperature for Sample 1100 of granodiorite.[54]

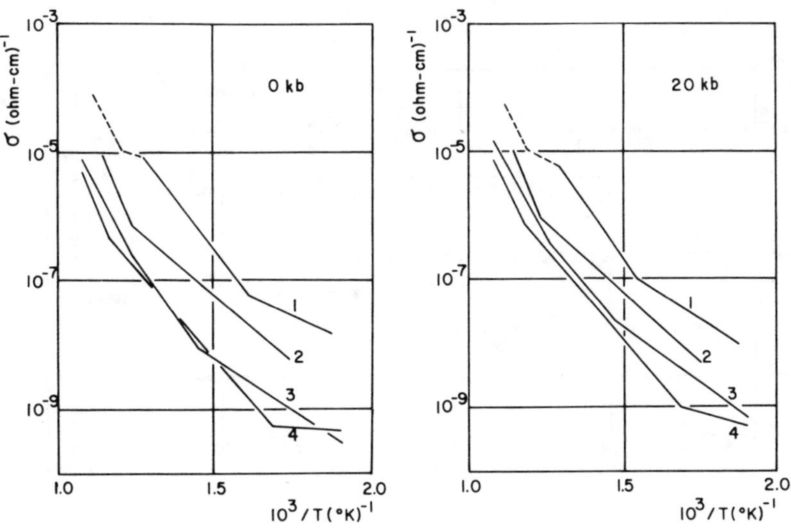

FIGURE 5. Conductivity versus 1/T for peridotites and olivinites: 1-6871 (broken line uncertain interpretation); 2-6013; 3-6014; 4-5375.

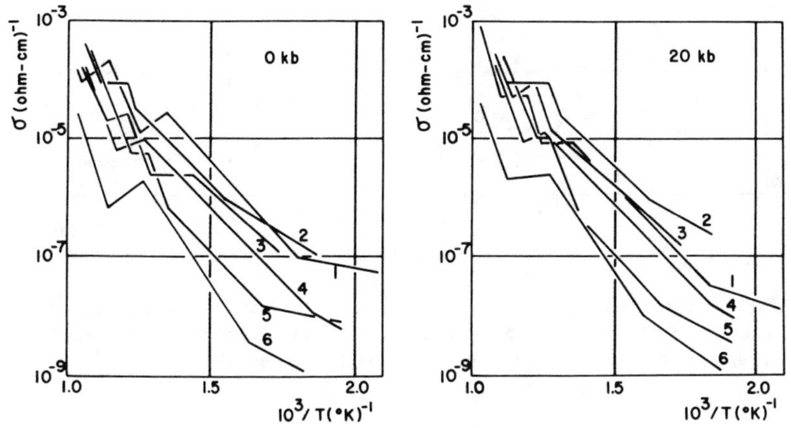

FIGURE 6. Conductivity versus 1/T for dunites: 1-4825; 2-7199; 3-6672; 4-5556; 5-7198; 6-5577.

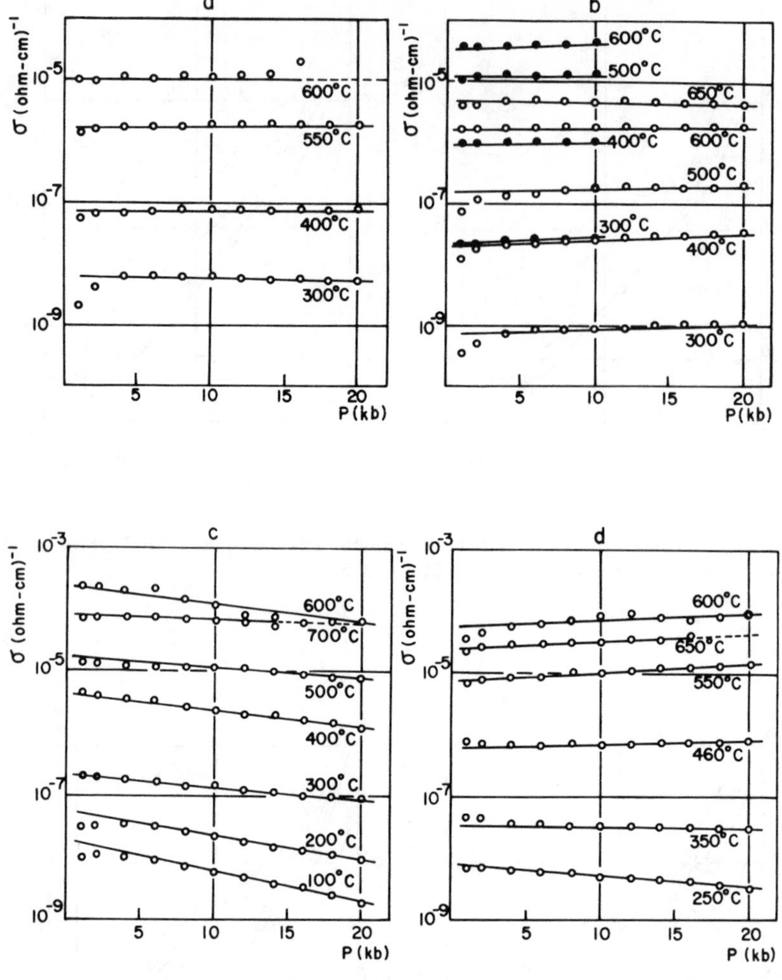

FIGURE 7. Variations of conductivity with pressure of several samples as a function of T: a-olivinite 6013; b-peridotite 5373 (open circles) with values taken during repeated heating on the same sample (full circles); c-Dunite 4825; d-dunite 7198.

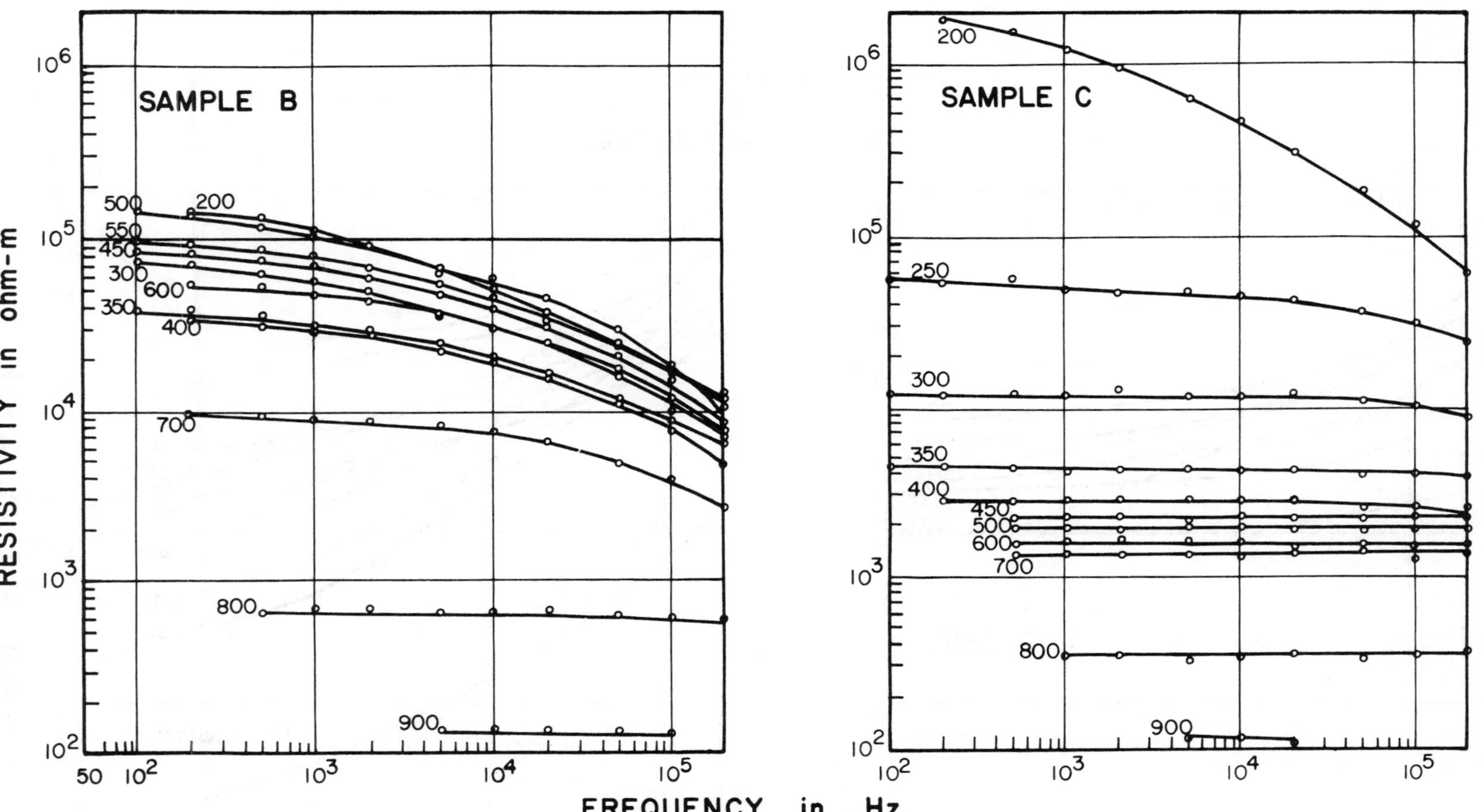

FIGURE 8. Resistivity as a function of frequency and temperature (°C) for Samples B, C, D, and E.[119]

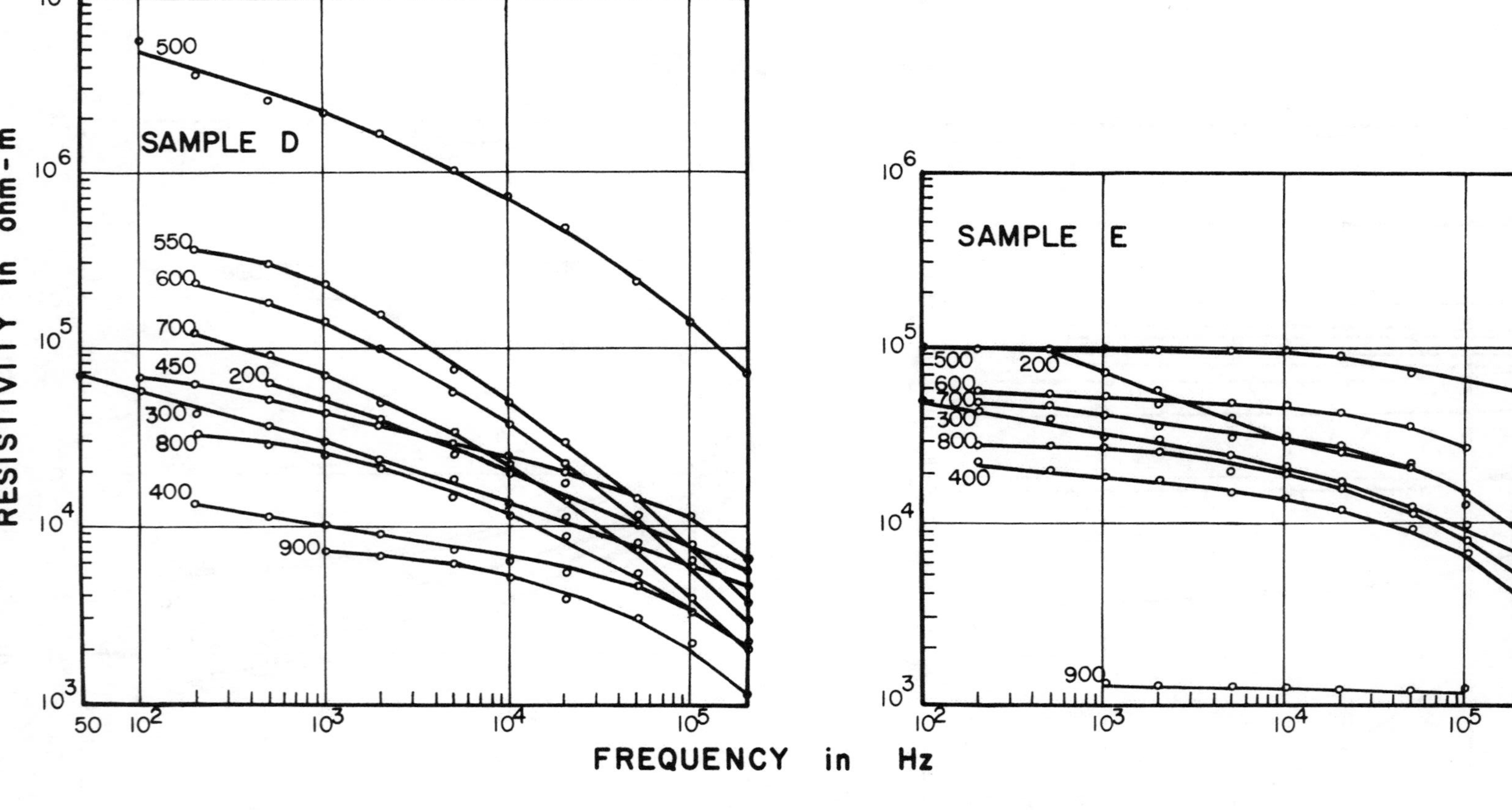

Figure 8 (continued)

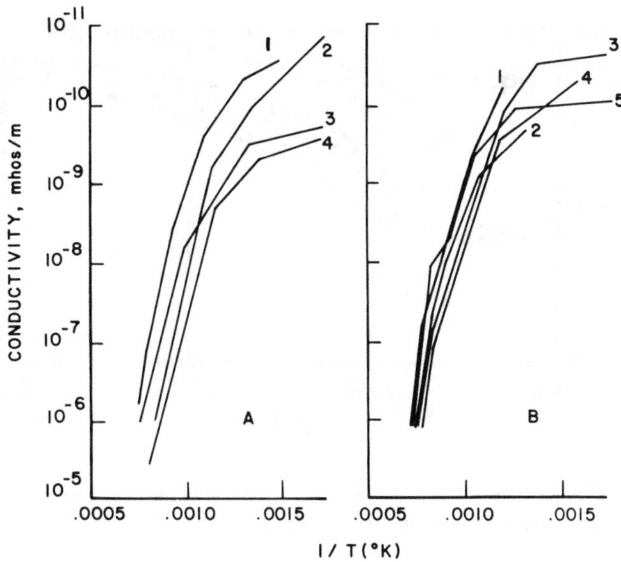

FIGURE 9. Relationship between conductivity and temperature in acidic and intermediate rocks. a. andesite (1,2) quartz diorite (3), and andesite basalt (4). b. Granite (2,3,4), quartzite (1), and perthite (5).

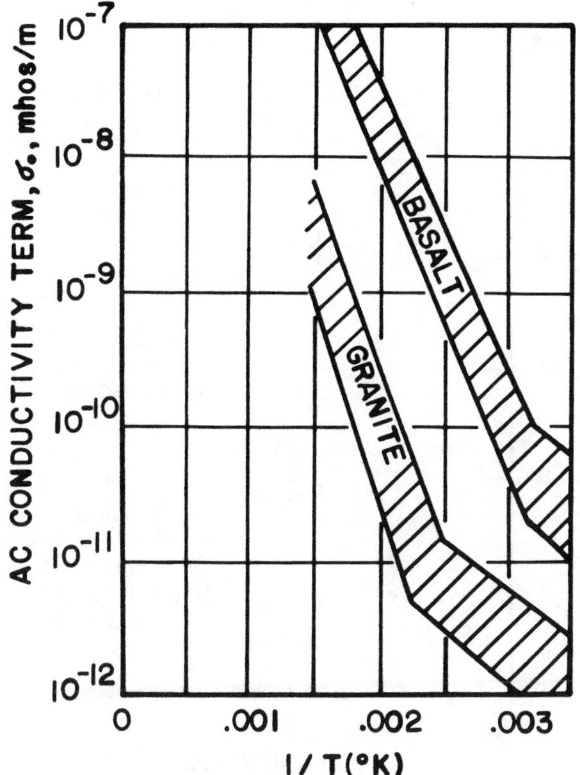

FIGURE 10. Behavior of the AC conduction term as a function of temperature and lithology.

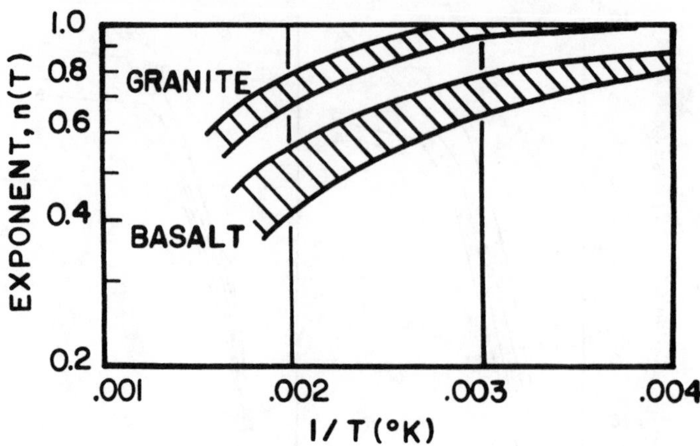

FIGURE 11. Variation of the exponent n(T) in the AC conduction term for granite and metabasalt.

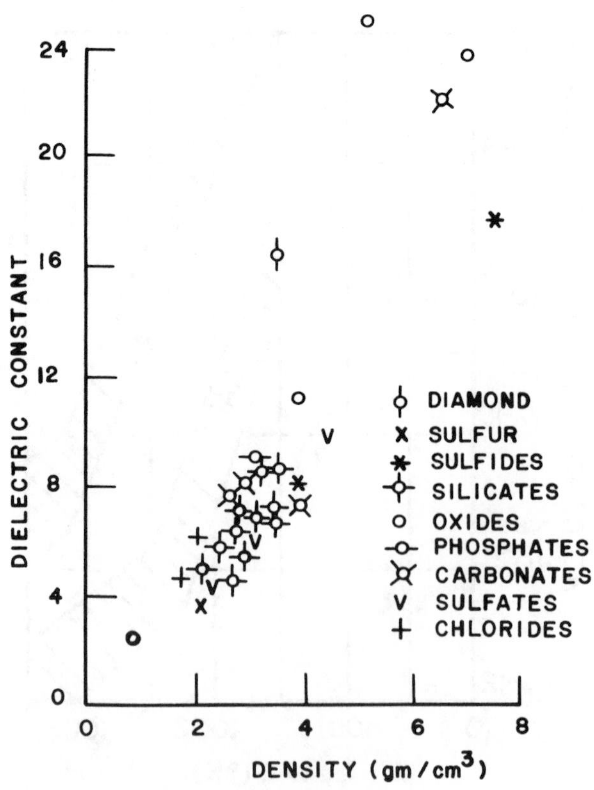

FIGURE 12A. Relationship between dielectric constant and density.[84]

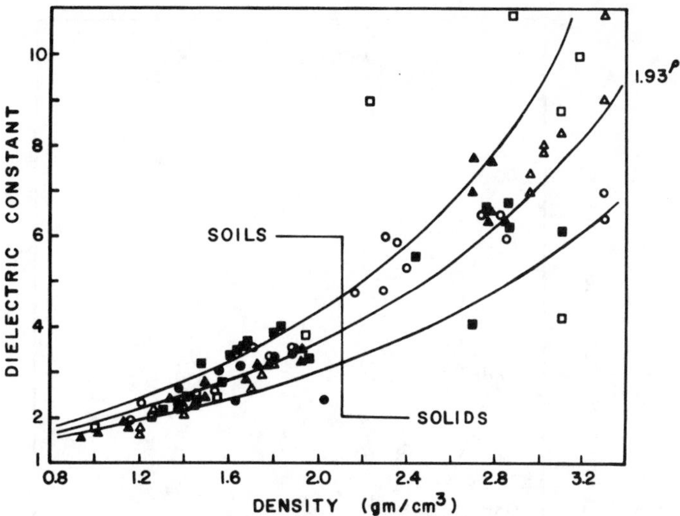

FIGURE 12B. Dielectric constant versus density with fitted equation from regression analysis (also shown are curves for plus or minus one standard deviation). Open squares, triangles, and circles are data from Apollo 11, 12, and 14 samples, respectively, and closed squares, triangles and circles are from Apollo 15, 16, and 17 samples, respectively.[82]

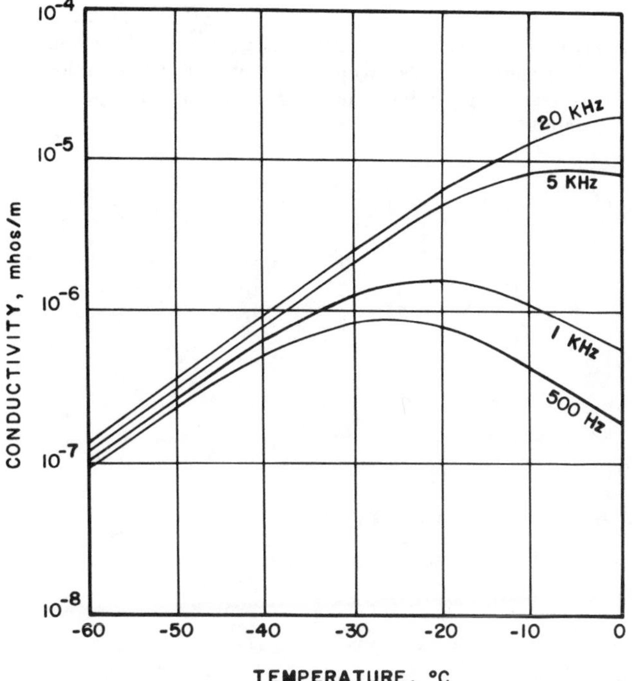

FIGURE 13A. The electrical conductivity of pure ice as a function of temperature and frequency.

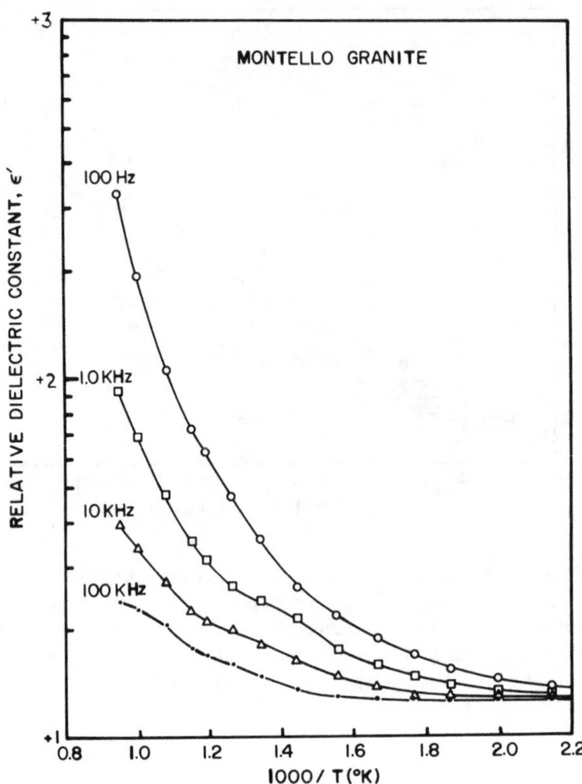

FIGURE 15. Relative dielectric constant versus 1/T for Montello granite at selected frequencies. Note convergence at low temperatures and rapid frequency-dependent rise at high temperatures.[41]

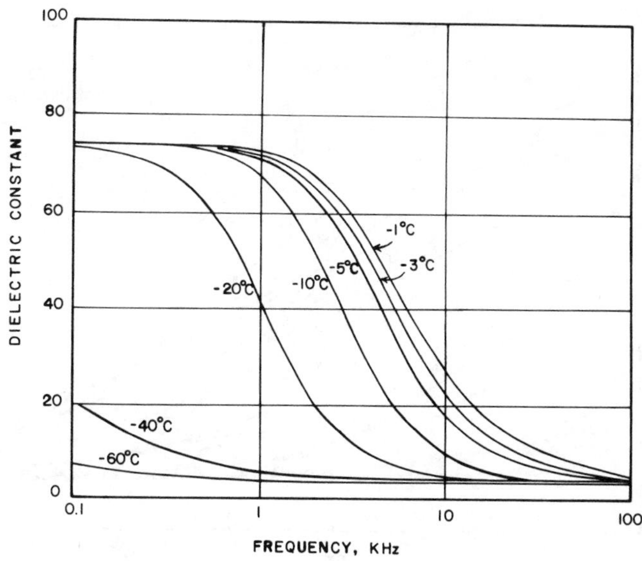

FIGURE 13B. The relative dielectric constant of ice as a function of temperature and frequency.

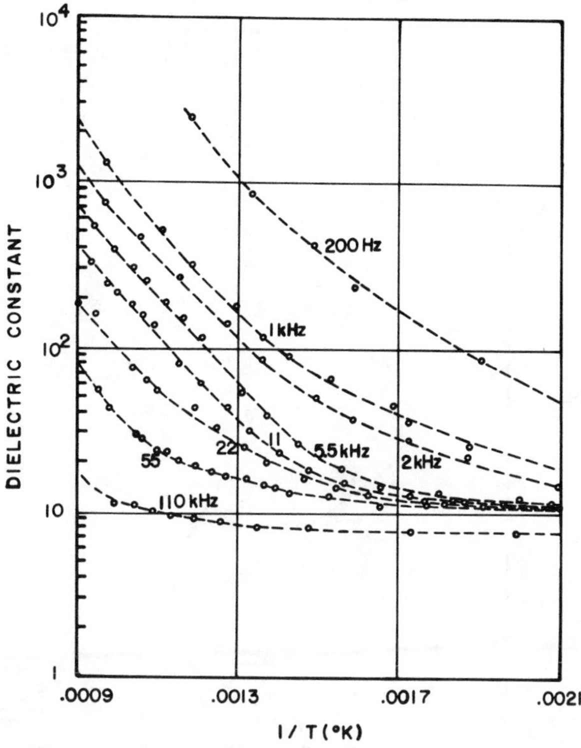

FIGURE 14. Dielectric constant for a sample of the mineral olivine measured as a function of temperature and frequency.[49]

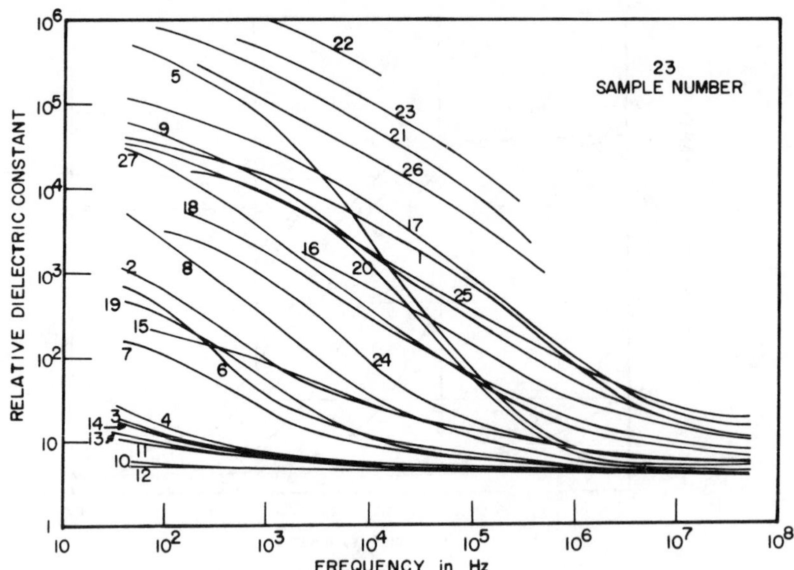

FIGURE 17. Dielectric constants of natural state Morrison cores.[51]

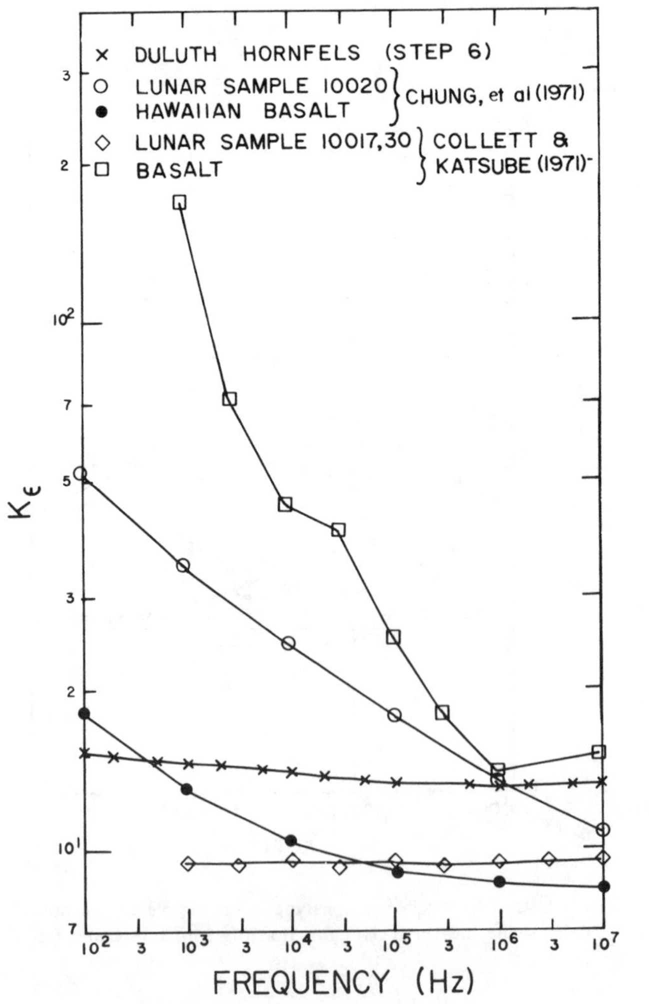

FIGURE 18. K_ϵ spectra of various samples.

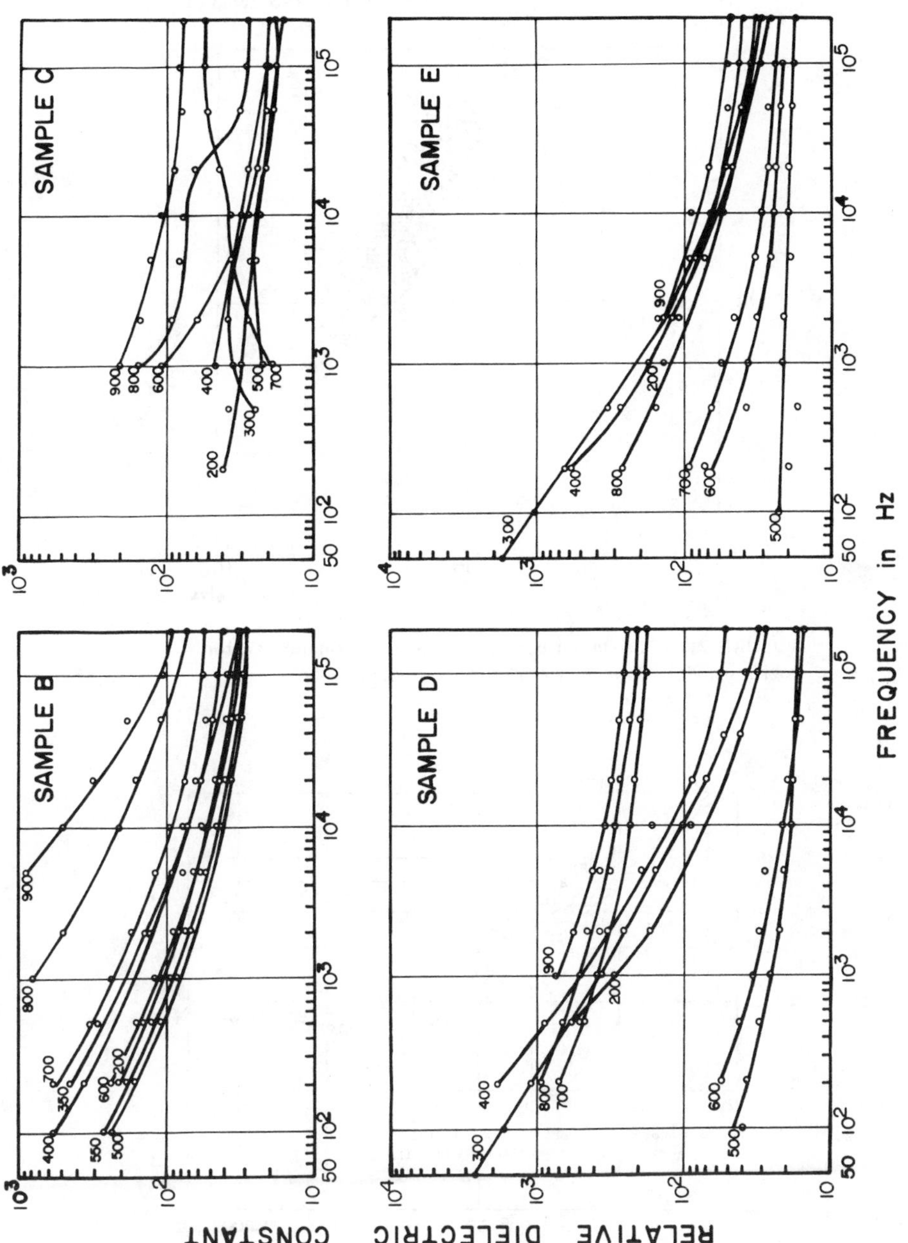

FIGURE 16. Relative dielectric constant as a function of frequency and temperature (°C) for Samples B, C, D, and E.[119]

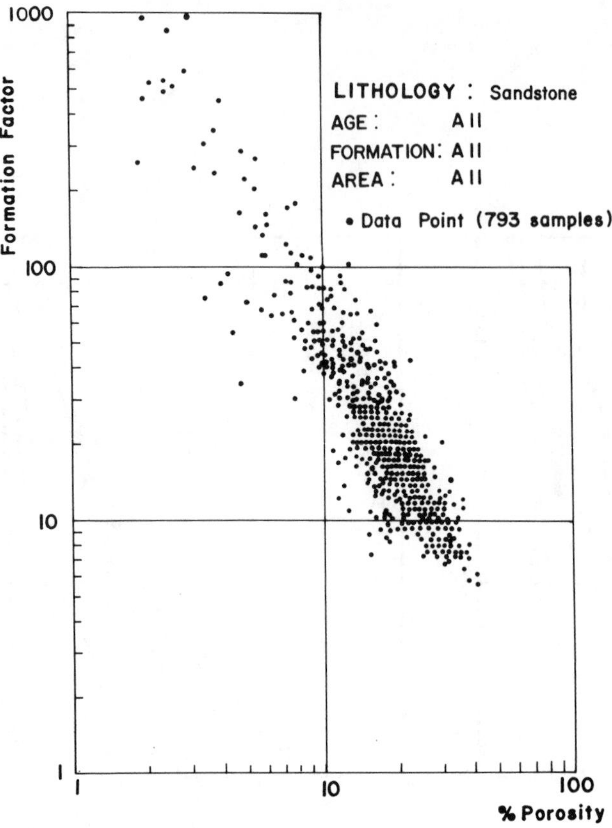

FIGURE 21. Correlation between resistivity (Formation Factor = rock resistivity/contained water resistivity) and porosity.[15]

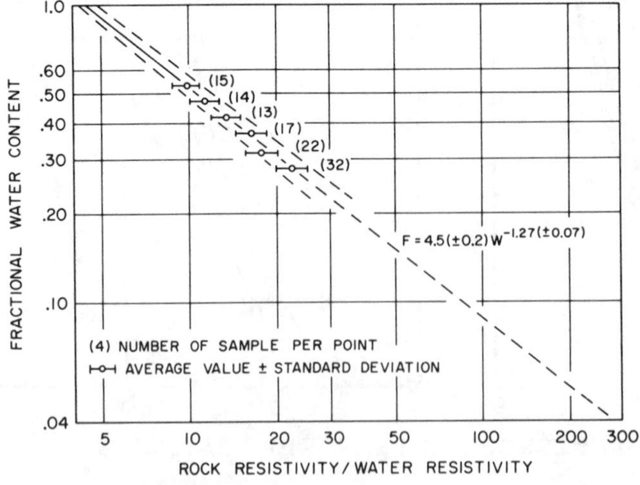

FIGURE 22. Correlation between average values of resistivity and porosity for samples of tuff.[53]

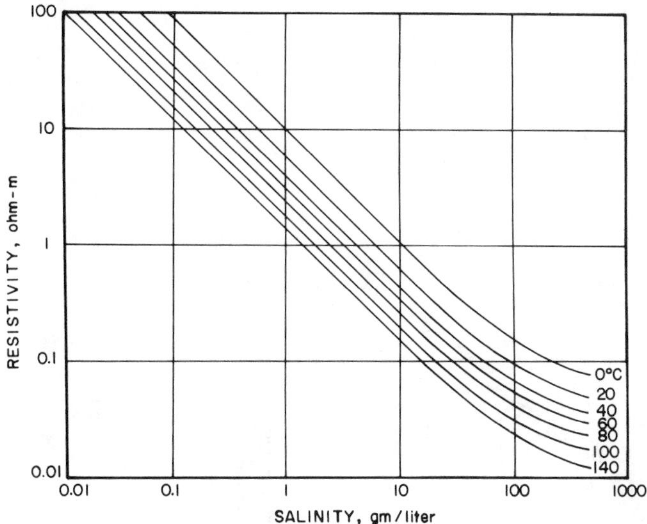

FIGURE 19. Resistivity of solutions of sodium chloride as a function of concentration and temperature.

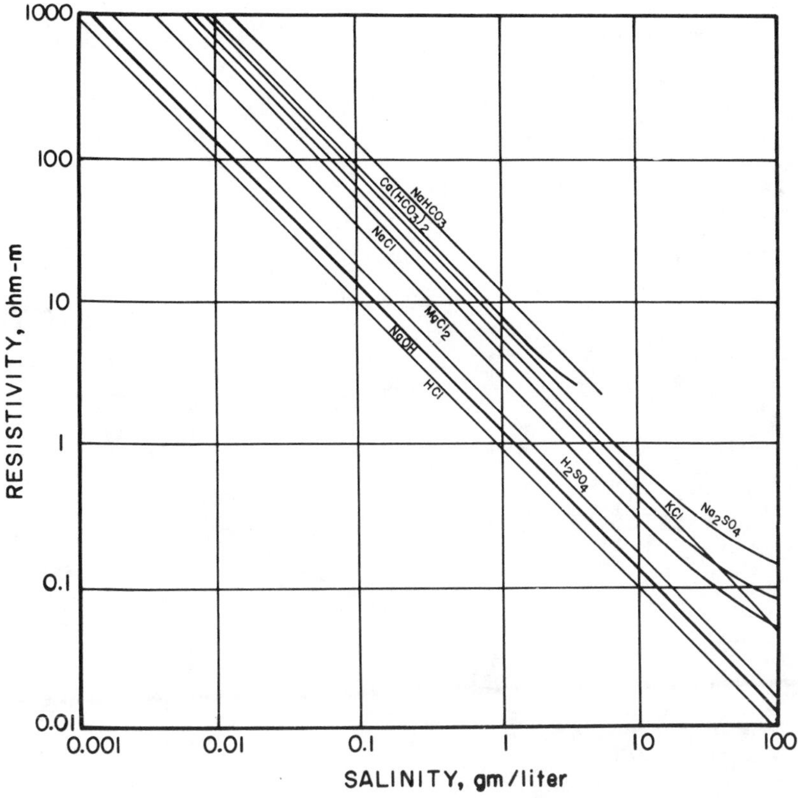

FIGURE 20. Relationship between resistivity and concentration for various salt solutions at a temperature of 18°C.

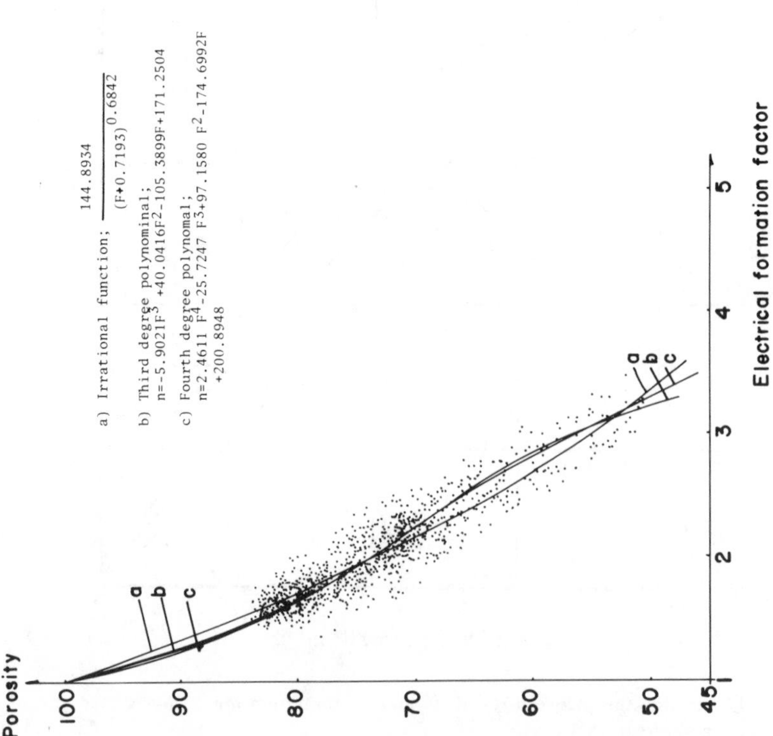

FIGURE 25. Three curves plotted on a set of points corresponding to 2500 measurements made on 21 cores.[66]

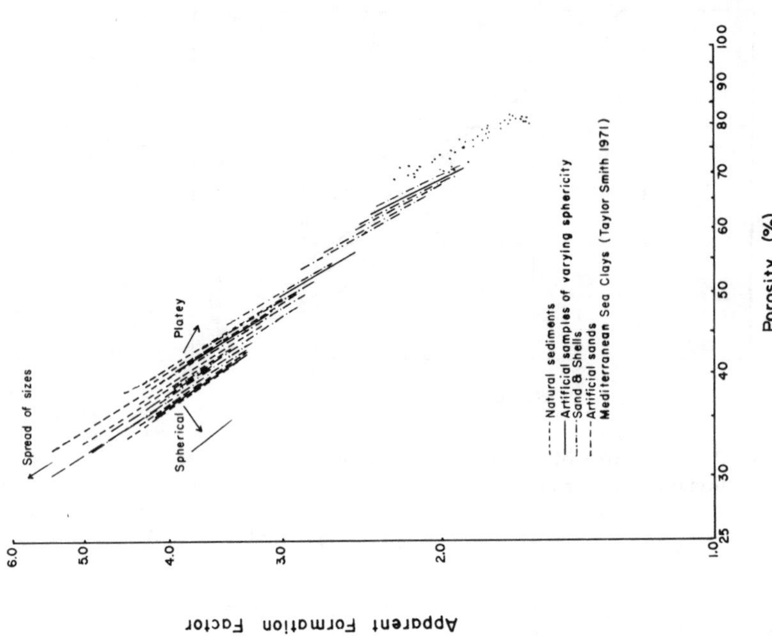

FIGURE 26. Best-fit Archie lines for the complete range of sediments considered, plus some individual values for marine clays.[108]

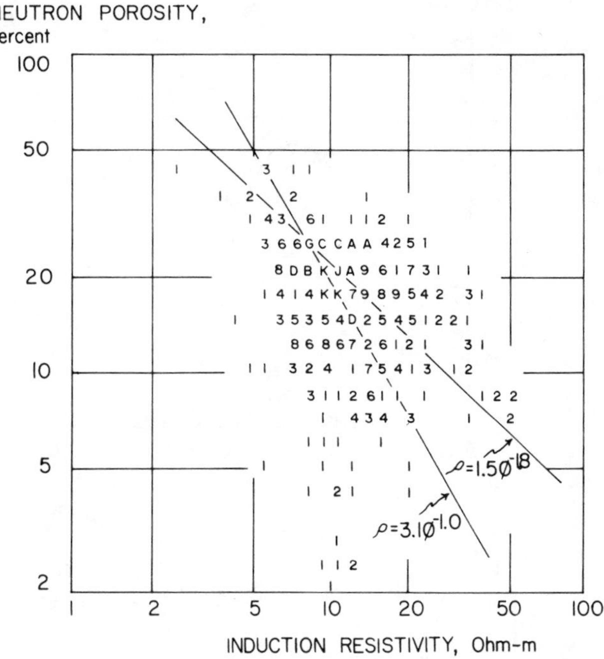

FIGURE 23. Porosity-resistivity correlation from logs run in basalt at Kilauea Volcano, Hawaii.[63]

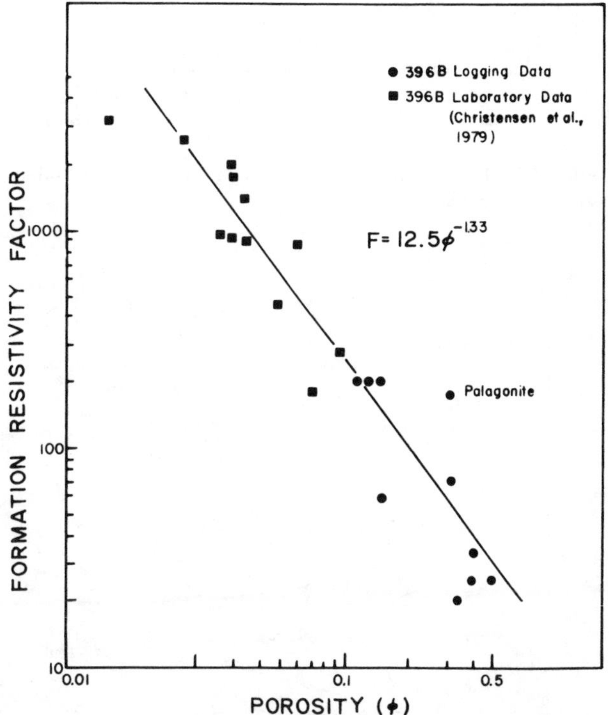

FIGURE 24. Formation factor (versus porosity rock resistivity/pore fluid resistivity) for hole 396B log data assuming a pore fluid resistivity of 0.4 ohm-m along with laboratory measurement of hole 396B samples assuming a fluid resistivity of 0.2 ohm-m.[69]

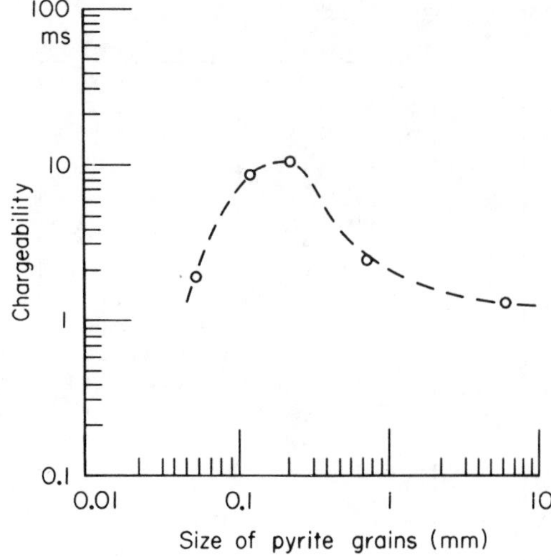

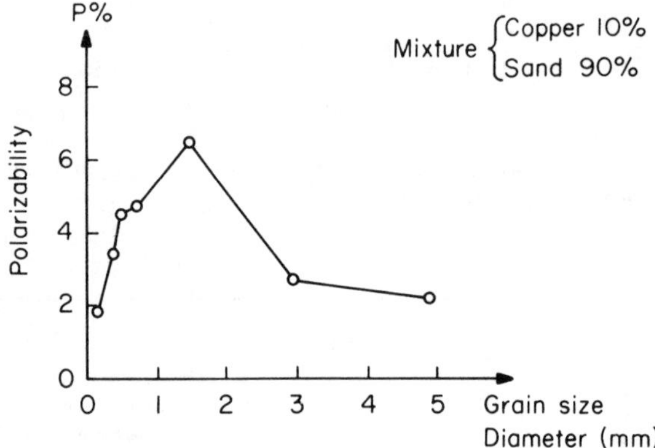

FIGURE 27. Variation of IP effects with grain size for sulfide grains in a sand mixture.[9]

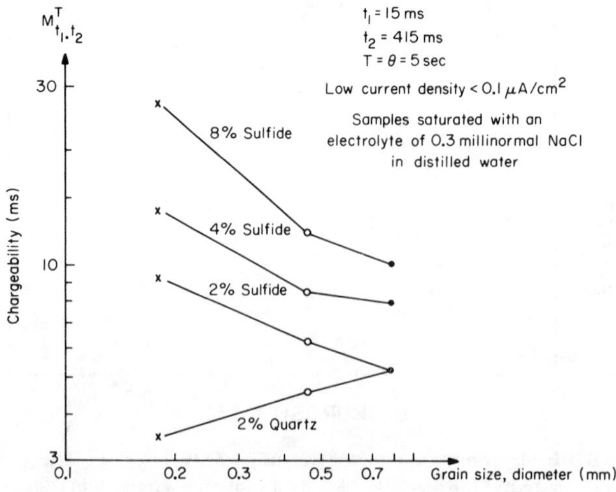

FIGURE 28. IP versus mineral grain size.[97]

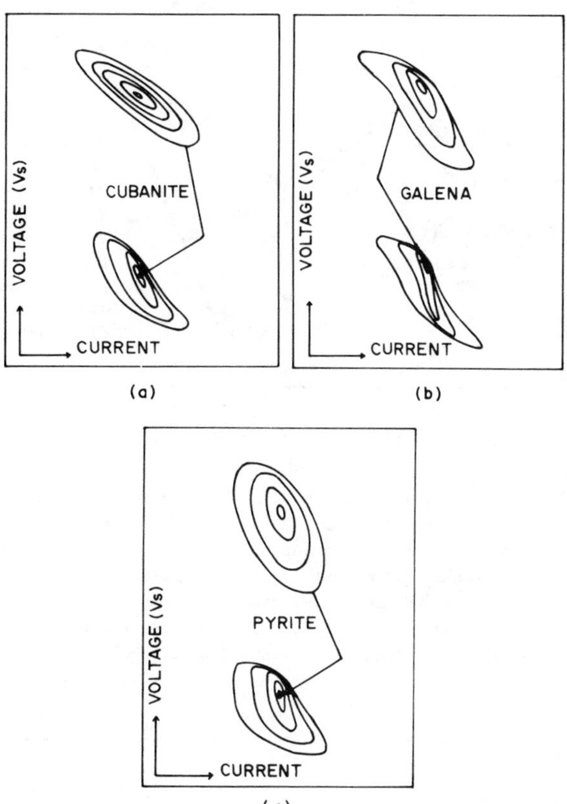

FIGURE 29. Nonlinear Lissajous figures for (a) cubanite, (b) galena, and (c) pyrite at 1.0 hz with progressively increasing current. In each case, the outermost trace of the upper Lissajous is repeated as the innermost trace of the lower Lissajous. The range of current densities is from 10^{-7} to 3×10^{-2} amp/cm^2.[18]

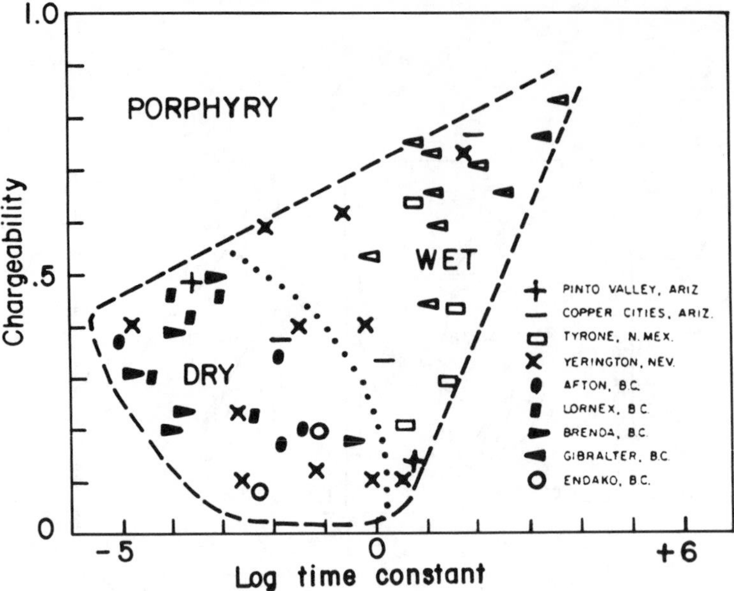

FIGURE 30. Summary of the spectral IP data obtained from porphyry deposits plotted in chargeability-time constant space. The dotted line roughly divides the measurement sites into two groups according to the type of mineralization observed at each site. "Dry" indicates that the total concentration of sulfide minerals is low and that the sulfides commonly occur as discrete disseminated grains. "Wet" indicates that the total concentration of sulfide minerals is high and that the sulfides typically occur as veins and veinlets.[86]

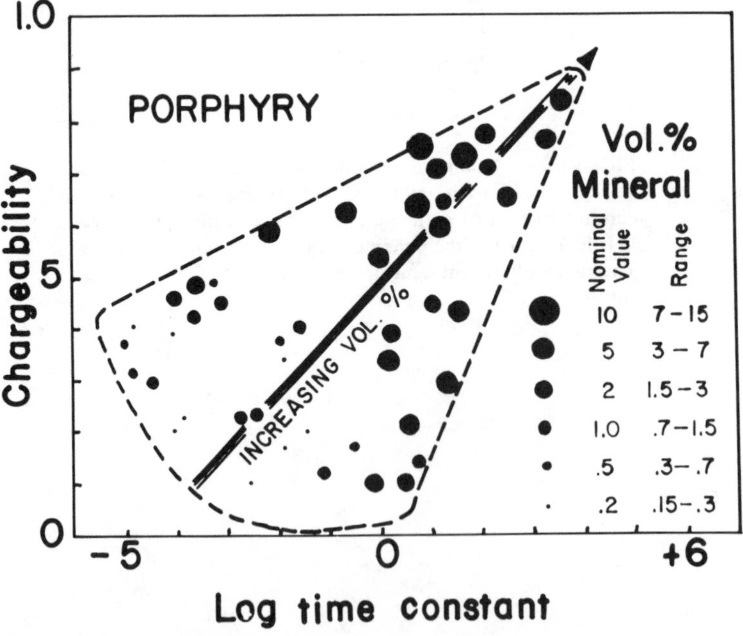

FIGURE 31. Data from porphyry deposits plotted in chargeability-time constant space. The larger dots indicate high sulfide concentration. Superimposed on the plot is the arrow from Figure 6 which indicates the trend due to increasing volume percent sulfides in artificial rocks.[86]

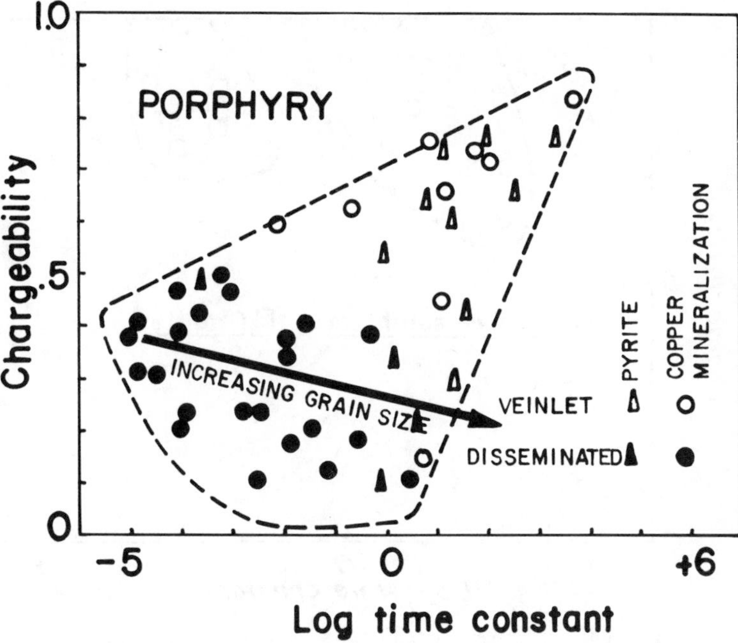

FIGURE 32. Data from porphyry deposits plotted in chargeability-time constant space. There is a grouping of veinlet mineralization (open symbols) versus discretely disseminated mineralization (closed symbols) but there is no distinct grouping of pyrite mineralization (triangles) versus copper mineralization (circles). Superimposed on the plot is the arrow from Figure 6 which indicates the trend due to increasing grain size in artificial rocks.[86]

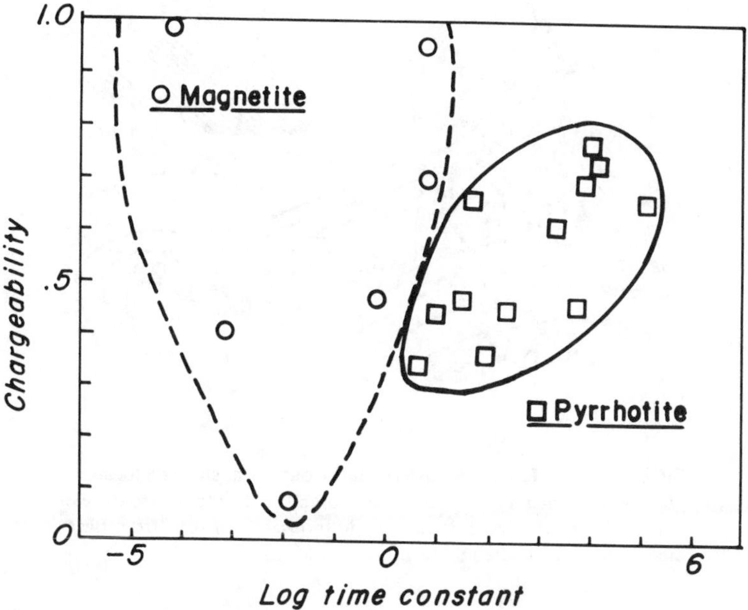

FIGURE 33. Summary of the spectral IP data obtained from magnetite and pyrrhotite deposits plotted in chargeability -time constant space.[86]

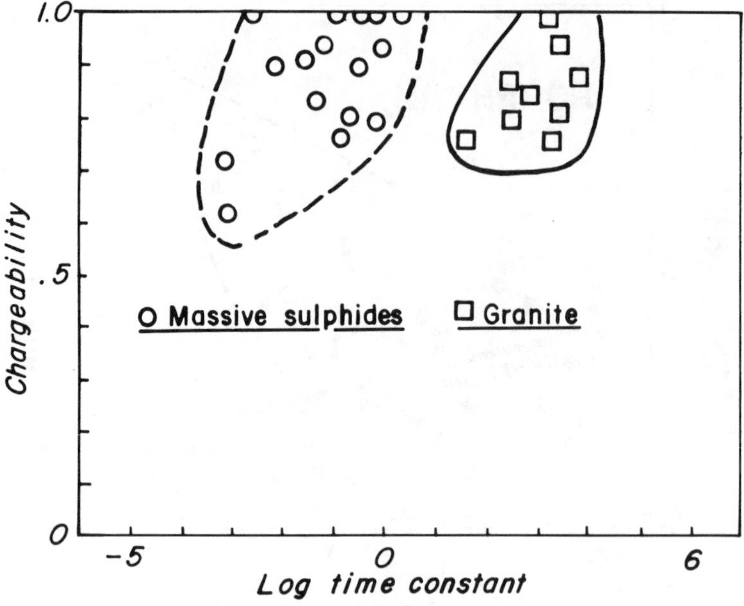

FIGURE 34. Summary of the spectral IP data obtained from massive sulfide and graphite deposits plotted in chargeability-time constant space.[86]

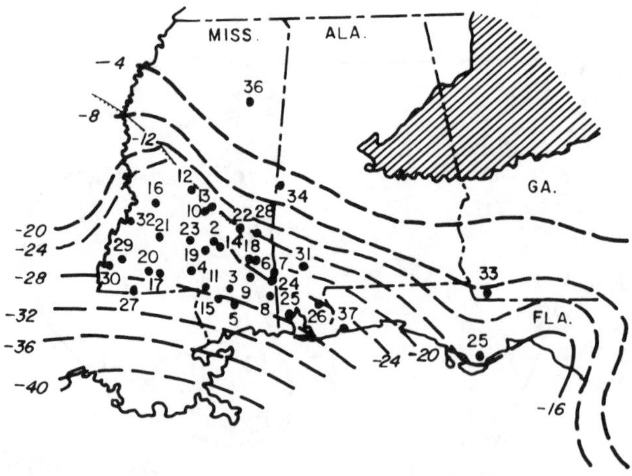

FIGURE 35. Map of the eastern Gulf Coast area, showing locations of wells for which electric logs were compiled. Contours are the elevation of the pre-Mesozoic surface, in thousands of feet (from Basement Map of the United States).

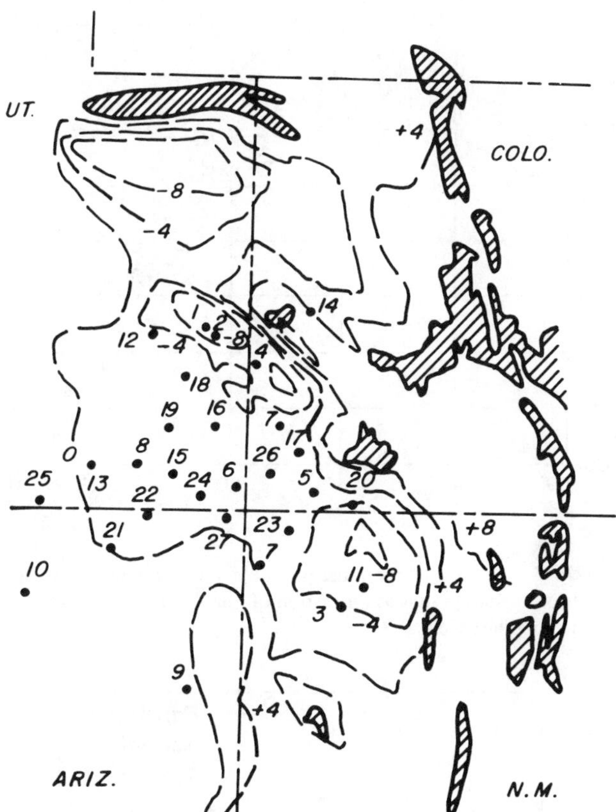

FIGURE 36. Map of the Four-Corners area in the states of Colorado, Utah, Arizona and New Mexico, with locations of wells for which electric logs were compiled. Contours represent the elevation of the Precambrian basement surface, in thousands of feet from sea-level (taken from USGS Basement Map of the United States). Shaded areas are of basement outcrop.

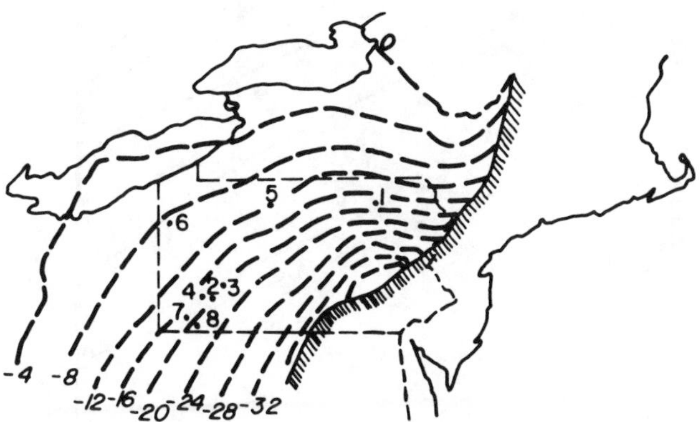

FIGURE 37. Locations of wells in Pennsylvania for which electric logs were compiled. Contours are elevations of the Precambrian surface in thousands of feet.

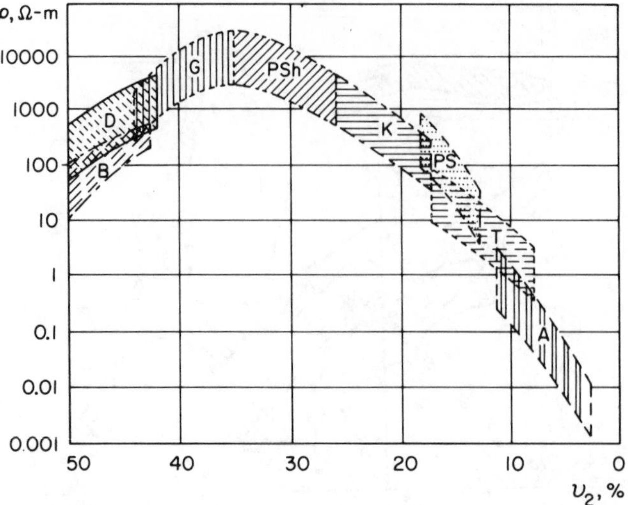

FIGURE 38. Graph of the most probable values for the resistivity of coals of various grades as a function of quality. The ash content is plotted along the abscissa.

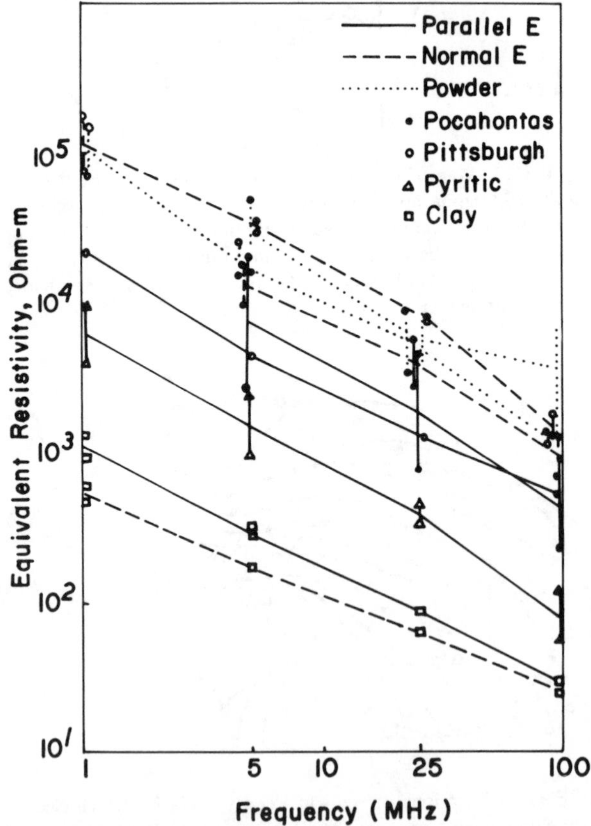

FIGURE 39A. RF resistivities of two bituminous coals and a coal clay.[19]

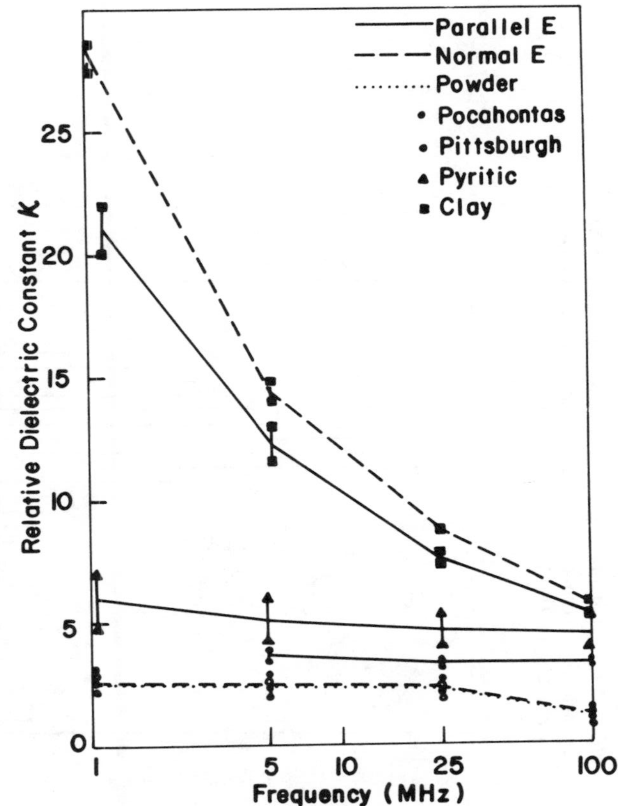

FIGURE 39B. **RF dielectric constants of two coals and a coal clay.**[19]

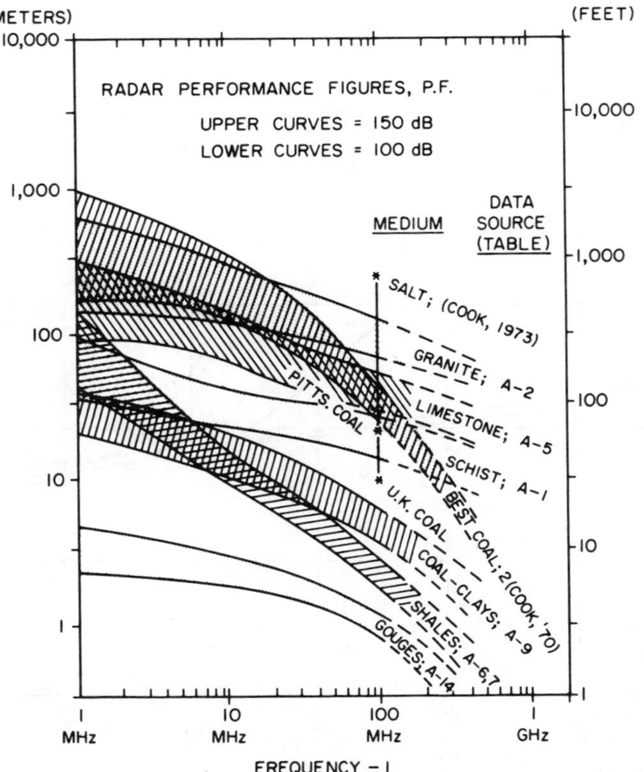

FIGURE 40. **Radar probing distances through some typical "rocks".**

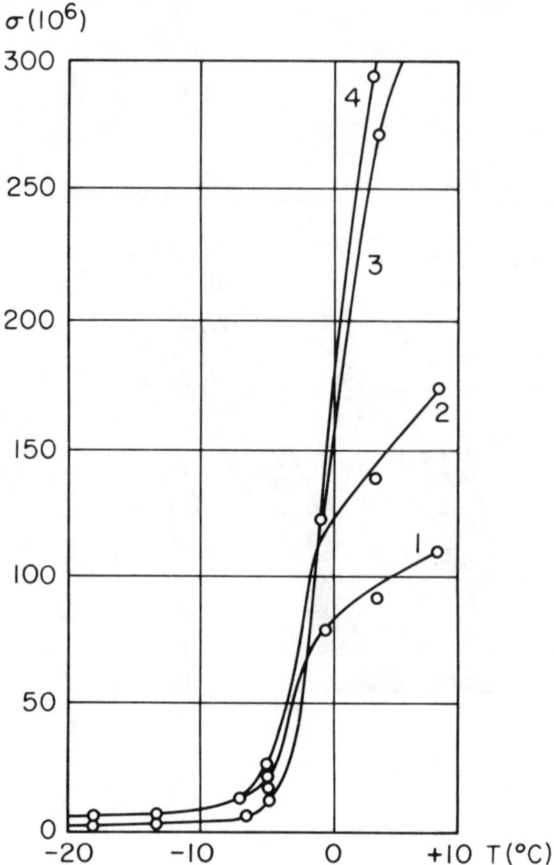

FIGURE 41. Relationship between resistivity and temperature near the freezing point as a function of water content. 1 to 1.8%; 2 to 14.2%; 3 to 24.4%; 4 to 34.2%.[10]

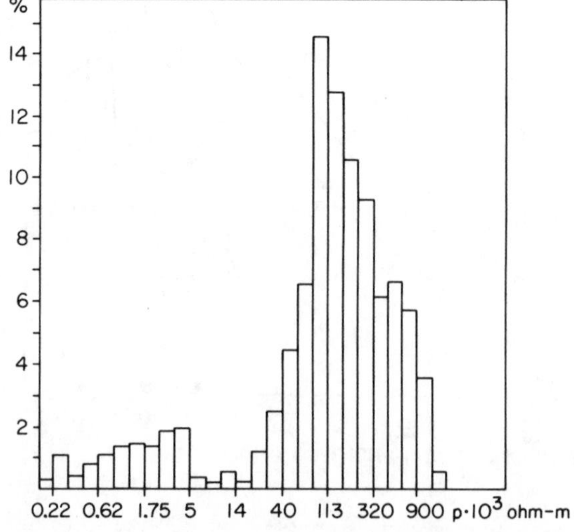

FIGURE 42. Histogram of observed resistivity in frozen rock, in thousands of ohm-meters.[117]

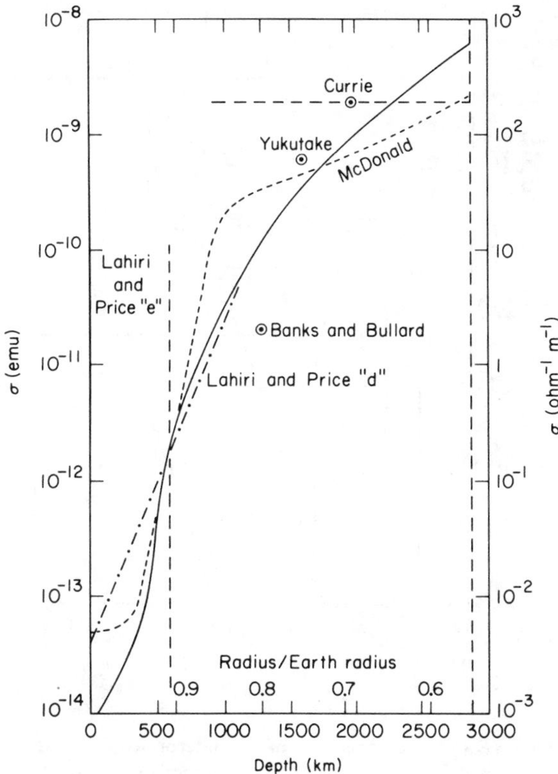

FIGURE 43. Electrical conductivity of the mantle. The author's preferred curve is shown as a solid line. Also shown are distributions by Lahiri and Price (1939) and McDonald (1957) and data by Yukutake (1965), Banks and Bullard (1966), and Currie (1968). Uncertainties are of the order of a factor of 10. (Stacey, 1969).

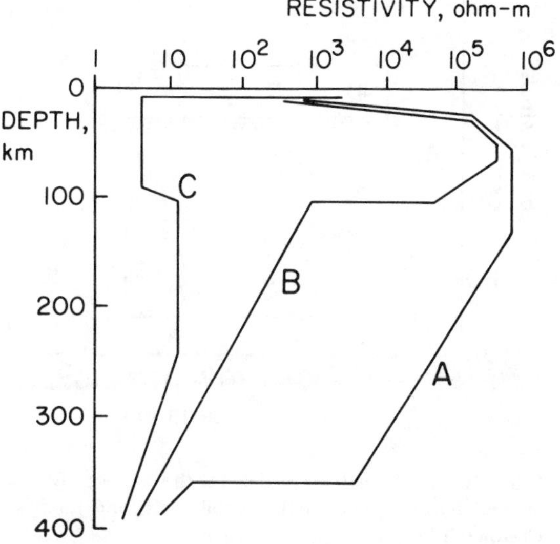

FIGURE 44. **Highly idealized resistivity profiles through the crust and upper mantle. Profile A pertains to a stable contental nucleus, profile B to a mobile crustal plate, and profile C to a volcanic rift area.**[57]

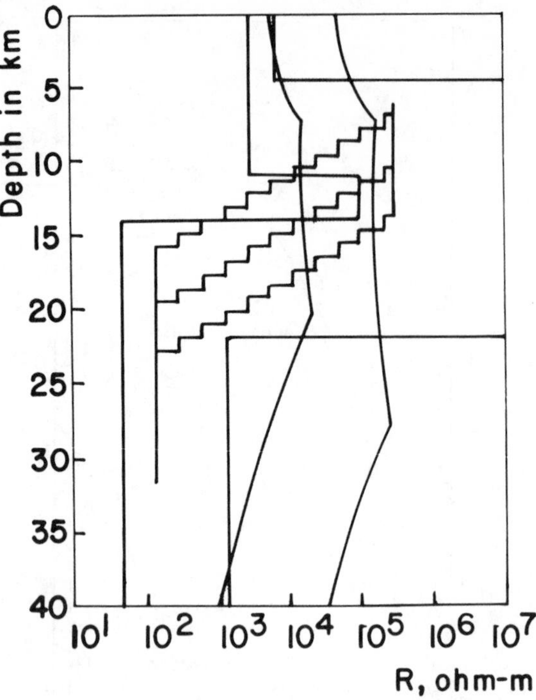

FIGURE 45. Comparison of our interpreted resistivity-depth profile for the lower crust with Brace's (1971) model. Brace's model for the lower crust was based on laboratory data and hypothesized temperatures and compositions for the lower crust. Rectangular boxed region (to $0.6 \times 10^2 - 10^3$ ohm-m at 40 km) represents our interpreted profile; vertical swath (to $10^3 - 5 \times 10^4$) represents Brace's profile. The step models represent the transition layers.[104]

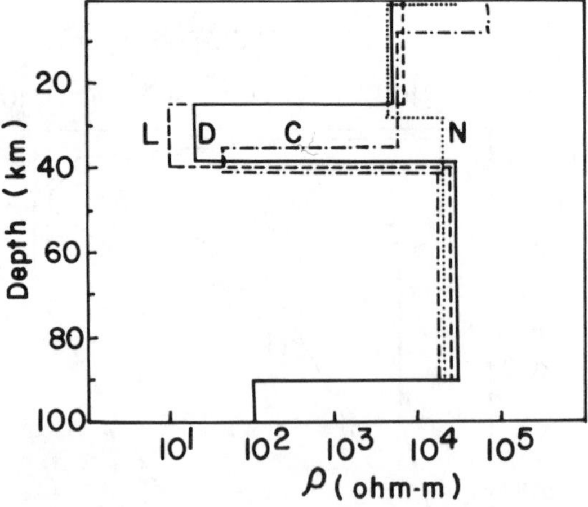

FIGURE 46. Resistivity models for the Limpop (L), Namaqua (N), and Damara (D) mobile belts and Kaapvaal Craton (C).[112]

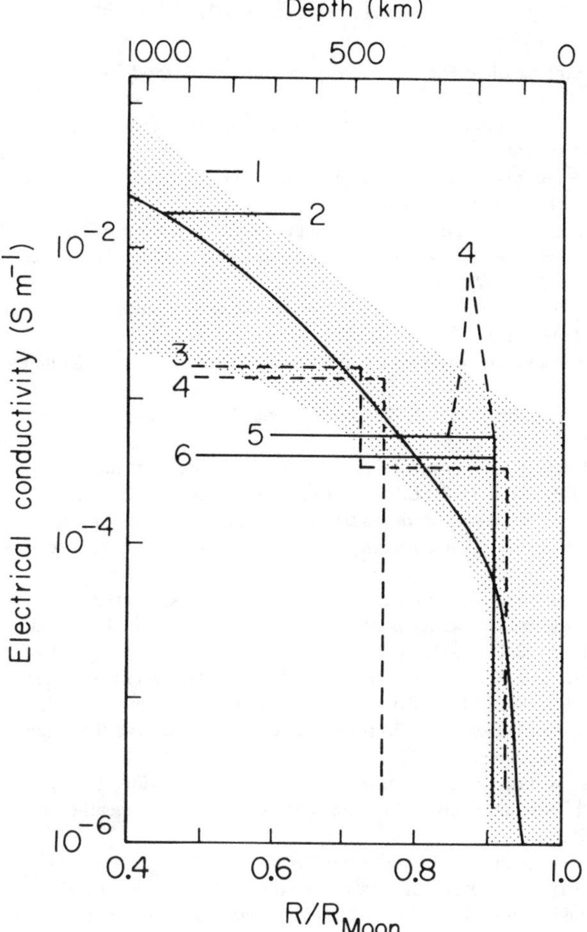

FIGURE 47. Lunar electrical conductivity profiles obtained by different techniques and investigators. (1) night-side transient analysis in the solar wind; (2) geomagnetic tail transient analysis; (3), (4) and (6) day-side harmonic analyses in the solar wind using different conductivity models; (5) day-side harmonic analysis using a two-layer model.[44]

REFERENCES

1. **Alvarez, R.**, Complex dielectric permittivity in rocks: a method for its measurement and analysis, *Geophysics*, 38(5), 920, 1973.
2. **Alvarez, R.**, Dielectric comparison of lunar and terrestrial fines at lunar conditions, *J. Geophys Res.*, 79(35), 5453, 1974.
3. **Anderson, L. A.**, Experimental deep resistivity probes in central and eastern United States, U.S. Geol. Surv. Tech. Lett., Crustal Studies 31, 1965.
4. **Anderson, L. A. and Keller, G. V.**, A study in induced polarization, *Geophysics*, 29(5), 848, 1964.
5. **Angoran, Y. and Madden, J. R.**, Induced polarization: a preliminary study of its chemical basis, *Geophysics*, 42(4), 788, 1977.
6. **Archie, G. E.**, The electrical resistivity log as an aid in determining some reservoir characteristics, *Trans. Am. Inst. Min. Eng.*, 146, 54, 1942.
7. **Banks, R. J.**, The overall conductivity distribution of the earth, *J. Geomagn. Geoelectr.*, 24, 337, 1972.
8. **Banks, R. J. and Bullard, E. C.**, The annual and 27 day magnetic variations, *Earth Planet. Sci. Lett.*, 1, 118, 1966.
9. **Bertin, J. and Loeb, J.**, *Experimental and Theoretical Aspects of Induced Polarization*, Vol. 1, Nostrand, V. R. and Parasins, D. S., Eds., International Publications Service, New York, 1976.
10. **Borovinskiy, B. A.**, *Elektro-i seismometricheskie issledovaniya mnogletnemerzlikh Gornikh Porod i Lednikov (Electrical and Seismic Investigations of Permafrost)*, Nauka, Moscow, 1969.
11. **Böttcher, C. J. F.**, *Theory of Polarization*, Elsevier, Amsterdam, 1952.
12. **Brace, W. F.**, Resistivity of saturated crustal rocks to 40 km based on laboratory studies, in *The Structure and Physical Properties of the Earth's Crust*, Heacock, J. G., Ed., American Geophysics Union, Washington, D.C., 1971, 243.
13. **Campbell, M. J.**, The electrical parameters of lunar surface rocks, in *Electromagnetic Exploration of the Moon*, Linlor, W. I., Ed., Mono Book Corporation, Baltimore, 1970, 91.
14. **Campbell, M. J. and Ulrichs, J.**, Electrical properties of rocks and their significance for lunar radar observations, *J. Geophys. Res.*, 75, 6524, 1969.
15. **Carothers, J. E.**, A statistical study of the formation factor relation, *Log Anal.*, 13, 1968.
16. **Chebotarev, I. I.**, Metamorphism of natural waters in the crust of weathering, *Geochim. Cosmochim. Acta*, 8, 53, 1955.
17. **Chung, D. H., Westphal, W. B., and Simmons, G.**, Dielectric properties of Apollo II lunar samples and their comparison with earth materials, *J. Geophys. Res.*, 75, 6524, 1970.
18. **Collett, L. S. and Katsube, T. J.**, Electrical parameters of rocks in developing geophysical techniques, *Geophysics*, 38(1), 76, 1973.
19. **Cook, J. C.**, R.F. electrical properties of bituminous coal samples, *Geophysics*, 35(6), 1079, 1970.
20. **Cook, J. C.**, Radar transparencies of mine and tunnel rocks, *Geophysics*, 40(5), 865, 1975.
21. **Currie, R. G.**, Geomagnetic spectrum of internal origin and lower mantle conductivity, *J. Geophys. Res.*, 73, 2779, 1968.
22. **Dortman, N. B., Vasileva, V. I., Veinberg, A. K., Dubinchik, E. Ya., Zhdanov, V. V., Zotova, I. F., Ilaev, M. G., Trunina, V. Ya., Khoreva, B. Ya., and Shoplo, L. E.**, *Fizicheskie Svoistva Gornikh Porod i Poleznikh Iskopaemikh SSR (Physical Properties of Rocks and Ores in the USSR)*, Nedra, Moscow, 1964.
23. **Duba, A.**, Electrical conductivity of olivine, *J. Geophys. Res.*, 77(14), 2483, 1972.
24. **Duba, A., Bolland, J. N., and Ringwood, A. E.**, The electrical conductivity of pyroxene, *J. Geol.*, 81(6), 727, 1973.
25. **Duba, A. G., Heard, H. C., and Schock, R. N.**, Electrical conductivity of olivine at high pressure and under controlled oxygen fugacity, *J. Geophys. Res.*, 79, 1667, 1974.
26. **Duba, A. G. and Nicholls, I. A.**, The influence of oxidation state on the electrical conductivity of olivine, *Earth Planet. Sci. Lett.*, 18, 59, 1973.
27. **Dvorak, Z.**, Electrical conductivity of several samples of olivinites, periodotites, and dunites as a function of pressure and temperature, *Geophysics*, 38(1), 14, 1973.
28. **Dvorak, Z. and Schoessin, H. H.**, On the anistropic electrical conductivity of enstatite as a function of pressure and temperature, *Geophysics*, 38(1), 25, 1973.
29. **Dyal, P. and Parkin, C. W.**, Global electromagnetic induction in the moon and planets, *Phys. Earth Planet. Inter.*, 7, 251, 1973.
30. **Eisenberg, D. and Kauzmann, W.**, *The Structure and Properties of Water*, Oxford University Press, Oxford, 1969, 296.
31. **Evjen, H. M.**, Theory and practice of low-frequency electromagnetic exploration, *Geophysics*, 13(4), 584, 1948.

32. **Fatt, I.**, The network model of porous media, *Trans. Am. Inst. Min. Eng.*, 207, 144, 1956.
33. **Fraser, D. C., Keevil, N. B., and Ward, S. H.**, Conductivity spectra of rocks from the Craigmont ore environment, *Geophysics*, 29, 832, 1964.
34. **Frolich, H.**, *Theory of Dielectrics*, Clarendon Press, Oxford, 1949.
35. **Grechukhin, V. V.**, *Geofizicheskie metodi issledovaniya ugolnikh skvashin (Geophysical Logging of Boreholes in Coal)*, Nedra, Moscow, 1965, 467.
36. **Hansen, W., Sill, W. R., and Ward, S. H.**, The dielectric properties of selected basalts, *Geophysics*, 38(1), 135, 1973.
37. **Harthill, N.**, The CSM test area for electrical surveying methods, *Geophysics*, 33(4), 675, 1968.
38. **Harvey, R. D.**, Electrical conductivity and polished mineral surfaces, *Econ. Geol.*, 23, 778, 1928.
39. **Hermance, J. F.**, An electrical model for the sub-Icelandic crust, *Geophysics*, 38(1), 3, 1973.
40. **Hill, D. G.**, A laboratory investigation of electrical anisotropy in precambrian rocks, *Geophysics*, 37(6), 1022, 1972.
41. **Housley, R. M. and Oliver, J. R.**, Electrical characteristics of igneous precambrian basement rocks of central North America, in *The Earth's Crust, Its Nature and Physical Properties*, Heacock, J. G., Ed., Am. Geophys. Un. Monogr. Ser., 20, 181, 1977.
42. **Hughes, H.**, The pressure effect on the electrical conductivity of peridot, *J. Geophys. Res.*, 60, 187, 1955.
43. **Hutt, J. E. and Berg, J. W., Jr.**, Thermal and electrical conductivities of sandstone rocks and ocean sediments, *Geophysics*, 33(3), 489, 1968.
44. **Hutton, V. R. S.**, The electrical conductivity of the earth and planets, *Rep. Prog. Phys.*, 39, 487, 1976.
45. **Jackson, D. B.**, Electrical properties of the sedimentary section in the high plains area, U.S. Geol. Surv. Tech. Lett. Rep., ARPA Order No. 193-61, 1962.
46. **Jackson, D. B.**, Deep resistivity probes in the southwestern United States, U.S. Geol. Surv. Tech. Lett., Crustal Studies 29, 1965.
47. **Jackson, D. B.**, Deep resistivity probes in the southwestern United States, *Geophysics*, 32(1), 1123, 1967.
48. **Jackson, D. B., Smith, D. T., and Stanford, P. N.**, Resistivity-porosity-particle shape relationships for marine sands, *Geophysics*, 43(6), 1250, 1978.
49. **Jacobson, J. J.**, The Electrical Conductivity of Olivine at High Temperatures, Master's thesis T-1002, Colorado School of Mines, Golden, 1964.
50. **Katsube, T. J., Ahrens, R. H., Collett, L. S.**, Electrical non-linear phenomena in rocks, *Geophysics*, 38(1), 106, 1973.
51. **Keller, G. V.**, Analysis of some electrical transient measurements on igneous, sedimentary, and metamorphic rocks, in *Overvoltage Research and Geophysical Applications*, Wait, J. R., Ed., Pergamon Press, New York, 1959.
52. **Keller, G. V.**, Electrical properties of sandstones of the Morrison formation, *U. S. Geol. Surv. Bull.*, 1052, 307, 1959.
53. **Keller, G. V.**, Electrical resistivity of rocks in the area 12 tunnels, Nevada Test Site, Nye County, Nevada, *Geophysics*, 27(2), 242, 1962.
54. **Keller, G. V.**, Electrical properties in the deep crust, *IEEE Trans. Antennas Propag.*, 11(3), 334, 1963.
55. **Keller, G. V.**, Electrical properties of rocks and minerals, in *Handbook of Physical Constants*, Clark, S. P., Ed., Geological Society of America, Boulder, Colo., 1966, 533.
56. **Keller, G. V.**, Electrical prospecting for oil, *Q. Colo. Sch. Mines*, 63(2), 1, 1968.
57. **Keller, G. V.**, Electrical characteristics of the earth's crust, in *Electromagnetic Probing in Geophysics*, Golem Press, Boulder, Colo., 1971, 13.
58. **Keller, G. V.**, Electrical properties of the earth's crust, a survey of the literature, ONR Contract N000-14-70-C-09290, Colorado School of Mines, Golden, 1971.
59. **Keller, G. V.**, Electrical studies of the crust and upper mantle, in *The Structure and Physical Properties of the Earth's Crust*, Heacock, J. G., Ed., American Geophysical Union, Washington, D.C., 1971, 107.
60. **Keller, G. V., Anderson, L. A., and Pritchard, J. I.**, Geological survey investigations of the electrical properties of the crust and upper mantle, *Geophysics*, 31(16), 1078, 1966.
61. **Keller, G. V. and Frischknecht, F. C.**, *Electrical Methods in Geophysical Prospecting*, Pergamon Press, New York, 1966, 517.
62. **Keller, G. V. and Licastro, P. H.**, Measurements of electrical resistivity and dielectric constant on some sandstone and siltstone cores, *Geophysics*, 21(2), 538, 1956.
63. **Keller, G. V., Grose, L. T., Murray, J. C., and Skokan, C. K.**, Results of an experimental drill hole at the summit of Kilauea Volcano, Hawaii, *Geophys. Res. Lett.*, 5(3), 345, 1979.

64. **Keller, G. V. and Licastro, P. H.**, Dielectric constant and electrical resistivity of natural-state cores, *U.S. Geol. Surv. Bull.*, 1052-H, 257, 1959.
65. **Keller, G. V., Murray, J. C., Skokan, J. J., and Skokan, C. J.**, CSM research drillhole at summit of Kilauea Volcano, *Mines Mag.*, 64, 14, 1974.
66. **Kermabom, A., Gehin, C., and Blavier, P.**, A deep sea electrical resistivity probe for measuring porosity and density of unconsolidated sediments, *Geophysics*, 34(4), 554, 1969.
67. **Khalafalla, G. S. and Maeglev, W. J.**, Low-frequency impedance parameters of basalt, granite, and quartzite, *Geophysics*, 38(1), 68, 1973.
68. **Khitarov, N. I., Slusky, A. B., and Pugin, V. A.**, Electrical conductivity of basalts at high T-P and phase transitions under upper mantle conditions, *Phys. Earth Planet. Inter.*, 3, 334, 1970.
69. **Kirkpatrick, R. J.**, The physical state of the oceanic crust: results of downhole geophysical logging in the mid-Atlantic ridge at 23° N, *J. Geophys. Res.*, 84(B1), 178, 1979.
70. **Klein, J. D. and Shuey, R. T.**, Non-linear impedance of mineral-electrolyte interface. I. Pyrite, *Geophysics*, 43(6), 1222, 1978.
71. **Klein, J. D. and Shuey, R. T.**, Non-linear impedance of mineral-electrolyte interface. II. Galena, chalcopyrite and graphite, *Geophysics*, 43, 1235, 1978.
72. **Kobranova, V. N.**, *Fizicheskie Svoistva Gornikh Popod (Physical Properties of Rocks)*, Gostoptekhizdat, Moscow, 1962, 490.
73. **Kwon, Bong-sung**, A Mathematical Pore Structure Model and Pore Structure Interrelationships, Ph.D. Thesis T-1717, Colorado School of Mines, Golden, 1975.
74. **Lahiri, B. N. and Price, A. T.**, Electromagnetic induction in non-uniform conductors, and the determination of the conductivity of the earth from terrestrial magnetic variations, *Philos. Trans. R. Soc. Ser. A*, 237, 509, 1939.
75. **Lebedev, E. B. and Khitarov, N. I.**, Beginning of melting in granite and the electrical conductivity of its melts in relation to pressure of pore water, *Geokhimiya*, 3, 195, 1964.
76. **Lidiard, A. B.**, Ionic conductivity, in *Handbuch der Physik*, Vol. 20, Electrische Leitungsphomene II, 1957, 246.
77. **Madden, T. R.**, Random networks and mixing laws, *Geophysics*, 41(6A), 1104, 1976.
78. **McDonald, K. L.**, Penetration of the geomagnetic secular field through a mantle with variable conductivity, *J. Geophys. Res.*, 62, 117, 1957.
79. **Olhoeft, G. R.**, Lunar sample electrical properties, *Geochim. Cosmochim. Acta*, 3, 3133, 1973.
80. **Olhoeft, G. R.**, Electrical properties of rocks, in *The Physics and Chemistry of Rocks and Minerals*, John Wiley & Sons, London, 1976, 261.
81. **Olhoeft, G. R., Frisillo, A. L., and Strangway, D. W.**, Electrical properties of lunar soil sample 15301, 39, *J. Geophys. Res.*, 79, 1599, 1974.
82. **Olhoeft, G. R. and Strangway, D. W.**, Dielectric properties of the first 100 meters of the moon, *Earth Planet. Sci. Lett.*, 24, 394, 1975.
83. **Olhoeft, G. R., Strangway, D. W., and Frisillo, A. L.**, Lunar sample electrical properties, in Proc. 4th Lunar Sci. Conf., (Supplement 4, *Geochim. Cosmochim. Acta)*, 3, 3133, 1973.
84. **Parkhomenko, E. I.**, *Electrical Properties of Rocks*, Plenum Press, New York, 1967, 314.
85. **Parkhomenko, E. I.**, *Electrification Phenomena in Rocks*, Plenum Press, New York, 1971, 285.
86. **Pelton, W. H., Ward, S. H., Halloff, P. G., Sill, W. R., and Nelson, P. H.**, Mineral discrimination and removal of inductive coupling with multifrequency IP, *Geophysics*, 43(3), 588, 1978.
87. **Pirson, S. J.**, *Handbook of Well Log Analysis*, Prentice Hall, Englewood Cliffs, N.J., 1963, 326.
88. **Porath, H.**, A review of the evidence on low resistivity layers in the earth's crust, in The Structure and Physical Properties of the Earth's Crust, Heacock, J. G., Ed., American Geophysics Union Washington, D.C., 1971, 127.
89. **Porter, C. R. and Carothers, J. E.**, Formation factor — porosity relation derived from well log data, *Log Anal.*, 16, 1971.
90. **Presnall, D. C., Simmons, G., and Porath, H.**, Changes in electrical conductivity of a synthetic basalt during melting, *J. Geophys. Res.*, 77, 5665, 1972.
91. **Quist, A. S. and Marshall, W. L.**, Electrical conductances of aqueous solutions at high temperatures and pressures. III. The conductances of potassium bisulfate solutions from 0 to 700° and at pressures to 4000 bars, *J. Phys. Chem.*, 73, 978, 1969.
92. **Quist, A. S. and Marshall, W. L.**, Electrical conductances of aqueous sodium chloride solutions from 0 to 800° and at pressures to 4000 bars, *J. Phys. Chem.*, 72, 684, 1968.
93. **Quist, A. S. and Marshall, W. L.**, The electrical conductances of some alkali metal halides in aqueous solutions from 0 to 800° and at pressures to 4000 bars, *J. Phys. Chem.*, 73, 978, 1969.
94. **Saint-Amant, M. and Strangway, D. W.**, Dielectric properties of dry, geologic materials, *Geophysics*, 35(4), 624, 1970.

95. **Schock, R. N., Duba, A. G., Heard, H. C., and Stromberg, H. D.,** The electrical conductivity of polycrystalline olivine and pyroxene under pressure, in *High Pressure Research Application in Geophysics,* Manghnani, M. and Akimoto, S., Eds., Academic Press, New York, 1977, 39.
96. **Scott, J. H., Carroll, R. D., and Cunningham, D. R.,** Dielectric constant and electrical conductivity measurements of moist rock: a new laboratory method, *J. Geophys. Res.,* 72, 5115, 1967.
97. **Scott, W. J. and West, G. F.,** Induced polarization of synthetic, high-resistivity rocks containing disseminated sulfides, *Geophysics,* 34(1), 87, 1969.
98. **Shankland, T. J.,** Electrical conduction in rock minerals: parameters for interpretation, *Phys. Earth Planet. Inter.,* 10, 209, 1975.
99. **Shankland, T. J. and Waff, H. S.,** Partial melting and electrical conductivity anomalies in the upper mantle, *J. Geophys. Res.,* 82(33), 5409, 1977.
100. **Shuey, R. T.,** *Semiconducting Ore Minerals,* Elsevier, Amsterdam, 1975, 415.
101. **Shuey, R. T. and Johnson, N.,** On the phenomenology of electrical relaxation in rocks, *Geophysics,* 38(1), 37, 1973.
102. **Stacey, F. D.,** *Physics of the Earth,* John Wiley & Sons, New York, 1969.
103. **Stanley, W. D., Boehl, J. E., Bostick, F. X., and Smith, H. W.,** Geothermal significance of magnetotelluric sounding in the eastern Snake River Plain-Yellowstone region, *J. Geophys. Res.,* 82(17), 2501, 1977.
104. **Sternberg, B. K.,** Electrical resistivity structure of the crust in the southern extension of the Canadian shield-layered earth models, *J. Geophys. Res.,* 84(B1), 212, 1979.
105. **Strangway, D. W.,** Moon: electrical properties of the uppermost layers, *Science,* 165, 1012, 1969.
106. **Strangway, D. W.,** Possible electrical and magnetic properties of near-surface lunar materials, in *Electromagnetic Exploration of the Moon,* Linlor, W. I., Ed., Mono Book Corp., Baltimore, 1970, 77.
107. **Sumner, J. S.,** *Principles of Induced Polarization for Geophysical Exploration,* Elsevier, Amsterdam, 1976, 277.
108. **Taylor-Smith, D.,** Acoustic and electric techniques for sea-floor sediment identification, Proc. Int. Symp. Engineering Properties of Sea Floor Soils and Their Geophysical Identification, Seattle, 1971.
109. **Tozer, D. C.,** The electrical properties of the earth's interior, *Phys. Chem. Earth,* 3, 414, 1959.
110. **Volarovich, M. P. and Parkhomenko, E. I.,** Electrical properties of rocks at high temperatures and pressures, in *Geoelectric and Geothermal Studies,* Akademiai Kiado, Budapest, 1976, 321.
111. **Voorhis, G. D., Nelson, P. H., and Drake, T. L.,** Complex resistivity spectra of porphyry copper mineralization, *Geophysics,* 38(1), 49, 1973.
112. **Van Zijl, J. S. V.,** Electrical studies of the deep crust in various tectonic provinces of southern Africa, in The Earth's Crust, Its Nature and Physical Properties, Heacock, J. G., Ed., American Geophysical Union, Washington, D.C., 1977, 470.
113. **von Hippel, A. R.,** *Dielectric Materials and Applications,* The MIT Press, Cambridge, Mass., 1954.
114. **von Hippel, A. R.,** *Dielectrics and Waves,* John Wiley & Sons, New York, 1954.
115. **Waff, H. S. and Weill, D. F.,** Electrical conductivity of magmatic liquids: effects of temperature, oxygen, fugacity, and composition, *Earth Planet. Sci. Lett.,* 28, 254, 1975.
116. **Wait, J. R.,** *Overvoltage Research and Geophysical Applications,* Pergamon Press, New York, 1959.
117. **Yakupov, V. S.,** *Elektroprovodnost i geoelektricheskiy razrez merzlikh tolshch (Electrical Conductivity and Geoelectrical Character of Permafrost),* Vol. 20, Nauka, Moscow, 1968, 179.
118. **Yukatake, T.,** The solar cycle contribution to the secular change in the geomagnetic field, *J. Geomagn. Geoelectr.,* 17, 287, 1965.
119. **Zablocki, C. J.,** Electrical properties of serpentinite from Mayaguez, Puerto Rico, in a study of serpentinite, *Natl. Acad. Sci. Rep.,* 1188, 107, 1964.
120. **Zablocki, C. J., Tilling, R. I., Peterson, D. W., Christianson, R. I., Keller, G. V., and Murray, J. C.,** A deep research drill hole at the summit of an active volcano, Kilauea, Hawaii, *Geophys. Res. Lett.,* 1, 323, 1974.

Section VI
Seismic Velocities

By
Nikolas I. Christensen

INTRODUCTION

Seismology has provided a wealth of evidence relating to the physical nature of the interior of the Earth. Most seismological studies using earthquakes or reflection and refraction techniques from artificially generated waves present layered models in which velocities and layer thicknesses are tabulated. Many significant results have emerged from these studies including:

1. The broad subdivision of the Earth into crust, mantle, and core
2. The recognition of seismic discontinuities within the core and mantle that are probably related to phase changes
3. The marked difference in overall structure of the oceanic and continental crust

Several intracrustal discontinuities have, in turn, been recognized which differ from one region to another, and as increasing data become available it is apparent that many regions of the Earth's interior are anisotropic and heterogeneous.

The information desired from seismic studies is not ultimately velocity-depth functions, but knowledge of the nature and distribution of materials with depth so we may understand the origin and evolution of the Earth. The velocities of elastic waves in materials for the interpretation of seismic data must be obtained through carefully controlled laboratory experiments which realistically simulate the physical conditions that exist within the Earth's interior.

In addition to being of interest to Earth scientists, velocities are of considerable significance to materials scientists since they yield information important in understanding forces between atoms and ions. Also, since velocities are related to the elastic properties of solids, they are important in describing the mechanical behavior of materials.

For homogeneous isotropic elastic materials, compressional (V_p) and shear (V_s) wave velocities, density (ϱ), and the elastic moduli are related by the following equations:

Bulk Modulus	$K = \varrho(V_p^2 - 4/3\ V_s^2)$
Shear Modulus	$\mu = \varrho V_s^2$
Poisson's Ratio	$\sigma = \dfrac{(r^2 - 2)}{2(r^2 - 1)}, \quad r = V_p/V_s$
Young's Modulus	$E = 2\mu(1 + \sigma)$
Compressional Wave Velocity	$V_p = \sqrt{[K + (4/3)\mu]/\varrho}$
Shear Wave Velocity	$V_s = \sqrt{\mu/\varrho}$

Laboratory studies of velocities in materials generally fall into three categories: (1) measurements of velocities in naturally occurring materials such as rocks, (2) studies of hot-pressed polycrystalline aggregates, and (3) velocity measurements in single crystals. The velocities in rocks and hot-pressed aggregates are commonly affected by porosity. Values useful for the interpretation of field measurements, except in near-surface studies, are obtained only after porosity has been reduced by application of a few kilobars pressure. Measurements of velocities in single crystals are useful in the interpretation of seismic anisotropy resulting from preferred mineral orientation. In addition, if the elastic constants have been completely determined it is possible to estimate the velocities of quasi-isotropic aggregates of single crystals.

The prediction of velocities of a quasi-isotropic rock containing a large number of randomly oriented, highly anisotropic crystals, from single-crystal data is complicated in many aspects. In theory, it is difficult to compromise between assumptions of uniform local strain and uniform local stress. Voigt[1] assumed that strain is uniform throughout the rock and averaged over solid angles the elastic constants (C_{ij}), whereas Reuss[2] assumed that uniform local stress was operative and averaged the elastic compliances (S_{ij}) over all directions. The appropriate relationships for the bulk moduli and shear moduli according to the two theories are as follows:

Voigt's Moduli

$$9K_v = (C_{11} + C_{22} + C_{33}) + 2(C_{12} + C_{23} + C_{31})$$

$$15\mu_v = (C_{11} + C_{22} + C_{33}) - (C_{12} + C_{23} + C_{31}) + 3(C_{44} + C_{55} + C_{66})$$

Reuss's Moduli

$$1/K_r = (S_{11} + S_{22} + S_{33}) + 2(S_{12} + S_{23} + S_{31})$$

$$15/\mu_r = 4(S_{11} + S_{22} + S_{33}) - 4(S_{12} + S_{23} + S_{31}) + 3(S_{44} + S_{55} + S_{66})$$

Calculated compressional and shear wave velocities for quasi-isotropic monomineralic rocks are obtained from the relationships $\varrho V_p^2 = K + 4\mu/3$ and $\varrho V_s^2 = \mu$ where ϱ is the density of the mineral.

Voigt's and Reuss's velocity averages frequently show considerable variance especially for the silicate minerals of low symmetry. Hill[3] has shown theoretically that the true values lie between the Voigt and Reuss Moduli and the Hill average is commonly taken as the mean of the Voigt and Reuss values.

The accuracies of seismic structure within the Earth depend to a large extent on the combination of field and analytical techniques used to identify the velocities and probably vary between 3 and 10% for most models. The accuracies of laboratory velocities in materials also depend on the specific technique employed, varying from 0.5% to 3% for the pulse-transmission method commonly used for rocks to approximately 0.01% with interferometric methods. The laboratory techniques typically use frequencies much higher than the field studies. However, several studies have demonstrated that dispersion in the frequency range of 10^{-1} to 10^7 Hz is negligible, thus allowing direct use of the laboratory data in the interpretation of field measurements.

The following tables list velocities in rocks, minerals, polycrystalline aggregates, strata, and other substances available in the published literature. When possible, data have been combined in common tables. To avoid inaccurate extrapolation of individual author's results, some similar tables exist, particularly for velocities at elevated temperatures and pressures.

ACKNOWLEDGMENTS

I am indebted to S. Blair, M. Patella, R. Carlson, D. Fountain, J. Hull, R. Prior, M. Salisbury, and R. Wilkens for assistance in preparation of the tables.

Table 1
COMPRESSIONAL WAVE VELOCITIES IN MARINE SEDIMENTS FROM SONOBUOY STATIONS[4,5]

Buoy number	Layer	V_p (km/s)	Water depth (km)	Latitude	Longitude
		Japan Sea and Tartar Strait			
J1	1	1.66	3.62	41°08′N	137°54′E
J2	1	1.84	3.52	43°43′N	138°01′E
J3	1	1.76	0.97	47°35′N	140°46′E
J4a	1	1.75	3.38	41°32′N	132°49′E
J4b	1	1.85	3.38	41°34′N	132°51′E
J5	1	1.59	2.26	37°00′N	131°02′E
		Bering Sea			
1	1	1.83	4.28	56°03′N	164°47′E
1	3	2.32	4.28	56°03′N	164°47′E
2	1	1.74	3.66	58°14′N	167°12′E
2	2	2.00	3.66	58°14′N	167°12′E
2	3	2.22	3.66	58°14′N	167°12′E
3	1	1.71	3.84	56°53′N	168°44′E
3	2	1.89	3.84	56°53′N	168°44′E
4	1	1.76	2.92	59°29′N	168°52′E
5	1	1.70	3.80	56°00′N	171°56′E
6	1	1.74	3.89	54°18′N	172°42′E
6	2	1.93	3.89	54°18′N	172°42′E
7	1	1.74	3.85	55°49′N	176°12′E
7	4	2.10	3.85	55°49′N	176°12′E
8	1	1.60	3.60	57°00′N	175°29′W
9	1	1.69	3.36	55°45′N	173°01′W
		Aleutian Trench			
A1	1	1.77	6.74	53°53′N	166°52′E
A2	2	2.05	7.26	51°05′N	171°13′W
A3	1	1.93	7.29	50°47′N	176°48′E
		Aleutian Abyssal Plain			
1	1	1.69	5.10	47°51′N	154°58′W
2	1	1.65	4.94	50°20′N	155°58′W
		California Continental Borderland			
A	1	1.63	4.07	29°06′N	116°42′W
B1	1	1.59	1.87	32°33′N	118°09′W
B2	1	1.76	1.77	33°02′N	119°00′W
C	1	1.47	1.16	32°40′N	117°38′W
C	2	1.79	1.16	32°40′N	117°38′W
D	1	1.51	2.84	30°00′N	116°35′W
		Northwest Pacific Pelagic Sediments			
P1	1	1.55	4.84	52°49′N	168°00′E
P2	1	1.74	6.32	52°29′N	162°00′E

Table 1 (continued)
COMPRESSIONAL WAVE VELOCITIES IN MARINE SEDIMENTS FROM SONOBUOY STATIONS[4,5]

Buoy number	Layer	V_p (km/s)	Water depth (km)	Latitude	Longitude
			Bengal Fan		
6A	1	1.79	3.57	10°06′N	91°23′E
6B	1	1.97	3.55	10°17′N	91°22′E
10B	1	1.87	3.53	10°09′N	84°50′E
11A	1	1.85	3.66	10°14′N	82°06′E
11B	1	1.96	3.67	10°15′N	81°47′E
	2	2.03	3.67	10°15′N	81°47′E
14A1	1	1.83	3.47	12°34′N	81°58′E
14A2	1	1.95	3.47	12°37′N	82°00′E
14D	1	1.80	3.46	12°40′N	82°03′E
14E	1	1.82	3.46	12°44′N	82°07′E
15	1	1.55	3.44	13°00′N	82°23′E
16A	1	1.52	3.35	13°37′N	82°36′E
16B	1	1.76	3.34	13°48′N	82°36′E
	2	2.19	3.34	13°48′N	82°36′E
18A	1	1.84	2.74	17°04′N	84°58′E
18B	1	2.07	2.72	17°04′N	85°05′E
18D	1	1.64	2.69	17°04′N	85°17′E
26A	1	1.73	2.29	18°11′N	92°49′E
33A	1	2.22	2.95	16°15′N	83°29′E
33B	1	1.73	2.89	16°11′N	83°24′E
35A	1	1.94	3.23	13°40′N	81°08′E
35C	1	1.90	3.29	13°37′N	81°22′E
35D	1	2.01	3.30	13°35′N	81°31′E
37A	1	1.78	3.10	13°32′N	86°39′E
	1	1.84	3.10	13°32′N	86°39′E
37B	1	1.68	3.09	13°31′N	86°44′E
	2	1.71	3.09	13°31′N	86°44′E
	3	2.17	3.09	13°31′N	86°44′E
37C	1	1.88	3.06	13°31′N	87°24′E
38A	1	1.97	3.02	13°36′N	88°20′E
39A	1	1.65	2.94	13°23′N	89°37′E
40C	1	1.57	2.95	13°12′N	90°42′E
40D	1	1.76	2.95	13°13′N	90°52′E
41A	1	1.95	2.83	14°06′N	91°27′E
41B	1	2.22	2.81	14°07′N	91°21′E
42A	1	1.82	2.80	14°16′N	91°42′E
42B	1	1.78	2.82	14°14′N	92°07′E
	2	2.30	2.82	14°14′N	92°07′E
46A	1	1.60	3.15	10°31′N	95°01′E
2A	1	1.70	2.69	07°04′N	94°55′E
	2	2.14	2.69	07°04′N	94°55′E
3A	1	1.61	2.66	07°58′N	95°16′E
	2	2.22	2.66	07°58′N	95°16′E
6B	1	1.74	2.58	08°46′N	95°19′E
9D	1	1.72	2.67	12°10′N	95°31′E
16A	1	1.75	2.56	11°40′N	95°50′E
	2	2.03	2.56	11°40′N	95°50′E
16D	1	1.94	2.49	11°25′N	95°47′E
16E	1	1.86	2.51	11°20′N	95°45′E
	2	1.96	2.51	11°20′N	95°45′E
18A	1	2.09	3.05	11°16′N	94°50′E
18C	1	1.99	3.05	11°12′N	94°43′E
19B	1	1.81	2.58	10°20′N	95°42′E

Table 1 (continued)
COMPRESSIONAL WAVE VELOCITIES IN MARINE SEDIMENTS FROM SONOBUOY STATIONS[4,5]

Buoy number	Layer	V_p (km/s)	Water depth (km)	Latitude	Longitude
20A	1	1.58	2.46	10°04′N	95°56′E
26A	1	2.02	4.44	03°38′N	93°19′E
27A	1	1.86	4.25	02°56′N	92°37′E
29A	1	1.80	4.22	02°25′N	92°18′E
29A	2	1.98	4.22	02°25′N	92°18′E
30B	1	1.66	4.34	01°38′N	92°31′E
	2	2.10	4.34	01°38′N	92°31′E
30E	1	1.89	4.35	01°22′N	92°35′E
	2	2.28	4.35	01°22′N	92°35′E
31A	1	2.24	4.40	00°42′N	92°52′E
32A	1	1.98	4.44	00°33′N	93°00′E
33A	1	1.90	4.39	00°49′N	93°13′E
33B	1	1.57	4.36	01°05′N	93°20′E
34B	1	2.03	4.39	01°50′N	93°49′E
35C	1	1.75	4.79	02°24′N	94°26′E
	2	2.19	4.79	02°24′N	94°26′E
39B	1	1.77	4.65	00°42′S	96°09′E
39D	1	1.91	4.89	04°04′S	94°32′E
45A	1	1.75	5.01	05°08′S	94°49′E
45D	1	2.07	5.04	05°13′S	94°57′E
A3	3	2.08	4.37	03°00′N	80°11′E
A4	1	1.62	4.64	00°05′N	80°35′E
	2	1.83	4.64	00°05′N	80°35′E
	3	1.89	4.64	00°05′N	80°35′E
A6	2	1.65	4.34	02°22′N	84°57′E
	3	1.80	4.34	02°22′N	84°57′E
A8	1	2.25	4.04	04°29′N	88°12′E
A9	1	2.20	3.92	05°44′N	87°03′E
A10	1	1.80	3.92	06°06′N	85°27′E
	2	2.16	3.92	06°06′N	85°27′E
A11	1	1.59	3.92	06°11′N	85°21′E
A12	1	1.57	3.86	06°43′N	84°55′E
	3	1.96	3.86	06°43′N	84°55′E
	4	2.10	3.86	06°43′N	84°55′E
	5	2.19	3.86	06°43′N	84°55′E
A13	1	1.82	4.03	05°56′N	83°18′E
A14	1	1.64	3.75	07°52′N	86°16′E
	3	2.07	3.75	07°52′N	86°16′E
A15	1	1.86	3.74	08°01′N	86°19′E
	2	1.94	3.74	08°01′N	86°19′E
A16	1	1.75	3.61	09°18′N	86°39′E
	2	1.98	3.61	09°18′N	86°39′E
	3	2.17	3.61	09°18′N	86°39′E
A17	1	1.60	3.40	10°40′N	87°08′E
A18	2	1.87	3.36	10°57′N	87°14′E
	3	1.96	3.36	10°57′N	87°14′E
	4	2.28	3.36	10°57′N	87°14′E
	5	2.36	3.36	10°57′N	87°14′E
A20	1	1.62	3.30	11°27′N	87°22′E
	3	1.92	3.30	11°27′N	87°22′E
	4	1.93	3.30	11°27′N	87°22′E
	5	2.11	3.30	11°27′N	87°22′E

Table 1 (continued)
COMPRESSIONAL WAVE VELOCITIES IN MARINE SEDIMENTS FROM SONOBUOY STATIONS[4,5]

Buoy number	Layer	V_p (km/s)	Water depth (km)	Latitude	Longitude
A23	1	2.21	2.82	15°09′N	87°53′E
A24	1	1.60	2.47	17°17′N	88°01′E
A25	1	1.58	2.26	18°08′N	88°13′E
A29	1	2.29	2.69	15°07′N	92°28′E
A30	1	1.94	2.65	15°10′N	92°03′E
A31	1	1.77	2.87	14°20′N	92°20′E
A33	1	1.76	3.33	11°25′N	91°32′E
	2	2.30	3.33	11°25′N	91°32′E
A35	1	1.56	4.39	04°29′N	92°36′E
A36	1	1.60	4.08	04°04′N	91°57′E
A38	1	1.65	4.32	01°31′N	93°13′E
A39	1	1.80	4.40	00°54′N	94°45′E
C2	1	1.86	3.15	15°15′N	83°31′E
C3	1	2.20	2.61	15°12′N	91°20′E
C4	1	1.68	1.70	19°11′N	90°19′E
C5	1	1.71	2.59	15°50′N	89°19′E
C6	1	1.94	3.31	12°12′N	85°27′E
C7	1	1.69	3.71	09°12′N	82°57′E
C10	1	2.00	4.62	01°00′S	83°54′E
C11	1	1.83	4.32	04°15′N	80°50′E

Table 2
SHEAR WAVE VELOCITIES IN MARINE SEDIMENTS[6]

Material	V_s (km/s)	Depth (m)
Medium sand	0.05	0.2
	0.12	0.4
	0.17	3.3
	0.21	5.2
	0.29	7.2
Silty sand	0.08	0.1
	0.15	3.7
Fine sand	0.09	0.2
	0.12	2.0
	0.17	4.2
Coral sand	0.14	0.4
	0.25	7.0
Silty clay	0.14	1.0
	0.17	4.0
Clayey silt	0.18	9.0
Silt	0.19	11.0
Firm clay	0.20	17.0
Sandy silt	0.24	21.0
Shaley clay	0.24	27.0

Table 3
REFRACTION VELOCITIES IN THE WEST ANTARCTIC ICE SHEET[7]

Station number	Temp (°C)	V_p (km/s)	V_s (km/s)
B-58	−28.4	3.86	1.93
S-150	−28.0	3.86	1.94
S-360	−26.6	3.85	1.94
S-858	−31.2	3.85	1.94
E-300	−28.1	3.86	—
E-348	−27.7	3.85	1.95
E-396	−25.8	3.85	1.91
E-444	−25.5	3.85	1.93
E-612	−27.7	3.86	1.94
E-708	−23.7	3.84	1.95
E-756	−23.0	3.85	—
E-828	−24.2	3.85	1.94
E-972	−25.9	3.84	1.96
E-1020	−25.0	3.83	1.91

Table 4
REFRACTION VELOCITIES IN THE GREENLAND ICE SHEET[8]

Station	V_p (km/s)	V_s (km/s)
Camp IV	3.80	1.91
Camp V	3.70	1.87
Camp VI	3.82	1.92
Point A34	3.72	1.91

Table 5
DETAILED COMPRESSIONAL WAVE VELOCITY SONOBUOY DATA FOR THE OCEANIC CRUST FROM THE NORTH AND EQUATORIAL PACIFIC[9]

Age (m.y.)	Layer 2A (km/s)	Layer 2B (km/s)	Layer 2C (km/s)	Layer 3 (km/s)	Layer 2A (km)	Layer 2B (km)	Layers 2B and 2C (km)	Water (km)	Sediment (km)	Sonobuoy	Latitude	Longitude
0.5		5.90		6.80	0.75		1.45	2.76	—	17C13	17°36.4′S	113°16.2′W
0.5		5.80		7.00	0.60		2.70	3.17	—	24C13	18°11.8′S	112°13.9′W
0.5		5.20	5.90		0.60	0.50		3.18	—	29C13	18°24.2′S	112°09.7′W
0.5		4.75	6.00		0.83	0.76		2.68	—	42C13	12°53.3′N	103°59.7′W
0.5	3.60	4.95	5.90		0.76	0.57		2.65	—	43C13	12°36.1′N	103°56.1′W
0.5	3.20	5.70	6.30		0.95	0.70		3.02	—	44C13	12°53.3′N	103°41.3′W
0.5	3.20	5.35	6.20		0.36	0.64		2.85	—	45C13	12°53.5′N	103°28.9′W
0.5	4.15	4.85	6.15		1.23	0.25		2.68	—	54C13	12°51.0′N	103°58.1′W
0.5		4.80	5.80	7.05	0.49	1.23	2.06	2.96	—	57C13	12°47.2′N	103°49.9′W
0.5		4.55	6.05	6.95	0.65	0.57	1.84	2.98	—	55C13	12°38.2′N	103°48.7′W
0.5		4.60	5.95	6.75	0.81	0.21	1.41	2.92	—	56C13	12°27.2′N	103°47.5′W
0.5	2.80	4.65	6.40		0.84	0.98		2.94	—	58C13	12°47.9′N	103°29.9′W
0.5	3.30	5.30	6.40		0.79	1.14		3.09	—	48C13	12°50.5′N	103°21.5′W
0.5		5.60			0.80			3.02	—	11C13	17°24.7′S	113°16.2′W
0.5		4.70	6.30		0.60	0.88		3.20	—	13C13	17°10.0′S	112°00.0′W
0.5		4.90			0.45			2.96	—	14C13	17°01.7′S	113°24.8′W
0.5		6.20		7.10	0.65		2.85	2.99	—	15C13	18°03.3′S	113°49.4′W
2	3.40	4.90	6.00		1.15	0.62		2.64	—	53C13	12°36.0′N	103°53.5′W
2	3.75	5.65			0.97			3.06	—	60C13	12°48.7′N	102°56.5′W
2		5.30	6.50	7.15	0.70	0.44	1.71	3.10	—	61C13	12°47.3′N	102°42.2′W
2	3.90	5.05	6.30	7.30	0.88	0.94	1.72	3.14	—	62C13	12°31.3′N	102°28.6′W
2		5.35		6.90	0.66		1.18	3.30	—	64C13	12°43.9′N	102°22.0′W
2	3.50	5.30		6.90	0.84		0.94	3.05	—	79C13	12°26.0′N	103°00.3′W
2		5.70			0.80			3.10	—	80C13	12°30.1′N	103°03.6′W
2		5.70	6.35	7.15	0.76	0.73	1.54	3.06	—	50C13	12°50.3′N	102°47.9′W
2		4.90	6.15	7.10	0.26	1.37	3.31	3.14	—	77C13	12°19.7′N	102°22.1′W
2	3.25	5.10	6.40		0.60	0.48		3.23	—	27C13	18°19.5′S	115°10.0′W
2		5.30		7.00	0.50		0.80	3.04	—	26C13	18°11.8′S	114°00.7′W

2	2.95							3.15	—	9C13	17°35.8′S	114°14.0′W
2		5.00			0.70			3.37	—	10C13	17°23.3′S	111°22.7′W
2	3.90							3.29	—	12C13	17°21.2′S	115°14.2′W
5		5.50	6.10		0.80	0.42		3.06	—	66C13	12°25.0′N	101°59.6′W
5	3.30	4.60	6.10	7.05	0.59	0.83	2.85	3.30	—	67C13	12°42.1′N	102°02.6′W
5		5.50		6.70	0.61		1.23	3.32	—	68C13	12°42.1′N	101°42.1′W
5		5.80	6.10		1.00	0.85		3.11	—	75C13	12°41.0′N	101°34.8′W
5	3.45	4.80	6.20	6.70	0.92	0.61	2.40	3.31	—	76C13	12°22.1′N	101°32.8′W
5		5.20			1.00			3.17	—	78C13	12°26.0′N	102°22.6′W
5		6.10		7.05	0.76		1.36	3.32	—	65C13	12°42.5′N	102°02.1′W
5		4.95	6.40		0.42	1.21		3.32	—	70C13	12°41.4′N	101°12.5′W
7.5		5.00	6.25	7.00	0.53	0.79	2.18	3.40	—	72C13	12°22.8′N	100°45.8′W
7.5		5.40	6.30		1.10	1.00		3.29	—	81C13	12°14.6′N	101°03.7′W
7.5		4.70	5.90	7.00	0.44	0.60	0.98	3.38	—	71C13	12°40.7′N	100°48.8′W
7.5		5.30	6.10		0.90	0.40		3.25	—	73C13	12°42.4′N	100°50.7′W
7.5	4.20	4.70	5.45		0.36	0.65		3.46	0.16	63V32	12°35.9′N	100°35.1′W
7.5		4.60	6.15	7.00	0.42	0.37	1.54	3.52	0.06	64V32	14°12.1′N	100°25.5′W
7.5		5.05			0.70			2.95	—	101V28	03°49.2′N	091°52.8′W
7.5	3.90	5.00	6.05		0.33	1.45		2.89		102V28	03°46.0′N	091°45.2′W
7.5		5.70			0.60			3.65		39C13	00°46.8′N	106°56.6′W
9.5	3.70	5.70			0.98			3.70	0.12	58V32	13°19.4′N	099°37.3′W
9.5		4.75	6.25		0.20	0.58		3.66	0.14	62V32	14°45.0′N	099°07.3′W
9.5		4.90	6.10	6.85	0.61	0.69	1.94	3.50		82C13	12°49.3′N	099°08.3′W
11		5.30		6.50	0.30		2.49	3.41	0.19	59V32	14°50.9′N	098°33.3′W
12	3.85	5.60			0.47			3.90	0.47	75V32	02°41.7′N	114°45.0′W
13		5.40			0.66			3.85	0.44	74V32	02°42.3′N	114°31.3′W
14		5.30		6.75	0.39		0.97	3.79	0.41	73V32	02°42.0′N	114°00.4′W
15	3.70	5.20			0.04			4.27	0.48	77V32	02°39.8′N	117°51.5′W
17		5.20			—			4.27	0.40	78V32	02°40.0′N	119°02.02′W
20.5		4.60	5.80	6.70	0.58	0.28	1.91	3.57		85C13	12°30.0′N	097°58.9′W
20.5		4.95			—			3.31		85C11	52°23.2′N	135°01.4′W
20.5		4.90	5.90		—	0.68		3.35		86C11	52°20.8′N	135°07.1′W
20.5		5.10			—			3.65		87C11	51°14.2′N	137°29.2′W
20.5		4.85		7.15	—		1.39	3.50		120C14	49°40.9′N	134°24.8′W
27		5.40		6.55	—		2.41	2.85	2.06	115C14	55°41.6′N	137°48.2′W
27		5.25	6.25		—	1.00		3.94	0.23	119C14	49°56.3′N	138°40.9′W
27		4.60	5.50	6.60	0.21	0.39	1.21	2.55		105V28	05°19.1′N	095°14.8′W
32		5.90		6.65	—		1.08	3.70	2.15	111C14	57°31.0′N	142°08.5′W

Table 5 (continued)
DETAILED COMPRESSIONAL WAVE VELOCITY SONOBUOY DATA FOR THE OCEANIC CRUST FROM THE NORTH AND EQUATORIAL PACIFIC[9]

Age (m.y.)	Layer 2A (km/s)	Layer 2B (km/s)	Layer 2C (km/s)	Layer 3 (km/s)	Layer 2A (km)	Layer 2B (km)	Layers 2B and 2C (km)	Water (km)	Sediment (km)	Sonobuoy	Latitude	Longitude
32		5.00			0.02			4.49	0.43	83V32	02°37.6′N	129°12.7′W
35		5.35	6.50		0.41	0.98		4.17	0.13	79C11	53°22.2′N	146°06.8′W
38		5.20			0.26			4.41	0.43	84V32	02°38.1′N	131°04.0′W
38		5.10	6.25	6.90	—	1.11	2.44	3.64	0.19	86C13	07°36.1′N	092°54.4′W
39		5.30	6.05	7.00	—	1.43	1.77	3.57	0.24	87C13	06°53.6′N	092°07.9′W
43		4.85	6.25		—	1.26		4.37	0.55	87V32	02°51.0′N	133°36.3′W
46		6.10			—			4.19	1.07	104C14	54°45.6′N	146°54.1′W
47		5.68		6.50	—		1.45	4.30	1.48	106C14	57°02.0′N	147°15.8′W
51		5.78	6.10		0.21	1.20		4.61	0.49	95C12	53°30.8′N	153°40.8′W
56		4.95	6.05		0.36	0.22		4.32	0.67	90V32	02°41.7′N	139°51.5′W
56		5.30			0.43			5.53	0.19	82C12	47°37.3′N	157°51.8′W
56		5.70			0.11			5.25	0.48	86C12	46°12.4′N	157°21.8′W
56		4.70	5.70		—	0.97		5.17	0.57	88C12	48°26.9′N	158°19.4′W
63		4.90	5.90		—	1.12		4.40	0.54	91V32	03°44.1′N	143°08.9′W
71		5.25		7.05	0.16		1.26	5.24	0.24	53C11	45°39.2′N	160°00.2′W
71		5.10	6.20		—	0.97		5.10	0.65	57C11	48°30.9′N	161°45.4′W
76		5.35		6.85	0.19		1.22	5.84	0.08	47C11	45°12.5′N	174°25.5′W
76		5.30			—			5.38	0.24	51C11	44°32.0′N	163°13.4′W
76		5.30	6.10		—			5.25	0.25	52C11	44°35.0′N	163°04.3′W
90		4.85	6.20	6.85	—	0.80	2.02	5.87	0.16	402C12	33°53.2′N	161°09.3′W
90			5.75	6.65	—	1.23	1.68	5.86	0.04	404C12	33°34.4′N	160°58.1′W
90		4.60	6.20		—	0.65		5.65	0.30	405C12	31°10.6′N	159°44.6′W
90		4.90	6.00	7.00	—	0.88	1.86	5.79	0.30	406C12	31°00.2′N	159°39.8′W
93		5.05	5.75		—	1.25		5.75	0.09	93V32	30°10.0′N	158°03.5′W
93		4.60	6.30		—	1.07		5.75	0.12	94V32	30°11.3′N	157°56.5′W
93		5.30	5.90	6.70	—	0.80	1.78	5.94	0.09	106V32	30°28.7′N	157°32.0′W
95		5.40	5.95	6.80	—	0.87	3.27	5.51	0.23	115C12	17°42.8′N	162°17.3′W

95		5.00	5.95		—	0.67		5.44	0.25	116C12	16°09.9′N	164°56.3′W
95		5.70	6.25	6.95	—	0.92	1.93	5.32	0.51	30V24	09°27.8′N	155°31.1′W
95		5.30	6.00		—	0.71		4.97	0.42	107V32	25°13.6′N	162°46.3′W
95		4.65	6.00	6.70	—	0.62	1.67	5.26	0.41	111V32	29°23.8′N	168°01.6′W
100		5.00	6.25		—	0.40		5.42	0.58	113V32	32°39.8′N	176°40.4′W
100		5.30			—			5.34	0.42	114V32	32°39.8′N	176°40.4′W
100		5.50			—			5.33	0.55	48C12	03°52.6′N	165°00.9′W
100		5.69	6.30		—	0.99		5.35	0.93	49C12	02°17.2′N	165°07.6′W
100		5.60	6.15		—	0.80		5.00	0.62	118C12	11°01.7′N	166°22.5′W
100		5.35		6.75	—		1.04	5.14	0.33	120C12	09°42.3′N	168°35.1′W
105				6.75	—		1.95	5.11	0.75	73C12	42°05.2′N	151°50.6′E
107		5.15	6.25	6.80	—	0.86	1.66	5.39	0.73	356C12	40°40.5′N	150°30.0′E
114		4.70	5.75		—	0.97		5.42	0.54	74C12	42°00.1′N	155°02.5′E
116		5.05	6.30	7.20	0.26	0.59	2.05	5.58	0.27	77C12	42°02.9′N	159°29.6′E
116		6.10		7.10	0.34		1.94	5.58	0.31	78C12	42°02.6′N	159°41.8′E
116		6.05		6.80	0.47		1.04	5.56	0.31	79C12	42°03.2′N	160°02.4′E
116		5.00	6.30		—	0.89		5.68	0.27	127V32	33°28.1′N	169°54.0′E
117		5.30	6.40		—	0.68	1.35	5.27	0.53	366C12	36°27.2′N	166°55.9′E
119		6.10		7.10	0.24		1.28	5.62	0.38	128V32	33°36.3′N	168°59.0′E
119		4.60	6.25	7.10	—	0.65	2.12	5.68	0.30	129V32	33°40.6′N	168°20.5′E
123		5.70		6.55	0.07		1.60	5.56	0.33	364C12	35°49.6′N	163°43.9′E
128		5.60			0.45			5.80	0.09	35V24	25°50.4′N	176°15.7′E
128		5.45	5.85	6.95	0.35	0.40	1.43	5.90	0.23	357C12	36°53.6′N	152°29.2′E
128		5.50	6.40	7.00	0.26	0.98	2.57	5.89	0.14	358C12	36°45.0′N	152°34.4′E
131		5.20	6.20		0.20	0.88		5.66	0.35	363C12	32°24.4′N	161°20.3′E
134		5.55	6.55		0.49	0.91		6.02	0.41	139V32	35°13.9′N	148°52.9′E
137					0.09			6.05	0.43	138V32	33°49.8′N	149°29.1′E
137		4.80	5.75		—	0.44		5.65	0.69	158V32	33°20.0′N	145°43.4′E
140		4.65	6.35		—	0.88		5.81	0.48	159V32	32°38.5′N	146°15.8′E
142		5.35	6.00	6.60	—	0.73	1.48	5.93	0.45	160V32	32°02.9′N	146°40.1′E
145		5.35	6.30		—	1.36		5.91	0.35	161V32	31°17.4′N	147°05.2′E
146		5.40		6.69	—		1.07	6.09	0.70	162V32	30°40.1′N	147°28.9′E
148		5.60		6.95	—		2.49	6.06	0.63	163V32	29°59.5′N	147°58.7′E
148	4.16	4.75			0.59	0.51		4.55	0.26	177V32	04°13.0′N	165°39.8′E
149		5.35	6.30		—	1.53		5.98	0.31	130V32	28°24.1′N	159°07.6′E
149		5.40			—			4.71	0.28	146C17	04°36.4′N	166°51.0′E
149		5.70			—			4.64	0.56	147C17	04°24.2′N	166°56.3′E
150		5.30	5.85	6.80	0.25	0.57	2.27	4.78	0.48	89C12	51°56.8′N	158°12.5′W

Table 5 (continued)
DETAILED COMPRESSIONAL WAVE VELOCITY SONOBUOY DATA FOR THE OCEANIC CRUST FROM THE NORTH AND EQUATORIAL PACIFIC[9]

Age (m.y.)	Layer 2A (km/s)	Layer 2B (km/s)	Layer 2C (km/s)	Layer 3 (km/s)	Layer 2A (km)	Layer 2B (km)	Layers 2B and 2C (km)	Water (km)	Sediment (km)	Sonobuoy	Latitude	Longitude
150		4.60	6.20		—	0.69		4.89	0.53	90C12	52°31.7′N	158°17.5′W
151		4.80	5.90	6.60	—	1.20	2.86	6.13	0.50	164V32	28°54.9′N	148°43.4′E
151	3.92				1.40			4.85	0.27	176V32	05°28.3′N	165°18.6′E
152	4.25	5.90		7.05	0.43		1.55	6.02	0.46	165V32	28°20.4′N	149°03.1′E
153		5.30	6.15	6.80	—	0.88	1.76	5.79	0.49	137V32	29°20.8′N	152°34.4′E
155		5.15	6.05		0.44	0.92		6.06	0.09	362C12	27°32.0′N	156°36.6′E
155		5.25	6.20		0.45	1.14		4.77	0.50	359C12	31°34.9′N	155°20.3′E
160			5.80	6.70	—	1.49	2.12	6.10	0.49	136V32	27°09.2′N	153°48.3′E
160		5.35			0.26			5.00	0.44	57C12	06°00.6′N	161°38.6′E
170		5.30	5.90	6.70	—	0.62	1.45	5.95	0.43	134V32	26°05.5′N	154°32.7′E
170		5.60	6.10	6.65	—	0.67	1.59	5.75	0.72	360C12	25°34.0′N	154°57.2′E
170		5.45		6.90	—		2.32	5.73	0.44	361C12	25°27.2′N	154°50.7′E
J		6.10		7.00	—		1.29	5.71	0.85	58C12	22°53.3′N	149°38.2′E
J		4.95	5.75		—	1.73		5.91	0.33	166V32	27°35.8′N	149°31.6′E
J		5.65	6.35		—	1.60		5.91	0.38	167V32	27°19.8′N	149°55.9′E
J		6.10			—			5.95	0.59	168V32	27°13.0′N	150°07.5′E
J		4.60		6.65	—		0.87	5.76	0.33	169V32	26°39.1′N	150°41.2′E
J	4.25	5.10	5.60	6.65	0.22	1.08	1.58	5.78	0.35	170V32	25°53.2′N	151°08.0′E
J	2.99	5.60			0.69			5.19	0.33	173V32	08°11.7′N	164°28.1′E

Note: J = Jurassic.

Table 6
COMPRESSIONAL AND SHEAR WAVE VELOCITIES IN THE CONTINENTAL CRUST AND UPPER MANTLE

Note: The velocities are grouped by continent and arranged from west to east longitude in bands of 5° increasing latitude away from the equator. The depths and velocities correspond to the top of the refracting layer. Azimuth begins at North equal to zero and increases in a clockwise direction. The type symbols are as follows: R is a refraction line, E is earthquake data, T is time term analysis, and O is other

Latitude	Longitude	Depth below surface (km)	V_p (km/s)	V_s (km/s)	Azimuth	Type	Ref.
			North America				
24° 03′N	104° 15′W	0.00	3.00	—	325°	R	63
		0.80	5.00	—			
		4.20	6.00	—			
		32.70	7.60	—			
		43.40	8.40	—			
29° 00′N	97° 00′W	0.00	2.30	—	76°	R	64
		2.00	3.94	—			
		7.30	5.38	—			
		19.80	6.92	—			
		33.00	8.18	—			
34° 00′N	118° 00′W	0.00	3.00	—	55°	R	65
		2.90	6.30	—			
		31.30	7.80	—			
34° 10′N	117° 49′W	0.00	5.70	—	302°	R	66
		11.30	7.20	—			
		43.90	8.10	—			
34° 10′N	117° 49′W	0.00	5.90	—	15°	R	67
		1.00	6.10	—			
		6.00	6.50	—			
		16.00	6.90	—			
		40.00	8.20	—			
34° 10′N	117° 49′W	0.00	5.00	—	297°	R	63
		0.50	5.90	—			
		6.00	6.10	—			
		26.00	7.00	—			
		32.00	8.20	—			
34° 30′N	117° 45′W	0.00	5.50	—		R	68
		4.00	6.30	—			
		27.40	6.80	—			
		32.40	7.80	—			
33° 00′N	116° 00′W	0.00	2.50	—		R	69
		0.40	5.10	—			
		2.90	6.00	—			
		14.00	7.10	—			
		25.00	7.90	—			
34° 10′N	109° 20′W	0.00	4.80	—		R	66
		4.30	6.10	—			
		26.00	7.40	—			
		48.10	7.30	—			
		56.90	8.20	—			
34° 00′N	103° 00′W	—	4.93	—	0°	R	70
		4.20	6.14	—			
		19.20	6.72	—			
		31.10	7.10	—			
		50.80	8.23	—			

Table 6 (continued)
COMPRESSIONAL AND SHEAR WAVE VELOCITIES IN THE CONTINENTAL CRUST AND UPPER MANTLE

Latitude	Longitude	Depth below surface (km)	V_p (km/s)	V_s (km/s)	Azimuth	Type	Ref.
34° 45′N	92° 18′W	0.00	4.60	—	43°	R	63
		2.00	5.20	—			
		10.20	6.60	—			
		41.20	8.20	—			
32° 00′N	90° 00′W	3.20	4.80	—	0°	R	71
		7.80	5.94	—			
		15.70	6.92	—			
		28.00	8.40	—			
34° 00′N	77° 00′W	—	1.70	—		R	72
		0.50	6.03	—			
		30.40	8.13	—			
38° 00′N	123° 00′W	—	6.00	—	57°	R	73
		20.00	7.90	—			
37° 40′N	122° 30′W	0.00	5.50	—	316°	R	63
		10.00	6.50	—			
		31.50	8.00	—			
37° 00′N	122° 00′W	0.00	6.20	3.40	317°	R	74
		22.00	8.00	—			
38° 00′N	122° 00′W	—	5.60	—		R	75
		12.00	6.70	—			
		27.00	8.00	—			
39° 47′N	121° 46′W	0.00	5.30	3.00		R	76
		10.00	6.30	3.80			
		25.00	7.80	4.00			
37° 00′N	121° 00′W	—	3.00	1.76	45°	E	77
		0.60	2.15	3.50			
		2.00	5.95	3.50			
		6.50	6.45	3.79			
		12.00	6.80	4.00			
37° 00′N	121° 00′W	0.00	3.00	1.69	45°	E	77
		0.80	4.90	2.75			
		4.00	5.40	3.03			
		8.50	5.70	3.20			
		12.00	6.80	4.00			
38° 00′N	121° 00′W	0.00	6.30	—	65°	R	78
		43.00	7.92	—			
38° 00′N	121° 00′W	0.00	3.50	—	53°	R	73
		1.30	6.00	—			
		20.30	7.90	—			
37° 00′N	120° 00′W	0.00	3.00	—	276°	R	79
		0.80	4.28	—			
		2.70	6.05	—			
		28.10	7.90	—			
38° 00′N	119° 00′W	0.00	6.15	—	302°	R	80
		30.00	7.10	—			
		41.10	7.80	—			
39° 00′N	119° 00′W	0.00	3.50	—	52°	R	73
		1.30	6.00	—			
		31.10	7.80	—			
39° 00′N	119° 00′W	0.00	3.50	—	52°	R	81
		1.30	6.00	—			
		19.90	6.60	—			
		33.70	7.80	—			
39° 55′N	118° 55′W	0.00	3.50	—	272°	R	73
		1.30	6.00	—			
		22.20	7.80	—			

Table 6 (continued)
COMPRESSIONAL AND SHEAR WAVE VELOCITIES IN THE CONTINENTAL CRUST AND UPPER MANTLE

Latitude	Longitude	Depth below surface (km)	V_p (km/s)	V_s (km/s)	Azimuth	Type	Ref.
		North America (continued)					
35° 00′N	117° 00′W	1.00	6.11	3.49		R	82
		23.00	7.66	—			
		49.00	8.11	—			
37° 07′N	116° 03′W	0.00	3.00	—		R	83
		1.40	6.00	—			
		21.70	6.70	—			
		25.50	7.90	—			
37° 07′N	116° 03′W	0.00	3.00	—		R	83
		1.40	6.00	—			
		24.30	7.90	—			
37° 07′N	116° 03′W	0.00	3.00	—		R	83
		1.40	6.00	—			
		24.30	6.50	—			
		25.90	7.60	—			
36° 30′N	116° 00′W	0.00	6.15	—	302°	R	80
		23.50	7.10	—			
		31.40	7.80	—			
39° 35′N	116° 00′W	0.00	6.00	—	0°	R	84
		19.00	6.70	—			
		31.00	7.90	—			
36° 00′N	115° 00′W	—	3.00	—	55°	R	65
		1.00	6.10	—			
		24.80	7.80	—			
35° 57′N	114° 46′W	0.00	3.80	—	25°	R	85
		2.50	6.20	—			
		36.00	8.20	—			
36° 45′N	111° 50′W	0.00	5.70	—	350°	R	86
		9.00	6.30	—			
		25.00	7.60	—			
		72.00	8.00	—			
38° 00′N	111° 00′W	0.00	3.15	1.81		0	87
		2.00	6.10	3.52			
		9.00	6.40	3.69			
		24.00	6.80	3.92			
		40.00	7.80	4.50			
37° 00′N	110° 00′W	0.00	6.20	—	336°	R	88
		25.00	6.80	—			
		40.00	7.80	—			
36° 04′N	109° 57′W	0.00	2.30	—	72°	R	89
		0.70	5.20	—			
		2.40	6.20	—			
		29.10	7.80	—			
35° 00′N	107° 00′W	0.00	6.15	—	0°	R	90
		18.60	6.50	—			
		39.90	8.12	—			
35° 30′N	98° 30′W	0.00	4.00	—		R	91
		0.60	6.05	—			
		1.00	5.50	—			
		3.10	6.04	—			
		6.10	6.12	—			
		13.40	6.48	—			
		16.40	6.20	—			

Table 6 (continued)
COMPRESSIONAL AND SHEAR WAVE VELOCITIES IN THE CONTINENTAL CRUST AND UPPER MANTLE

Latitude	Longitude	Depth below surface (km)	V_p (km/s)	V_s (km/s)	Azimuth	Type	Ref.
		17.90	6.68	—			
		26.10	7.01	—			
		35.20	7.33	—			
36° 00′N	97° 00′W	0.00	5.96	—		R	92
		13.70	6.66	—			
		29.60	7.20	—			
37° 00′N	92° 00′W	—	6.00	—	58°	R	93
		40.00	6.10	—			
		30.00	7.30	—			
		42.00	8.20	—			
38° 00′N	91° 00′W	—	6.00	—	58°	R	93
		5.00	6.30	—			
		25.00	6.90	—			
		40.00	8.00	—			
36° 05′N	82° 10′W	0.00	6.00	—		R	63
		5.30	6.30	—			
		13.70	6.70	—			
		45.30	8.10	—			
38° 05′N	76° 26′W	0.00	6.10	—		R	94
		5.00	6.30	—			
		10.00	6.40	—			
		16.00	6.60	—			
		21.00	6.80	—			
		26.00	6.90	—			
		31.00	8.00	—			
38° 05′N	76° 20′W	0.00	6.00	—		R	94
		24.00	6.00	—			
		27.00	7.50	—			
		30.00	8.00	—			
38° 00′N	75° 00′W	—	2.10	—		R	72
		1.60	5.78	—			
		9.90	6.34	—			
		26.30	7.97	—			
40° 00′N	118° 00′W	0.00	3.50	—	347°	R	73
		1.30	6.00	—			
		32.40	7.80	—			
40° 00′N	118° 00′W	0.00	3.50	—	347°	R	73
		1.30	6.00	—			
		20.30	6.60	—			
		35.80	7.80	—			
40° 00′N	117° 00′W	—	6.00	—	90°	R	73
		20.00	6.60	—			
		25.00	7.80	—			
43° 40′N	116° 00′W	0.00	5.20	—	0°	R	84
		10.00	6.70	—			
		42.50	7.90	—			
40° 00′N	113° 00′W	0.00	3.20	1.85		0	87
		1.50	5.90	3.41			
		5.50	6.05	3.49			
		12.00	5.95	3.39			
		15.00	5.80	3.26			
		18.00	6.40	3.59			
		29.00	7.70	4.32			

Table 6 (continued)
COMPRESSIONAL AND SHEAR WAVE VELOCITIES IN THE CONTINENTAL CRUST AND UPPER MANTLE

Latitude	Longitude	Depth below surface (km)	V_p (km/s)	V_s (km/s)	Azimuth	Type	Ref.
		North America (continued)					
40° 00′N	112° 00′W	0.00	3.40	2.00		R	95
		1.70	6.00	3.50			
		8.40	5.50	2.90			
		14.70	6.50	3.50			
		24.70	7.40	4.00			
41° 22′N	111° 57′W	0.00	5.20	—		R	66
		5.90	5.80	—			
		17.60	6.30	—			
		38.90	7.30	—			
		47.70	8.20	—			
41° 30′N	111° 00′W	0.00	3.57	2.06		R	96
		2.20	6.06	3.50			
		9.40	5.80	3.00			
		14.60	6.40	3.50			
		19.60	6.90	3.80			
		28.60	7.60	—			
40° 00′N	110° 00′W	0.00	3.40	1.96		0	87
		2.00	4.50	2.59			
		5.00	5.30	3.05			
		8.00	6.10	3.52			
		12.00	6.40	3.69			
		17.00	6.70	3.86			
		22.00	6.90	3.98			
		40.00	7.80	4.50			
44° 00′N	109° 00′W	0.00	3.70	2.13		0	87
		2.00	4.95	2.85			
		5.00	5.70	3.29			
		8.00	6.10	3.52			
		17.00	6.80	3.95			
		34.00	7.80	4.50			
44° 55′N	106° 58′W	0.00	2.60	—	5°	R	63
		0.30	3.70	—			
		3.70	6.10	—			
		23.30	7.00	—			
		40.20	7.60	—			
		50.30	8.10	—			
44° 46′N	105° 27′W	0.00	3.70	—		R	97
		3.08	6.00	—			
		14.15	6.90	—			
		43.08	8.00	—			
44° 55′N	105° 00′W	0.00	3.60	—	90°	R	63
		3.10	6.10	—			
		16.70	6.90	—			
		27.70	7.30	—			
		46.00	8.00	—			
40° 00′N	92° 00′W	—	6.30	—		R	98
		21.70	6.97	—			
		49.20	8.13	—			
41° 32′N	90° 37′W	0.00	4.50	—		R	63
		0.60	6.00	—			
		5.00	6.10	—			
		10.00	6.20	—			
		15.00	6.40	—			

Table 6 (continued)
COMPRESSIONAL AND SHEAR WAVE VELOCITIES IN THE CONTINENTAL CRUST AND UPPER MANTLE

Latitude	Longitude	Depth below surface (km)	V_p (km/s)	V_s (km/s)	Azimuth	Type	Ref.
		20.00	6.50	—			
		25.00	6.70	—			
		30.00	6.90	—			
		43.60	8.20	—			
44° 00′N	89° 10′W	0.00	5.60	—	334°	R	99
		0.70	6.10	—			
		22.70	6.40	—			
		39.10	8.10	—			
44° 02′N	89° 07′W	0.00	5.40	—	333°	R	63
		1.00	6.10	—			
		12.30	6.50	—			
		37.50	8.00	—			
44° 51′N	87° 20′W	0.00	4.60	—	0°	R	63
		1.40	5.70	—			
		7.50	6.20	—			
		37.50	8.20	—			
40° 43′N	77° 33′W	0.00	5.60	—	84°	R	100
		1.40	6.00	—			
		32.70	8.20	—			
44° 04′N	74° 03′W	0.00	6.40	—		R	100
		5.00	6.60	—			
		10.00	6.70	—			
		15.00	6.80	—			
		20.00	6.90	—			
		25.00	7.00	—			
		30.00	7.20	—			
		36.00	8.10	—			
44° 00′N	64° 45′W	0.00	5.72	—	74°	R	101
		2.50	6.10	—			
		36.30	8.11	—			
44° 35′N	63° 40′W	0.00	5.44	—		R	101
		8.30	6.10	—			
		32.80	8.11	—			
45° 00′N	123° 00′W	0.00	5.50	3.00	5°	E	102
		10.00	6.70	3.90			
		16.00	8.00	4.60			
		21.00	7.90	4.50			
		22.50	7.80	4.20			
		24.50	7.70	3.70			
		26.00	7.50	3.50			
		27.50	7.20	3.30			
		29.50	6.90	—			
		32.00	6.60	—			
47° 00′N	123° 00′W	0.00	2.00	1.00	5°	E	103
		1.20	3.00	1.50			
		2.40	4.50	2.30			
		4.90	6.80	3.90			
		8.90	7.40	4.20			
		13.40	7.80	4.30			
		40.70	6.50	3.10			
		47.70	5.50	2.90			
		55.70	8.00	—			

Table 6 (continued)
COMPRESSIONAL AND SHEAR WAVE VELOCITIES IN THE CONTINENTAL CRUST AND UPPER MANTLE

Latitude	Longitude	Depth below surface (km)	V_p (km/s)	V_s (km/s)	Azimuth	Type	Ref.
		North America (continued)					
48° 05′N	122° 30′W	0.00	5.60	—		R	104
		0.70	6.20	—			
		5.00	6.30	—			
		10.00	6.40	—			
		15.00	6.60	—			
		20.00	6.70	—			
		25.00	6.80	—			
		30.00	6.90	—			
		32.50	8.10	—			
49° 00′N	121° 00′W	—	6.30	—	223°	R	105
		—	6.93	—			
		28.00	7.96	—			
48° 20′N	114° 18′W	0.00	5.00	—	334°	R	63
		3.70	6.00	—			
		29.90	7.40	—			
		35.40	7.90	—			
49° 00′N	114° 00′W	—	5.90	—		R	106
		4.00	6.00	—			
		15.00	4.80	—			
		21.00	6.40	—			
		37.00	8.00	—			
47° 00′N	113° 15′W	0.00	3.80	—		R	97
		1.54	6.00	—			
		23.38	7.40	—			
		46.15	7.60	—			
		49.85	8.20	—			
45° 45′N	112° 26′W	0.00	5.00	—	334°	R	63
		2.30	6.00	—			
		22.20	7.40	—			
		46.10	7.90	—			
48° 11′N	112° 12′W	0.00	5.00	—		R	97
		1.23	5.80	—			
		12.62	6.60	—			
		23.08	7.30	—			
		44.62	8.20	—			
48° 15′N	111° 58′W	0.00	3.60	—	273°	R	63
		0.70	5.60	—			
		15.50	6.70	—			
		29.00	7.20	—			
		54.70	7.90	—			
45° 40′N	110° 18′W	0.00	3.60	—	273°	R	63
		2.80	6.10	—			
		15.70	6.90	—			
		40.30	8.20	—			
48° 08′N	108° 14′W	0.00	2.80	—		R	97
		1.23	6.10	—			
		7.69	6.40	—			
		23.08	6.60	—			
		50.15	8.20	—			
47° 51′N	106° 37′W	0.00	2.60	—	5°	R	63
		0.30	3.70	—			
		2.40	6.10	—			

Table 6 (continued)
COMPRESSIONAL AND SHEAR WAVE VELOCITIES IN THE CONTINENTAL CRUST AND UPPER MANTLE

Latitude	Longitude	Depth below surface (km)	V_p (km/s)	V_s (km/s)	Azimuth	Type	Ref.
		17.50	7.00	—			
		34.40	7.60	—			
		57.20	8.10	—			
46° 39′N	106° 15′W	0.00	3.00	—		E	107
		3.30	6.10	—			
		18.00	6.70	—			
		50.00	8.20	—			
48° 00′N	105° 00′W	0.00	2.80	—		R	97
		2.31	6.10	—			
		8.46	6.40	—			
		37.85	7.80	—			
		53.85	8.30	—			
46° 42′N	100° 12′W	0.00	2.80	—		R	97
		3.08	6.20	—			
		17.23	6.70	—			
		35.69	7.30	—			
		60.00	8.40	—			
47° 30′N	94° 00′W	0.00	5.60	—		R	108
		1.00	6.10	—			
		5.00	6.00	—			
		10.00	6.30	—			
		15.00	6.40	—			
		20.00	6.50	—			
		25.00	6.60	—			
		30.00	6.90	—			
		42.50	8.10	—			
47° 00′N	91° 15′W	0.30	3.50	—	334°	R	99
		2.10	5.60	—			
		5.70	6.90	—			
		12.00	6.40	—			
		39.60	8.10	—			
46° 53′N	91° 12′W	0.00	5.40	—	333°	R	63
		3.70	6.10	—			
		15.90	6.50	—			
		42.40	8.00	—			
47° 00′N	90° 00′W	0.00	5.17	—		R	109
		4.70	6.23	—			
		21.80	7.17	—			
		42.20	8.13	—			
47° 00′N	89° 00′W	—	5.00	—		T	110
		7.50	6.70	—			
		19.50	7.10	—			
		52.50	8.16	—			
47° 00′N	89° 00′W	—	5.50	—	72°	R	111
		10.00	6.63	—			
		—	8.10	—			
45° 40′N	88° 40′W	0.00	4.20	—	45°	R	63
		2.70	6.10	—			
		10.00	6.20	—			
		15.00	6.40	—			
		20.00	6.50	—			
		25.00	6.60	—			

Table 6 (continued)
COMPRESSIONAL AND SHEAR WAVE VELOCITIES IN THE CONTINENTAL CRUST AND UPPER MANTLE

Latitude	Longitude	Depth below surface (km)	V_p (km/s)	V_s (km/s)	Azimuth	Type	Ref.
		North America (continued)					
		30.00	6.70	—			
		35.00	6.80	—			
		40.70	8.20	—			
47° 29′N	87° 46′W	0.00	4.80	—	60°	R	63
		1.90	6.40	—			
		15.70	6.70	—			
		37.20	8.10	—			
45° 20′N	61° 05′W	0.00	5.26	—	74°	R	101
		2.10	6.10	—			
		32.60	8.11	—			
46° 40′N	60° 50′W	0.00	5.90	—	286°	R	101
		14.70	6.35	—			
		25.30	7.35	—			
		43.30	8.50	—			
46° 10′N	59° 40′W	1.00	3.10	—		R	101
		2.30	3.80	—			
		4.60	5.40	—			
		13.80	6.25	—			
		35.00	8.00	—			
48° 40′N	59° 10′W	0.00	4.76	—	28°	R	101
		5.10	6.14	—			
		23.30	7.08	—			
		34.70	7.98	—			
47° 20′N	57° 00′W	0.00	5.40	—	92°	R	101
		2.40	5.87	—			
		7.20	6.38	—			
		30.90	7.85	—			
49° 10′N	56° 15′W	0.00	6.00	—	49°	R	101
		7.50	6.42	—			
		36.70	8.51	—			
49° 10′N	53° 40′W	0.00	6.01	—		R	101
		12.30	6.70	—			
		22.80	7.52	—			
		41.50	8.69	—			
54° 00′N	130° 00′W	0.00	6.03	—	300°	R	112
		12.90	6.41	—			
		17.60	6.70	—			
		29.90	8.11	—			
50° 00′N	125° 00′W	0.00	6.00	—		R	113
		6.20	6.73	—			
		51.20	7.74	—			
53° 00′N	124° 00′W	0.00	5.60	—	80°	T	114
		2.90	6.10	—			
		34.30	8.03	—			
51° 00′N	120° 00′W	0.00	6.40	—		T	114
		36.00	7.83	—			
51° 28′N	113° 35′W	0.00	3.60	—	341°	R	115
		2.00	6.10	—			
		3.00	6.20	—			
		29.00	7.20	—			
		43.00	8.20	—			

Table 6 (continued)
COMPRESSIONAL AND SHEAR WAVE VELOCITIES IN THE CONTINENTAL CRUST AND UPPER MANTLE

Latitude	Longitude	Depth below surface (km)	V_p (km/s)	V_s (km/s)	Azimuth	Type	Ref.
50° 00′N	113° 00′W	0.00	6.16	—	90°	R	116
		14.00	6.50	—			
		30.00	7.16	—			
		44.00	8.33	—			
50° 12′N	112° 35′W	3.00	6.10	—		R	117
		14.00	6.50	—			
		36.00	7.20	—			
		47.00	8.20	—			
50° 00′N	112° 00′W	—	6.40	—		R	118
		22.50	6.10	—			
		33.70	7.32	—			
		47.30	8.25	—			
50° 30′N	111° 51′W	0.00	3.20	—	90°	R	119
		1.50	5.90	—			
		2.70	6.40	—			
		22.70	6.10	—			
		34.20	7.30	—			
		47.50	8.30	—			
50° 00′N	109° 00′W	0.00	6.00	—		R	118
		9.90	6.51	—			
		35.80	7.23	—			
		48.20	8.01	—			
50° 00′N	96° 00′W	—	6.11	—		R	120
		28.00	6.80	—			
		25.50	7.10	—			
		34.00	7.90	—			
50° 00′N	94° 00′W	0.00	6.05	3.46		R	121
		18.30	6.85	4.00			
		34.30	7.92	4.60			
53° 00′N	90° 00′W	0.00	6.30	—	0°	R	122
		30.00	8.05	—			
58° 00′N	155° 00′W	—	5.50	3.37		R	123
		12.00	6.50	3.91			
		32.00	8.10	4.44			
59° 35′N	135° 15′W	0.00	5.50	—		R	124
		4.00	6.10	—			
		14.00	6.70	—			
		38.00	8.10	—			
56° 00′N	133° 00′W	0.00	5.90	—	344°	R	112
		9.30	6.30	—			
		13.10	6.90	—			
		26.50	7.86	—			
60° 00′N	155° 00′W	—	5.50	3.33		R	123
		15.00	6.50	3.67			
		38.00	8.10	4.35			
64° 00′N	149° 00′W	0.00	3.67	2.31		R	125
		2.60	5.40	3.30			
		4.80	5.80	3.45			
		11.60	6.43	3.66			
		40.00	8.10	4.50			
61° 25′N	147° 50′W	0.00	5.70	—		R	124
		6.10	6.60	—			
		16.70	7.30	—			
		48.90	8.30	—			

Table 6 (continued)
COMPRESSIONAL AND SHEAR WAVE VELOCITIES IN THE CONTINENTAL CRUST AND UPPER MANTLE

Latitude	Longitude	Depth below surface (km)	V_p (km/s)	V_s (km/s)	Azimuth	Type	Ref.
		North America (continued)					
63° 00′N	144° 00′W	—	5.84	—	32°	R	126
		10.20	6.79	—			
		43.50	6.98	—			
		53.00	8.07	—			
62° 30′N	115° 00′W	1.00	6.00	—	135°	R	127
		8.00	5.80	—			
		10.00	6.40	—			
		30.00	8.00	—			
66° 00′N	135° 00′W	0.00	6.20	3.57		E	128
		36.00	8.20	4.75			
78° 00′N	98° 00′W	0.00	2.60	—		R	129
		0.90	4.00	—			
		2.30	4.50	—			
		3.00	5.20	—			
		9.30	6.00	—			
		25.10	7.30	—			
		37.80	8.20	—			
		Europe					
37° 00′N	8° 00′W	0.00	4.40	—	315°	R	130
		10.00	5.30	—			
		20.00	7.07	—			
		38.00	8.15	—			
40° 00′N	23° 00′E	—	5.78	3.48		E	131
		16.00	6.14	3.71			
		31.00	6.88	4.00			
		42.00	7.87	4.55			
42° 00′N	30° 00′E	—	3.00	—		R	132
		5.00	4.00	—			
		14.00	6.70	—			
		19.00	8.10	—			
43° 00′N	30° 00′E	10.00	6.20	—		R	132
		17.00	6.80	—			
		25.00	8.20	—			
44° 00′N	30° 00′E	—	5.60	—	0°	R	132
		8.00	6.20	—			
		18.00	7.20	—			
		30.00	8.20	—			
45° 00′N	30° 00′E	—	5.60	—		R	132
		10.00	6.20	—			
		20.00	7.00	—			
		35.00	8.20	—			
44° 02′N	33° 58′E	0.00	4.50	—		R	133
		10.00	6.00	—			
		23.00	8.10	—			
42° 21′N	42° 59′E	0.00	4.30	—		R	134
		7.00	5.60	—			
		24.00	6.50	—			
		48.00	8.00	—			
48° 00′N	36° 00′W	8.00	6.40	—		R	135
		20.00	6.80	—			
		40.00	—	—			

Table 6 (continued)
COMPRESSIONAL AND SHEAR WAVE VELOCITIES IN THE CONTINENTAL CRUST AND UPPER MANTLE

Latitude	Longitude	Depth below surface (km)	V_p (km/s)	V_s (km/s)	Azimuth	Type	Ref.
48° 00′N	34° 00′W	6.00	6.40	—		R	135
		15.00	7.00	—			
		50.00	—	—			
48° 28′N	8° 52′W	0.00	4.00	—		R	136
		0.98	6.00	—			
		10.15	5.50	—			
		19.12	6.80	—			
		24.39	7.60	—			
		36.49	8.00	—			
48° 10′N	7° 20′W	0.00	6.00	—		R	136
		7.80	5.50	—			
		18.93	6.80	—			
		23.41	7.60	—			
		37.85	8.00	—			
46° 00′N	7° 00′E	—	5.30	—		R	137
		5.00	5.90	—			
		11.00	5.80	—			
		23.50	6.50	—			
		36.00	7.80	—			
48° 07′N	7° 19′E	0.00	3.20	—		R	136
		0.57	6.00	—			
		8.19	5.50	—			
		16.19	6.90	—			
		19.43	6.20	—			
		25.90	7.60	—			
		40.95	8.20	—			
50° 00′N	8° 00′E	0.00	3.20	—		R	138
		2.00	4.10	—			
		3.80	6.00	—			
		10.00	5.50	—			
		19.00	6.90	—			
		23.00	6.20	—			
		26.00	7.60	—			
		40.00	8.20	—			
47° 58′N	8° 17′E	0.00	6.00	—		R	136
		10.93	5.50	—			
		19.71	6.80	—			
		26.73	7.60	—			
		40.98	8.00	—			
48° 33′N	8° 22′E	0.00	5.60	—	8°	R	63
		2.40	6.00	—			
		20.10	6.50	—			
		30.20	8.20	—			
48° 26′N	9° 03′E	0.00	3.20	—		R	136
		0.57	5.90	—			
		14.29	5.50	—			
		21.71	6.70	—			
		23.81	6.20	—			
		26.67	7.60	—			
		40.95	8.20	—			
46° 24′N	9° 40′E	0.00	—	—		R	139
		3.10	6.00	—			
		7.20	6.20	—			
		11.60	6.40	—			
		17.60	6.20	—			

Table 6 (continued)
COMPRESSIONAL AND SHEAR WAVE VELOCITIES IN THE CONTINENTAL CRUST AND UPPER MANTLE

Latitude	Longitude	Depth below surface (km)	V_p (km/s)	V_s (km/s)	Azimuth	Type	Ref.
		North America (continued)					
		22.80	6.00	—			
		27.00	5.60	—			
		32.00	6.00	—			
		36.20	6.40	—			
		47.60	7.00	—			
46° 00′N	10° 00′E	0.00	5.00	—	273°	R	141
		2.00	5.90	—			
		9.00	6.50	—			
		20.00	7.30	—			
		57.00	8.20	—			
48° 42′N	10° 00′E	0.00	4.00	—		R	136
		1.17	6.00	—			
		7.41	5.50	—			
		12.49	6.30	—			
		16.59	6.20	—			
		22.24	6.70	—			
		29.85	8.20	—			
47° 36′N	11° 30′E	0.00	6.00	—		R	140
		0.80	6.80	—			
		3.30	5.90	—			
		10.90	5.80	—			
		15.60	5.90	—			
		20.70	6.50	—			
		36.70	7.00	—			
		38.20	8.40	—			
47° 36′N	12° 30′E	0.00	5.80	—		R	140
		1.50	6.20	—			
		3.30	4.90	—			
		8.90	6.10	—			
		10.00	5.90	—			
		14.20	5.90	—			
		18.50	6.50	—			
		36.70	7.00	—			
		38.20	8.40	—			
46° 00′N	13° 00′E	0.00	5.00	—	78°	R	141
		1.50	5.90	—			
		10.00	6.10	—			
		20.00	6.50	—			
		41.00	8.20	—			
49° 51′N	15° 30′E	0.00	5.60	—	291°	R	63
		10.90	6.40	—			
		30.80	8.20	—			
47° 05′N	20° 53′E	0.00	2.40	—	71°	R	63
		2.00	5.90	—			
		19.10	6.70	—			
		23.70	8.10	—			
50° 10′N	5° 30′W	0.00	5.90	—		R	142
		12.00	6.50	—			
		27.00	8.00	—			

Table 6 (continued)
COMPRESSIONAL AND SHEAR WAVE VELOCITIES IN THE CONTINENTAL CRUST AND UPPER MANTLE

Latitude	Longitude	Depth below surface (km)	V_p (km/s)	V_s (km/s)	Azimuth	Type	Ref.
54° 00′N	4° 00′W	—	4.00	—		R	143
		4.50	6.10	—			
		25.00	7.28	—			
		30.00	8.10	—			
54° 00′N	3° 00′W	0.00	6.12	—		R	144
		25.00	7.99	—			
54° 00′N	0° 00′W	—	4.00	—		R	145
		5.00	5.80	—			
		22.00	6.50	—			
		31.00	8.00	—			
55° 00′N	3° 00′E	—	4.00	—	330°	R	145
		8.00	6.15	—			
		25.00	6.50	—			
		31.00	8.15	—			
60° 00′N	7° 00′E	0.00	5.20	—		R	146
		1.00	6.00	—			
		18.00	6.51	—			
		34.00	8.05	—			
52° 52′N	8° 26′E	0.00	3.60	—	324°	R	147
		6.00	5.40	—			
		13.50	6.50	—			
		27.40	8.20	—			
57° 00′N	9° 00′E	0.00	3.00	—	343°	R	148
		5.00	6.10	—			
		8.00	6.60	—			
		28.00	8.10	—			
51° 00′N	12° 00′E	—	5.75	—		R	149
		18.70	6.24	—			
		29.10	8.02	—			
51° 00′N	12° 00′E	0.00	5.86	—	354°	R	149
		16.30	7.00	—			
		32.00	8.48	—			
58° 00′N	14° 00′E	—	6.10	—	306°	R	150
		11.00	6.50	—			
		24.50	6.90	—			
		37.00	8.00	—			
59° 46′N	22° 57′E	0.00	5.70	—	73°	R	151
		2.20	6.00	—			
		18.40	6.40	—			
		26.40	8.20	—			
60° 00′N	24° 00′E	—	6.07	—	90°	R	152
		15.00	6.60	—			
		41.00	8.00	—			
60° 00′N	24° 00′E	—	5.73	—	90°	R	152
		—	5.95	3.52			
		21.00	6.37	3.72			
		29.00	8.23	4.67			
60° 00′N	27° 00′E	—	5.89	3.30	90°	R	152
		—	6.21	3.55			
		21.00	6.65	3.96			
		38.00	8.23	4.67			
57° 25′N	35° 25′E	0.00	3.50	—		R	153
		1.60	4.90	—			
		3.80	5.40	—			
		11.90	6.31	—			

Table 6 (continued)
COMPRESSIONAL AND SHEAR WAVE VELOCITIES IN THE CONTINENTAL CRUST AND UPPER MANTLE

Latitude	Longitude	Depth below surface (km)	V_p (km/s)	V_s (km/s)	Azimuth	Type	Ref.
		North America (continued)					
		17.80	6.70	—			
		26.70	8.00	—			
61° 00′N	9° 00′E	0.00	5.17	—	110°	R	154
		4.70	6.22	—			
		21.80	7.14	—			
		42.20	8.06	—			
62° 00′N	10° 00′E	0.00	6.20	—	0°	R	155
		17.00	6.55	—			
		27.00	7.10	—			
		34.00	8.20	—			
61° 00′N	11° 00′E	—	6.20	—		R	156
		17.00	6.60	—			
		24.00	7.40	—			
		31.50	8.20	—			
64° 00′N	14° 00′E	—	6.13	3.59		R	157
		12.00	6.65	3.79			
		35.00	7.84	4.57			
63° 00′N	16° 00′E	—	6.06	—	285°	R	158
		13.20	6.66	—			
		28.60	7.17	—			
		45.40	8.13	—			
64° 00′N	16° 00′E	0.00	6.00	3.58		R	159
		12.30	6.42	3.83			
		38.80	8.42	4.75			
64° 00′N	17° 00′E	0.00	6.27	3.54		R	160
		10.40	6.69	3.69			
		36.70	7.87	4.55			
62° 00′N	19° 00′E	—	6.32	3.58		R	157
		12.00	6.65	3.64			
		35.00	8.12	4.58			
64° 00′N	20° 00′E	—	6.25	—		T	161
		14.00	6.64	—			
		42.00	8.12	—			
61° 00′N	23° 00′E	—	6.07	3.51	300°	R	152
		18.00	6.51	3.76			
		42.00	8.03	4.64			
61° 05′N	23° 25′E	—	6.06	3.54		R	162
		20.00	6.65	3.88			
		44.00	7.96	4.96			
60° 22′N	26° 58′E	0.00	5.70	—	73°	R	151
		2.20	6.00	—			
		21.50	6.40	—			
		29.70	8.20	—			
65° 00′N	40° 00′E	0.00	5.40	—		R	163
		10.00	6.70	—			
		20.00	7.00	—			
		35.00	8.10	—			
61° 00′N	21° 00′E	21.00	6.85	—	50°	R	152
		38.00	8.20	—			

Table 6 (continued)
COMPRESSIONAL AND SHEAR WAVE VELOCITIES IN THE CONTINENTAL CRUST AND UPPER MANTLE

Latitude	Longitude	Depth below surface (km)	V_p (km/s)	V_s (km/s)	Azimuth	Type	Ref.
62° 29′N	24° 00′E	—	5.80	3.28	355°	R	152
		—	6.17	3.52			
		18.00	6.77	3.73			
		36.00	8.39	4.69			
69° 00′N	15° 00′E	0.00	6.00	—	48°	R	146
		17.00	6.66	—			
		31.00	8.26	—			
69° 00′N	21° 00′E	—	5.95	—	315°	R	152
		14.00	6.70	—			
		33.00	8.18	—			
68° 00′N	22° 00′E	0.00	5.95	—		R	146
		14.00	6.70	—			
		33.00	8.18	—			
66° 00′N	24° 00′E	—	6.23	3.55	320°	R	152
		24.00	6.72	3.85			
		40.00	8.23	4.62			
			South America				
10° 00′N	84° 00′W	0.00	5.10	—		E	164
		8.20	6.20	—			
		21.10	6.60	—			
		43.40	7.90	—			
0° 00′N	78° 00′W	0.00	5.50	—	40°	R	165
		6.00	6.20	—			
		2.90	6.70	—			
		53.00	7.40	—			
		66.00	7.80	—			
		95.00	9.10	—			
3° 00′N	75° 00′W	0.00	5.50	—	40°	R	165
		6.00	6.10	—			
		30.00	6.80	—			
		52.00	7.30	—			
		66.00	7.80	—			
		100.00	9.10	—			
16° 00′S	72° 50′W	0.00	5.30	—	287°	R	166
		4.10	6.20	—			
		25.30	6.70	—			
		51.70	8.00	—			
15° 00′S	71° 00′W	—	4.50	—		R	167
		6.50	6.04	—			
		9.10	5.00	—			
		12.10	6.10	—			
		31.50	6.75	—			
		36.60	6.15	—			
		45.50	6.90	—			
		73.00	8.00	—			
18° 00′S	67° 00′W	0.00	4.50	—		R	167
		6.50	6.04	—			
		9.10	5.00	—			
		12.00	6.10	—			
		28.00	6.80	—			
		31.10	6.15	—			
		39.00	6.90	—			

Table 6 (continued)
COMPRESSIONAL AND SHEAR WAVE VELOCITIES IN THE CONTINENTAL CRUST AND UPPER MANTLE

Latitude	Longitude	Depth below surface (km)	V_p (km/s)	V_s (km/s)	Azimuth	Type	Ref.
			South America (continued)				
17° 15′S	66° 30′W	0.00	5.30	—	283°	R	166
		4.10	6.20	—			
		25.30	6.70	—			
		64.90	8.00	—			
21° 58′S	68° 52′W	0.00	5.30	—	282°	R	166
		6.00	6.40	—			
		34.40	7.00	—			
		56.60	8.00	—			
21° 58′S	68° 52′W	0.00	5.30	—	317°	R	166
		6.00	6.40	—			
		34.40	7.00	—			
		70.30	8.00	—			
			Japan				
34° 35′N	133° 38′E	0.00	5.50	—	60°	R	168
		2.50	6.00	—			
		36.00	7.70	—			
35° 20′N	134° 42′E	0.00	5.50	—	90°	T	169
		4.20	6.10	—			
		17.50	6.65	—			
		34.70	7.79	—			
36° 07′N	136° 54′E	0.00	5.50	—	60°	R	168
		1.80	6.00	—			
		27.70	7.70	—			
36° 07′N	136° 54′E	0.00	5.60	—	88°	R	168
		3.10	6.00	—			
		38.00	7.70	—			
36° 43′N	138° 38′E	0.00	2.70	—	291°	R	170
		0.10	5.50	—			
		6.10	6.10	—			
		25.10	7.70	—			
40° 00′N	140° 00′E	—	2.50	—	308°	T	171
		3.00	5.90	—			
		16.30	6.60	—			
		28.00	7.50	—			
39° 33′N	140° 13′E	0.00	5.50	—		R	172
		3.80	6.00	—			
		13.80	6.60	—			
		31.00	8.00	—			
36° 08′N	140° 34′E	0.00	1.70	—	16°	R	173
		0.90	5.50	—			
		5.20	6.20	—			
		26.70	7.90	—			
36° 08′N	140° 34′E	0.00	1.80	—	291°	R	170
		0.70	5.50	—			
		6.70	6.10	—			
		25.70	7.70	—			
39° 06′N	140° 54′E	0.00	2.50	—		R	174
		0.50	5.80	—			
		9.50	6.20	—			
		25.00	7.80	—			

Table 6 (continued)
COMPRESSIONAL AND SHEAR WAVE VELOCITIES IN THE CONTINENTAL CRUST AND UPPER MANTLE

Latitude	Longitude	Depth below surface (km)	V_p (km/s)	V_s (km/s)	Azimuth	Type	Ref.
39° 18′N	141° 42′E	0.00	1.70	—	16°	R	173
		0.90	5.50	—			
		5.90	6.20	—			
		18.70	7.50	—			
			Australia				
8° 00′S	147° 30′E	0.00	4.40	—		T	175
		4.60	6.98	—			
		20.00	7.96	—			
9° 00′S	149° 00′E	0.00	2.80	—		T	175
		1.70	3.70	—			
		4.60	5.66	—			
		10.00	6.86	—			
		24.00	7.96	—			
15° 00′S	143° 00′E	—	5.82	—	60°	R	176
		15.00	6.71	—			
		35.00	8.09	—			
21° 00′S	145° 00′E	—	5.94	—	45°	R	176
		17.00	6.62	—			
		35.00	7.84	—			
30° 13′S	131° 24′E	0.00	6.00	—	84°	R	177
		37.40	8.30	—			
36° 08′S	148° 37′E	0.00	4.50	—	40°	R	178
		1.50	6.00	—			
		36.30	8.00	—			
40° 00′S	147° 00′E	—	6.00	—	15°	R	176
		15.00	6.60	—			
		35.00	7.90	—			
41° 08′S	174° 55′E	0.00	3.50	—	57°	R	179
		0.60	5.50	—			
		3.70	6.10	—			
		7.00	6.20	—			
		17.60	8.00	—			
			Africa				
9° 00′N	38° 40′E	0.00	6.00	—		R	180
		24.46	6.95	—			
		40.22	8.10	—			
5° 10′N	38° 20′E	0.00	6.00	3.50	12°	R	180
		20.00	6.80	4.00			
		48.00	7.95	4.50			
		78.00	8.20	4.05			
50° 40′N	8° 35′E	0.00	3.60	—	345°	R	−0
		2.50	5.40	—			
		9.00	6.60	—			
		24.00	8.30	—			
0° 30′N	35° 00′E	0.00	5.80	—		R	181
		26.00	6.50	—			
		44.00	8.00	—			
2° 00′N	36° 00′E	—	3.00	1.80	0°	R	182
		2.80	6.38	3.53			
		18.50	7.48	4.53			

Table 6 (continued)
COMPRESSIONAL AND SHEAR WAVE VELOCITIES IN THE CONTINENTAL CRUST AND UPPER MANTLE

Latitude	Longitude	Depth below surface (km)	V_p (km/s)	V_s (km/s)	Azimuth	Type	Ref.
2° 20′S	27° 30′E	0.00	3.40	—		R	180
		1.63	6.00	—			
		13.04	6.80	—			
		35.87	8.10	—			
1° 20′S	36° 50′E	0.00	6.00	—		R	180
		16.30	6.80	—			
		44.35	8.10	—			
26°15′S	28° 08′E	0.00	5.40	—	31°	R	183
		1.30	6.10	—			
		36.00	8.40	—			
26° 15′S	28° 08′E	0.00	6.00	—	282°	R	184
		28.80	7.20	—			
		36.60	8.00	—			
26° 15′S	28° 08′E	0.00	5.40	—	22°	R	183
		1.30	6.20	—			
		37.90	8.20	—			
26° 15′S	28° 08′E	0.00	5.70	—	77°	R	185
		4.50	6.10	—			
		27.20	6.80	—			
		42.70	8.30	—			
			Asia				
15° 00′N	75° 00′E	0.00	6.00	3.50		E	186
		16.00	6.60	3.98			
		52.00	8.24	4.73			
13° 00′N	77° 00′E	0.00	5.67	3.46		R	187
		15.80	6.51	3.96			
		34.70	7.98	4.61			
15° 00′N	77° 00′E	0.00	5.57	3.35	0°	E	186
		12.00	6.64	3.92			
		41.00	8.24	4.73			
17° 00′N	74° 00′E	—	5.78	3.42		R	188
		19.80	6.58	3.92			
		38.60	8.19	4.62			
28° 00′N	77° 00′E	0.00	2.70	—		E	189
		3.70	5.64	—			
		18.90	6.49	—			
		40.30	8.06	—			
26° 00′N	80° 00′E	—	2.70	—		R	190
		6.00	6.20	—			
		14.00	6.90	—			
		28.00	8.20	—			
28° 00′N	84° 00′E	0.00	5.92	3.54		E	191
		13.40	6.80	3.92			
		30.10	8.00	4.54			
33° 00′N	56° 00′E	0.00	3.49	2.00		R	192
		6.00	6.00	3.53			
		30.00	7.00	4.05			
		52.00	8.10	4.65			
35° 00′N	80° 00′E	—	5.48	3.33		E	193
		22.70	6.00	3.56			
		39.00	6.45	3.90			
		57.70	8.07	4.57			

Table 6 (continued)
COMPRESSIONAL AND SHEAR WAVE VELOCITIES IN THE CONTINENTAL CRUST AND UPPER MANTLE

Latitude	Longitude	Depth below surface (km)	V_p (km/s)	V_s (km/s)	Azimuth	Type	Ref.
32° 00′N	85° 00′E	0.00	4.50	2.60	0°	E	194
		3.50	5.98	3.45			
		12.00	5.98	3.42			
		28.00	5.80	3.37			
		38.00	6.30	3.64			
		68.00	7.70	4.45			
39° 52′N	55° 33′E	0.00	3.50	—		R	195
		8.30	5.00	—			
		18.30	5.50	—			
		27.30	6.30	—			
		46.30	8.00	—			
39° 52′N	55° 33′E	0.00	4.00	—		R	195
		6.40	5.50	—			
		20.00	6.30	—			
		39.00	8.00	—			
39° 52′N	55° 33′E	0.00	3.50	—		R	195
		3.90	5.50	—			
		13.80	6.30	—			
		29.80	8.00	—			
		99.00	—	—			
37° 19′N	60° 36′E	0.00	3.70	—	65°	R	196
		6.00	6.10	—			
		15.00	6.70	—			
		30.00	7.20	—			
		47.00	8.20	—			
37° 38′N	61° 51′E	0.00	3.70	—	65°	R	196
		5.00	6.10	—			
		15.00	6.70	—			
		27.00	7.20	—			
		44.00	8.20	—			
39° 35′N	66° 48′E	0.00	3.70	—	65°	R	196
		5.00	6.10	—			
		10.00	6.70	—			
		23.00	7.50	—			
		42.00	8.20	—			
38° 35′N	72° 33′E	0.00	5.50	—	0°	R	197
		33.00	6.40	—			
		51.00	8.10	—			
38° 33′N	72° 56′E	0.00	5.50	—	90°	R	197
		43.00	6.40	—			
		72.00	8.10	—			
38° 33′N	73° 00′E	0.00	5.50	—	0°	R	197
		32.00	6.40	—			
		57.00	8.10	—			
40° 16′N	69° 40′E	0.00	3.70	—	77°	R	196
		7.00	6.00	—			
		14.00	6.60	—			
		28.00	7.40	—			
		42.00	8.20	—			
40° 45′N	72° 20′E	0.00	3.70	—	77°	R	196
		6.50	6.20	—			
		15.50	6.80	—			
		32.00	7.50	—			
		50.00	8.30	—			

Table 6 (continued)
COMPRESSIONAL AND SHEAR WAVE VELOCITIES IN THE CONTINENTAL CRUST AND UPPER MANTLE

Latitude	Longitude	Depth below surface (km)	V_p (km/s)	V_s (km/s)	Azimuth	Type	Ref.
40° 30′N	73° 00′E	0.00	5.50	—	0°	R	197
		20.00	6.40	—			
		44.00	8.10	—			
43° 00′N	75° 00′E	0.00	5.50	—	275°	R	198
		10.00	6.40	—			
		50.00	8.10	—			
42° 45′N	75° 30′E	0.00	5.50	—	282°	R	199
		10.00	6.40	—			
		46.00	8.10	—			
44° 00′N	76° 00′E	0.00	5.50	—	334°	R	198
		10.00	6.40	—			
		50.00	8.10	—			
42° 08′N	77° 00′E	0.00	5.50	—	14°	R	200
		10.50	6.40	—			
		51.50	8.10	—			
42° 46′N	77° 18′E	0.00	5.80	—	56°	R	199
		30.00	6.40	—			
		50.00	7.90	—			
42° 40′N	77° 20′E	0.00	5.50	—	11°	R	198
		10.00	6.40	—			
		50.00	8.10	—			
46° 30′N	78° 20′E	0.00	5.50	—	11°	R	198
		20.00	6.40	—			
		40.00	8.10	—			
46° 35′N	78° 36′E	0.00	5.50	—	14°	R	200
		11.00	6.40	—			
		47.00	8.10	—			
54° 47′N	60° 20′E	0.00	5.60	—	90°	R	201
		9.90	6.00	—			
		16.70	6.30	—			
		29.00	7.20	—			
		38.00	8.00	—			
52° 30′N	104° 00′W	0.00	5.10	—	327°	R	202
		2.86	6.00	—			
		18.57	6.10	—			
		23.13	6.40	—			
		37.50	8.10	—			
52° 00′N	108° 00′E	0.00	5.80	—	300°	R	202
		6.25	6.00	—			
		17.86	6.40	—			
		37.50	7.80	—			
51° 30′N	110° 00′E	0.00	5.75	—	300°	R	202
		5.50	6.00	—			
		13.70	6.40	—			
		44.50	8.10	—			
51° 30′N	113° 30′E	0.00	5.80	—	16°	R	202
		5.92	5.90	—			
		14.64	6.00	—			
		18.08	6.40	—			
		38.40	8.10	—			

Table 6 (continued)
COMPRESSIONAL AND SHEAR WAVE VELOCITIES IN THE CONTINENTAL CRUST AND UPPER MANTLE

Latitude	Longitude	Depth below surface (km)	V_p (km/s)	V_s (km/s)	Azimuth	Type	Ref.
63° 00′N	12° 00′E	—	6.05	—	285°	R	74
		4.20	6.24	—			
		20.30	6.68	—			
		30.30	7.10	—			
		41.50	8.13	—			
60° 40′N	151° 40′E	0.00	5.30	—	8°	R	203
		4.50	6.00	—			
		21.00	6.70	—			
		32.50	8.10	—			
61° 45′N	152° 00′E	0.00	5.30	—	8°	R	203
		6.00	6.00	—			
		21.00	6.70	—			
		38.00	8.10	—			
			Antarctica				
60° 30′S	46° 10′W	0.10	5.10	—		R	204
		3.50	6.30	—			
61° 00′S	46° 00′W	0.20	2.50	—		R	204
		2.00	3.40	—			
		4.40	5.50	—			
61° 40′S	46° 10′W	0.50	1.90	—		R	204
		3.00	5.00	—			
		6.25	6.70	—			
70° 47′S	12° 10′E	0.00	6.20	—		R	205
		4.00	6.00	—			
		9.50	6.10	—			
		17.00	6.35	—			
		27.00	6.42	—			
		38.00	7.80	—			
74° 27′S	67° 08′W	0.00	—	—		R	206
		1.15	4.40	—			
		1.40	5.30	—			
		4.75	6.00	—			
75° 17′S	116° 24′W	0.00	—	—		R	206
		1.50	4.60	—			
		4.50	6.10	—			
76° 22′S	9° 32′E	0.00	—	—		R	207
		3.00	5.73	—			
77° 24′S	100° 24′W	0.00	—	—		R	206
		3.10	2.40	—			
		3.70	4.10	—			
		6.00	6.00	—			
77° 46′S	87° 12′W	0.00	—	—		R	206
		0.60	5.20	—			
		1.80	6.10	—			
78° 00′S	55° 00′E	0.00	—	—		E	208
		3.00	5.77	3.33			
		5.00	6.25	3.59			
		15.00	6.35	3.64			
		27.00	6.60	3.74			
		42.00	7.85	4.44			

Table 6 (continued)
COMPRESSIONAL AND SHEAR WAVE VELOCITIES IN THE CONTINENTAL CRUST AND UPPER MANTLE

Latitude	Longitude	Depth below surface (km)	V_p (km/s)	V_s (km/s)	Azimuth	Type	Ref.
78° 01′S	158° 25′E	0.00	—	—		R	206
		1.00	4.30	—			
		1.60	5.40	—			
		2.50	6.50	—			
78° 10′S	162° 13′W	0.00	2.40	—		R	206
		1.40	4.20	—			
		2.00	6.40	—			
80° 00′S	120° 00′W	0.00	—	—		R	206
		2.60	4.30	—			
		4.00	5.90	—			
82° 09′S	109° 07′W	0.00	—	—		R	206
		2.25	5.20	—			
		2.95	5.80	—			
		5.75	7.00	—			
84° 41′S	114° 02′W	0.00	—	—		R	206
		2.20	4.40	—			
		2.40	5.50	—			
		2.70	7.00	—			
87° 09′S	112° 06′W	0.00	—	—		R	206
		1.90	5.70	—			
		7.25	6.70	—			
87° 55′S	126° 12′W	0.00	—	—		R	206
		2.25	5.70	—			
		6.80	6.70	—			
89° 30′S	103° 54′W	0.00	—	—		R	206
		3.10	5.30	—			
		4.30	6.10	—			
			Iceland				
64° 55′N	23° 00′W	0.00	—	—	25°	R	209
		0.97	6.50	—			
		13.95	7.20	—			
64° 30′N	22° 38′W	0.00	—	—	302°	R	209
		2.51	6.50	—			
		8.85	7.20	—			
63° 53′N	22° 04′W	0.00	3.70	—		R	210
		2.10	6.70	—			
		17.80	7.40	—			
		27.80	8.00	—			
64° 12′N	21° 50′W	0.00	—	—	25°	R	209
		2.56	6.50	—			
		8.97	7.20	—			
63° 58′N	20° 26′W	0.00	—	—	302°	R	209
		2.88	6.50	—			
65° 20′N	19° 00′W	0.00	4.70	2.70	40°	R	211
		4.94	6.30	3.60			
		10.21	7.40	4.30			
63° 24′N	18° 14′W	0.00	—	—	302°	R	209
		9.01	6.50	—			
		14.66	7.20	—			

Table 6 (continued)
COMPRESSIONAL AND SHEAR WAVE VELOCITIES IN THE CONTINENTAL CRUST AND UPPER MANTLE

Latitude	Longitude	Depth below surface (km)	V_p (km/s)	V_s (km/s)	Azimuth	Type	Ref.
			Hawaii				
21° 20N	158° 10′W	0.00	1.50	—		R	212
		0.50	4.80	—			
		5.00	6.40	—			
		19.50	8.50	—			
21° 43′N	157° 57′W	0.00	3.00	—		R	212
		1.50	4.20	—			
		4.00	5.70	—			
		5.67	7.80	—			
21° 25′N	157° 42′W	0.00	1.50	—		R	212
		0.67	3.60	—			
		5.00	4.70	—			
		9.83	6.70	—			
		13.33	8.30	—			
20° 50′N	157° 06′W	0.00	4.80	—		R	212
		5.83	6.90	—			
		14.00	7.80	—			
19° 28′N	155° 53′W	0.00	2.50	—		R	212
		0.83	4.70	—			
		3.83	6.00	—			
		9.67	7.20	—			
		16.67	8.20	—			
20° 10′N	155° 50′W	0.00	3.50	—		R	212
		0.83	5.20	—			
		3.33	7.00	—			
19° 00′N	155° 41′W	0.00	3.00	—		R	212
		0.83	5.30	—			
		7.17	7.00	—			
		12.00	8.20	—			
19° 20′N	155° 19′W	0.00	1.80	—		R	212
		0.33	3.10	—			
		1.83	5.10	—			
		8.33	7.10	—			
		12.83	8.10	—			
19° 44′N	155° 08′W	0.00	3.00	—		R	212
		1.25	5.30	—			
		5.00	7.00	—			
		12.17	8.20	—			
19° 33′N	154° 54′W	0.00	2.50	—		R	212
		0.83	3.50	—			
		1.67	5.10	—			
		8.67	7.10	—			
		11.67	8.20	—			

Table 7
VELOCITIES, DENSITY, PRESSURE, AND GRAVITY IN THE EARTH AS A FUNCTION OF RADIUS AND DEPTH[213]

Index	Radius (km)	Depth (km)	V_p (km/s)	V_s (km/s)	ϱ (g/cm³)	Pressure (kb)	Gravity g (cm/sec²)
1	1	6370	11.17	3.50	12.58	3617	0
2	100	6271	11.18	3.50	12.57	3614	52
3	300	6071	11.19	3.50	12.53	3592	122
4	400	5971	11.20	3.50	12.53	3575	150
5	600	5771	11.20	3.50	12.52	3529	218
6	800	5571	11.20	3.48	12.52	3466	286
7	1000	5371	11.17	3.47	12.48	3385	355
8	1215	5156	10.89	3.46	12.30	3281	427
9	1215	5156	10.33	0.0	12.12	3281	427
10	1300	5071	10.31	0.0	12.09	3236	454
11	1400	4971	10.28	0.0	12.05	3179	486
12	1500	4871	10.24	0.0	11.99	3116	517
13	1600	4771	10.21	0.0	11.92	3055	549
14	1700	4671	10.14	0.0	11.85	2988	580
15	1800	4571	10.06	0.0	11.77	2917	611
16	1900	4471	9.98	0.0	11.69	2844	641
17	2100	4271	9.79	0.0	11.52	2688	702
18	2200	4171	9.71	0.0	11.44	2606	731
19	2300	4071	9.61	0.0	11.36	2520	760
20	2400	3971	9.51	0.0	11.28	2433	789
21	2500	3871	9.41	0.0	11.20	2342	818
22	2600	3771	9.30	0.0	11.10	2249	846
23	2700	3671	9.19	0.0	11.00	2154	874
24	2800	3571	9.07	0.0	10.89	2057	901
25	2900	3471	8.95	0.0	10.76	1958	928
26	3000	3371	8.82	0.0	10.63	1858	953
27	3200	3171	8.55	0.0	10.37	1652	1003
28	3300	3071	8.39	0.0	10.23	1548	1027
29	3400	2971	8.18	0.0	10.09	1442	1050
30	3485	2886	7.98	0.0	9.96	1352	1069
31	3485	2886	13.64	7.23	5.53	1352	1069
32	3510	2861	13.63	7.24	5.50	1337	1065
33	3550	2821	13.62	7.23	5.50	1314	1060
34	3625	2746	13.59	7.22	5.47	1270	1050
35	3700	2671	13.53	7.21	5.45	1227	1042
36	3775	2596	13.45	7.18	5.43	1185	1034
37	3850	2521	13.37	7.14	5.40	1143	1028
38	3925	2446	13.28	7.11	5.36	1102	1022
39	4000	2371	13.20	7.07	5.31	1061	1017
40	4075	2296	13.12	7.04	5.27	1021	1012
41	4150	2221	13.04	7.01	5.22	981	1008
42	4225	2146	12.97	6.98	5.16	942	1005
43	4300	2071	12.89	6.95	5.12	903	1002
44	4375	1996	12.80	6.92	5.07	865	999
45	4450	1921	12.71	6.89	5.03	827	997
46	4525	1846	12.63	6.85	4.99	789	995
47	4600	1771	12.54	6.81	4.96	752	994
48	4675	1696	12.44	6.77	4.92	715	993
49	4750	1621	12.36	6.72	4.88	679	992
50	4825	1546	12.26	6.67	4.84	643	992
51	4900	1471	12.16	6.63	4.81	607	991
52	4975	1396	12.06	6.59	4.77	571	991
53	5050	1321	11.95	6.56	4.74	536	991
54	5125	1246	11.83	6.52	4.71	501	992

Table 7 (continued)
VELOCITIES, DENSITY, PRESSURE, AND GRAVITY IN THE EARTH AS A FUNCTION OF RADIUS AND DEPTH[213]

Index	Radius (km)	Depth (km)	V_p (km/s)	V_s (km/s)	ϱ (g/cm³)	Pressure (kb)	Gravity g (cm/sec²)
55	5200	1171	11.71	6.45	4.67	466	992
56	5275	1096	11.58	6.38	4.64	431	993
57	5350	1021	11.45	6.37	4.61	397	994
58	5425	946	11.31	6.36	4.58	362	995
59	5500	871	11.17	6.31	4.54	328	997
60	5550	821	11.07	6.23	4.51	306	998
61	5573	798	11.03	6.17	4.50	295	998
62	5602	769	10.96	6.13	4.46	282	999
63	5625	746	10.95	6.09	4.43	272	999
64	5643	728	10.95	6.08	4.41	264	999
65	5660	711	10.91	6.06	4.40	257	1000
66	5675	696	10.86	6.04	4.38	250	1000
67	5700	671	10.64	5.90	4.36	239	1000
68	5700	671	10.25	5.60	4.07	239	1000
69	5725	646	10.11	5.45	4.03	229	1000
70	5750	621	10.09	5.43	4.00	219	1000
71	5775	596	10.08	5.42	3.98	209	999
72	5800	571	10.07	5.40	3.95	199	999
73	5825	546	9.93	5.34	3.90	189	999
74	5850	521	9.70	5.26	3.76	180	998
75	5875	496	9.51	5.12	3.74	170	997
76	5900	471	9.51	5.10	3.76	161	997
77	5925	446	9.50	5.07	3.79	152	996
78	5950	421	9.46	5.04	3.80	142	996
79	5967	404	9.12	4.86	3.69	136	995
80	5983	388	8.79	4.71	3.62	130	995
81	6000	371	8.64	4.64	3.59	124	994
82	6025	346	8.60	4.59	3.53	115	994
83	6050	321	8.58	4.57	3.47	106	993
84	6075	296	8.55	4.57	3.41	98	992
85	6100	271	8.52	4.59	3.37	89	991
86	6125	246	8.40	4.62	3.34	81	990
87	6150	221	8.19	4.57	3.34	73	989
88	6175	196	7.98	4.45	3.35	64	988
89	6200	171	7.78	4.30	3.37	56	987
90	6225	146	7.75	4.22	3.39	48	986
91	6250	121	7.79	4.18	3.40	39	985
92	6270	101	7.93	4.36	3.44	33	984
93	6290	81	8.08	4.62	3.48	26	984
94	6310	61	8.38	4.72	3.52	19	984
95	6330	41	8.38	4.73	3.51	12	983
96	6350	21	8.38	4.71	3.49	5	983
97	6350	21	6.50	3.72	2.80	5	983
98	6368	3	6.50	3.72	2.80	3	982
99	6368	3	1.45	0.0	1.02	3	982
100	6371	0	1.45	0.0	1.02	0	981

Table 8
COMPRESSIONAL WAVE VELOCITIES IN AIR AND CO_2 AT 0°C[214]

Frequency (cps)	V_p (km/sec)
Air	
41,009	0.33245
42,071	0.33237
50,701	0.33247
56,319	0.33232
70,118	0.33198
88,585	0.33197
98,183	0.33177
205,620	0.33167
610,220	0.33181
1,034,060	0.33176
1,479,900	0.33164
CO_2	
42,071	0.25882
98,183	0.25894
205,620	0.26015
1,034,060	opaque

Table 9
COMPRESSIONAL WAVE VELOCITIES IN SEA WATER AS A FUNCTION OF PRESSURE (P), TEMPERATURE (T), AND SALINITY (S)[215]

P = O S = 35‰		P = O T = 22°C		S = 35‰ T = O°C	
Temp (°C)	V_p (km/s)	Salinity (‰)	V_p (km/s)	Pressure (bars)	V_p (km/s)
0	1.45	37	1.5245	90.5	1.463
2.6	1.46	36	1.5225	181.0	1.488
5.2	1.47	35	1.521	271.5	1.493
8	1.48	34	1.519	362.0	1.509
11	1.49	33	1.5175	452.5	1.525
14.3	1.50	32	1.516	543.0	1.540
18	1.51	31	1.5145	633.5	1.555
22	1.52			724.0	1.570
				814.5	1.585

Table 10
LABORATORY MEASUREMENTS OF COMPRESSIONAL WAVE VELOCITIES AND CARBONATE CONTENTS OF MARINE SEDIMENTS[219,220]

Sample number	Sediment type	Density (g/cm^3)	Porosity (%)	V_p (km/s)	Carbonate (wt %)
1	Red clay	1.49	—	1.52	—
2	Siltite	1.60	65.0	1.59	83.0
3	Lutite	1.74	52.0	1.68	94.0
4	Lutite	1.77	50.0	1.73	91.0
5	Lutite	1.77	46.0	1.69	94.0
6	Lutite	1.77	55.0	1.68	94.0
7	Siltite	1.71	57.0	1.68	95.0
8	Arenite	1.62	61.0	1.70	95.0
9	Lutite	1.57	62.0	1.60	93.0
10	Arenite	1.77	57.0	1.80	93.0
11	Arenite	1.73	55.0	1.83	94.0
12	Arenite	1.74	58.0	1.77	93.0
13	Arenite	1.50	59.0	2.04	96.0
14	Lutite	1.89	47.0	2.77	96.0
15	Lutite	1.80	52.0	2.37	92.0
16	Siltite	1.71	57.0	1.74	95.0
17	Siltite	1.72	57.0	1.73	95.0
18	Siltite	1.77	54.0	1.71	97.0
19	Siltite	1.71	57.0	1.72	94.0
20	Arenite	1.57	63.0	1.75	88.0
21	Arenite	1.61	60.0	2.02	88.0
22	Red clay	1.56	68.0	1.51	49.0
23	Gray clay	1.86	48.0	1.62	32.0
24	Gray clay	1.90	47.0	1.61	33.0
25	Red clay	1.77	55.0	1.61	68.0
26	Lutite	1.78	54.0	1.69	61.0
27	Gray clay	1.77	55.0	1.64	—
28	Gray clay	1.79	54.0	1.63	74.0
29	Gray clay	1.48	70.0	1.50	35.0
30	Gray clay	1.48	71.0	1.48	32.0
31	Gray clay	1.61	63.0	1.54	29.0
32	Gray clay	1.82	51.0	1.59	—
33	Gray clay	1.46	74.0	1.49	20.0
34	Lutite	1.55	65.0	1.59	—
35	Gray clay	1.54	69.0	1.53	33.0
36	Gray clay	1.61	64.0	1.51	33.0
37	Gray clay	1.62	61.0	1.57	24.0
38	Gray clay	1.49	69.0	1.49	27.0
39	Gray clay	1.57	64.0	1.55	28.0
40	Silt and sand	1.91	43.0	1.71	—
41	Gray clay	1.52	68.0	1.51	25.0
42	Gray clay	1.55	66.0	1.51	23.0
43	Gray silt	1.84	46.0	1.67	21.0
44	Gray clay	1.58	66.0	1.52	26.0
45	Siltite	1.63	61.0	1.65	98.0
46	Siltite	1.61	62.0	1.69	96.0
2992	Terrigenous mud	1.61	—	1.51	21.0
2993	Terrigenous mud	1.62	—	1.54	27.0
2994	Globerigina ooze	1.62	—	1.63	54.0
2995	Globerigina ooze	1.55	—	1.54	54.0
2996	Terrigenous mud	1.54	—	1.47	17.0
2997	Terrigenous mud	1.67	—	1.50	26.0

Table 11
COMPRESSIONAL AND SHEAR WAVE VELOCITIES VERSUS COMPACTION PRESSURE IN GLOBIGERINA OOZE[220]

Pressure (kg/cm²)	V_p (km/s)	V_s (km/s)
512.0	2.68	1.20
768.0	2.89	1.42
1024.0	3.06	1.57
512.0	2.98	1.51
256.0	2.77	1.40
128.0	2.63	1.30
64.0	2.48	1.22

Table 12
COMPRESSIONAL AND SHEAR WAVE VELOCITIES IN POLYCRYSTALLINE ICE[221]

Sample #1 V_p (km/sec)	Sample #2 V_p (km/sec)	Sample #3 V_p (km/sec)	Sample #4 V_s (km/sec)
3.11	3.15	3.07	1.93
3.11	3.23	3.19	1.92
3.14	3.20	3.16	1.94
3.18	3.23	3.09	1.88
3.11	3.22	3.14	1.91
3.23	3.28		
3.16			
3.17			
3.11			

Table 13
COMPRESSIONAL WAVE VELOCITIES IN PERMAFROST SAMPLES

Sample type	Moisture content (%)	V_p @ −1°C (km/sec)	V_p@−7°C (km/sec)
Sandy clay	22.2	3.32	3.55
Silty clay	18.1	3.10	3.34
Clay	58.9	2.47	3.65
Clay	10.2	2.02	3.11
Clay	13.7	2.29	4.10
Clay	—	3.09	3.62
Clay	35.2	3.87	4.11
Peat	488.0	2.51	2.79
Peat	437.6	2.77	2.88

Table 14
COMPRESSIONAL WAVE VELOCITIES IN FROZEN ROCKS[223]

Rock type	Porosity (%)	V_p @ 25°C (km/sec)	V_p, frozen (km/sec)
Dry Berea sandstone	18.1	3.82	3.86
Wet Berea sandstone	18.1	3.93	5.15
Boise sandstone	26.4	3.22	4.88
Navajo sandstone	13.0	4.36	5.35
Spergen limestone	14.6	4.38	5.74
Duvernay dolomite	16.9	4.72	5.83
Porous porcelain	41.6	2.89	5.36
Volcanic sandstone	24.4	3.10	4.51
Black shale	3.5	3.56	3.85
Green River shale	9.6	3.98	5.44

Table 15
COMPRESSIONAL WAVE VELOCITIES IN ROCKS AS A FUNCTION OF SATURATION[224]

Rock type	Density (g/cm^3)	Porosity (%)	V_p dry (km/sec)	V_p air-dry (km/sec)	V_p saturated (km/sec)
Granite	2.62	—	3.76	3.85	5.10
	2.65	—	3.20	3.25	5.10
	2.66	—	4.20	4.30	5.20
	2.65	1.1	3.35	3.45	5.45
	2.67	1.6	3.65	3.75	5.55
	2.65	1.8	3.50	3.60	5.60
	2.64	1.1	4.00	4.10	5.65
	2.63	0.6	4.30	4.40	5.75
	2.64	0.6	4.10	4.20	5.80
	2.62	—	5.25	5.30	5.90
	2.62	—	5.35	5.35	6.00
	2.62	—	5.30	5.40	6.30
Nepheline Syenite	2.68	0.8	3.95	4.10	5.40
Nepheline Syenite	2.67	0.7	4.45	4.60	5.70
Syenite	2.74	1.1	5.20	5.45	6.45
Gneiss	2.68	0.9	3.50	3.65	5.50
	2.75	0.3	4.75	4.85	5.90
	2.72	0.4	4.95	5.10	5.90
	2.72	0.9	5.00	5.20	6.00
	2.79	0.7	4.00	4.15	5.45
	2.74	0.4	4.55	4.65	5.75
	2.76	0.4	4.30	4.35	5.80
	2.76	0.5	4.20	4.30	5.95
Granulite	2.83	0.4	4.60	4.90	5.45
	2.86	0.2	5.25	5.35	5.85
	2.84	0.3	5.60	5.60	6.00
Diorite	2.79	1.2	4.55	4.65	5.95
	2.79	0.6	5.20	5.30	6.20
	2.80	0.7	5.30	5.30	6.20
	2.81	0.6	5.85	5.90	6.40
	2.86	0.6	5.70	5.70	6.60
Gabbro-Norite	2.93	0.2	6.15	6.15	6.70
Norite	3.16	0.2	6.40	6.40	6.75
	3.14	0.2	6.65	6.65	6.95
	3.16	0.2	6.75	6.80	7.10
	3.16	0.2	7.00	7.00	7.10
Pyroxenite	3.22	0.2	7.10	7.10	7.70
	3.29	—	7.55	7.60	8.10
	3.32	0.2	7.65	7.70	8.15
	3.33	—	8.10	8.15	8.40

Table 16
VELOCITIES IN LOW-DENSITY ROCKS[225]

Rock type	Density (g/cm^3)	V_p (km/s)	V_s (km/s)
Tuff San Luis Obispo, Calif.	1.38	1.43	0.87
Kaolin Dry Branch, Colo.	1.58	1.44	0.93
Rhyolite Castle Rock, Colo.	2.05	3.27	1.98
Volcanic Breccia Park County, Colo.	2.19	4.22	2.49
Basaltic Scoria Klamath Falls, Ore.	2.23	4.33	2.51
Latite Chaffee County, Colo.	2.45	3.77	2.21
Graphite, Ceylon	2.16	3.06	1.86
Tremolite, New York	2.86	6.17	3.70
Limonite, Alabama	3.55	5.36	2.97
Pyrrhotite, Ontario	4.55	4.69	2.76

Table 17
LABORATORY MEASUREMENTS OF SEISMIC ANISOTROPY IN VOLCANIC ROCKS[226]

Rock type	V_p (km/sec)		Anisotropy (%)	Porosity (%)
	Perp. foliation	Par. foliation		
Basalt tuff Cape Ann, Mass.	1.98	1.40	34.3	41.1
Trachyte San Pedro, Mex.	5.12	5.30	3.5	3.8
Rhyolite Castle Rock, Colo.	2.94	3.48	16.8	32.0
Vesicular Rhyolite Castle Mtns., Calif.	3.78	4.90	25.8	18.0
Massive pumice Mono Craters, Calif.	1.34	3.03	77.3	76.0

Table 18
COMPRESSIONAL WAVE VELOCITIES IN ROCKS AS A FUNCTION OF PRESSURE

Note: Compressional wave velocities in rocks as a function of pressure. The samples are arranged according to increasing velocity at 2 kilobars pressure. The type symbols refer to dry (0) and wet (1) samples. Anistropy is measured by the percent difference in maximum and minimum velocity compared to the mean for two or three mutually perpendicular cores. If no value is given, velocity was measured in only one direction.

	P (Kbar)								Wet		
	0.1	0.5	1	2	4	6	10	ϱ (g/cm³)	or dry	Anis (%)	Ref.
Lunar powder 175081	—	1.81	2.04	2.20	—	—	—	1.50	0	—	227
Mudstone 249-28-2 (56-59)	2.05	2.16	2.24	2.38	2.64	2.89	—	1.49	1	—	228
Terrestrial powdered basalt	—	2.04	2.36	2.59	—	—	—	1.50	0	—	227
Lunar powder 172201	—	2.22	2.58	2.84	—	—	—	1.50	0	—	227
Terrestrial volcanic ash	—	2.33	2.63	2.99	—	—	—	1.50	0	—	227
Lunar powder 170051	—	2.38	2.79	3.17	—	—	—	1.50	0	—	227
Tuff 70-4	2.61	2.97	3.12	3.28	3.51	3.70	4.01	1.76	1	—	229
Lunar powder 172161	—	2.53	2.91	3.31	—	—	—	1.50	0	—	227
Lunar breccia 14313	—	2.61	2.88	3.39	4.12	4.61	5.16	2.39	0	—	230
Tuff 70-1	3.22	3.26	3.32	3.43	3.62	3.77	4.05	1.83	1	—	229
Tuff 70-2	3.26	3.31	3.37	3.48	3.68	3.85	4.14	1.98	1	—	229
Tuff 71-2	3.20	3.30	3.37	3.50	3.72	3.89	4.15	1.91	1	—	229
Tuff 70-3	3.04	3.17	3.29	3.52	3.84	3.99	4.08	1.80	1	—	229
Tuff 71-1	3.25	3.39	3.47	3.62	3.85	4.03	4.33	1.96	1	—	229
Tuff 71-4	3.29	3.37	3.46	3.63	3.88	4.07	4.35	1.93	1	—	229
Tuff 71-3	3.28	3.38	3.50	3.69	3.92	4.05	4.45	1.85	1	—	229
Serpentinized peridotite V25-09-8, Kane F. Z.	—	3.55	3.60	3.70	3.86	4.02	—	2.32	0	—	231
Basalt 259-36-2, 66-70	3.35	3.47	3.56	3.70	3.93	4.14	—	2.08	1	—	232
Altered basalt 137-17-1 (397-401)	—	3.52	3.63	3.78	3.99	4.14	—	2.33	0	1.3	233
Basalt 335 6-CC	3.63	3.76	3.80	3.82	—	—	—	2.75	1	—	234
Basalt 249-33-2 (27-30)	3.50	3.61	3.70	3.84	4.03	4.21	—	2.32	1	—	228
Basalt 5-34-18	3.74	3.78	3.83	3.89	4.01	4.13	4.34	2.15	1	4.5	235
Basalt 249-33-2 (126-129)	3.80	3.84	3.88	3.97	4.14	4.31	—	2.19	1	—	228
Tuff 296-56-6 (10-13)	3.92	3.96	3.98	4.00	4.02	—	—	1.99	1	—	236

Basalt 259-35-1, 128-131	3.50	3.78	3.89	4.03	4.24	4.45	—	2.26	1	—	232
Noritic breccia 61175,22	—	2.94	3.46	4.03	4.66	5.04	—	2.25	0	—	237
Lunar microbreccia 10065 Sea of Tranquillity	—	2.90	3.50	4.05	4.30	—	—	2.34	0	—	238
Basalt 236-33-3A	3.16	3.68	3.84	4.08	4.45	4.74	—	2.35	0	—	239
Sandstone 246-10-CC	3.69	3.86	3.98	4.12	4.32	4.51	—	2.10	1	—	228
Altered basalt 141-10-1 (295-298)	—	3.81	3.94	4.15	4.44	4.63	—	2.53	0	—	233
Basalt 7-66.0-11	3.98	4.03	4.08	4.16	4.29	4.42	4.62	2.34	1	2.6	235
Halite 134-10-2 (95-108)	—	3.80	4.13	4.22	4.31	4.36	—	2.16	0	2.3	240
Sandstone 73-2	3.38	3.72	3.95	4.22	4.59	4.87	5.21	2.03	0	—	229
Serpentinized peridotite V25-09-5, kane F.Z.	—	3.95	4.12	4.25	4.47	4.66	—	2.42	0	—	231
Lunar gabbro 77017,24	—	2.69	3.20	4.25	5.16	5.73	—	2.48	0	—	237
Serpentinite famous AII 77-52-24	—	3.75	3.97	4.26	4.62	4.84	—	2.33	0	—	241
Basalt 136-9-1 (308-313)	—	3.97	4.11	4.31	4.51	4.63	—	2.52	0	0.7	233
Altered basalt 138-7-1 (437-442)	—	4.06	4.18	4.31	4.48	4.61	—	2.58	0	22.3	233
Serpentinite famous AII 77-52-6	—	4.05	4.16	4.33	4.59	4.78	—	2.31	0	—	241
Basalt 151-14-1 (109-111)	—	4.17	4.23	4.34	4.55	4.73	—	2.58	0	—	242
Basalt 152-24-2 (136-150)	—	3.95	4.11	4.35	4.61	4.79	—	2.60	0	3.0	242
Basalt 155-1-1, 143-150	4.15	4.21	4.27	4.36	4.48	4.56	4.64	2.45	1	—	243
Serpentinite famous AII 77-52-54	—	4.14	4.22	4.38	4.68	4.99	—	2.42	0	—	241
Basalt 259-37-2, 105-109	4.20	4.27	4.32	4.38	4.49	4.61	—	2.43	1	—	232
Basalt 249-33-3 (128-131)	4.15	4.23	4.30	4.39	4.54	4.68	—	2.39	1	—	228
Basalt 152-24-2 (47-55)	—	4.03	4.21	4.43	4.67	4.80	—	2.63	0	—	242
Basalt 151-14-1 (130-137)	—	4.13	4.28	4.45	4.66	4.78	—	2.58	0	—	242
Basalt 152-23-1 (71-76)	—	4.24	4.34	4.47	4.61	4.73	—	2.56	0	—	242
Basalt 152-23-1 (118-127)	—	4.24	4.36	4.51	4.69	4.81	—	2.62	0	6.0	242
Lunar breccia 15015, 18	3.68	4.13	4.38	4.54	4.74	4.96	—	—	0	—	244
Basalt 151-15-1 (138-153)	—	4.29	4.41	4.55	4.77	4.94	—	2.68	0	6.9	242
Quartzite 69-5	4.33	4.45	4.48	4.56	4.71	4.84	5.10	2.50	0	—	229
Basalt 261-35-3, 84-87	4.42	4.47	4.52	4.59	4.73	4.87	—	2.60	1	—	232
Dolerite 150-11-2 (74-81)	—	4.26	4.43	4.63	4.87	5.04	—	2.66	0	3.3	242
Metabasalt V25-06-95, Kane F.Z.	—	4.46	4.60	4.67	4.80	4.90	—	2.63	0	—	231
Basalt 152-24-1 (11-19)	—	4.53	4.60	4.71	4.85	4.95	—	2.72	0	—	242
Massive anhydrite 124-10-1 (72-75)	—	4.48	4.57	4.71	4.89	5.03	—	2.54	0	6.1	240
Basalt 4-23-7	4.48	4.56	4.62	4.71	4.84	4.95	5.11	2.54	1	3.9	245
Altered basalt 138-7-1 (437-442)	—	4.49	4.58	4.73	4.96	5.13	—	2.50	0	2.6	233

Table 18 (continued)
COMPRESSIONAL WAVE VELOCITIES IN ROCKS AS A FUNCTION OF PRESSURE

	P (Kbar)								Wet		
	0.1	0.5	1	2	4	6	10	ϱ (g/cm^3)	or dry	Anis. (%)	Ref.
Serpentinite Burro Mountain, Calif.	4.30	4.54	4.64	4.75	4.91	5.02	5.23	2.52	0	9.4	246
Metabasalt V25-06-30, Kane F.Z.	—	4.41	4.56	4.75	4.90	5.05	—	2.51	0	—	231
Basalt 233-9-1A	4.25	4.42	4.55	4.75	5.01	5.13	—	2.67	0	—	239
Basalt 3-19-12	4.59	4.66	4.72	4.80	4.94	5.07	5.29	2.45	1	4.9	245
Biomicrite 249-27-2 (133-136)	4.22	4.44	4.58	4.80	5.07	5.20	—	2.25	1	—	228
Basalt 259-39-1, 135-138	4.58	4.66	4.72	4.81	4.96	5.11	—	2.55	1	—	232
Basalt 153-20-1 (126-132)	—	4.45	4.62	4.82	5.06	5.23	—	2.72	0	2.4	242
Sandstone 70-11	4.42	4.68	4.73	4.82	4.94	5.01	5.08	2.29	1	—	229
Bedford limestone	3.04	—	4.38	4.83	—	—	—	2.62	0	—	247
Serpentinite AII-42-2-4, Mar	4.15	4.44	4.64	4.83	5.09	5.26	5.45	2.47	0	1.6	248
Sandstone 70-5	3.84	4.58	4.75	4.85	4.94	5.01	5.09	2.40	0	—	229
Basalt 294-7-1 (116-119)	4.60	4.71	4.78	4.88	5.04	5.14	—	2.46	1	—	236
Quartzite 69-6	3.85	4.20	4.50	4.88	5.20	5.34	5.50	2.63	0	—	229
Serpentinite 3 Mayaguez	—	4.68	4.75	4.88	5.06	—	—	2.51	0	—	249
Trondjhemite AII 60-12-27	—	4.67	4.78	4.88	5.02	5.12	—	2.57	1	—	250
Basalt 236-34-2	4.64	4.80	4.84	4.89	5.01	5.13	—	2.68	0	—	239
Dolerite 150-12-1 (119-131)	—	4.72	4.75	4.92	5.07	5.17	—	2.74	0	1.8	242
Hawaiite 7IH, Hawaii	4.07	4.41	4.67	4.93	5.18	5.32	5.51	2.47	0	—	251
Sandstone 72-1	3.60	4.51	4.84	4.95	5.04	5.08	5.09	2.44	0	—	229
Bedford limestone	4.68	—	4.90	4.96	—	—	—	2.62	1	—	247
Serpentinite AII-20-26-118, Mar	4.64	4.74	4.84	4.96	5.18	5.33	5.47	2.55	0	9.7	248
Basalt 153-20-2 (6-16)	—	4.67	4.79	4.96	5.10	5.18	—	2.72	0	—	242
Massive anhydrite 124-10-2 (114-124)	—	4.83	4.90	4.97	5.04	5.08	—	2.63	0	1.8	240
Serpentinite 2 Canyon Mountain, Ore.	4.76	4.84	4.90	4.99	5.11	5.21	5.38	2.55	1	—	252
Sandstone 72-2	4.02	4.75	4.88	4.99	5.24	5.40	5.67	2.47	1	—	229
Gabbro V25-06-28, Kane F.Z.	—	4.55	4.75	5.00	5.25	5.40	—	2.66	0	—	231
Dolerite 146-41R-2 (104-113)	—	4.68	4.80	5.00	5.23	5.38	—	2.74	0	—	242
Basalt 322-13-2, 56-62	4.86	4.90	4.94	5.01	5.09	5.14	—	2.55	1	—	253
Serpentinite 6 Mayaguez	—	4.86	4.91	5.01	5.17	—	—	2.54	0	—	249
Pyrophyllite unknown	—	—	4.73	5.02	5.38	5.58	5.89	2.66	0	—	245

Nodular anhydrite 124-11-2 (114-124)	—	4.81	4.90	5.02	5.20	5.33	—	2.74	0	1.3	240
Serpentinite Mt. Boardman, Calif.	4.89	4.93	4.96	5.03	5.15	5.25	5.42	2.51	1	—	252
Dolerite 146-41R-2 (19-27)	—	4.63	4.81	5.03	5.29	5.45	—	2.81	0	7.2	242
Gypsum 132-27-2 (0-5)	—	4.88	4.95	5.04	5.14	5.20	—	2.29	0	3.5	240
Sandstone 70-10	4.34	4.92	4.99	5.04	5.12	5.14	5.16	2.27	0	—	229
Metabasalt V25-06-105, Kane F.Z.	—	4.70	4.85	5.05	5.16	5.25	—	2.62	0	—	231
Basalt 236-36-1	4.81	4.93	4.98	5.05	5.18	5.31	—	2.73	0	—	239
Basalt 259-41-3, 53-56	4.89	4.95	4.99	5.06	5.17	5.27	—	2.66	1	—	232
Basalt 2-10-20	4.76	4.87	4.95	5.06	5.21	5.33	5.49	2.62	1	1.5	245
Basalt 292-41-2 (37-40)	4.59	4.75	4.89	5.06	5.28	—	—	2.57	1	—	236
Basalt 231-64-1	4.59	4.75	4.85	5.06	5.21	5.34	—	2.74	0	—	239
Serpentinite Paskenta, Calif.	4.87	4.95	5.00	5.07	5.19	5.31	5.49	2.52	1	—	252
Trachyte 32IH, Hawaii	4.94	4.98	5.01	5.07	5.16	5.25	5.42	2.45	0	—	251
Basalt 19-192A-5-4, 133-142	4.71	4.85	4.95	5.09	5.26	5.38	5.55	2.56	1	0.5	255
Metabasalt V25-08-27, Kane F.Z.	—	4.77	5.00	5.10	5.23	5.34	—	2.66	0	—	231
Basalt 261-33-1, 55-59	4.90	4.97	5.03	5.10	5.21	5.32	—	2.59	1	—	232
Dolerite 150-11-2 (11-29)	—	4.84	4.95	5.10	5.27	5.40	—	2.77	0	3.4	242
Breccia 14321, 93	—	4.68	4.86	5.10	5.40	5.61	—	2.40	0	—	256
Metabasalt V25-08-10, Kane F.Z.	—	4.90	5.03	5.13	5.17	5.22	—	2.60	0	—	231
Basalt 254-35-1, 107	4.79	4.97	5.07	5.13	—	—	—	2.75	1	—	257
Serpentinite AII-32-8-4, Mar	4.71	4.91	5.00	5.14	5.32	5.44	5.57	2.47	0	3.4	248
Alkalic basalt C205A, Hawaii	4.65	4.80	4.95	5.15	5.33	5.50	5.77	2.88	0	—	251
Basalt V25-013-33, Kane F.Z.	—	4.95	5.05	5.16	5.35	5.45	—	2.75	0	—	231
Dolerite V25-06-62, Kane F.Z.	—	4.85	4.95	5.16	5.50	5.70	—	2.80	0	—	231
Basalt 153-19-1 (140-150)	—	4.79	4.95	5.16	5.39	5.51	—	2.95	0	—	242
Basalt 153-20-2 (123-129)	—	4.91	5.02	5.16	5.32	5.42	—	2.81	0	—	242
Basalt V25-013-6, Kane F.Z.	—	4.85	5.00	5.20	5.40	5.60	—	2.72	0	—	231
Basalt 7-61.0-2	4.83	4.97	5.08	5.20	5.34	5.43	5.53	2.71	1	1.7	235
Dolerite 150-12-2 (26-42)	—	4.95	5.05	5.20	5.39	5.50	—	2.78	0	3.3	242
Sandstone W2	—	4.85	5.02	5.20	5.41	5.57	—	2.55	1	—	258
Basalt 292-41-5 (40-43)	4.86	4.95	5.04	5.21	5.39	—	—	2.61	1	—	236
Basalt 7-61.1-2	5.06	5.11	5.16	5.22	5.31	5.38	5.49	2.60	1	4.0	235
Basalt 332B 22-2 (140-142)	4.75	5.06	5.14	5.23	—	—	—	2.61	1	—	234
Basalt V25-06-40, Kane F.Z.	—	5.05	5.15	5.26	5.40	5.46	—	2.43	0	—	231
Basalt 292-43-4 (40-43)	4.95	5.07	5.16	5.26	5.36	—	—	2.68	1	—	236
Basalt V25-06-29, Kane F.Z.	—	5.08	5.16	5.28	5.44	5.55	—	2.75	0	—	231
Basalt 292-42-5 (29-32)	5.02	5.11	5.19	5.29	5.41	—	—	2.61	1	—	236

Table 18 (continued)
COMPRESSIONAL WAVE VELOCITIES IN ROCKS AS A FUNCTION OF PRESSURE

	P (Kbar)								Wet		
	0.1	0.5	1	2	4	6	10	ϱ (g/cm³)	or dry	Anis. (%)	Ref.
Basalt 231-64-3A	5.00	5.11	5.16	5.29	5.39	5.52	—	2.75	0	—	239
Basalt V25-08-6, Kane F.Z.	—	5.16	5.22	5.30	5.45	5.55	—	2.78	0	—	231
Basalt V30-RD8-31, oceanographer F.Z.	—	4.86	5.12	5.30	5.48	5.56	—	—	0	—	259
Basalt RD8-P31, oceanographer F.Z.	4.72	4.98	5.12	5.30	5.46	5.56	—	2.61	0	—	260
Metagabbro (greenschist) 120-8-1 (86-90	—	4.97	5.14	5.33	5.52	5.67	—	2.61	0	—	240
Basalt AM2, Hawaii	4.72	4.95	5.12	5.34	5.61	5.79	6.01	2.64	0	—	251
Dolerite 150-12-1 (0-15)	—	5.06	5.19	5.34	5.52	5.62	—	2.85	0	2.2	242
Basalt 292-44-4 (48-51)	5.07	5.19	5.27	5.34	5.45	—	—	2.69	1	—	236
Basalt 233-13-1	4.59	4.96	5.12	5.34	5.57	5.71	—	2.73	0	—	239
Serpentinite 1 Mayaguez	—	5.16	5.24	5.34	5.46	—	—	2.43	0	—	249
Serpentinite 1 Canyon Mountains, Ore.	5.25	5.28	5.31	5.35	5.42	5.50	5.62	2.54	1	—	252
Dolerite 146-43R-4 (28-36)	—	5.03	5.18	5.35	5.53	5.62	—	2.84	0	2.5	242
Dolerite V25-06-23, Kane F.Z.	—	5.05	5.20	5.36	5.60	5.70	—	2.80	0	—	231
Basalt 332B 44-1 (75-77)	4.88	5.22	5.30	5.36	—	—	—	2.61	1	—	234
Serpentinite 4 Mayaguez	—	5.24	5.28	5.36	5.48	—	—	2.59	0	—	249
Basalt 320B-3-1, 64-67	5.24	5.28	5.31	5.37	5.48	5.61	—	2.73	1	—	261
Dolerite 146-42R-1 (27-33)	—	4.97	5.17	5.37	5.54	5.62	—	2.81	0	—	242
Dolerite 150-12-2 (107-120)	—	5.12	5.23	5.37	5.56	5.69	—	2.85	0	1.8	242
Dolerite 146-43R-3 (119-131)	—	5.10	5.22	5.38	5.55	5.65	—	3.04	0	—	242
Dolerite 146-43R-1 (67-82)	—	5.15	5.26	5.39	5.53	5.61	—	2.88	0	0.5	242
Basalt 257-14-2, 95	5.15	5.27	5.33	5.39	—	—	—	2.75	1	—	257
Metabasalt V25-05-10, Kane F.Z.	—	5.05	5.28	5.40	5.55	5.65	—	2.68	0	—	231
Basalt 335 6-4 (23-25)	—	5.35	5.35	5.40	—	—	—	2.52	1	—	234
Metabasalt V25-05-37, Kane F.Z.	—	5.10	5.28	5.41	5.49	5.55	—	2.61	0	—	231
Basalt 323-20-CC	5.29	5.32	5.35	5.41	5.48	5.54	—	2.72	1	—	253
Dolerite 146-43R-3 (0-10)	—	5.09	5.23	5.41	5.58	5.67	—	2.84	0	2.9	242
Basalt 261-37-2, 92-95	5.31	5.35	5.38	5.42	5.49	5.56	—	2.72	1	—	232

Basalt 235-20-2	4.88	5.26	5.32	5.42	5.57	5.70	—	2.71	0	—	239
Metasandstone SR70-1	—	5.08	5.26	5.42	5.65	5.78	—	2.64	1	—	258
Basalt 333A 10-2 (112-115)	5.15	5.28	5.35	5.43	—	—	—	2.76	1	—	234
Greenstone W-2-3, Mar	4.81	5.07	5.24	5.43	5.66	5.79	5.93	2.72	1	5.4	262
Sandstone Catskill, N.Y.	—	5.00	5.27	5.44	5.63	5.75	5.85	2.66	0	6.2	245
Basalt 321-13-4, 104-107	5.20	5.28	5.35	5.44	5.57	5.70	—	2.82	1	—	261
Basalt 235-20-5	5.09	5.31	5.36	5.44	5.52	5.72	—	2.67	0	—	239
Basalt 9-77B-54	5.34	5.37	5.40	5.45	5.52	5.58	5.65	2.71	1	1.3	235
Mugearite C211A, Hawaii	5.37	5.39	5.42	5.45	5.51	5.54	5.58	2.63	0	—	251
Dolerite 146-42R-1 (117-123)	—	5.20	5.30	5.45	5.68	5.85	—	2.88	0	—	242
Dolerite 146-42R-2 (10-19)	—	5.17	5.29	5.45	5.65	5.75	—	2.90	0	0.7	242
Basalt 257-11-2, 74	5.06	5.21	5.31	5.45	—	—	—	2.74	1	—	257
Shale BCS57	—	5.13	5.29	5.45	5.64	5.75	—	2.65	1	—	258
Basalt 153-20-1 (10-23)	—	5.22	5.34	5.47	5.62	5.68	—	2.87	0	3.9	242
Quartzite 69-4	5.37	5.42	5.43	5.47	5.53	5.59	5.68	2.58	0	—	229
Diorite HU-159-34	—	5.20	5.33	5.47	5.72	5.86	—	2.64	1	—	250
Andesite 69-3	5.08	5.17	5.27	5.48	5.57	5.58	5.71	2.64	0	—	229
Basalt Famous Cyp 74-31-38	—	5.14	5.29	5.49	5.66	5.73	—	2.78	0	—	241
Dolerite 146-43R-1 (144-151)	—	5.14	5.30	5.49	5.64	5.72	—	2.88	0	2.7	242
Basalt Cyp 74-31-38, Famous	—	5.14	5.29	5.49	5.61	5.73	—	—	0	—	259
Basalt 248-15-1 (35-38)	5.33	5.39	5.44	5.49	5.60	5.70	—	2.68	1	—	228
Metagabbro V25-05-17, Kane F.Z.	—	5.20	5.32	5.50	5.61	5.73	—	2.64	0	—	231
Basalt 238-57-3	5.19	5.33	5.40	5.50	5.67	5.81	—	2.82	0	—	239
Basalt 239-20-1 (125-128)	5.37	5.43	5.47	5.50	5.57	5.63	—	2.72	1	—	228
Dolerite 146-42R-3 (28-37)	—	5.23	5.37	5.51	5.68	5.82	—	2.79	0	—	242
Dolerite 146-42R-2 (93-101)	—	5.28	5.38	5.52	5.71	5.84	—	2.87	0	—	242
Basalt 235-20-4	5.10	5.40	5.45	5.52	5.63	5.75	—	2.75	0	—	239
Lithographic limestone	—	—	5.59	5.54	5.56	—	—	2.54	0	3.6	245
Basalt 33B 31-1 (108-111)	5.31	5.45	5.50	5.54	—	—	—	2.74	1	—	234
Basalt W-2-8, Mar	4.79	5.09	5.29	5.54	5.83	5.97	6.14	2.76	1	1.0	262
Basalt 238-55-2A	5.21	5.32	5.41	5.54	5.72	5.85	—	2.81	0	—	239
Basalt 239-21-1 (46-49)	5.41	5.47	5.50	5.54	5.60	5.66	—	2.76	1	—	228
Basalt 323-18-6, 110-120	5.43	5.46	5.49	5.55	5.62	5.67	—	2.73	1	—	253
Basalt 333A 1-3 (74-76)	5.36	5.39	5.45	5.55	—	—	—	2.85	1	—	234
Basalt 254-36-3, 105	5.24	5.36	5.48	5.55	—	—	—	2.82	1	—	257
Basalt 261-38-2, 23-26	5.47	5.50	5.53	5.56	5.62	5.68	—	2.75	1	—	232
Dolerite 146-42R-3 (135-145)	—	5.29	5.42	5.56	5.73	5.81	—	2.89	0	0.9	242

Table 18 (continued)
COMPRESSIONAL WAVE VELOCITIES IN ROCKS AS A FUNCTION OF PRESSURE

	P (Kbar)								Wet		
	0.1	0.5	1	2	4	6	10	ϱ (g/cm³)	or dry	Anis. (%)	Ref.
Basalt V30-RD8-27, Oceanographer F.Z.	—	5.38	5.48	5.56	5.65	5.73	—	—	0	—	259
Basalt RD8-P27B, Oceanographer F.Z.	5.15	5.38	5.48	5.56	5.67	5.73	—	2.67	0	—	260
Metagabbro RD20-P10, Oceanographer F.Z.	4.87	5.29	5.41	5.56	5.75	5.90	—	2.89	0	—	260
Basalt V30-RD8-29, Oceanographer F.Z.	—	5.28	5.41	5.57	5.69	5.80	—	—	0	—	259
Basalt RD8-P29B, Oceanographer F.Z.	5.09	5.28	5.41	5.57	5.73	5.80	—	2.71	0	—	260
Basalt 332A 16-1 (33-36)	5.36	5.45	5.51	5.58	—	—	—	2.72	1	—	234
Metagabbro RD18-P19A, Oceanographer F.Z.	4.55	5.25	5.40	5.58	5.76	5.82	—	2.95	0	—	260
Basalt 42IH, Hawaii	4.55	5.16	5.44	5.60	5.75	5.88	6.04	2.44	0	—	251
Basalt 332B 36-4 (131-133)	5.36	5.53	5.56	5.60	—	—	—	2.70	1	—	234
Chlorite-quartz rock V25-05-36, Kane F.Z.	—	5.30	5.46	5.61	5.66	5.70	—	2.65	0	—	231
Basalt 322-12-1, piece 7	4.49	5.53	5.56	5.61	5.69	5.74	—	2.73	1	—	253
Basalt 292-46-CC	5.30	5.41	5.50	5.61	5.73	—	—	2.79	1	—	236
Lunar basalt 10057 Sea of Tranquillity	—	3.80	4.65	5.62	6.52	—	—	2.88	0	—	238
Serpentinite 2 Black Mountain, Calif.	5.51	5.54	5.57	5.62	5.70	5.76	5.83	2.63	1	—	252
Basalt 236-37-1A	5.34	5.53	5.59	5.62	5.65	5.69	—	2.76	0	—	239
Metagabbro RD18-P16, Oceanographer F.Z.	4.52	5.38	5.49	5.62	5.77	5.89	—	2.77	0	—	260
Lunar basalt 15065, 27	—	3.90	4.70	5.62	6.52	6.84	7.10	2.86	0	—	263
Serpentinite 1 Black Mountain, Calif.	5.50	5.53	5.56	5.63	5.71	5.78	5.89	2.62	1	—	252
Basalt 261-33-1, 131-134	5.51	5.55	5.58	5.63	5.71	5.80	—	2.79	1	—	232
Basalt 1EIH, Hawaii	5.34	5.50	5.57	5.63	5.70	5.75	5.79	2.69	0	—	251
Basalt 332B 14-2 (81-83)	5.41	5.52	5.58	5.63	—	—	—	2.69	1	—	234
Basalt Famous 523-2	—	5.40	5.50	5.64	5.82	5.92	—	2.59	0	—	241

Schist (meta-sandstone series) 133-3-1	—	5.36	5.49	5.64	5.80	5.88	—	2.69	0	3.1	240
Basalt 19-191-16-1, 21-25	5.49	5.55	5.59	5.64	5.70	5.76	5.86	2.79	1	3.7	255
Basalt 523-2, Famous	—	5.40	5.50	5.64	5.78	5.92	—	—	0	—	259
Dolerite 146-43R-2 (32-39)	—	5.43	5.53	5.66	5.81	5.93	—	2.93	0	—	242
Basalt 238-61-4A	5.41	5.54	5.59	5.66	5.76	5.86	—	2.82	0	—	239
Spilite 2 Black Mountain, Calif.	4.74	5.15	5.43	5.67	5.86	5.99	6.19	2.70	1	—	252
Metagabbro RD19-P12, Oceanographer F.Z.	5.07	5.44	5.52	5.67	5.88	6.01	—	2.92	0	—	260
Sandstone W5	—	5.40	5.54	5.67	5.83	5.92	—	2.63	1	—	258
Metasandstone SP7402	—	5.33	5.51	5.68	5.84	5.93	—	2.66	1	—	258
Basalt V25-08-46, Kane F.Z.	—	5.50	5.60	5.70	5.81	5.85	—	2.84	0	—	231
Basalt 1DIH, Hawaii	5.10	5.44	5.57	5.70	5.80	5.86	5.94	2.67	0	—	251
Basalt 332B 6-2 (122-123)	—	5.52	5.61	5.70	—	—	—	2.66	1	—	234
Metagabbro V25-05-51, Kane F.Z.	—	5.45	5.60	5.71	5.90	6.06	—	2.84	0	—	231
Basalt 332B 28-2 (23-26)	5.45	5.62	5.66	5.71	—	—	—	2.83	1	—	234
Basalt 238-64-1A	5.30	5.57	5.62	5.71	5.89	6.04	—	2.71	0	—	239
Sandstone 73-1	4.14	5.04	5.46	5.71	5.91	6.05	6.15	2.56	0	—	229
Solenhafen limestone	5.62	—	5.68	5.72	—	—	—	2.66	0	—	247
Basalt 5-32-13	5.63	5.67	5.69	5.72	5.77	5.81	5.86	2.83	1	—	235
Serpentinite Thetford, Quebec	—	—	5.67	5.73	5.80	5.87	6.00	2.60	0	4.3	245
Basalt 332A 12-1 (122-125)	5.61	5.66	5.69	5.73	—	—	—	2.80	1	—	234
Basalt JF-1 cross, Juan de Fuca rise	3.37	4.45	5.11	5.73	6.17	6.31	6.40	2.87	0	—	264
Basalt 254-31-1, 111	5.20	5.52	5.65	5.73	—	—	—	2.74	1	—	257
Basalt 257-13-5, 15	5.46	5.60	5.67	5.73	—	—	—	2.82	1	—	257
Basalt V30-RD8-2, Oceanographer F.Z.	—	5.38	5.54	5.73	5.87	6.00	—	—	0	—	259
Basalt RD8-P2A, Oceanographer F.Z.	4.86	5.38	5.54	5.73	5.92	6.00	—	2.67	0	—	260
Serpentinized peridotite H7, Linhorka, CSSR	5.63	5.67	5.72	5.74	5.83	—	—	2.60	0	—	265
Basalt 1BIH, Hawaii	4.58	5.17	5.42	5.74	6.02	6.18	6.35	2.62	0	—	251
Greenstone A-11-42-1-3, Mar	5.19	5.47	5.61	5.75	5.92	6.03	6.13	2.69	1	8.1	262
Basalt 7-63. 0-11	5.60	5.69	5.73	5.76	5.80	5.84	5.91	2.79	1	12.1	235
Basalt 3-14-10	5.62	5.66	5.69	5.76	5.82	5.87	5.97	2.77	1	1.2	245
Basalt 251A-31-2, 84	5.55	5.61	5.68	5.76	—	—	—	2.82	1	—	257
Greenstone A-II-42-1-4, Mar	5.32	5.54	5.66	5.77	5.90	5.99	6.07	2.71	1	3.7	262
Metabasalt RD12-P22A, Oceanographer F.Z.	5.28	5.42	5.56	5.77	5.93	6.01	—	2.80	0	—	260

Table 18 (continued)
COMPRESSIONAL WAVE VELOCITIES IN ROCKS AS A FUNCTION OF PRESSURE

	P (Kbar)								Wet		
	0.1	0.5	1	2	4	6	10	ϱ (g/cm^3)	or dry	Anis. (%)	Ref.
Obsidian Modoc	—	—	—	5.78	5.73	5.70	5.62	2.38	0	—	245
Solenhafen limestone	5.64	—	5.75	5.78	—	—	—	2.66	1	—	247
Nephelenite KC6, Hawaii	5.29	5.46	5.63	5.78	5.91	6.00	6.14	2.72	0	—	251
Basalt 332A 27-1 (128-131)	5.57	5.65	5.71	5.78	—	—	—	2.71	1	—	234
Basalt 332B 20-1 (56-59)	5.38	5.55	5.61	5.78	—	—	—	2.83	1	—	234
Basalt V30-RD8-1, Oceanographer F.Z.	—	5.59	5.66	5.78	5.90	6.01	—	—	0	—	259
Basalt RD8-P1B, Oceanographer F.Z.	5.15	5.59	5.66	5.78	5.94	6.01	—	2.72	0	—	260
Basalt Famous Arp 74-14-28	—	5.44	5.60	5.79	5.96	6.00	—	2.68	0	—	241
Alkalic basalt AM4, Hawaii	4.72	5.10	5.49	5.79	5.93	6.11	6.28	2.70	0	—	251
Basalt C200A, Hawaii	5.46	5.55	5.64	5.79	5.91	5.97	6.03	2.82	0	—	251
Basalt 332B 9-1 (112-115)	5.43	5.63	5.71	5.79	—	—	—	2.81	1	—	234
Basalt 333A 8-6 (58-60)	5.59	5.67	5.73	5.79	—	—	—	2.77	1	—	234
Basalt ARP 74-14-28, Famous	—	5.44	5.60	5.79	5.90	6.00	—	—	0	—	259
Actinolite rock V25-09-9, Kane F.Z.	—	5.41	5.62	5.80	6.01	6.21	—	2.77	0	—	231
Basalt 3IH, Hawaii	4.96	5.31	5.56	5.80	6.03	6.13	6.26	2.79	0	—	251
Basalt 1CIH, Hawaii	4.90	5.40	5.57	5.80	5.93	6.00	6.14	2.68	0	—	251
Basalt 332B 9-2 (104-106)	5.42	5.60	5.70	5.80	—	—	—	2.80	1	—	234
Hawaiite 83IH, Hawaii	5.24	5.48	5.62	5.81	6.02	6.16	6.37	2.88	0	—	251
Basalt 1AIH, Hawaii	4.00	5.07	5.64	5.82	6.01	6.15	6.38	2.70	0	—	251
Basalt 332B 25-2 (91-93)	5.51	5.60	5.69	5.82	—	—	—	2.88	1	—	234
Granite Westerly, R.I.	4.82	—	5.71	5.83	6.00	—	6.21	2.65	0	—	266
Basalt 261-38-5, 94-97	5.74	5.77	5.80	5.83	5.89	5.96	—	2.80	1	—	232
Metagabbro RD12-P38, Oceanographer F.Z.	5.21	5.49	5.64	5.84	6.13	6.32	—	3.10	0	—	260
Aragonite	—	—	5.82	5.85	5.90	5.93	5.97	2.92	0	—	245
Basalt 332A 7-2 (41-42)	—	5.70	5.79	5.85	—	—	—	2.81	1	—	234
Metagabbro RD18-P15, Oceanographer F.Z.	5.13	5.61	5.74	5.85	6.01	6.12	—	2.81	0	—	260
Basalt 15058, 57	—	5.33	5.54	5.85	6.31	6.57	—	2.99	0	—	256

Diorite HU-159-38	—	5.60	5.72	5.85	6.06	6.19	—	2.63	1	—	250
Basalt V25-01-4, Kane F.Z.	—	5.66	5.75	5.86	6.00	6.05	—	2.83	0	—	231
Basalt 321-14-4, 51-54	5.65	5.72	5.78	5.86	5.95	6.00	—	2.92	1	—	261
Dolerite 146-43R-4 (131-138)	—	5.64	5.73	5.86	5.98	6.02	—	2.94	0	—	242
Basalt 332A 31-2 (32-35)	5.59	5.72	5.79	5.86	—	—	—	2.73	1	—	234
Quartz schist 703 H-N, Shikoku, Japan	—	5.37	5.68	5.86	5.98	6.04	—	2.75	0	12.0	267
Metagabbro V25-05-18, Kane F.Z.	—	5.50	5.70	5.87	6.10	6.27	—	2.72	0	—	231
Graywacke New Zealand	—	5.63	5.76	5.87	5.98	6.04	6.13	2.68	0	0.2	245
Limestone 261-33-1, 67-71	5.46	5.61	5.73	5.87	6.00	6.07	—	2.68	1	—	232
Alkalic basalt 351H, Hawaii	4.85	5.24	5.57	5.87	6.12	6.24	6.37	2.70	0	—	251
Basalt JF-2, Juan de Fuca rise	3.44	4.51	5.26	5.87	6.22	6.34	6.43	2.86	0	0.8	264
Metasandstone LO71-5	—	5.45	5.68	5.87	6.05	6.15	—	2.70	1	—	258
Metasandstone BCSS55	—	5.51	5.73	5.87	5.99	6.08	—	2.71	1	—	258
Basalt Famous KN 42-77-28	—	5.78	5.81	5.88	6.03	6.12	—	2.77	0	—	241
Marble Rutland, Vt.	—	—	5.74	5.88	6.10	6.24	6.46	2.72	0	5.5	268
Basalt 332A 28-3 (28-31)	—	5.75	5.81	5.88	—	—	—	2.79	1	—	234
Basalt 332A 31-3 (79-82)	5.67	5.80	5.83	5.88	—	—	—	2.78	1	—	234
Basalt 332A 34-2 (104-107)	5.65	5.79	5.83	5.88	—	—	—	2.81	1	—	234
Basalt KN 42-77-28, Famous	—	5.78	5.81	5.88	6.00	6.12	—	—	0	—	259
Epidosite RD20-P15, Oceanographer F.Z.	5.24	5.43	5.60	5.88	6.21	6.48	—	3.19	0	—	260
Basalt 7-63, 0-10	5.78	5.83	5.85	5.89	5.93	5.96	6.01	2.83	1	0.2	235
Basalt 320B-4-1, 144-147	5.60	5.69	5.77	5.89	6.03	6.14	—	2.84	1	—	261
Basalt 332A 28-1 (83-86)	5.66	5.76	5.81	5.89	—	—	—	2.79	1	—	234
Spilite 1 Black Mountain, Calif.	5.48	5.63	5.77	5.90	6.02	6.11	6.22	2.70	1	—	252
Basalt 1IH, Hawaii	5.52	5.56	5.65	5.90	6.07	6.13	6.19	2.72	0	—	251
Basalt 335 6-3 (44-49)	5.85	5.88	5.88	5.90	—	—	—	2.80	1	—	234
Serpentinite 2 Mayaguez	—	5.82	5.86	5.90	5.99	—	—	2.60	0	—	249
Slate Medford, Mass.	—	—	5.79	5.91	6.02	6.10	6.22	2.73	0	8.6	245
Basalt 9-84-30	5.79	5.84	5.87	5.91	5.95	6.01	6.09	2.81	1	1.2	235
Basalt Famous Cyp 74-29-31A	—	5.78	5.82	5.91	6.05	6.15	—	2.91	0	—	241
Limestone Solenhofen	—	—	5.70	5.91	6.19	6.27	6.10	2.54	0	2.8	268
Basalt 332B 10-2 (125-128)	5.69	5.74	5.82	5.91	—	—	—	2.84	1	—	234
Basalt Cyp 74-29-31A, Famous	—	5.78	5.82	5.91	6.03	6.15	—	—	0	—	259
Basalt 332B 36-5 (40-42)	5.72	5.77	5.84	5.92	—	—	—	2.78	1	—	234
Basalt 332B 47-2 (145-147)	5.64	5.77	5.85	5.92	—	—	—	2.89	1	—	234

Table 18 (continued)
COMPRESSIONAL WAVE VELOCITIES IN ROCKS AS A FUNCTION OF PRESSURE

	P (Kbar)								Wet		
	0.1	0.5	1	2	4	6	10	ϱ (g/cm³)	or dry	Anis. (%)	Ref.
Basalt W-4-115, Mar	5.60	5.67	5.76	5.92	6.07	6.15	6.23	2.78	1	1.2	262
Kinzigite gneiss IV-21	—	—	5.65	5.92	6.13	6.26	6.37	2.73	0	—	269
Metasandstone IV5	—	—	5.80	5.92	6.03	6.09	—	2.66	1	—	258
Basalt 9-83-9	5.76	5.81	5.85	5.93	6.02	6.08	6.15	2.83	1	—	235
Basalt 240-7-1 (120-123)	5.83	5.87	5.89	5.93	6.00	6.07	—	2.80	1	—	228
Actinolite rock V25-09-10A, Kane F.Z.	—	5.47	5.70	5.94	6.18	6.40	—	2.96	0	—	231
Basalt 332A 33-2 (92-95)	5.85	5.90	5.92	5.94	—	—	—	2.81	1	—	234
Basalt 332B 24-1 (125-127)	5.53	5.76	5.84	5.94	—	—	—	2.73	1	—	234
Metagabbro RD18-P17A, Oceanographer F.Z.	4.97	5.63	5.78	5.94	6.15	6.28	—	2.87	0	—	260
Schist PT212	—	5.61	5.80	5.94	6.11	6.22	—	2.74	1	—	258
Metagabbro V25-06-58, Kane F.Z.	—	5.61	5.75	5.95	6.11	6.26	—	2.78	0	—	231
Spilite Canyon Mountain, Ore.	5.72	5.79	5.88	5.95	6.05	6.13	6.28	2.71	1	—	252
Basalt 335 6-6 (75-77)	5.80	5.90	5.93	5.95	—	—	—	2.79	1	—	234
Basalt JF-1, Juan de Fuca rise	3.85	4.19	5.40	5.95	6.29	6.40	6.45	2.86	0	1.0	264
Metagabbro RD18-P20, Oceanographer F.Z.	5.21	5.79	5.85	5.95	6.05	6.09	—	2.89	0	—	260
Granite 69-1	5.26	5.66	5.80	5.95	6.09	6.14	6.14	2.63	0	—	229
Basalt 9-79-17	5.85	5.88	5.92	5.96	6.02	6.07	6.14	2.73	1	3.2	235
Basalt 332B 21-1 (99-102)	5.73	5.83	5.89	5.96	—	—	—	2.90	1	—	234
Basalt 3-18-7	5.78	5.86	5.90	5.96	6.04	6.09	6.13	2.82	1	0.5	245
Granite Westerly, R.I.	—	5.63	5.84	5.97	6.10	6.16	6.23	2.62	0	3.1	245
Basalt Famous AII 77-76-42	—	5.49	5.68	5.97	6.30	6.44	—	2.74	0	—	241
Basalt 332A 37-1 (116-119)	5.88	5.92	5.94	5.97	—	—	—	2.83	1	—	234
Basalt AII 77-76-42, Famous	—	5.49	5.68	5.97	6.25	6.44	—	—	0	—	259
Serpentinized peridotite 1 Burro Mountain, Calif.	5.70	5.87	5.91	5.98	6.05	6.14	6.30	2.75	0	8.3	246
Basalt 332B 19-1 (104-107)	5.63	5.82	5.90	5.98	—	—	—	2.87	1	—	234
Basalt 335 6-1 (5-7)	5.80	5.94	5.95	5.98	—	—	—	2.82	1	—	234

Metagabbro RD18-P1, Oceanographer F.Z.	5.15	5.71	5.83	5.98	6.12	6.22	—	2.81	0	—	260
Slate Poultney, Vt.	5.84	5.90	5.94	5.99	6.08	6.15	6.29	2.76	0	19.6	270
Serpentinized peridotite Ehime Prep., Japan	—	—	5.92	5.99	6.09	6.19	6.33	2.73	0	6.2	271
Basalt 248-17-2 (122-125)	5.77	5.88	5.93	5.99	6.06	6.13	—	2.76	1	—	228
Quartz schist 609 H-N, Shikoku, Japan	—	5.54	5.80	5.99	6.13	6.17	—	2.70	0	14.5	267
Gneiss 6 Goshen, Conn.	4.90	5.50	5.79	6.00	6.20	6.29	6.42	2.76	0	13.4	270
Basalt V25-04-14, Kane F.Z.	—	5.65	5.81	6.00	6.15	6.20	—	2.77	0	—	231
Dolerite V25-06-9, Kane F.Z.	—	5.65	5.80	6.00	6.20	6.30	—	2.91	0	—	231
Serpentinite 1 Stonyford, Calif.	5.73	5.85	5.92	6.00	6.10	6.18	6.31	2.63	1	—	252
Basalt 332B 8-3 (22-24)	5.58	5.87	5.95	6.00	—	—	—	2.73	1	—	234
Basalt 334 16-2 (88-91)	5.92	5.95	5.96	6.00	—	—	—	2.82	1	—	234
Greenstone W-2-1, Mar	5.76	5.86	5.92	6.00	6.09	6.16	6.27	2.81	1	2.5	262
Metagabbro RD18-P24, Oceanographer F.Z.	5.50	5.79	5.91	6.00	6.11	6.18	—	2.85	0	—	260
Quartzite Rutland, Vt.	5.63	—	5.96	6.01	6.10	—	6.24	2.64	0	—	266
Basalt Famous Arp 73-10-2	—	5.52	5.78	6.01	6.20	6.25	—	2.73	0	—	241
Greenstone Famous KN 42-146-10	—	5.84	5.90	6.01	6.15	6.25	—	2.85	0	—	241
Metabasalt breccia Famous KN 42-146-98	—	5.92	5.95	6.01	6.08	6.09	—	2.75	0	—	241
Basalt EPR5	5.22	5.87	5.93	6.01	6.10	6.16	6.21	2.82	1	2.6	272
Basalt arp 73-10-2, Famous	—	5.52	5.78	6.01	6.13	6.25	—	—	0	—	259
Quartz schist 508 H-K, Shikoku, Japan	—	5.93	5.97	6.01	6.06	6.10	—	2.68	0	1.5	267
Basalt GF1, Hawaii	5.62	5.74	5.86	6.02	6.21	6.31	6.43	2.82	0	—	251
Basalt JF-3, Juan de Fuca rise	3.78	4.83	5.52	6.02	6.29	6.39	6.44	2.86	0	2.5	264
Metagabbro RD18-P4, Oceanographer F.Z.	5.17	5.66	5.82	6.02	6.20	6.34	—	3.01	0	—	260
Quartz schist 610 H-N, Shikoku, Japan	—	5.51	5.77	6.02	6.20	6.28	—	2.70	0	14.8	267
Basalt 14053, 32	—	4.54	5.32	6.02	6.53	6.79	—	3.18	0	—	256
Granite Stone Mountain, Ga.	4.34	—	5.77	6.03	6.19	—	6.37	2.63	0	1.6	266
Spilite Mt. Boardman, Calif.	5.62	5.75	5.87	6.03	6.14	6.20	6.28	2.74	1	—	252
Nephelenite C186B, Hawaii	5.84	5.92	6.00	6.03	6.08	6.13	6.22	2.67	0	—	251
Basalt 332B 22-1 (57-60)	5.87	5.95	5.98	6.03	—	—	—	2.85	1	—	234
Basalt 332B 36-3 (76-78)	5.54	5.94	5.98	6.03	—	—	—	2.78	1	—	234

Table 18 (continued)
COMPRESSIONAL WAVE VELOCITIES IN ROCKS AS A FUNCTION OF PRESSURE

	P (Kbar)								Wet		
	0.1	0.5	1	2	4	6	10	ϱ (g/cm³)	or dry	Anis. (%)	Ref.
Metagabbro RD18-P23, Oceanographer F.Z.	4.79	5.49	5.81	6.03	6.23	6.35	—	2.95	0	—	260
Graywacke Quebec	—	—	5.92	6.04	6.14	6.20	6.28	2.71	0	4.4	245
Basalt Famous 527-6-2	—	5.46	5.73	6.04	6.29	6.46	—	2.65	0	—	241
Nephelenite KC2AM, Hawaii	5.81	5.91	5.96	6.04	6.15	6.23	6.35	2.74	0	—	251
Basalt 332B 36-3 (143-145)	5.81	5.91	5.98	6.04	—	—	—	2.81	1	—	234
Basalt 332B 37-3 (54-56)	5.71	5.84	5.94	6.04	—	—	—	2.88	1	—	234
Basalt 527-6-2, Famous	—	5.46	5.73	6.04	6.25	6.46	—	—	0	—	259
Basalt 238-59-2A	5.71	5.96	6.00	6.04	6.12	6.20	—	2.90	0	—	239
Diorite AII 60-12-26	—	5.75	5.91	6.04	6.19	6.26	—	2.57	1	—	250
Gabbro V25-05-3, Kane F.Z.	—	5.68	5.85	6.05	6.33	6.53	—	2.96	0	—	231
Basalt 9-82-7	5.63	5.78	5.91	6.05	6.20	6.29	6.37	2.80	1	—	235
Metagabbro (greenschist) 120-8-1 (46-50)	—	5.62	5.55	6.05	6.20	6.27	—	2.71	0	6.5	240
Basalt 332A 8-2 (6-9)	5.73	5.83	5.95	6.05	—	—	—	2.81	1	—	234
Basalt 332B 36-1 (77-79)	5.73	5.93	6.00	6.05	—	—	—	2.79	1	—	234
Gneiss 2 Torrington, Conn.	4.50	5.05	5.50	6.06	6.18	6.24	6.33	2.65	0	2.1	270
Gneiss Pelham, Mass.	—	5.67	5.91	6.06	6.18	6.27	6.31	2.64	0	3.4	245
Anhydrite	—	—	6.00	6.06	6.15	6.19	6.27	2.93	0	0.3	245
Westerly granite	4.98	—	5.85	6.06	—	—	—	2.65	0	—	247
Basalt 319A-2-3, 46-48	5.91	5.95	6.00	6.06	6.19	6.32	—	2.86	1	—	261
Basalt 332B 25-4 (47-49)	5.77	5.88	5.95	6.06	—	—	—	2.83	1	—	234
Quartz monzonite Porterville, Calif.	—	—	5.95	6.07	6.22	6.28	6.37	2.64	0	1.3	245
Granodiorite gneiss N.H.	—	—	5.95	6.07	6.16	6.21	6.30	2.76	0	4.7	245
Basalt 319A-7-1, 65-68	5.91	5.99	6.03	6.07	6.13	6.18	—	2.85	1	—	261
Basalt Famous 523-4	—	5.74	5.91	6.07	6.17	6.21	—	2.70	0	—	241
Basalt 332B 33-1 (62-65)	—	5.95	6.01	6.07	—	—	—	2.84	1	—	234
Basalt JF-1 star, Juan de Fuca rise	4.97	5.47	5.77	6.07	6.30	6.39	6.47	2.90	1	—	264
Basalt 523-4, Famous	—	5.74	5.91	6.07	6.14	6.21	—	—	0	—	259
Serpentinite Calif.	—	—	6.02	6.08	6.15	6.21	6.31	2.71	0	1.0	245

Basalt 333A 6-2 (58-61)	5.90	5.97	6.03	6.08	—	—	—	2.87	1	—	234
Basalt 1, Hartford, Conn.	5.80	5.95	6.02	6.08	6.16	6.21	6.30	2.90	0	3.2	273
Greenstone W-2-2, Mar	5.90	5.97	6.01	6.08	6.19	6.23	6.34	2.86	1	6.6	262
Limestone Solenhafen, Germany	6.00	6.03	6.05	6.08	6.12	6.14	—	2.66	0	—	274
Granite Chelmsford, Mass.	—	5.64	5.91	6.09	6.22	6.28	6.35	2.63	0	1.8	245
Basalt Famous arp 73-7-1	—	5.30	5.67	6.09	6.36	6.46	—	2.72	0	—	241
Basalt 332B 6-1 (100-103)	5.93	5.99	6.03	6.09	—	—	—	2.84	1	—	234
Basalt 332B 29-1 (86-89)	5.75	5.96	6.01	6.09	—	—	—	2.88	1	—	234
Basalt W-7-1, Mar	5.37	5.66	5.87	6.09	6.28	6.35	6.39	2.83	1	1.1	262
Basalt arp 73-7-1, Famous	—	5.30	5.67	6.09	6.28	6.46	—	—	0	—	259
Quartz schist 10, Shikoku, Japan	—	5.79	5.96	6.09	6.20	6.28	—	2.73	0	15.5	267
Metagabbro V25-05-14, Kane F.Z.	—	5.66	5.92	6.10	6.25	6.40	—	2.82	0	—	231
Basalt Famous 525-5-2	—	5.76	5.93	6.10	6.26	6.29	—	2.81	0	—	241
Basalt 332A 21-1 (131-134)	6.00	6.06	6.08	6.10	—	—	—	2.81	1	—	234
Basalt 332B 27-2 (93-95)	5.89	6.05	6.11	6.10	—	—	—	2.84	1	—	234
Basalt 525-5-2, Famous	—	5.76	5.93	6.10	6.20	6.29	—	—	0	—	259
Basalt 2, Hartford, Conn.	5.80	6.04	6.08	6.11	6.16	6.21	6.29	2.91	0	1.8	273
Lunar basalt 14310, 72	4.17	4.94	5.53	6.11	6.55	6.93	—	2.88	0	—	244
Gabbroic anorthosite 14310, 72	—	4.94	5.53	6.11	6.70	—	—	2.88	0	—	275
Quartzite Clarendon Springs, Va.	5.50	5.90	6.05	6.12	6.19	6.24	6.30	2.63	0	1.5	270
Gneiss 1 Torrington, Conn.	4.80	5.70	5.97	6.12	6.22	6.27	6.35	2.64	0	2.2	270
Hornblende granodiorite 49 Iwate Pref., Japan	—	—	5.92	6.12	6.26	6.34	6.48	2.71	0	—	271
Basalt 321-14-1, 76-79	6.02	6.06	6.08	6.12	6.19	6.26	—	2.90	1	—	261
Basalt 332B 15-1 (129-131)	5.86	6.01	6.07	6.12	—	—	—	2.88	1	—	234
Basalt 256-10-2, 68	5.67	5.88	6.04	6.12	—	—	—	2.96	1	—	257
Basalt W-2-4, Mar	5.93	5.99	6.04	6.12	6.25	6.33	6.42	2.85	1	3.0	262
Dolerite W-10-3, Mar	5.84	5.93	6.01	6.12	6.27	6.36	6.42	2.89	1	0.2	262
Granite Westerly, R.I.	—	5.98	6.06	6.12	6.18	6.26	—	2.62	0	—	276
Quartz schist 402 H-K, Shikoku, Japan	—	5.64	5.93	6.12	6.26	6.32	—	2.71	0	19.0	267
Serpentinite 8 Mayaguez	—	6.07	6.09	6.12	6.19	—	—	2.73	0	—	249
Westerly granite	5.70	—	6.06	6.13	—	—	—	2.65	1	—	247
Basalt 6-54-8	5.95	6.03	6.07	6.13	6.20	6.26	6.37	2.87	1	0.2	235
Basalt 3M, Hawaii	4.63	5.29	5.75	6.13	6.39	6.47	6.52	2.70	0	—	251
Basalt 332A 40-2 (95-98)	5.52	5.95	6.02	6.13	—	—	—	2.74	1	—	234
Metagabbro RD18-P22, Oceanographer F.Z.	5.24	5.80	5.95	6.13	6.29	6.37	—	2.73	0	—	260

Table 18 (continued)
COMPRESSIONAL WAVE VELOCITIES IN ROCKS AS A FUNCTION OF PRESSURE

	P (Kbar)							ϱ	Wet or	Anis.	
	0.1	0.5	1	2	4	6	10	(g/cm³)	dry	(%)	Ref.
Quartz schist 615 H-N, Shikoku, Japan	—	5.83	6.00	6.13	6.25	6.31	—	2.78	0	11.5	267
Gneiss 5, Hull, Quebec	—	5.85	6.03	6.13	6.21	6.25	—	2.65	0	2.9	267
Kreep rock 60315, 33	—	4.81	5.40	6.13	6.68	6.96	—	3.05	0	—	256
Gneiss 5 Torrington, Conn.	5.50	5.90	6.05	6.14	6.27	6.38	6.51	2.85	0	5.0	270
Metabasalt V25-06-71, Kane F.Z.	—	5.96	6.08	6.14	6.20	6.25	—	2.87	0	—	231
Basalt 320B-5, CC	6.00	6.06	6.10	6.14	6.19	6.24	—	2.83	1	—	261
Hawaiite 811H, Hawaii	5.87	5.98	6.05	6.14	6.30	6.43	6.65	2.86	0	—	251
Metagabbro RD12-P4, Oceanographer F.Z.	5.16	5.77	6.00	6.14	6.34	6.45	—	3.01	0	—	260
Metagabbro RD12-P8, Oceanographer F.Z.	5.20	5.77	6.01	6.14	6.28	6.37	—	2.79	0	—	260
Schist IV-13	—	—	5.98	6.14	6.28	6.38	6.50	2.74	0	—	269
Metasandstone P3	—	5.79	5.98	6.14	6.30	6.41	—	2.85	1	—	258
Quartz schist 914 S-O, Shikoku, Japan	—	6.03	6.09	6.14	6.20	6.24	—	2.72	0	0.6	267
Gneiss 4, Mattawa, Ontario	—	5.76	5.97	6.14	6.23	6.29	—	2.66	0	4.8	267
Metabasalt V25-06-36, Kane F.Z.	—	5.95	6.05	6.15	6.25	6.30	—	2.80	0	—	231
Gabbro V25-06-38, Kane F.Z.	—	5.80	5.98	6.15	6.33	6.51	—	2.89	0	—	231
Quartzite Montana	—	—	6.11	6.15	6.22	6.26	6.35	2.65	0	2.9	245
Granite Barre, Vt.	—	5.86	6.06	6.15	6.25	6.32	6.39	2.66	0	1.1	245
Basalt 319A-4-1, 137-140	5.99	6.05	6.09	6.15	6.23	6.31	—	2.91	1	—	261
Basalt W-4-427, Mar	5.67	5.83	5.98	6.15	6.32	6.40	6.44	2.84	1	0.6	262
Granite Stone Mountain, Ga.	—	5.42	5.94	6.16	6.27	6.33	6.40	2.63	0	1.3	245
Basalt 332B 22-3 (39-41)	5.80	6.00	6.09	6.16	—	—	—	2.86	1	—	234
Basalt W-8-2, Mar	5.49	5.76	5.95	6.16	6.33	6.40	6.44	2.79	1	0.5	262
Greenstone W-2-10, Mar	5.98	6.03	6.08	6.16	6.27	6.34	6.43	2.84	1	1.6	262
Basalt W-2-6, Mar	5.74	5.90	6.02	6.16	6.31	6.40	6.48	2.84	1	1.7	262
Dolerite W-10-1, Mar	5.90	5.98	6.05	6.16	6.30	6.38	6.48	2.89	1	1.0	262
Granulite Ontario	—	—	6.06	6.17	6.31	6.41	6.48	2.71	0	0.8	277

Feldspathic mica quartzite Thomaston, Conn.	5.20	6.00	6.10	6.17	6.24	6.30	6.38	2.67	0	6.7	270
Basalt Famous arp 74-7-9	—	5.89	6.01	6.17	6.30	6.31	—	2.77	0	—	241
Basalt Famous arp 74-9-13	—	5.53	5.92	6.17	6.35	6.40	—	2.78	0	—	241
Basalt arp 74-7-9, Famous	—	5.89	6.01	6.17	6.24	6.31	—	—	0	—	259
Basalt arp 74-9-13, Famous	—	5.53	5.92	6.17	6.29	6.40	—	—	0	—	259
Biotite granodiorite Iwate Pref., Japan	—	—	6.07	6.18	6.28	6.36	6.45	2.69	0	—	271
Hornblende granodiorite 50 Iwate Pref., Japan	—	—	5.37	6.18	6.46	6.54	6.65	2.77	0	—	271
Basalt 332A 30-1 (127-130)	6.13	6.14	6.15	6.18	—	—	—	2.82	1	—	234
Metabasalt RD18-P12A, Oceanographer F.Z.	5.66	5.93	6.04	6.18	6.31	6.36	—	2.85	0	—	260
Metagabbro RD18-P13, Oceanographer F.Z.	5.29	5.88	6.05	6.18	6.31	6.38	—	2.86	0	—	260
Limestone 69-2	6.04	6.09	6.13	6.18	6.26	6.31	6.15	2.66	0	—	229
Quartz schist 603 H-N, Shikoku, Japan	—	5.92	6.08	6.18	6.26	6.30	—	2.73	0	7.5	267
Actinolite V25-09-10B, Kane F.Z.	—	5.62	5.86	6.19	6.65	6.87	—	2.97	0	—	231
Serpentinized peridotite 1 Mt. Boardman, Calif.	6.06	6.10	6.14	6.19	6.26	6.31	6.38	2.84	1	—	252
Basalt 6-54-9	6.01	6.08	6.13	6.19	6.26	6.31	6.39	2.87	1	3.2	235
Basalt 319A-6-1, 145-148	6.10	6.13	6.16	6.19	6.23	6.26	—	2.88	1	—	261
Basalt 332A 29-1 (74-77)	6.08	6.14	6.16	6.19	—	—	—	2.85	1	—	234
Metabasalt RD12-P18B, Oceanographer F.Z.	5.62	5.87	6.04	6.19	6.35	6.44	—	2.90	0	—	260
Gray granite Llano County, Tex.	—	5.96	6.10	6.19	6.25	6.30	—	2.61	0	—	276
Quartz schist 701 H-N, Shikoku, Japan	—	5.84	6.09	6.19	6.35	—	—	2.75	0	19.7	267
Serpentinized peridotite 2 Burro Mountain, Calif.	6.00	6.09	6.14	6.20	6.28	6.37	6.50	2.84	0	3.0	246
Granite Quincy, Mass.	—	6.04	6.11	6.20	6.30	6.37	6.45	2.62	0	1.7	245
Granite Chelmsford, Mass.	5.56	5.89	6.04	6.20	—	—	—	2.62	1	—	278
Basalt Famous cyp 74-31-39	—	5.65	5.92	6.21	6.45	6.51	—	2.78	0	—	241
Basalt 332A 33-2 (128-131)	6.14	6.18	6.20	6.21	—	—	—	2.86	1	—	234
Basalt 332B 3-2 (116-119)	6.08	6.13	6.17	6.21	—	—	—	2.75	1	—	234
Basalt W-9-78, Mar	5.38	5.69	5.93	6.21	6.40	6.48	6.55	2.82	1	0.2	262
Basalt cyp 74-31-39, Famous	—	5.65	5.92	6.21	6.36	6.51	—	—	0	—	259

Table 18 (continued)
COMPRESSIONAL WAVE VELOCITIES IN ROCKS AS A FUNCTION OF PRESSURE

	P (Kbar)								Wet or	Anis.	
	0.1	0.5	1	2	4	6	10	ϱ (g/cm³)	dry	(%)	Ref.
Basalt 245-19-1 (37-40)	6.06	6.11	6.16	6.21	6.28	6.35	—	2.89	1	—	228
Biotite granite Woodbury, Vt.	—	6.05	6.16	6.22	6.29	6.33	—	2.63	0	—	276
Gneiss Hells Gate, N.Y.	—	6.06	6.13	6.23	6.33	6.37	6.50	2.68	0	6.6	245
Basalt 332A 34-1 (88-91)	6.02	6.09	6.16	6.23	—	—	—	2.82	1	—	234
Metasandstone 21RGC60	—	5.99	6.11	6.23	6.37	6.45	—	2.82	1	—	258
Greywacke U2-U7, Pribram, Czechoslovakia	—	6.14	6.19	6.23	6.28	—	—	2.69	0	—	279
Serpentinite 5 Mayaguez	—	6.19	6.20	6.23	6.30	—	—	2.74	0	—	249
Diabase Paskenia, Calif.	5.91	6.05	6.15	6.24	6.36	6.42	6.49	2.86	1	—	252
Basalt 250A-26-6, 58	6.02	6.17	6.24	6.24	—	—	—	2.82	1	—	257
Basalt 3, Hartford, Conn.	6.00	6.11	6.17	6.24	6.30	6.35	6.42	2.94	0	1.1	273
Granodiorite 69-11	6.07	6.15	6.19	6.24	6.31	6.36	6.40	2.69	0	—	229
Lunar basalt 12065	4.44	5.21	5.80	6.24	6.61	6.80	6.96	3.26	0	—	280
Quartz schist 1007, Saitama Pref., Japan	—	6.16	6.20	6.24	6.28	6.32	—	2.76	0	6.7	267
Gneiss 4 Torrington, Conn.	4.80	5.70	6.03	6.25	6.40	6.47	6.58	2.82	0	4.5	270
Granite Latchford, Ontario	—	6.13	6.19	6.25	6.30	6.34	6.41	2.68	0	1.0	245
Basalt Famous arp 74-11-18	—	5.49	5.90	6.25	6.56	6.69	—	2.80	0	—	241
Serpentinized peridotite H5, Linhorka, CSSR	5.95	6.13	6.19	6.25	6.33	—	—	2.86	0	—	265
Basalt 74-11-18, Famous	—	5.49	5.90	6.25	6.47	6.69	—	—	0	—	259
Basalt Famous 525-2	—	5.55	5.89	6.26	6.61	6.80	—	2.77	0	-0	241
Basalt 525-2, Famous	—	5.55	5.89	6.26	6.53	6.80	—	—	0	—	259
Serpentinite 7 Mayaguez	—	6.18	6.21	6.26	6.33	—	—	2.73	0	—	249
Basalt 319A-3-2, 114-117	6.05	6.11	6.17	6.27	6.38	6.44	—	2.92	1	—	261
Limestone Oak Hall quarry	—	—	5.76	6.27	6.83	6.85	6.84	2.71	0	1.2	268
Basalt W-14-442, Mar	5.85	6.00	6.11	6.27	6.44	6.51	6.54	2.86	1	2.6	262
Greywacke 6, Pribaum, Czechoslovakia	—	6.21	6.23	6.27	6.33	—	—	2.75	0	—	279

Quartz schist 909 S-O, Shikoku, Japan	—	6.12	6.21	6.27	6.33	—	—	2.75	0	17.4	267
Granite Sacred Heart, Minn.	—	—	6.24	6.28	6.34	6.38	6.45	2.66	0	3.0	245
Basalt 319A-1-1, 32-35	6.04	6.13	6.20	6.28	6.38	6.41	—	2.92	1	—	261
Basalt 19-183-39-1, 148-150	6.12	6.17	6.22	6.28	6.36	6.41	6.43	2.84	1	—	255
Serpentinized peridotite 2 Mt. Boardman, Calif.	6.08	6.18	6.24	6.29	6.37	6.43	6.51	2.87	1	—	252
Granite Rockport, Mass.	—	5.96	6.18	6.29	6.39	6.43	6.51	2.62	0	0.9	245
Serpentinite 240, Urals, Serebry, USSR	5.86	6.06	6.20	6.29	6.37	—	—	2.66	0	—	265
Basalt 332A 36-2 (33-36)	6.21	6.25	6.27	6.29	—	—	—	2.83	1	—	234
Basalt 335 5-2 (36-38)	6.18	6.22	6.26	6.29	—	—	—	2.87	1	—	234
Quartz schist 604 H-N, Shikoku, Japan	—	6.04	6.18	6.29	6.37	6.41	—	2.73	0	17.0	267
Quartz schist 704 H-N, Shikoku, Japan	—	5.93	6.15	6.29	6.39	6.45	—	2.74	0	8.6	267
Lunar anorthosite 10020 Sea of Tranquillity	—	4.80	5.55	6.30	7.00	—	—	3.18	0	—	238
Charnockite Pallavaram, India	—	—	6.24	6.30	6.36	6.40	6.46	2.74	0	1.6	245
Basalt 319-13-1, 52-55	6.11	6.18	6.23	6.30	6.38	6.46	—	2.92	1	—	261
Basalt Famous CH 31-DR3-356	—	5.25	5.89	6.30	6.60	6.66	—	2.75	0	—	241
Basalt CH 31-DR3-356, Famous	—	5.25	5.89	6.30	6.48	6.66	—	—	0	—	259
Quincy granite Massachusetts	—	6.06	6.21	6.30	6.38	6.42	—	2.63	0	—	276
Quartz schist 605 H-N, Shikoku, Japan	—	6.06	6.19	6.30	6.39	6.45	—	2.71	0	17.1	267
Albitite	—	6.18	6.24	6.31	6.40	6.45	6.52	2.62	0	2.5	281
Basalt 332B 22-4 (11-13)	6.16	6.23	6.27	6.31	—	—	—	2.88	1	—	234
Basalt 257-15-1, 133	6.04	6.13	6.22	6.31	—	—	—	2.89	1	—	257
Metagabbro RD22-P11, Oceanographer F.Z.	5.27	6.00	6.12	6.31	6.48	6.56	—	2.77	0	—	260
Schist IV-22	—	—	6.17	6.31	6.44	6.52	6.61	2.75	0	—	269
Garnet schist Thomaston, Conn.	5.20	6.00	6.20	6.32	6.43	6.50	6.59	2.76	0	11.7	270
Gneiss Torrington, Conn.	5.10	5.80	6.15	6.32	6.43	6.49	6.57	2.76	0	0.3	270
Gabbro V25-06-163, Kane F.Z.	—	5.95	6.15	6.32	6.56	6.70	—	2.84	0	—	231
Basalt 250A-26-2, 140	6.13	6.22	6.32	6.32	—	—	—	2.85	1	—	257
Dolerite W-10-2, Mar	6.10	6.19	6.24	6.32	6.43	6.50	6.56	2.88	1	0.5	262
Lunar basalt 12052	4.90	5.55	5.93	6.32	6.68	6.88	7.01	3.27	0	—	280

Table 18 (continued)
COMPRESSIONAL WAVE VELOCITIES IN ROCKS AS A FUNCTION OF PRESSURE

	P (Kbar)							ϱ	Wet or	Anis.	
	0.1	0.5	1	2	4	6	10	(g/cm³)	dry	(%)	Ref.
Serpentinized peridotite Vallecitos, Calif.	6.21	6.25	6.28	6.33	6.41	6.46	6.55	2.72	1	—	252
Basalt 251A-31-5, 105	5.93	6.06	6.24	6.33	—	—	—	2.94	1	—	257
Metadiabase 293-18-1 (90-93)	6.20	6.24	6.27	6.33	6.44	6.53	—	2.83	1	—	236
Metagabbro RD20-P12, Oceanographer F.Z.	5.34	6.02	6.16	6.33	6.53	6.67	—	2.79	0	—	260
Gabbroic anorthosite 65015,9	—	5.58	5.98	6.33	6.72	—	—	2.97	0	—	275
Basalt 5-36-14	6.16	6.21	6.26	6.34	6.38	6.42	6.48	2.91	1	0.5	235
Basalt 319A-3-4, 85-88	6.12	6.21	6.27	6.34	6.43	6.47	—	2.94	1	—	261
Basalt Famous arp 73-10-3C	—	5.20	5.78	6.34	6.80	6.94	—	2.92	0	—	241
Basalt 332B 36-6 (44-46)	—	6.18	6.24	6.34	—	—	—	2.87	1	—	234
Basalt arp 73-10-3C, Famous	—	5.20	5.78	6.34	6.64	6.94	—	—	0	—	259
Schist 1404 B-T, Shikoku, Japan	—	5.82	6.06	6.34	6.59	6.69	—	3.04	0	11.1	267
Metabasalt V25-06-63, Kane, F.Z.	—	6.00	6.20	6.35	6.55	6.65	—	2.86	0	—	231
Granulite 5 Adirondacks	6.04	6.20	6.27	6.35	6.44	6.48	6.54	2.74	0	0.2	282
Granite Barriefield, Ontario	—	6.21	6.29	6.35	6.42	6.46	6.51	2.67	0	1.6	245
Granodiorite Butte, Mont.	—	—	6.27	6.35	6.43	6.48	6.56	2.71	0	1.1	245
Basalt 257-12-1, 130	5.97	6.11	6.25	6.35	—	—	—	2.73	1	—	257
Basalt W-4-15, Mar	5.33	5.82	6.10	6.35	6.51	6.57	6.61	2.86	1	1.1	262
Metagabbro RD20-P5B, Oceanographer F.Z.	5.68	6.08	6.21	6.35	6.48	6.57	—	2.81	0	—	260
Metagabbro V25-05-11, Kane F.Z.	—	6.10	6.26	6.36	6.50	6.60	—	2.91	0	—	231
Lunar breccia 14311	—	6.02	6.18	6.36	6.52	6.58	6.62	2.86	0	—	230
Trondjhemite 1 Trinity Complex, Calif.	5.94	6.16	6.28	6.37	6.48	6.56	6.71	2.65	1	—	252
Granite Englehart, Ontario	—	6.28	6.33	6.37	6.43	6.48	6.57	2.68	0	3.0	245
Alkalic basalt 2M, Hawaii	5.93	6.10	6.21	6.37	6.50	6.56	6.64	2.89	0	—	251
Basalt 3-15-10	6.23	6.29	6.32	6.37	6.45	6.50	6.58	2.91	1	1.6	245
Basalt 251A-31-4, 48	5.94	6.13	6.25	6.37	—	—	—	2.93	1	—	257
Limestone Irving, Tex.	5.60	5.98	6.13	6.37	—	—	—	2.77	1	—	278

Schist 1401 B-T, Shikoku, Japan	—	6.18	6.28	6.37	6.45	6.84	—	2.89	0	12.7	267
Serpentinized peridotite 53 Iwate Pref., Japan	—	—	6.25	6.38	6.55	6.64	6.75	2.92	0	—	271
Granite Hyderabad, India	—	6.26	6.31	6.38	6.44	6.49	6.56	2.65	0	3.4	245
Basalt 332B 1-5 (120-123)	—	6.24	6.31	6.38	—	—	—	2.81	1	—	234
Granite Barrefield, Ontario	5.88	6.22	6.34	6.38	6.43	6.47	—	2.67	0	—	274
Metasandstone P5	—	6.05	6.19	6.38	6.59	6.70	—	2.93	1	—	258
Staurolite-garnet schist Litchfield, Conn.	5.50	6.07	6.27	6.39	6.52	6.58	6.68	2.75	0	23.5	270
Basalt 319A-5-1, 80-83	6.22	6.29	6.34	6.40	6.46	6.49	—	2.95	1	—	261
Metagabbro RD17-P4, Oceanographer F.Z.	5.16	5.98	6.18	6.40	6.55	6.63	—	3.11	0	—	260
Gneiss 6, Renfrew, Ontario	—	6.32	6.36	6.40	6.45	6.47	—	2.89	0	13.3	267
Lunar anorthosite 15415,57	—	5.00	5.60	6.40	6.70	6.83	6.87	2.70	0	—	263
Granulite 2 Adirondacks	5.79	6.23	6.34	6.41	6.51	6.57	6.64	2.70	0	2.0	282
Granulite 3 Adirondacks	5.97	6.25	6.34	6.41	6.48	6.53	6.60	2.70	0	3.2	282
Magnetite ore Transvaal	—	—	6.32	6.41	6.52	6.58	6.67	4.54	0	3.2	245
Marble Danby, Vt.	—	—	6.27	6.41	6.62	6.76	6.92	2.71	0	10.5	268
Basalt 334 18-2 (12-14)	6.34	6.40	6.41	6.41	—	—	—	2.89	1	—	234
Granodiorite 71-5	5.91	6.24	6.34	6.41	6.47	6.51	6.56	2.67	0	—	229
Gabbroic anorthosite 62295,18	—	5.54	5.98	6.42	6.79	—	—	2.83	0	—	275
Metagabbro 1 Point Sal, Calif.	6.30	6.34	6.38	6.43	6.50	6.54	6.60	2.72	1	—	252
Granulite 4 Adirondacks	6.16	6.31	6.36	6.43	6.49	6.54	6.59	2.73	0	1.7	282
Tonalite Val Verde, Calif.	—	—	6.33	6.43	6.49	6.54	6.60	2.76	0	1.1	245
Basalt 257-12-3, 35	5.88	6.13	6.28	6.43	—	—	—	2.73	1	—	257
Basalt EPR4	5.90	6.08	6.25	6.43	6.58	6.65	6.69	2.88	1	1.2	272
Pink Granite Llano County, Tex.	—	6.29	6.34	6.43	6.50	6.54	—	2.64	0	—	276
Schist 1005 M-O, Shikoku, Japan	—	6.10	6.26	6.43	6.59	6.69	—	2.90	0	8.6	267
Anorthosite 15415,96	—	6.20	6.28	6.43	6.66	6.82	—	2.76	0	—	256
Granulite New Jersey	—	—	6.36	6.44	6.52	6.57	6.63	2.68	0	2.6	277
Kyanite Schist 2 Torrington, Conn.	5.10	6.02	6.24	6.44	6.60	6.70	6.82	2.77	0	15.4	270
Basalt 334 18-1 (84-87)	6.40	6.40	6.42	6.44	—	—	—	2.95	1	—	234
Basalt 163-29-4, 67-74	6.34	6.36	6.41	6.45	6.50	6.54	6.60	2.94	1	0.3	243
Metadiabase Marin County, Calif.	5.90	—	6.35	6.46	6.59	6.70	—	2.88	0	2.3	283
Metagabbro 2 Point Sal, Calif.	6.30	6.35	6.40	6.46	6.54	6.59	6.64	2.85	1	—	252
Granite Hyderabad, India	—	—	6.42	6.46	6.51	6.55	6.61	2.68	0	1.5	245
Dunite Webster, N.C.	—	—	6.37	6.46	6.55	6.64	6.79	2.98	0	6.4	245
Basalt 6-57-3	6.34	6.38	6.41	6.46	6.51	6.55	6.61	2.98	1	0.6	235

Table 18 (continued)
COMPRESSIONAL WAVE VELOCITIES IN ROCKS AS A FUNCTION OF PRESSURE

	P (Kbar)								Wet		
	0.1	0.5	1	2	4	6	10	ϱ (g/cm³)	or dry	Anis. (%)	Ref.
Basalt 332B 2-2 (86-89)	6.21	6.35	6.39	6.46	—	—	—	2.84	1	—	234
Granulite 8 Adirondacks	6.22	6.37	6.43	6.47	6.53	6.55	6.60	2.83	0	1.1	282
Diabase Holyoke, Mass.	—	6.40	6.43	6.47	6.52	6.56	6.63	2.98	0	0.8	245
Basalt 332B 9-3 (80-82)	6.33	6.44	6.44	6.47	—	—	—	2.87	1	—	234
Basalt 334 16-4 (104-107)	6.39	6.47	6.46	6.47	—	—	—	2.93	1	—	234
Mica schist Woodsville, Vt.	—	—	6.43	6.48	6.53	6.57	6.64	2.80	0	10.3	245
Basalt 332B 1-5 (37-39)	6.28	6.36	6.41	6.48	—	—	—	2.80	1	—	234
Metagabbro Goshen, Conn.	5.90	6.18	6.35	6.49	6.64	6.72	6.85	2.99	0	3.2	270
Trondjhemite 2 Trinity Complex, Calif.	6.06	6.25	6.38	6.49	6.60	6.67	6.80	2.73	1	—	252
Ijolite C192B, Hawaii	6.26	6.34	6.39	6.49	6.54	6.57	6.61	2.97	0	—	251
Metagabbro RD20-P6A, Oceanographer F.Z.	5.99	6.25	6.36	6.49	6.62	6.72	—	2.90	0	—	260
Schist 1201 M-O, Shikoku, Japan	—	5.81	6.15	6.49	6.74	6.84	—	3.01	0	14.4	267
Talc Schist Chester, Vt.	—	—	6.30	6.50	6.71	6.82	6.97	2.91	0	2.7	245
Metagabbro Famous AII 73-47-25	—	6.19	6.33	6.50	6.69	6.79	—	3.02	0	—	241
Basalt 332A 6-2 (117-120)	—	6.39	6.45	6.50	—	—	—	2.79	1	—	234
Lunar breccia 15418,43	5.00	5.50	6.02	6.50	6.69	6.81	—	2.80	0	—	244
Gabbroic anorthosite 15418,43	—	5.50	6.02	6.50	6.70	—	—	2.80	0	—	275
Metagabbro RD22-P4, Oceanographer F.Z.	5.44	6.22	6.34	6.51	6.65	6.72	—	2.89	0	—	260
Quartz diorite Calif.	—	—	6.43	6.52	6.60	6.64	6.71	2.80	0	1.1	245
Schist 601 H-N, Shikoku, Japan	—	6.00	6.25	6.52	6.72	6.86	—	2.98	0	18.5	267
Lunar basalt 60015,29	—	5.50	6.00	6.52	6.86	6.94	7.02	2.76	0	—	263
Granulite Saranac Lake, N.Y.	—	—	6.40	6.53	6.65	6.70	6.76	2.85	0	3.0	277
Limestone Oak Hull Quarry, Pa.	6.35	—	6.50	6.53	6.58	—	6.60	2.71	0	—	266
Quartz diorite Dedham, Mass.	—	—	6.46	6.53	6.60	6.65	6.71	2.91	0	0.5	245
Schist 1105 M-O, Shikoku, Japan	—	6.16	6.36	6.53	6.70	6.78	—	2.90	0	14.5	267
Granulite Saranac Lake, N.Y.	—	—	6.40	6.54	6.66	6.71	6.77	2.83	0	0.9	277
Metagabbro V25-05-26, Kane F.Z.	—	6.31	6.44	6.54	6.65	6.73	—	2.78	0	—	231

Serpentinite 2 Stonyford, Calif.	6.43	6.47	6.49	6.54	6.60	6.64	6.69	2.66	1	—	252
Casco granite	6.02	—	6.48	6.54	—	—	—	2.63	1	—	247
Basalt 332A 40-3 (40-43)	6.45	6.49	6.51	6.54	—	—	—	2.86	1	—	234
Pyroxene granulite IV-8	—	—	6.32	6.54	6.70	6.78	6.83	2.79	0	—	269
Gabbro V25-06-41, Kane F.Z.	—	6.22	6.40	6.55	6.70	6.80	—	2.90	0	—	231
Casco granite	5.05	—	6.46	6.55	—	—	—	2.63	0	—	247
Basalt EPR3	6.19	6.32	6.42	6.55	6.67	6.72	6.74	2.87	1	5.8	272
Granulite Santa Lucia Mountains, Calif.	—	—	6.40	6.56	6.64	6.69	6.76	2.73	0	2.0	277
Peridotite HA-01 Higashi-Akaishi-Yama	—	6.50	6.53	6.56	6.62	—	6.72	3.37	0	7.7	284
Basalt EPR1	6.33	6.39	6.46	6.56	6.66	6.71	6.75	2.95	1	0.9	272
Granite 72-3	5.54	6.41	6.52	6.56	6.63	6.65	6.68	2.61	0	—	229
Gabbro V25-05-52, Kane F.Z.	—	6.30	6.44	6.57	6.72	6.83	—	2.93	0	—	231
Serpentinite Ludlow, Vt.	—	—	6.51	6.57	6.67	6.74	6.84	2.80	0	3.4	245
Basalt 332A 7-1 (66-69)	6.53	6.56	6.59	6.58	—	—	—	2.81	1	—	234
Basalt 332B 3-4 (17-20)	6.42	6.47	6.52	6.58	—	—	—	2.89	1	—	234
Pyriclasite IV-9	—	—	6.31	6.58	6.85	6.99	7.09	2.94	0	—	269
Peridotite Miye Pref., Japan	—	—	6.50	6.59	6.70	6.79	6.90	3.06	0	1.6	271
Serpentinite Ludlow, Vt.	—	6.33	6.46	6.59	6.70	6.75	6.82	2.61	0	12.7	245
Basalt 261-34-3, 69-73	6.47	6.50	6.55	6.59	6.65	6.68	—	3.00	1	—	232
Andesite IBA8, Hawaii	6.42	6.50	6.55	6.59	6.63	6.66	6.72	2.98	0	—	251
Dolomitized Micrite Limestone 127-18-1	—	6.50	6.58	6.59	6.54	6.66	—	2.73	0	1.5	240
Gabbroic anorthosite 68415,54	—	5.78	6.18	6.59	6.89	—	—	2.78	0	—	275
Metagabbro V25-06-43, Kane F.Z.	—	6.40	6.50	6.60	6.70	6.80	—	2.87	0	—	231
Metagabbro V25-06-17, Kane F.Z.	—	6.35	6.48	6.60	6.70	6.80	—	2.90	0	—	231
Serpentinized peridotite Miye Pref., Japan	—	—	6.55	6.60	6.65	6.74	6.88	3.04	0	—	271
Basalt EPR2	6.23	6.36	6.47	6.60	6.71	6.76	6.79	2.95	1	1.0	272
Amphibolite IV-16	—	—	6.06	6.60	7.04	7.21	7.32	3.04	0	—	269
Lunar basalt 61016,34	—	5.60	6.20	6.60	6.87	6.96	7.02	2.79	0	—	263
Basalt 332B 3-4 (10-13)	—	6.52	6.57	6.61	—	—	—	2.81	1	—	234
Lunar basalt 74275,25	—	5.20	5.98	6.61	7.01	7.19	—	3.36	0	—	237
Basalt Famous 527-1-2A	—	6.00	6.35	6.62	6.84	6.89	—	2.82	0	—	241
Basalt 527-1-2A, Famous	—	6.00	6.35	6.62	6.76	6.89	—	—	0	—	259
Granulite 9 Adirondacks	6.42	6.53	6.58	6.63	6.68	6.70	6.75	2.95	0	3.3	282

Table 18 (continued)
COMPRESSIONAL WAVE VELOCITIES IN ROCKS AS A FUNCTION OF PRESSURE

	P (Kbar)								Wet		
	0.1	0.5	1	2	4	6	10	ϱ (g/cm^3)	or dry	Anis. (%)	Ref.
Augite syenite Ontario	—	—	6.58	6.63	6.70	6.73	6.79	2.78	0	1.5	245
Gabbro 293-19-1 (108-111)	6.50	6.54	6.58	6.63	6.70	6.74	—	2.85	1	—	236
Basalt 15545,24	—	6.10	6.37	6.63	6.87	6.94	—	2.56	0	—	256
Serpentinized peridotite 52 Iwate Pref., Japan	—	—	6.52	6.64	6.70	6.78	6.82	2.85	0	—	271
Basalt 332B 2-5 (115-117)	—	6.60	6.64	6.64	—	—	—	2.83	1	—	234
Schist 1407 B-T, Shikoku, Japan	—	6.09	6.38	6.64	6.85	6.95	—	2.95	0	16.1	267
Metabasalt 2 Luray, Va.	6.50	—	6.58	6.65	6.71	6.75	—	2.93	0	1.3	283
Metagabbro V25-05-32, Kane F.Z.	—	6.42	6.54	6.65	6.75	6.88	—	2.89	0	—	231
Albitite Sylmar, Pa.	—	—	6.62	6.65	6.68	6.72	6.76	2.69	0	2.4	245
Basalt 256-10-3, 84	6.23	6.47	6.56	6.65	—	—	—	2.96	1	—	257
Schist 1102 M-O, Shikoku, Japan	—	6.12	6.42	6.65	6.83	6.91	—	2.95	0	14.2	267
Schist 1109 M-O, Shikoku, Japan	—	6.47	6.56	6.65	6.76	6.84	—	2.98	0	7.1	267
Marble Danby, Vt.	—	—	6.61	6.66	6.72	6.74	6.76	2.70	0	3.4	245
Basalt 332B 2-5 (69-72)	6.46	6.55	6.60	6.66	—	—	—	2.75	1	—	234
Lunar basalt 15555,88	—	5.60	6.10	6.66	7.02	7.25	7.42	3.10	0	—	263
Diabase Cobalt, Ontario	—	—	6.64	6.67	6.71	6.75	6.82	2.96	0	0.3	245
Basalt 332B 2-1 (60-63)	6.41	6.59	6.61	6.67	—	—	—	2.78	1	—	234
Schist 1406 B-T, Shikoku, Japan	—	6.33	6.51	6.67	6.80	6.86	—	2.97	0	12.1	267
Diorite Aichi Pref., Japan	—	—	6.61	6.68	6.78	6.85	6.95	2.94	0	4.1	271
Peridotite HA-05 Higashi-Akaishi-Yama	—	6.54	6.62	6.68	6.75	—	6.85	3.67	0	1.9	284
Metagabbro 1 Canyon Mountain, Ore.	6.57	6.60	6.64	6.69	6.73	6.80	6.94	2.82	1	—	252
Peridotite HA-13 Higashi-Akaishi-Yama	—	6.57	6.61	6.69	6.75	—	6.90	3.05	0	9.0	284
Schist 418 H-K, Shikoku, Japan	—	6.16	6.45	6.69	6.84	6.94	—	2.97	0	2.6	267
Metagabbro 2 Canyon Mountain, Ore.	6.60	6.64	6.66	6.70	6.75	6.78	6.84	2.87	1	—	252
Basalt 332B 35-1 (66)	—	6.51	6.61	6.70	—	—	—	2.81	1	—	234

Lunar basalt 70215,30	—	5.77	6.23	6.70	7.00	7.11	—	3.37	0	—	237
Gabbro 1D Canyon Mountain, Ore.	6.54	6.61	6.65	6.71	6.79	6.85	6.95	2.84	1	5.3	252
Nephelenite IHAM2, Hawaii	6.36	6.50	6.59	6.71	6.84	6.93	7.01	2.99	0	—	251
Schist 1101 M-O, Shikoku, Japan	—	6.15	6.46	6.71	6.90	6.94	—	2.98	0	15.9	267
Schist 1405 B-T, Shikoku, Japan	—	6.31	6.55	6.72	6.85	6.93	—	3.02	0	12.0	267
Lunar Metabreccia 73235,18	—	6.02	6.39	6.72	6.98	7.10	—	2.93	0	—	237
Granulite Santa Lucia Mountains, Calif.	—	—	6.51	6.73	6.91	6.99	7.12	2.98	0	5.1	277
Anorthosite New Glasgow, Quebec	—	6.64	6.69	6.73	6.78	6.81	6.85	2.71	0	1.5	281
Amphibolite Canyon Mountain, Ore.	6.63	6.66	6.70	6.74	6.80	6.83	6.86	2.93	1	—	252
Granulite 1 Adirondacks	5.96	6.41	6.58	6.74	6.83	6.89	6.95	2.71	0	2.3	282
Stronalite gneiss IV-23	—	—	6.60	6.74	6.87	6.96	7.05	2.95	0	—	269
Magnetite ore Tahawus, N.Y.	—	—	6.65	6.75	6.85	6.92	6.98	4.53	0	—	245
Anorthosite Whiteface, N.Y.	—	6.61	6.69	6.75	6.82	6.85	6.91	2.71	0	1.5	281
Diabase Centreville, Va.	—	—	6.70	6.76	6.82	6.86	6.93	2.98	0	0.4	245
Diabase Sudbury, Ontario	—	6.67	6.72	6.76	6.81	6.84	6.91	3.00	0	2.2	245
Schist 420 H-K, Shikoku, Japan	—	6.20	6.55	6.76	6.88	6.94	—	2.99	0	7.3	267
Schist 1502 B-T, Shikoku, Japan	—	6.30	6.53	6.76	6.94	7.00	—	2.97	0	17.9	267
Basalt 251A-31-3, 50	6.31	6.49	6.62	6.77	—	—	—	2.86	1	—	257
Granulite IV-11	—	—	6.58	6.77	6.91	6.98	7.05	2.92	0	—	269
Diabase 7, Pribram, Czechoslovakia	—	6.65	6.72	6.77	6.84	—	—	2.88	0	—	279
Schist 920 S-O, Shikoku, Japan	—	6.36	6.63	6.77	6.90	6.94	—	2.95	0	16.3	267
Kyanite Schist 1 Torrington, Conn.	5.10	6.10	6.48	6.78	7.05	7.20	7.41	3.00	0	3.5	270
Serpentine Middlefield, Mass.	6.66	6.71	6.74	6.79	6.84	6.90	6.97	2.79	0	9.9	285
Diabase Frederick, Md.	—	—	6.77	6.80	6.84	6.88	6.92	3.01	0	0.1	245
Metagabbro 293-20-1 (100-103)	6.46	6.59	6.71	6.80	6.86	—	—	2.85	1	—	236
Chlorite schist Chester Quarry, Vt.	—	—	6.75	6.82	6.92	6.98	7.07	2.84	0	9.2	245
Schist 509 H-K, Shikoku, Japan	—	6.45	6.67	6.82	6.91	6.95	—	3.04	0	11.7	267
Schist 812 K-O, Shikoku, Japan	—	6.57	6.72	6.82	6.90	6.92	—	3.00	0	11.2	267
Granulite Santa Lucia Mountains, Calif.	—	—	6.75	6.83	6.91	6.95	7.02	2.90	0	3.2	277
Diabase Frederick, Md.	6.67	—	6.78	6.83	6.91	—	7.02	3.02	0	—	266
Granulite 7 Adirondacks	6.35	6.62	6.75	6.83	6.90	6.94	6.98	2.79	0	1.2	282
Granulite 13 Adirondacks	6.40	6.62	6.74	6.83	6.90	6.94	7.03	3.09	0	5.2	282
Schist 702 H-N, Shikoku, Japan	—	6.47	6.68	6.83	6.93	6.97	—	2.97	0	16.2	267
Schist 921 S-O, Shikoku, Japan	—	6.52	6.70	6.83	6.91	6.97	—	2.90	0	11.6	267
Schist 1104 M-O, Shikoku, Japan	—	6.64	6.73	6.83	6.93	6.99	—	3.02	0	13.0	267
Gabbro V25-06-39A. Kane F.Z.	—	6.60	6.75	6.84	7.00	7.06	—	2.90	0	—	231

Table 18 (continued)
COMPRESSIONAL WAVE VELOCITIES IN ROCKS AS A FUNCTION OF PRESSURE

	P (Kbar)								Wet		
	0.1	0.5	1	2	4	6	10	ϱ (g/cm³)	or dry	Anis. (%)	Ref.
Granulite 6 Adirondacks	6.20	6.66	6.77	6.85	6.94	7.00	7.09	2.77	0	1.9	282
Granulite 12 Adirondacks	5.92	6.59	6.72	6.85	6.95	7.02	7.12	3.07	0	0.3	282
Amphibolite IV-1	—	—	6.61	6.85	7.03	7.08	7.18	3.06	0	—	269
Schist 918 S-O, Shikoku, Japan	—	6.68	6.78	6.86	6.93	6.99	—	2.89	0	13.9	267
Serpentinized Peridotite 4 Burro Mountain, Calif.	6.40	6.69	6.78	6.87	6.98	7.05	7.19	3.07	0	2.4	246
Amphibolite 2 Bantam, Conn.	5.50	6.30	6.63	6.87	7.04	7.10	7.18	3.03	0	14.6	270
Peridotite HA-14 Higashi-Akaishi-Yama	—	6.80	6.86	6.87	6.94	—	6.95	3.05	0	1.7	284
Schist 602 H-N, Shikoku, Japan	—	6.47	6.69	6.87	7.01	7.07	—	3.04	0	15.4	267
Metagabbro 3 Trinity Complex, Calif.	6.75	6.79	6.83	6.88	6.96	7.01	7.09	3.02	1	—	252
Diabase 3, Pribram, Czechoslovakia	—	6.69	6.81	6.88	6.92	—	—	2.90	0	—	279
Schist 916 S-O, Shikoku, Japan	—	6.48	6.74	6.88	6.96	7.02	—	2.99	0	13.4	267
Granulite Adirondack Mountains, N.Y.	—	—	6.82	6.89	6.97	7.01	7.07	2.93	0	0.9	277
Serpentinized Peridotite Iwate Pref., Japan	—	—	6.78	6.89	7.04	7.10	7.16	2.82	0	3.3	271
Metabasalt 1 Yreka, Calif.	6.70	—	6.84	6.90	6.96	6.99	—	2.91	0	3.2	283
Anorthosite Tahawus, N.Y.	—	—	6.86	6.90	6.94	6.97	7.02	2.77	0	4.2	245
Magnetite ore Port Henry, N.Y.	—	—	6.77	6.90	6.99	7.04	7.11	4.87	0	0.1	245
Brecciated Dolomitized Limestone 127-19-1	—	6.73	6.82	6.91	6.97	7.01	—	2.81	0	0.4	240
Schist 410 H-K, Shikoku, Japan	—	6.63	6.78	6.91	7.03	7.09	—	3.04	0	16.8	267
Diopside Hornblendite Iwate Pref.,Japan	—	—	6.34	6.92	7.08	7.16	7.25	3.20	0	—	271
Schist 912 S-O, Shikoku, Japan	—	6.59	6.78	6.92	7.02	7.08	—	2.95	0	13.8	267
Siderite Roxbury, Conn.	—	—	—	6.93	7.04	7.09	7.15	3.75	0	1.3	286
Metagabbro 1 Trinity Complex, Calif.	6.82	6.85	6.88	6.93	7.01	7.07	7.17	2.91	1	—	252
Metagabbro 4 Trinity Complex, Calif.	6.81	6.85	6.88	6.93	6.99	7.04	7.12	3.04	1	—	252
Gabbro 334 24-4 (86-88)	6.71	6.85	6.90	6.93	—	—	—	2.85	1	—	234

Schist 406A H-K, Shikoku, Japan	—	6.72	6.83	6.93	7.04	7.06	—	3.02	0	15.8	267
Metabasalt 3 Mt. Vernon, Wash.	6.80	—	6.89	6.94	7.00	7.04	—	2.94	0	5.7	283
Idiocrase Crestmore, Calif.	5.62	6.10	6.54	6.95	7.27	7.40	7.54	3.14	0	1.2	285
Gabbro V25-06-39B, Kane F.Z.	—	6.72	6.86	6.95	7.05	7.18	—	2.89	0	—	231
Gabbro 293-21-1 (5-8)	6.79	6.84	6.88	6.95	7.05	7.10	—	2.94	1	—	236
Schist 412 H-K, Shikoku, Japan	—	6.65	6.82	6.95	7.06	7.12	—	3.02	0	9.3	267
Hornblende Gabbro Kyoto, Japan	—	—	6.78	6.96	7.02	7.06	7.15	3.11	0	—	271
Peridotite Miye Pref., Japan	—	—	6.87	6.96	7.09	7.16	7.26	3.15	0	1.8	271
Granulite 10 Adirondacks	6.66	6.79	6.86	6.96	7.07	7.14	7.22	2.99	0	1.6	282
Serpentinite 334 22-2 (43-45)	6.15	6.75	6.90	6.96	—	—	—	2.84	1	—	234
Amphibolite 1 Bantam, Conn.	6.50	6.77	6.88	6.97	7.08	7.14	7.22	3.04	0	10.9	270
Schist 401 H-K, Shikoku, Japan	—	6.87	6.91	6.97	7.03	7.07	—	3.01	0	21.1	267
Schist 906 S-O, Shikoku, Japan	—	6.69	6.87	6.98	7.05	7.08	—	3.04	0	7.7	267
Gabbro Pegmatite Papua, New Guinea	6.60	6.86	6.93	6.98	7.02	—	7.09	2.78	0	1.0	287
Schist 503 H-K, Shikoku, Japan	—	6.40	6.76	6.99	7.15	7.23	—	3.02	0	14.5	267
Granulite 15 Adirondacks	6.83	6.93	6.96	7.00	7.04	7.08	7.14	3.23	0	0.6	282
Webatuck Dolomite	6.40	—	6.94	7.00	—	—	—	2.87	0	—	247
Gabbro 293-21-1 (32-35)	6.75	6.84	6.91	7.00	7.07	—	—	2.94	1	—	236
Dolomite 72-4	5.32	5.90	6.68	7.00	7.27	7.36	7.42	2.85	0	—	229
Schist 811 K-O, Shikoku, Japan	—	6.74	6.89	7.00	7.09	7.15	—	3.08	0	9.2	267
Rhodochrosite Argentina	—	—	—	7.01	7.09	7.14	7.20	3.57	0	4.4	286
Microcline Labrador	6.49	6.84	6.95	7.01	7.06	7.09	7.15	2.57	0	43.2	285
Gabbro 1A Canyon Mountain, Ore.	6.80	6.87	6.93	7.01	7.13	7.19	7.30	2.90	1	1.7	252
Anorthosite Stillwater, Mont.	—	—	6.97	7.01	7.05	7.07	7.10	2.77	0	4.1	245
Serpentinized Peridotite 3 Burro Mountain, Calif.	6.40	6.76	6.90	7.02	7.12	7.20	7.35	3.05	0	9.1	246
Gabbro French Creek, Pa.	—	6.74	6.93	7.02	7.11	7.17	7.23	3.05	0	0.8	245
Peridotite HA-15 Higashi-Akaishi-Yama	—	6.95	6.97	7.02	7.08	—	7.22	3.56	0	2.7	284
Gabbro 334 22-1 (69-71)	6.95	6.97	6.99	7.02	—	—	—	3.01	1	—	234
Gabbro 293-20-1 (136-139)	6.78	6.87	6.95	7.02	7.10	7.16	—	2.83	1	—	236
Schist 1503 B-T, Shikoku, Japan	—	6.56	6.85	7.02	7.12	7.16	—	3.05	0	8.4	267
Epidosite Luray, Va.	6.10	—	6.77	7.03	7.16	7.23	—	3.17	0	0.4	283
Dolomite Vt.	—	—	6.98	7.03	7.09	7.14	7.22	2.84	0	2.1	245
Schist 504B H-K, Shikoku, Japan	—	6.45	6.85	7.03	7.13	7.21	—	3.05	0	11.2	267
Schist 706 K-O, Shikoku, Japan	—	6.85	6.94	7.03	7.11	7.17	—	2.99	0	19.1	267
Anorthosite West Greenland	—	6.64	6.84	7.03	7.12	7.15	—	2.08	0	—	256

Table 18 (continued)
COMPRESSIONAL WAVE VELOCITIES IN ROCKS AS A FUNCTION OF PRESSURE

	P (Kbar)								Wet		
	0.1	0.5	1	2	4	6	10	ϱ (g/cm³)	or dry	Anis. (%)	Ref.
Anorthosite Bushveld Complex	—	6.92	6.98	7.05	7.13	7.16	7.21	2.81	0	3.2	245
Webatuck Dolomite	6.73	—	6.99	7.05	—	—	—	2.87	1	—	247
Granulite Valle D Ossola, Italy	—	—	6.68	7.06	7.29	7.40	7.48	3.09	0	3.2	277
Dolomite Williamstown, Mass.	6.30	6.77	6.93	7.06	7.17	7.23	7.36	2.85	0	4.7	285
Gabbro V25-06-6, Kane F.Z.	—	6.85	6.95	7.06	7.20	7.30	—	2.96	0	—	231
Granulite 11 Adirondacks	6.87	6.98	7.03	7.06	7.11	7.15	7.20	3.04	0	2.1	282
Granulite 14 Adirondacks	5.84	6.80	6.95	7.06	7.17	7.22	7.30	3.17	0	3.8	282
Hornblende-Pyroxene granofels IV-6	—	—	6.67	7.06	7.27	7.37	7.45	3.07	0	—	269
Pyriclasite IV-25	—	—	6.68	7.06	7.29	7.40	7.48	3.09	0	—	269
Gabbro 2 Canyon Mountain, Ore.	6.82	6.90	6.97	7.07	7.18	7.26	7.40	3.01	1	—	252
Schist 904 S-O, Shikoku, Japan	—	6.72	6.96	7.07	7.15	7.19	—	3.10	0	10.8	267
Pyriclasite IV-17	—	—	6.90	7.08	7.21	7.28	7.35	2.91	0	—	269
Granulite 16 Adirondacks	6.03	6.86	7.00	7.09	7.18	7.23	7.29	3.24	0	1.9	282
Gabbro Mellen, Wisc.	—	7.04	7.07	7.09	7.13	7.16	7.21	2.93	0	3.9	245
Metagabbro 3 Canyon Mountain, Ore.	6.90	6.98	7.03	7.10	7.18	7.23	7.33	3.03	1	—	252
Stronalite gneiss IV-24	—	—	6.79	7.10	7.40	7.56	7.73	3.00	0	—	269
Norite Pretoria, Transvaal	—	7.02	7.07	7.11	7.16	7.20	7.28	2.98	0	1.4	245
Eclogite ME-1-15(P), Colorado Plateau	—	6.03	6.65	7.11	7.47	—	7.67	3.28	0	—	284
Gabbro 293-21-2 (11-14)	6.98	7.02	7.06	7.11	7.17	7.23	—	2.93	1	—	236
Schist 501A H-K, Shikoku, Japan	—	6.76	6.94	7.11	7.21	7.27	—	3.06	0	16.6	267
Gneiss IV-7	—	—	6.99	7.12	7.27	7.35	7.43	3.10	0	—	269
Pyriclasite IV-20	—	—	6.69	7.12	7.38	7.48	7.57	3.05	0	—	269
Serpentinized Peridotite 6 Burro Mountain, Calif.	6.70	7.00	7.09	7.16	7.23	7.30	7.42	3.14	0	2.1	246
Granulite 17 Adirondacks	6.84	7.01	7.10	7.16	7.22	7.27	7.32	3.72	0	1.4	282
Dunite TW-13, Wash.	6.78	6.96	7.05	7.18	7.30	7.41	7.51	3.07	0	1.7	288
Magnesite Unknown	6.97	7.06	7.11	7.19	7.27	7.33	7.45	2.80	0	4.3	285
Eclogite SL002, Hawaii	5.51	6.29	6.77	7.19	7.58	7.70	7.85	3.23	0	—	251

Schist 413 H-K, Shikoku, Japan	—	6.83	7.03	7.19	7.32	7.38	—	3.05	0	7.2	267
Schist 805 K-O, Shikoku, Japan	—	7.07	7.15	7.19	7.25	7.29	—	3.01	0	16.1	267
Peridotite Ehime, Japan	—	—	7.07	7.20	7.39	7.49	7.59	3.16	0	5.1	271
Plagioclase peridotite Miye Pref., Japan	—	—	7.07	7.20	7.30	7.40	7.62	3.13	0	—	271
Eclogite 1 Sittampundi, India	—	—	7.08	7.21	7.42	7.53	—	3.56	0	—	289
Hortonolite dunite Mooihoek Mine, Transvaal	6.70	7.13	7.16	7.21	7.27	7.30	7.36	3.74	0	6.1	290
Amphibolite Madison County, Mont.	—	—	7.17	7.21	7.27	7.31	7.35	3.12	0	12.0	245
Dunite Mooihoek Mine, Transvaal	—	7.13	7.16	7.21	7.27	7.30	7.36	3.74	0	6.1	245
Granulite Wind River Mountains, Wyo.	—	—	7.11	7.23	7.32	7.37	7.43	3.04	0	4.8	277
Metagabbro 3 Point Sal, Calif.	6.90	7.04	7.13	7.23	7.30	7.35	7.39	2.94	1	—	252
Plagioclase IV-18	—	—	7.09	7.24	7.39	7.48	7.57	2.96	0	—	269
Serpentinized Peridotite 5 Burro Mountain, Calif.	6.80	7.07	7.15	7.25	7.37	7.46	7.62	3.13	0	3.5	246
Schist 417 H-K, Shikoku, Japan	—	7.00	7.17	7.25	7.32	7.34	—	3.06	0	5.9	267
Hornblende-Pyroxene Granofels IV-15	—	—	7.12	7.26	7.38	7.45	7.51	3.08	0	—	269
Gabbro 1C Canyon Mountain, Ore.	7.11	7.17	7.21	7.27	7.35	7.41	7.50	2.99	1	2.6	252
Gabbro 334 23-1 (76-78)	—	7.23	7.25	7.28	—	—	—	3.03	1	—	234
Eclogite 2 Sittampundi, India	—	—	7.17	7.29	7.44	7.55	—	3.58	0	—	289
Gabbro 1B Canyon Mountain, Ore.	7.15	7.19	7.24	7.29	7.35	7.41	7.50	2.99	1	0.5	252
Gabbro 334 21-1 (78-82)	7.02	7.17	7.22	7.29	—	—	—	2.97	1	—	234
Pyriclasite IV-14	—	—	7.15	7.29	7.41	7.48	7.57	3.08	0	—	269
Schist 419 H-K, Shikoku, Japan	—	6.51	6.96	7.29	7.47	7.55	—	3.12	0	8.3	267
Albitite 3204, Sugajima, Mie Pref., Japan	—	7.17	7.21	7.29	7.32	7.34	—	2.87	0	7.2	267
Hortonolite Dunite N.Y.	7.20	7.24	7.27	7.30	7.35	7.39	7.46	3.93	0	3.4	290
Eclogite Sunnmore, Norway	—	—	7.13	7.30	7.46	7.54	7.69	3.38	0	3.9	245
Serpentinized Dunite 337, Urals, Uktus, USSR	6.91	7.10	7.23	7.30	7.34	—	—	3.02	0	—	265
Monticellite Crestmore, Calif.	7.13	7.22	7.27	7.31	7.36	7.40	7.50	3.01	0	2.4	285
Metagabbro 2 Trinity Complex, Calif.	7.13	7.20	7.25	7.31	7.40	7.46	7.56	3.00	1	—	252
Epidote Amphibolite 2 Litchfield, Conn.	6.20	6.80	7.09	7.32	7.52	7.60	7.67	3.26	0	5.6	270
Dunite Miye Pref., Japan	—	—	7.16	7.32	7.46	7.52	7.65	3.20	0	—	271

Table 18 (continued)
COMPRESSIONAL WAVE VELOCITIES IN ROCKS AS A FUNCTION OF PRESSURE

	P (Kbar)							ϱ	Wet or	Anis.	
	0.1	0.5	1	2	4	6	10	(g/cm^3)	dry	(%)	Ref.
Actinolite Schist Chester, Vt.	—	—	7.20	7.32	7.41	7.47	7.54	3.19	0	14.2	245
Gabbro 334 21-1 (39-41)	—	7.29	7.30	7.32	—	—	—	3.00	1	—	234
Epidote Amphibolite 1 Litchfield, Conn.	6.40	7.00	7.20	7.39	7.56	7.66	7.75	3.13	0	1.2	270
Pyroxenite 457, Kola Kumuzja, USSR	7.26	7.32	7.40	7.40	7.40	—	—	3.24	0	—	265
Dunite TW-7, Wash.	7.28	7.32	7.35	7.41	7.48	7.53	7.59	3.13	0	1.2	288
Granular gabbro 2612, Papua, New Guinea	7.07	7.25	7.33	7.41	7.48	—	7.59	3.03	0	2.6	287
Granular gabbro 2613, Papua, New Guinea	7.21	7.31	7.36	7.41	7.46	—	7.56	2.99	0	3.9	287
Harzburgite 2603, Papua, New Guinea	7.20	7.32	7.46	7.41	7.50	—	7.63	3.22	0	9.6	287
Wollastonite Unknown	5.67	6.85	7.21	7.42	7.56	7.64	7.71	2.87	0	—	285
Eclogite SL001, Hawaii	5.98	6.45	6.94	7.42	7.81	7.90	7.94	3.39	0	—	251
Gabbro 334 24-1 (63-65)	6.87	7.27	7.35	7.42	—	—	—	2.87	1	—	234
Hornblendite IV-19	—	—	7.14	7.42	7.62	7.72	7.81	3.23	0	—	269
Serpentinized dunite Addie, N.C.	7.00	7.22	7.32	7.44	7.57	7.65	7.78	3.19	0	7.1	246
Eclogite Tanzania	—	7.30	7.38	7.46	7.57	7.62	7.71	3.33	0	4.5	245
Harzburgite 2604, Papua, New Guinea	7.28	7.38	7.43	7.49	7.56	—	7.71	3.24	0	6.4	287
Harzburgite 2601, Papua, New Guinea	7.08	7.39	7.44	7.50	7.40	—	7.58	3.17	0	13.0	287
Harzburgite 2607, Papua, New Guinea	7.24	7.39	7.44	7.50	7.58	—	7.73	3.16	0	3.0	287
Peridotite HA-H12 Higashi-Akaishi-Yama	—	7.45	7.48	7.53	7.57	—	7.67	3.28	0	4.6	284
Dunite TW-6, Wash.	7.34	7.43	7.47	7.55	7.66	7.73	7.81	3.17	0	9.9	288
Dunite ON003, Hawaii	6.06	6.72	7.14	7.57	7.90	8.03	8.12	3.32	0	—	251
Plagioclase Peridotite Hokkaido, Japan	—	—	7.35	7.58	7.70	7.78	7.90	2.99	0	—	271

Dunite Webster, N.C.	—	—	7.54	7.59	7.65	7.69	7.78	3.24	0	3.1	245
Bronzitite Bushveld Complex	—	7.40	7.49	7.60	7.75	7.85	8.02	3.29	0	2.8	245
Eclogite ME-1-1(P), Colorado Plateau	—	7.22	7.44	7.62	7.78	—	7.94	3.35	0	—	284
Peridotite 1 Kailua, Hawaii	5.40	6.60	7.21	7.63	8.00	8.09	8.21	3.29	0	4.1	246
Eclogite GR-3(P), Colorado Plateau	—	7.16	7.43	7.63	7.81	—	8.01	3.37	0	5.9	284
Bronzitite Stillwater Complex, Mont.	—	—	7.62	7.65	7.72	7.75	7.83	3.28	0	1.2	245
Eclogite Kimberley, South Africa	—	7.49	7.56	7.65	7.79	7.85	7.92	3.34	0	—	245
Pyroxenite IV-12	—	—	7.57	7.65	7.74	7.82	7.89	3.28	0	—	269
Bronzitite Stillwater, Mont.	—	7.58	7.62	7.67	7.74	7.80	7.90	3.26	0	5.8	291
Eclogite GR-1-008(C), Colorado Plateau	—	7.42	7.55	7.69	7.88	7.99	8.09	3.40	0	8.9	284
Dunite ON002, Hawaii	6.23	6.88	7.29	7.69	8.00	8.07	8.13	3.37	0	—	251
Peridotite Hokkaido, Japan	—	—	7.44	7.70	7.86	7.94	8.02	3.30	0	—	271
Peridotite 455, Kola Moncegorsk, Nittis, USSR	7.48	7.64	7.70	7.70	7.70	—	—	3.28	0	—	265
Eclogite 7 Kimberley, South Africa	7.31	7.53	7.62	7.71	7.79	7.84	7.90	3.42	0	1.5	282
Eclogite Kimberley, South Africa	—	7.65	7.68	7.73	7.79	7.82	7.87	3.38	0	2.4	245
Hematite	—	—	7.72	7.73	7.74	7.76	7.80	5.00	0	—	245
Peridotite HD-8 Horoman Hidaka	—	7.69	7.72	7.75	7.79	—	7.88	3.28	0	11.3	284
Eclogite 3 Tasmania	7.38	7.58	7.66	7.77	7.91	7.97	8.07	3.41	0	—	282
Eclogite GR-3A(P), Colorado Plateau	—	7.34	7.59	7.78	7.94	—	8.07	3.34	0	4.4	284
Pyroxenite Sonoma County, Calif.	—	—	7.73	7.79	7.88	7.93	8.01	3.25	0	5.8	245
Eclogite HA-04, Higashi-Akaishi-Yama	—	7.34	7.55	7.79	7.98	—	8.10	3.43	0	2.0	284
Dunite ON001, Hawaii	5.90	6.84	7.37	7.79	8.08	8.18	8.24	3.30	0	—	251
Peridotite 2 Kailua, Hawaii	5.40	6.44	7.18	7.80	8.13	8.22	8.36	3.29	0	3.1	246
Dunite Mt. Dun, New Zealand	—	7.69	7.75	7.80	7.86	7.92	8.00	3.26	0	9.7	245
Harzburgite Bushveld Complex	—	7.74	7.78	7.81	7.85	7.90	7.95	3.37	0	2.9	245
Eclogite Healdsburg, Calif.	—	—	7.69	7.81	7.89	7.94	8.01	3.44	0	1.9	245
Eclogite ME-1-17(P), Colorado Plateau	—	7.64	7.74	7.82	7.91	7.96	8.01	3.52	0	2.7	284
Eclogite 1 Norway	7.10	7.51	7.69	7.84	8.00	8.07	8.14	3.27	0	2.9	282
Eclogite HA-04, Higashi-Akaishi-Yama	—	7.42	7.63	7.84	8.02	—	8.27	3.48	0	1.6	284
Eclogite 4 Sonoma, Calif.	6.48	7.34	7.63	7.85	8.00	8.07	8.15	3.42	0	2.4	282
Eclogite 6 Tasmania	7.49	7.63	7.74	7.85	7.95	8.00	8.08	3.42	0	—	282
Pyroxenite 1 Canyon Mountain, Ore.	7.72	7.77	7.81	7.86	7.91	7.95	8.03	3.21	1	—	252

Table 18 (continued)
COMPRESSIONAL WAVE VELOCITIES IN ROCKS AS A FUNCTION OF PRESSURE

	P (Kbar)							ϱ	Wet or	Anis.	
	0.1	0.5	1	2	4	6	10	(g/cm³)	dry	(%)	Ref.
Eclogite 2 Valley Ford, Calif.	7.58	7.69	7.77	7.88	7.99	8.03	8.09	3.36	0	2.5	282
Pyroxenite 2 Canyon Mountain, Ore.	7.72	7.79	7.83	7.89	7.96	8.00	8.08	3.27	1	—	252
Eclogite 12 Norway	7.05	7.64	7.78	7.91	8.04	8.12	8.21	3.52	0	—	282
Garnet	—	—	7.81	7.91	7.99	8.01	8.07	3.95	0	—	245
Eclogite ME-1-11(P), Colorado Plateau	—	7.75	7.84	7.91	7.98	8.03	8.07	3.37	0	—	284
Orthoenstatite 69-8 (C-AXIS)	7.85	7.87	7.89	7.91	7.95	7.99	8.03	3.27	0	—	229
Dunite TW-10, Wash.	7.71	7.80	7.85	7.92	8.00	8.05	8.11	3.24	0	6.8	288
Eclogite ME-1-12(P), Colorado Plateau	—	7.84	7.89	7.93	7.98	—	8.05	3.23	0	—	284
Eclogite 9 Russian River, Calif.	7.32	7.69	7.81	7.94	8.06	8.12	8.22	3.44	0	0.7	282
Eclogite 5 Healdsburg, Calif.	7.68	7.79	7.86	7.95	8.03	8.09	8.17	3.42	0	1.6	282
Websterite 2605, Papua, New Guinea	7.82	7.88	7.92	7.96	8.01	—	8.12	3.27	0	3.4	287
Pyroxenite 469, Kola Moncegorsk, Sopca, USSR	7.70	7.75	7.87	7.97	8.00	—	—	3.29	0	—	265
Harzburgite Pyroxenite 2602, Papua, New Guinea	7.75	7.87	7.93	7.98	8.04	—	8.17	3.25	0	2.5	287
Eclogite 10 Norway	7.10	7.63	7.83	7.99	8.14	8.24	8.31	3.46	0	—	282
Eclogite GR-33(P), Colorado Plateau	—	7.71	7.86	8.00	8.10	—	8.18	3.36	0	2.0	284
Peridotite 462, Kola Moncegorsk, Nittis, USSR	7.08	7.65	7.83	8.00	8.03	—	—	3.21	0	—	265
Dunite Balsam Gap, N.C.	—	7.82	7.89	8.01	8.13	8.19	8.28	3.27	0	8.9	245
Harzburgite Pyroxenite 2608, Papua, New Guinea	7.77	7.89	7.96	8.01	8.10	—	8.25	3.23	0	4.7	287
Eclogite GR-34(P), Colorado Plateau	—	6.79	7.96	8.03	8.10	—	8.15	3.37	0	3.7	284
Dunite Addie, N.C.	—	—	7.99	8.05	8.14	8.20	8.28	3.30	0	8.4	245
Dunite TW-1, Wash.	7.87	7.95	7.99	8.05	8.13	8.18	8.24	3.24	0	10.6	288
Eclogite 13 Norway	7.68	7.91	7.98	8.06	8.13	8.17	8.23	3.54	0	2.5	282
Pyroxenite 2606, Papua, New Guinea	7.85	7.97	8.02	8.07	8.13	—	8.24	3.34	0	0.2	287

Eclogite HA-04, Higashi-Akaishi-Yama	—	7.94	8.00	8.09	8.16	—	8.17	3.48	0	2.7	284
Eclogite Ehime Pref., Japan	—	—	7.55	8.10	8.31	8.40	8.50	3.51	0	2.9	271
Dunite TW-3, Wash.	7.97	8.03	8.05	8.10	8.15	8.18	8.21	3.24	0	8.5	288
Magnesite Chewelah, Wash.	—	—	—	8.12	8.24	8.31	8.41	2.97	0	0.7	286
Eclogite Ehime Pref., Japan	—	—	7.86	8.12	8.24	8.34	8.44	3.49	0	1.5	271
Dunite A Twin Sisters, Wash.	—	—	8.08	8.12	8.18	8.21	8.25	3.26	0	10.1	292
Eclogite 11 Kimberley, South Africa	7.85	7.99	8.06	8.13	8.21	8.25	8.29	3.50	0	2.9	282
Clinopyroxenite 2611, Papua, New Guinea	7.87	8.04	8.09	8.13	8.17	—	8.29	3.28	0	1.8	287
Dunite Hokkaido, Japan	—	—	7.94	8.16	8.36	8.45	8.52	3.30	0	—	271
Eclogite 15 Norway	7.66	7.92	8.05	8.16	8.28	8.35	8.43	3.58	0	1.3	282
Dunite TW-14, Wash.	7.91	8.04	8.09	8.18	8.29	8.35	8.40	3.29	0	8.0	288
Dunite TW-5, Wash.	8.03	8.10	8.14	8.18	8.25	8.29	8.34	3.29	0	5.6	288
Dunite TW-2, Wash.	7.68	8.02	8.12	8.19	8.28	8.33	8.39	3.28	0	5.1	288
Dunite Twin Sisters, Wash.	—	8.11	8.15	8.20	8.26	8.30	8.38	3.29	0	15.0	291
Peridotite HD-8 Horoman Hidaka	—	8.14	8.17	8.20	8.24	—	8.33	3.31	0	3.0	284
Eclogite 14 Norway	7.82	8.05	8.14	8.22	8.30	8.35	8.42	3.57	0	3.5	282
Jadeite Japan	—	—	8.21	8.22	8.23	8.24	8.28	3.18	0	1.3	245
Eclogite MR-61-A(C), Colorado Plateau	—	8.02	8.11	8.23	8.31	8.42	8.46	3.28	0	2.6	284
Eclogite H3, Bernartice, Czech Massif, CSSR	8.00	8.15	8.18	8.23	8.30	—	—	3.52	0	1.8	265
Dunite TW-12, Wash.	8.00	8.14	8.20	8.25	8.33	8.38	8.43	3.28	0	11.0	288
Dunite Twin Sisters, Wash.	—	8.11	8.19	8.27	8.32	8.35	8.42	3.31	0	10.9	245
Eclogite MR-B-10(C), Colorado Plateau	—	—	8.18	8.27	8.37	8.43	8.47	3.31	0	1.5	284
Dunite TW-9, Wash.	8.06	8.25	8.27	8.29	8.34	8.38	8.42	3.29	0	6.4	288
Eclogite 8 Norway	7.57	7.97	8.14	8.31	8.45	8.53	8.61	3.44	0	3.1	282
Dunite B Twin Sisters, Wash.	—	—	8.25	8.31	8.37	8.41	8.47	3.32	0	15.7	292
Eclogite H4X, Bernartice, Czech Massif, CSSR	8.01	8.09	8.20	8.33	8.40	—	—	3.39	0	—	265
Dunite TW-11, Wash.	8.10	8.22	8.28	8.35	8.43	8.48	8.52	3.31	0	0.5	288
Dunite TW-4, Wash.	8.25	8.30	8.33	8.39	8.43	8.46	8.51	3.30	0	6.8	288
Dunite TW-8, Wash.	8.29	8.34	8.37	8.41	8.47	8.51	8.56	3.30	0	8.1	288
Eclogite HA-03, Higashi-Akaishi-Yama	—	8.08	8.29	8.45	8.57	—	8.62	3.71	0	10.5	284

Table 18 (continued)
COMPRESSIONAL WAVE VELOCITIES IN ROCKS AS A FUNCTION OF PRESSURE

	P (Kbar)								Wet		
	0.1	0.5	1	2	4	6	10	ϱ (g/cm^3)	or dry	Anis. (%)	Ref.
Dunite Twin Sisters, Wash.	8.40	8.43	8.49	8.52	8.55	8.59	8.66	3.33	0	8.3	246
Grossularite Conn.	—	—	8.41	8.55	8.72	8.83	8.99	3.56	0	3.1	245
Polycrystalline Forsterite 69-9	7.04	8.14	8.44	8.64	8.71	—	8.77	3.13	0	—	229
Jadeite Burma	—	—	8.67	8.69	8.72	8.75	8.78	3.33	0	0.1	245
Dunite Twin Sisters, Wash.	8.69	8.74	8.87	8.93	8.96	8.99	—	3.16	0	—	274
Sillimanite Australia	9.43	9.51	9.55	9.60	9.65	9.68	9.73	3.19	0	4.7	285

Table 19
COMPRESSIONAL WAVE VELOCITIES IN ROCKS TO PRESSURES OF 30 KILOBARS[295]

	Velocity (km/sec)				
Pressure (bars)	Pyroxenite, Stillwater, Mont. (ϱ = 3.311 g/cm^3)	Pyroxenite, Twin Sisters, Wash. (ϱ = 3.286 g/cm^3)	Dunite, Twin Sisters, Wash. (ϱ = 3.309 g/cm^3)	Eclogite, Sunnmore, Norway (ϱ = 3.504 g/cm^3)	Eclogite, Nove Dvory, Czechoslovakia (ϱ = 3.559 g/cm^3)
10	7.651	7.623	7.842	7.501	8.248
2,000	7.895	7.816	8.275	7.972	8.324
4,000	7.967	7.894	8.372	8.112	8.375
6,000	8.010	7.930	8.434	8.173	8.402
8,000	8.052	7.962	8.470	8.225	8.430
10,000	8.081	8.000	8.498	8.270	8.453
12,000	8.117	8.029	8.527	8.292	8.475
14,000	8.151	8.061	8.548	8.320	8.502
16,000	8.180	8.090	8.576	8.342	8.525
18,000	8.212	8.118	8.605	8.365	8.545
20,000	8.248	8.150	8.632	8.390	8.572
22,000	8.280	8.181	8.655	8.418	8.591
24,000	8.311	8.209	8.684	8.435	8.617
26,000	8.341	8.239	8.708	8.462	8.642
28,000	8.376	8.272	8.732	8.485	8.663
30,000	8.408	8.301	8.761	8.508	8.690

Table 20
COMPRESSIONAL AND SHEAR WAVE VELOCITIES IN SANDSTONES AS A FUNCTION OF EXTERNAL PRESSURE (P ext) AND PORE PRESSURE (P pore)[296]

Sandstone 1

V_p(km/sec) P ext (bars)	0	68	136	204	272	340	408	544
P pore (bars)								
0 (wet)	2.82	3.14	3.32	3.42	3.48	3.52	3.57	3.61
68 (wet)	—	2.88	—	3.34	3.42	3.48	—	—
136 (wet)	—	—	2.90	—	—	—	3.48	—
204 (wet)	—	—	—	—	—	3.34	3.42	—
272 (wet)	—	—	—	—	2.90	3.22	3.22	3.34
340 (wet)	—	—	—	—	—	—	—	3.42
408 (wet)	—	—	—	—	—	—	—	3.34
476 (wet)	—	—	—	—	—	—	—	3.22
544 (wet)	—	—	—	—	—	—	—	2.90

Sandstone 2

V_p (km/sec) P ext (bars)	34	68	136	204	272	340	408	476
P pore (bars)								
0 (dry)	3.08	3.38	3.73	3.87	3.96	4.02	4.05	—
0 (wet)	3.63	3.75	3.95	4.04	4.08	4.11	4.13	—
68 (dry)	—	—	3.35	—	—	—	4.02	—
68 (wet)	—	—	3.73	—	—	—	4.11	—
136 (dry)	—	—	—	3.32	—	—	—	4.01
136 (wet)	—	—	—	3.72	—	—	—	4.11
204 (dry)	—	—	—	—	3.29	—	—	—
204 (wet)	—	—	—	—	3.72	—	—	—
272 (dry)	—	—	—	—	—	3.26	—	—
272 (wet)	—	—	—	—	—	3.70	—	—
340 (dry)	—	—	—	—	—	—	3.23	—
340 (wet)	—	—	—	—	—	—	3.70	—

V_s (km/sec) P ext (bars)	34	68	136	204	272	340	408	476
P pore (bars)								
0 (dry)	1.95	2.10	2.30	2.42	2.50	2.53	2.56	—
0 (wet)	1.89	2.03	2.19	2.29	2.33	2.36	2.39	—
68 (dry)	—	—	2.09	—	—	—	2.53	—
68 (wet)	—	—	2.01	—	—	—	2.38	—
136 (dry)	—	—	—	2.07	—	—	—	2.51
136 (wet)	—	—	—	2.01	—	—	—	2.38
204 (dry)	—	—	—	—	2.06	—	—	—
204 (wet)	—	—	—	—	2.00	—	—	—
272 (dry)	—	—	—	—	—	2.04	—	—
272 (wet)	—	—	—	—	—	2.00	—	—
340 (dry)	—	—	—	—	—	—	2.03	—
340 (wet)	—	—	—	—	—	—	1.98	—

Table 21
COMPRESSIONAL AND SHEAR WAVE VELOCITIES AS A FUNCTION OF LITHOSTATIC PRESSURE, PORE PRESSURE AND TEMPERATURE FOR SEDIMENTARY ROCKS[297]

Sample	Porosity (%)	P lithostatic (bars)	P pore (bars)	T (°C)	V_p (km/sec)	V_s (km/sec)	$\partial V_p/\partial T$ (km/sec - °C)	$\partial V_s/\partial T$ (km/sec - °C)
Sandstone-1	17.1	1380	600	16	4.40	2.60	$-5.5 \cdot 10^{-4}$	$-7.8 \cdot 10^{-6}$
	17.1	345	150	21	4.25	2.51	$-6.8 \cdot 10^{-4}$	$-3.0 \cdot 10^{-4}$
	17.1	138	60	17	4.13	2.34	$-4.7 \cdot 10^{-4}$	$-2.9 \cdot 10^{-4}$
Sandstone-2	30.8	690	345	18	3.35	1.81	$-8.6 \cdot 10^{-4}$	$-4.4 \cdot 10^{-4}$
Carbonate-1	2.9	915	398	26	5.96	3.30	$-8.3 \cdot 10^{-4}$	$-1.7 \cdot 10^{-4}$
Carbonate-2	3.6	916	398	22	5.84	3.41	$-9.8 \cdot 10^{-4}$	$-1.2 \cdot 10^{-4}$
Carbonate-3	1.3	917	399	24	6.26	3.24	$-1.1 \cdot 10^{-3}$	$-2.3 \cdot 10^{-4}$
Carbonate-4	6.4	920	400	26	5.94	3.13	$-7.7 \cdot 10^{-4}$	$-5.6 \cdot 10^{-4}$
Carbonate-5	9.9	921	400	17	5.61	2.93	$-1.1 \cdot 10^{-3}$	$-1.6 \cdot 10^{-4}$
Carbonate-6	8.7	953	414	24	5.59	2.96	$-1.1 \cdot 10^{-3}$	$-8.8 \cdot 10^{-4}$
Carbonate-7	8.5	953	414	24	5.49	2.92	$-1.4 \cdot 10^{-3}$	$-5.4 \cdot 10^{-4}$

Note: The change in velocity with temperature is linear for $T \leqslant 180°C$.

Table 22
COMPRESSIONAL AND SHEAR WAVE VELOCITIES AS A FUNCTION OF TEMPERATURE FOR ROCKS[299]

T(°C)	V_p (km/sec)	V_s (km/sec)	T(°C)	V_p (km/sec)	V_s (km/sec)
		Quartzite (P = 2.1 kbar)			
20	6.23	4.00	560	5.70	—
125	6.20	4.00	600	5.58	4.00
220	6.17	4.00	625	5.21	3.96
290	6.08	4.00	650	6.14	3.98
375	6.01	4.00	675	6.51	3.98
420	5.95	—	700	6.70	3.96
470	5.92	4.00	725	6.79	3.96
520	5.83	4.00			
		Granite (P = 4.2 kbar)			
60	6.11	3.50	625	5.73	3.41
110	6.10	3.50	655	5.64	3.38
185	6.08	3.50	675	5.53	3.38
275	6.04	3.48	685	5.80	3.39
390	5.98	3.46	720	6.11	3.40
500	5.90	3.44	740	6.19	3.40
		Gabbro (P = 4.1 kbar)			
80	—	3.90	360	6.70	3.86
100	6.80	3.90	425	6.68	3.85
140	6.78	3.90	475	6.66	3.83
200	6.77	3.89	555	6.58	3.81
260	6.75	3.88	655	6.50	3.78
315	6.73	3.87	—	—	—
		Eclogite (P = 4.1 kbar)			
110	7.90	4.60	500	7.73	4.50
200	7.88	4.59	630	7.63	4.44
290	7.84	4.56	655	7.62	4.43
335	7.82	4.55	700	7.58	4.42
400	7.78	4.53			
		Peridotite (P = 4.1 kbar)			
50	7.82	4.52	460	—	4.37
100	7.80	4.50	530	7.61	4.34
200	7.76	4.46	550	7.60	4.33
290	7.72	4.44	610	7.57	4.32
350	7.69	4.41	650	7.53	4.30
400	—	4.39	700	7.52	4.28
435	7.65	4.37			

Table 23
COMPRESSIONAL AND SHEAR WAVE VELOCITIES AS A FUNCTION OF TEMPERATURE AND PRESSURE FOR ROCKS[300-302]

	25°C		100°C		200°C		300°C	
P (kbar)	V_p (km/sec)	V_s (km/sec)	V_p (km/sec)	V_s (km/sec)	V_p (km/sec)	V_s (km/sec)	V_p (km/sec)	V_s (km/sec)
			Woodbury Biotite Granite (ϱ = 2.634 g/cm³)					
0.2	5.77	3.31	—	—	—	—	—	—
0.5	6.05	3.46	6.04	3.41	6.00	—	5.77	—
1.0	6.16	3.56	6.13	3.54	6.06	3.45	5.87	—
1.5	6.20	3.61	6.18	3.58	6.13	3.55	5.92	—
2.0	6.22	3.63	6.21	3.61	6.15	3.60	5.95	—
3.0	6.26	3.66	6.22	3.64	6.18	3.62	6.01	—
4.0	6.29	3.67	6.26	3.66	6.20	3.63	6.04	—
5.0	6.31	3.68	6.29	3.68	6.22	3.65	6.08	—
			Texas Pink Granite (ϱ = 2.636 g/cm³)					
0.2	6.14	3.27	6.01	3.16	—	—	—	—
0.5	6.29	3.35	6.23	3.28	5.86	3.13	5.57	—
1.0	6.34	3.35	6.32	3.32	6.17	3.26	5.94	—
1.5	6.41	3.38	6.35	3.33	6.26	3.31	6.07	—
2.0	6.43	3.39	6.38	3.34	6.30	3.32	6.12	—
3.0	6.47	3.38	6.42	3.35	6.35	3.34	6.18	—
4.0	6.50	3.36	6.47	3.36	6.38	3.35	6.25	—
5.0	6.52	3.37	6.49	3.36	6.42	3.36	6.33	—
6.0	6.54	3.39	6.52	3.36	6.46	3.36	6.37	—
7.0	6.56	3.39	6.53	3.37	6.49	3.36	6.43	—
			Texas Gray Granite (ϱ = 2.609 g/cm³)					
0.2	5.78	3.42	—	—	—	—	—	—
0.5	5.96	3.55	5.94	3.54	5.80	3.41	5.27	—
1.0	6.10	3.58	6.02	3.57	5.89	3.55	5.53	—
1.5	6.15	3.59	6.08	3.58	5.96	3.55	5.67	—
2.0	6.19	3.59	6.11	3.59	6.00	3.57	5.75	—
3.0	6.23	3.60	6.15	3.59	6.06	3.58	5.90	—
4.0	6.25	3.61	6.18	3.60	6.14	3.59	6.00	—
5.0	6.28	3.61	6.22	3.60	6.14	3.59	6.08	—
6.0	6.30	3.61	6.24	3.60	6.18	3.59	6.10	—
7.0	6.32	3.62	6.27	3.61	—	—	6.15	—
8.0	6.34	3.62	—	—	—	—	6.18	—
			Dunite (ϱ = 3.160 g/cm³)					
0.1	8.69	4.24	8.43	4.29	8.16	4.13	7.70	4.02
0.3	8.75	4.41	8.64	4.39	8.45	4.29	8.03	—
0.7	8.82	4.44	8.75	4.43	8.69	4.44	8.40	4.32
1.0	8.87	4.46	8.79	4.44	8.71	4.51	8.57	4.37
1.7	8.91	4.52	8.82	4.47	8.78	4.47	8.65	4.44
2.4	8.93	4.55	8.88	4.48	8.79	4.44	8.70	4.40
3.1	8.94	4.55	8.86	4.50	8.83	4.47	8.74	4.42
4.1	8.96	4.53	8.89	4.53	8.89	4.49	8.78	4.43
5.2	8.98	4.53	8.99	4.56	8.90	4.51	8.79	4.46

Table 23 (continued)
COMPRESSIONAL AND SHEAR WAVE VELOCITIES AS A FUNCTION OF TEMPERATURE AND PRESSURE FOR ROCKS[300-302]

	25°C		100°C		200°C		300°C	
P (kbar)	V_p (km/sec)	V_s (km/sec)	V_p (km/sec)	V_s (km/sec)	V_p (km/sec)	V_s (km/sec)	V_p (km/sec)	V_s (km/sec)
				Barriefield Granite				
0.1	5.88	2.96	6.01	3.05	5.20	2.87	—	—
0.25	6.08	3.05	6.13	3.11	5.52	2.98	4.49	—
0.5	6.22	3.11	6.24	3.16	5.96	3.08	4.84	—
0.75	6.29	3.16	6.28	3.20	6.14	3.16	5.10	—
1.0	6.34	3.16	6.33	3.20	6.24	3.17	5.22	—
1.5	6.37	3.17	6.35	3.20	6.30	3.18	5.40	—
2.0	6.38	3.21	6.37	3.20	6.32	3.19	5.53	—
2.5	6.40	3.23	6.39	3.21	6.34	3.18	5.62	—
3.0	6.41	3.22	6.41	3.23	6.35	3.19	5.68	—
4.0	6.43	3.23	6.43	3.24	6.39	3.20	5.80	—
5.0	5.45	3.23	6.45	3.24	6.42	3.22	5.87	—
				Solenhofen Limestone (ϱ = 2.656 g/cm³)				
0.1	6.00	2.95	5.89	2.95	5.76	2.89	5.55	—
0.25	6.01	2.98	5.91	2.96	5.76	2.91	5.63	—
0.5	6.03	2.99	5.93	2.96	5.78	2.93	5.69	—
0.75	6.04	3.00	5.93	2.97	5.80	2.93	5.72	—
1.0	6.05	3.01	5.95	2.97	5.81	2.94	5.74	—
1.5	6.06	3.01	5.97	2.97	5.84	2.94	5.77	—
2.0	6.08	3.02	5.99	2.99	5.85	2.94	5.80	—
2.5	6.10	3.02	6.01	2.99	5.88	2.95	5.82	—
3.0	6.11	3.03	6.02	3.00	5.89	2.95	5.84	—
4.0	6.12	3.05	6.02	3.00	5.91	2.96	5.87	—
5.0	6.13	3.04	6.04	3.00	5.93	2.97	5.89	—
				San Marcos Gabbro (ϱ = 2.993 g/cm³)				
0.2	6.69	3.47	6.63	3.43	—	—	—	—
0.5	6.79	3.48	6.78	3.50	6.75	3.48	6.53	3.35
1.0	6.88	3.50	6.87	3.51	6.87	3.50	6.72	3.46
1.5	6.93	3.50	6.92	3.52	6.92	3.51	6.80	3.48
2.0	6.95	3.51	6.94	3.53	6.94	3.52	6.86	3.50
3.0	6.98	3.51	6.97	3.53	6.97	3.53	6.93	3.51
4.0	7.01	3.52	6.98	3.53	6.99	3.53	6.95	3.52
5.0	7.03	3.53	7.00	3.53	7.00	3.53	6.97	3.52
6.0	7.05	3.54	7.02	3.54	7.01	3.54	7.00	3.52
				Bytownite Gabbro (ϱ = 2.885 g/cm³)				
0.2	6.45	3.42	6.42	—	—	—	—	—
0.5	6.61	3.45	6.56	3.40	6.37	3.35	5.96	3.25
1.0	6.69	3.47	6.60	3.42	6.47	3.39	6.13	3.32
1.5	6.72	3.51	6.64	3.43	6.53	3.41	6.27	3.36
2.0	6.76	3.52	6.68	3.45	6.57	3.44	6.35	3.39
3.0	6.78	3.52	6.72	3.46	6.62	3.45	6.45	3.42
4.0	6.81	3.53	6.73	3.48	6.64	3.46	6.52	3.44
5.0	6.83	3.53	6.76	3.49	6.67	3.47	6.57	3.45
6.0	6.84	3.54	6.79	3.50	—	—	—	—

Table 23 (continued)
COMPRESSIONAL AND SHEAR WAVE VELOCITIES AS A FUNCTION OF TEMPERATURE AND PRESSURE FOR ROCKS[300-302]

	25°C		100°C		200°C		300°C	
P (kbar)	V_p (km/sec)	V_s (km/sec)	V_p (km/sec)	V_s (km/sec)	V_p (km/sec)	V_s (km/sec)	V_p (km/sec)	V_s (km/sec)
			Hornblende Gabbro (ϱ = 2.933 g/cm^3)					
0.2	6.60	3.56	6.56	3.57	—	—	—	—
0.5	6.67	3.59	6.63	3.62	6.49	3.55	6.14	3.44
1.0	6.74	3.65	6.69	3.64	6.61	3.60	6.44	3.51
1.5	6.78	3.66	6.74	3.67	6.69	3.63	6.56	3.55
2.0	6.80	3.69	6.77	3.69	6.73	3.65	6.64	3.59
3.0	6.84	3.71	6.81	3.70	6.77	3.67	6.70	3.64
4.0	6.86	3.71	6.85	3.71	6.82	3.71	6.76	3.68
5.0	6.88	3.71	6.87	3.72	6.83	3.73	6.78	3.71
6.0	6.89	3.72	6.88	3.71	6.84	3.74	6.81	3.71
			Analcime (ϱ = 2.712 g/cm^3)					
0.2	5.41	3.05	5.35	3.01	—	—	—	—
0.5	5.48	3.08	5.42	3.05	5.33	3.03	5.10	2.86
1.0	5.55	3.09	5.50	3.06	5.42	3.06	5.25	2.92
1.5	5.62	3.10	5.56	3.07	5.49	3.07	5.37	2.97
2.0	5.67	3.10	5.59	3.08	5.55	3.08	5.44	2.99
3.0	5.73	3.11	5.66	3.10	5.65	3.08	5.55	3.02
4.0	5.76	3.11	5.69	3.11	5.68	3.10	5.62	3.05
5.0	5.78	3.11	5.72	3.11	5.71	3.11	5.65	3.07
			Basalt (ϱ = 2.586 g/cm^3)					
0.2	5.41	3.21	—	—	—	—	—	—
0.5	5.57	3.23	5.59	3.20	5.50	3.18	5.40	3.16
1.0	5.66	3.25	5.67	3.21	5.59	3.20	5.44	3.17
1.5	5.73	3.26	5.71	3.22	5.65	3.21	5.50	3.18
2.0	5.75	3.26	5.75	3.23	5.71	3.22	5.58	3.20
3.0	5.79	3.26	5.78	3.24	5.77	3.23	5.69	3.22
4.0	5.80	3.27	5.79	3.24	5.80	3.24	5.79	3.23
5.0	5.81	3.27	5.80	3.23	5.81	3.24	5.82	3.24
6.0	5.82	3.27	5.81	3.23	5.82	3.24	5.83	3.24
			Dunite (ϱ = 3.198 g/cm^3)					
0.2	7.40	3.79	7.15	3.69	—	—	—	—
0.5	7.54	3.88	7.38	3.83	6.78	3.50	—	—
1.0	7.63	3.99	7.54	3.88	6.97	3.54	—	—
1.5	7.70	4.09	7.63	3.97	7.06	3.56	—	—
2.0	7.77	4.10	7.69	4.05	7.13	3.63	—	—
3.0	7.82	4.13	7.77	4.07	7.28	3.72	—	—
4.0	7.86	4.15	7.83	4.11	7.40	3.79	—	—
5.0	7.91	4.17	7.86	4.12	7.48	3.83	—	—
			Dry Sandstone (ϱ = 2.543 g/cm^3, ϕ = 5.1%)					
	0.1	4.04	2.51	3.82	—	3.30	—	
	0.25	4.37	2.59	4.09	2.47	3.80	—	
	0.5	4.58	2.70	4.38	—	4.14	2.60	
	0.75	4.76	2.81	4.59	2.69	4.41	2.64	

Table 23 (continued)
COMPRESSIONAL AND SHEAR WAVE VELOCITIES AS A FUNCTION OF TEMPERATURE AND PRESSURE FOR ROCKS[300-302]

	25°C		100°C		200°C		300°C	
P (kbar)	V_p (km/sec)	V_s (km/sec)	V_p (km/sec)	V_s (km/sec)	V_p (km/sec)	V_s (km/sec)	V_p (km/sec)	V_s (km/sec)
1.0	4.87	2.85	4.74	2.75	4.58	2.68		
1.5	5.02	2.88	4.93	2.83	4.85	2.80		
2.0	5.11	2.93	5.06	2.88	4.99	2.85		
2.5	5.17	2.94	5.15	2.92	5.10	2.88		
3.0	5.13	2.96	5.21	2.97	5.16	2.92		
4.0	5.28	2.98	5.30	2.97	5.27	2.94		
5.0	5.36	2.97	5.35	2.97	5.31	2.98		
Wet Sandstone (ϱ = 2.606 g/cm³, ϕ = 5.1%)								
0.1	4.46	2.69	4.20	—	3.79	—		
0.25	4.47	2.76	4.22	—	3.80	—		
0.5	4.51	2.90	4.23	—	3.87	—		
0.75	4.54	3.01	4.27	—	3.93	—		
1.0	4.55	3.10	4.31	—	3.96	—		
1.5	4.61	3.26	4.32	—	4.07	—		
2.0	4.65	3.38	4.37	—	4.18	—		
2.5	4.69	3.47	4.41	—	4.30	—		
3.0	4.71	3.60	4.44	—	4.40	—		
4.0	4.78	3.65	4.54	—	4.60	—		
5.0	4.89	3.95	4.66	—	4.80	—		
Marble								
0.14	6.06	3.07	4.94	2.67	4.65	2.55		
0.35	6.25	3.10	5.82	2.95	5.33	2.75		
0.69	6.50	3.15	6.29	3.06	5.94	2.97		
1.0	6.55	3.17	6.41	3.10	6.24	3.06		
1.7	6.62	3.20	6.54	3.13	6.40	3.09		
2.4	6.65	3.12	6.58	3.14	6.44	3.10		
3.1	6.67	3.21	6.59	3.15	6.46	3.11		
4.1	6.67	3.21	6.60	3.15	6.47	3.11		
5.2	6.66	3.25	6.62	3.13	6.47	3.10		
Argillaceous Limestone (ϱ = 2.739)								
0.1	5.74	3.06	5.65	—	5.47	2.94		
0.25	5.81	3.10	5.71	—	5.48	2.96		
0.5	5.90	3.13	5.78	—	5.67	3.01		
0.75	5.95	3.12	5.84	—	5.73	3.03		
1.0	5.98	3.13	5.89	—	5.77	3.06		
1.5	6.05	3.16	5.93	3.20	5.84	3.10		
2.0	6.06	3.19	5.79	3.34	5.88	3.12		
2.5	6.09	3.21	6.00	3.39	5.92	3.12		
3.0	6.11	3.22	6.03	3.14	5.95	3.13		
4.0	6.14	3.22	6.03	3.18	5.99	3.18		
5.0	6.17	3.24	6.10	3.18	6.04	3.18		

Table 23 (continued)
COMPRESSIONAL AND SHEAR WAVE VELOCITIES AS A FUNCTION OF TEMPERATURE AND PRESSURE FOR ROCKS[300-302]

	25°C		100°C		200°C		300°C	
P (kbar)	V_p (km/sec)	V_s (km/sec)	V_p (km/sec)	V_s (km/sec)	V_p (km/sec)	V_s (km/sec)	V_p (km/sec)	V_s (km/sec)
	Argillaceous Limestone (ϱ = 2.731 g/cm^3)							
0.1	6.05	3.13	6.02	3.18	6.12	3.21		
0.25	6.09	3.18	6.06	3.19	6.16	3.23		
0.5	6.13	3.21	6.10	3.21	6.19	3.25		
0.75	6.16	3.23	6.13	3.22	6.22	3.25		
1.0	6.17	3.24	6.16	3.23	6.24	3.26		
1.5	6.24	3.26	6.18	3.25	6.25	3.28		
2.0	6.28	3.28	6.21	3.26	6.27	3.28		
2.5	6.30	3.29	6.22	3.26	6.29	3.28		
3.0	6.32	3.29	6.23	3.27	6.33	3.30		
4.0	6.34	3.30	6.26	3.27	6.35	3.29		
5.0	6.37	3.29	6.29	3.28	6.37	3.30		

Table 24
COMPRESSIONAL WAVE VELOCITIES AS A FUNCTION OF TEMPERATURE FOR ROCKS FROM THE PAPUAN OPHIOLITE BELT (P = 3 kbar)[287]

T (°C)	Gabbro V_p (km/sec)	Orthopyroxenite V_p (km/sec)	Pyroxenite V_p (km/sec)
25	7.51	7.98	8.00
50	7.50	7.95	7.98
75	7.48	7.92	7.95
100	7.46	7.87	7.93
125	7.45	7.84	7.91
150	7.42	7.81	7.88
175	7.35	7.76	7.85
200	7.31	7.74	7.81
225	7.27	7.70	7.75
250	7.23	7.66	7.71
275	7.20	7.62	7.69
300	7.16	7.56	7.64
325	7.12	—	7.59
350	7.08	—	—
Plagioclase	46%	—	—
Orthopyroxene	3%	62%	14%
Clinopyroxene	50%	—	77%
Olivine	—	35%	8%
Density (g/cm^3)	3.03	3.30	3.28

Table 25
PRESSURE AND TEMPERATURE DERIVATIVES OF COMPRESSIONAL AND SHEAR WAVE VELOCITIES FOR POLYCRYSTALLINE AGGREGATES RELEVANT TO GEOPHYSICS

Composition	Structure	$\partial V_p/\partial P$ (km/sec-kb)	$\partial V_s/\partial P$ (km/sec-kb)	$\partial V_p/\partial T$ (km/sec-°C)	$\partial V_s/\partial T$ (km/sec-°C)	Ref.
MgO	Rocksalt	$7.71 \cdot 10^{-3}$	$4.35 \cdot 10^{-3}$	$-4.3 \cdot 10^{-4}$	$-3.6 \cdot 10^{-4}$	344
		$8.66 \cdot 10^{-3}$	$4.23 \cdot 10^{-3}$			307
		$7.71 \cdot 10^{-3}$	$4.35 \cdot 10^{-3}$	$-5.0 \cdot 10^{-4}$	$-4.8 \cdot 10^{-4}$	306
		—	—	$-8.25 \cdot 10^{-4}$	$-6.75 \cdot 10^{-4}$	308
		$7.80 \cdot 10^{-3}$	$3.75 \cdot 10^{-3}$			309
CaO	Rocksalt	$10.43 \cdot 10^{-3}$	$2.90 \cdot 10^{-3}$	$-5.15 \cdot 10^{-4}$	$-3.68 \cdot 10^{-4}$	311
$NiFe_2O_4$	Spinel	$4.41 \cdot 10^{-3}$	$-0.03 \cdot 10^{-3}$			320
TiO_2	Rutile	$21.0 \cdot 10^{-3}$	$-0.6 \cdot 10^{-3}$			328
		$7.6 \cdot 10^{-3}$	$0.9 \cdot 10^{-3}$			328
SiO_2	α	$14.2 \cdot 10^{-3}$	$-3.3 \cdot 10^{-3}$			328
Al_2O_3	α-Al_2O_3	—	—	$-6.18 \cdot 10^{-4}$	$-4.64 \cdot 10^{-4}$	308
		$5.18 \cdot 10^{-3}$	$2.21 \cdot 10^{-3}$			331
		$5.35 \cdot 10^{-3}$	$2.20 \cdot 10^{-3}$	$-3.7 \cdot 10^{-4}$	$-2.9 \cdot 10^{-4}$	332
Fe_2O_3		$4.67 \cdot 10^{-3}$	$0.63 \cdot 10^{-3}$			334
Mg_2SiO_4	Olivine	$10.3 \cdot 10^{-3}$	$2.45 \cdot 10^{-3}$			340, 345
		$10.3 \cdot 10^{-3}$	$3.8 \cdot 10^{-3}$			341
$(Mg_{95}, Fe_5)_2SiO_4$	Olivine	$10.3 \cdot 10^{-3}$	$3.7 \cdot 10^{-3}$			341
$(Mg_{90}, Fe_{10})_2SiO_4$	Olivine	$10.3 \cdot 10^{-3}$	$3.7 \cdot 10^{-3}$			341
$(Mg_{85}, Fe_{15})_2SiO_4$	Olivine	$10.2 \cdot 10^{-3}$	$3.6 \cdot 10^{-3}$			341
$(Mg_{80}, Fe_{20})_2SiO_4$	Olivine	$10.1 \cdot 10^{-3}$	$3.3 \cdot 10^{-3}$			341
$(Mg_{50}, Fe_{50})_2SiO_4$	Olivine	$9.5 \cdot 10^{-3}$	$2.4 \cdot 10^{-3}$			341
Fe_2SiO_4	Olivine	$8.8 \cdot 10^{-3}$	$0.6 \cdot 10^{-3}$			341

Table 26
VELOCITIES IN ROCK FORMING MINERALS

Mineral	Propagation direction	Displacement direction	Velocity (km/sec)	Ref.
Natrolite	[001]	[001]	7.80	
ϱ = 2.25 g/cm³		[100]	3.27	
		[010]	2.96	
V_p = 6.11 km/sec	[010]	[010]	5.35	
V_s = 3.53 km/sec		[001]	2.94	
		[100]	4.10	
	[100]	[100]	5.64	
		[010]	4.34	
		[001]	3.26	
	[110]	[110]	6.46	
		[110]	2.97	
		[001]	3.12	
	[101]	[101]	6.45	
		[101]	3.93	
		[010]	3.73	
	[011]	[011]	6.56	
		[011]	3.49	
		[100]	3.84	
Perthite	[001]	[001]	6.40	
$Or_{75}Ab_{22}An_0$		[100]	2.74	
ϱ = 2.54 g/cm³		[010]	2.33	
	[010]	[010]	7.64	
V_p = 5.56 km/sec		[001]	2.34	
V_s = 3.06 km/sec		[100]	3.52	
	[100]	[100]	4.90	
		[010]	3.56	
		[001]	2.56	
	[110]	[110]	6.53	
		[110]	3.31	
		[001]	2.51	
	[101]	[101]	4.76	
		[101]	2.73	
		[010]	2.82	
	[011]	[011]	5.96	
		[011]	4.45	
		[100]	3.15	
Perthite	[001]	[001]	6.30	347
$Or_{67}Ab_{29}An_0$		[100]	2.55	
		[010]	2.22	
ϱ = 2.54 g/cm³	[010]	[010]	7.60	
		[001]	2.20	
V_p = 5.58 km/sec		[100]	3.68	
V_s = 3.04 km/sec	[100]	[100]	4.90	
		[010]	3.68	
		[001]	2.50	
	[110]	[110]	6.58	
		[110]	3.30	
		[001]	2.45	
	[101]	[101]	4.60	
		[101]	2.80	
		[010]	2.83	
	[011]	[011]	5.91	
		[011]	4.45	
		[100]	3.00	

Table 26 (continued)
VELOCITIES IN ROCK FORMING MINERALS

Mineral	Propagation direction	Displacement direction	Velocity (km/sec)	Ref.
Perthite	[001]	[001]	6.95	347
$Or_{79}Ab_{19}An_2$		[100]	2.93	
ϱ = 2.56 g/cm³		[010]	2.37	
	[010]	[010]	8.15	
V_p = 5.91 km/sec		[001]	2.14	
V_s = 3.25 km/sec		[100]	3.83	
	[100]	[100]	5.10	
		[010]	3.75	
		[001]	3.04	
	[110]	[110]	7.14	
		[1$\bar{1}$0]	3.44	
		[001]	2.88	
	[101]	[101]	5.20	
		[$\bar{1}$01]	3.55	
		[010]	3.04	
	[011]	[011]	6.30	
		[0$\bar{1}$1]	4.96	
		[100]	3.20	
Perthite	[001]	[001]	6.30	347
$Or_{74}Ab_{19}An_2$		[100]	2.63	
ϱ = 2.57 g/cm³		[010]	2.34	
	[010]	[010]	7.85	
V_p = 5.79 km/sec		[001]	2.32	
V_s = 3.11 km/sec		[100]	3.72	
	[100]	[100]	5.00	
		[010]	3.74	
		[001]	2.58	
	[110]	[110]	6.83	
		[1$\bar{1}$0]	3.33	
		[001]	2.39	
	[101]	[101]	4.76	
		[$\bar{1}$01]	2.80	
		[010]	2.98	
	[011]	[011]	6.08	
		[0$\bar{1}$1]	4.45	
		[100]	3.15	
Perthite	[001]	[001]	6.45	347
$Or_{65}Ab_{27}An_4$		[100]	2.66	
ϱ = 2.57 g/cm³		[010]	2.33	
	[010]	[010]	7.84	
V_p = 5.74 km/sec		[001]	2.32	
V_s = 3.13 km/sec		[100]	3.83	
	[100]	[100]	4.96	
		[010]	3.80	
		[001]	2.54	
	[110]	[110]	6.76	
		[1$\bar{1}$0]	3.40	
		[001]	2.54	
	[101]	[101]	4.75	
		[$\bar{1}$01]	2.78	
		[010]	2.98	
	[011]	[011]	6.16	
		[0$\bar{1}$1]	4.45	
		[100]	3.14	

Table 26 (continued)
VELOCITIES IN ROCK FORMING MINERALS

Mineral	Propagation direction	Displacement direction	Velocity (km/sec)	Ref.
Perthite	[001]	[001]	6.86	347
$Or_{61}Ab_{36}An_2$		[100]	2.86	
ϱ = 2.57 g/cm³		[010]	2.31	
	[010]	[010]	7.81	
V_p = 5.65 km/sec		[001]	2.15	
V_s = 3.19 km/sec		[100]	3.62	
	[100]	[100]	5.06	
		[010]	3.65	
		[001]	2.53	
	[110]	[110]	6.68	
		[1$\bar{1}$0]	3.42	
		[001]	2.50	
	[101]	[101]	4.88	
		[$\bar{1}$01]	2.88	
		[010]	2.94	
	[011]	[011]	6.10	
		[0$\bar{1}$1]	4.75	
		[100]	3.20	
Perthite	[001]	[001]	6.90	347
$Or_{54}Ab_{35}An_9$		[100]	2.91	
ϱ = 2.57 g/cm³		[010]	1.98	
	[010]	[010]	7.68	
V_p = 5.88 km/sec		[001]	1.92	
V_s = 3.05 km/sec		[100]	3.70	
	[100]	[100]	5.10	
		[010]	3.71	
		[001]	2.96	
	[110]	[110]	6.69	
		[1$\bar{1}$0]	3.44	
		[001]	2.55	
	[101]	[101]	4.98	
		[$\bar{1}$01]	2.70	
		[010]	2.79	
	[011]	[011]	6.16	
		[0$\bar{1}$1]	4.50	
		[100]	3.20	
Plagioclase	[001]	[001]	7.13	348
An_9		[100]	3.19	
ϱ = 2.61 g/cm³		[010]	2.56	
	[010]	[010]	7.26	
V_p = 6.07 km/sec		[001]	2.58	
V_s = 3.40 km/sec		[100]	3.56	
	[100]	[100]	5.42	
		[010]	5.45	
		[001]	3.30	
	[110]	[110]	6.38	
		[1$\bar{1}$0]	3.69	
		[001]	2.74	
	[101]	[101]	5.31	
		[$\bar{1}$01]	3.40	
		[010]	2.95	
	[011]	[011]	6.20	
		[0$\bar{1}$1]	4.63	
		[100]	3.10	

Table 26 (continued)
VELOCITIES IN ROCK FORMING MINERALS

Mineral	Propagation direction	Displacement direction	Velocity (km/sec)	Ref.
Plagioclase	[001]	[001]	6.88	349
An_{16}		[100]	3.19	
ϱ = 2.64 g/cm³		[010]	2.58	
	[010]	[010]	7.87	
V_p = 6.22 km/sec		[001]	2.71	
V_s = 3.23 km/sec		[100]	3.66	
	[100]	[100]	5.68	
		[010]	3.70	
		[001]	2.95	
	[110]	[110]	6.81	
		$[1\bar{1}0]$	3.72	
		[001]	2.78	
	[101]	[101]	6.60	
		$[\bar{1}01]$	3.07	
		[101]	3.13	
	[011]	[011]	6.45	
		$[0\bar{1}1]$	4.53	
		[100]	3.36	
Plagioclase	[001]	[001]	7.18	348
An_{24}		[100]	3.24	
		[010]	2.59	
ϱ = 2.64 g/cm³	[010]	[010]	7.41	
		[001]	2.84	
V_p = 6.22 km/sec		[100]	3.58	
V_s = 3.34 km/sec	[100]	[100]	5.62	
		[010]	3.55	
		[001]	3.35	
	[110]	[110]	8.55	
		$[1\bar{1}0]$	3.72	
		[001]	2.86	
	[101]	[101]	5.48	
		$[\bar{1}01]$	3.48	
		[010]	3.06	
	[011]	[011]	6.32	
		$[0\bar{1}1]$	4.62	
		[100]	3.26	
Plagioclase	[001]	[001]	7.17	348
An_{29}		[100]	3.27	
ϱ = 2.64 g/cm³		[010]	2.65	
	[010]	[010]	7.55	
V_p = 6.30 km/sec		[001]	2.70	
V_s = 3.44 km/sec		[100]	3.61	
	[100]	[100]	5.70	
		[010]	3.60	
		[001]	3.37	
	[110]	[110]	6.67	
		$[1\bar{1}0]$	3.76	
		[001]	2.90	
	[101]	[101]	5.50	
		$[\bar{1}01]$	3.52	
		[010]	3.10	
	[011]	[011]	6.35	
		$[0\bar{1}1]$	4.70	
		[100]	3.29	

Table 26 (continued)
VELOCITIES IN ROCK FORMING MINERALS

Mineral	Propagation direction	Displacement direction	Velocity (km/sec)	Ref.
Quartz	[$\bar{1}2\bar{1}0$]	[$\bar{1}2\bar{1}0$]	5.749	350
ϱ = 2.649 g/cm^3	[$\bar{1}2\bar{1}0$]	[$10\bar{1}0$]	3.297	
	[$\bar{1}2\bar{1}0$]	[0001]	5.114	
V_p = 6.05 km/sec	[$10\bar{1}0$]	[$10\bar{1}0$]	6.006	
V_s = 4.09 km/sec	[$10\bar{1}0$]	[$\bar{1}2\bar{1}0$]	4.323	
	[$10\bar{1}0$]	[0001]	3.918	
$\partial V_p/\partial P$ = 13.9 km/sec^{-1}Mb^{-1}	[0001]	[0001]	6.319	
$\partial V_s/\partial P$ = −3.28 km/sec^{-1}Mb^{-1}	[0001]	in (0001) plane	4.687	
Plagioclase	[001]	[001]	7.30	348
An_{53}		[100]	3.40	
ϱ = 2.68 g/cm^3		[010]	2.70	
	[010]	[010]	7.80	
V_p = 6.57 km/sec		[001]	2.78	
V_s = 3.53 km/sec		[100]	3.63	
	[100]	[100]	6.06	
		[010]	3.72	
		[001]	3.44	
	[110]	[110]	6.98	
		[$1\bar{1}0$]	3.84	
		[001]	2.95	
	[101]	[101]	7.25	
		[$\bar{1}01$]	3.65	
		[010]	3.14	
	[011]	[011]	6.55	
		[$0\bar{1}1$]	4.76	
		[100]	3.44	
Plagioclase	[001]	[001]	7.53	349
An_{58}		[100]	3.49	
ϱ = 2.68 g/cm^3		[010]	2.83	
	[010]	[010]	7.71	
V_p = 6.70 km/sec		[001]	2.76	
V_s = 3.55 km/sec		[100]	3.53	
	[100]	[100]	6.10	
		[010]	3.72	
		[001]	3.59	
	[110]	[110]	7.38	
		[$1\bar{1}0$]	4.14	
		[001]	2.89	
	[101]	[101]	7.05	
		[$\bar{1}01$]	3.65	
		[010]	2.96	
	[011]	[011]	6.48	
		[$0\bar{1}1$]	4.89	
		[100]	3.63	

Table 26 (continued)
VELOCITIES IN ROCK FORMING MINERALS

Mineral	Propagation direction	Displacement direction	Velocity (km/sec)	Ref.
Plagioclase	[001]	[001]	7.33	348
An_{56}		[100]	3.40	
ϱ = 2.69 g/cm³		[010]	2.72	
	[010]	[010]	8.00	
V_p = 6.62 km/sec		[001]	2.80	
V_s = 3.75 km/sec		[100]	3.65	
	[100]	[100]	6.10	
		[010]	3.74	
		[001]	3.50	
	[110]	[110]	7.10	
		$[1\bar{1}0]$	3.90	
		[001]	3.02	
	[101]	[101]	7.38	
		$[\bar{1}01]$	3.67	
		[010]	3.17	
	[011]	[011]	6.64	
		$[0\bar{1}1]$	4.80	
		[100]	3.48	
Calcite	$[\bar{1}2\bar{1}0]$	$[\bar{1}2\bar{1}0]$	7.30	351
ϱ = 2.712 g/cm³	$[\bar{1}2\bar{1}0]$	$[10\bar{1}0]$	4.71	
	$[10\bar{1}0]$	$[10\bar{1}0]$	7.35	
V_p = 6.53 km/sec	$[10\bar{1}0]$	$[\bar{1}2\bar{1}0]$	4.01	
V_s = 3.36 km/sec	$[10\bar{1}0]$	[0001]	3.26	
	[0001]	[0001]	5.54	
	[0001]	in (0001) plane	3.47	
Muscovite	[001]	[001]	4.44	352
ϱ = 2.79 g/cm³		[100]	2.03	
		[010]	2.05	
V_p = 5.78 km/sec	[010]	[010]	8.03	
V_s = 3.33 km/sec		[001]	2.06	
		[100]	5.01	
	[100]	[100]	7.90	
		[010]	4.95	
		[001]	2.19	
	[110]	[110]	8.06	
		$[1\bar{1}0]$	4.86	
		[001]	2.16	
	[101]	[101]	4.70	
Phlogopite	[001]	[001]	4.30	352
ϱ = 2.80 g/cm³		[100]	1.42	
		[010]	1.44	
V_p = 5.55 km/sec	[010]	[010]	7.95	
V_s = 2.88 km/sec		[100]	5.20	
	[100]	[100]	7.94	
		[010]	5.06	
		[001]	1.50	
	[110]	[110]	8.03	
		$[1\bar{1}0]$	5.14	
		[001]	1.44	
	[011]	[011]	5.85	
		[100]	3.66	

Table 26 (continued)
VELOCITIES IN ROCK FORMING MINERALS

Mineral	Propagation direction	Displacement direction	Velocity (km/sec)	Ref.
Phlogopite	[001]	[001]	4.26	352
ϱ = 2.82 g/cm³		[100]	1.50	
		[010]	1.51	
V_p = 5.44 km/sec	[010]	[010]	7.97	
V_s = 2.99 km/sec		[001]	1.53	
		[100]	5.19	
	[100]	[100]	7.94	
		[010]	5.05	
		[001]	1.52	
	[110]	[110]	7.95	
		[$1\bar{1}0$]	5.11	
		[001]	1.56	
	[101]	[101]	5.76	
	[011]	[011]	5.81	
		[100]	3.5	
Biotite	[001]	[001]	4.21	352
ϱ = 3.05 g/cm³		[100]	1.38	
		[010]	1.38	
V_p = 5.26 km/sec	[010]	[010]	7.78	
V_s = 2.87 km/sec		[001]	1.34	
		[100]	5.06	
	[100]	[100]	7.87	
		[010]	5.06	
		[001]	1.40	
	[110]	[110]	7.83	
		[$1\bar{1}0$]	5.08	
		[001]	1.29	
	[101]	[101]	4.26	
Hornblende	[001]	[001]	7.85	353
ϱ = 3.12 g/cm³		[100]	3.16	
		[010]	4.29	
V_p = 6.81 km/sec	[010]	[010]	7.16	
V_s = 3.72 km/sec		[001]	4.52	
		[100]	3.53	
	[100]	[100]	6.11	
		[010]	3.43	
		[001]	3.18	
	[110]	[110]	6.50	
		[$1\bar{1}0$]	3.56	
		[001]	3.92	
	[101]	[101]	7.11	
		[$\bar{1}01$]	3.65	
		[010]	3.62	
	[011]	[011]	7.55	
		[$0\bar{1}1$]	4.20	
		[100]	3.30	

Table 26 (continued)
VELOCITIES IN ROCK FORMING MINERALS

Mineral	Propagation direction	Displacement direction	Velocity (km/sec)	Ref.
Hornblende	[001]	[001]	8.13	353
ϱ = 3.15 g/cm³		[100]	3.03	
		[010]	4.40	
V_p = 7.04 km/sec	[010]	[010]	7.54	
V_s = 3.81 km/sec		[001]	4.45	
		[100]	3.72	
	[100]	[100]	6.45	
		[010]	3.78	
		[001]	3.46	
	[110]	[110]	7.01	
		[1$\bar{1}$0]	3.87	
		[001]	3.77	
	[101]	[101]	6.18	
		[$\bar{1}$01]	3.98	
		[010]	4.05	
	[011]	[011]	7.80	
		[0$\bar{1}$1]	4.48	
		[100]	3.48	
Forsterite	[001]	[001]	8.565	354
ϱ = 3.22 g/cm³		[100]	5.029	
		[010]	4.569	
V_p = 8.59 km/sec	[010]	[010]	7.889	
V_s = 5.03 km/sec		[001]	4.569	
$\partial V_p/\partial P$ = 9.84 km sec⁻¹Mb⁻¹	[100]	[100]	5.022	
$\partial V_s/\partial P$ = 3.64 km sec⁻¹Mb⁻¹		[100]	10.110	
$\partial V_p/\partial T = -5.28 \times 10^{-4}$ km sec⁻¹deg⁻¹		[010]	5.019	
$\partial V_s/\partial T = -3.49 \times 10^{-4}$ km sec⁻¹deg⁻¹		[001]	5.028	
Forsterite	[001]	[001]	8.544	355
ϱ = 3.224 g/cm³		[100]	5.019	
		[010]	4.519	
V_p = 8.57 km/sec	[010]	[010]	7.872	
V_s = 5.02 km/sec		[001]	4.521	
		[100]	5.010	
$\partial V_p/\partial P$ = 10.7 km sec⁻¹Mb⁻¹	[100]	[100]	10.093	
$\partial V_s/\partial P$ = 3.58 km sec⁻¹Mb⁻¹		[010]	5.008	
		[001]	5.019	
$\partial V_p/\partial T = -4.80 \times 10^{-4}$ km sec⁻¹deg⁻¹	[110]	[110]	8.819	
		[1$\bar{1}$0]	5.350	
		[001]	4.753	
$\partial V_s/\partial T = -3.40 \times 10^{-4}$ km sec⁻¹deg⁻¹	[101]	[101]	8.959	
		[$\bar{1}$01]	5.617	
		[010]	4.784	
	[011]	[011]	8.125	
		[0$\bar{1}$1]	4.727	
		[100]	5.016	

Table 26 (continued)
VELOCITIES IN ROCK FORMING MINERALS

Mineral	Propagation direction	Displacement direction	Velocity (km/sec)	Ref.
Diallage	[001]	[001]	8.00	356
ϱ = 3.30 g/cm³		[100]	4.30	
		[010]	4.39	
V_p = 7.03 km/sec	[010]	[010]	6.72	
V_s = 4.26 km/sec		[001]	4.41	
		[100]	3.95	
	[100]	[100]	6.87	
		[010]	3.98	
		[001]	4.25	
	[110]	[110]	6.99	
		[1$\bar{1}$0]	4.00	
		[001]	4.13	
	[101]	[101]	7.78	
		[$\bar{1}$01]	4.65	
		[010]	3.86	
	[011]	[011]	7.20	
		[0$\bar{1}$1]	4.67	
		[100]	4.16	
Diopside	[001]	[001]	8.60	356
ϱ = 3.31 g/cm³		[100]	3.95	
		[010]	4.51	
V_p = 7.70 km/sec	[010]	[010]	7.25	
V_s = 4.38 km/sec	[100]	[100]	7.90	
		[010]	4.60	
		[001]	4.10	
	[110]	[110]	7.98	
		[1$\bar{1}$0]	3.94	
		[001]	4.35	
	[101]	[101]	6.94	
		[$\bar{1}$01]	4.45	
		[010]	4.17	
	[011]	[011]	7.82	
		[0$\bar{1}$1]	4.83	
		[100]	4.19	
Olivine	[001]	[001]	8.427	357
$Fo_{93}Fa_7$		[100]	4.874	
		[010]	4.418	
ϱ = 3.311 g/cm³	[010]	[010]	7.725	
		[001]	4.418	
V_p = 8.42 km/sec		[100]	4.886	
V_s = 4.89 km/sec	[100]	[100]	9.887	
$\partial V_p/\partial P$ = 10.2 km sec⁻¹Mb⁻¹		[010]	4.886	
		[001]	4.874	
$\partial V_s/\partial P$ = 3.60 km sec⁻¹Mb⁻¹	[110]	[110]	8.658	
		[1$\bar{1}$0]	5.203	
		[001]	4.644	
$\partial V_p/\partial T = -4.86 \times 10^{-4}$ km sec⁻¹deg⁻¹	[101]	[101]	8.826	
		[$\bar{1}$01]	5.530	
$\partial V_s/\partial T = -3.40 \times 10^{-4}$ km sec⁻¹deg⁻¹		[010]	4.662	
	[011]	[011]	7.976	
		[0$\bar{1}$1]	4.569	
		[100]	4.876	

Table 26 (continued)
VELOCITIES IN ROCK FORMING MINERALS

Mineral	Propagation direction	Displacement direction	Velocity (km/sec)	Ref.
Augite	[001]	[001]	8.15	356
ϱ = 3.32 g/cm^3		[100]	3.82	
		[010]	4.58	
V_p = 7.22 km/sec	[010]	[010]	6.81	
V_s = 4.18 km/sec		[001]	4.66	
		[100]	4.32	
	[100]	[100]	7.48	
		[010]	4.10	
		[001]	3.81	
	[110]	[110]	7.34	
		[1$\bar{1}$0]	3.86	
		[001]	4.09	
	[101]	[101]	8.36	
		[$\bar{1}$01]	4.34	
		[010]	4.49	
	[011]	[011]	7.44	
		[0$\bar{1}$1]	4.72	
		[100]	3.87	
Olivine	[001]	[001]	8.65	357
ϱ = 3.324 g/cm^3		[100]	5.00	
		[010]	4.54	
V_p = 8.48 km/sec	[010]	[010]	7.73	
V_s = 4.93 km/sec		[001]	4.42	
		[100]	4.88	
	[100]	[100]	9.87	
		[010]	4.88	
		[001]	4.87	
Orthopyroxene	[001]	[001]	7.853	358
$En_{85}Fs_{15}$		[100]	4.791	
		[010]	4.990	
ϱ = 3.335 g/cm^3	[010]	[010]	7.043	
V_p = 7.85 km/sec		[001]	4.992	
V_s = 4.76 km/sec		[100]	4.857	
	[100]	[100]	8.303	
		[010]	4.847	
		[001]	4.780	
	[110]	[110]	7.991	
		[1$\bar{1}$0]	4.270	
		[001]	4.893	
	[101]	[101]	8.015	
		[$\bar{1}$01]	4.898	
		[010]	4.918	
	[011]	[011]	7.806	
		[0$\bar{1}$1]	4.549	
		[100]	4.815	

Table 26 (continued)
VELOCITIES IN ROCK FORMING MINERALS

Mineral	Propagation direction	Displacement direction	Velocity (km/sec)	Ref.
Bronzite	[001]	[001]	7.920	359
$Mg_{0.8}Fe_{0.2}SiO_3$		[100]	4.741	
ϱ = 3.354 g/cm³		[010]	4.936	
	[010]	[010]	6.918	
V_p = 7.78 km/sec		[001]	4.937	
V_s = 4.72 km/sec		[100]	4.812	
	[100]	[100]	8.254	
$\partial V_p/\partial P$ = 20.6 km $sec^{-1}Mb^{-1}$		[010]	4.812	
$\partial V_s/\partial P$ = 5.16 km $sec^{-1}Mb^{-1}$		[001]	4.745	
$\partial V_p/\partial T$ = -9.08×10^{-4} km $sec^{-1}deg^{-1}$				
$\partial V_s/\partial T$ = -4.86×10^{-4} km $sec^{-1}deg^{-1}$				
Epidote	[001]	[001]	7.75	346
ϱ = 3.40 g/cm³		[100]	3.47	
		[010]	3.39	
V_p = 7.43 km/sec	[010]	[010]	8.38	
V_s = 4.24 km/sec		[001]	3.39	
		[100]	4.86	
	[100]	[100]	7.89	
		[010]	4.78	
		[001]	3.56	
	[110]	[110]	8.14	
		[1$\bar{1}$0]	4.85	
		[001]	3.62	
	[101]	[101]	7.35	
		[$\bar{1}$01]	4.87	
		[010]	4.25	
	[011]	[011]	7.24	
		[0$\bar{1}$1]	5.11	
		[100]	4.30	
Aegirite-augite	[001]	[001]	7.99	356
ϱ = 3.42 g/cm³		[100]	3.60	
		[010]	3.41	
V_p = 7.32 km/sec	[010]	[010]	6.66	
V_s = 4.09 km/sec		[001]	3.69	
		[100]	3.93	
	[100]	[100]	6.86	
		[010]	3.79	
		[001]	3.47	
	[110]	[110]	7.06	
		[1$\bar{1}$0]	3.18	
		[001]	3.47	
	[101]	[101]	7.99	
		[$\bar{1}$01]	4.17	
		[010]	3.77	
	[011]	[011]	7.05	
		[0$\bar{1}$1]	4.24	
		[100]	3.41	

Table 26 (continued)
VELOCITIES IN ROCK FORMING MINERALS

Mineral	Propagation direction	Displacement direction	Velocity (km/sec)	Ref.
Aegirite	[001]	[001]	8.21	356
ϱ = 3.50 g/cm^3		[100]	3.72	
		[010]	4.23	
V_p = 7.32 km/sec	[010]	[010]	7.20	
V_s = 4.09 km/sec		[001]	3.97	
		[100]	3.48	
	[100]	[100]	7.30	
		[010]	3.68	
		[001]	3.78	
	[110]	[110]	7.60	
		$[1\bar{1}0]$	4.07	
		[001]	3.95	
	[101]	[101]	6.75	
		$[\bar{1}01]$	4.43	
		[010]	3.68	
	[011]	[011]	8.30	
		$[0\bar{1}1]$	4.65	
		[100]	3.86	
Garnet				
3 $(Mn_0Fe_1Mg_2Ca_{97})$ O · Al_2O_3 · $3SiO_2$				
ϱ = 3.60 g/cm^3				
	[100]	[100]	8.86	360
V_p = 8.72 km/sec		[010]	5.00	
V_s = 5.07 km/sec		[001]	5.00	
	[110]	[110]	8.70	
		$[1\bar{1}0]$	5.16	
		[001]	5.02	
Spinel	[001]	[001]	9.083	361
MgO · 2.6 Al_2O_3		in (001) plane	6.598	
	[110]	[110]	10.296	
ϱ = 3.619 g/cm^3		$[1\bar{1}0]$	4.473	
		[001]	6.598	
V_p = 9.93 km/sec				
V_s = 5.66 km/sec				
$\partial V_p/\partial P$ = 4.74 km sec^{-1}Mb^{-1}				
$\partial V_s/\partial P$ = 0.32 km sec^{-1}Mb^{-1}				
Spinel	[100]	[100]	9.10	357
MgO · 3.5(Al_2O_3)		in (100) plane	6.61	
ϱ = 3.63 g/cm^3	[110]	[110]	10.30	
V_p = 9.93 km/sec				
V_s = 5.66 km/sec				
Garnet				
3(Mn,$Fe_{17}Mg_{72}Ca_{11}$)O · Al_2O_3 · $3SiO_2$				
ϱ = 3.67 g/cm^3	[110]	[110]	8.60	360
		$[1\bar{1}0]$	4.90	
V_p = 8.55 km/sec		[001]	4.90	
V_s = 4.86 km/sec				

Table 26 (continued)
VELOCITIES IN ROCK FORMING MINERALS

Mineral	Propagation direction	Displacement direction	Velocity (km/sec)	Ref.
Spinel	[001]	[001]	8.393	362
$Mg_{0.75}Fe_{0.36}Al_{1.90}O_4$		in (001) plane	6.124	
	[110]	[110]	9.203	
ϱ = 3.826 g/cm³		[1$\bar{1}$0]	3.725	
		[001]	6.124	
V_p = 9.25 km/sec				
V_s = 5.01 km/sec				
$\partial V_p/\partial P$ = 5.13 km $sec^{-1}Mb^{-1}$				
$\partial V_s/\partial P$ = 0.58 km $sec^{-1}Mb^{-1}$				
Garnet				
$3(Mn, Fe_{63}Mg_{29}Ca_8)O \cdot Al_2O_3 \cdot 3SiO_2$				
ϱ = 4.01 g/cm³				
	[110]	[110]	8.34	360
		[1$\bar{1}$0]	4.82	
V_p = 8.17 km/sec		[001]	4.82	
V_s = 4.71km/sec				
Garnet				
$3(Mn_2Fe_{64}Mg_{23}Ca_{11})O \cdot Al_2O_3 \cdot 3SiO_2$				
ϱ = 4.06 g/cm³				
	[100]	[100]	8.45	360
V_p = 8.45 km/sec		[010]	4.82	
V_s = 4.85 km/sec		[001]	4.82	
	[110]	[110]	8.45	
		[1$\bar{1}$0]	4.92	
		[001]	4.87	
Garnet				
$3(Mn_4Fe_{77}Mg_{12}Ca_8)O \cdot Al_2O_3 \cdot 3SiO_2$				
ϱ = 4.06 g/cm³				
	[100]	[100]	8.01	360
V_p = 8.04 km/sec		[010]	4.55	
V_s = 4.54 km/sec		[001]	4.52	
	[110]	[110]	8.12	
		[1$\bar{1}$0]	4.50	
		[001]	4.50	
Garnet				
$3(Mn_6Fe_{74}Mg_{13}Ca_7)O \cdot Al_2O_3 \cdot 3SiO_2$				
ϱ = 4.16 g/cm³				
	[100]	[100]	8.36	360
V_p = 8.22 km/sec		[010]	4.62	
V_s = 4.67 km/sec		[001]	4.62	
	[110]	[110]	8.24	
		[1$\bar{1}$0]	4.77	
		[001]	4.67	

Table 26 (continued)
VELOCITIES IN ROCK FORMING MINERALS

Mineral	Propagation direction	Displacement direction	Velocity (km/sec)	Ref.
Garnet				
$3(Fe_{76}Mg_{21}Ca_3)O \cdot Al_2O_3 \cdot 3SiO_2$	[001]	[001]	8.579	363
		[$1\bar{1}0$]	4.725	
ϱ = 4.160 g/cm³	[110]	[110]	8.520	
		[$1\bar{1}0$]	4.825	
		[001]	4.714	
V_p = 8.53 km/sec				
V_s = 4.76 km/sec				
$\partial V_p/\partial P$ = 7.84 km sec^{-1}Mb^{-1}				
$\partial V_s/\partial P$ = 2.17 km sec^{-1}Mb^{-1}				
$\partial V_p/\partial T$ = -3.93×10^{-4} km sec^{-1}deg^{-1}				
$\partial V_s/\partial T$ = -2.18×10^{-4} km sec^{-1}deg^{-1}				
Garnet				
$3(Fe_{81}Mg_{14}Mn,Ca_4)O \cdot Al_2O_3 \cdot 3SiO_2$				
ϱ = 4.183 g/cm³				
	[100]	[100]	8.54	357
V_p = 8.52 km/sec		in (100) plane	4.75	
V_s = 4.77 km/sec	[110]	[110]	8.51	
Garnet				
$3(Mn_{55}Fe_{43.5}Mg_{0.2}Ca_{1.3})O \cdot Al_2O_3 \cdot 3SiO_2$				
ϱ = 4.247 g/cm³				
	[100]	[100]	8.51	357
V_p = 8.47 km/sec		in (100) plane	4.74	
V_s = 4.77 km/sec	[110]	[110]	8.47	
Garnet	[001]	[001]	8.521	364
$3(Mn_{54}Fe_{46})O \cdot Al_2O_3 \cdot 3SiO_2$		in (001) plane	4.723	
	[110]	[110]	8.475	
ϱ = 4.249 g/cm³		[$1\bar{1}0$]	4.850	
		[001]	4.723	
V_p = 8.48 km/sec				
V_s = 4.76 km/sec				
$\partial V_p/\partial P$ = 7.14 km sec^{-1}Mb^{-1}				
$\partial V_s/\partial P$ = 2.22 km sec^{-1}Mb^{-1}				
Rutile	[001]	[001]	10.659	365, 366
ϱ = 4.260 g/cm³		in (001) plane	5.404	
	[100]	[100]	7.982	
V_p = 9.26 km/sec		[010]	6.762	
V_s = 5.14 km/sec	[110]	[110]	9.899	
$\partial V_p/\partial P$ = 7.64 km sec^{-1}Mb^{-1}		[$1\bar{1}0$]	3.312	
$\partial V_s/\partial P$ = 5.20 km sec^{-1}Mb^{-1}		[001]	5.405	
$\partial V_p/\partial T$ = -8.9×10^{-4} km sec^{-1}deg^{-1}	[101]	[101]	9.667	
$\partial V_s/\partial T$ = -3.9×10^{-4} km sec^{-1}deg^{-1}		[$\bar{1}01$]	6.116	
Spinel	[001]	[001]	7.883	362
$FeAl_2O_4$		in (001) plane	5.585	
	[110]	[110]	9.143	
ϱ = 4.280 g/cm³		[$1\bar{1}0$]	3.123	
		[001]	5.585	
V_p = 8.67 km/sec				
V_s = 4.42 km/sec				

Table 26 (continued)
VELOCITIES IN ROCK FORMING MINERALS

Mineral	Propagation direction	Displacement direction	Velocity (km/sec)	Ref.
Zircon				
ϱ = 4.70 g/cm³	[001]	[001]	9.00	346
		[100]	3.89	
		[010]	3.89	
V_p = 8.06 km/sec	[010]	[010]	7.41	
V_s = 3.97 km/sec		[001]	4.87	
		[100]	3.90	
	[110]	[110]	8.39	
		[1$\bar{1}$0]	2.94	
		[001]	4.05	
	[011]	[011]	8.36	
		[0$\bar{1}$1]	3.16	
		[100]	4.64	

Note: V_p and V_s refer to Hill[3] averages for polycrystalline aggregates. Propagation and displacement directions refer to an orthogonal set of coordinates. Because of twinning, the feldspars have been treated as monoclinic with crystallographic a and b corresponding to the [100] and [010] propagation directions, respectively. For the monoclinic amphiboles and pyroxenes [010] and [001] correspond to crystallographic b and c, respectively. For the micas [100] and [010] refer to crystallographic a and b, respectively.

Table 27
COMPRESSIONAL AND SHEAR WAVE VELOCITIES IN GLASSY SPHERES, USING THE RESONANCE TECHNIQUE[367,368]

Sample	Density (g/cm³)	V_p (km/sec)	V_s (km/sec)
Lunar Sphere			
LG-102	2.79	6.48	3.69
LG-103	3.09	6.47	3.52
LG-104	3.01	6.31	3.54
LG-107	2.79	6.30	3.61
LG-108	3.15	6.42	3.52
LG-109	2.98	6.43	3.61
LG-110	3.03	6.44	3.56
LG-112	3.05	6.36	3.57
LG-116	3.03	6.40	3.54
LG-117	3.09	6.33	3.50
Moldavite, Bohemian Tektite	2.37	5.92	3.63
Indochinite, Thailand Tektite	2.42	6.00	3.64

Table 28
COMPRESSIONAL AND SHEAR WAVE VELOCITIES IN MISCELLANEOUS MATERIALS[342-344]

Material	V_p (km/sec)	V_s (km/sec)
Aluminum	7.05	2.94
Window glass	6.79	3.26
Aluminum	6.38	3.10
Aluminum	6.32	3.10
Steel	6.15—6.30	2.72—2.83
Aluminum	6.26	3.04
Fused silica	5.97	—
Steel	5.94	3.23
Iron	5.92	3.23
Steel	5.92	—
CR Steel II	5.89	3.21
CR Steel I	5.88	3.20
Iron	5.84	3.26
Copper	4.82—5.96	2.30
Glass	5.80	3.35
Magnesium	5.78	3.06
Birch	5.00	0.76
Copper	4.66	2.32
Brass	4.28	2.03
Bakelite®	3.46	1.99
Cellulose	3.59	1.71
Concrete	3.56	2.16
Polymethyl methacrylate	2.87	1.44
Lucite®	2.64	1.27
Plexiglas®	2.55	1.28
Resin	2.44	1.02
Plastic	2.34	1.46—1.56
Grout	2.31	1.18
Sapsago cheese	2.12	—
Gjetost cheese	1.83	—
Provolone cheese	1.75	—
Romano cheese	1.75	—
Cheddar cheese	1.72	—
Emmenthal cheese	1.65	—
Muenster cheese	1.57	—
Mercury	1.45	—
Rubber	1.04	0.03
Agar-agar	0.10	0.03

Table 29
COMPRESSIONAL AND SHEAR WAVE VELOCITIES VERSUS PRESSURE IN THREE MATERIALS[375]

	Polystyrene (ϱ = 1.05 g/cm³)		Lucite® (ϱ = 1.17 g/cm³)		Polyethylene (ϱ = 0.91 g/cm³)
Pressure (bars)	V_p (km/sec)	V_s (km/sec)	V_p (km/sec)	V_s (km/sec)	V_p (km/sec)
0	2.30	1.14	—	—	1.98
171	2.34	1.15	2.62	1.30	2.06
342	2.38	1.16	2.70	1.32	2.13
513	2.41	1.17	2.74	1.36	2.19
684	2.45	1.18	2.78	1.38	2.26
855	2.48	1.19	2.82	1.39	2.32
1026	2.52	1.19	2.85	1.40	2.37

REFERENCES

1. **Voigt, W.**, *Lehrbuch der Krystallphysik,* Teubner, Berlin, 1910.
2. **Reuss, A.**, Berechnung der Fliessgrenze von Meschkristallen auf Grund der Plastizitatsbedingung für Einkristalle, *Z. Angew. Math. Mech.*, 9, 49, 1929.
3. **Hill, R.**, The elastic behavior of a crystalline aggregate, *Proc. Phys. Soc. London, Sect. A*, 65, 349, 1952.
4. **Hamilton, E. L., Moore, D. G., Buffington, E. C., Sherrer, P. L., and Curray, J. R.**, Sediment velocities from sonobuoys: Bay of Bengal, Bering Sea, Japan Sea, and North Pacific, *J. Geophys. Res.*, 79, 2653, 1974.
5. **Hamilton, E. L., Bachman, R. T., Curray, J. R., and Moore, D. G.**, Sediment velocities from sonobuoys: Bengal Fan, Sunda Trench, Andaman Basin and Nicobar Fan, *J. Geophys. Res.*, 82, 3003, 1977.
6. **Cunny, R. W. and Fry, Z. B.**, Vibratory in-situ and laboratory soil moduli compared, *J. Soil. Mech. Fdn. Div., Am. Soc. Civil Eng.*, 99, 1055, 1973.
7. **Bentley, C. R.**, Seismic anistropy in the West Antarctic Ice Sheet, snow and ice studies. II, *Antarct. Res. Ser.*, 16, 131, 1971.
8. **Joset, A. and Holtzscherer, J. J.**, Étude des vitesses de propagation des ondes séismiques sur l'inlandsis de Groenland, *Ann. Geophys.*, 9, 330, 1953.
9. **Houtz, R. E.**, Seismic properties of Layer 2A in the Pacific, *J. Geophys. Res.*, 81, 6321, 1976.
10. **Ross, D. A. and Schlee, J.**, Shallow structure and geologic development of the southern Red Sea, *Geol. Soc. Am. Bull.*, 84, 3827, 1973.
11. **Naini, B. R. and Leyden, R.**, Ganges Cone: a wide angle seismic reflection and refraction study, *J. Geophys. Res.*, 78, 8711, 1973.
12. **Laughton, A. S. and Tramotini, C.**, Recent studies of the crustal structure in the Gulf of Aden, *Tectonophysics,* 8, 359, 1969.
13. **Francis, T. J. G. and Shor, G. C.**, Seismic refraction measurements in the northwest Indian Ocean, *J. Geophys. Res.*, 71, 427, 1966.
14. **Curray, J. R., Shor, G. C., Raitt, R. W., and Henry, M.**, Seismic refraction and reflection studies of crustal structure of the eastern Sunda and western Banda Arcs, *J. Geophys. Res.*, 82, 2479, 1977.
15. **Francis, T. J. G. and Raitt, R. W.**, Seismic refraction measurements in the southern Indian Ocean, *J. Geophys. Res.*, 72, 3015, 1967.
16. **Ludwig, W. J., Nafe, J. E., Simpson, E. S. W., and Sacks, S.**, Seismic refraction measurements on the southeast African continental margin, *J. Geophys. Res.*, 73, 3707, 1968.
17. **König, M. and Talwani, M.**, A geophysical study of the southern continental margin of Australia: Great Australian Bight and western sections, *Geol. Soc. Am. Bull.*, 88, 1000, 1977.
18. **Houtz, R. E., Ludwig, W. J., Milliman, J. D., and Grow, J. A.**, Structure of the northern Brazilian continental margin, *Geol. Soc. Am. Bull.*, 88, 711, 1977.
19. **Le Pichon, X., Houtz, R. E., Drake, C. L., and Nafe, J. W.**, Crustal structure of the mid-ocean ridges. I. Seismic refraction measurements, *J. Geophys. Res.*, 70, 319, 1965.
20. **Dash, B. P., Ball, M. M., King, G. A., Butler, L. W., and Rona, P. A.**, Geophysical investigation of the Cape Verde Archipelago, *J. Geophys. Res.*, 81, 5249, 1976.
21. **Bunce, E. T., Fahlquist, D. A., and Clough, J. W.**, Seismic refraction and reflection measurements — Puerto Rico outer ridge, *J. Geophys. Res.*, 74, 3082, 1969.
22. **Bosshard, E. and MacFarlane, D. J.**, Crustal structure of the western Canary Islands from seismic refraction and gravity data, *J. Geophys. Res.*, 75, 4901, 1970.
23. **Keen, C. and Loncarevic, B. D.**, Crustal structure on the eastern seaboard of Canada: studies on the continental margin, *Can. J. Earth Sci.*, 66, 65, 1966.
24. **Matthews, D. H., Laughton, A. S., Pugh, D. T., Jones, E. T. W., Sunderland, J., Takin, M., and Bacon, M.**, Crustal structure and origin of Peake and Freen Deep, N.E. Atlantic, *Geophys. J. R. Astron. Soc.*, 18, 517, 1969.
25. **Grau, G., Fail, J. P., Montadert, L., and Patriat, Ph.**, A seismic study in the Bay of Biscay, *Earth Planet. Sci. Lett.*, 23, 357, 1974.
26. **Fenwick, D. K. B., Keen, M. J., Keen, C., and Lambert, A.**, Geophysical studies of the continental margin northeast of Newfoundland, *Can. J. Earth Sci.*, 68, 483, 1968.
27. **Talwani, M., Windisch, C. C., and Langseth, M. G.**, Reykjanes Ridge crest: a detailed geophysical study, *J. Geophys. Res.*, 76, 473, 1971.
28. **Talwani, M. and Eldholm, O.**, Continental margin off Norway: a geophysical study, *Geol. Soc. Am. Bull.*, 83, 3575, 1972.
29. **Sundvor, E.**, Seismic refraction measurements on the Norwegian continental shelf between Andoya and Fugloybanken, *Mar. Geophys. Res.*, 1, 303, 1971.

30. **Houtz, R. and Windisch, C.**, Barents Sea continental margin sonobuoy data, *Geol. Soc. Am. Bull.*, 88, 1030, 1977.
31. **Eldholm, O. and Talwani, M.**, Sediment distribution and structural framework of the Barents Sea, *Geol. Soc. Am. Bull.*, 88, 1015, 1977.
32. **Van Andel, T., Rea, D. K., Von Herzen, R. P., and Hoskins, H.**, Ascension Fracture Zone, Ascension Island and the Mid-Atlantic Ridge, *Geol. Soc. Am. Bull.*, 84, 1527, 1973.
33. **Goslin, J., Mascle, J., Sibuet, J., and Hoskins, H.**, Geophysical study of easternmost Walvis Ridge, South Atlantic: morphology and shallow structure, *Geol. Soc. Am. Bull.*, 85, 619, 1974.
34. **Goslin, J. and Sibuet, J. C.**, Geophysical study of the easternmost Walvis Ridge, South Atlantic: deep structure, *Geol. Soc. Am. Bull.*, 86, 1713, 1975.
35. **Ewing, J., Ludwig, W. J., Ewing, M., and Eittreim, S. L.**, Structure of the Scotia Sea and Falkland Plateau, *J. Geophys. Res.*, 76, 7118, 1971.
36. **Houtz, R., Ewing, J., and Buhl, P.**, Seismic data from sonobuoy stations in the northern and equatorial Pacific, *J. Geophys. Res.*, 75, 5093, 1970.
37. **Raitt, R. W.**, Refraction studies of the Pacific Ocean basin, Part 1: Crustal thickness of the central equatorial Pacific, *Geol. Soc. Am. Bull.*, 67, 1623, 1956.
38. **Shor, G. G., Menard, H. W., and Raitt, R. W., II.**, Regional observations. I. Structure of the Pacific Basin, in *The Sea*, Maxwell, A. E., Ed., John Wiley & Sons, New York, 1971, 3.
39. **Murachi, S., Ludwig, W. J., Den, N., Hotta, H., Asanuma, T., Yoshii, T., Kubotera, A., and Hagiwara, K.**, Structure of the Sulu Sea and Celebes Sea, *J. Geophys. Res.*, 78, 3437, 1973.
40. **Den, N., Ludwig, W. J., Murauchi, S., Ewing, M., Hotta, H., Asanuma, T., Yoshii, T., Kubotera, A., and Hagiwara, K.**, Sediments and structure of the Eauripile-New Guinea Rise, *J. Geophys. Res.*, 76, 4711, 1971.
41. **Murauchi, S., Den, N., Asano, S., Hotta, H., Yoshii, T., Asanuma, T., Hagiwara, K., Ichikawa, K., Sato, T., Ludwig, W. J., Ewing, J. I., Edgar, N. T., and Houtz, R. E.**, Crustal structure of the Philippine Sea, *J. Geophys. Res.*, 73, 3143, 1968.
42. **Sutton, G. H., Maynard, G. L., and Hussong, D. M.**, Widespread occurrence of a high velocity basal layer in the Pacific crust found with repetitive sources and sonobuoys, in *The Structure and Physical Properties of the Earth's Crust*, Heacock, J. G., Ed., American Geophysical Union, Washington, D.C., 1971, 14.
43. **Furumoto, A. S., Woollard, G. P., Campbell, J. F., and Hussong, D. M.**, Variation in the thickness of the crust in the Hawaiian archipelago, in *The Crust and Upper Mantle of the Pacific Area*, Knopoff, L., Drake, C. L., and Hart, P. J., Eds., American Geophysical Union, Washington, D.C., 1968, 12.
44. **Furumoto, A. S., Campbell, J. F., and Hussong, D. M.**, Seismic refraction surveys along the Hawaiian ridge, Kauai to Midway, *Bull. Seismol. Soc. Am.*, 61, 147, 1971.
45. **Helmberger, D. V. and Morris, G. B.**, A travel time and amplitude interpretation of a marine refraction profile: primary waves, *J. Geophys. Res.*, 74, 483, 1969.
46. **Ludwig, W. J., Murauchi, S., Den, N., Buhl, P., Hotta, H., Ewing, M., Asanuma, T., Yoshii, T., and Sakajiri, N.**, Structure of the East China Sea — West Philippine Sea margin off southern Kyushu, Japan, *J. Geophys. Res.*, 78, 2526, 1973.
47. **Ludwig, W. J., Murauchi, S., and Houtz, R. E.**, Sediments and structure of the Japan Sea, *Geol. Soc. Am. Bull.*, 86, 651, 1975.
48. **Yoshii, T., Ludwig, W. J., Den, N., Murauchi, S., Ewing, M., Hotta, H., Buhl, P., Asanuma, T., and Sakajiri, N.**, Structure at Southwest Japan margin off Shikoku, *J. Geophys. Res.*, 78, 2517, 1973.
49. **Ludwig, W. J., Ewing, J. I., Ewing, M., Murauchi, S., Den, N., Asano, S., Hotta, H., Hayakawa, M., Asanuma, T., Ichikawa, K., and Ichikawa, I.**, Sediments and structure of the Japan Trench, *J. Geophys. Res.*, 71, 2121, 1966.
50. **Den, N., Ludwig, W. J., Murauchi, S., Ewing, J., Hotta, H., Edgar, T. N., Yoshii, T., Asanuma, T., Hagiwara, K., Sato, T., and Ando, S.**, Seismic refraction measurements of the northwest Pacific Basin, *J. Geophys. Res.*, 74, 1421, 1969.
51. **Shor, G. G., Dehlinger, P., Kirk, H. K., and French, W. S.**, Seismic refraction studies off Oregon and northern California, *J. Geophys. Res.*, 73, 2175, 1968.
52. **Clowes, R. M. and Malecek, S. J.**, Preliminary interpretation of a marine deep seismic sounding survey in the region of Explorer Ridge, *Can. J. Earth Sci.*, 13, 1545, 1976.
53. **Ludwig, W. J., Murauchi, S., Den, N., Ewing, M., Hotta, H., Houtz, R., Yoshii, T., Asanuma, T., Hagiwara, K., Sato, T., and Ando, S.**, Structure of Bowers Ridge, Bering Sea, *J. Geophys. Res.*, 76, 6350, 1971.
54. **Shor, G. G. and Fornari, D. J.**, Seismic refraction measurements in the Kamchatka Basin, western Bering Sea, *J. Geophys. Res.*, 81, 5260, 1976.

55. **Murauchi, S., Ludwig, W. J., Den, N., Hotta, H., Asanuma, T., Yoshii, T., Kubotera, A., and Hagiwara, K.**, Seismic refraction measurements on the Ontong Java Plateau, northeast of New Ireland, *J. Geophys. Res.*, 78, 8653, 1973.
56. **Ewing, M., Hawkins, L. V., and Ludwig, W. J.**, Crustal structure of the Coral Sea, *J. Geophys. Res.*, 75, 1953, 1970.
57. **Shor, G. G., Kirk, H. K., and Menard, H. W.**, Crustal structure of the Melanesian area, *J. Geophys. Res.*, 76, 2562, 1971.
58. **Houtz, R. E. and Markl, R. G.**, Seismic profiler data between Antarctica and Australia, in *Antarctic Oceanology. II. The Australian-New Zealand Sector*, Hayes, D. E., Ed., American Geophysical Union, Washington, D.C., 1972, 19.
59. **Houtz, R. E.**, South Tasman basin and borderlands: a geophysical summary, in Initial Reports of the Deep Sea Drilling Project, Kennett, J. P., Houtz, R. E., et al., Eds., U.S. Government Printing Office, Washington, D.C., 1975, 29.
60. **Houtz, R. E. and Davey, F. J.**, Seismic profiles and sonobuoy measurements in Ross Sea, Antarctica, *J. Geophys. Res.*, 78, 3448, 1973.
61. **Lort, J. M., Limond, W. Q., and Gray, F.**, Preliminary seismic studies in the eastern Mediterranean, *Earth Planet Sci. Lett.*, 21, 355, 1974.
62. **Ludwig, W. J., Houtz, R. E., and Ewing, J. I.**, Profiler — sonobuoy measurements in Columbia and Venezuela Basins, Caribbean, *Am. Assoc. Pet. Geol. Bull.*, 59, 115, 1975.
63. **Steinhart, J. S. and Meyer, R. P.**, in *Explosion Studies of Continental Structure*, Carnegie Institution of Washington, Washington, D.C., 1961.
64. **Cram, I. H.**, A crustal structure refraction survey in South Texas, *Geophysics*, 26, 560, 1961.
65. **Roller, J. C. and Healy, J. H.**, Seismic refraction measurements of crustal structure between Santa Monica Bay and Lake Mead, *J. Geophys. Res.*, 68, 5837, 1963.
66. **Tatel, H. E. and Tuve, M. A.**, Seismic exploration of a continental crust, *Geol. Soc. Am. Spec. Pap.*, 62, 35, 1955.
67. **Gutenberg, B.**, Waves from blasts recorded in southern California, *Am. Geophys. Union Trans.*, 33, 427, 1952.
68. **Kanamori, H. and Hadley, D.**, Crustal structure and temporal velocity change in southern California, *Pure Appl. Geophys.*, 113, 257, 1975.
69. **Hamilton, R. M.**, Time term analysis of explosion data from the vicinity of the Borrego Mtn., California, earthquake of April 9, 1968, *Bull. Seismol. Soc. Am.*, 60, 367, 1970.
70. **Stewart, S. W. and Pakiser, L. C.**, Crustal structure in eastern New Mexico interpreted from the Gnome explosion, *Bull. Seismol. Soc. Am.*, 52, 1017, 1962.
71. **Warren, D. H., Healey, J. H., and Jackson, W. H.**, Crustal seismic measurements in southern Mississippi, *J. Geophys. Res.*, 71, 3437, 1966.
72. **Hales, A. L., Helsley, C. E., Dowling, J. J., and Nation, J. B.**, The East Coast onshore-offshore experiment. I. The first arrival phases, *Bull. Seismol. Soc. Am.*, 58, 757, 1968.
73. **Eaton, J. P.**, Crustal structure from San Francisco to Eureka, Nevada, from seismic refraction measurements, *J. Geophys. Res.*, 68, 5789, 1963.
74. **Hamilton, R. M., Ryall, A., and Berg, E.**, Crustal structure southwest of the San Andreas Fault from quarry blasts, *Bull. Seismol. Soc. Am.*, 54, 67, 1964.
75. **Filson, J.**, S velocities at near distances in western central California, *Bull. Seismol. Soc. Am.*, 60, 901, 1970.
76. **Lomnitz, C. and Bolt, B. A.**, Evidence on crustal structure in California from the Chase V explosion and the Chico earthquake of May 24, 1966, *Bull. Seismol. Soc. Am.*, 57, 1093, 1967.
77. **Healey, J. H. and Peake, L. G.**, Seismic velocity structure along a section of the San Andreas Fault near Bear Valley, California, *Bull. Seismol. Soc. Am.*, 65, 1177, 1975.
78. **Mikumo, T.**, Crustal structure in central California in relation to the Sierra Nevada, *Bull. Seismol. Soc. Am.*, 55, 65, 1965.
79. **Carder, D. S., Qamar, A., and McEvilly, T. V.**, Trans-California seismic profile — Pahute Mesa to San Francisco, *Bull. Seismol. Soc. Am.*, 60, 1829, 1970.
80. **Johnson, L. R.**, Crustal structure between Lake Mead, Nevada, and Mono Lake, California, *J. Geophys. Res.*, 70, 2863, 1965.
81. **Healey, J. H.**, Crustal structure along the coast of California from seismic refraction measurements, *J. Geophys. Res.*, 68, 5789, 1963.
82. **Press, F.**, Crustal structure in the California-Nevada region, *J. Geophys. Res.*, 65, 1039, 1960.
83. **Ryall, A. and Stuart, D. J.**, Travel times and amplitudes from nuclear explosions, Nevada Test Site to Ordway, Colorado, *J. Geophys. Res.*, 68, 5821, 1963.
84. **Hill, D. P. and Pakiser, L. C.**, Crustal structure between the Nevada Test site and Boise, Idaho, from seismic refraction measurements, in *The Earth Beneath the Continents*, Steinhart, D. H. and Smith, R. B., Eds., American Geophysical Union, Washington, D. C., 1966, 391.

85. **Carder, D. S. and Bailey, L. F.**, Seismic wave travel times from nuclear explosions, *Bull. Seismol. Soc. Am.*, 48, 377, 1958.
86. **Berg, J. W., Cook, K. L., Narans, H. D., and Dolan, W. M.**, Seismic investigation of crustal structure in the eastern part of the Basin and Range Province, *Bull. Seismol. Soc. Am.*, 50, 511, 1960.
87. **Keller, G. R., Smith, R. B., Braile, L. W., Heaney, R., and Shurbet, D. H.**, Upper crustal structure of the eastern Basin and Range, northern Colorado Plateau, and middle Rocky Mountains from Rayleigh wave dispersion, *Bull. Seismol. Soc. Am.*, 66, 869, 1976.
88. **Roller, J. C.**, Crustal structure in the eastern Colorado Plateaus province from seismic refraction measurements, *Bull. Seismol. Soc. Am.*, 55, 107, 1965.
89. **Diment, W. H., Stewart, S. W., and Roller, J. C.**, Crustal structure from the Nevada Test site to Kingman, Arizona, from seismic and gravity observations, *J. Geophys. Res.*, 66, 201, 1961.
90. **Toppozada, T. R. and Sanford, A. R.**, Crustal structure in central New Mexico interpreted from the Gasbuggy explosion, *Bull. Seismol. Soc. Am.*, 66, 877, 1976.
91. **Mitchell, B. J. and Landisman, M.**, Interpretation of a crustal section across Oklahoma, *Geol. Soc. Am. Bull.*, 81, 2647, 1970.
92. **Tryggvason, E. and Qualls, B. R.**, Seismic refraction measurements of crustal structure in Oklahoma, *J. Geophys. Res.*, 72, 3738, 1967.
93. **Stewart, S. W.**, Crustal structure in Missouri by seismic refraction methods, *Bull. Seismol. Soc. Am.*, 58, 291, 1968.
94. **Tatel, H. E., Tuve, M. A., and Hart, P. J.**, The Earth's crust seismic studies, *Carnegie Inst. Washington Yearb.*, 53, 43, 1954.
95. **Keller, G. R., Smith, R. B., and Braile, L. W.**, Crustal structure along the Great Basin-Colorado Plateau transition from seismic refraction studies, *J. Geophys. Res.*, 810, 1093, 1975.
96. **Braile, L. W., Smith, R. B., Keller, G. R., Welch, R. M., and Meyer, R. P.**, Crustal structure across the Wasatch Front from detailed seismic refraction surveys, *J. Geophys. Res.*, 79, 2669, 1974.
97. **McCamy, K. and Meyer, R. P.**, A correlation method of apparent velocity measurements, *J. Geophys. Res.*, 69, 691, 1964.
98. **Green, R. W. E. and Hales, A. L.**, The travel times of P waves to 30° in the central United States and upper mantle structure, *Bull. Seismol. Soc. Am.*, 58, 267, 1968.
99. **Ocola, L. C. and Meyer, R. P.**, Central North American rift system. I. Structure of the axial zone from seismic and gravimetric data, *J. Geophys. Res.*, 78, 5173, 1973.
100. **Katz, S.**, Seismic study of crustal structure in Pennsylvania and New York, *Bull. Seismol. Soc. Am.*, 45, 303, 1955.
101. **Dainty, A. M., Keen, C. E., Keen, M. J., and Blanchard, J. E.**, Review of geophysical evidence on crust and upper mantle structure on the eastern seaboard of Canada, in *The Earth Beneath the Continents*, Steinhart, J. S. and Smith, T. J., Eds., American Geophysical Union, Washington, D.C., 1966, 349.
102. **Langston, C. A.**, Corvallis, Oregon, crustal and upper mantle receiver structure from teleseismic P and S waves, *Bull. Seismol. Soc. Am.*, 67, 713, 1977.
103. **Langston, C. A. and Blum, D. E.**, The April 29, 1965, Puget Sound earthquake and the crustal and upper mantle structure of western Washington, *Bull. Seismol. Soc. Am.*, 67, 693, 1977.
104. **Tuve, M. A.**, Annual report of the Director of the Department of Terrestrial Magnetism, *Carnegie Inst. Washington Yearb.*, 50, 65, 1951.
105. **Johnson, S. H. and Couch, R. W.**, Crustal structure in the North Cascade Mountains of Washington and British Columbia from seismic refraction measurements, *Bull. Seismol. Soc. Am.*, 60, 1259, 1970.
106. **Hales, A. L. and Nation, J. B.**, A seismic refraction survey in the northern Rocky Mountains: more evidence for an intermediate crustal layer, *Geophys. J. R. Astron. Soc.*, 35, 381, 1973.
107. **Capon, J.**, Characterization of crust and upper mantle structure under LASA as a random medium, *Bull. Seismol. Soc. Am.*, 64, 235, 1974.
108. **Tatel, H. E. and Tuve, M. A.**, The Earth's crust-seismic studies, *Carnegie Inst. Washington Yearb.*, 52, 103, 1953.
109. **Massé, R. P.**, Compressional wave velocity distribution beneath central and eastern North America, *Bull. Seismol. Soc. Am.*, 63, 911, 1973.
110. **O'Brien, P. N. S.**, Lake Superior crustal structure — a reinterpretation of the 1963 seismic experiment, *J. Geophys. Res.*, 73, 2669, 1968.
111. **Berry, M. J. and West, G. F.**, An interpretation of the first arrival data of the Lake Superior experiment by the time-term method, *Bull. Seismol. Soc. Am.*, 56, 141, 1966.
112. **Johnson, S. H., Couch, R. W., Gemperle, M., and Banks, E. R.**, Seismic refraction measurements in southeast Alaska and western British Columbia, *Can. J. Earth Sci.*, 9, 1756, 1972.
113. **White, W. R. H. and Savage, J. C.**, A seismic refraction and gravity study of the earth's crust in British Columbia, *Bull. Seismol. Soc. Am.*, 55, 463, 1965.

114. **Berry, M. J. and Forsyth, D. A.**, Structure of the Canadian Cordillera from seismic refraction and other data, *Can. J. Earth Sci.*, 12, 182, 1975.
115. **Richard, T. C. and Walker, J. D.**, Measurement of the thickness of the Earth's crust in the Albertan Plains of Western Canada, *Geophysics*, 24, 262, 1959.
116. **Chandra, N. N. and Cumming, G. L.**, Seismic refraction studies in western Canada, *Can. J. Earth Sci.*, 9, 1099, 1972.
117. **Cumming, G. L. and Chandra, N. N.**, Further studies of reflections from the deep crust in southern Alberta, *Can. J. Earth Sci.*, 12, 539, 1975.
118. **Kanasewich, E. R. and Cumming, G. L.**, Near-vertical-incidence seismic reflections from the Conrad discontinuity, *J. Geophys. Res.*, 70, 3441, 1965.
119. **Cumming, G. L., Garland, G. D., and Vozoff, K.**, in *Seismological Measurement in Southern Alberta*, University of Alberta Physics Dept., Edmonton, 1962.
120. **Gurbuz, B. M.**, A study of the earth's crust and upper mantle using travel times and spectrum characteristics of body waves, *Bull. Seismol. Soc. Am.*, 60, 1921, 1970.
121. **Hall, D. H. and Hajnal, Z.**, Crustal structure of northeastern Ontario. Refraction seismology, *Can. J. Earth Sci.*, 6, 81, 1969.
122. **Mereu, R. F. and Hunter, J. A.**, Crustal and upper mantle structure under the Canadian Shield from Project Early Rise data, *Bull. Seismol. Soc. Am.*, 59, 147, 1969.
123. **Berg, E., Kubota, S., and Kienle, J.**, Preliminary determination of crustal structure in the Katmai National Monument, Alaska, *Bull. Seismol. Soc. Am.*, 57, 1367, 1963.
124. **Tatel, H. E. and Tuve, M. A.**, The Earth's crust — seismic studies, *Carnegie Inst. Washington Yearb.*, 55, 69, 1956.
125. **Hanson, K., Berg, E., and Gedney, L.**, A seismic refraction profile and crustal structure in central interior Alaska, *Bull. Seismol. Soc. Am.*, 58, 1657, 1968.
126. **Berg, E.**, Crustal structure in Alaska, *Tectonophysics*, 20, 165, 1973.
127. **Clee, T. E., Barr, K. G., and Berry, M. J.**, Fine structure of the crust near Yellowknife, *Can. J. Earth Sci.*, 11, 1534, 1974.
128. **LeBlanc, G. and Wetmiller, R. J.**, An evaluation of seismological data available for the Yukon Territory and the Mackenzie Valley, *Can. J. Earth Sci.*, 11, 1435, 1974.
129. **Sander, G. W. and Overton, A.**, Deep seismic refraction investigation in the Canadian Arctic archipelago, *Geophysics*, 30, 87, 1965.
130. **Mueller, S., Prodehl, C., Mendes, A. S., and Moreira, V. S.**, Crustal structure in the southwestern part of the Iberian Peninsula, *Tectonophysics*, 20, 307, 1973.
131. **Papazachos, B. C., Comninakis, P. S., and Drakopoulos, J. C.**, Preliminary results of an investigation of crustal structure in southeastern Europe, *Bull. Seismol. Soc. Am.*, 56, 1241, 1966.
132. **Neprochov, Y. P., Kosminskaya, I. P., and Malovitsky, Y. P.**, Structure of the crust and upper mantle of the Black and Caspian Seas, *Tectonophysics*, 10, 517, 1970.
133. **Neprochov, Y. P.**, The deep structure of the Earth's crust under the Black Sea southwest of the Crimea according to seismic data, *Dokl. Akad. Nauk SSSR*, 125, 1119, 1959.
134. **Balavadze, B. K. and Tvaltvadze, G. K.**, Structure of the Earth's crust in Georgia according to geophysical data, *Bull. Acad. Sci. USSR Geophys. Ser.*, 9, 623, 1958.
135. **Sollogub, V. B.**, On certain regularities of crustal structure associated with the major features of southeastern Europe, *Tectonophysics*, 10, 549, 1970.
136. **Mueller, S., Peterschmitt, E., Fuchs, K., Emter, D., and Ansorge, J.**, Crustal structure of the Rhinegraben area, in *Developments in Geotectonics*, Mueller, S., Ed., Elsevier, Amsterdam, 1974, 381.
137. **Miller, H.**, A lithospheric seismic profile along the axis of the Alps, 1975, *Pure Appl. Geophys.*, 114, 1109, 1976.
138. **Mueller, S., Peterschmitt, E., Fuchs, K., and Ansorge, J.**, Crustal structure beneath the Rhinegraben from seismic refraction and reflection, *Tectonophysics*, 8, 529, 1969.
139. **Giese, P. and Prodehl, C.**, Main features of crustal structure in the Alps, in *Explosion Seismology in Central Europe*, Giese, P., Prodehl, C., and Stein, A., Eds., Springer-Verlag, Berlin, 1976, 347.
140. **Will, M.**, Calculation of travel times and ray paths for lateral inhomogeneous media, in *Explosion Seismology in Central Europe*, Giese, P., Prodehl, C., and Stein, A., Eds., Springer-Verlag, Berlin, 1976, 168.
141. **Behnke, C. L., Giese, P., Prodehl, C. L., and DeVisintini, G.**, Seismic refraction investigations in the Dolomites for the exploration of the Earth's crust in the eastern Alpine area, *Boll. Geofis. Teor. Appl.*, 4, 110, 1962.
142. **Bott, H. P., Holder, A. P., Long, R. E., and Lucas, A. L.**, Crustal structure beneath the granites of southwest England, *Geol. J.* (Spec. Iss.), 2, 93, 1970.
143. **Blundell, D. J. and Parks, R.**, A study of crustal structure beneath the Irish Sea, *Geophys. J. R. Astron. Soc.*, 17, 45, 1969.

144. **Agger, H. E. and Carpenter, E. W.**, A crustal study in the vicinity of the Eskdalemuir seismological array station, *Geophys. J. R. Astron. Soc.*, 9, 69, 1964.
145. **Collette, B. J., Lagaay, R. A., Ritsema, A. R., and Schouter, J. A.**, Seismic investigations in the North Sea, *Geophys. J. R. Astron. Soc.*, 19, 183, 1970.
146. **Kanestrøm, R.**, Seismic investigations of the crust and upper mantle in Norway, in Deep Seismic Structure in Northern Europe, Vogel, A., Ed., Swedish National Science Research Council, 1971, 17.
147. **Reich, H., Foertsch, O., and Schulze, G. A.**, Results of seismic observations in Germany on the Heligoland explosion of April 18, 1947, *J. Geophys. Res.*, 56, 147, 1951.
148. **Hjelme, J.**, Review of seismic sounding of the crust below Denmark, in Deep Seismic Structure in Northern Europe, Vogel, A., Ed., Swedish National Science Research Council, 1971, 28.
149. **Knothe, C. and Walther, K.- F.**, Deep seismic sounding in the German Democratic Republic, in Deep Seismic Structure in Northern Europe, Vogel, A., Ed., Swedish National Science Research Council, 1971, 43.
150. **Gregersen, S.**, Profile section 4-5, in Deep Seismic Structure in Northern Europe, Vogel, A., Ed., Swedish National Science Research Council, 1971, 92.
151. **Penttila, A., Karros, M., Normia, M., Siirola, A., and Vesanen, E.**, Report on the 1959 explosion seismic investigation in southern Finland, Univ. Helsinki Publ. Seism., 35, 1960.
152. **Pentilla, E.**, Seismic investigations on the earth's crust in Finland, in Deep Seismic Structure in Northern Europe, Vogel, A., Ed., Swedish National Science Research Council, 1971, 9.
153. **Grubbe, K.**, Seismic-refraction measurements along two crossing profiles in northern Germany and their interpretation by a ray tracing method, in *Explosion Seismology in Central Europe*, Giese, P., Prodehl, C., and Stein, A., Eds., Springer-Verlag, Berlin, 1976, 268.
154. **Massé, R. P. and Alexander, S. S.**, Compressional velocity distribution beneath Scandinavia and western Russia, *Geophys. J. R. Astron. Soc.*, 39, 587, 1974.
155. **Kanestrøm, R. and Haugland, K.**, Profile section 3-4, in Deep Seismic Structure in Northern Europe, Vogel, A., Ed., Swedish National Science Research Council, 1971, 76.
156. **Kanestrøm, R.**, A crust-mantle model for the NORSAR area, *Pure Appl. Geophys.*, 105, 729, 1973.
157. **Båth, M.**, Average crustal structure of Sweden, *Pure Appl. Geophys.*, 88, 75, 1971.
158. **Vogel, A. and Lund, C.- E.**, Profile section 2-3, in *Deep Seismic Structure in Northern Europe*, Vogel, A., Ed., Swedish National Science Research Council, 1971, 62.
159. **Dahlman, O.**, Deep Seismic sounding in Sweden, in *Deep Seismic Structure in Northern Europe*, Vogel, A., Ed., Swedish National Science Research Council, 1971, 14.
160. **Wahlström, R.**, Seismic wave velocities in the Swedish crust, *Pure Appl. Geophys.*, 113, 673, 1975.
161. **Leong, L. S.**, Crustal structure of the Baltic Shield beneath Umeå, Sweden, from the spectral behavior of long period P waves, *Bull. Seismol. Soc. Am.*, 65, 113, 1975.
162. **Penttila, E.**, Profile section 1-2, in *Deep Seismic Structure in Northern Europe*, Vogel, A., Ed., Swedish National Science Research Council, 1971, 58.
163. **Anderson, A. J.**, Deep seismic sounding in north European part of U.S.S.R., in *Deep Seismic Structure of Northern Europe*, Vogel, A., Ed., Swedish National Science Research Council, 1971, 50.
164. **Matumoto, T., Ohtake, M., Latham, G., and Umana, J.**, Crustal structure in southern Central America, *Bull. Seismol. Soc. Am.*, 67, 121, 1977.
165. **Ocola, L. C., Aldrich, L. T., Gettrust, J. F., Meyer, R. P., and Ramirez, J. E.**, Project Narino. I. Crustal structure under southern Colombian-northern Ecuador Andes from seismic refraction data, *Bull. Seismol. Soc. Am.*, 65, 1681, 1975.
166. **Woollard, G. P.**, Seismic crustal studies during the IGY. II. Continental program, *IGY Bull.*, 34, 1960.
167. **Ocola, L. C. and Meyer, R. P.**, Crustal low-velocity zone under the Peru-Bolivia Altiplano, *Geophys. J. R. Astron. Soc.*, 30, 199, 1972.
168. Research Group for Explosion Seismology, Crustal structure in Central Japan as derived from the Miboro explosion-seismic observations. II. On the crustal structure, *Bull. Earthquake Res. Inst.*, 39, 327, 1961.
169. **Yoshii, T., Sasaki, Y., Tada, T., Okada, H., Asano, S., Muramatu, I., Hashizume, M., and Moriya, T.**, The third Kurayosi explosion and the crustal structure in the western part of Japan, *J. Phys. Earth*, 22, 109, 1974.
170. Research Group for Explosion Seismology, Crustal structure in northern Kwanto district by explosion-seismic observations, *Bull. Earthquake Res. Inst.*, 36, 329, 1958.
171. **Yoshii, T. and Asano, S.**, Time term analysis of explosion seismic data, *J. Phys. Earth*, 20, 47, 1972.
172. **Hashizume, M., Oike, K., Asano, S., Hamaguchi, H., Okada, A., Murauchi, S., Shima, E., and Nogoshi, M.**, Crustal structure in the profile across the northeastern part of Honshu, Japan, as derived from explosion seismic observations. II. Crustal structure, *Bull. Earthquake Res. Inst.*, 46, 607, 1968.

173. Research Group for Explosion Seismology, A: Observations of seismic waves from the second Hodaka explosion. B: On the crustal structure derived from observations of the second Hodaka explosion, *Bull. Earthquake Res. Inst.*, 37, 495, 1959.
174. Research Group for Explosion Seismology, The third explosion seismic observations in the northeastern Japan, *Bull. Earthquake Res. Inst.*, 31, 281, 1953.
175. **Finlayson, D. M., Muirhead, K. J., Webb, J. P., Gibson, G., Furomoto, A. S., Cooke, R. J. S., and Russell, A. J.**, Seismic investigation of the Papuan ultramafic belt, *Geophys. J. R. Astron. Soc.*, 44, 45, 1976.
176. **Cleary, J.**, Australian crustal structure, *Tectonophysics*, 20, 241, 1973.
177. **Bolt, B. A., Doyle, A. A., and Sutton, D. J.**, Seismic observations from the 1956 atomic explosions in Australia, *Geophys. J. R. Astron. Soc.*, 1, 135, 1958.
178. **Doyle, H. A., Everingham, I. B., and Hogan, T. K.**, Seismic recordings of large explosions in southeastern Australia, *Austr. J. Phys.*, 12, 222, 1959.
179. **Eiby, G. A.**, Crustal structure project, the Wellington profile, Wellington, New Zealand, *Geophys. Mem. N.Z. Dept. Sci. Ind. Res.*, 5, 1, 1957.
180. **Mueller, S. and Bonjer, K.- P.**, Average structure of the crust and upper mantle in East Africa, in *Developments in Geotectonics*, Mueller, S., Ed., Elsevier, Amsterdam, 1974, 283.
181. **Maguire, P. K. H. and Long, R. E.**, The structure on the western flank of the Gregory Rift (Kenya). I. The crust, *Geophys. J. R. Astron. Soc.*, 44, 661, 1976.
182. **Griffiths, D. H.**, Some comments on the results of a seismic refraction experiment in the Kenya Rift, *Tectonophysics*, 15, 151, 1972.
183. **Gane, P. G., Atkins, A. R., Sellschop, J. P. F., and Seligman, P.**, Crustal structure in the Transvaal, *Bull. Seismol. Soc. Am.*, 46, 293, 1956.
184. **Hales, A. L. and Sacks, I. S.**, Evidence for an intermediate layer from crustal structure studies in the eastern Transvaal, *Geophys. J. R. Astron. Soc.*, 2, 15, 1959.
185. **Willmore, P. L., Hales, A. L., and Gane, P. G.**, A seismic investigation of crustal structure in the western Transvaal, *Bull. Seismol. Soc. Am.*, 42, 53, 1952.
186. **Bhattacharya, S. N.**, The crust-mantle structure of the Indian Peninsula from surface wave dispersion, *Geophys. J. R. Astron. Soc.*, 36, 273, 1944.
187. **Arora, S. K.**, A study of the earth's crust near Gauribinaur in southern India, *Bull. Seismol. Soc. Am.*, 61, 671, 1971.
188. **Dube, R. K., Bhayana, J. C., and Choudhury, H. M.**, Crustal structure of the peninsular India, *Pure Appl. Geophys.*, 109, 1718, 1973.
189. **Dube, R. K. and Bhayana, J. C.**, Crustal structure in the Gangetic Plains of the Indian subcontinent from body waves, *Bull. Seismol. Soc. Am.*, 64, 571, 1974.
190. **Kaila, K. L., Reddy, P. R., and Narain, H.**, Crustal structure in the Himalayan foothills area north of India, from P-wave data of shallow earthquakes, *Bull. Seismol. Soc. Am.*, 58, 597, 1968.
191. **Verma, G. S.**, Structure of the foothills of the Himalayas, *Pure Appl. Geophys.*, 112, 18, 1974.
192. **Moazami-Goudarzi, P. K.**, La vitesse de phase des ondes de Rayleigh et les structures de la croûte et du manteau superiéur entre Machhad et Chiraz (Iran), *Pure Appl. Geophys.*, 112, 675, 1974.
193. **Tandon, A. N. and Dube, R. K.**, A study of the crustal structure beneath the Himalayas from body waves, *Pure Appl. Geophys.*, 111, 2207, 1973.
194. **Chun, K. Y. and Yoshii, T.**, Crustal structure of the Tibetan Plateau: a surface-wave study by a moving window analysis, *Bull. Seismol. Soc. Am.*, 67, 735, 1977.
195. **Kosminskaia, I. P. and Tulina, Y. Y.**, An experimental application of the seismic depth-sounding method to the investigation of the structure of the earth's crust in parts of western Turkmenia, *Bull. Acad. Sci. USSR Geophys. Ser.*, 7, 38, 1957.
196. **Godin, Y. N., Volvovski, B. S., Volvovski, I. S., and Fomenko, K. E.**, Determination of the structure of the earth's crust by means of regional seismic investigation on the Russian platform and in central Asia, *Bull. Acad. Sci. USSR Geophys. Ser.*, 10, 955, 1961.
197. **Kosminskaia, I. P., Mikhota, C. G., and Tulina, Y. V.**, Crustal structure in the Pamir-Alai Zone according to deep seismic sounding, *Bull. Acad. Sci. USSR Geophys. Ser.*, 10, 673, 1958.
198. **Gambortsev, G. A.**, Deep seismic crustal probing, *Trans. Geophys. Inst. Acad. Sci. USSR*, 25, 124, 1954.
199. **Ulomov, V. U.**, Some special features in the structure of the earth's crust in central Asia according to records of high power explosions, *Bull. Acad. Sci. USSR Geophys. Ser.*, 1, 83, 1960.
200. **Gambortsev, G. A., Vietsman, P. A., and Tulina, Y. V.**, The structure of the earth's crust in the northern Tienshan region according to seismic depth-sounding data, *Dokl. Akad. Nauk SSSR*, 105, 83, 1955.
201. **Demenitskaya, R. M.**, Basic features of the earth's crustal structure on geophysical data, *Trans. Sci. Res. Inst. Arctic Geol.*, 115, 1, 1961.

202. **Puzyrev, N. N., Mandelbaum, M. M., Krylov, S. V., Mishenkin, B. P., Krupskaya, G. V., and Petrick, G. V.,** Deep seismic investigations in the Baikal Rift zone, in *Developments in Geotectonics,* Mueller, S., Ed., Elsevier, Amsterdam, 1974, 85.
203. **Rezanov, I. A.,** The geological interpretation of the Magadan-Kolyna seismic depth-sounding profile, *Bull. Acad. Sci. USSR,* P, 555, 1963.
204. **Harrington, P. K., Barker, P. F. and Griffiths, D. H.,** Crustal structure of the South Orkney Islands area from seismic refraction and magnetic measurements, in *Antarctic Geology and Geophysics,* Adie, R. J., Ed., Universitetsforlaget, Oslo, 1972, 27.
205. **Kogan, A. L.,** Results of deep seismic sounding of the earth's crust in East Antarctica, in *Antarctic Geology and Geophysics,* Adie, R. J., Ed., Universitetsforlaget, Oslo, 1972, 485.
206. **Bentley, C. R. and Clough, J. W.,** Antarctic subglacial structure from seismic refraction measurements, in *Antarctic Geology and Geophysics,* Adie, R. J., Ed., Universitetsforlaget, Oslo, 1972, 683.
207. **Bentley, C. R. and Clough, J. W.,** Seismic refraction shooting in Ellsworth and Dronning Maud Lands, in *Antarctic Geology and Geophysics,* Adie, R. J., Ed., Universitetsforlaget, Oslo, 1972, 169.
208. **Dewart, G. and Toksöz, M. N.,** Crustal structure in East Antarctica from surface wave dispersion, *Geophys. J. R. Astron. Soc.,* 10, 127, 1965.
209. **Pålmason, G.,** *Crustal Structure of Iceland from Explosion Seismology,* Societas Scientiarum Islandica, Reykjavik, 1971.
210. **Båth, M.,** Crustal structure of Iceland, *J. Geophys. Res.,* 65, 1793, 1960.
211. **Tryggvason, E.,** Crustal structure of the Iceland region from dispersion of surface waves, *Bull. Seismol. Soc. Am.,* 52, 359, 1962.
212. **Furumoto, A. S., Weibenga, W. A., Webb, J. P., and Sutton, G. H.,** Crustal structure of the Hawaiian Archipelago, northeastern Melanesia, and the Central Pacific Basin by seismic refraction methods, in *Developments in Geotectonics,* Mueller, S., Ed., Elsevier, Amsterdam, 1974, 153.
213. **Anderson, D. L. and Hart, R. S.,** An earth model based on free oscillations and body waves, *J. Geophys. Res.,* 81, 1461, 1976.
214. **Pierce, G. W.,** Piezoelectric crystal oscillators applied to the precision measurement of the velocity of sound in air and CO_2 at high frequencies, *Proc. Am. Acad. Arts Sci.,* 60, 271, 1925.
215. **Heck, N. H. and Service, J. H.,** Velocity of Sound in Sea Water, *Spec. Publ. 108, Coast and Geodetic Survey, U.S. Department of Commerce,* 1924, 1.
216. **Sutton, G. H., Berckhemer, H., and Nafe, J. E.,** Physical analysis of deep-sea sediments, *Geophysics,* 22, 779, 1957.
217. **Hamilton, E. L.,** Sound speed and related properties of sediments from the experimental Mohole (Guadalupe site), *Geophysics,* 30, 257, 1965.
218. **Hamilton, E. L.,** Variations of density and porosity with depth in deep sea sediments, *J. Sediment. Petrol.,* 40, 280, 1976.
219. **Morton, R. W.,** Sound velocity in carbonate sediments from the Whitney Basin, Puerto Rico, *Mar. Geol.,* 19, 1, 1975.
220. **Laughton, A. S.,** Sound propagation in compacted ocean sediments, *Geophysics,* 22, 233, 1957.
221. **Ewing, M., Crary, A. P., and Thorne, A. M.,** Propagation of elastic waves in ice. I, *Physics,* 5, 165, 1934.
222. **Kurfurst, P. J.,** Ultrasonic wave measurements on frozen soils at permafrost temperatures, *Can. J. Earth Sci.,* 13, 1571, 1976.
223. **Timur, A.,** Velocity of compressional waves in porous media at permafrost temperatures, *Geophysics,* 33, 584, 1968.
224. **Dortman, N. B. and Magid, M. S.,** Velocity of elastic waves in crystalline rocks and its dependence on moisture content, *Dokl. Akad. Nauk SSSR Geophys. Ser.,* 179, 76, 1968.
225. **Woeber, A. F., Katz, S. and Ahrens, T. J.,** Elasticity of selected rocks and minerals, *Geophysics,* 28, 658, 1963.
226. **Watkins, J. S., Walters, L. A., and Godson, R. H.,** Dependence of in situ compressional wave velocities on porosity in undersaturated rocks, *Geophysics,* 37, 29, 1972.
227. **Mizutani, H. and Osako, M.,** Elastic wave velocities and thermal diffusivities of Apollo 17 rocks and their geophysical implications, *Proc. Fifth Lunar Sci. Conf.,* 5-3, 2891, 1974.
228. **Christensen, N. I., Fountain, D. M., Carlson, R. H., and Salisbury, M. H.,** Velocities and elastic modul of volcanic and sedimentary rocks recovered on DSDP Leg 25, in Initial Reports of the Deep Sea Drilling Project, Simpson, E. S. W., Schlich, R., et al., Eds., U.S. Government Printing Office, Washington, D.C., 1974, 25.
229. **Schock, R. N., Bonner, B. P., and Louis, H.,** Collection of ultrasonic velocity data as a function of pressure for polycrystalline solids, *Lawr. Live. Lab. Tech. Rept.,* UCRL-51508, 1, 1974.
230. **Mizutani, H., Fujii, N., Hamano, Y., and Osako, M.,** Elastic wave velocities and thermal diffusivities of Apollo 14 rocks, *Proc. Third Lunar Sci. Conf.,* 3-3, 2557, 1972.

231. **Fox, P. J., Schreiber, E., and Peterson, J. J.**, The geology of the oceanic crust: compressional wave velocities of oceanic rocks, *J. Geophys. Res.*, 78, 5155, 1973.
232. **Christensen, N. I., Salisbury, M. H., Fountain, D. M., and Carlson, R. L.**, Velocities of compressional and shear waves in DSDP Leg 27 basalts, in Initial Reports of the Deep Sea Drilling Project, Veevers, J. J., Heirtzler, J. R., et al., Eds., U.S. Government Printing Office, Washington, D.C., 1974, 27.
233. **Fox, P. J., Schreiber, E., and Peterson, J.**, Compressional wave velocities in basalt and altered basalt recovered during Leg 14, in Initial Reports of the Deep Sea Drilling Project, Hayes, D. E., Pimm, A. C., et al., Eds., U.S. Government Printing Office, Washington, D.C., 1972, 14, 773.
234. **Hyndman, R. D.**, Seismic velocity measurements of basement rocks from DSDP Leg 37, in Initial Reports of the Deep Sea Drilling Project, Aumento, F. and Melson, W. G., Eds., U.S. Government Printing Office, Washington, D.C., 1976, 37, 373.
235. **Christensen, N. I. and Salisbury, M. H.**, Velocities, elastic moduli and weathering-age relations for Pacific layer 2 basalts, *Earth Planet. Sci. Lett.*, 19, 461, 1973.
236. **Christensen, N. I., Carlson, R. L., Salisbury, M. H., and Fountain, D. M.**, Elastic wave velocities in volcanic and plutonic rocks recovered on DSDP Leg 31, in Initial Reports of the Deep Sea Drilling Project, Karig, D. E., Ingle, J. C., et al., Eds., U.S. Government Printing Office, Washington, D.C., 1975, 31.
237. **Talwani, P., Nur, A., and Kovach, R. L.**, Implications of elastic wave velocities for Apollo 17 rock powders, *Proc. Fifth Lunar Sci. Conf.*, 5-3, 2919, 1974.
238. **Kanamori, H., Nur, A., Chung, D., and Simmons, G.**, Elastic wave velocities of lunar samples at high pressures and their geophysical implications, *Proc. First Lunar Sci. Conf.*, 1-3, 2289, 1970.
239. **Schreiber, E., Perfit, M., and Cernock, P. J.**, Compressional wave velocities in samples recovered by DSDP Leg 24, in Initial Reports of the Deep Sea Drilling Project, Fisher, R. L., et al., Eds., U.S. Government Printing Office, Washington, D.C., 1974, 24, 787.
240. **Schreiber, E., Fox, P. J., and Peterson, J. J.**, Compressional wave velocities in selected samples of gabbro, schist, limestone, anhydrite, gypsum and halite, in Initial Reports of the Deep Sea Drilling Project, Ryan, W. B. E., Hsu, K. J., et al., Eds., U.S. Government Printing Office, Washington, D.C., 1972, 13, 595.
241. **Schreiber, E. and Fox, P. J.**, Density and P-wave velocity of rocks from the FAMOUS region and their implication to the structure of the oceanic crust, *Geol. Soc. Am. Bull.*, 88, 600, 1977.
242. **Fox, P. J. and Schreiber, E.**, Compressional wave velocities in basalt and dolerite samples recovered during Leg 15, in Initial Reports of the Deep Sea Drilling Project, Edgar, N. T., Saunders, J. B., et al., Eds., U.S. Government Printing Office, Washington, D.C., 1973, 15, 1013.
243. **Christensen, N. I.**, Compressional and shear wave velocities in basaltic rocks, DSDP Leg 16, in Initial Reports of the Deep Sea Drilling Project, van Andel, T. H., Heath, G. R., et al., Eds., U.S. Government Printing Office, Washington, D.C., 1973, 16, 647.
244. **Todd, T., Wang, H., Baldridge, W. S., and Simmons, G.**, Elastic properties of Apollo 14 and 15 rocks, *Proc. Third Lunar Sci. Conf.*, 3-3, 2577, 1972.
245. **Christensen, N. I. and Salisbury, M. H.**, Sea floor spreading, progressive alteration of layer 2 basalts, and associated changes in seismic velocities, *Earth Planet. Sci. Lett.*, 15, 367, 1972.
246. **Christensen, N. I.**, Elasticity of ultrabasic rocks, *J. Geophys. Res.*, 71, 5921, 1966.
247. **Nur, A. and Simmons, G.**, The effect of saturation on velocity in low porosity rocks, *Earth Planet. Sci. Lett.*, 7, 183, 1969.
248. **Christensen, N. I.**, The abundance of serpentinites in the oceanic crust, *J. Geol.*, 80, 709, 1972.
249. **Birch, F.**, Velocity of compressional waves in serpentine from Mayaguez, Puerto Rico, in *A Study of Serpentine*, Burk, C. A., Ed., National Academy of Science — National Research Council, Washington, D.C., 1964, 132.
250. **Christensen, N. I.**, The geophysical significance of oceanic plagiogranite, *Earth Planet. Sci. Lett.*, 36, 297, 1977.
251. **Manghnani, M. H. and Woollard, G. P.**, Elastic wave velocities in Hawaiian rocks at pressures to ten kilobars, in *The Crust and Upper Mantle of the Pacific Area*, Knopoff, L., Drake, C. L., and Hart, P. J., Eds., American Geophysical Union, Washington, D.C., 1968, 12, 501.
252. **Christensen, N. I.**, Ophiolites, seismic velocities, and oceanic crustal structure, *Tectonophys.*, 47, 131, 1978.
253. **Christensen, N. I.**, Seismic velocities, densities and elastic constants of basalts from DSDP Leg 35, in Initial Reports of the Deep Sea Drilling Project, Hollister, C. D., Craddock, C., et al., Eds., U.S. Government Printing Office, Washington, D.C., 1976, 35, 335.
254. **Birch, F.**, The velocity of compressional waves in rocks to 10 kilobar, 1, *J. Geophys. Res.*, 65, 1083, 1960.
255. **Christensen, N. I.**, Compressional and shear wave velocities and elastic moduli of basalts, DSDP Leg 19, in *Initial Reports of the Deep Sea Drilling Project*, Creager, J. S., Scholl, D. W., et al., Eds., U.S. Government Printing Office, Washington, D.C., 1973, 19, 657.

256. **Mizutani, H. and Newbigging, D. F.**, Elastic wave velocities of Apollo 14, 15 and 16 rocks, *Proc. Fourth Lunar Sci. Conf.*, 4-3, 2601, 1973.
257. **Hyndman, R. D.**, Seismic velocities of basalts from DSDP Leg 26, in Initial Reports of the Deep Sea Drilling Project, Davies, T. A., Luyendyk, B. P., et al., Eds., U.S. Government Printing Office, Washington, D.C., 1974, 26, 509.
258. **Stewart, R. and Peselnick, L.**, Velocity of compressional waves in dry Franciscan rocks to 8 kilobar and 300°C, *J. Geophys. Res.*, 82, 2027, 1977.
259. **Schreiber, E. and Fox, P. J.**, Compressional wave velocities and mineralogy of fresh basalts from the FAMOUS area and the Oceanographer Fracture Zone and the texture of Layer 2a of the oceanic crust, *J. Geophys. Res.*, 81, 4071, 1976.
260. **Fox, P. J., Schreiber, E., Rowlett, H., and McKamy, K.**, The geology of the Oceanographer Fracture Zone: a model for fracture zones, *J. Geophys. Res.*, 81, 4117, 1976.
261. **Salisbury, M. H. and Christensen, N. I.**, Sonic velocities and densities of basalts from the Nazca Plate, DSDP Leg 34, in Initial Reports of the Deep Sea Drilling Project, Yeats, R. S., Hart, S. R., et al., Eds., U.S. Government Printing Office, Washington, D.C., 1976, 34, 543.
262. **Christensen, N. I. and Shaw, G. H.**, Elasticity of mafic rocks from the mid-Atlantic Ridge, *Geophys. J. R. Astron. Soc.*, 20, 271, 1970.
263. **Chung, D. H.**, Elastic wave velocities in anorthosite and anorthositic gabbros from Apollo 15 and 16 landing sites, *Proc. Fourth Lunar Sci. Conf.*, 4-3, 2591, 1973.
264. **Christensen, N. I.**, Compressional wave velocities in basalts from the Juan de Fuca Ridge, *J. Geophys. Res.*, 75, 2773, 1970.
265. **Bajuk, E. I., Volarovich, M. P., Klima, K., Pros, Z., and Vanek, J.**, Velocity of longitudinal waves in eclogite and ultrabasic rocks under pressures to 4 kilobars, *Stud. Geophys. Geod.*, 11, 271, 1957.
266. **Simmons, G. and Brace, W. F.**, Comparison of static and dynamic measurements of compressibility of rocks, *J. Geophys. Res.*, 70, 5649, 1965.
267. **Iida, K., Sugino, T., Furuhashi, H., and Kumazawa, M.**, Elastic dilational wave velocity in crystalline schists from Sanbagawa metamorphic terrain, Shikoku, Japan, *J. Earth Sci. Nagoya Univ.*, 15, 112, 1967.
268. **Wang, C.**, Velocity of compressional waves in limestones, marbles and a single crystal of calcite to 20 kilobars, *J. Geophys. Res.*, 71, 3543, 1966.
269. **Fountain, D. M.**, The Ivrea-Verbano and Strona-Ceneri Zones, northern Italy: a cross-section of the continental crust — new evidence from seismic velocities of rock samples, *Tectonophysics*, 33, 145, 1976.
270. **Christensen, N. I.**, Compressional wave velocities in metamorphic rocks at pressures to 10 kilobar, *J. Geophys. Res.*, 70, 6147, 1965.
271. **Kanamori, H. and Mizutani, H.**, Ultrasonic measurements of elastic constants of rocks under high pressures, *Bull. Earthquake Res. Inst.*, 43, 173, 1965.
272. **Christensen, N. I.**, Compressional and shear wave velocities at pressures to 10 kilobars for basalts from the East Pacific Rise, *Geophys. J. R. Astron. Soc.*, 28, 425, 1972.
273. **Christensen, N. I.**, Compressional wave velocities in basic rocks, *Pac. Sci.*, 22, 41, 1968.
274. **Hughes, D. S. and Cross, J. H.**, Elastic wave velocities in rocks at high pressure and temperature, *Geophysics*, 16, 577, 1951.
275. **Wang, H., Todd, T., Richter, D., and Simmons, G.**, Elastic properties of plagioclase aggregates and seismic velocities in the moon, *Proc. 4th Lunar Sci. Conf.*, 4-3, 2663, 1973.
276. **Hughes, D. S. and Maurette, C.**, Variation of elastic wave velocities in granites with pressure and temperature, *Geophysics*, 21, 277, 1956.
277. **Christensen, N. I. and Fountain, D. M.**, Constitution of the lower continental crust based on experimental studies of seismic velocities in granulites, *Geol. Soc. Am. Bull.*, 86, 227, 1975.
278. **Todd, T. and Simmons, G.**, Effect of pore pressure on the velocity of compressional waves in low porosity rocks, *J. Geophys. Res.*, 77, 3731, 1972.
279. **Pros, Z., Vanek, J., and Klima, K.**, The velocity of elastic waves in diabase and greywacke under pressures up to 4 kilobars, *Stud. Geophys. Geod.*, 6, 347, 1962.
280. **Kanamori, H., Mizutani, H., and Hamano, Y.**, Elastic wave velocities of Apollo 12 rocks at high pressures, *Proc. 2nd Lunar Sci. Conf.*, 2-3, 2323, 1971.
281. **Birch, F.**, The velocity of compressional waves in rocks to 10 kilobars, 2, *J. Geophys. Res.*, 66, 2199, 1961.
282. **Manghnani, M. H., Ramananantoandro, R., and Clark, S. P.**, Compressional and shear wave velocities in granulite facies rocks and eclogites to 10 kilobars, *J. Geophys. Res.*, 79, 5427, 1974.
283. **Christensen, N. I.**, Possible greenschist facies metamorphism of the oceanic crust, *Geol. Soc. Am. Bull.*, 81, 905, 1970.
284. **Kumazawa, M. H., Helmstaedt, H., and Masaki, K.**, Elastic properties of eclogite xenoliths from diatremes of the east Colorado plateau and their implications to the upper mantle structure, *J. Geophys. Res.*, 76, 1231, 1971.

285. **Simmons, G.**, Velocity of compressional waves in various minerals at pressures to 10 kilobars, *J. Geophys. Res.*, 69, 1117, 1964.
286. **Christensen, N. I.**, Elastic properties of polycrystalline magnesium, iron, and manganese carbonates to 10 kilobars, *J. Geophys. Res.*, 77, 369, 1972.
287. **Kroenke, I. W., Manghnani, M. H., Rai, C. S., Fryer, P., and Ramananantoandro, R.**, Elastic properties of selected ophiolitic rocks from Papua, New Guinea: nature and composition of oceanic lower crust and upper mantle, in *The Geophysics of the Pacific Ocean Basin and its Margins*, Sutton, G. H., Manghnani, M. H., and Moberly, R., Eds., American Geophysical Union, Washington, D.C., 1976, 19, 407.
288. **Christensen, N. I.**, Fabric, seismic anisotropy and tectonic history of the Twin Sisters dunite, *Geol. Soc. Am. Bull.*, 82, 1681, 1971.
289. **Rao, M., Ramana, Y. V., and Gogte, B. S.**, Dependence of compressional velocity on the mineral chemistry of eclogites, *Earth Planet. Sci. Lett.*, 23, 15, 1974.
290. **Mao, N.- H., Ito, J., Hays, J. F., Drake, J., and Birch, F.**, Composition and elastic constants of hortonolite dunite, *J. Geophys. Res.*, 75, 4071, 1970.
291. **Babuska, V.**, Elasticity and anisotropy of dunite and bronzitite, *J. Geophys. Res.*, 77, 6955, 1972.
292. **Christensen, N. I. and Ramananantoandro, R.**, Elastic moduli and anisotropy of dunite to 10 kilobars, *J. Geophys. Res.*, 76, 4003, 1971.
293. **Simmons, G.**, The velocity of shear waves in rocks to 10 kilobar, 1, *J. Geophys. Res.*, 69, 1123, 1964.
294. **Christensen, N. I.**, Shear wave velocities in metamorphic rocks at pressures to 10 kilobars, *J. Geophys. Res.*, 71, 3549, 1966.
295. **Christensen, N. I.**, Compressional wave velocities in possible mantle rocks to pressures of 30 kilobars, *J. Geophys. Res.*, 79, 407, 1974.
296. **Wyllie, M. R. J., Gregory, A. R., and Gardner, G. H. F.**, An experimental investigation of factors affecting elastic wave velocities in porous media, *Geophysics*, 23, 459, 1958.
297. **King, M. S.**, Wave velocities in rocks as a function of changes in overburden pressure and pore fluid saturants, *Geophysics*, 31, 50, 1966.
298. **Timur, A.**, Temperature dependence on compressional and shear wave velocities in rocks, *Geophysics*, 42, 950, 1977.
299. **Fielitz, K.**, Elastic wave velocities in different rocks at high pressure and temperatures up to 750°C, *Z. Geophys.*, 37, 943, 1971.
300. **Hughes, D. S. and Maurette, C.**, Variation of elastic wave velocities in granites with pressure and temperature, *Geophysics*, 21, 277, 1956.
301. **Hughes, D. S. and Cross, J. H.**, Elastic wave velocities in rocks at high pressures and temperatures, *Geophysics*, 16, 577, 1951.
302. **Hughes, D. S. and Maurette, C.**, Variation of elastic wave velocities in basic igneous rocks with pressure and temperature, *Geophysics*, 22, 23, 1957.
303. **Birch, F.**, Elasticity of igneous rocks at high temperatures and pressures, *Geol. Soc. Am. Bull.*, 54, 263, 1943.
304. **Peselnick, L. and Stewart, R. M.**, A sample assembly for velocity measurements of rocks at elevated temperatures and pressures, *J. Geophys. Res.*, 80, 3765, 1975.
305. **Jones, L. E. A. and Liebermann, R. C.**, Elastic and thermal properties of fluoride and oxide analogues in the rocksalt, fluorite, rutile and perovskite structures, *Phys. Earth Planet. Inter.*, 9, 101, 1974.
306. **Schreiber, E. and Anderson, O. L.**, Temperature dependence of the velocity derivatives of periclase, *J. Geophys. Res.*, 71, 3007, 1966.
307. **Schreiber, E. and Anderson, O. L.**, Revised data on polycrystalline magnesium oxide, *J. Geophys. Res.*, 73, 2837, 1968.
308. **Soga, N. and Anderson, O. L.**, High temperature elastic properties of polycrystalline MgO and Al_2O_3, *J. Am. Ceram. Soc.*, 49, 355, 1966.
309. **Chung, D. H. and Simmons, G.**, Elastic properties of polycrystalline periclase, *J. Geophys. Res.*, 74, 2133, 1969.
310. **Soga, N.**, New measurements on the sound velocity of calcium oxide and its relation to Birch's law, *J. Geophys. Res.*, 72, 5157, 1967.
311. **Soga, N.**, Elastic properties of CaO under pressure and temperature, *J. Geophys. Res.*, 73, 5385, 1968.
312. **Akimoto, S.**, The system MgO-FeO-Sio_2 at high pressures and temperatures: phase equilibria and elastic properties, *Tectonophysics*, 13, 161, 1972.
313. **Notis, M. R., Spriggs, R. M., and Hahn, W. C.**, Elastic moduli of pressure-sintered nickel oxide, *J. Geophys. Res.*, 76, 7052, 1971.
314. **Liebermann, R. C., Jackson, I., and Ringwood, A. E.**, Elasticity and phase equilibria of spinel disproportionation reactions, *Geophys. J. R. Astron. Soc.*, 50, 553, 1977.

315. **Chung, D. H.**, Elasticity of high pressure phases, EOS, 54, 475, 1973.
316. **Liebermann, R. C.**, Elastic properties of germanate analogues of olivine, spinel, and β-polymorphs of $(Mg,Fe)_2SiO_4$, *Nature (London) Phys. Sci.*, 244, 105, 1973.
317. **Mizutani, H., Hamano, Y., Iida, Y., and Akimoto, S.**, Compressional-wave velocities in fayalite, Fe_2SiO_4 spinel, and coesite, *J. Geophys. Res.*, 75, 2741, 1970.
318. **Liebermann, R. C.**, Elasticity of olivine (α), beta (β), and spinel (γ) polymorphs of germanates and silicates, *Geophys. J. R. Astron. Soc.*, 42, 899, 1975.
319. **Syono, Y., Fukai, Y., and Ishikawa, Y.**, Anomalous elastic properties of Fe_2TiO_4, *J. Phys. Soc. Jpn.*, 31, 471, 1971.
320. **Liebermann, R. C.**, Pressure and temperature dependence of the elastic properties of polycrystalline trevorite ($NiFe_2O_4$), *Phys. Earth Planet. Intern.*, 6, 360, 1973.
321. **Chung, D. H.**, General relationships among sound speeds. 1. New experimental information, *Phys. Earth Planet. Inter.*, 8, 113, 1974.
322. **Shaw, G.**, Phase transitions, elasticity-density relations and the univalent halides, *J. Geophys. Res.*, 79, 2635, 1974.
323. **Liebermann, R. C., Jones, L., and Ringwood, A. E.**, Elasticity of aluminate, titanate, stannate and germanate compounds with the perovskite structure, *Phys. Earth Planet. Inter.*, 14, 165, 1977.
324. **Liebermann, R. C.**, Elasticity of pyroxene-garnet and pyroxene-ilmenite phase transformations in germanates, *Phys. Earth Planet. Inter.*, 8, 361, 1974.
325. **Mizutani, H., Hamano, Y., and Akimoto, S.- I.**, Elastic wave velocities of polycrystalline stishovite, *J. Geophys. Res.*, 77, 3744, 1972.
326. **Chung, D. H. and Buessem, W. R.**, The Voigt-Reuss-Hill approximation and the elastic moduli of polycrystalline Zno, TiO_2 (rutile) and α-Al_2O_3, *J. Appl. Phys.*, 39, 2777, 1968.
327. **Liebermann, R. C.**, Compressional velocities of polycrystalline olivine, spinel and rutile minerals, *Earth Planet. Sci. Lett.*, 17, 263, 1972.
328. **Chung, D. H. and Simmons, G.**, Pressure derivatives of the elastic properties of polycrystalline quartz and rutile, *Earth Planet. Sci. Lett.*, 6, 134, 1969.
329. **Soga, N.**, Sound velocity of some germanate compounds and its relation to the law of corresponding states, *J. Geophys. Res.*, 76, 3983, 1971.
330. **Liebermann, R. C.**, Elastic properties of polycrystalline SnO_2 and GeO_2: comparison with stishovite and rutile data, *Phys. Earth Planet. Inter.*, 7, 461, 1973.
331. **Schreiber, E. and Anderson, O. L.**, The pressure derivatives of the sound velocities of polycrystalline alumina, *J. Am. Ceram. Soc.*, 49, 184, 1966.
332. **Chung, D. H. and Simmons, G.**, The pressure and temperature dependences of the isotropic elastic moduli of polycrystalline alumina, *J. Appl. Phys.*, 39, 5316, 1968.
333. **Rossi, L. R. and Lawrence, W. G.**, Elastic properties of oxide solid solutions: the system Al_2O_3-Cr_2O_3, *J. Am. Ceram. Soc.*, 53, 604, 1970.
334. **Liebermann, R. C. and Schreiber, E.**, Elastic constants of polycrystalline hematite as a function of pressure to 3 kilobars, *J. Geophys. Res.*, 73, 6585, 1968.
335. **Soga, N.**, Elastic constants of BeO as a function of pressure and temperature, *J. Am. Ceram. Soc.*, 52, 246, 1969.
336. **Bentle, G. G.**, Some elastic properties of BeO at room temperature, *J. Nucl. Mater.*, 6, 336, 1962.
337. **Soga, N. and Anderson, O. L.**, Anomalous behavior of the shear sound velocity under pressure for polycrystalline ZnO, *J. Appl. Phys.*, 38, 2985, 1967.
338. **Liebermann, R. C.**, Elasticity of ilmenites, *Phys. Earth Planet. Inter.*, 12, 5, 1976.
339. **Liebermann, R. C.**, Elasticity of the ilmenite-perovskite phase transformation in $CdTiO_3$, *Earth Planet. Sci. Lett.*, 29, 326, 1976.
340. **Schreiber, E. and Anderson, O. L.**, Pressure derivatives of the sound velocities of polycrystalline forsterite with 6% porosity, *J. Geophys. Res.*, 72, 762, 1967.
341. **Chung, D. H.**, Elasticity and equations of state of olivines in the Mg_2SiO_4—Fe_2SiO_4 system, *Geophys. J. R. Astron. Soc.*, 25, 511, 1971.
342. **Liebermann, R. C. and Mayson, D. S.**, Elastic properties of polycrystalline diopside ($CaMgSi_2O_6$), in press, 1977.
343. **Liebermann, R. C. and Mayson, D. J.**, Elastic properties of polycrystalline anorthite ($CaAl_2Si_2O_8$), in press, 1977.
344. **Anderson, O. L. and Schreiber, E.**, The pressure derivatives of the sound velocities of polycrystalline magnesia, *J. Geophys. Res.*, 70, 5241, 1965.
345. **Schreiber, E. and Anderson, O. L.**, Correction to paper by E. Schreiber and O. L. Anderson, 'Pressure derivatives of sound velocities of polycrystalline forsterite with 6% porosity', *J. Geophys. Res.*, 72, 3751, 1967.
346. **Ryzhova, T. V., Aleksandrov, K. S., and Korobkova, V. M.**, The elastic properties of rock-forming minerals. V. Additional data on silicates, *Izv. Acad. Sci. USSR Phys. Solid Earth*, 2, 111, 1966.

347. **Ryzhova, T. V. and Aleksandrov, K. S.**, The elastic properties of potassium-sodium feldspars, *Bull. Acad. Sci. USSR Geophys. Ser.*, 7, 53, 1965.
348. **Ryzhova, T. V.**, Elastic properties of plagioclase, *Bull. Acad. Sci. USSR Geophys. Ser.*, 7, 633, 1964.
349. **Alexandrov, K. S. and Ryzhova, T. V.**, Elastic properties of rock-forming minerals. 3. Feldspars, *Bull. Acad. Sci. USSR Geophys. Ser.*, 2, 1129, 1962.
350. **McSkimin, H. J., Andreatch, P., and Thurston, R. W.**, Elastic moduli of quartz versus hydrostatic pressure at 25° and −195.8°C, *J. Appl. Phys.*, 36, 1624, 1965.
351. **Dandekar, D. P.**, Pressure dependence of the elastic constants of calcite, *Phys. Rev.*, 172, 873, 1968.
352. **Aleksandrov, K. S. and Ryzhova, T. V.**, Elastic properties of rock-forming minerals. 2. Layered silicates, *Bull. Acad. Sci. USSR Geophys. Ser.*, 9, 1165, 1961.
353. **Aleksandrov, K. S. and Ryzhova, T. V.**, The elastic properties of rock-forming minerals. 1. Pyroxenes and amphiboles, *Bull. Acad. Sci. USSR Geophys. Ser.*, 9, 871, 1961.
354. **Graham, E. K. and Barsch, G. R.**, Elastic constants of single-crystal forsterite as a function of temperature and pressure, *J. Geophys. Res.*, 74, 5949, 1969.
355. **Kumazawa, M. and Anderson, O. L.**, Elastic moduli, pressure derivatives, and temperature derivatives of single-crystal olivine and single-crystal forsterite, *J. Geophys. Res.*, 74, 5961, 1969.
356. **Aleksandrov, K. S., Ryzhova, T. V., and Belikov, B. P.**, The elastic properties of pyroxenes, *Sov. Phys. Crystallogr.*, 8, 589, 1964.
357. **Verma, R. K.**, Elasticity of some high density crystals, *J. Geophys. Res.*, 65, 757, 1960.
358. **Kumazawa, M.**, The elastic constants of single-crystal orthopyroxene, *J. Geophys. Res.*, 74, 5973, 1969.
359. **Frisillo, A. L. and Barsch, G. R.**, Measurement of single-crystal elastic constants of bronzite as a function of pressure and temperature, *J. Geophys. Res.*, 77, 6360, 1972.
360. **Ryzhova, T. V., Reshchikova, L. M., and Aleksandrov, K. S.**, Elastic properties of rock-forming minerals. 6. Garnets, *Bull. Acad. Sci. USSR Geophys. Ser.*, 7, 447, 1966.
361. **Schreiber, E.**, Elastic moduli of single crystal spinel at 25°C and to 2 kilobar, *J. Appl. Phys.*, 38, 2508, 1967.
362. **Wang, H. and Simmons, G.**, Elasticity of some mantle crystal structures. 1. Pleonaste and hercynite spinel, *J. Geophys. Res.*, 77, 4379, 1972.
363. **Soga, N.**, Elastic constants of garnet under pressure and temperature, *J. Geophys. Res.*, 72, 4227, 1967.
364. **Wang, H. and Simmons, G.**, Elasticity of some mantle crystal structures. 3. Spessartite-almandine garnet, *J. Geophys. Res.*, 79, 2607, 1974.
365. **Manghnani, M. H.**, Elastic constants of single-crystal rutile under pressures to 7.5 kilobars, *J. Geophys. Res.*, 74, 4317, 1969.
366. **Manghnani, M. H., Fisher, E. S., and Brower, W. S.**, Temperature dependence of the elastic constants of single-crystal rutile between 4° and 583°K, *J. Phys. Chem. Solids*, 33, 2149, 1972.
367. **Anderson, O. L., Scholz, C., Soga, N., Warren, N., and Schreiber, E.**, Elastic properties of a microbreccia, igneous rock and lunar fines from Apollo 11 mission, *Proc. Apollo 11 Lunar Sci. Conf.*, 1-3, 1959, 1970.
368. **Soga, N. and Anderson, O. L.**, Elastic properties of tektites measured by resonant sphere technique, *J. Geophys. Res.*, 72, 1733, 1967.
369. **Spinner, S.**, Elastic moduli of glasses by a dynamic method, *J. Am. Ceram. Soc.*, 37, 229, 1954.
370. **Manghnani, M. H.**, Pressure and temperature dependence of the elastic moduli of Na_2O-TiO_2-SiO_2 glasses, *J. Am. Ceram. Soc.*, 55, 360, 1972.
371. **Sokolowski, T. J. and Manghnani, M. H.**, Adiabatic elastic moduli of vitreous calcium aluminates to 3.5 kilobar, *J. Am. Ceram. Soc.*, 52, 539, 1969.
372. **Molotova, L. V. and Vassil'ev, Y. I.**, Velocity ratio of longitudinal and transverse waves in rocks, 2, *Bull. Acad. Sci. USSR Geophys. Ser.*, 8, 731, 1960.
373. **Hughes, D. S., Pondrom, W. L., and Mims, R. L.**, Transmission of elastic pulses in metal rods, *Phys. Rev.*, 75, 1552, 1949.
374. **Schreiber, E. and Anderson, O. L.**, Properties and composition of lunar materials: earth analogies, *Science*, 168, 1579, 1970.
375. **Hughes, D. S., Blankenship, E. B., and Mims, R. L.**, Variation of elastic wave moduli with pressure and temperature in plastics, *J. Appl. Phys.*, 21, 294, 1950.

Section VII
Seismic Attenuation

By
Marius Vassiliou, Carlos A. Salvado, and Bernhard R. Tittmann

INTRODUCTION

The attenuation of seismic waves in rock has been the subject of considerable attention, both experimental and theoretical, in recent years. We may conveniently divide much of the work that has been done into two broad classes.

One class comprises work done with a view toward understanding attenuation in the interior of the earth. There is a broad connection here with the larger questions of geodynamics, involving mantle flow and global tectonic processes. Seismologically, one deals here generally with teleseismic data, seismograms recorded on the global array of WWSSN instruments. Attenuation measurements have been made both for traveling waves and for free oscillations. Laboratory experimental data relevant to this problem are those conducted at high temperature-pressure and low frequency. Unfortunately, although some work has been done, very few such data have been obtained. Most of the work in this class has been theoretical and seismological.

The second class comprises work done with a view toward understanding attenuation in the shallow crust of the earth. Here, laboratory data are relatively much more numerous. Since the focus of many of the studies has been to help develop attenuation as a tool for the exploration for oil and gas, much work has been done on sedimentary rocks, especially sandstones. The physical mechanisms operating here, where temperatures are low compared to the melting temperature, and where pores and pore fluids are important, are different from the mechanisms likely to be operating in the mantle of the earth. We might note that much laboratory work performed on nonsedimentary rock falls more into this class than the first. Field seismic studies in this second class are at considerably higher frequency than those in the first class (100 Hz as opposed to 1 Hz or less), and are much more local in area/coverage.

In the next Section, we review definitions and terminology in the field of seismic attenuation. We then review methods of laboratory and seismological measurement. The section following presents seismological and some of the few laboratory data for Class 1, and the last section presents laboratory and field seismic data for Class 2. We present much of the laboratory data in the form of figures rather than tables, for the reason that there are enough parameters influencing attenuation that a tabular presentation would be both confusing and impractical. For each rock type, one would need a different table for different strain amplitude, frequency, saturation condition. etc. The original publications themselves rarely display data in tabular form.

In preparing for this work, we drew heavily on many previous reviews and compilations. We owe a special debt of gratitude to the following papers: Minster,[1] Johnston and Toksoz,[2] Anderson and Hart,[3,4] Anderson and Given,[5] and Johnston.[6] The reader is enthusiastically referred to these works.

MEASURES OF ATTENUATION

We provide here a concise summary of the various definitions of attenuation appearing in the literature, and of the relations between them.

One of the more common measures, and the one we shall refer to here, is Q, the quality factor''. This was defined[7] following electrical engineering practice as

$$Q = \frac{2\pi E_{peak}}{\Delta E} \tag{1}$$

where ΔE is the energy dissipated per cycle, and E_{peak} is the peak stored elastic energy. Q^{-1} is often referred to as the internal friction. Q has also been defined[8,9] as

$$Q = \frac{4\pi\langle E\rangle}{\Delta E} \tag{2}$$

where E is the average stored energy. Equation 2 is preferable to Equation 1 for a variety of reasons,[9] one of them being that given Equation 2, and monochromatic stress strain histories

$$\begin{aligned} \sigma(t) &= \mathrm{Re}[\sigma_O \exp(i\omega t)] \\ \epsilon(t) &= \mathrm{Re}[\epsilon_O \exp(i\omega t)] \end{aligned} \tag{3}$$

it can be shown easily[1] that

$$Q(\omega) = \frac{M_1(\omega)}{M_2(\omega)} = \frac{J_1(\omega)}{J_2(\omega)} \tag{4}$$

where

$$J^*(i\omega) = J_1(\omega) - iJ_2(\omega) \tag{5}$$

and

$$M^*(i\omega) = M_1(\omega) + iM_2(\omega) \tag{6}$$

are complex compliance and modulus, respectively. Equation 4 is only valid for Q as defined by Equation 1 in the limit of low loss, i.e., $Q \to \infty$.[9] $Q^{-1}(\omega)$ is also related[1] to the phase difference ϕ between stress and strain

$$\frac{1}{Q} = \tan\phi \tag{7}$$

Other common definitions of attenuation are given as follows:[2,10]

1. The logarithmic decrement, $\delta = \ell n\,[A_1/A_2]$ where A_1 and A_2 are amplitudes of two successive maxima or minima in an exponentially decaying free vibration.
2. Attenuation coefficient, α, in the expression

$$A(x,t) = A_0\, e^{-\alpha x}\, e^{i(k_R x - \omega t)}$$

for the amplitude of a plane wave in an unbounded attenuating medium (k_R is the real part of the wavenumber).
3. Resonance peak bandwidth ($\Delta f/f_R$), where f_R is the frequency of a resonance peak and Δf is the half-power bandwidth.

The various measures of attenuation are related, in the low loss approximation, by[1,2]

$$\frac{1}{Q} = \tan\phi = \frac{M_2}{M_1} = \frac{\Delta f}{f} = \frac{\delta}{\pi} = \frac{\alpha V}{\pi f} = \frac{\alpha}{8.686\pi}$$

where in the last term α is in dB/wavelength (this is a common measure in the exploration literature dealing with field seismic studies); V is elastic wave velocity.

We note finally that when one sees "Q" reported in the literature, one must be careful to note to which wave the value pertains. In teleseismic seismology, in field exploration seismic measurements, and in ultrasonic wave propagation laboratory methods, one will generally encounter Q_P and Q_S, for P and S waves, respectively. In some laboratory measurements, Q_P and Q_S are not directly measured; instead one obtains the Q for the wave corresponding to a given elastic modulus. Commonly, since many experiments are conducted using forced longitudinal resonance, one encounters Q_E, for the Young's modulus wave. One also hears sometimes of Q_K, the bulk loss. These are related to Q_S and Q_P by[2,11]

$$\frac{1 + \nu}{Q_E} = \frac{(1 - \nu)(1 - 2\nu)}{Q_P} + \frac{2\nu(2 - \nu)}{Q_S}$$

and

$$\frac{1 + \nu}{Q_K} = \frac{3(1 - \nu)}{Q_P} - \frac{2(1 - 2\nu)}{Q_S}$$

where ν is Poisson's ratio.

MEASUREMENT METHODS

Laboratory Measurements

Laboratory measurements are reviewed in several monographs.[12-15] An excellent recent treatment is that of Johnston and Toksoz,[16] from which we have drawn heavily to make the summary Table 1.

Seismological Methods

Traveling Waves

The Parameter t*

The attenuation of seismic body waves is usually given in terms of the parameter t*,[35,38,52] defined by

$$t^* = \pi \int_{path} \frac{dx}{QV} \simeq \frac{T}{Q_{av}}$$

where V is wave velocity, T is elastic travel time, and Q_{av} is average Q. Q is presumed independent of frequency. Q_p and Q_s are connected by[4]

$$Q_p^{-1} = L\, Q_s^{-1} + (1 - L)\, Q_K^{-1}$$

where $L = (4/3)(V_s/V_p)^2$. If all losses are in shear ($Q_K^{-1} = 0$), and we assume that $\lambda = \mu$, then, given that the ratio of travel times of direct P and S (at 33° distance) is $T_s/T_p = 1.8$ we obtain[53] $t^*_s = 4t^*_p$. In the case where Q is frequency dependent, the parameter t* and its application must be modified.[1,39]

The Spectral Ratio Method

This is one of the most commonly applied methods, used in teleseismic body wave

Table 1
LABORATORY METHODS FOR MEASURING ATTENUATION

Type of method	Form of measurement	Remarks and difficulties	Ref.
Free vibration	Logarithmic decrement δ	Frequency depends on sample size; can be modified to make measurements at high temperature and low frequencies (~ 1 Hz) relevant to mantle of the earth.[17-21] Must take special care not to fracture sample	17—22
Forced vibration	Resonance peak bandwidth Δf/f; get Q_E or Q_S	Must take special care to account for extraneous losses in the case of longitudinal oscillation (damping along surface, radiation from ends) especially in high Q materials; jacket on sample may change resonance frequency slightly,[23] may penetrate porous rocks. Frequency depends on sample size, one example is 5—1.5 kHz for 1-in. long cylinder. Can end-load sample to lower resonant frequency (50 Hz has been achieved)[24]	23,24
Wave propagation		In general, ultrasonic (MHz) frequency range. Cannot go as low as with other techniques, must assume plane wave behavior. Losses may occur in transducer or transducer bond. Ideally, large samples are needed.	25
Pulse echo method	Attenuation coefficient α; get Q_P and Q_S	Observe amplitude decay of multiple reflections from a free surface; must assume reflection is loss free; this limits range of pressures to which technique is applicable (assumption not as valid at high pressure)	26
Through transmission	$\alpha, t^* \rightarrow Q_P, Q_S$	May use spectral ratio technique to get differential attenuation between two receivers. Very similar to teleseismic body-wave seismology	27—29
Observation of pulse shape	Rise time of width of waveform, leading to Q_P, Q_s	Very large sample, or massive rock	30, 31; Pertinent theory in 32—40
Observation of stress-strain curves	Energy lost per cycle,	Should be far off resonance frequency; can observe nonlinearity in mechanism	41—44
Transient creep	Creep function, phenomenologically related to attenuation as a function of frequency	An indirect measurement; ideally, if one can determine transient creep, one can determine Q(ω), and vice-versa; high-temperature transient creep experiments[48-51] can yield information about igneous rock Q(ω) under conditions relevant to mantle of the Earth; has been so used very sparsely[51]	45—51

seismology,[54] exploration-oriented studies, and reduction of laboratory ultrasonic data.[16] We write the noise-free spectral amplitude of the propagating wave as[1,16]

$$A(f,x) = S(f)\ P(f)\ R(f)\ e^{-t^*f}$$

S(f) is the source spectum; R(f) is the receiver response; and P(f) is a propagation operator. Usually it is assumed that P is in fact not a function of frequency, but rather is a constant G incorporating geometric spreading and possibly reflection and transmission.

Consider now two amplitude spectra from the same source

$$A_1(f,x) = G_1S(f)\ R_1(f)\ e^{-t_1^*f}$$

$$A_2(f,x) = G_2S(f)\ R_2(f)\ e^{-t_2^*f}$$

Taking logarithms and forming the ratio,

$$\ell n \frac{A_1(f)}{A_2(f)} = \ell n \frac{G_1}{G_2} + \ell n \frac{R_1(f)}{R_2(f)} + (t_2^* - t_1^*)f$$

In the case of two rays arriving at different stations from the same source, one obtains the differential attenuation $\delta t^* = (t^*_2 - t^*_1)$ as the slope of the best fit straight line in log spectral frequency space.[54,55] The spectral ratio method is also used for the case when A_1 and A_2 pertain to the same seismic phase recorded by different stations along the same path. This is the case for surface waves[55-57] and some multiply reflected phases.[58] The technique has also been used for two different phases recorded at the same station, namely multiple ScS phases, by Sipkin and Jordan,[59,60] who used a phase equalization and stacking procedure to estimate the attenuation operator.

Waveform Fitting

Attenuation has also been estimated using the synthetic seismogram technique, adjusting the attenuation operator to obtain best fits to observed waveforms.[53] A well-known source is essential to the success of the method.

Free Oscillations

We list here some basic approaches to the problem of determining Q from free oscillations.[1, 61]

1. Study of decay of power in a given peak as Q function of time.[61]
2. Study peak halfwidths.
3. Compare narrow-band filtered traces with synthetic traces in the time domain; this has been shown effective in studying the modes split by the rotation of the earth.[62]
4. Multiply the time series representing the free oscillation record by $e^{wt/2Q}$, which removes the effect of attenuation, and vary Q until the peak under study is narrowest.[63]
5. Use a phase equalization and stacking procedure, developed by Gilbert and Dziewonski.[64] We do not summarize the procedure here; a succinct summary is given in Reference 1.

SEISMOLOGICAL DATA AND THE ANELASTIC STRUCTURE OF THE EARTH

Tables 2 to 9 present seismological Q measurements. Tables 2 and 3 are for body waves, and are taken from the compilation of Anderson and Hart.[3] Tables 4 to 9 are for free oscillations and are synthesized from the tables in Anderson and Hart[4] and Anderson and Given.[5]

The available data may be used[4,5,66] to construct models of the Q structure of the earth. Tables 9 to 11 show important examples of such models, which tend to have the following important features:[1]

1. All the data taken together seem to suggest the following gross structure:[4,66]

    a. A moderate to high-Q lithosphere (0 to 80 km, $Q_s \cong 200$)
    b. A low-Q upper mantle (80 to 670 km, $Q_s \cong 100$)
    c. A high-Q lower mantle (670 to 2885 km, $Q_s \cong 400$)
    d. A nonattenuating outer core

Table 2
SHEAR WAVE ATTENUATION IN THE MANTLE[3]

Region	Depth (km)	Period (sec)	Q	Ref.
—	Whole	12	700	73
—	Whole	24	400	73
South America	Whole	11	500	72
South America	Whole	25	508	52
South America	Whole	25	440	52
South America	Whole	14—67	600	77
South America	Whole	25	330	82
South America	Whole	25	360	82
South America — North America	Whole	30	690	78
South America — North America	Whole	40	590	78
South America — North America	Whole	50	500	78
South America — North America	Whole	90	230	78
Southwestern U.S.	Whole	1.5—5.0	230	79
Japan	Whole	2—20	260	74
Japan	Whole	2—20	280	74
Japan	Whole	5	300	75
Sea of Japan	Whole	1.25—66	290	81
Southwestern Pacific	Whole	16—160	156—178	59
Hawaii	Whole	—	300	83
Tonga-Albuquerque	Whole	25	380	82
Tonga-Hawaii	Whole	25	230	82
Tonga-Guam	Whole	25	365	82
Tonga-Solomon Ils.	Whole	25	300	82
Celebes Ils.-Solomon Ils.	Whole	25	230	82
Kurile Ils.-Dugway	Whole	25	270	82
Kurile Ils.-Manila	Whole	25	270	82
Tasman Sea — South Pole	Whole	25	325	82
Tasman Sea — South Pole	Whole	10	380	80
South America	<600	—	160	76
South America	<600	25	151	52
South America	<600	25	185	52
South America	<600	14—67	200	77
Japan	<600	2—20	110	74
Sea of Japan	<600	28—67	150	81
Sea of Japan	<600	10—28	220	81
Sea of Japan	<600	1.25—3.3	260	81
Japan	<1000	2—20	260	74
Japan	<1000	5—50	180	84
Japan	<2000	5—50	200	84
South America	>600	—	500	76
South America	>600	25	1430	52
South America	>600	14—67	2200	77
Japan	>1100	2—20	350	74

2. Most data are consistent with zero bulk loss, which implies, for a Poisson solid, that $t^*_s \cong 4t^*_p$ (see "Measurement Methods"). However, the assumption of zero bulk loss leads to predictions of Q for radial modes that are ~ 30% higher than the observed values.[66] A region in the Earth where bulk loss is nonzero (though still relatively small) seems to be required. One model[66] places the region in the upper mantle; another places it in the inner core.[4]
3. "Most seismic data do not require a frequency dependence of Q and can be explained entirely by a variation of Q with depth."[1] There is, however, some evidence for a rapid increase in Q with frequency around 1 Hz.[67] Theoretically, a frequency de-

Table 3
OBSERVED AVERAGE MANTLE P-WAVE Q's[3]

Depth interval (km)	Q_p			
	(1)[a]	(2)[b]	(3)[c]	(4)[d]
0—100	220	100		
0—760	530 ± 150	150		166—272
0—900			180—240	
0—2900	$845\begin{cases}+420\\-260\end{cases}$	375	410—630	300—412
100—760	710 ± 150	165		
100—2900	$1080\begin{cases}+420\\-285\end{cases}$	420		
760—2900	$1260\begin{cases}+950\\-365\end{cases}$	1210		2050—3650
900—2900			1600-6000	

[a] 0.6—5 sec[85]
[b] 1 sec[79]
[c] 1 sec[86]
[d] 8—33 sec[87]

Table 4
LOW ORDER FUNDAMENTAL MODE Q VALUES[4,5]

Mode	Period, secs.	1[66]	2[88,89]	3[90]	4[62]	5[a]
$_0S_2$	3232	589	500		425—550	500—589
$_0S_3$	2134	460	450		325—450	450—520
$_0S_4$	1546	411	400		275—400	400—411
$_0S_5$	1190	352	300		300—325	300—400
$_0S_6$	964	343	270			343—399
$_0S_7$	812	373	460			373—460
$_0S_8$	708	357	230			295—357
$_0S_9$	634	326	366			328
$_0S_{10}$	580	329	320			320
$_0S_{11}$	537		254			
$_0S_{12}$	503	335	280			308
$_0S_{13}$	474	305	310			
$_0S_{14}$	448	298	403			294
$_0T_2$	2631		250			250—400
$_0T_3$	1703		370		325	325—400
$_0T_4$	1304		290	138	425	290—425
$_0T_5$	1076		280	185		185—280
$_0T_6$	926		280	357		266—357
$_0T_7$	818			125		125—141
$_0T_8$	736		170	200		170—295
$_0T_9$	672		180			157—180
$_0T_{10}$	619		200	188		188—250

[a] Range chosen by Anderson and Given.[5]

pendence of Q is definitely expected.[46] For a solid characterized by a single relaxation time τ, Q^{-1} is a Debye function with maximum absorption at $\omega\tau = 1$. When the solid is characterized by a spectrum of relaxation times (e.g., arising from a distribution of dislocation lengths[68-70]) the band of maximum absorption is widened.[71] Within the

Table 5
FUNDAMENTAL SPHEROIDAL MODE Q[4,5]

Mode	Period (sec)	1[89,91-93]	2[63,94]	3[56]	4[55]	5[66,88]	6[a]
$_0S_{15}$	426	288				227	
$_0S_{16}$	407	224	278			300	276
$_0S_{17}$	390	215				316	
$_0S_{18}$	374	219				173	282
$_0S_{19}$	360	167				251	
$_0S_{20}$	348	185		146		250	240
$_0S_{21}$	356	188				222	
$_0S_{22}$	325	207		167		200	228
$_0S_{23}$	315	200				210	
$_0S_{24}$	306	201				210	210
$_0S_{25}$	298	185	213	178	198	200	
$_0S_{26}$	290						198
$_0S_{28}$	275						188
$_0S_{29}$	269	175	203	182		164	
$_0S_{30}$	262						179
$_0S_{40}$	212	149	155	172	177	149	151
$_0S_{50}$	178	137	155			113	137
$_0S_{51}$	175	135	152	137			
$_0S_{57}$	160	132	140	123			
$_0S_{60}$	153	122				110	122
$_0S_{65}$	143	116				137	
$_0S_{70}$	134	122					120
$_0S_{76}$	125	127					122

[a] Values selected by Anderson and Given.[5]

"absorption band" Q may be approximately frequency independent, although more likely it is proportional to ω^α where $\alpha \sim 0.3$,[46] (see also Figure 2) as suggested by transient creep experiments.[48-51] Outside the band, in the neighborhood of the long and short period cutoffs (τ_2 and τ_1, respectively) to the relaxation spectrum, Q, depends strongly on frequency. Anderson and Given[5] approximate the absorption band as

$$Q = Q_m\,(f\tau_2)^{-1},\ f < 1/\tau_2$$
$$Q = Q_m\,(f\tau_2)^{\alpha},\ 1/\tau_2 < f < 1/\tau_1$$

$$Q = Q_m(\tau_2/\tau_1)^{\alpha}\,(f\tau_1),\ f > 1/\tau_1$$

Fixing τ_2/τ_1 and α, they use the available data to obtain values of Q_m and τ_1. Despite the limited resolving power of the data, their results (Figure 1B and Tables 10, 11) are extremely important because they are based on the most realistic and complete physical model yet applied to the problem. Figure 1A and Tables 10, 11 also present Q model SL8,[4] where Q is assumed independent of frequency. As mentioned before, the laboratory data relevant to Q in the interior of the earth are few. Figure 2 shows internal friction as a function of frequency at high temperature for a mantle peridotite (lherzolite from Ivrea, Italy), derived from transient creep experiments by Berckhemer et al.[51] Figures 3A and 3B show Q^{-1} as a function of temperature for forsterite, enstatite, and peridotite, as measured by Woirgard and Gueguen.[21]

Table 6
FUNDAMENTAL TOROIDAL MODE Q[4,5]

Mode	Period (sec)	1[92]	2[91]	3[88,90,93,95,96]	4[55,56,66]	5[a]
$_0T_{11}$	575	258	270	260		
$_0T_{12}$	538		189	150—220		189—220
$_0T_{13}$	505		204	190—260	118—258	
$_0T_{14}$	477	220	200	135—200		200—270
$_0T_{15}$	452	240	157	172	158—186	
$_0T_{16}$	430	174	149	215	185—245	168—215
$_0T_{17}$	410	163	127	126		
$_0T_{18}$	391	116	111	188—270	172—204	
$_0T_{19}$	375	81—94	105	281		
$_0T_{20}$	360	81		249	97	175—249
$_0T_{21}$	346	86—111	105	170		
$_0T_{22}$	333	95—207	108		114	
$_0T_{23}$	321	91—111	115		123—160	
$_0T_{24}$	310	123—153	116			
$_0T_{25}$	300	123	116		104—149	110—149
$_0T_{26}$	290	97—209				
$_0T_{27}$	281	93—146	115			
$_0T_{28}$	273	92—221	116		110	
$_0T_{29}$	265	95—147	114			
$_0T_{30}$	258	120—184	115		111—142	111—142
$_0T_{31}$	250	96—104	110		105	
$_0T_{32}$	244	106—176	114		133	
$_0T_{33}$	237	94	108			
$_0T_{34}$	231	101—167	114			
$_0T_{35}$	226	105	108		102	
$_0T_{36}$	220	139—162	115		131	
$_0T_{37}$	215	93—142				
$_0T_{38}$	210	111—169				
$_0T_{39}$	205	95—137				
$_0T_{40}$	201		118		102—133	102—133
$_0T_{45}$	181	104—123	115		117	
$_0T_{50}$	164	93—130	108			108—130
$_0T_{55}$	151	114—122	108		114—118	
$_0T_{60}$	139	103—112	104			104—116
$_0T_{65}$	129		108			
$_0T_{70}$	121	100	99		116	100
$_0T_{75}$	114	95—140	113			
$_0T_{80}$	107	86—121	109			86—121
$_0T_{85}$	101				108	108
$_0T_{110}$	79				109	109

[a] Range selected by Anderson and Given.[5]

ATTENUATION IN SEDIMENTARY ROCKS

Laboratory Work

Much of the work on attenuation in sedimentary rocks is to be found in the geophysical literature, particularly the exploration geophysical literature. Considerable attention has been focused on sandstones, which are important because of their role as oil and gas reservoirs.

It is difficult to quote a ''ballpark figure'' for attenuation in sedimentary rocks, because, as we shall see, Q is a complicated function of a number of parameters: rock type (which includes such factors as permeability and porosity), temperature, pressure, strain amplitude,

Table 7
Q OF TOROIDAL OVERTONES[4,5]

		Q					Q		
Mode	**Period (sec)**	**1[66,88,91,92,96]**	**2[a]**	**3[b]**	**Mode**	**Period (sec)**	**1[66,88,91,92,96]**	**2[a]**	**3[b]**
${}_1T_7$	475			238	${}_1T_{43}$	140	161	122—193	
${}_1T_9$	407	178			${}_1T_{44}$	138	161	145—173	
${}_1T_{11}$	359	176	176—249		${}_1T_{45}$	136	167	151—202	
${}_1T_{12}$	339			195	${}_1T_{50}$	126	152	122—193	
${}_1T_{19}$	250	141	151—208	195	${}_1T_{55}$	117	135		
${}_1T_{22}$	225	133	133—172		${}_1T_{59}$	111	134		
${}_1T_{23}$	218	149	139—182		${}_1T_{61}$	109	138		
${}_1T_{24}$	212	164	124—164		${}_1T_{62}$	107			138
${}_1T_{25}$	206		192—286	192	${}_2T_2$	448	320		
${}_1T_{26}$	200	227	202—227		${}_2T_{36}$	133			183
${}_1T_{27}$	195	222	222—300		${}_2T_{37}$	131	172		
${}_1T_{28}$	190	208	155—284		${}_2T_{46}$	112	189		
${}_1T_{29}$	185	208	172—208		${}_2T_{47}$	110	182		
${}_1T_{30}$	181	189	165—207		${}_2T_{49}$	107			207
${}_1T_{31}$	177	179			${}_2T_{51}$	104	178		
${}_1T_{32}$	173	182	173—191		${}_2T_{52}$	103	162		
${}_1T_{33}$	169	192			${}_2T_{54}$	100	144		
${}_1T_{34}$	166	192	155—192		${}_2T_{62}$	91	143		
${}_1T_{35}$	162	192	172—255		${}_3T_{26}$	147	264		
${}_1T_{36}$	159	200	145—249		${}_3T_{27}$	144	296		
${}_1T_{37}$	156	196	147—239		${}_3T_{30}$	134			215
${}_1T_{38}$	153	185	169—310		${}_3T_{53}$	91	236		
${}_1T_{39}$	150	179	134—245		${}_4T_{11}$	200	208		
${}_1T_{40}$	148	175	163—180		${}_4T_{17}$	170	204		
${}_1T_{41}$	145	161	145—179		${}_5T_9$	175	243		
${}_1T_{42}$	143	159	152—209						

[a] Range selected by Anderson and Hart.[6]
[b] Values selected by Anderson and Given.[5]

frequency, and the presence of fluids (saturation condition). Figure 4, taken from Johnston et al.,[98] plots Q values as a function of porosity. The data, taken from Bradley and Fort,[10] show a definite dependence of Q on rock type, with more porous types such as sandstone showing lower Q values. The scatter in the data is easily understood because a broad range of frequency and saturation conditions is represented.

Underlying much of the work on sedimentary rocks has been the search for a physical mechanism, or combination of mechanisms, to explain the observed attenuation phenomena. This search, which is still incomplete, has spawned several theoretical studies and has definitely guided the experimental studies beyond merely making measurements only for measurements' sake. There has been a real effort to study systematically the factors influencing Q. The mass of papers on the subject often makes it difficult to keep all these factors in mind; we present below a critical summary of some of the pertinent trends, which any proposed combination of physical mechanisms must satisfactorily explain.

Factors Influencing Q

Strain Amplitude

This is an extremely important parameter: essentially, strain amplitude dependence tells us whether or not the mechanism we are dealing with is linear, which of course has a great bearing on the tractability of its mathematical formulation. The essential facts are as follows:

Table 8
Q OF SPHEROIDAL OVERTONES[4,5]

Mode	Period (sec)	Q 1[a]	Q 2[b]	Mode	Period (sec)	Q 1[a]	Q 2[b]
${}_1S_7$	604		484	${}_4S_{23}$	170	312	
${}_2S_2$	1049		546	${}_4S_{25}$	161	260	
${}_2S_4$	726		350—546	${}_4S_{26}$	157	306	
${}_2S_{15}$	309		244	${}_4S_{31}$	139		264
${}_2S_{23}$	202	514		${}_4S_{32}$	136	299	
${}_2S_{26}$	179	158—275		${}_4S_{34}$	130	219—234	234
${}_2S_{30}$	161	150—207		${}_4S_{35}$	127		246
${}_2S_{31}$	157	143		${}_4S_{39}$	118	193	
${}_2S_{39}$	131		179	${}_5S_7$	304		496
${}_2S_{57}$	98	174		${}_5S_{22}$	154	306—346	
${}_2S_{60}$	94		151	${}_5S_{24}$	147	308—339	
${}_3S_1$	1061		1020	${}_5S_{25}$	143	231—433	
${}_3S_{12}$	297	179—239		${}_5S_{26}$	140	236—299	
${}_3S_{13}$	285	227—271		${}_5S_{30}$	129		248
${}_3S_{14}$	273	274		${}_5S_{38}$	111		223
${}_3S_{15}$	262	163—394		${}_6S_1$	505	613—700	
${}_3S_{16}$	252	259—285		${}_6S_8$	268	286	
${}_3S_{18}$	233	157—232		${}_6S_9$	252		292
${}_3S_{20}$	217		229	${}_6S_{13}$	191	291	
${}_3S_{42}$	111	180		${}_6S_{23}$	138	186—299	
${}_4S_3$	488	560		${}_6S_{26}$	129	368	
${}_4S_{14}$	225		288	${}_6S_{31}$	116	391	
${}_4S_{19}$	192		291	${}_6S_{36}$	106	342	
${}_4S_{21}$	181	275		${}_6S_{47}$	89	276	

[a] Range selected by Anderson and Hart.[4]
[b] Values selected by Anderson and Given.[5]

1. Q^{-1} is in general roughly independent of strain amplitude provided that this amplitude is low enough. Beyond a threshold amplitude, which we shall denote as ϵ_{NL}, Q^{-1} becomes amplitude dependent, sometimes very strongly.[22,47,99] Figures 5, 6, and 7 show some representative curves. As the reader will note, ϵ_{NL} is not a well-defined point, but denotes a range of amplitude over which the transition from roughly linear to nonlinear behavior takes place. Figure 6 shows that this general behavior is observed in igneous as well as sedimentary rocks at the low temperatures and pressures pertinent here.
2. Materials which are known to be free of microcracks exhibit no amplitude dependence,[99,100] suggesting that sliding along crack faces may help explain the observations in 1.
3. Material which has been thermally cycled shows a lower ϵ_{NL}.[100] We note that thermal cycling has the effect of opening cracks.
4. ϵ_{NL} depends on the structure of the rock. Available evidence (Figure 7) shows that rocks which are well indurated (i.e., have well-cemented grains) have a higher ϵ_{NL} than rocks which are poorly indurated. Whatever nonlinear processes are operating seem to operate with greater ease when the rock is poorly cemented.
5. Increasing the confining pressure increases ϵ_{NL}[99,101] (Figures 5, 7). Moreover, this pressure effect depends on rock induration.[101] When the rock is poorly indurated, the effect is greater (Figure 7). Thus we again observe a general trend that somehow forcing the grains together seems to inhibit the nonlinear processes.

Table 9
HIGH Q MODES[4,5]

Radial modes			Other high Q modes		
Mode	Period (sec)	Q	Mode	Period (sec)	Q
${}_0S_0$	1230	12,000	${}_1S_7$	604	484
		7,500	${}_2S_2$	1049	546
		7,470	${}_2S_4$	725	350
		4,229	${}_2S_{23}$	202	514
		3,996	${}_3S_1$	1,061	1020
		900	${}_4S_3$	488	560
${}_1S_0$	614	5,160	${}_4S_{23}$	170	312
		1,970	${}_4S_{26}$	157	306
${}_2S_0$	399	1,170	${}_5S_7$	304	496
		1,059	${}_6S_1$	505	613—700
		870	${}_8S_1$	348	704
		704	${}_8S_9$	192	483
		672	${}_{10}S_2$	248	870
${}_3S_0$	306	992	${}_{10}S_{28}$	96	399
		874	${}_{11}S_3$	224	368—696
${}_4S_0$	244	1,264	${}_{11}S_4$	210	652
		1,173	${}_{13}S_1$	228	574—1,573
		1,156	${}_{13}S_2$	207	1125
		989	${}_{16}S_{20}$	89	463
		790	${}_{17}S_{24}$	80	435
		750	${}_{19}S_{13}$	96	496
${}_5S_0$	205	1,570	${}_{20}S_{18}$	83	630
		942			
		938			
		927			
		824			
${}_6S_0$	174	933			

6. The presence of water decreases ϵ_{NL}.[99,101] Apparently, also, the more water there is, the greater the effect, although changes are more dramatic with the initial addition. Figure 8 shows that ϵ_{NL} is lower for partially saturated Berea sandstone than it is for the dry rock, and it is even lower for the fully saturated rock. Again, this effect on ϵ_{NL} is induration dependent: well-indurated sandstone shows the effect less than poorly indurated sandstone.[101]

Pressure Dependence

1. In general Q^{-1} decreases with confining pressure, until it reaches a roughly constant value[11,23,29,41,102—104] (Figures 9 to 11). This appears to be true regardless of saturation condition.
2. $|dQ^{-1}/dP|$ (at low pressures, when Q^{-1} is still changing significantly) is greater for water-saturated rocks than it is for air-dry rocks. This effect is more apparent when the rock is poorly indurated[101] (Figure 10).
3. $|dQ^{-1}/dP|$ is greater at higher frequencies than at lower ones, when the rock is not well indurated[101,105] (Figure 11).

Frequency Dependence

1. Dry sedimentary rocks have attenuation values which are frequency independent over a wide band.[22,107-110] (Figures 12, 13).

Table 10
Q AS A FUNCTION OF DEPTH IN THE EARTH[4,5]

(Approximate values)[97]			Q_s for frequency-dependent absorption band model					Q_p for frequency-dependent absorption band model				
Depth (km)	**Density (g/cc)**	**Pressure (Kbar)**	**1 sec**	**10 sec**	**100 sec**	**1000 sec**	**Q_s for a frequency-independent Q model**	**1 sec**	**10 sec**	**100 sec**	**1000 sec**	**Q_p for a frequency independent Q model**
11	2.6	3	500	500	500	500		487	767	1168	1232	
11	2.9	3	200	141	100	181	500	287	262	207	377	1047
200	3.4	63	157	111	90	900	105	270	256	237	2365	279
421	3.5	140	190	134	95	254		302	296	244	659	
421	3.7	140	5691	569	330	234	140	741	840	819	603	364
671	4.0	239	8919	892	353	250	230	2921	2060	866	615	566
2200	5.2	984	11350	1135	366	259	515	2938	2687	942	668	1330
2400	5.3	1100	184	130	92	315	515	427	345	247	846	1358
2843	5.5	1340	184	130	92	315	100	427	345	247	846	269
2887	5.6	1354	—	—	—	—	—	7530	753	600	6000	10^6
4044	11.4	2522	—	—	—	—	—	4518	493	1000	10^4	10^6
5142	12.1	3280	100	1000	10000	10^5	425	511	454	3322	3.3×10^4	425

Table 11
AVERAGE MANTLE Q VALUES[4,5]

		Q_s, ABM[a]					Q_p, ABM[a]				
Region	**Depth (km)**	**1 sec**	**10 sec**	**100 sec**	**1000 sec**	**Q_s, SL_8[b]**	**1 sec**	**10 sec**	**100 sec**	**1000 sec**	**Q_p, SL_8[b]**
Upper mantle	0—671	267	173	127	295	130	362	354	311	727	328
Lower mantle	671—2886	721	382	211	266	360	1228	979	550	671	912
Whole mantle	0—2886	477	280	176	274	235	713	639	446	687	593

[a] Frequency-dependent absorption band model.
[b] Frequency-independent model.

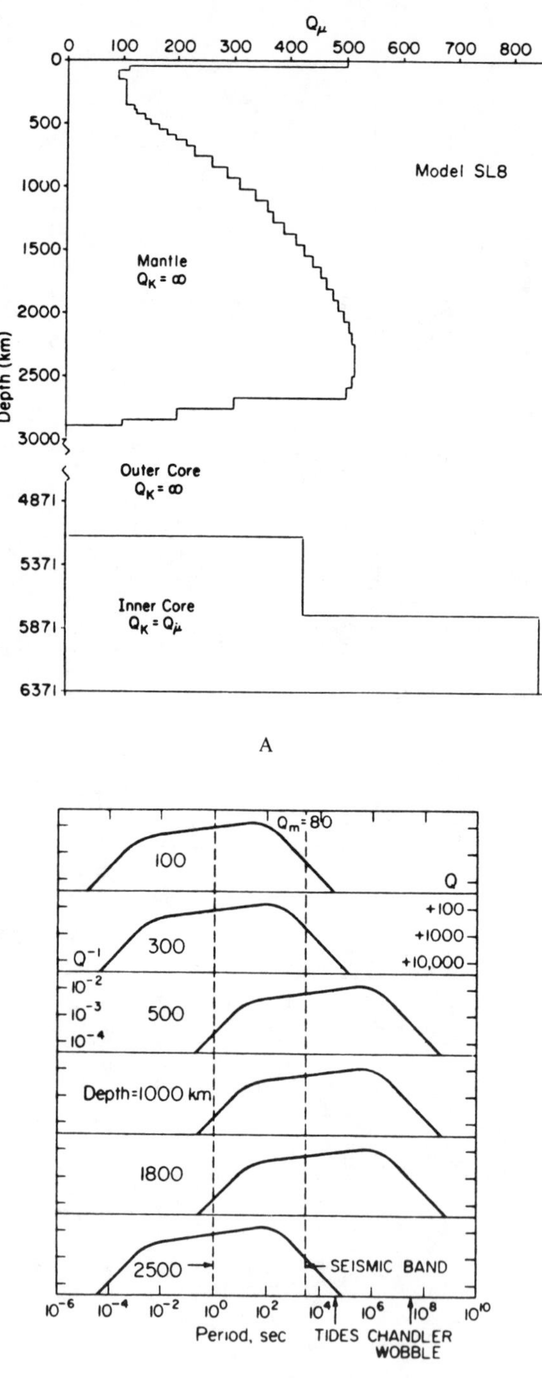

FIGURE 1. (A) Shear attenuation structure of the earth according to Anderson and Hart's[4] model SL8, where Q is assumed independent of frequency. (B) Shear attenuation structure of the earth for a frequency dependent absorption band model.[5]

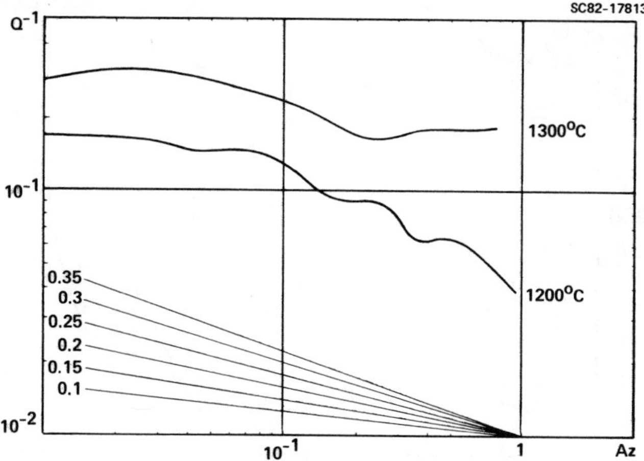

FIGURE 2. Internal friction (Young's modulus) as a function of log frequency at high temperature for a mantle periodotite (lherzolite from Ivrea, Italy). The Q^{-1} function was derived from transient creep experiments by Berckhemer et al.[51] (From Berckhemer, H., Auer, F., and Drisler, J., *Phys. Earth Planet. Inter.*, 20, 48, 1979. With permission.)

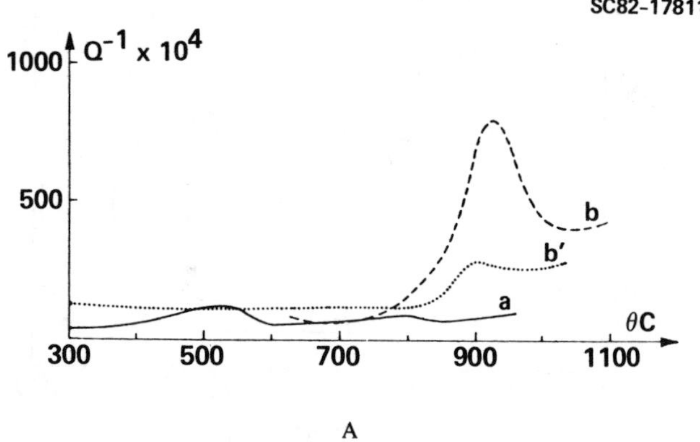

A

FIGURE 3. Shear attenuation vs. temperature at 2 to 8 Hz. (A) Results for synthetic forsterite (a), peridotite (b), and annealed peridotite (b′). (B) Results for single crystal enstatite. (From Woirgard, J. and Gueguen, Y., *Phys. Earth Planet. Inter.*, 17, 140, 1978. With permission.)

2. Water-saturated rocks have a frequency dependent Q^{-1},[110,111] with some evidence for relaxation peaks (Figures 13, 14). Figure 15 shows some frequency dependence in granite saturated with glycerine.
3. Sandstones which are not well indurated appear to have more pronounced frequency dependence in Q^{-1} (Figure 11).[101]
4. The frequency dependence of Q^{-1} for saturated sandstone is decreased by the application of confining pressure especially when such rocks are not well indurated[101] (Figure 11).

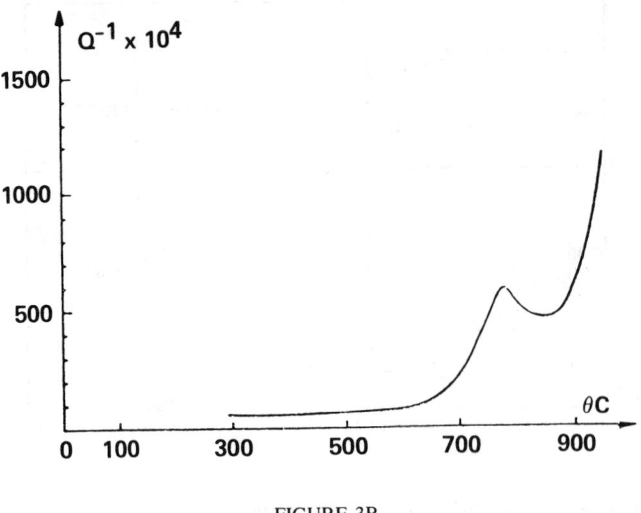

FIGURE 3B

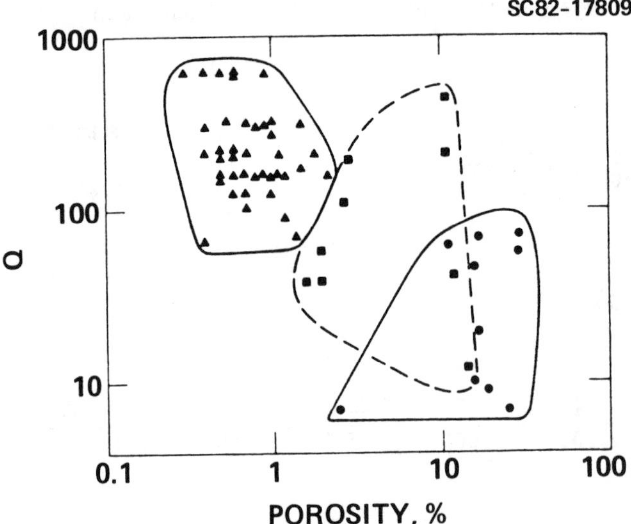

FIGURE 4. From Johnston et al.,[98] plotting data from the compilation of Bradley and Fort.[10] Igneous and metamorphic rocks are denoted by triangles, limestones by squares, and sandstones by circles. Depite the large scatter caused by the wide ranges of frequency and saturation represented, a general inverse trend of Q with porosity is visible.

Saturation Dependence

In general, the presence of fluids increases Q^{-1} values.[6,11,23,28,29,98,105,110,111] As we have seen, the presence of fluids affects the dependence of Q^{-1} on other parameters, such as strain amplitude, pressure, and frequency. We extend our discussion now to include the effects of fluid type, and also some specific effects of the degree of saturation on the Q_p/Q_s ratio, a possibly important diagnostic in oil and gas exploration.

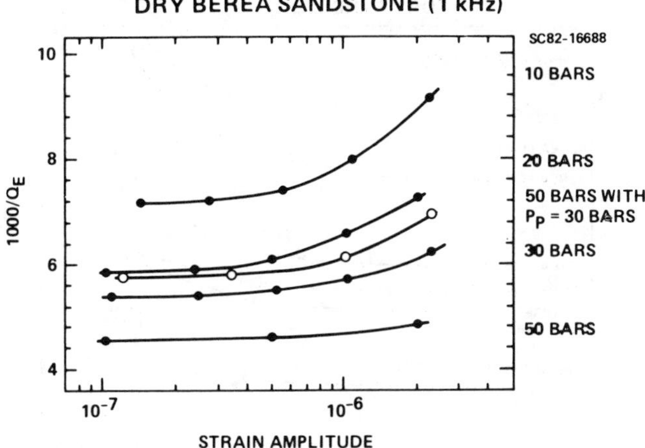

FIGURE 5. Threshold amplitude for onset of nonlinear behavior shifts upward with increasing pressure in dry Berea sandstone. (From Winkler, K. W., Nur, A., and Gladwin, M., *Nature (London)*, 227, 528, 1979. With permission.)

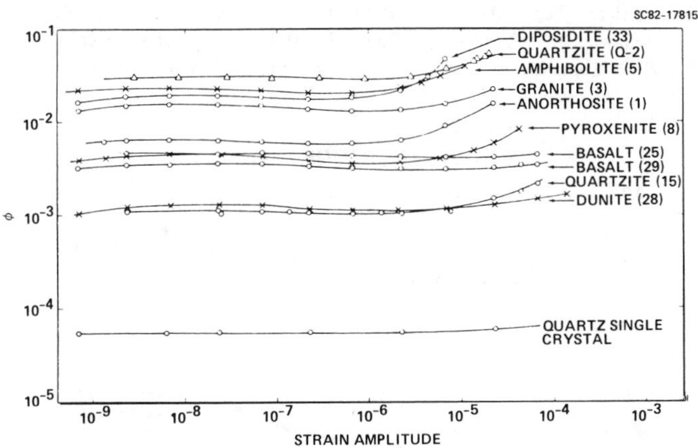

FIGURE 6. Internal friction (shear) vs. strain amplitude for several rock types. Linear to nonlinear transition is observed in igneous as well as sedimentary rocks. (From Gordon, R. B. and Davis L. A., *J. Geophys. Res.*, 73, 3917, 1968. With permission.)

1. The first few monolayers of fluid have an extremely important effect. A vacuum dry rock has a much lower Q^{-1} than an air dry rock, which in general has already managed to adsorb some volatiles.[112,113] Q^{-1} increases roughly linearly with absorbed mass for the first one or two monolayers[114] (Figure 16). After this effect, Q^{-1} tends to remain fairly constant until the effects of bulk fluid behavior are felt. Pandit and King[115] estimated 110Å as the thickness of the first layer where bulk behavior is important. (We note that there may be some problems with this estimate. The authors claim to see a transition from frequency independent to frequency dependent Q^{-1} and interpret this as indicating the onset of bulk behavior. The transition, however, is not particularly well resolved by their data.)
2. Fluids composed of polar molecules appear to be more effective than those composed of nonpolar molecules in increasing Q^{-1} (Figure 17).[116]
3. The relative attenuation of P and S waves varies with degree of saturation. For air dry

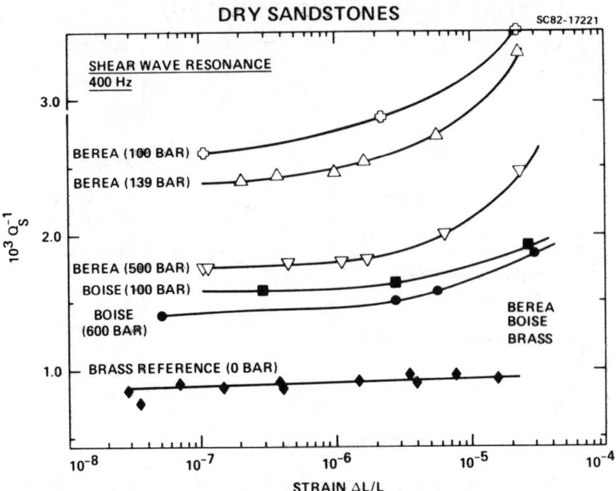

FIGURE 7. Showing the upward shift with pressure of the threshold amplitude for nonlinear behavior. The effect is greater in the poorly indurated Berea than in the better indurated Boise. (From Tittmann, B. R., Abdel-Gawad, M., Salvado, C., Bulau, J., Ahlberg, L., and Spencer, T. W., *Proc. Lunar Planet. Sci.*, 12B, 1737, 1981. With permission.)

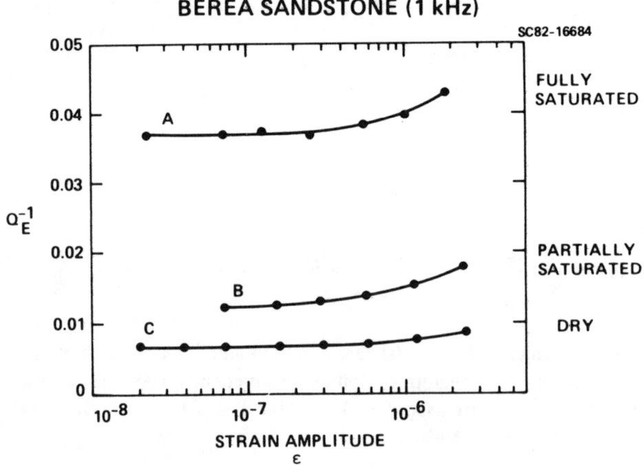

FIGURE 8. Showing the downward shift with degree of water saturation of the threshold amplitude for nonlinear behavior. (From Winkler, K. W., Nur, A., and Gladwin, M., *Nature (London)*, 227, 528, 1979. With permission.)

rocks, $Q^{-1}{}_p \cong Q^{-1}{}_s$. For fully saturated rocks, $Q^{-1}{}_p$ can range from somewhat less than $Q^{-1}{}_s$ to about the same value. For partially saturated rocks $Q^{-1}{}_p$ is higher than $Q^{-1}{}_s$[11,110] (Figures 18 to 20).

Mechanisms of Attenuation in Sedimentary Rocks

Several mechanisms have been proposed to explain attenuation patterns in sedimentary rocks. These fall naturally into two categories: those involving the solid rock and those involving pore fluids.

Most of the discussion of solid rock mechanisms has centered around frictional sliding

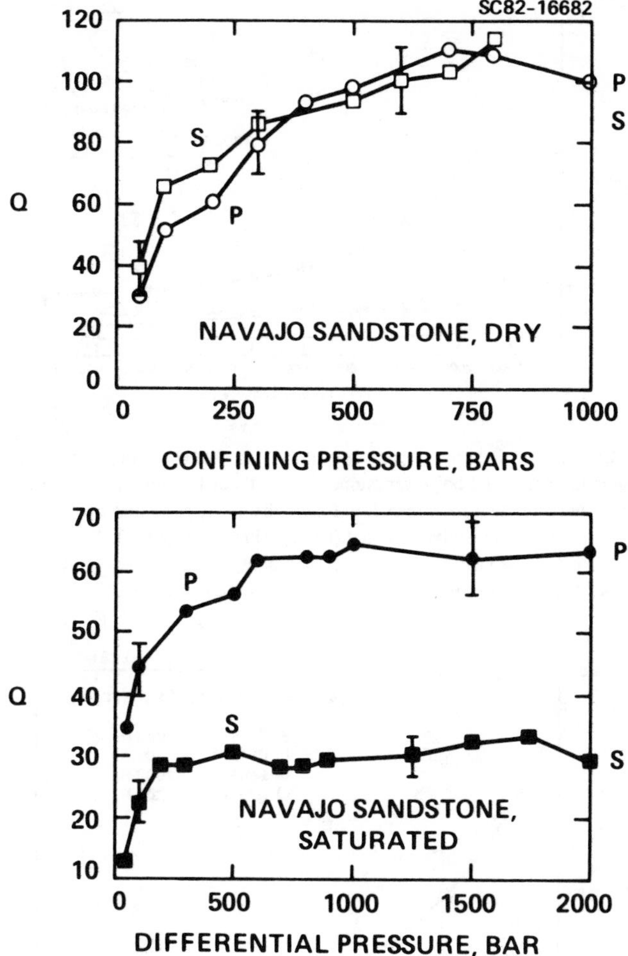

FIGURE 9. Showing the increase of Q with pressure. (From Johnston, D. H. and Toksoz, M. N., *J. Geophys. Res.*, 85, 925, 1980. With permission.)

along crack faces and grain boundaries. Models have been developed both for thin elliptical cracks and for spheres in contact; more recently, more generalized crack geometries have been considered. The formulation of Walsh has received considerable attention. Walsh[117] developed some mathematically quite complicated expressions for internal friction as a function of elastic moduli, coefficient of friction, and "critical" crack is one whose faces are barely touching: this kind of crack can contribute to dissipation, whereas cracks which are completely shut and completely open do not, presumably because the former will not slip and the latter will slip with no resistance. For a Q of 100, walsh requires a rather high critical crack density of one per grain.

The frictional mechanism is intuitively appealing and some variation of it can probably explain, or at least be consistent with, several pertinent facts. There is much evidence attesting to the importance of the presence of cracks. For one thing, rocks, which are polycrystals, have Q values which are much lower than single crystals of their constituent minerals.

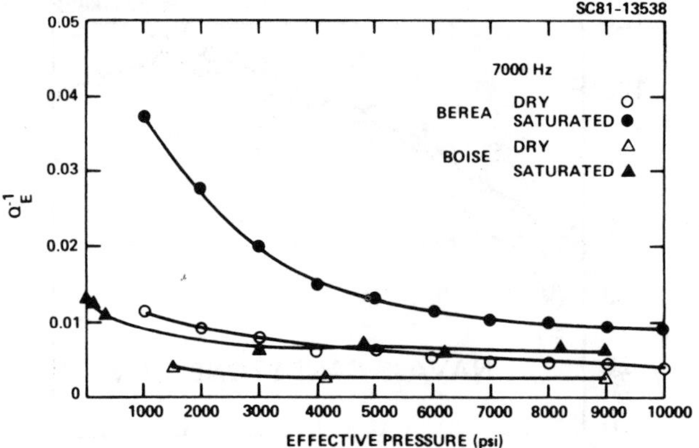

FIGURE 10. Pressure dependence of Q_E^{-1} at 7 kHz for dry and water-saturated berea and boise sandstone. $|dQ^{-1}/dP|$ at low pressure is greater for water saturated rock than it is for air-dry rock, particularly when the rock is relatively poorly indurated (berea). (From Tittman, B. R., Abdel-Gawad, M., Salvado, C., Bulau, J., Ahlberg, L., and Spencer, T. W., *Proc. Lunar Plant. Sci.*, 12B, 1737, 1981. With permission.)

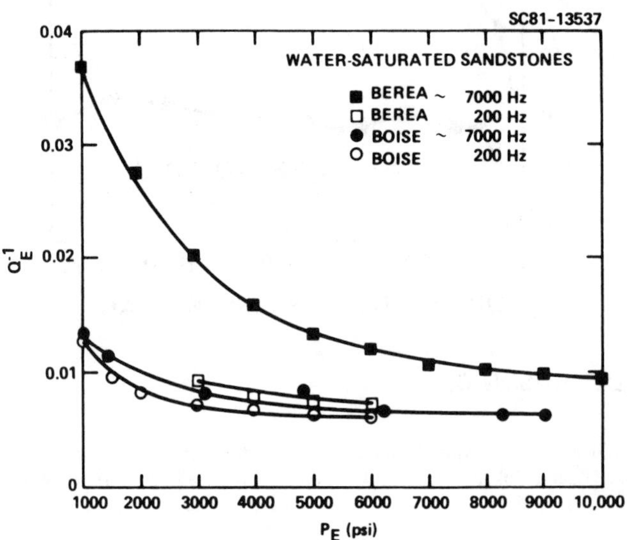

FIGURE 11. Pressure dependence of Q_E^{-1} for water-saturated berea and boise sandstone at two different frequencies. The frequency dependence of Q^{-1} is decreased markedly by the application of confining pressure in the case of the poorer-indurated berea sandstone. (From Tittmann, B. R., Abdel-Gawad, M., Salvado, C., Bulau, J., Ahlberg, L., and Spencer, T. W., *Proc. Lunar Planet. Sci.*, 12B, 1737, 1981. With permission.)

Clark[118] offers an instructive specific example: the Q of the Sioux quartzite, which is greater than 99% quartz in composition and has less than 5% porosity, is greater than the Q of a single crystal of quartz by three orders of magnitude. Another important fact is the increase of Q with pressure discussed in the previous section. A frictional mechanism is consistent with this; Johnston et al.,[98] have, in fact, extracted from Walsh's formulation an exponential increase of Q with pressure and have fit it successfully to their data for dry and saturated

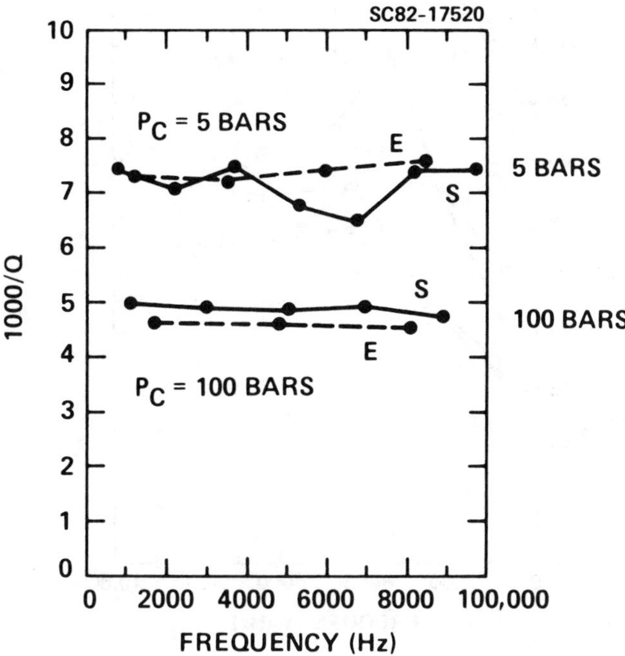

FIGURE 12. Showing frequency independence of Q_E^{-1} and Q_S^{-1} in dry berea sandstone. (From Winkler K. W. and Nur, A., *Geophysics*, 47, 1, 1982. With permission.)

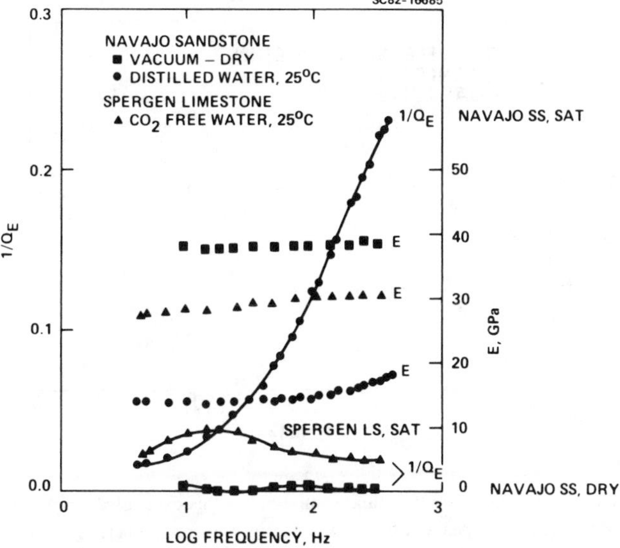

FIGURE 13. Showing frequency independence of Q_E^{-1} for dry Navajo sandstone compared to the frequency dependent behavior observed when this rock is water saturated. Frequency dependent behavior for saturated Spergen limestone is also shown. (From Spencer, J. W., *J. Geophys. Res.*, 86, 1803, 1981. With permission.)

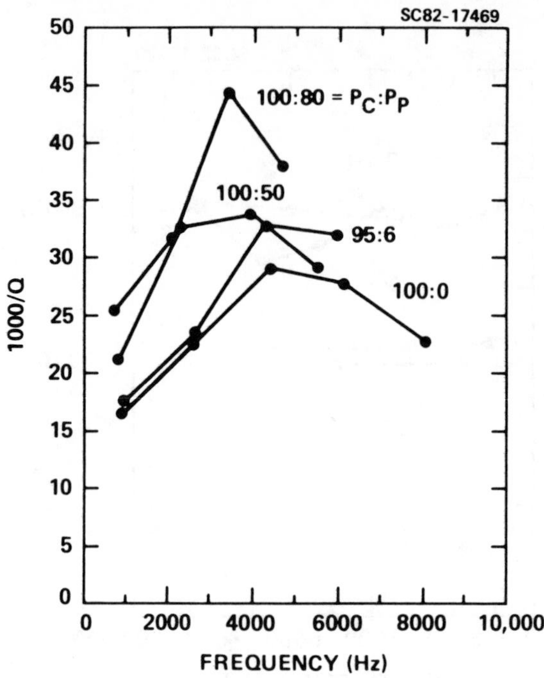

FIGURE 14. Frequency dependence of Q_S^{-1} in water-saturated berea sandstone for different ratios of confining pressure (P_c) to pore pressure (P_p). From Winkler, K. W. and Nur, A., *Geophysics*, 47, 1, 1982. With permission.)

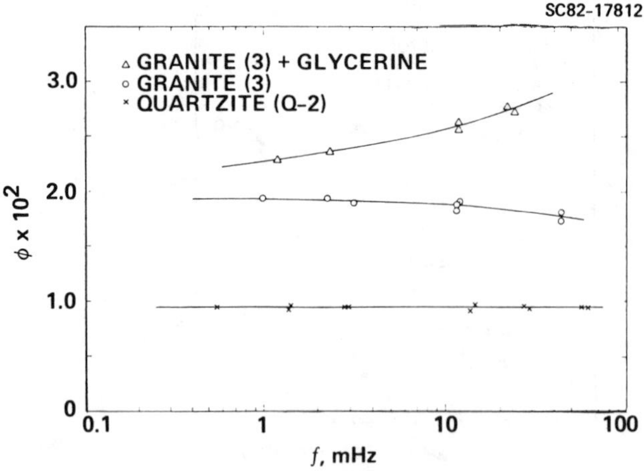

FIGURE 15. Showing that fluid saturation (in this case, glycerine) can cause frequency dependent behavior in an igneous rock where Q is otherwise more or less frequency dependent. The granite here is at low temperature and pressure compared to conditions in the mantle of the earth so that the discussion in the previous section of the text does not apply, and behavior is similar to that observed in sedimentary rocks. Measurements were made here by the stress-strain curve method. (From Gordon, R. B. and Davis, L. A., *J. Geophys. Res.*, 73, 3917, 1968. With permission.)

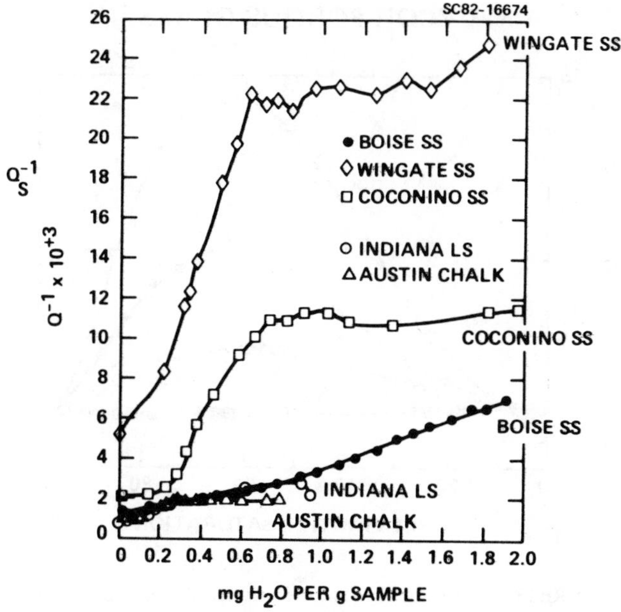

FIGURE 16. Shear attenuation vs. amount of adsorbed water for a variety of rocks. Attenuation increases linearly with mass for the first one or two monolayers. After that, there is little effect (until one enters the regime of bulk fluid behavior). (From Clark, V. A., Spencer, T. W., Tittmann, B. R., Ahlberg, L.A., and Coombe, L. T., *J. Geophys. Res.*, 85, 5190, 1980. With permission.)

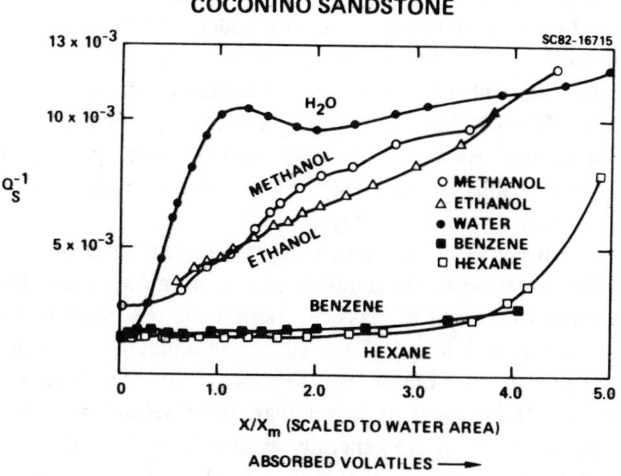

FIGURE 17. Shear attenuation vs. amount of adsorbed volatiles for coconino sandstone. Volatiles composed of polar molecules appear more effective in increasing Q^{-1}. (From Tittmann, B. R., Clark, V. A., Richardson, J., and Spencer, T. W., *J. Geophys. Res.*, 85, 5199, 1980. With permission.)

Berea sandstone, as well as to the granite data of Gordon and Davis.[41] Johnston et al. point out that friction, being a nonlinear process, is also consistent with the observation that Q in dry rocks is independent of frequency.

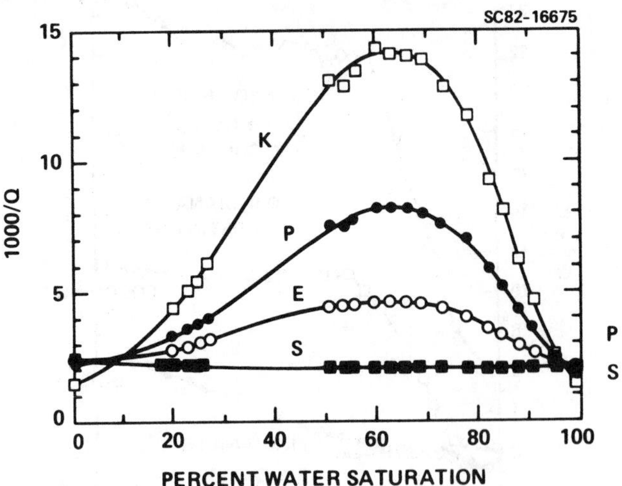

FIGURE 18. Attenuation as a function of water saturation in vycor porous glass. Extensional (E) and shear (S) attenuation are measured, while P wave and bulk compressional attenuation are computed. Shows that $Q_p^{-1} > Q_s^{-1}$ at partial saturation. Figures 19 and 20 show this effect for Massilon sandstone, and Figure 21 shows a simple conceptual model which might explain the observation. (From Winkler, K. W. and Nur, A., *Geophysics*, 47, 1, 1982. With permission.)

A causal linear process cannot have both Q and velocity constant with frequency. Clark,[118] however, argues that dispersion might be present within the resolution of experimental measurements, and hence that one cannot take the independence of Q on frequency as being good evidence in itself for friction (or a nonlinear mechanism in general).

In fact the intrinsic nonlinearity of a classical frictional mechanism has been cited by some as its downfall. Mavko[119] has shown that a frictional mechanism implies a Q^{-1} proportional to strain amplitude. As we have seen, Q^{-1} is roughly independent of strain amplitude below ϵ_{NL}, and seismic strains are generally below observed values of ϵ_{NL} for sandstones. Friction may be important beyond ϵ_{NL} only.

At low strains, not only is Q^{-1} amplitude independent, but the displacements are so small that they become of the same order of magnitude as interatomic spacings, and the applicability of the large displacement concepts of classical friction can be called in question.[120] Winkler et al.[99] calculate that for a crack length (L) of 10^{-2} cm (which they consider an upper bound) and a displacement (d) of 10^{-8} cm across the crack face, the strain $\epsilon = (d/L)$ is 10^{-6}. Hence they argue that strains must be larger than 10^{-6} before one can think of invoking friction, and in fact, values of ϵ_{NL} are generally greater than 10^{-6}. They explain the patterns shown in Figures 18 to 20 by arguing that adding a relatively small amount of water to the rock wets most of the internal surfaces and lubricates them, making it easier for sliding to occur and hence lowering ϵ_{NL}. The addition of more fluid does not have a significant extra lubricating effect and hence does not influence ϵ_{NL} as much, but influences the linear region more via flow losses. Winkler et al. are then faced with the problem of accounting for the nonzero attenuation in dry rock, if one is not willing to accept friction as a possibility. They mention as possibilities grain boundary relaxation[121,122] or a dislocation mechanism.[123,124]

A good qualitative review of the possible mechanisms of attenuation by pore fluids is given by Winkler and Nur.[110] The type of mechanism thought to be operating differs according

MASSILON SANDSTONE

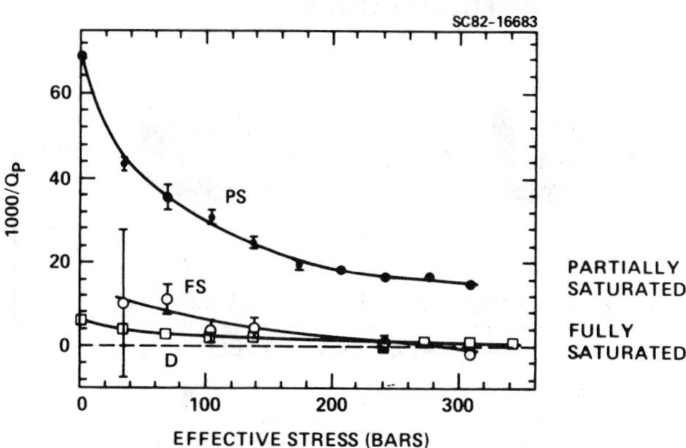

FIGURE 19. P wave attenuation (computed from Q_E and Q_s) in dry (D), fully staurated (FS), and partially saturated (PS) Massilon sandstone. Comparison with Figure 20 shows that P waves are more attenuated than S waves at partial saturation. (From Winkler, K. W. and Nur, A., *Geophysics,* 47, 1, 1982. With permission.)

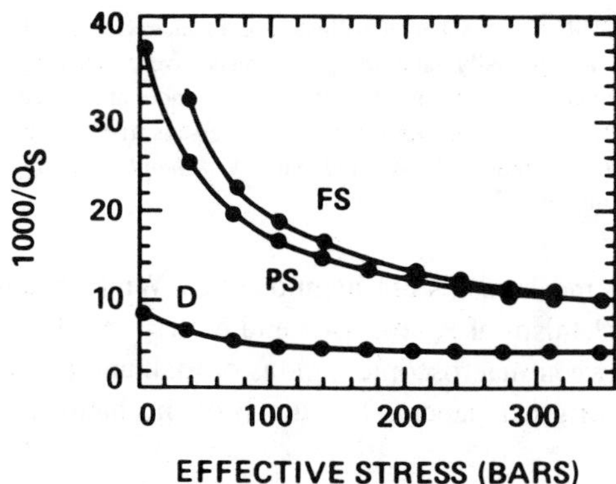

FIGURE 20. S wave attenuation in dry (D), fully saturated (FS), and partially saturated (PS) Massilon sandstone. Comparison with Figure 19 shows that S waves are less attenuated than P waves at partial saturation. (From Winkler, K. W. and Nur, A., *Geophysics,* 47, 1, 1982. With permission.)

to whether the rock is fully[125-127] or partially[128] saturated. Walsh[125,126] and O'Connell and Budiansky[127] studied viscous shear relaxation in fully saturated rocks, and Mavko and Nur[128] did the same for partially saturated rocks. As Winkler and Nur[110] point out, such shear relaxation is probably unimportant in water-bearing rocks below very high (megahertz) frequencies. At seismic frequencies, attenuation may be caused by wave-induced (or squirt) flow of fluid between cracks (intercrack flow) in the case of full saturation, or flow within cracks (intracrack flow) in the case of partial saturation. Figure 21 shows schematically how in partially saturated rock bulk compression may induce more intracrack flow than will pure shear, while in fully saturated rock, pure shear may induce more intercrack flow than will bulk compression. This kind of model may explain why compressional waves are more attenuated then shear waves in partially saturated rocks, while this is not true for fully saturated rocks (see "Attenuation in Sedimentary Rocks").

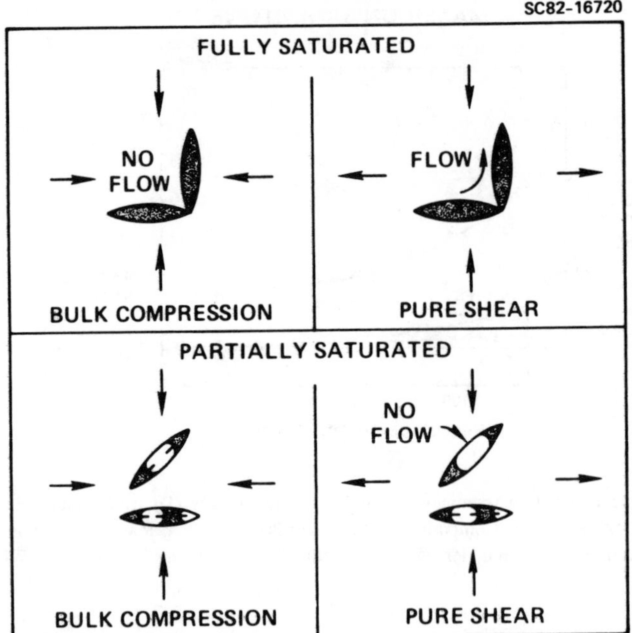

FIGURE 21. Schematic illustration of intercrack and intracrack flow in full and partially saturated pores, respectively. This type of model may explain the observation that P waves are more attenuated than S waves in partially saturated rocks, while the reverse is true in fully saturated rocks. (From Winkler, K. W. and Nur, A., *Geophysics*, 47, 1, 1982. With permission.)

Other fluid-type mechanisms are mentioned by Winkler and Nur,[110] such as the thermal relaxation and mechanism of Kjartansson and Nur,[129] the Biot mechanism,[130,131] and models involving flow between macroscopic regions of total and partial saturation.[132-135] The reader is referred to the original papers. For the Biot mechanism, the paper by Stoll[136] is also useful.

Field Measurements of Attenuation in Sedimentary Sequences

As mentioned previously, a primary stimulus for the experimental studies of elastic wave attenuation in rocks, especially sedimentary rocks, has been the prospect of using attenuation as a possible indicator in the seismic exploration for oil and gas. Certainly, in order to make such a program operative, one must be able to measure attenuation somehow in the field. In this section, we review some of the progress which has been made in effecting field measurements of seismic attenuation in the crust of the earth.

Generally, field measurements have been made by appropriately reducing data obtained by vertical seismic profiling (VSP). In VSP one places geophones (that is, seismometers) at various depths in a borehole, and records waves from a near-surface artificial source. In some of the earlier work carried out by McDonal et al.[137] and Tullos and Reid,[138] several geophones were cemented at different depths in the borehole, so that records at different depths were obtained from one experiment (i.e., one shot of the source). More recently, some important work has been done by Hauge[139] using only one downhole geophone placed at different depths for different experiments. In this later work, reproducibility of the source was a key issue. The earlier investigators used explosive sources, while Hauge used vibrators with much less high frequency energy (a top frequency of about 80 Hz, as opposed to one in excess of 400 Hz).

The procedure generally used to obtain estimates of attenuation from VSP is that of spectral ratios (see "Seismological Methods"). The spectral ratio technique must be applied with some care: one should look at the amplitude spectra of waves which are the direct arrivals to the receiver. If one's waveforms contain interference from, e.g., near-receiver reflections, one can incur serious errors. The problem is similar to that in earthquake seismology, where one must worry about free surface reflections when modeling shallow earthquakes.[140] Tullos and Reid[138] attempted to deal with this problem by applying a tapered time domain window to their waveforms from the various geophones, in order to "gate out," to some measure, the effect of interference from reflections.

It is important to note that a Q measurement of this kind is not a direct measurement of anelasticity, but rather of an aggregate apparent loss, of which anelasticity may be only a part. Other parts of the loss may be due to various forms of scattering. Many of these processes have been exhaustively discussed ih papers by Spencer et al.[141, 142] An example of an important nonabsorptive loss mechanism is the effect of the layering in the rock sequence, as is often found in sand-shale sequences. Where fine layering exists, and there may be an impedance structure which is oscillatory with depth, the system is referred to as "cyclic". Systems without fine layering and with more gradual variation of rock properties (and impedance) with depth are referred to as "transitional". In cyclic systems, the lower frequencies (that is, those with wavelengths long compared to the scale of the layering) will effectively see a homogeneous system and not be affected by the rapid impedance variation. Higher frequencies, on the other hand, will suffer multiple reflections within the cyclic systems. The multiple reflections delay, and hence, effectively remove energy from the perceived transmitted pulse. In the frequency domain, the cyclic system acts as a low pass filter, similar to the filter representing anelastic absorption. Schoenberger and Levin[143, 144] attempted to assess the effect of multiple reflections in cases where sonic and density logs were available and theoretical seismograms could be computed taking multiple reflections into account. They found Q values due to multiples alone ranging from about 125 to 3000, with multiples generally accounting for from 1/7 to 1/2 the total loss, but sometimes as high as 4/5.

Table 12 shows the results of field measurements from a number of investigators. Despite the problems and uncertainties associated with the measurements, some important patterns emerge. The field measurements appear able to discern high Qs in intervals rich in limestone and seem to be able to see the difference between shale/sand sections with high sand percentage (generally low Q) and low sand percentage (generally higher Q).

Table 12
FIELD MEASUREMENTS OF Q IN SEDIMENTARY SEQUENCES

Location	Material	Depth (m)	Frequency (Hz)	Apparent Q_p	Corrected Q_p[a]	Remarks	Ref.
Limon, Colorado	Pierre Shale	0—225	50—450	32		Q_s = 10, 20—125 Hz	137
Gulf Coast, 20 mi. So. of Houston	Loam-sand-clay	0—3	50—400	2			138
Gulf Coast, 20 mi. So. of Houston	Clay-sand	3—30	50—400	181			138
Gulf Coast, 20 mi. So. of Houston	Sandy clay	30—150	50—400	75			138
Gulf Coast, 20 mi. So. of Houston	Clay- sand	150—300	50—400	136			138
Offshore Louisiana	Pleistocene sands and shales	1170—1770	≤125	67	67		139
Offshore Louisiana	Pleistocene sand and shales, predominantely shale	1770—2070	≤125	>273	>273		139
Offshore Louisiana	Pleistocene sand and shales, mostly shale, with highest % of sand in the section	2070—2850	≤125	28	31		139
S.E. Texas 1	19% sand localized in ≤30m strips, the rest silts and silty shales	900—1560	≤80	52	109		139
S.E. Texas 1	Mostly shale	1560—1800	≤80	>273	>273		139
S.E. Texas 1	23% sand, rest shale	1800—2100	≤80	30	37		139
S.E. Texas 2	20% sand, rest shale	600—1560	≤80	41	46		139
S.E. Texas 2	Late Cretaceous limestone and chalk	1590—1755	≤80	>273	>273		139
S.E. Texas 3	Miocene, 45% sand, rest shale	660—1320	15—40	28	34		139
S.E. Texas 4	24% sand (localized), rest shale	1020—1620	40—70	55	94		139
Beaufort Sea N. Canada		549—1193	125		43		145
Beaufort Sea N. Canada		945—1311	125		67		145

[a] Corrected for multiple reflections.

REFERENCES

1. **Minster, J. B.,** Anelasticity and attenuation, in *Physics of the Earth's Interior,* Dziewonski, A. M. and Boschi, E., Eds., North-Holland, New York, 1980.
2. **Johnston, D. H. and Toksoz, M. N.,** Definitions and terminology, in Seismic Wave Attenuation, Toksoz, M. N. and Johnston, D. H., Eds., Society of Exploration Geophysicists, Tulsa, Okla., 1981.
3. **Anderson, D. L. and Hart, R. S.,** Attenuation models of the Earth, *Phys. Earth Planet. Inter.,* 16, 289, 1978.
4. **Anderson, D. L. and Hart, R. S.,** Q of the Earth, *J. Geophys. Res.,* 83, 5869, 1978.
5. **Anderson, D. L. and Given, J. W.,** Absorption band Q model for the Earth, *J. Geophys. Res.,* 87, 3893, 1982.
6. **Johnston, D. H.,** Attenuation: a state of the art summary, in Seismic Wave Attenuation, Toksoz, M. N. and Johnston, D. H., Eds., Society of Exploration Geophysicists, Tulsa, Okla., 1981.
7. **Knopoff, L. and MacDonald, G. J. F.,** *Rev. Mod. Phys.,* 30, 1958.
8. **Borcherdt, R. D.,** *J. Geophys. Res.,* 78, 2442, 1973.
9. **O'Connell, R. J., and Budiansky, B.,** Measures of dissipation in viscoelastic media, *Geophys. Res. Lett.,* 5, 5, 1978.
10. **Bradley, J. J. and Fort, A. N.,** Internal friction in rocks, in *Handbook of Physical Constants,* Clark, S. P. Jr., Ed., Geological Society of America, Boulder, Co., 1966, 97.
11. **Winkler, K. and Nur, A.,** Pore fluids and seismic attenuation in rocks, *Geophys. Res. Lett.,* 6, 1, 1979.
12. **Zener, C.,** *Elasticity and Anelasticity of Metals,* University of Chicago Press, Chicago, 1948.
13. **Kolsky, H.,** *Stress Waves in Solids,* Dover Publications, New York, 1963.
14. **Schreiber, E., Anderson, O. L., and Soga, N.,** *Elastic Constants and Their Measurement,* McGraw-Hill, New York, 1973.
15. **Nowick, A. S. and Berry, B. S.,** *Anelastic Relaxation in Crystalline Solids,* Academic Press, New York, 1972.
16. **Toksoz, M. N. and Johnston, D. H.,** Laboratory measurements of attenuation, in Seismic Wave Attenuation, Toksoz, M. N. and Johnston, D. H., Eds., Society of Exploration Geophysicists, Tulsa, Okla., 1981.
17. **Ke, T. S.,** Experimental evidence on the viscoelastic behavior of grain boundaries in metals, *Phys. Rev.,* 25, 533, 1947.
18. **Kingery, W. D.,** *Property Measurements at High Temperature,* John Wiley & Sons, New York, 1959.
19. **Jackson, D. D.,** Grain Boundary Relaxations and the Attenuation of Seismic Waves, Ph.D. thesis, Massachusetts Institute of Technology, Cambridge, 1969.
20. **Woirgard, J., Amirault, J. P., Chaumet, H., and Fouquet, J., de,** Appareil de mesure du module d'elasticite et du frottement interieur en flexion, a basse frequence, sous vide entre 20 et 800°C, *Rev. Phys. Appl.,* 6, 355, 1971.
21. **Woirgard, J. and Gueguen, Y.,** Elastic modulus and internal friction in enstatite, forsterite, and peridotite at seismic frequencies and high temperatures, *Phys. Earth Planet. Inter.,* 17, 140, 1978.
22. **Peselnick, L. and Outerbridge, W. F.,** Internal friction in shear and shear modulus of solenhofen limestome over a frequency range of 10^7 cycles per second, *J. Geophys Res.,* 66, 581, 1961.
23. **Gardner, G. H. F., Wyllie, M. R. J., and Droschak, D. M.,** Effects of pressure and fluid saturation on the attenuation of elastic waves in sands, *J. Petr. Tech.,* 189, 1964.
24. **Tittmann, B. R.,** Internal friction measurements and their implications in seismic Q structure models of the crust, in The Earth's Crust, Geophysical Monograph 20, American Geophysical Union, Washington, D.C., 1977.
25. **Truell, R., Elbaum, C., and Chick, B.,** *Ultrasonic Methods in Solid State Physics,* Academic Press, New York, 1969.
26. **Peselnick, L. and Zietz, X.,** Internal friction of fine grained limestones at ultrasonic frequencies, *Geophysics,* 24, 285, 1959.
27. **McSkimin, H. J.,** Propagation of longitudinal waves and shear waves in cylindrical rods at high frequencies, *J. Acoust. Soc. Am.,* 28, 484, 1956.
28. **Watson, T. H. and Wuenschel, P. C.,** An Experimental Study of Attenuation in Fluid Saturated Porous Media, Compressional Waves and Interfacial Waves, 43rd Annu. Society of Exploration Geophysicists Meet., Tulsa, Ok., 1973.
29. **Toksoz, M. N., Johnston, D. H., and Timur, A.,** Attenuation of seismic waves in dry and saturated rocks. I. Laboratory measurements, *Geophysics,* 44, 681, 1979.
30. **Gladwin, M. T. and Stacey, F. D.,** Anelastic degradation of acoustic pulses in rock, *Phys. Earth Planet. Inter.,* 8, 332, 1974.
31. **Ramana, Y. V. and Rao, M. V.,** Q by pulse broadening in rocks under pressure, *Phys. Earth Planet. Inter.,* 8, 337, 1974.

32. **Ricker, N. H.,** The form and laws of propagation of seismic wavelets, *Geophysics,* 18, 10, 1953.
33. **Kolsky, H.,** The propagation of stress pulses in viscoelastic solids, *Philos. Mag.*, 1, 693, 1956.
34. **Futterman, W. I.,** Dispersive body waves, *J. Geophys. Res.*, 67, 5257, 1962.
35. **Carpenter, E. W.,** Absorption of Elastic Waves — An Operator for A Constant Q Mechanism, U.K. Atomic Energy Authority AWRE Rep. 0-43/66, 1966; reprinted in *Toksoz, M. N. and Johnston, D. H., Eds.*, Seismic Wave Attenuation, Society of Exploration Geophysicists, Tulsa, Okla., 1981.
36. **Strick, E.,** A predicted pedestal effect for pulse propagation in constant Q solids, *Geophysics,* 35, 387, 1970.
37. **Minster, J. B.,** Transient and impulse responses of a one dimensional linearly attenuating medium. I. Analytical results, *Geophys. J. R. Astr. Soc.*, 52, 479, 1978.
38. **Minster, J. B.,** Transient and impulse responses of a one dimensional linearly attenuating medium. II. A parametric study, *Geophys. J. R. Astron. Soc.*, 52, 503, 1978.
39. **Minster, J. B. and Vassiliou, M. S.,** Pulse Propagation in a Frequency Dependent Linearly Attenuating Medium, Vol. 17, Stanford University Publications, Geological Sciences, 1979.
40. **Brennan, B. J.,** Pulse propagation in media with frequency dependent Q, *Geophys. Res. Lett.*, 7, 211, 1980.
41. **Gordon, R. B. and Davis, L. A.,** Velocity and attenuation of seismic waves in imperfectly elastic rock, *J. Geophys. Res.*, 73, 3917, 1968.
42. **McKavanagh, B. and Stacey, F. D.,** Mechanical hysteresis in rocks at low strain amplitudes and seismic frequencies, *Phys. Earth Planet. Inter.* , 8, 246, 1974.
43. **Brennan, B. J. and Stacey, F. D.,** Frequency dependence of elasticity of rock — test of seismic velocity dispersion, *Nature (London)*, 268, 220, 1977.
44. **Peselnick, L., Liu, H. P., and Harper, K. R.,** Observations of details of hysteresis loops in westerly granite, *Geophys. Res. Lett.*, 6, 693, 1979.
45. **Gross, B.,** *Mathematical Structure of the Theories of Viscoelasticity,* Hermann, Paris, 1953.
46. **Anderson, D. L. and Minster, J. B.,** The frequency dependence of Q in the earth and implications for mantle rheology and Chandler wobble, *Geophys. J. R. Astron. Soc.*, 58, 431, 1979.
47. **Jeffreys, H.,** Rock Creep, *Mon. Not. R. Astron. Soc.*, 118, 14, 1958.
48. **Goetze, C.,** High temperature rheology of westerly granite, *J. Geophys. Res.*, 76, 1223, 1971.
49. **Goetze, C. and Brace, W. F.,** Laboratory observations of high temperature rheology of rocks, *Tectonophysics,* 13, 583, 1972.
50. **Murrell S. A. F. and Chakravarty, S.,** Some new rheological experiments on igneous rocks at temperatures up to 1120°C, *Geophys. J. R. Astron. Soc.*, 34, 211, 1973.
51. **Berckhemer, H., Auer, F., and Drisler, J.,** High temperature elasticity and anelasticity of mantle peridotite, *Phys. Earth Planet. Inter.*, 20, 48, 1979.
52. **Anderson, D. L. and Kovach, R. L.,** Attenuation of the mantle and rigidity of the core from multiply reflected core phases, *Proc. Natl. Acad. Sci. U.S.A.*, 51, 168, 1964.
53. **Burdick, L. J.,** Broad Band Seismic Studies of Body Waves, Ph.D. thesis, California Institute of Technology, Pasadena, 1977.
54. **Der, Z. A. and McElfresh, T. W.,** *Bull. Seismol. Soc. Am.*, 67, 1303, 1977.
55. **Ben Menahem, A.,** Observed attenuation and Q values of seismic waves in the upper mantle, *J. Geophys. Res.*, 70, 4641, 1965.
56. **Kanamori, H.,** Velocity and Q of mantle waves, *Phys. Earth Planet. Inter.*, 2, 259, 1970.
57. **Nakanishi, I.,** *Geophys. J. R. Astron. Soc.*, 58, 35, 1979.
58. **Butler, R. G.,** Seismological Studies Using Observed and Synthetic Waveforms, Ph.D. thesis, California Institute of Technology, Pasadena, 1979.
59. **Jordan, T. H. and Sipkin, S. A.,** Estimation of the attenuation for multiple ScS waves, *Geophys. Res. Lett.*, 4, 167, 1977.
60. **Sipkin, S. A. and Jordan, T. H.,** *Bull. Seismol. Soc. Am.*, 69, 1055, 1979.
61. **Dziewonski, A. M.,** *Rev. Geophys. Space Phys.*, 17, 303, 1979.
62. **Stein, S. and Geller, R. J.,** *Bull. Seismol. Soc. Am.*, 68, 325, 1978.
63. **Buland, R. and Gilbert, F.,** Improved resolution of complex eigenfrequencies in analytically continued seismic spectra, *Geophys. J. R. Astron. Soc.*, 52, 457, 1978.
64. **Gilbert, F. and Dziewonski, A. M.,** An application of normal mode theory to the retrieval of structural parameters and source mechanisms from seismic spectra, *Philos. Trans. R. Soc. London, Ser. A*, 278, 187, 1975.
65. **Dziewonski, A. and Anderson, D. L.,** Preliminary reference Earth model, *Phys. Earth Planet. Inter.*, 25, 297, 1981.
66. **Sailor, R. V. and Dziewonski, A.,** Measurements and interpretation of normal mode attenuation, *Geophys. J. R. Astron. Soc.*, 53, 559, 1978.
67. **Lay, T. and Helmberger, D. V.,***Geophys. J. R. Astron. Soc.*, 66, 691, 1981.

68. **Anderson, D. L. and Minster, J. B.,** Seismic velocity, attenuation, and rheology of the upper mantle, in Source Mechanism and Earthquake Prediction, Allegre, C. J., Ed., Centre National de la Recherche Scientifique, Paris, 1980.
69. **Minster, J. B. and Anderson, D. L.,** Dislocations and nonelastic processes in the mantle, *J. Geophys. Res.*, 85, 6347, 1980.
70. **Minster, J. B. and Anderson, D. L.,** A model of dislocation-controlled rheology for the mantle, *Philos. Trans. R. Soc. London, Ser, A.*, 299, 319, 1981.
71. **Anderson, D. L., Kanamori, H., Hart, R. S., and Liu, H. P.,** The Earth as a seismic absorption band, *Science,* 196, 1104, 1977.
72. **Press, F.,** Rigidity of the Earth's core, *Science,* 124, 1204, 1956.
73. **Gutenberg, B.,** Attenuation of seismic waves in the Earth's mantle, *Bull. Seismol. Soc. Am.*, 48, 269, 1958.
74. **Otsuka, M.,** On the Forms of the S and ScS waves of some deep earthquakes, *Zisin,* 2, 169, 1962.
75. **Otsuka, M.,** Some considerations on the waveforms of ScS phases, *Spec. Contrib. Geophys. Inst., Kyoto Univ.*, 2, 415, 1963.
76. **Steinhart, H. S., Smith, T. J., Sacks, I. S., Sumner, R., Suzuki, Z., Rodriguez, A., Lomnitz, C., Tuve, M. A., and Aldrich, L. T.,** Explosion seismology, *Carnegie Inst. Wash. Yearb.*, 62, 286, 1964.
77. **Kovach, R. L. and Anderson, D. L.,** Attenuation of shear waves in the upper and lower mantle, *Bull. Seismol. Soc. Am.*, 54, 1855, 1964.
78. **Sato, R. and Espinosa, A. F.,** Dissipation in the Earth's mantle and rigidity and viscosity of the Earth's core determined from waves multiply reflected from the mantle-core boundary, *Bull. Seismol. Soc. Am.*, 57, 829, 1967.
79. **Kanamori, H.,** Spectrum of P and PcP in relation to the mantle-core boundary and attenuation in the mantle, *J. Geophys. Res.*, 72, 559, 1967.
80. **Choudbury, M. A. and Dorel, J.,** Spectral ratio of short-period ScP and ScS phases in relation to the attenuation in the mantle beneath the tasman sea and Antarctic region, *J. Geophys. Res.*, 78, 462, 1973.
81. **Yoshida, M. and Tsujiura, M.,** Spectrum and attenuation of multiply reflected corephases, *J. Phys. Earth,* 23, 31, 1975.
82. **Okal,** unpublished data.
83. **Best, W. J., Johnson, L. R., and McEvilly, T. V.,** ScS and the mantle beneath Hawaii, *EOS,* 56, 1147, 1974.
84. **Sima, H.,** On the attenuation of SS and SSS waves, *Q. J. Seismol. (Tokyo),* 29, 109, 1965.
85. **Berzon, I. S., Passechnik, I. P., and Polikarpov,** *Geophys. J. R. Astron. Soc.*, 39, 603, 1974.
86. **Kanamori, H.,** Attenuation of P waves in the upper and lower mantle, *Bull. Earthquake Res. Inst. Tokyo Univ.*, 45, 299, 1967.
87. **Mikumo, T. and Kurita, T.,** Q distribution for long period P waves in the mantle, *J. Phys. Earth,* 16, 1968.
88. **Smith, S. W.,** The anelasticity of the mantle, *Tectonophysics,* 13, 601, 1972.
89. **Nowroozi, A. A.,** Characteristic periods and Q for oscillations of the Earth following an intermediate earthquake, *J. Phys. Earth,* 22, 1, 1974.
90. **Bolt, B. A. and Brillinger, D. R.,** Estimation of uncertainties in fundamental frequencies of decaying geophysical time series, *EOS,* 56, 403, 1975.
91. **Deschamps, A.,** Inversion of the attenuation data of free oscillations of the Earth (fundamental and first higher modes), *Geophys. J. R. Astr. Soc.*, 50, 699, 1977.
92. **Roult, G.,** Attenuation of seismic waves of very low frequency, *Phys. Earth Planet. Inter.*, 10, 159, 1975.
93. **Nowroozi, A. A.,** Measurement of Q values from the free oscillations of the Earth, *J. Geophys. Res.*, 73, 1407, 1968.
94. **Wu, F. T.,** Mantle rayleigh wave dispersion and tectonic provinces, *J. Geophys. Res.*, 77, 6445, 1972.
95. **Smith, S. W.,** An Investigation of the Earth's Free Oscillations, Ph.D. thesis, California Institute of Technology, Pasadena, 1961.
96. **Jobert, N. and Roult, G.,** Periods and damping of free oscillations observed in France after sixteen earthquakes, *Geophys. J. R. Astron. Soc.*, 45, 155, 1976.
97. **Jacobs, J. A.,** *The Earth's Core,* Academic Press, New York, 1975.
98. **Johnston, D. H., Toksoz, M. N., and Timur, A.,** Attenuation of seismic waves in dry and saturated rocks. II. Mechanisms, *Geophysics,* 44, 691, 1979.
99. **Winkler, K., Nur, A., and Gladwin, M.,** Friction and seismic attenuation in rocks, *Nature (London),* 227, 528, 1979.
100. **Johnston, D. H. and Toksoz, M. N.,** Thermal cracking and amplitude dependent attenuation, *J. Geophys. Res.*, 85, 937, 1980.
101. **Tittmann, B. R., Abdel-Gawad, M., Salvado, C., Bulau, J., Ahlberg, L., and Spencer, T. W.,** A brief note on the effect of interface bonding on seismic dissipation, *Proc. Lunar Planet. Sci.*, 12B, 1737, 1981.

102. **Klima, K., Vanek, J., and Pros. Z.,** The attenuation of longitudinal waves in diabase and greywacke under pressures up to 4 kilobars, *Studia Geophys. Geod.*, 8, 247, 1964.
103. **Levykin, A. I.,** Longitudinal and transverse wave absorption and velocity in rock specimens at multilateral pressures up to 4000 kg/cm^2, *Izv. Phys. Solid Earth U.S.S.R. Acad. Sci.*, 1, 94, 1965.
104. **Johnston, D. H. and Toksoz, M. N.,** Ultrasonic P and S wave attenuation in dry and saturated rocks under pressure, *J. Geophys. Res.*, 85, 925, 1980.
105. **Tittmann, B. R., Nadler, H., Clark, V. A., and Ahlberg, L. A.,** Frequency dependence of seismic dissipation in saturated rocks, *Geophys. Res. Lett.*, 8, 36, 1981.
106. **Attewell, P. B. and Ramana, Y. V.,** Wave attenuation and internal friction as functions of frequency in rocks, *Geophysics*, 31, 1049, 1966.
107. **Birch, F. and Bancroft, D.,** Elasticity and internal friction in a long column of granite, *Bull. Seismol Soc. Am.*, 28, 243, 1938.
108. **Born, W. T.,** Attenuation constant of Earth materials, *Geophysics*, 6, 132, 1941.
109. **Pandit, B. I. and Savage, J. C.,** An experimental test of Lomnitz's theory of internal friction in rocks, *J. Geophys. Res.*, 78, 6097, 1973.
110. **Winkler, K. W. and Nur, A.,** Seismic attenuation: effects of pore fluids and frictional sliding, *Geophysics*, 47, 1, 1982.
111. **Spencer, J. W.,** Stress relaxations at low frequencies in fluid saturated rocks, attenuation and modulus dispersion, *J. Geophys. Res.*, 86, 1803, 1981.
112. **Pandit, B. I. and Tozer, D. C.,** Anomalous propagation of elastic energy within the moon, *Nature (London)*, 226, 335, 1970.
113. **Warren, N., Trice, R., and Stephens, J.,** Ultrasonic attenuation, Q measurements on 70215,29, *Geochim. Cosmochim. Acta*, Suppl. 5, 2927, 1974.
114. **Clark, V. A., Spencer, T. W., Tittmann, B. R., Ahlberg, L. A., and Coombe, L. T.,** Effect of volatiles on attenuation (Q^{-1}) and velocity in sedimentary rocks, *J. Geophys. Res.*, 85, 5190, 1980.
115. **Pandit, B. I. and King, M. S.,** The Variation of elastic wave velocities and quality factor of a sandstone with moisture content, *Can. J. Earth Sci.*, 16, 2187, 1979.
116. **Tittmann, B. R., Clark, V. A., Richardson, J., and Spencer, T. W.,** Possible mechanism for seismic attenuation in rocks containing small amounts of volatiles, *J. Geophys. Res.*, 85, 5199, 1980.
117. **Walsh, J. B.,** Seismic wave attenuation in rock due to friction, *J. Geophys. Res.*, 71, 2591, 1966.
118. **Clark, V. A.,** Effects of Volatiles on Seismic Attenuation and Velocity in Sedimentary Rocks, PhD. thesis, Texas A & M University, 1980
119. **Mavko, G. M.,** Frictional attenuation: an inherent amplitude dependence, *J. Geophys, Res.*, 84, 4769, 1979.
120. **Savage, J. C.,** *J. Geophys. Res.*, **74, 726, 1969.**
121. **Gordon, R. B. and Nelson, C. W.,** *Rev. Geophys.*, 4, 457, 1966.
122. **Jackson D. D. and Anderson, D. L.,** Physical mechanisms of seismic wave attenuation, *Rev. Geophys. Space Phys.*, 8, 1, 1970.
123. **Mason, W. P., Beshers, D. N., and Kuo, J. T.,** Internal friction in westerly granite: relation to dislocation theory, *J. Appl. Phys.*, 41, 5206, 1970.
124. **Mason, W. P., Marfurt, K. J., Beshers, D. N., and Kuo, J. T.,** Internal friction in rocks, *J. Acoust. Soc. Am.*, 63, 1596, 1978.
125. **Walsh, J. B.,** Attenuation in partially melted material, *J. Geophys. Res.*, 73, 2209, 1968.
126. **Walsh, J. B.,** New analysis of attenuation in partially melted Rock, *J. Geophys. Res.*, 74, 2209, 1969.
127. **O'Connell, R. J. and Budiansky, B.,** Viscoelastic properties of fluid saturated cracked solids, *J. Geophys. Res.*, 82, 5719, 1977.
128. **Mavko, G. M. and Nur, A.,** Wave attenuation in partially saturated rocks, *Geophysics*, 44, 161, 1979.
129. **Kjartansson, E. and Nur, A.,** Attenuation due to thermal relaxation in porous rocks, submitted to Geophysics, 1982.
130. **Biot, M. A.,** Theory of propagation of elastic waves in a fluid saturated, porous solid. I. Low frequency range, *J. Acoust. Soc. Am.*, 28, 168, 1956.
131. **Biot, M. A.,** Theory of propagation of elastic waves in a fluid saturated, porous solid. II. Higher frequency range, *J. Acoust. Soc. Am.*, 28, 179, 1956.
132. **White, J. E.,** Computed seismic speeds and attenuation in rocks with partial gas saturation, *Geophysics*, 40, 224, 1975.
133. **Dutta, N. C. and Ode, H.,** Attenuation and dispersion of compressional waves in fluid-filled rocks with partial gas saturation (White model). I. Biot theory, *Geophysics*, 44, 1806, 1979.
134. **Dutta, N. C. and Ode, H.,** Attenuation and dispersion of compressional waves in fluid-filled rocks with partial gas saturation (White model). II. Results, *Geophysics*, 44, 1789, 1979.
135. **Dutta, N. C. and Seriff, A. J.,** On White's model of attenuation in rocks with partial gas saturation, *Geophysics*, 44, 1806, 1979.

136. **Stoll, R. D.,** Acoustic waves in saturated sediments, in *Physics of Sound in Marine Sediments,* Hampton, L., Ed., Plenum Press, New York, 1974.
137. **McDonal, F. J., Angona, F. A., Mills, R. L., Sengbush, R. L., Van Nostrand, R. G., and White, J. E.,** Attenuation of shear and compressional waves in Pierre shale, *Geophysics,* 23, 421, 1958.
138. **Tullos, F. N. and Reid, A. C.,** Seismic attenuation of Gulf coast sediments, *Geophysics,* 34, 516, 1969.
139. **Hauge, P. S.,** Measurements of attenuation from vertical seismic profiles, *Geophysics,* 46, 1548, 1981.
140. **Langston, C. A.,** Moments, corner frequencies, and the free surface, *J. Geophys. Res.,* 83, 3422, 1978.
141. **Spencer, T. W., Edwards, C. M., and Sonnad, J. R.,** Seismic wave attenuation in nonresolvable cyclic stratification, *Geophysics,* 42, 939, 1977.
142. **Spencer, T. W., Sonnad, J. R., and Butler, T. M.,** Seismic Q — stratigraphy or dissipation, *Geophysics,* 47, 16, 1982.
143. **Schoenberger, M. and Levin, F. K.,** Apparent attenuation due to intrabed multiples. I, *Geophysics,* 39, 278, 1974.
144. **Schoenberger, M. and Levin, F. K.,** Apparent attenuation due to intrabed multiples. II, *Geophysics,* 43, 730, 1978.
145. **Ganley, D. C. and Kanasewich, E. R.,** Measurement of absorption and dispersion from check-shot surveys, *J. Geophys. Res.,* 85, 5219, 1980.

Section VIII
Radioactivity Properties of Minerals and Rocks

By
William Randall Van Schmus

INTRODUCTION

Radioactive elements found in rocks, minerals, and other crustal materials (e.g., sediment, soil) form the basis for several major applications in geochemistry and geophysics. Direct use of various elements and their decay products can be grouped into three main categories: (1) geochronology and cosmochronology, (2) radioactivity surveying and well logging, and (3) radiogenic heat production. In addition, various natural and artificial radionuclides or their decay products have found wide application in geochemical analysis and as tracers in various geologic processes. The information presented here is chosen primarily to relate to the naturally occurring properties and their application.

GEOCHRONOLOGY AND COSMOCHRONOLOGY

Radioactive nuclides that are, or once were, naturally occurring form the basis of measuring the time parameter for a wide variety of geologic and cosmologic processes. A summary of the radionuclides that have found significant application or have future potential application is given in Table 1. The decay constants given are those that are generally used by geochemists or geophysicists. In several instances there are alternate values available from the literature, but no attempt has been made to review this data. A few of the radionuclides are generally considered "extinct", and their decay products are found only in primitive meteorites. In one case, however (^{26}Al), it is also found as a cosmic-ray produced component in modern sediment.

Radiometric "clocks" generally belong to one of two main categories: decay clocks and accumulation clocks. In a simple decay clock, best represented by the familiar ^{14}C method, the elapsed time being measured is computed from the decay of the original radioactive isotope:

$$t = (1/\lambda)\ \ln(N_o/N) \tag{1}$$

where N_o is the starting quantity or radioactivity of the radionuclide in the sample, N is the present quantity, and λ is the decay constant. In such applications it is necessary to know, or be able to evaluate, N_o from independent arguments and to assume that there has been no net loss or gain of that isotope by other than normal decay processes (i.e., a "closed system"). In many instances it is better to use pairs of radionuclides, and for these techniques there is a wide variety of assumptions inherent to obtaining meaningful results. Many of these are summarized elsewhere.[8] In accumulation clocks, elapsed time is measured by use of the build-up of decay products of the radionuclide in question; the decay products may include intermediate, unstable decay products or radiation damage (e.g., "fission tracks") in the crystal as well as stable decay products. For simple accumulation, such as the ^{87}Rb-^{87}Sr or ^{147}Sm-^{143}Nd system, ages can be calculated from a basic equation such as:

$$t = \frac{1}{\lambda} \ln\left(1 + \frac{D - D_o}{N}\right) \tag{2}$$

in which D represents the total abundance of the decay-product isotope, D_o represents the amount of that isotope that was present at the start of the time interval being measured (the difference is that attributable to *in situ* decay of the radionuclide), and N is the amount of the radionuclide in the sample at present. As with the simple decay clock, it is necessary to assume a "closed system" behavior for all isotopes involved in the computation. Fur-

Table 1
RADIOMETRIC SYSTEMS USED IN GEOCHRONOLOGY AND COSMOCHRONOLOGY

Parent isotope	Half-life (years)	Decay constant (years^{-1})	Decay mode[a]	Daughter product(s)	Ref.
^{3}H	12.33	5.62×10^{-2}	β^-	^{3}He	1
^{10}Be	1.6×10^{6}	4.33×10^{-7}	β^-	^{10}B	1
^{14}C	5730[b]	1.210×10^{-4}	β^-	^{14}N	2
^{26}Al	7.2×10^{5}	9.6×10^{-7}	β^+	^{26}Mg	1
^{40}K	1.25×10^{9}	4.962×10^{-10}	β^-	^{40}Ca (89.5%)	3
		0.581×10^{-10}	EC	^{40}Ar (10.5%)	3
^{87}Rb	4.88×10^{10}	1.42×10^{-11}	β^-	^{87}Sr	3
^{107}Pd	6.5×10^{6}	1.07×10^{-7}	β^-	^{107}Ag	4
^{129}I	1.6×10^{7}	4.33×10^{-8}	β^-	^{129}Xe	1
^{147}Sm	1.06×10^{11}	6.54×10^{-12}	α	^{143}Nd	5
^{176}Lu	3.53×10^{10}	1.96×10^{-11}	β^-	^{176}Hf	6
^{187}Re	4.3×10^{10}	1.61×10^{-11}	β^-	^{187}Os	7
^{210}Pb	22.26	3.11×10^{-2}	β^-	^{210}Bi	8
^{226}Ra	1.62×10^{3}	4.27×10^{-4}	α	^{222}Rn	8
^{230}Th	7.52×10^{4}	9.22×10^{-6}	α	^{226}Ra	8
^{231}Pa	3.25×10^{4}	2.134×10^{-5}	α	^{227}Ac	8
^{232}Th	1.40×10^{10}	4.9475×10^{-11}	α,β^-	$^{208}Pb + 6\ ^{4}He$[c]	3
^{234}U	2.45×10^{5}	2.794×10^{-6}	α	^{230}Th	8
^{235}U	7.04×10^{8}	9.8485×10^{-10}	α,β^-	$^{207}Pb + 7\ ^{4}He$[c]	3
^{238}U	4.47×10^{9}	1.5513×10^{-10}	α,β^-	$^{206}Pb + 6\ ^{4}He$[c]	3
		8.46×10^{-17}[d]	SF	Various	9
^{244}Pu	8.2×10^{7}	8.47×10^{-9}	α	$^{232}Th + 3\ ^{4}He$	10
		1.06×10^{-11}	SF	Various	10

[a] Decay mode includes loss of electron (beta particle, β^-), or of positron (β^+), or of alpha particle (α); or EC (electron capture), or SF (spontaneous fission).

[b] Most accurate value. However, by convention most ^{14}C dating labs report ages based on a half-life of 5568 years.

[c] Decay series. See Tables 2, 3, or 4 for details.

[d] Many workers use 6.85×10^{-17} for this decay constant. See discussion by Faure[8] about this problem.

thermore, it is necessary to be able to evaluate D_o in some independent, precise manner. In many instances it is possible to use other derivatives from Equation 2 and multiple samples to solve for D_o; in other cases (e.g., U–Pb systems) ages can be calculated from the systematic behavior of coupled systems. Details on these and other methods may be found elsewhere.[8,11]

NATURAL RADIOACTIVITY MEASUREMENTS

In refering to ''natural radioactivity'', most users mean those radionuclides that contribute the largest portion of observed natural radiation. These are the isotopes ^{40}K, ^{232}Th, ^{235}U, and ^{238}U. The first undergoes branching decay to ^{40}Ca and ^{40}Ar, both of which are stable (Table 1). The U and Th isotopes, however, do not decay directly to a stable product, but achieve ultimate stability through a succession of α and β^- decays, including several minor closed branches (e.g., branches that converge a step later to the same subsequent product). The U and Th decay series are summarized in Tables 2 to 4 and Figure 1. Only the major branching decay in the ^{232}Th series at ^{212}Bi has been included. Full details on each series can be derived from more detailed references.[1]

The major penetrative natural radioactivity is γ-radiation that accompanies many α and

Table 2
PRINCIPAL STEPS IN THE ^{238}U DECAY SERIES

Step	Parent	Half-life	Decay mode	Daughter
1	^{238}U	4.47×10^9 years	Alpha	^{234}Th
2	^{234}Th	24.1 days	Beta	^{234}Pa
3	^{234}Pa	1.17 minutes	Beta	^{234}U
4	^{234}U	2.44×10^5 years	Alpha	^{230}Th
5	^{230}Th	7.7×10^4 years	Alpha	^{226}Ra
5	^{226}Ra	1.60×10^3 years	Alpha	^{222}Rn
7	^{222}Rn	3.82 days	Alpha	^{218}Po
8	^{218}Po	3.05 minutes	Alpha	^{214}Pb
9	^{214}Pb	26.8 minutes	Beta	^{214}Bi
10	^{214}Bi	19.8 minutes	Beta	^{214}Po
11	^{214}Po	1.64×10^{-4} sec	Alpha	^{210}Pb
12	^{210}Pb	22.3 years	Beta	^{210}Bi
13	^{210}Bi	5.01 days	Beta	^{210}Po
14	^{210}Po	138.4 days	Alpha	^{206}Pb (stable)

Table 3
PRINCIPAL STEPS IN THE ^{235}U DECAY SERIES

Step	Parent	Half-life	Decay mode	Daughter
1	^{235}U	7.04×10^8 years	Alpha	^{231}Th
2	^{231}Th	25.52 hr	Beta	^{231}Pa
3	^{231}Pa	3.28×10^4 years	Alpha	^{227}Ac
4	^{227}Ac	21.77 years	Beta	^{227}Th
5	^{227}Th	18.72 days	Alpha	^{223}Ra
6	^{223}Ra	11.43 days	Alpha	^{219}Rn
7	^{219}Rn	3.96 sec	Alpha	^{215}Po
8	^{215}Po	1.78×10^{-3} sec	Alpha	^{211}Pb
9	^{211}Pb	36.1 minutes	Beta	^{211}Bi
10	^{211}Bi	2.14 minutes	Alpha	^{207}Tl
11	^{207}Tl	4.77 minutes	Beta	^{207}Pb (stable)

Table 4
PRINCIPAL STEPS IN THE ^{232}Th DECAY SERIES

Step	Parent	Half-life	Decay mode	Daughter
1	^{232}Th	1.40×10^{10} years	Alpha	^{228}Ra
2	^{228}Ra	5.75 years	Beta	^{228}Ac
3	^{228}Ac	6.13 hr	Beta	^{228}Th
4	^{228}Th	1.913 years	Alpha	^{224}Ra
5	^{224}Ra	3.66 days	Alpha	^{220}Rn
6	^{220}Rn	55.6 sec	Alpha	^{216}Po
7	^{216}Po	0.15 sec	Alpha	^{212}Pb
8	^{212}Pb	10.64 hr	Beta	^{212}Bi
9a	^{212}Bi	60.6 minutes	Beta	^{212}Po (64%)
9b	^{212}Bi	60.6 minutes	Alpha	^{208}Tl (36%)
10a	^{212}Po	2.98×10^{-7} sec	Alpha	^{208}Pb (stable)
10b	^{208}Tl	3.053 minutes	Beta	^{208}Pb (stable)

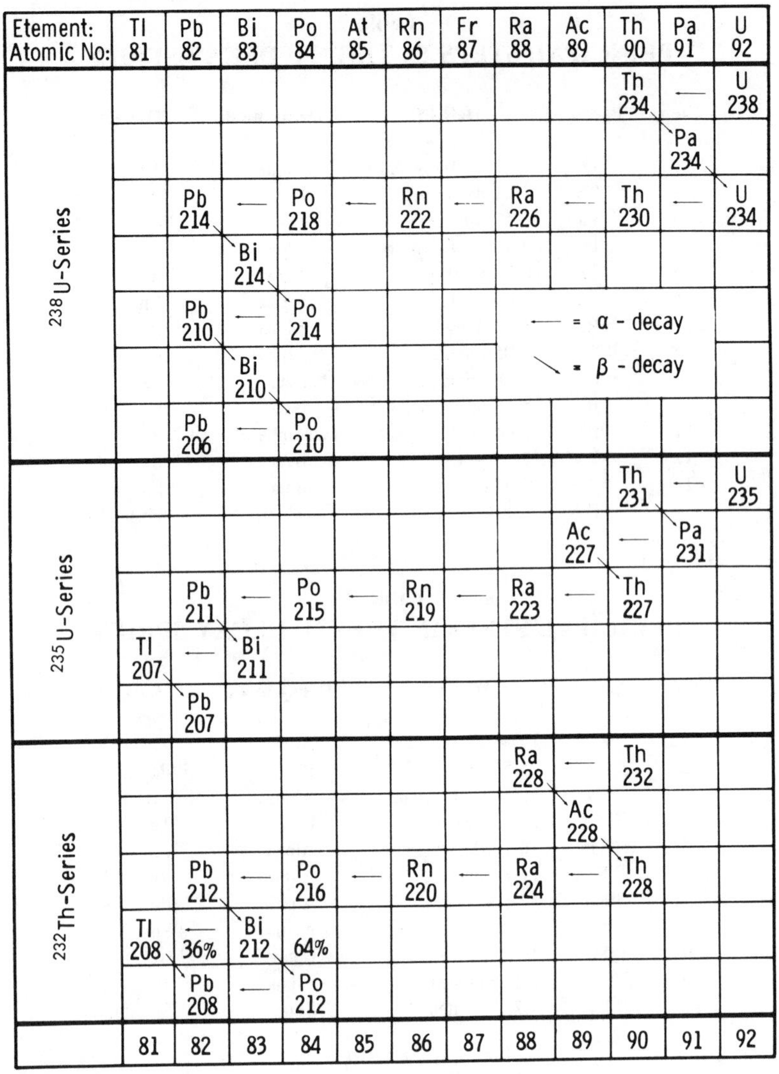

Etement: Atomic No:	Tl 81	Pb 82	Bi 83	Po 84	At 85	Rn 86	Fr 87	Ra 88	Ac 89	Th 90	Pa 91	U 92
^{238}U-Series										Th 234	—	U 238
											Pa 234	
		Pb 214	—	Po 218	—	Rn 222	—	Ra 226	—	Th 230	—	U 234
			Bi 214									
		Pb 210	—	Po 214				— = α - decay				
			Bi 210					↘ = β - decay				
		Pb 206	—	Po 210								
^{235}U-Series										Th 231	—	U 235
									Ac 227	—	Pa 231	
		Pb 211	—	Po 215	—	Rn 219	—	Ra 223	—	Th 227		
	Tl 207	—	Bi 211									
		Pb 207										
^{232}Th-Series								Ra 228	—	Th 232		
									Ac 228			
		Pb 212	—	Po 216	—	Rn 220	—	Ra 224	—	Th 228		
	Tl 208	— 36%	Bi 212	64%								
		Pb 208	—	Po 212								
	81	82	83	84	85	86	87	88	89	90	91	92

FIGURE 1. Summary of major steps in radioactive decay series of ^{238}U, ^{235}U, and ^{232}Th. Minor cases of branching decay have been omitted. See Tables 2—4 for half-lives of intermediate isotopes.

β^- transitions of the major natural radionuclides. This radiation is widely used in geophysical and geochemical prospecting, including well logging. The major γ-radiations from K, U, and Th are summarized in Table 5 and Figure 2. No γ-rays for ^{235}U are given because its low natural abundance ($^{238}U/^{235}U$ = 137.88[3]) means that the intensity of its major γ-rays would still be minor compared to those from ^{238}U, and hence are not detectable for all practical purposes. Of the γ-rays shown in Table 5 and Figure 2, the main ones used as diagnostic radiations are 1.461 MeV (^{40}K), 1.765 MeV (^{238}U), and 2.615 MeV (^{232}Th); most of the others are either too low in abundance for easy measurement by routine methods or overlap radiations from another element (e.g., 0.58 and 0.61 MeV), although in certain applications they can be used.

There are many minerals that contain major to significant trace amounts of K, U, and Th. Table 6 lists the more common or otherwise significant minerals that contain K, U, and

Table 5
PRINCIPAL GAMMA RAYS FROM MAJOR NATURALLY OCCURRING NUCLIDES[a]

Parent	Nuclides	Energy (MeV)	Frequency[b]	
^{40}K	K-40	1.461	11	
^{232}Th	Ac-228	0.210	4	
	Pb-212	0.239	43	48
	Ra-224	0.241	5	
	Tl-208	0.277	3	
	Pb-212	0.300	3	
	Ac-228	0.339	12	
	Ac-228	0.463	5	
	Tl-208	0.511	8	
	Tl-208	0.583	31	
	Bi-212	0.727	6	
	Tl-208	0.860	4	
	Ac-228	0.911	27	
	Ac-228	0.964	5	21
	Ac-228	0.969	16	
	Ac-228	1.588	3	
	Tl-208	2.615	36	
^{238}U	Ra-226	0.186	3	
	Pb-214	0.242	6	
	Pb-214	0.295	19	
	Pb-214	0.352	37	
	Bi-214	0.609	46	
	Bi-214	0.768	5	
	Bi-214	0.934	3	
	Bi-214	1.120	15	
	Bi-214	1.238	6	
	Bi-214	1.378	4	
	Bi-214	1.408	3	
	Bi-214	1.730	3	
	Bi-214	1.765	16	
	Bi-214	2.204	5	

[a] Compiled from Reference 1; only events with energies greater than 100 keV are tabulated.
[b] Events per 100 decays of primary parent; secular equilibrium assumed.

Th. Specific abundance ranges have not been given for minor or trace levels, since they can vary over a few orders of magnitude. Instead, approximate ranges are indicated.

In the broader context, it is not individual minerals, but their host rocks, that are studied. Table 7 lists K, U, Th, plus Rb and Sm abundances for a variety of common rock types or rock suites. As for minerals, the abundances of a given trace element in rocks can vary over a couple of orders of magnitude. The abundances given in Table 7 can therefore be considered as representative of rocks and rock types in general, but caution must be used in attaching too much significance to a single example. For example, the high Th contents of G-1 and GSP-1 represent the higher end of the abundance spectrum, and the mode is probably about 10 to 20 ppm Th for felsic rocks. Results for average rock types are probably more representative, but even in some of these cases uncertainties of a factor of 2 are possible for trace constituents.

RADIOGENIC HEAT PRODUCTION

In terms of the geophysical behavior of the earth, one of the most important applications of radionuclides is in modeling of the thermal state and thermal history of the earth and planets. In this context it is necessary to know the specific heat production of the major heat-producing radionuclides: ^{40}K, ^{232}Th, ^{235}U. In the radioactive decay process, a portion of the mass of each decaying nuclide is converted to energy. Most of this energy is the kinetic energy of emitted particles or of electromagnetic radiation (γ-rays). For β^- decays, however, part of the energy is carried away by neutrinos. All decay energy other than that carried away by neutrinos is absorbed within the earth and converted to heat. Because neutrinos are not easily captured, they pass completely out of the earth and that portion of the decay energy is lost to space. In determining how much radiogenic heat production a

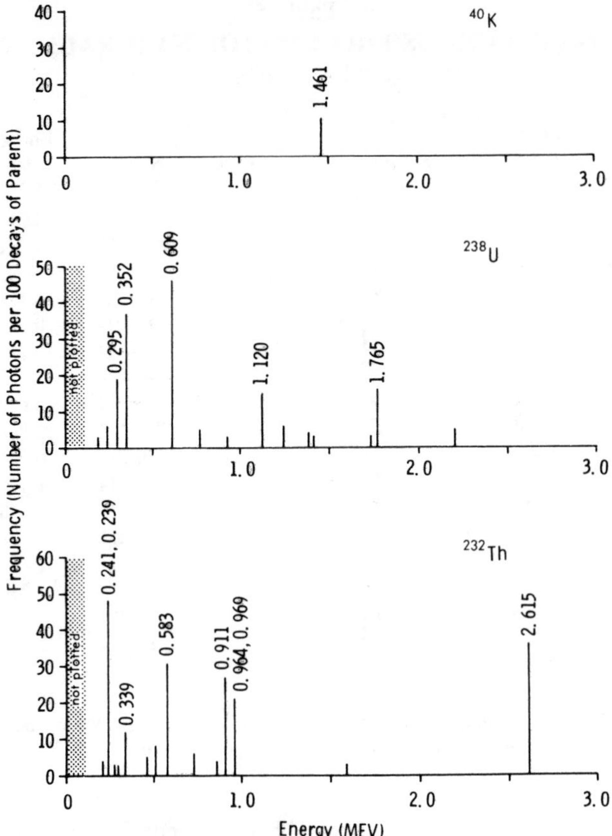

FIGURE 2. Summary of decay energies and relative intensities for major naturally occurring X-ray emitters. Intensities less than 3% not plotted. Energies less than 100 KeV not plotted for ^{238}U or ^{232}Th since they are more properly classed X-rays. Spectra for ^{235}U not presented because low natural abundance of ^{235}U relative to ^{238}U means all intensities of ^{235}U γ-rays are a minor part of the uranium spectrum.

particular radionuclide has, it is therefore necessary to correct for the loss of neutrino energy by β^- emitters.

All four of the major heat producing isotopes involve β^- decay. For ^{40}K there is the β^- decay branch to ^{40}Ca. In this case, the evaluation of the mean β^- energy was based on the β^- spectrum measured by Kelly et al.[19] All the β^- emitters in the ^{232}Th, ^{235}U, and ^{238}U decay series (Tables 2 to 4) constitute only a small fraction of the total decay energy, so individual determinations of β^- spectra were not used to determine neutrino energy loss. Instead, the general relationship that, on average, neutrinos represent 2/3 of the decay energy for β^- decay[20] was used. Table 8 summarizes some of the information needed to determine specific heat production and the results. This is a new compilation using recent data for individual nuclides.[1] Total decay energy was based on the mass difference between starting isotope and stable end products (Table 1).

Perhaps one of the most interesting results of this determination of specific heat production is that the refined values listed in Table 8 do not differ significantly from those reported nearly 30 years ago[21] and which are still widely used. The results are given in the traditional c.g.s. units, although many authors are now tending to report data in SI units (e.g., J/Kg-year, mW/cm^2, etc.).

Table 6
MAJOR NATURALLY RADIOACTIVE MINERALS

Mineral	Nominal composition	Radioactive element abundances[a] K	U	Th
Adularia	$KAlSi_3O_8$	14.0	—	—
Allanite	$(Ca,X)_2(Al,Fe,Mg)_3Si_3O_{12}(OH)$	—	*	***
Alunite	$KAl_3(SO_4)_2(OH)_6$	9.4	—	—
Apatite	$Ca_5(PO_4)_3(F,Cl,OH)$	—	*	*
Apophyllite	$KCa_4(Si_4O_{10})_2F{\cdot}8H_2O$	4.1	—	—
Autunite	$Ca(UO_2)_2(PO_4)_2{\cdot}10–12H_2O$	—	48–50	—
Biotite	$K(Mg,Fe)_3(AlSi_3O_{10})(OH)_2$	8–9	—	—
Carnallite	$KMgCl_3{\cdot}6H_2O$	14.1	—	—
Carnotite	$K_2(UO_2)_2(VO_4)_2{\cdot}3H_2O$	7.2	53	—
Glauconite	Complex sheet silicate	4.6–6.2	—	—
Hornblende	$NaCa_2(Mg,Fe,Al)_5(Si,Al)_8O_{22}(OH)_2$	***	—	—
Lepidolite	Lithium mica	7.1–8.3	—	—
Leucite	$KAlSi_2O_6$	17.9	—	—
Microcline	$KAlSi_3O_8$	14.0	—	—
Monazite	$(Ce,La,Y,Th)PO_4$	—	**	2–20
Muscovite	$KAl_2(AlSi_3O_{10})(OH)_2$	9.8	—	—
Nepheline	$(Na,K)AlSiO_4$	3–10	—	—
Orthoclase	$KAlSi_3O_8$	14.0	—	—
Phlogopite	$KMg_3(AlSi_3O_{10})(OH)_2$	9.4	—	—
Pitchblende	Massive UO_2	—	88	—
Polyhalite	$K_2Ca_2Mg(SO_4)_4{\cdot}2H_2O$	13.0	—	—
Sanidine	$KAlSi_3O_8$	14.0	—	—
Sphene	$CaTiSiO_5$	—	*	*
Sylvite	KCl	52.4	—	—
Thorianite	ThO_2	—	—	88
Thorite	$ThSiO_4$	—	***	72
Torbernite	$Cu(UO_2)\ (PO_4)_2{\cdot}8–12H_2O$	—	32–36	—
Tyuyamunite	$Ca(UO_2)_2(VO_4)_2{\cdot}5–8^1/_2H_2O$	—	45–48	—
Uraninite	UO_2	—	88	—
Xenotime	YPO_4	—	***	**
Zircon	$ZrSiO_4$	—	**	**

[a] Abundances in percents except as noted: ***, 0.5—3% range, **, 0.1—0.5% range, *, 0.001—0.1% range.

Table 9 presents some typical ranges of specific heat production for various rock types, ranging from U, Th-rich granite (atypical) through intermediate to mafic and ultramafic rock types. One point that should be kept in mind with regard to Tables 7 and 9 is that attempts to model the thermal history of terrestrial components are only approximations; the abundances of K, U, and Th for any specific case (e.g., "continental crust") are not known well enough to permit precise interpretations.

Table 10 and Figure 3 summarize past heat production in the whole earth for *one* assumed bulk composition. For the model used, the Th/U ratio is probably accurate to better than 10%, but the K/U ratio could be uncertain to a factor of 2, and the absolute concentrations could also be uncertain to a factor of 2. Thus, the data in Table 10 and Figure 3 should only be considered as representative of possible past histories and not accepted as the best model. However, they do give a good indication of the relative contributions of K, U, Th to heat production in the earth now and in the past.

Table 7
ABUNDANCES OF RADIOACTIVE ELEMENTS IN USGS* ROCK STANDARDS AND OTHER SELECTED AVERAGE ROCK TYPES

Rock/rock type	K (%)	Rb (ppm)	Sm (ppm)	Th (ppm)	U (ppm)	Ref.
G-1 granite	4.45	220	8	50	3.4	12
Av Lo-Ca granite	4.20	170	7.1	20	4.7	13
G-2 Granite	3.67	168	7	24	2.0	12
Av metalum. granite	—	—	—	16.5	3.7	14
Av peralum. granite	—	—	—	19.0	4.5	14
RGM-1 rhyolite	3.49	154	—	13	5.8	15, 16
STM-1 neph. syenite	3.54	113	—	27	9.1	15, 16
GSP-1 granodiorite	4.50	254	27	104	2.0	12
QLO-1 quartz latite	2.90	68	—	13	5.8	15, 16
AGV-1 andesite	2.35	67	6	6.4	1.9	12
BCR-1 basalt	1.38	47	6.7	6.0	1.7	12
W-1 diabase	0.52	21	4	2.4	0.6	12
BHVO-1 basalt	0.43	9	—	0.9	0.5	15, 16
Av basalt	0.83	30	6.9	2.7	0.9	13
PCC-1 peridotite	0.001	0.063	0.008	0.01	0.005	12
DTS-1 dunite	0.001	0.053	0.004	0.01	0.004	12
Av ultramafic	0.003	0.13	1.1	0.004	0.001	13
MAG-1 marine mud	2.96	186	—	12.2	2.8	15, 16
SCo-1 shale	2.20	122	—	9.5	3.1	15, 16
SDC-1 mica schist	2.71	129	—	11.4	3.1	15, 16
Av shale	2.66	140	7.0	12	3.7	13
Av sandstone	1.07	60	1.9	5.5	1.7	13
Av carbonate	0.27	3	0.6	1.7	2.2	13
Av upper continental crust	2.7	110	5.6	10.5	2.5	17
Av continental crust	1.25	50	3.7	2.5	1.0	17
Bulk Earth	0.0200	—	—	0.074	0.020	18

* U.S. Geological Survey.

Table 8
INFORMATION ON MAJOR HEAT PRODUCING ISOTOPES

Element / Isotope	Potassium (^{40}K)	Thorium (^{232}Th)	Uranium ^{235}U	Uranium ^{238}U
Isotopic abundance (Wt %)	0.0119	100	0.71	99.28
Decay constant, λ (year^{-1})	5.54×10^{-10}	4.95×10^{-11}	9.85×10^{-10}	1.551×10^{-10}
Total decay energy (MeV/decay)	1.34[a]	42.66[b]	46.40[b]	51.70[b]
Beta decay energy (MeV/decay)	1.19[a]	3.5	3.0	6.3
Beta energy lost as neutrinos (MeV/decay)	0.65[c]	2.3[d]	2.0[d]	4.2[d]
Total energy retained in earth (MeV/decay)	0.69	40.4	44.4	47.5
Specific isotopic heat production (cal/g-year)	0.220	0.199	4.29	0.714
Present elemental heat production (cal/g-year)	26×10^{-6}	0.199		0.740

[a] Averaged for branching decay; $\beta^- = 1.32$ MeV.

[b] Summed for entire decay chain.

[c] Based on mean decay energy for β^- of 0.60 MeV (19).

[d] Assumed average neutrino loss = 2/3 total β^- energy (20).

Table 9
PRESENT RADIOGENIC HEAT PRODUCTION IN SELECTED ROCK UNITS

	Annual heat production[b]			
Rock unit[a]	**Due to K**	**Due to Th**	**Due to U**	**Total**
GSP-1 "granodiorite"	1.17	20.70	1.48	23.35[c]
G-1 "granite"	1.16	9.95	2.52	13.63
Av upper continental crust	0.70	2.09	1.85	4.64
AGV-1 "andesite"	0.61	1.27	1.41	3.29
Av continental crust	0.33	0.50	0.74	1.56
BHVO-1 "oceanic basalt"	0.11	0.18	0.37	0.66
PCC-1 "peridotite"	0.0003	0.0020	0.0037	0.0060
Bulk Earth	0.0052	0.0147	0.0148	0.0347

[a] Elemental abundances from Table 7.
[b] In μcal/g-year; multiply × 0.004184 for J/kg-year.
[c] Not a typical granodiorite, but it illustrates natural range due to granitic rocks high in Th or U.

Table 10
PAST HEAT PRODUCTION IN THE BULK EARTH[a]

	K		Th		U		Total	
Time[b]	**Abs.[c]**	**Rel.[d]**	**Abs.**	**Rel.**	**Abs.**	**Rel.**	**Abs.**	**Rel.**
0.0	5.2	0.15	14.7	0.42	14.8	0.43	34.7	1.00
0.5	6.9	0.20	15.1	0.44	16.3	0.47	38.3	1.10
1.0	9.1	0.26	15.5	0.45	18.2	0.52	42.8	1.23
1.5	12.0	0.35	15.9	0.46	20.6	0.59	48.5	1.40
2.0	15.9	0.46	16.3	0.47	23.7	0.68	55.9	1.61
2.5	20.9	0.60	16.7	0.48	28.0	0.81	65.6	1.89
3.0	27.6	0.80	17.1	0.49	34.3	0.99	79.0	2.28
3.5	36.4	1.05	17.5	0.50	43.5	1.25	97.4	2.81
4.0	48.1	1.39	17.9	0.52	57.7	1.66	123.7	3.56
4.5	63.4	1.83	18.4	0.53	79.7	2.30	161.5	4.65

[a] Assumed present abundances: K = 200 ppm, Th = 74 ppb, U = 20 ppb (K : U : Th = 10,000 : 1 : 3.7).
[b] Billions of years ago.
[c] Abs. = absolute in 10^{-9} cal/g-year.
[d] Rel. = relative to present total

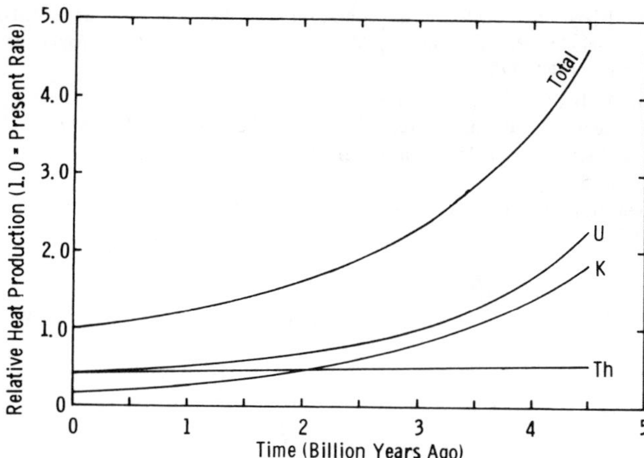

FIGURE 3. Curves showing relative heat production due to K, Th, and U during the geologic past. Normalized to current total heat production = 1.00. Numerical data summarized in Table 10.

REFERENCES

1. **Lederer, C. M. and Shirley, V. S., Ed.,** *Table of Isotopes,* 7th ed., John Wiley & Sons, New York, 1978.
2. **Godwin, H.,** Half-life of radiocarbon, *Nature (London),* 195, 984, 1962.
3. **Steiger, R. H. and Jager, E.,** Subcommission on Geochronology: convention on the use of decay constants in geo- and cosmochronology, *Earth Planet. Sci. Lett.,* 36, 359, 1977.
4. **Flynn, K. F. and Glendenin, L. E.,** Half-life of ^{107}Pd, *Phys. Rev.,* 185, 1591, 1969.
5. **Lugmair, G. W. and Marti, K.,** Lunar initial ^{143}Nd/^{144}Nd: differential evolution of the Lunar crust and mantle, *Earth Planet. Sci. Lett.,* 39, 349, 1978.
6. **Patchett, P. J. and Tatsumoto, M.,** Lu–Hf isotope systematics of the eucrite meteorites, *Meteoritics,* 15, 349, 1980.
7. **Hirt, B., Tilton, G. R., Herr, W., and Hoffmeister, W.,** The half-life of ^{187}Re, in *Earth Science and Meteorites,* Geiss, J., and Goldberg, E. D., Eds., North-Holland, Amsterdam, 1963, 273.
8. **Faure, G.,** *Principles of Isotope Geology,* John Wiley & Sons, New York, 1977.
9. **Galliker, D., Hugentobler, E., and Hahn, B.,** Spontane Kerspaltung von U-238 und Am-241, *Helv. Phys. Acta,* 43, 593, 1970.
10. **Fields, P. R., Friedman, A. M., Milsted, J., Lerner, J., Stevens, C. M., Metta, D., and Sabine, W. K.,** Decay properties of plutonium-244 and comments on its existence in nature, *Nature (London),* 212, 131, 1966.
11. **O'Nions, R. K., Carter, S. R., Evensen, N. M., and Hamilton, P. J.,** Geochemical and cosmochemical applications of Nd isotope analysis, *Ann. Rev. Earth Planet. Sci.,* 7, 11, 1979.
12. **Flanagan, F. J.,** 1972 compilation of data on USGS standards, in Descriptions and Analyses of Eight New USGS Rock Standards, Prof. Pap. 840, U.S. Geological Survey, Reston, Va., 1976, 131.
13. **Turekian, K. K.,** *Chemistry of the Earth,* Holt, Rinehart & Winston, New York, 1972, 84.
14. **Stuckless, J.,** personal communication, 1981.
15. **Fabbi, B. B. and Espos, L. F.,** X-ray fluorescence analysis of 21 selected major, minor, and trace elements in eight new USGS standard rocks, in Descriptions and Analyses of Eight New USGS Rock Standards, Prof. Pap. 840, U.S. Geological Survey, Reston, Va., 1976, 89.
16. **Millard, H. T., Jr.,** Determination of uranium and thorium in USGS standard rocks by the delayed neutron technique, in Descriptions and Analyses of Eight New USGS Rock Standards, Prof. Pap. 840, U.S. Geological Survey, Reston, Va., 1976, 61.

17. **Taylor, S. R.,** Island arc models and the composition of the continental crust, in Island Arcs, Deep Sea Trenches, and Back-Arc Basins, Talwani, M. and Pitman, W. C., Eds., American Geophysical Union, Washington, D.C., 1977, 325.
18. **Van Schmus, W. R.,** unpublished estimate, 1981.
19. **Kelly, W. H., Beard, G. B., and Peters, R. A.,** The beta decay of K^{40}, *Nucl. Phys.*, 11, 492, 1959.
20. **Friedlander, G., Kennedy, J. W., and Miller, J. M.,** *Nuclear and Radiochemistry*, John Wiley & Sons, New York, 1964, 52.
21. **Birch, F.,** Heat from radioactivity, in *Nuclear Geology*, Faul, H., Ed., John Wiley & Sons, New York, 1954, chap. 5.

Section IX

Spectroscopic Properties of Rocks and Minerals

By
Graham R. Hunt

INTRODUCTION

Spectroscopy may be defined as the study of the interaction of matter with electromagnetic radiation (EMR).

EMR may be expressed on a wavelength scale, typically calibrated in angstrom (Å), nanometer (nm), or micrometer (μm) units; a frequency scale in wavenumber (cm^{-1}) units; or an energy scale in calorie (cal) or electron volt (eV) units; where 10,000 Å = 1,000 nm = 1 μm = 10^{-4} cm = $10^4 cm^{-1}$ = 2.859×10^4 cal = 1.24 eV.

The radiation wavelength range considered covers the visible and infrared, from 0.35 to 40 μm. Within this range, intrinsic interactions are caused by electronic and/or vibrational processes.

For solids, radiation is absorbed or emitted as a consequence of changes in the total energy content of the material. These changes, called "transitions", take place between the "discrete" energy levels or energy bands of the material. These may be electronic energy levels, in which case evidence (in the form of spectral features) for the transitions typically appears in the visible (0.35 to 0.70 μm) and near infrared (0.7 to 2.5 μm), or between vibrational energy levels, in which case evidence appears in the infrared range (1.2 to 40+ μm).

The fundamental properties that determine spectroscopic behavior of a solid in absorption (≡transmission), reflection, or emission are the optical constants, namely the refractive index (n) and extinction (or absorption) coefficient (k). Other parameters such as surface condition, particle size and distribution, illumination angle, temperature, and pressure have their effects on the appearance of a spectrum, but all the compositionally characteristic information is contained in n and k. The latter are wavelength-dependent because of optically active transitions between the energy levels (Figure 1).

For geologic materials, electronic processes provide information about the presence and nature of a particular ion and its environment, about adjacent ions, or about defects in specific crystal locations. Thus, information concerning bulk composition is of an indirect nature. Vibrational processes are governed by three characteristic properties of the material: the chemical composition, the geometry and equilibrium positions of the constituent atoms, and the potential field of the interatomic forces representative of interatomic bonds and angles. Consequently the information available is directly related to bulk composition.

A spectrum is a plot of the absorption, transmission, reflection, or emission intensity as a function of wavelength, frequency, or energy.

Rocks are basically assemblages of minerals, so much of the data presented is for pure minerals since the spectrum of a rock is typically some composite of the spectra of its constituent minerals.

The rationale for igneous rock classification is shown in Figure 2 and a simplified igneous rock classification system is shown in Table 1.

VISIBLE AND NEAR INFRARED (0.35 TO 2.5 μm) REGION

Spectra in this region are typically recorded in the transmission mode for thin sections, for polished single crystals, finely ground particles embedded in a matrix or suspended in a transparent fluid; or for particulate (crushed and ground) samples in the bidirectional, diffuse, or integrating sphere reflection mode. In all these cases, the spectral information available in the form of bands and slope changes is a consequence of the absorption process.

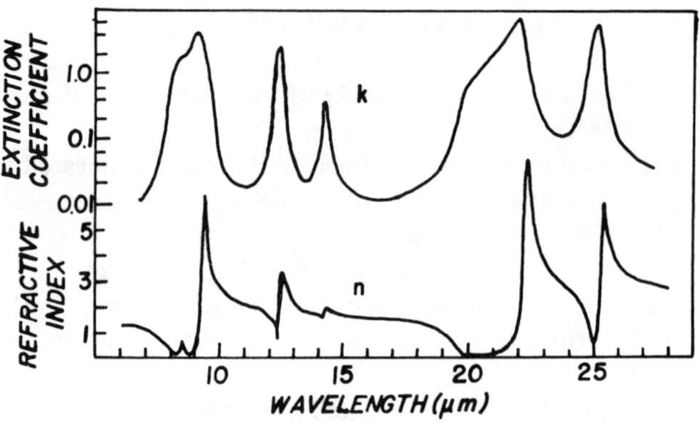

FIGURE 1. Optical constants of quartz derived from the data of Spitzer and Kleinman (1961).[7]

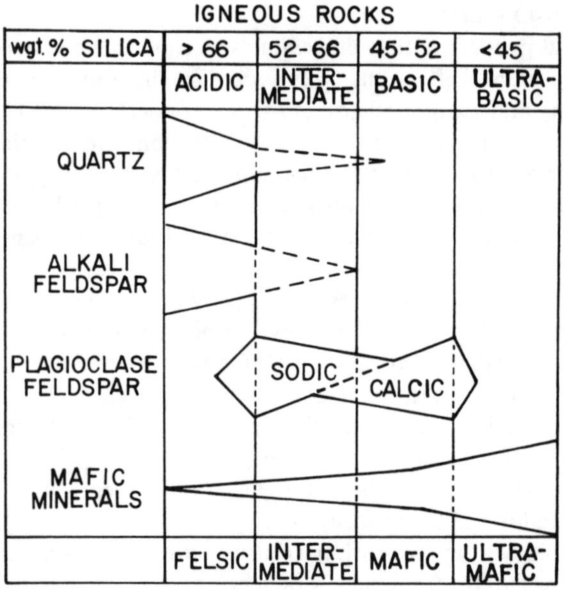

IGNEOUS ROCKS

wgt.% SILICA	> 66	52-66	45-52	<45
	ACIDIC	INTER-MEDIATE	BASIC	ULTRA-BASIC
QUARTZ				
ALKALI FELDSPAR				
PLAGIOCLASE FELDSPAR		SODIC	CALCIC	
MAFIC MINERALS				
	FELSIC	INTER-MEDIATE	MAFIC	ULTRA-MAFIC

FIGURE 2. Rationale for igneous rock classification.[12]

Mineral Spectra

Electronic Processes

There are four distinct electronic processes that produce spectral features in geologic materials.

Crystal Field Effects

When an ion, especially that of a transition element, is embedded in a crystal structure, the crystal field perturbs the energy levels of the ion so that they differ from those of the isolated ion. This provides new energy differences between the levels. When transitions take place between them, the resultant spectrum reflects the type of field in which the ion is located. An energy level diagram for an ion in different crystal fields is shown in Figure 3.

Table 1
SIMPLIFIED IGNEOUS ROCK CLASSIFICATION SYSTEM[12]

Characterizing parameters	Felsic rocks	Intermediate rocks					Mafic rocks	Ultramafic rocks
Minerals								
Alkali feldspar	>⅔ Total feldspar	>⅔ Total feldspar	>⅔ Total feldspar	⅓—⅔ Total feldspar	¼—⅓ Total feldspar	<10% Total feldspar	<10% Total feldspar	None
Plagioclase feldspar	<⅓ Total feldspar	<⅓ Total feldspar	<⅓ Total feldspar	⅔ — ⅓ Total feldspar	>⅔ Total feldspar	⅔ Total feldspar (sodic)	>⅔ Total feldspar (calcic)	<⅓ Total minerals (calcic)
Quartz	>10%	<10%	<10%	<10%	>10%	<10%	<10%	None
Feldspathoids	<10%	<10%	>10%	<10%	<10%	<10%	<10%	<20%
Mafic minerals	Minor accessory	Accessory	Accessory	Accessory	Accessory	Accessory	Major accessory	60—100%
Texture								
Coarse grained	Granite	Syenite	Nepheline syenite	Monzonite	Granodiorite	Diorite	Gabbro Diabase	Peridotite
Fine grained	Rhyolite	Trachyte	Phonolite	Latite	Dacite	Andesite	Basalt	Limburgite

Note: Glassy varieties (tuff, obsidian, pitchstone) are almost all felsic.

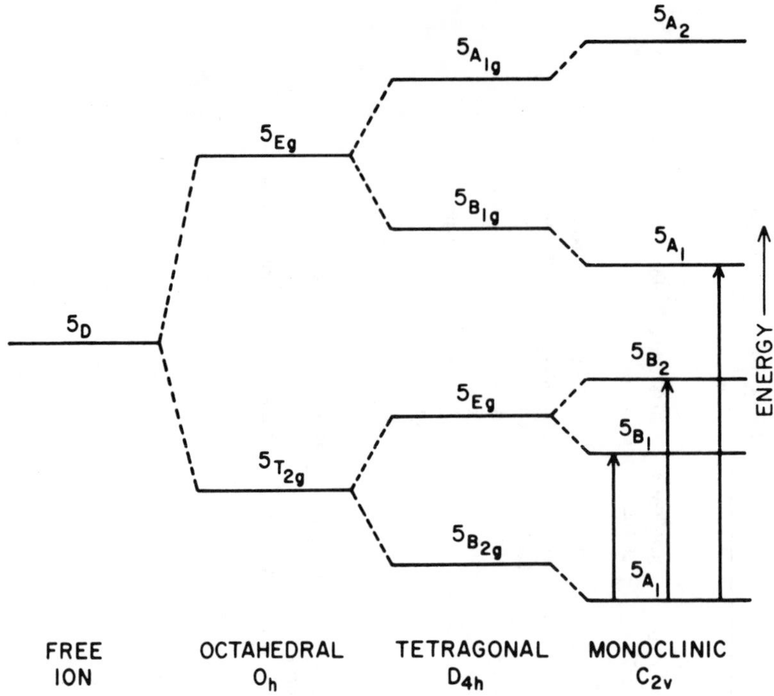

FIGURE 3. Partial energy level diagram representative of a transition ion located in different crystal fields.

Polarized Absorption Spectra

The spectra shown in Figure 4 are recorded from oriented single crystals and display intense features in the 10,000 Å and 20,000 Å (1.0 and 2.0 μm) regions due to crystal field effect absorption in the ferrous ion located in olivines, pyroxenes, and amphiboles. The variations in absorption of the polarized light in different yield additional information concerning the precise site location of the ion.

Low Temperature Absorption Spectra

The spectra shown in Figure 5 illustrate the effect of low temperature, which sharpens the features in the absorption spectra of some minerals.

Bidirectional (Diffuse) Reflection Spectra of Particulate Samples

1. The effect of particle size on the appearance of a spectrum: Typically, for transparent materials, the smaller the grain size, the greater the overall reflection and the smaller the spectral contrast of the spectral features. For opaque materials, the smaller the particle size the lower the reflectance. The spectra shown in Figure 6 also illustrate that regardless of the overall reflectance and contrast, the wavelength positions of the bands remain constant.
2. Spectra due to the presence of different (transition element) ions in various minerals: The features due to the particular ion listed are indicated by a vertical line to the center of the band and attached to a shaded area indicating the half width of the band (Figure 7). The positions of bands in the spectra of chromium (Cr^{3+}) bearing minerals and Fe^{3+} oxide and silicate minerals are shown in Tables 2 and 3.

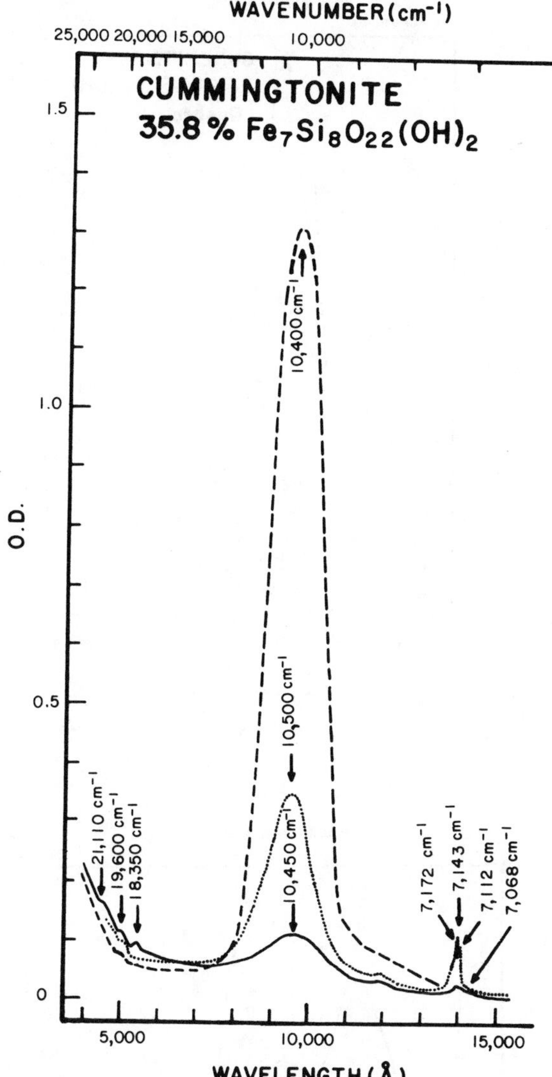

FIGURE 4. Polarized absorption spectra of minerals containing the ferrous ion spectra.[4] Dotted line is α spectrum, dashed line is β spectrum, and solid line is γ spectrum. (Continued next four pages).

3. The effect of different crystal fields in various minerals on the spectrum of the same ion (Fe^{2+}) is shown in Figure 8.
4. Uniqueness of mineral crystal field absorption bands is shown in Figure 9.

Charge — Transfer Processes

Charge transfer refers to the process whereby absorbed energy causes an electron to migrate between neighboring ions, or between ions and ligands (Figure 10 and Table 4). Although the electron is actually transfered, it remains localized in its new position.

Color Centers

This phenomenon, due to lattice defects, can produce discrete energy levels of excited electrons which have become bound to the defect (Figure 11).

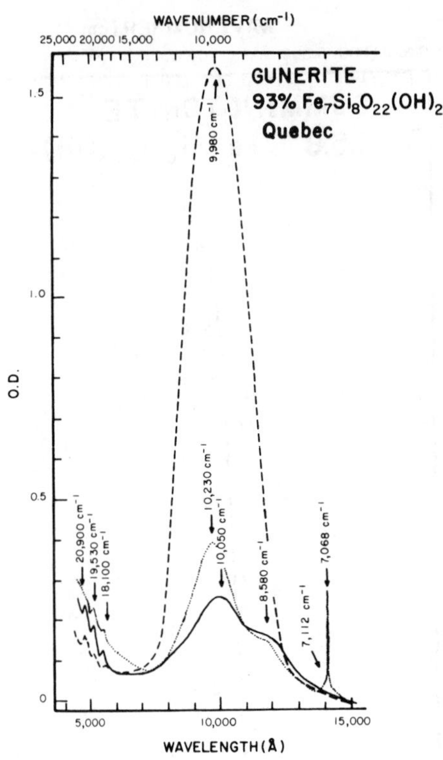

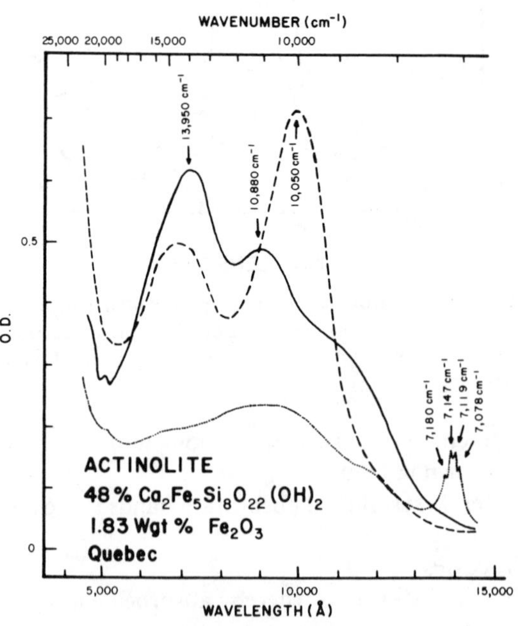

FIGURE 4 (continued)

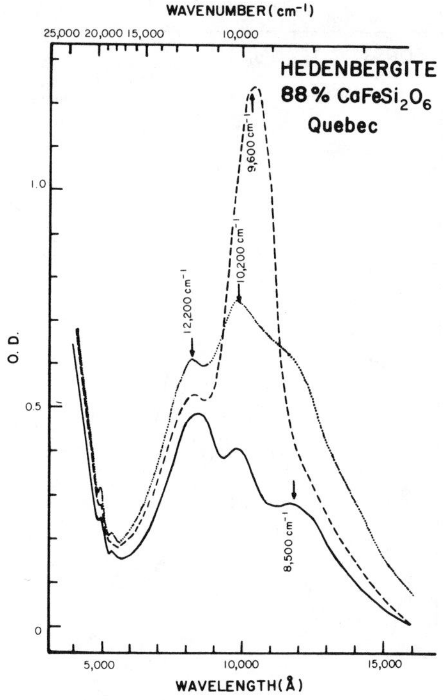

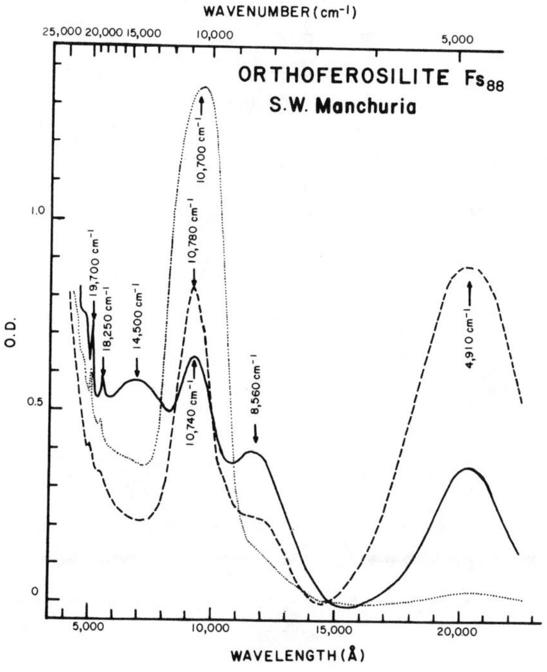

FIGURE 4 (continued)

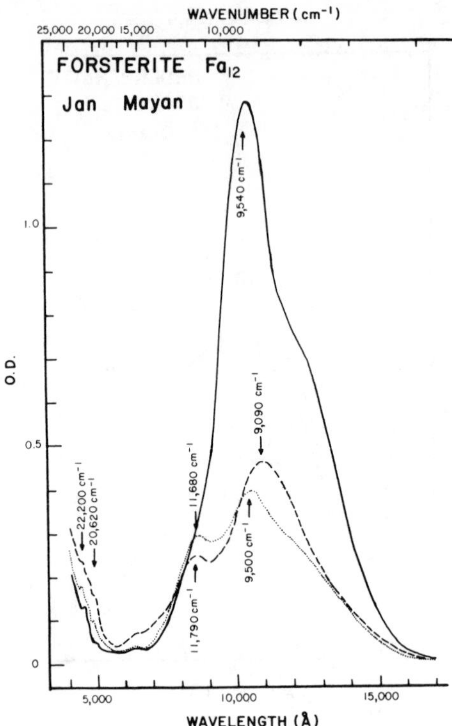

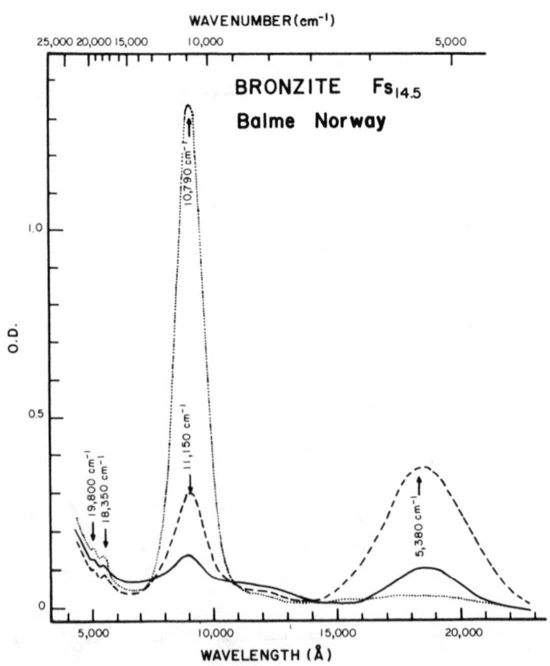

FIGURE 4 (continued)

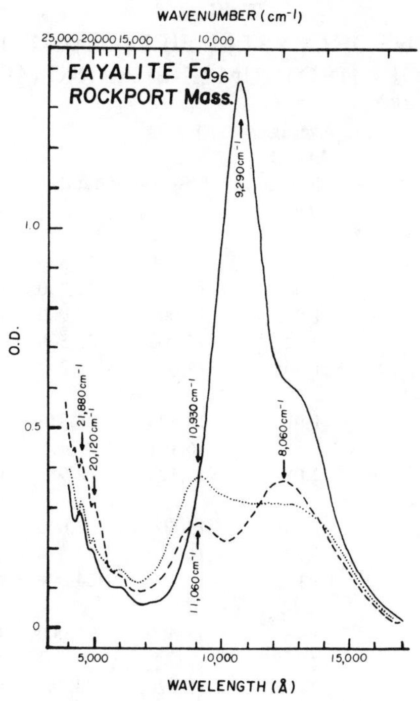

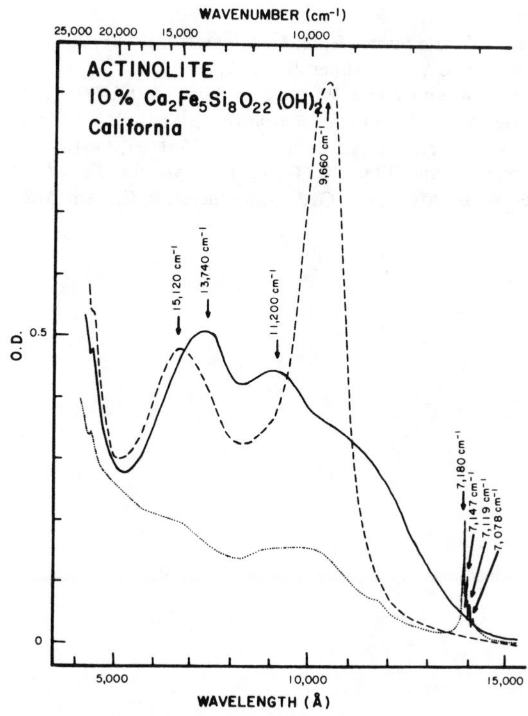

FIGURE 4 (continued)

Table 2
POSITIONS OF CRYSTAL FIELD BANDS IN THE SPECTRA OF CHROMIUM III-BEARING MINERALS[4]

Mineral	Average M—O Distance (Å)	Absorption bands (cm^{-1})		Color	Ref.
Ruby (Al_2O_3/Cr^{3+})	1.91	17,990	24,390	Red	5
Eskolaite (Cr_2O_3)	2.00	16,670	21,750	Green	5
Spinel ($MgAl_2O_4/Cr^{3+}$)	1.91	18,000	25,000	Pink	6
Olivine (Mg_2SiO_4/Cr^{3+})	2.12	16,900	23,500	Green	7
Pyrope ($Mg_3Al_2(SiO_4)_3/Cr^{3+}$)	1.905	17,606	24,272	Red	4
Uvarovite	1.985	16,529	24,814	Green	5
$Ca_3Cr_2(SiO_4)_3$		16,191	22,676	Green	4
Chrome diopside ($CaMgSi_2O_6/Cr^{3+}$)	2.08	16,129	22,989	Green	5
Ureyite ($NaCrSi_2O_6$)	1.998	15,600	22,000	Green	8
Cr-tremolite	2.07	16,310	23,530	Green	1, 5
Cr-epidote	1.93 or 2.05	≃16,300	≃24,000	Green	2
Cr-chlorite	—	18,450	25,000	Red-violet	1, 5
Cr-mica	—	15,820	23,580	Green	1, 5
Cr-tourmaline	1.93 or 2.05	17,000	24,000	Green	3

References

1. **Burns, R. G.**, unpublished data.
2. **Burns, R. G. and Strens, R. G. J.**, *Mineral. Mag.*, 36, 204, 1967.
3. **Manning, P. G.**, *Can. Mineral.*, 10, 57, 1969.
4. **Moore, R. K. and White, W. B.**, *Can. Mineral.*, 11, 791, 1972.
5. **Neuhaus, A.**, *Z. Kristallogr. Kristallgeom. Kristallphys. Kristallchem.*, 113, 195, 1960.
6. **Poole, C. P., Jr.**, *J. Phys. Chem. Solids*, 25, 1169, 1964.
7. **Scheetz, B. E. and White, W. B.**, *Contrib. Mineral. Petrol.*, 37, 221, 1972.
8. **White, W. B., McCarthy, G. J., and Scheetz, B. E.**, *Am. Mineral.*, 56, 72, 1971.

Table 3
POSITIONS OF CRYSTAL FIELD BANDS IN SPECTRA OF FERRIC IRON OXIDE AND SILICATE MINERALS[4]

Mineral	Absorption bands (cm^{-1})					Ref.
	$^4T_{1g}$	$^4T_{2g}$	$^4E_g, ^4A_{1g}$	$^4T_{2g}$	^4E_g	
Sapphire (Al_2O_3/Fe^{3+})	9,700	18,700	22,200	25,600	26,700	8
Kyanite (Al_2SiO_3/Fe^{3+})	(16,000-17,000)		22,400	26,500	27,000	7
			23,400			
Pyroxene	14,000	16,700	22,200	–	–	2, 4
Epidote	9,470	16,480	21,200	24,800		3
			22,000			
Vesuvianite			21,600	26,000		9
Andradite	12,453	16,650	22,701	24,000	27,000	
			22,999			10
Grossularite	13,111		22,865	23,592	27,040	10
			23,121			
Grossularite[a]	18,000	19,700	21,758	26,400		10
Orthoclase[a]		20,700	24,000	26,500		5
			22,650			
Phlogopite[a]	19,200	20,300	22,700	25,000		6

[a] Tetrahedral Fe^{3+}

References

1. Bell, P. M. and Mao, H. K., *Annu. Rep. Geophys. Lab. Yearb.*, 68, 253, 1969.
2. Bell, P. M. and Mao, H. K., *Proc. Lunar Sci. Conf., Geochim. Cosmochim. Acta,* 1972.
3. Burns, R. G. and Strens, R. G. J., *Mineral. Mag.*, 36, 204, 1967.
4. Burns, R. G., unpublished data.
5. Faye, G. H., *Can. Mineral.*, 10, 112, 1969.
6. Faye, G. H. and Hogarth, D. D., *Can. Mineral.*, 10, 25, 1969.
7. Faye, G. H. and Nickel, E. H., *Can. Mineral.*, 10, 35, 1969.
8. Ferguson, J. and Fielding, P. E., *Chem. Phys. Lett.*, 10, 262, 1971.
9. Manning, P. G., *Can. Mineral.*, 9, 348, 1968.
10. Moore, R. K. and White, W. B., *Can. Mineral.*, 11, 791, 1972.

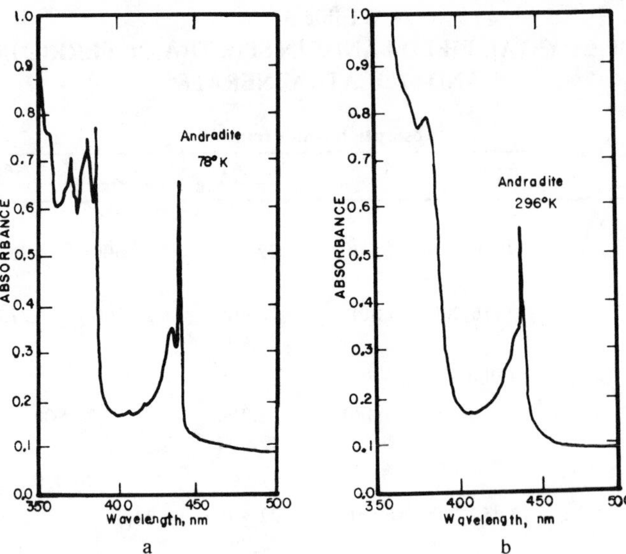

FIGURE 5. Expanded scale absorption spectrum of Val Malenco andradite garnet at (a) 78°K and (b) 296°K.[2]

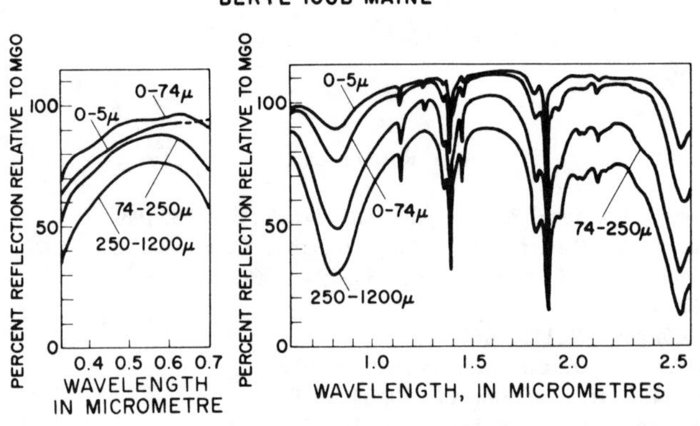

FIGURE 6. Bidirectional reflection spectra of four particle size range samples of beryl.[8]

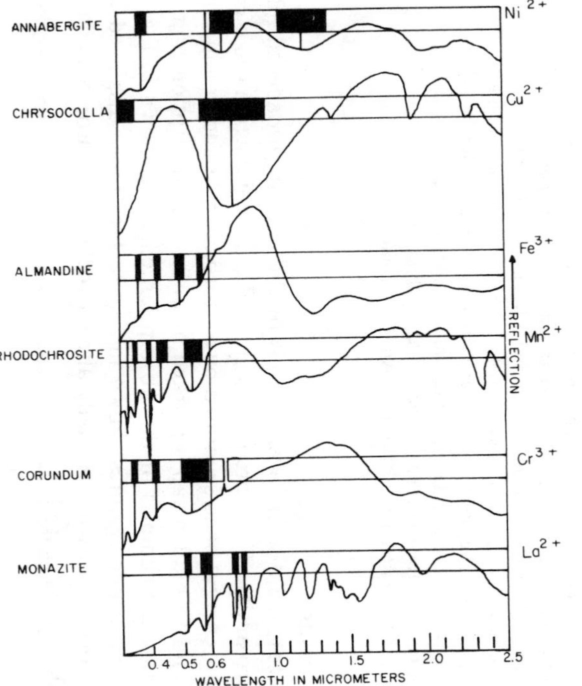

FIGURE 7. Spectra of seven minerals, the top six of which contain a different transition element and the lowest contains a rare-earth element. Solid areas show wavelength range of halfwidth of spectral bands due to crystal field transitions, and vertical lines locate the minimum of earth band.[8]

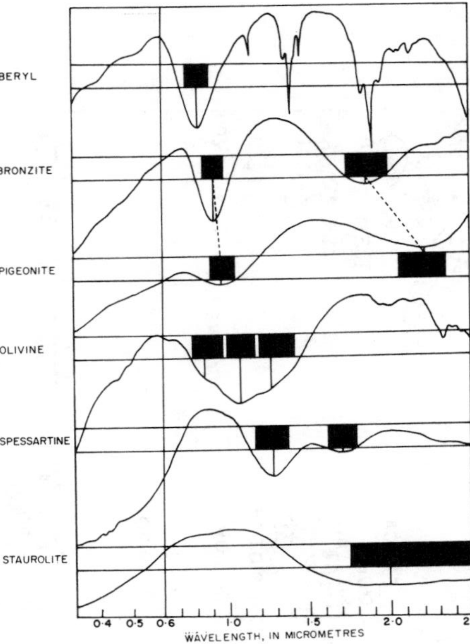

FIGURE 8. Spectra of six minerals containing ferrous ions, either in different crystal fields or in different sites. All features indicated in these spectra are due to spin-allowed transitions in the ferrous ions.[8]

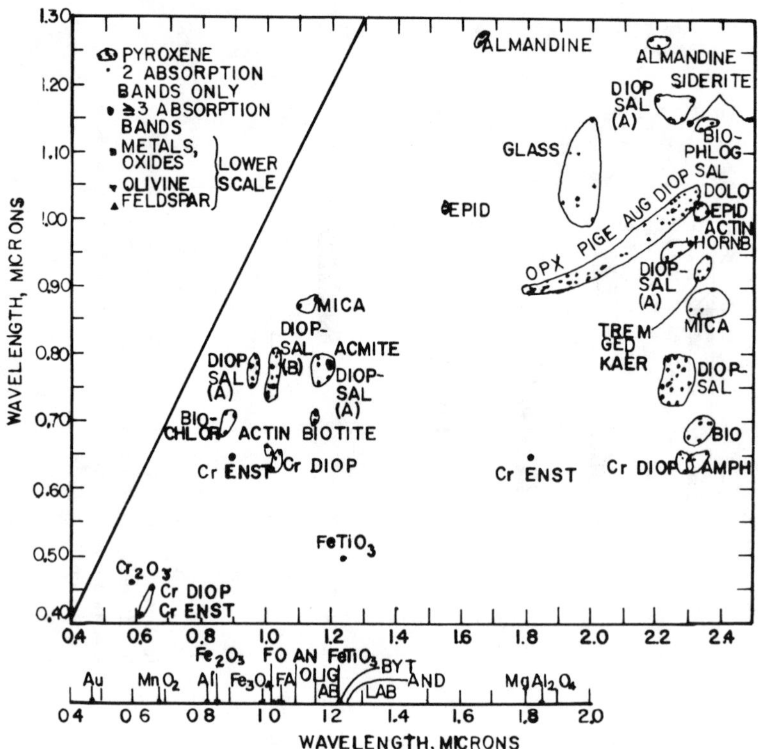

FIGURE 9. Plot of enters of absorption bands in the diffuse reflectance spectra of several common minerals. Minerals having only one main band are plotted on the lower scale. Minerals with two or more bands in their spectra are designed by points in the main upper field. Hydroxyl and water bands at 1.4 and 1.9 μm are not used. Bands for mineral species and for glasses plot in separate fields. Diop-sal = diopside-salite; ortho-pyroxene; dolo = dolomite; epid = epidote; actin-hornb = actinolite-hornblende; trem-ged-kaer = tremolite-gedrite-kaersutite; amph = amphibole; enst = enstatite; chlor = chlorite; Fo = forsterite; Fa = fayalite; Ab = albite; olig = oligoclase; and = andesine; lab = labradorite; byt = bytownite; An = anorthite.[1]

Conduction Band Transitions

In some periodic lattices, the discrete energy levels of the ions are broadened into energy bands by their proximity. There are two such bands, the conduction (high energy) and the valence (low energy) band separated by a "forbidden" band or gap. The edge of the conduction and forbidden gap is marked by an intense absorption edge in the visible and near infrared. Examples of this absorption edge are shown in Figure 12a. In addition to the electronic features shown in Figure 12a, those displayed by minerals that frequently occur in hydrothermally altered rocks — the iron oxide and sulfate minerals — are shown in Figure 12b.

Vibrational Processes

The apparent random motions of any vibrating system are made up of from a restricted number of simple motions, called the "normal modes" or "fundamentals". For most materials the fundamentals produce spectral features at wavelengths longer than 2.7 μm. When a fundamental mode is excited with two or more quanta of energy, an "overtone" occurs producing a spectral band at or near twice (or some multiple integral value) the fundamental frequency (υ_1), i.e., at $2\upsilon_1$, $3\upsilon_1$, $4\upsilon_1$, etc. When two or more fundamental or overtone vibrations interact, a "combination tone" feature appears located at or near the sum of all the fundamental or overtone frequencies involved. In geological materials only features due to overtone or combinations occur in the near infrared, and they are predominantly due to the presence of the hydroxyl group or water, or occasionally to other anions such as carbonate, sulfate, or phosphate.

Spectra due to the presence of molecular water in various minerals are shown in Figure 13. Characteristic spectra due to the fundamental vibrations of the hydroxyl group in different environments in two mineral series are shown in Figure 14.

Spectra due to the overtone and combination tone vibrations of the hydroxyl group in different environments in several minerals are shown in Figure 15. The characteristic spectra of minerals containing other anionic groups and shown in Figure 16. The locations of the most commonly occurring electronic and vibrational features and the cations and anions specifically responsible for their appearance in the spectra of minerals are shown in Figure 17. The near infrared spectra of those minerals that commonly occur in hydrothermally altered rocks are shown in Figure 18.

Rock Spectra

Because each rock is a combination of several minerals, it is not expected that features in its spectrum will be as well-defined as those in its constituents. Generally, the features appear muted and in some cases the contributions from minor constituents dominate the spectrum.

Igneous Rocks

The principal constituents of igneous rocks, the silicon-oxygen and aluminum-oxygen tetrahedra, display no features in the visible and near infrared range. Spectral features occur as a consequence of other components which are present constitutionally, substitutionally, or as impurities.

Felsic Igneous Rocks

Of the felsic rocks, granites — composed principally of quartz and orthoclase — should not display features. However, quartz normally contains water and some orthoclases contain exsolved ferric ion. Those felsic volcanic glasses that are water rich or have abundant pore spaces weather rapidly to form clay minerals. Rhyolites are the least altered but frequently contain water. Tuffs are sometimes composed of fragments of widely varying chemistry, and are often altered. The reflection spectra of suites of granites, volcanic glasses, rhyolites, and tuffs are shown in Figures 19 to 22. The superimposed spectra of unaltered felsic rocks are shown in Figure 23. The superimposed spectra of altered felsic rocks are shown in Figure 24.

Intermediate Igneous Rocks

The mafic accessory minerals of intermediate rocks offer the potential for spectral features due to the presence of ferrous and ferric ions, but the best displayed bands are the water and hydroxyl features of alteration products. In most cases, the presence of magnetite reduces the overall reflectance and decreases contrast. For igneous rocks, coarsely crystalline intrusives typically display more prominent bands than do their extrusive equivalents.

The reflection spectra of typical diorites and andesites are shown in Figure 25, and superimposed spectra of suites of syenites, trachytes, nepheline syenites, phonolites, monzonites, latites, granodiorites, dacites, diorites, and andesites are shown in Figure 26. The latter collected spectra of unaltered felsic rocks are given in Figure 27, in relation to the region occupied by intermediate rocks designated by the shaded area.

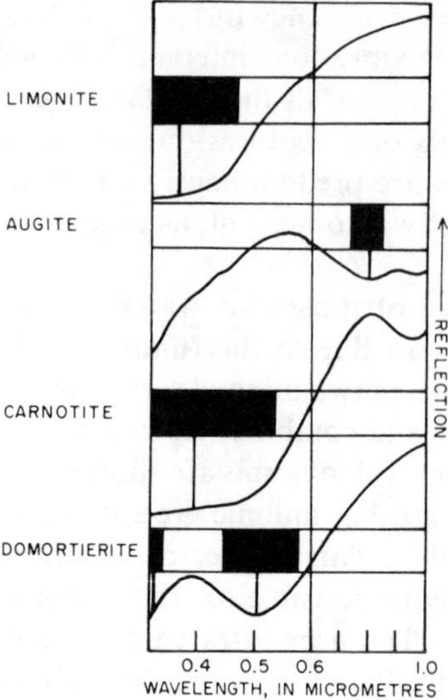

FIGURE 10. Spectra of four minerals, the top three of which illustrate various types of charge transfer spectral features. The bottom spectrum illustrates bands due to π-π transitions.[8]

Table 4
POSITIONS OF SUGGESTED FE²⁺—FE³⁺ CHARGE TRANSFER BANDS IN IRON-BEARING MINERALS[5]

Mineral	M—M distance (Å)	Intervalence transition (cm^{-1})	Ref.
Vivianite	2.85	15,200	4,9,10,12
Sapphire	2.65, 2.79	16,400	5,6
Kyanite	2.75	16,700	7
Orthopyroxene	3.20, 3.25	15,500—16,500	2
Augite	3.15, 3.25	13,000	2
Crocidolite	3.09, 3.10,	15,000 and	1,8,10,11
	3.22, 3.21	18,000	
Cordierite	2.85	17,500	9
Tourmaline	3.00	14,000	3,9
		17,000	
Calculated[a]	3.20	13,600	13

[a] MO Calculation on Fe^{3+} and Fe^{2+} in regular octahedra of oxygens (Fe^{3+} − O = 2.06 Å; Fe^{2+} − O = 2.17 Å; Fe^{3+} − Fe^{2+} = 3.20 Å). References on next page.

Table 4 (continued)
POSITIONS OF SUGGESTED FE^{2+}—FE^{3+} CHARGE TRANSFER BANDS IN IRON-BEARING MINERALS[5]

References

1. Bancroft, G. M. and Burns, R. G., *Mineral. Soc. Am. Spec. Paper,* 2, 137, 1969.
2. Burns, R. G., *Mineralogical Applications of Crystal Field Theory,* Cambridge University Press, London, 1970.
3. Burns, R. G. and Simon, H. F., *Abstr. Geol. Soc. Am. Annu. Meet.*, 5, 563, 1973.
4. Faye, G. H., *Can. Mineral.*, 9, 403, 1968.
5. Faye, G. H., *Can. Mineral.*, 10, 889, 1971.
6. Faye, G. H., *Am. Mineral.*, 56, 344, 1971.
7. Faye, G. H. and Nickel, E. H., *Can. Mineral.*, 10, 35, 1969.
8. Faye, G. H. and Nickel, E. H., *Can. Mineral.*, 10, 616, 1970.
9. Faye, G. H., Manning, P. G., and Nickel, E. H., *Am. Mineral.*, 53, 1174, 1968.
10. Hush, N. S., *Progr. Inorg. Chem.*, 8, 357, 1967.
11. Littler, J. G. F. and Williams, R. J. P., *J. Chem. Soc. London,* 6368, 1965.
12. Townsend, M. G. and Faye, G. H., *Phys. Status Solid,* 38, K57, 1970.
13. Tossell, J. A., Burns, R. G., Vaughan, D. J., and Johnson, K. H., unpublished data.

Mafic Igneous Rocks

With the exception of some norite gabbros, the only spectral features commonly found in mafic rocks are those due to iron near 1.0 μm and they are typically weak and broad. With the exception of some light-colored anorthositic gabbros, mafic rocks have the lowest reflectivities of any rocks because of the presence of large amounts of dark mafic minerals and particularly the presence of magnetite and other opaques. The reflection spectra of suites of gabbros, basalts, and diabases are shown in Figures 28 and 29. the collected superimposed reflection spectra of mafic rocks are shown in Figure 30.

Ultramafic Igneous Rocks

The spectra of ultramafic rocks always display a well-defined band near 1.0 μm and quite frequently an accompanying feature near 1.8 μm. Unlike mafic rocks, they contain relatively little opaque material. Their spectra are shown in Figure 31. The relative areas occupied by felsic, intermediate, mafic, and ultramafic rocks are indicated in Figures 32a and 32b.

Sedimentary Rocks

Well-defined features due to carbonates or clay as well as iron oxides are common in these spectra, except when they are masked by the presence of opaque carbonaceous materials. The specra of suites of limestones, sandstones, and shales are shown in Figures 33 to 35.

Metamorphic Rocks

These rocks typically display well-defined features due to the presence of the carbonate, hydroxyl, and occasionally borate group, and to the iron and chromium ions. The spectra of suites of marbles, quartzite, gneisses, slates and phyllites, and schistose rocks are shown in Figures 36 to 37.

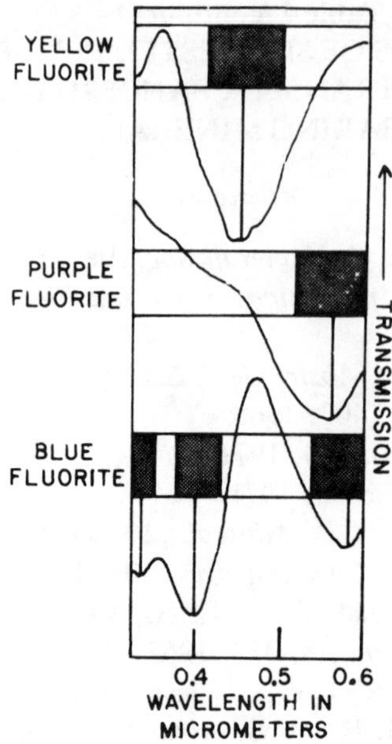

FIGURE 11. Spectra of three different colored samples of fluorite, illustrating features due to the presence of color centers.[8]

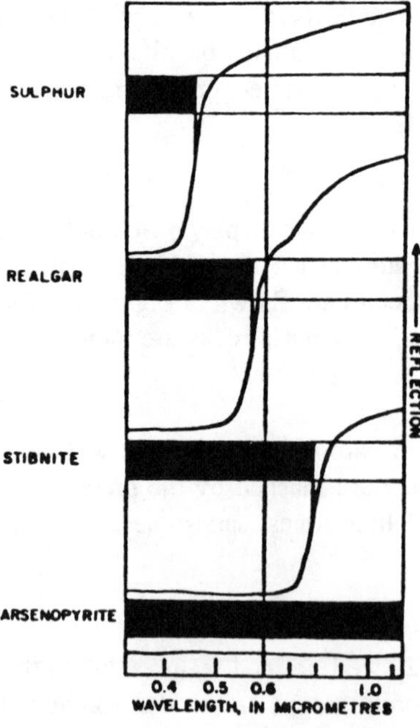

FIGURE 12a. Spectra of four minerals that illustrate the sharp transition between intense absorption and transparency in materials that display features due to the presence of a forbidden gap and a condution band. In the bottom spectrum, the absorption edge occurs in the mid-infrared.[8]

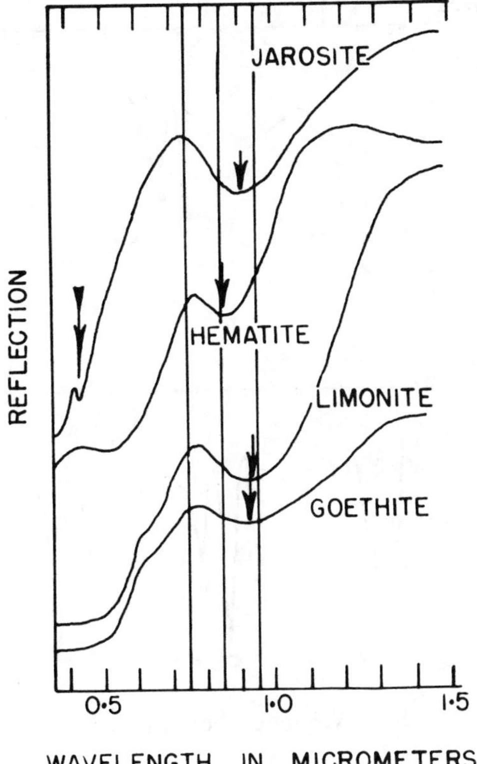

FIGURE 12b. Visible and near-infrared spectra of ferric oxide and sulfate minerals that commonly occur in hydrothermally altered rocks.[14]

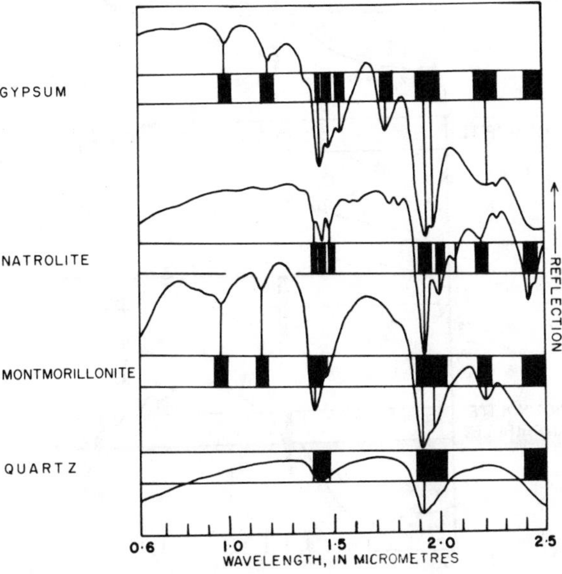

FIGURE 13. Spectra displaying features due to the vibrational combinations and overtones of molecular water, present in the various minerals in various forms, i.e., constitutionally, substitutionally, adsorbed, and included.[8]

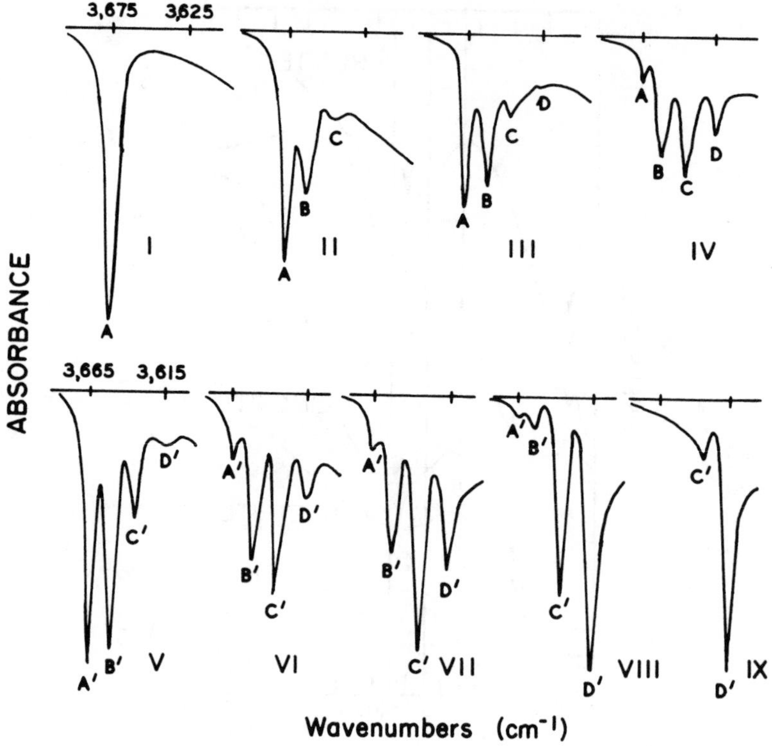

FIGURE 14. Spectra due to the fundamental O-H stretching modes in Fe^{2+}-Mg^{2+} amphiboles. I-IV, the tremolite-ferroactinolite series with 0 to 48% $Ca_2Fe_5Si_8O_{22}(OH)_2$; V to IX, the cummingtonite-grunerite series with 35.4 to 95.3% $Fe_7Si_8O_{22}(OH)_2$.[3]

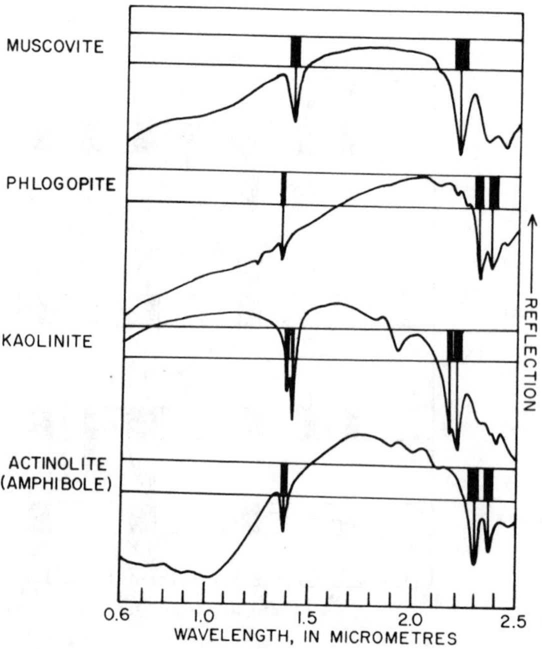

FIGURE 15. Spectra displaying features due to overtone and combination tones of the hydroxyl group present in different environments in various minerals.[8]

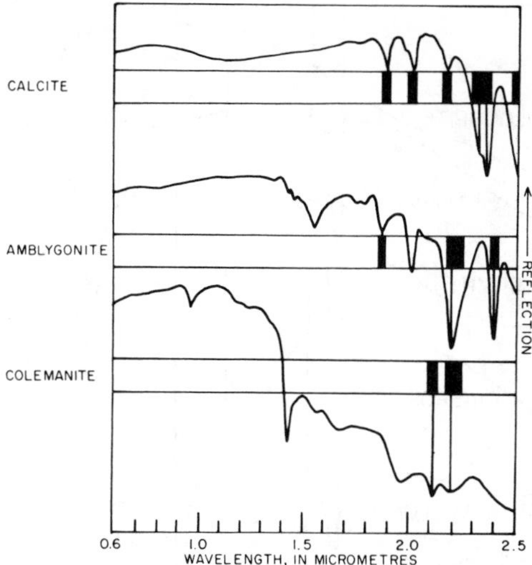

FIGURE 16. Spectra displaying features due to overtone and combination tones of internal vibrations of the carbonate (top), phosphate (middle), and borate anions present in various minerals.[8]

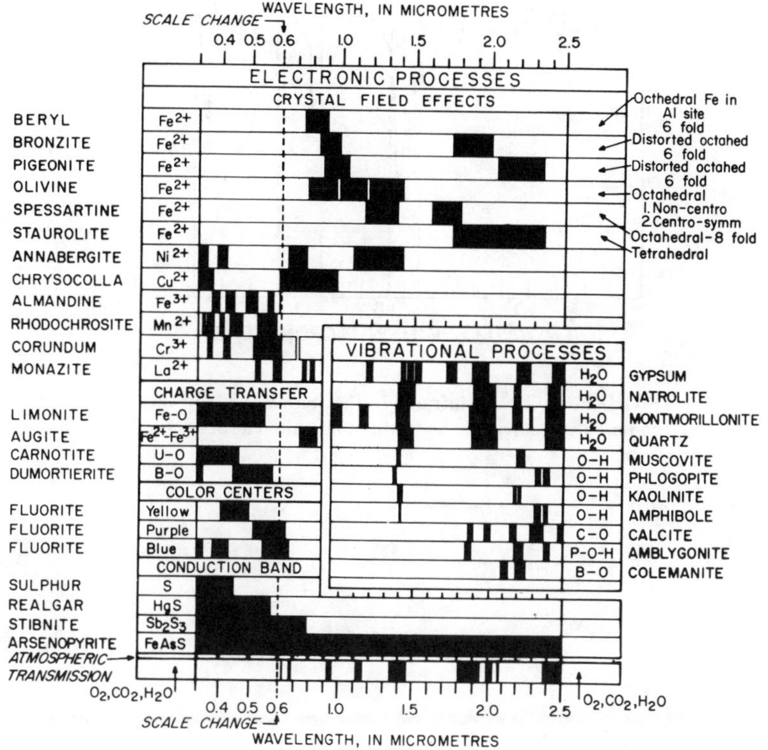

FIGURE 17. Composite diagram showing the location of the spectral features (wavelength of the center of black areas) and half-width (width of black areas) of most common minerals.[8]

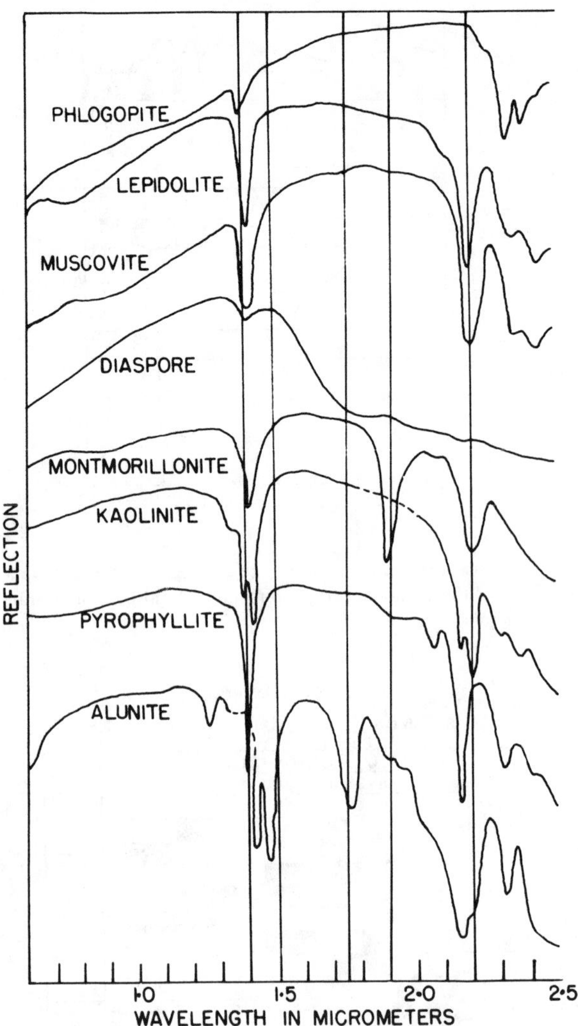

FIGURE 18. Near-infrared reflection spectra of minerals commonly occurring in hydrothermally-altered rocks.[14]

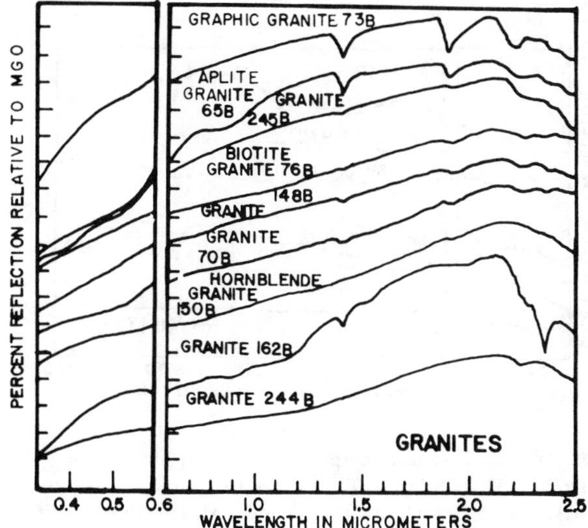

FIGURE 19. Reflection spectra of 74-250 μm particle-size granites. 73, graphic, Grand Canyon, Arizona; 65, aplite, Boulder, Colo.; 245, Wisconsin; 76, biotite, Rhode Island; 148, Westerly, R.I.; 70, Pikes Peak, Colo.; 150, hornblende, Rockport, Mass.; 162, Lowell Lakes, Ontario; 244, Georgia. Spectra displaced vertically.[10]

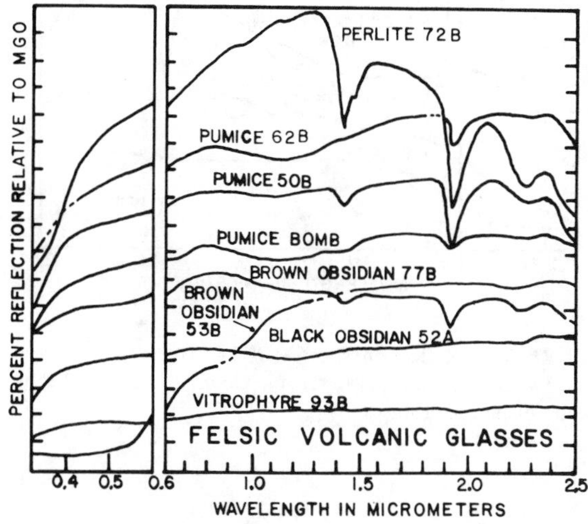

FIGURE 20. Reflection spectra of 74-250 μm particle-size felsic volcanic glasses. 72, Chaffee Co., Colo.; 62, U.S.B.M.; 50, Millard Co., Utah; Pumice Bomb, N. Crater, Mono craters, Calif.; 53, Lake Co., Oregon; 77, Custer Co., Colo.; 52, Lake Co., Oregon; 93, New Mexico. Spectra displaced vertically.[10]

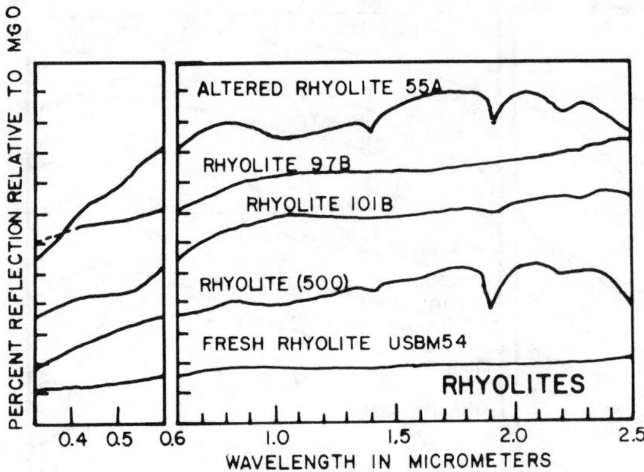

FIGURE 21. Reflection spectra of 74-250 μm particle-size rhyolites: 55, U.S.B.M.; 97, Chaffee Co., Colo.; 101, Castle Rock, Colo.; 500, Devil's Punch Bowl, Mono Craters Calif.; 54, U.S.B.M. Spectra displaced vertically.[10]

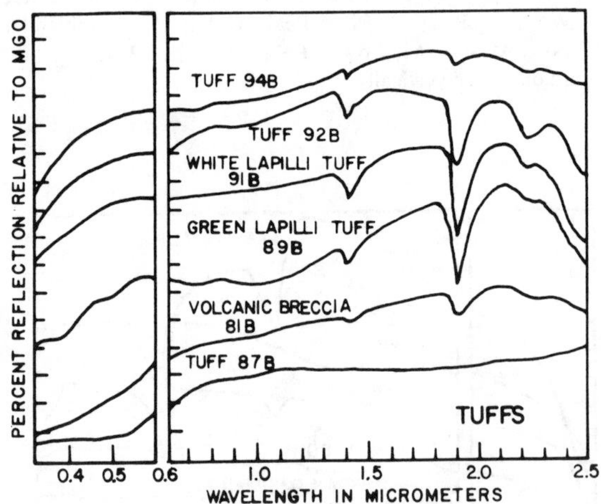

FIGURE 22. Reflection spectra of 74-250 μm particle-size tuffs. 94, Butte, Mont.; 81, Guffy, Colo.; 87, Ennis, Mont. Spectra displaced vertically.[10]

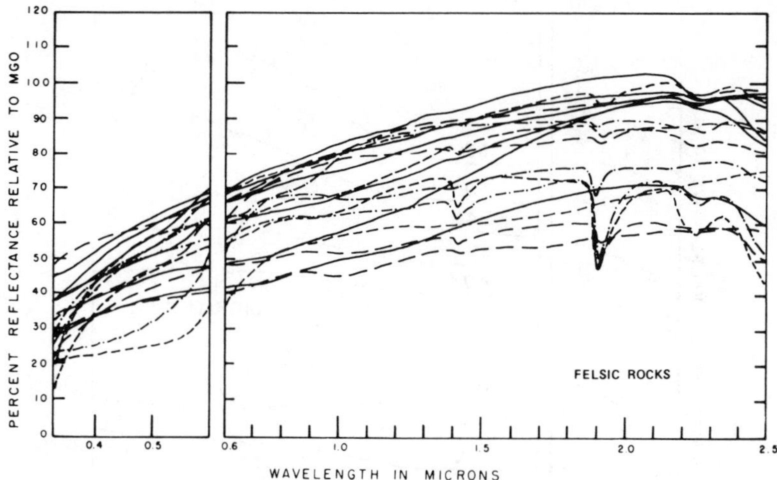

FIGURE 23. Collected, superimposed spectra of felsic rocks. ——granites;— —rhyolites; —·—tuffs;——felsic volcanic glasses.[10]

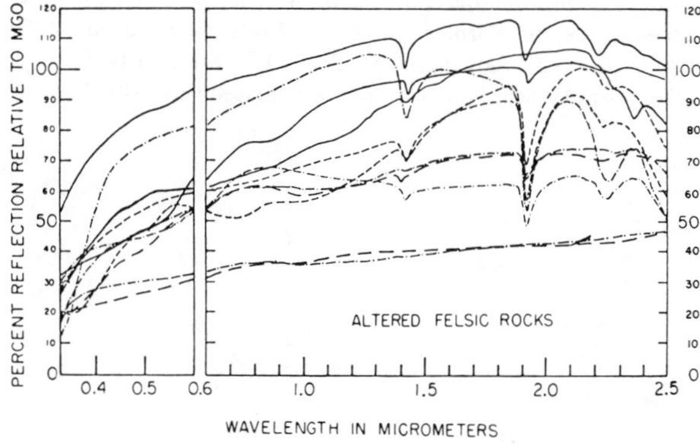

FIGURE 24. Collected superimposed spectra of altered felsic rocks. ——granites;— —rhyolites; —·—tuffs;—— felsic volcanic glasses.[10]

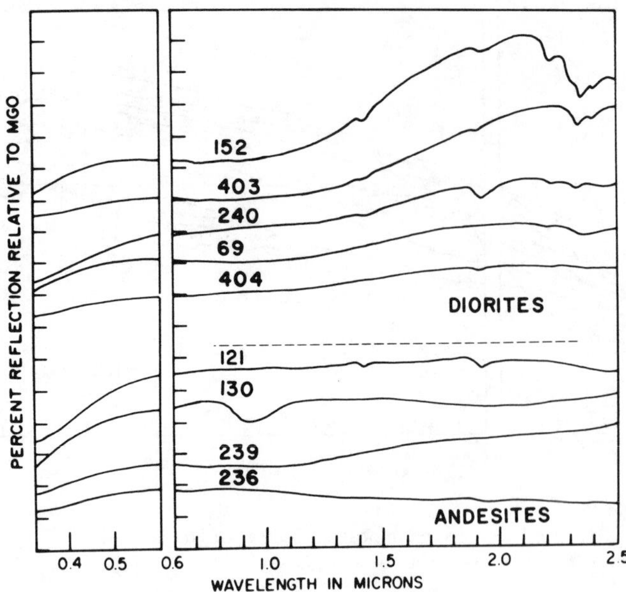

FIGURE 25. Reflection spectra of 74-250 μm particle-size diorites: 152, hornblende, Fremont Co., Colo.; 403, hornblende, Calif.; 240, hornblende, Tx; 69, hornblende, Salem, Mass.; 404, porphyry, Quebec. Andesites: 121, San Juan Co., Colo.; 130, Mt. Shasta, Clif.; 239, Tx; 236, hornblende, Colo. Spectra are displaced vertically.[10]

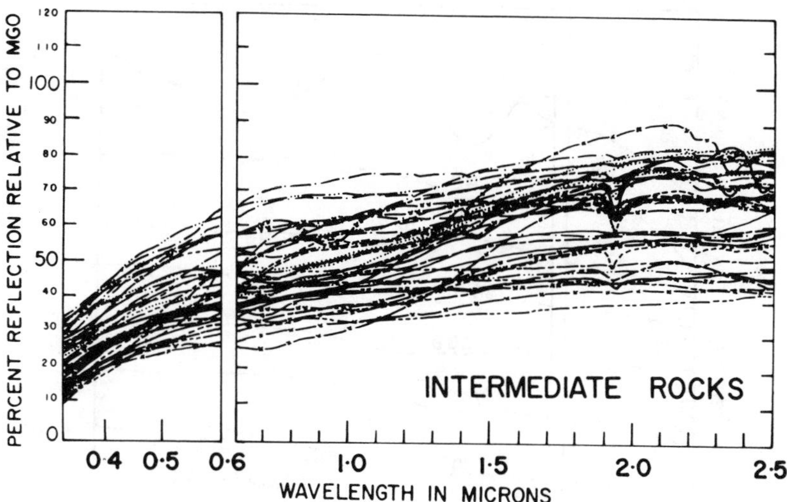

FIGURE 26. Collected superimposed spectra of intermediate rocks: ——syenites;—·—trachytes;—··—nepheline syenites;———phonolites;—— — —— monzonites;— — —quartz monzonites:—— -- ——latites;······granodiorites; ··· — ··· dacites;—x—diorites;—xx—andesites.[10]

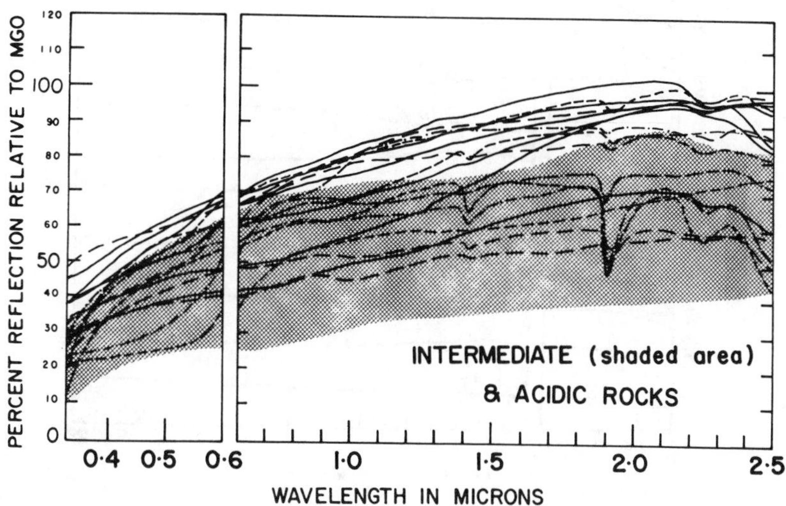

FIGURE 27. Collected spectra of unaltered felsic rocks (individual spectra) with the area (shaded) occupied by all the intermediate igneous rocks given above.[10]

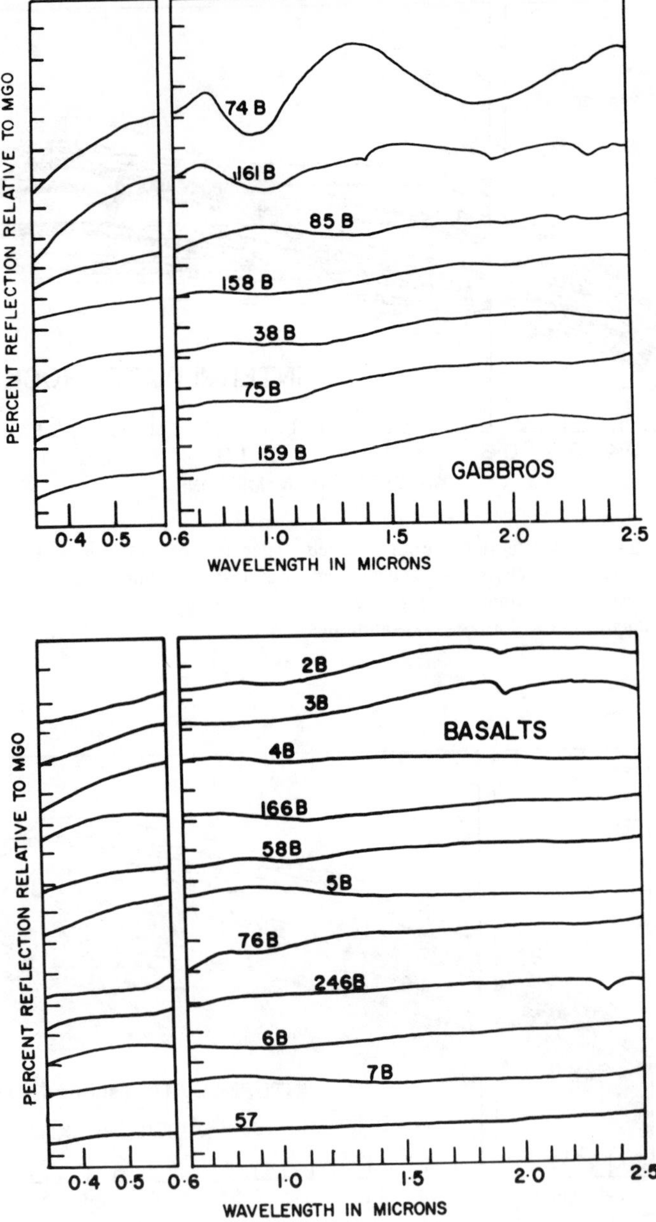

FIGURE 28. Reflection spectra of 74-250 μm particle-size gabbros and basalts: 74, norite, S. Africa; 161, norite porphyry, Butte, Mont.; 85, anorthositic, Colo.; 158, olivine, Wichita Mt., Colo.; 38, bytownite, Duluth, Minn.; 75, hypersthene, Ontario; 159, syenite, Butte, Mont. Basalts: 2, olivine, Jefferson Co., Colo.; 3, Boulder Co., Colo.; 4, Somerset Co., N. J.; 166, olivine, Hawaii, 58, flood, U.S.B.M.; 5, Chaffee Co., Colo.; 76, scoria, Ubehelhe Crater; 246, amygdaloidal, Michigan, 6, Germany; 7, Chaffee Co., Colo.; 57, vesicular, U.S.B.M. Spectra are displaced vertically.[10]

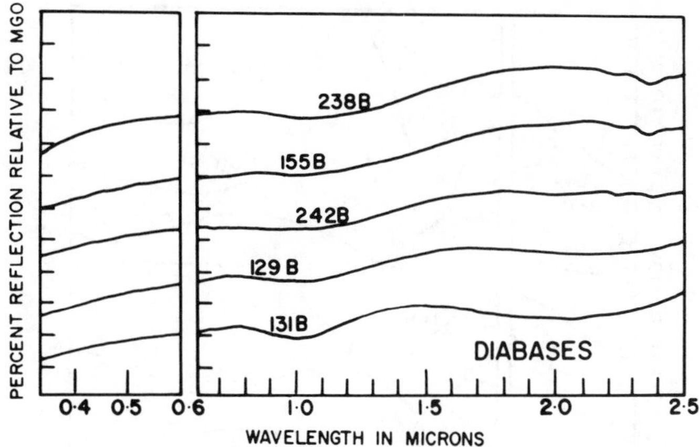

FIGURE 29. Reflection spectra of 74-250 μm particle-size diabases: 238, Finland; 155, Mt. Tom, Mass.; 242, Colo.; 129, Jersey City, N.J.; 131, St. Peters, Pa. Spectra are displaced vertically.[10]

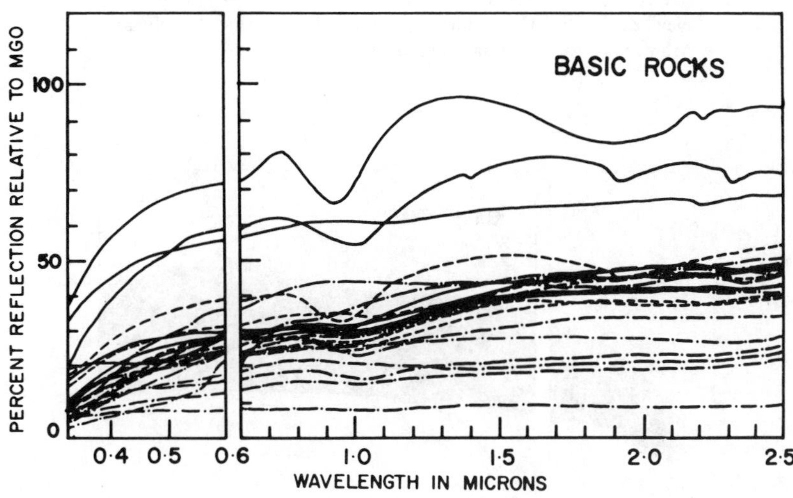

FIGURE 30. Collected superimposed reflection spectra of mafic rocks.[10]

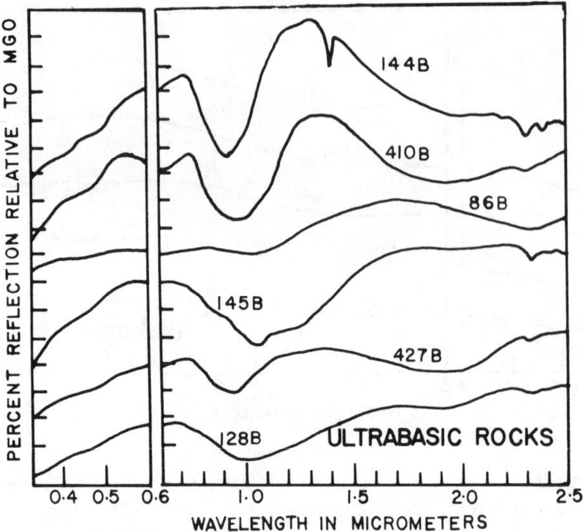

FIGURE 31. Reflection spectra of 74-250 μm particle-size ultramafic rocks: 144, peridotite, Webster, N.C.; 410, pyroxenite, N.C.; 86, pyroxenite, McClure Mt., Colo.; 145, dunite, New Zealand; 247, peridotite, Mont.; 128, peridotite, Nye, Mont. Spectra are displaced vertically.[10]

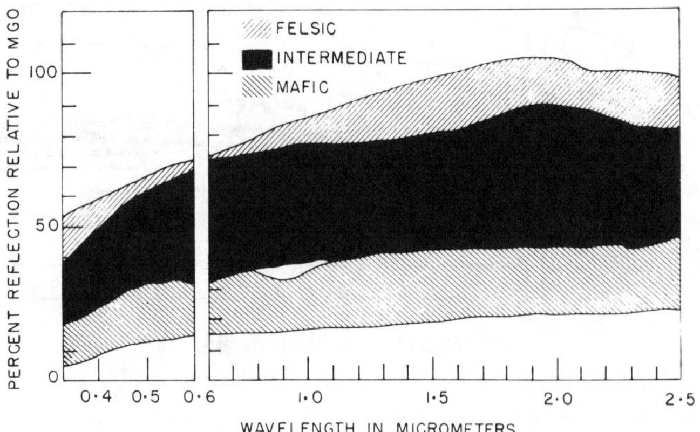

FIGURE 32a. Reflectance vs. wavelength. The three coded shaded areas indicate the locations into which most felsic (top), intermediate (middle), and mafic (bottom) igneous rock spectra fall.[10]

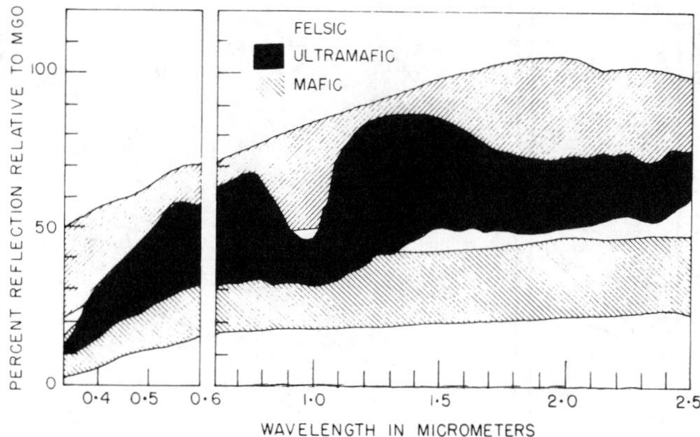

FIGURE 32b. Reflectance vs. wavelength. The three coded shaded areas indicate the locations into which most felsic (top), ultramafic (middle), and mafic (bottom) igneous rock spectra fall.[10]

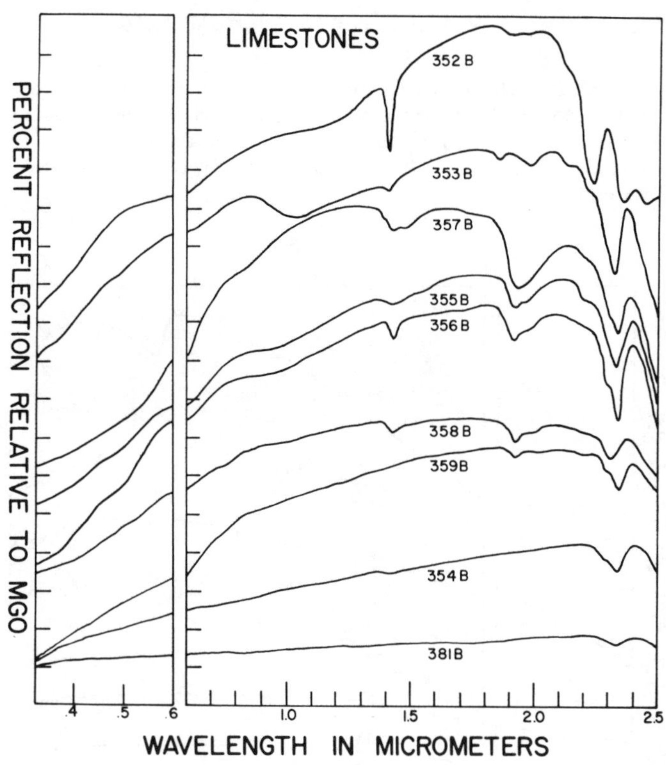

FIGURE 33. Reflection spectra of 74-250 μm particle-size limestones: 352; gray, Pa; 353, dolomitic, Colo.; 357; travertine, N. Mex.; 355, fossiliferous, Colo.; 356, lithographic, Germany; 358, oolitic, Indiana; 359, argillaceous, Colo.; 354, oolitic, Indiana; 381, argillaceous, Colo. Spectra displaced vertically.[12]

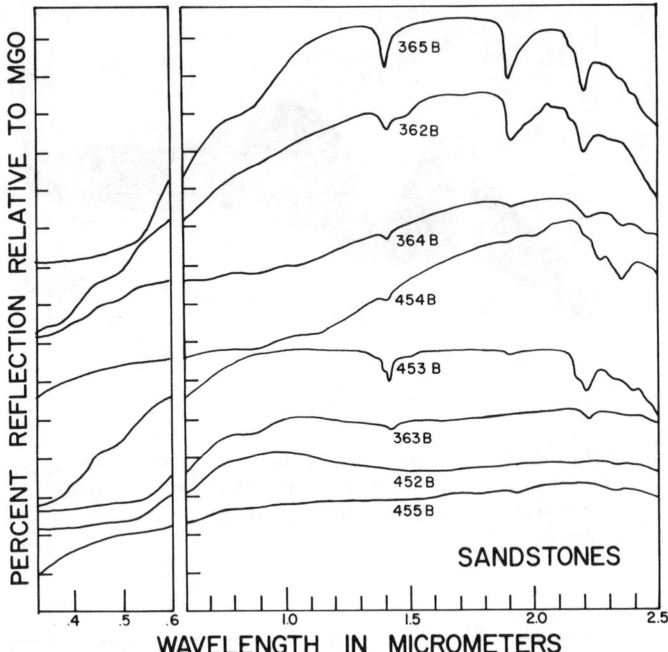

FIGURE 34. Reflection spectra of 74-250 μm particle-size sandstones: 365, red, Colo.; 362, arkosic, Colo.; 364, glauconitic, S. Dak.; 454, graywacke, Ontario; 453, S. Dak.; 363, micaceous red, Colo.; 452, ferruginous, N.Y.; 455, siltstone, Calif. Spectra are displaced vertically.[12]

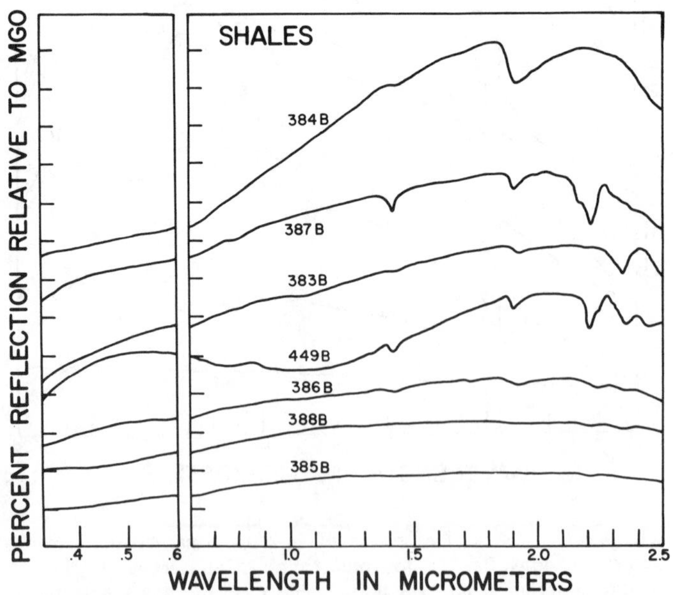

FIGURE 35. Reflection spectra of 74-250 μm particle-size shales: 384, phosphatic, Wyo.; 387, arenaceous, Colo.; 383, calcareous, Colo.; 449, illite-bearing, N.Y.; 386, argillaceous, Colo.; 388, carbonaceous, Colo.; 385, black, S. Dak. Spectra are displaced vertically.[12]

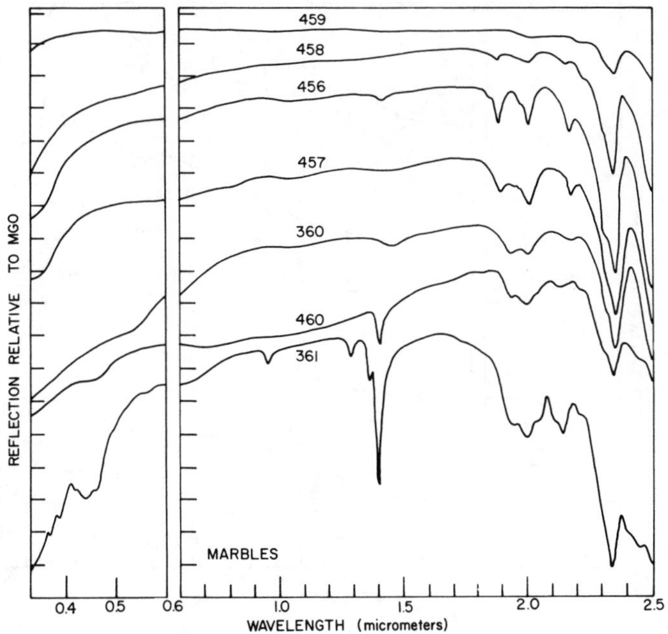

FIGURE 36. Reflection spectra of 74-250 μm particle-size marbles: 459, dolomitic, Mass.; 458, dolomitic, N.Y.; 456, Vermont; 457, Georgia; 360, pink, Colo.; 460, serpentine, N.Y.; 361, serpentine, Vermont. Spectra are displaced vertically.[12]

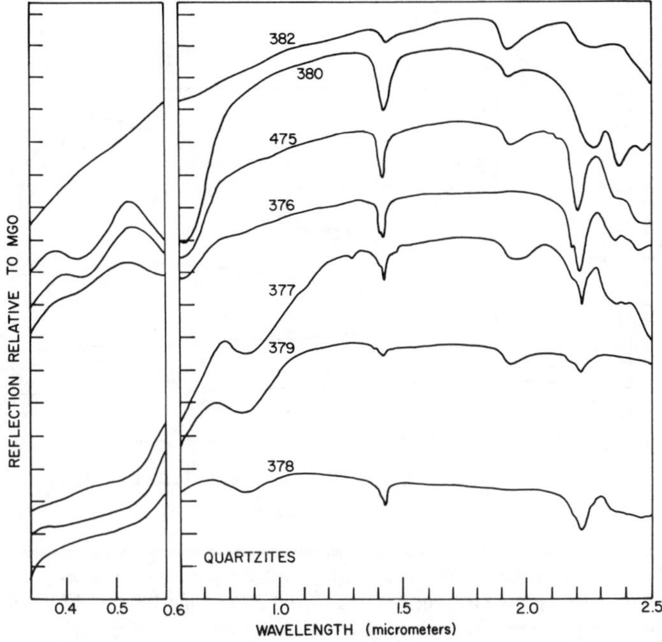

FIGURE 37. Reflection spectra of 74-250 μm particle-size quartzites: 382, gray, Colo.; 380, chloritic, S. Africa; 475, green, Ontario; 376, green, Colo.; 377, red, Norway; 379, red, Colo.; 378, purple, Colo. Spectra are displaced vertically.[12]

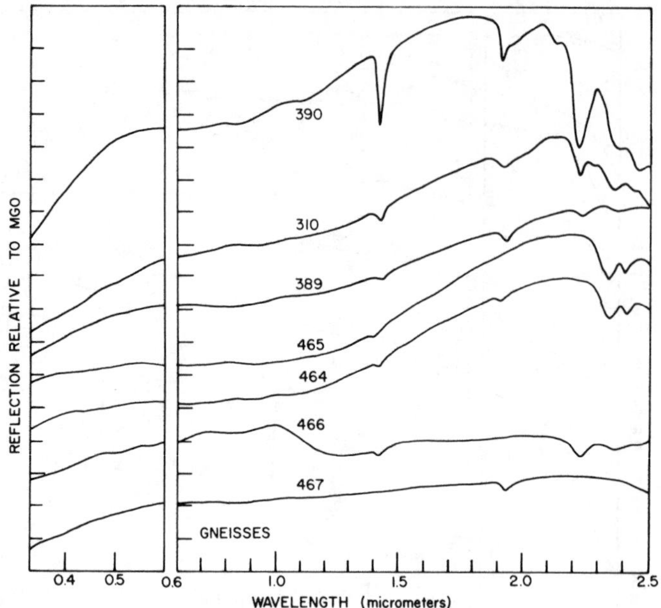

FIGURE 38. Reflection spectra of 74-250 μm particle-size gneisses: 390, albite, Colo.; 310, felsic, Colo.; 389, augen, Colo.; 465, hornblende, Essex Co., N.Y.; 464, diorite, Calif.; 466, sillimanite-garnet, Warren Co., N.Y.; 467, syenite, Wisc. Spectra are displaced vertically.[12]

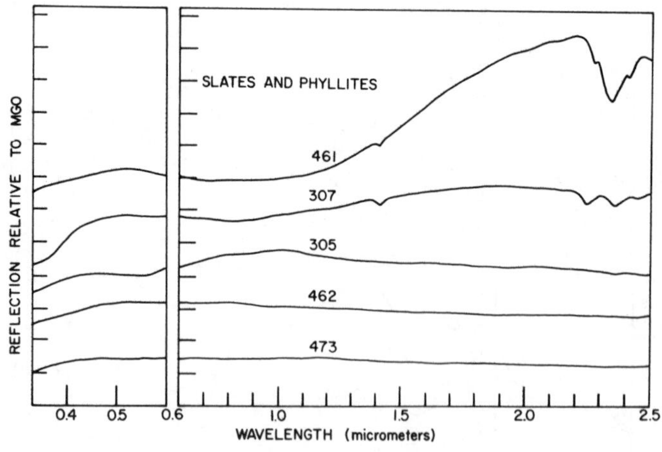

FIGURE 39. Reflection spectra of 74-250 μm particle-size slates: 461, Mont.; 307, gray, Oregon; 305, red, Colo.; 462, chiastolic, Calif. And phyllite; 473, Vermont. Spectra displaced vertically.[12]

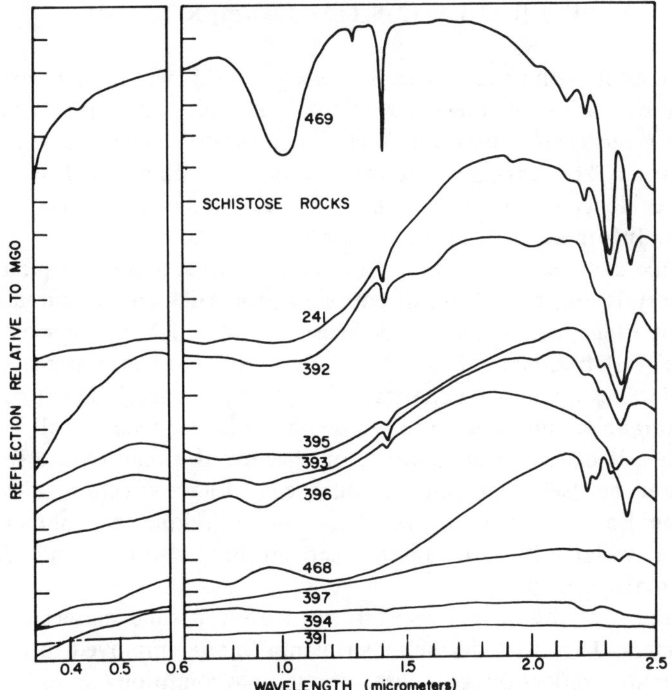

FIGURE 40. Reflection spectra of 74-250 μm particle-size schistose rocks: 469, tremolite schist, N.Y.; 392, green schist, Norway; 395, chlorite schist, Colo.; 393, hornblende schist, Colo.; 396, anthophyllite mica schist, Colo.; 468, tourmaline schist, S. Dak.; 397, graphite schist, Canada; 394, mica schist, Colo.; 391, graphite schist, Colo. Spectra displaced vertically.[12]

INFRARED (2.5 TO 40.0 μm) REGION

Information about solid substances is available in this wavelength range as a consequence of the excitation of internal vibrations of their constituent atoms relative to each other, of groups (molecular units) relative to the crystal structure, or of vibrations of the lattice itself. The vibrations are commonly classified as bond stretching or angle bending, and evidence for fundamentals, overtones, and combination tones may occur in this region. Information regarding the nature of these vibrations, in the form of bands and slope changes, is available from conventional spectroscopic techniques. In these, the transmission, reflection, or emission properties are measured at each wavelength over the range, or by Raman spectroscopy, in which monochromatic light (the "exciting line") of frequency υ, to which the material is transparent, is scattered by molecules or lattices involving a change in frequency to $\upsilon \pm \upsilon^1$. These new frequencies (υ^1) are the Raman frequencies and are due to the lattice (acoustical) and vibrational (optical) modes. Because the appearance or absence of a feature in a spectrum is primarily governed by "selection rules", and because these selection rules are different for conventional and Raman spectra, these two techniques compliment each other. Some data are available from a Raman spectrum that is "forbidden" from appearing in a conventional spectrum.

The appearance of the infrared spectrum of a given solid material is governed by many parameters. They include the experimental mode employed to record the spectrum (transmission, reflection, emission), the sample condition (smooth or rough plane surface, particulate, etc.), and the environmental conditions (temperature, gradients, pressure, insolation angle, etc.). Because varying these parameters will alter the appearance of a spectrum, comparing spectral information requires that the data used be collected using the same set of experimental conditions or else that the effects of varying any of the parameters be fully understood and accounted for.

In the infrared, large variations occur in the value of the extinction coefficient, k, for solid materials, and it reaches very high values in the region of the resonance frequencies, particularly for silicates near 10 and 20 μm. These variations are accompanied by and related to large variations in the refractive index, n, a phenomenon referred to as anomalous dispersion (see Figure 1). The high values of n and k produce very high specular reflectances at small angles of incidence to smooth surfaces (referred to as the "reststrahlen effect") and for rough surfaces and particulates, as well as absorption effects. Maxima and minima appear in their spectra that are associated with, but are not concident with, the resonance frequencies. They are due to the Christiansen effect, and "Christiansen frequencies" can be particularly useful diagnostically, especially when dealing with the emission spectra of particulate samples.

Spectra shown in Figures 41 and 42 illustrate two transmission spectra (at the top) and a corresponding reflection spectrum recorded from a polished surface (at the bottom). The reflection spectra are roughly the inverse of the transmission spectra: for the transmission spectra, the dashed curve represents transmission through a particulate samples in air, while the solid curve is for transmission through a particulate sample embedded in KBr. The major difference between the transmission spectra occurs at just shorter wavelengths than the shortest wavelength major minimum where the dashed curve rises to a maximum, called the principal Christiansen frequency, which is not apparent in the solid curve (for the KBr-embedded sample). Additional Christiansen effect maxima are apparent on the short wavelength side of other absorption features.

Mineral Spectra

Transmission and Absorption Spectra

Infrared absorption bands are typically so intense that it is difficult to prepare sections sufficiently thin to allow complete penetration at the resonance frequencies. Consequently, the usual technique for recording spectra is to grind the samples to very small particle sizes and record their spectra in air or suspended in a KBr matrix.

Spectra of particulate minerals in air — The spectra of particulate mineral samples recorded by suspending the samples on a mirror (thus allowing the beam to traverse the sample twice) are shown in Figures 43 to 52. All these spectra display principal Christiansen maxima.

Spectral signatures of silicates — The characteristic features in the absorption spectra in air of different silicate minerals are illustrated in Figure 53. In the 7 to 9-μm region, the Christiansen maximum is obvious, and its location migrates systematically to longer wavelengths as the nature of the material progresses from felsic to intermediate to mafic to ultramafic. From 8.5 to 12 μm, the most intense absorption due to Si-O stretching vibration occurs and its position also migrates in parallel with the Christiansen peak. From 12 to 15 μm, features occur indicating (Si, Al)-O-(Si, Al) bridges have formed. All feldspars display bands here, but their spectra vary from showing a single feature to a set of four well-defined bands. From 15 to 20 μm constitutes a feature gap for most silicates, but all feldspars show a group of three similar and characteristic bands. From 20 to 40 μm all silicates display features due to deformation and bending modes and some (Al, Si)-O-Metal valence stretching modes. The information available from the infrared spectra of silicates is summarized in Figure 54.

Reflection Spectra

Specular reflection from smooth plane surfaces — The spectra recorded from these surfaces typically display well-defined intense features (the "reststrahlen bands") with reflection maxima at or close to the resonance frequencies. The reflection (R) at any given wavelength can be expressed in terms of the optical constants n and k by

$$R = \frac{(n - 1)^2 + n^2k^2}{(n + 1)^2 + n^2k^2}$$

The intensity of the maxima is governed by the degree of flatness and smoothness of the surface. The better the polish, the more intense the maxima. Figure 55 shows the spectra of four mafic mineral specimens, and the top halves of Figures 56 and 57 show the specular reflectance of calcite and gypsum, respectively, compared with their transmission spectra recorded in KBr.

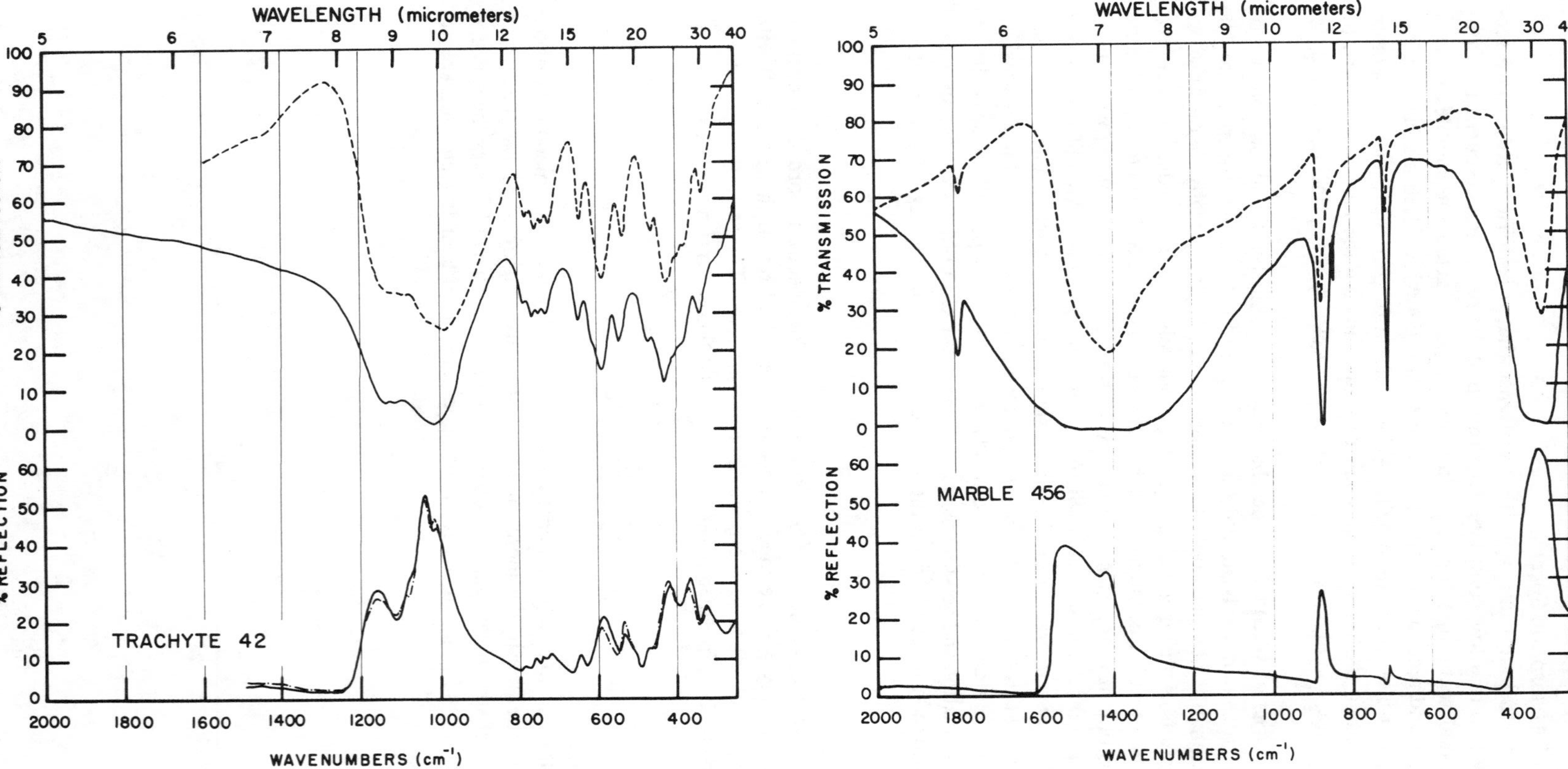

FIGURE 41. Spectra of a trachyte recorded under different experimental conditions. Top:———transmission in air; top:——transmission in KBr pellet. Bottom: ——specular reflection from polished plate.[12]

FIGURE 42. Spectra of a pure white Vermont marble recorded under different conditions. Top:———transmission in air. Top:——transmission in KBr pellet. Bottom:——specular reflection from polished plate.[13]

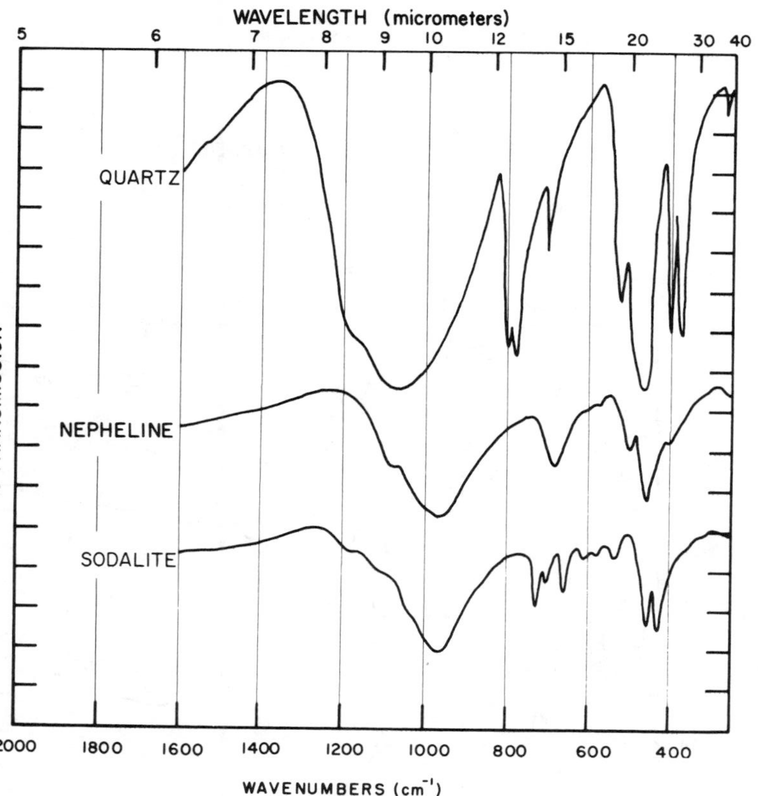

FIGURE 43. Spectra of quartz and feldspathoids.[12]

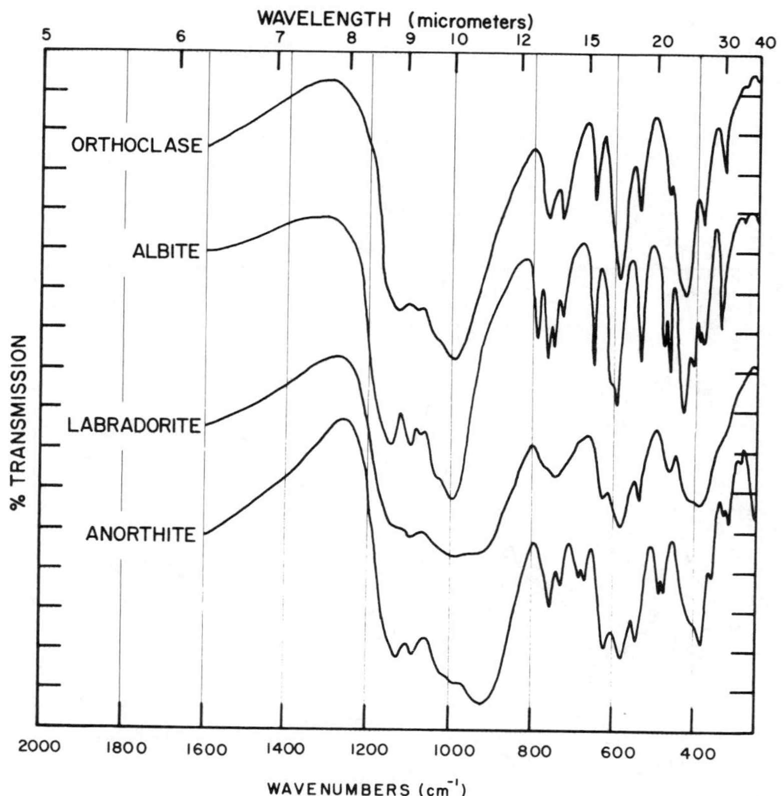

FIGURE 44. Spectra of feldspars.[12]

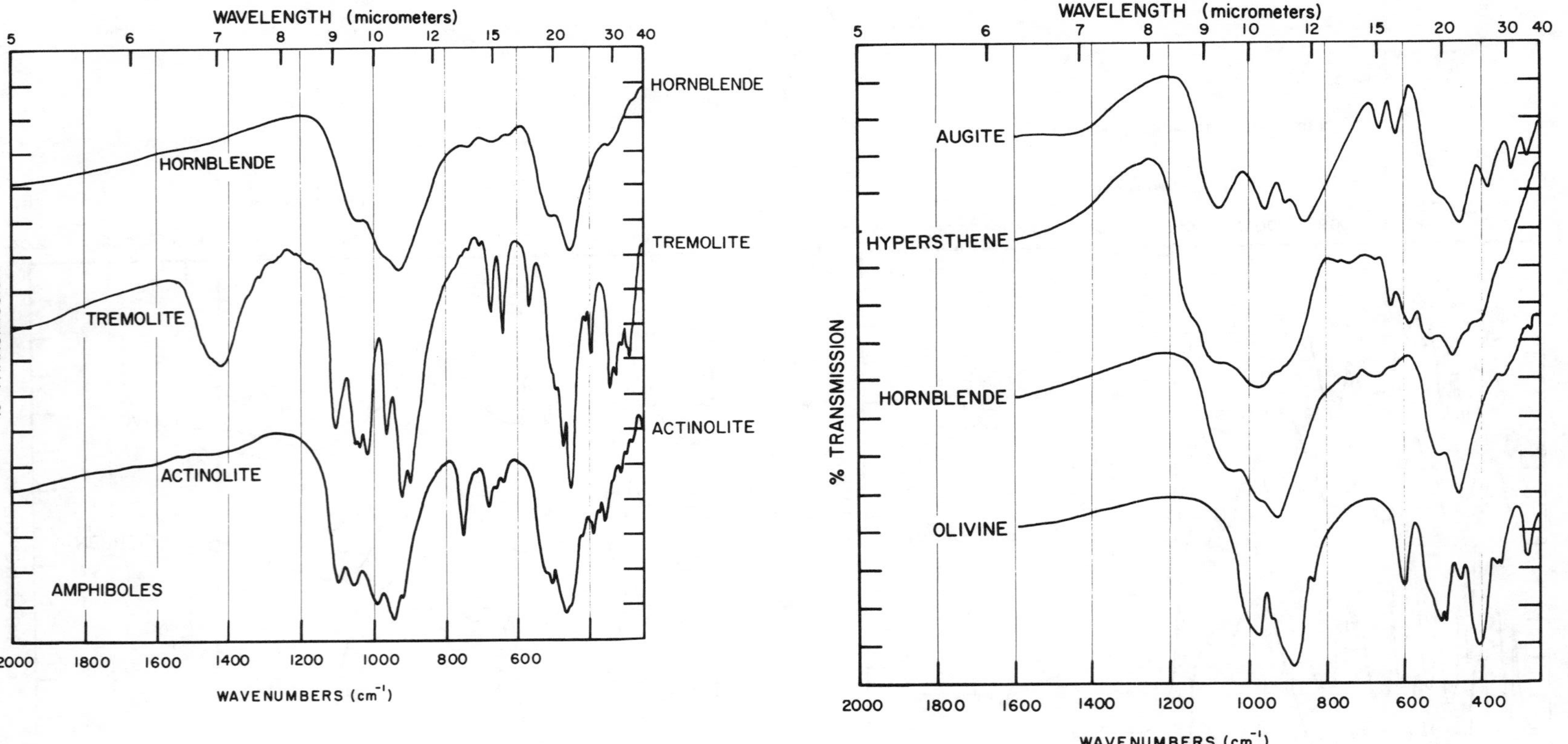

FIGURE 45. Spectra of amphiboles.[12]

FIGURE 46. Spectra of mafic minerals.[12]

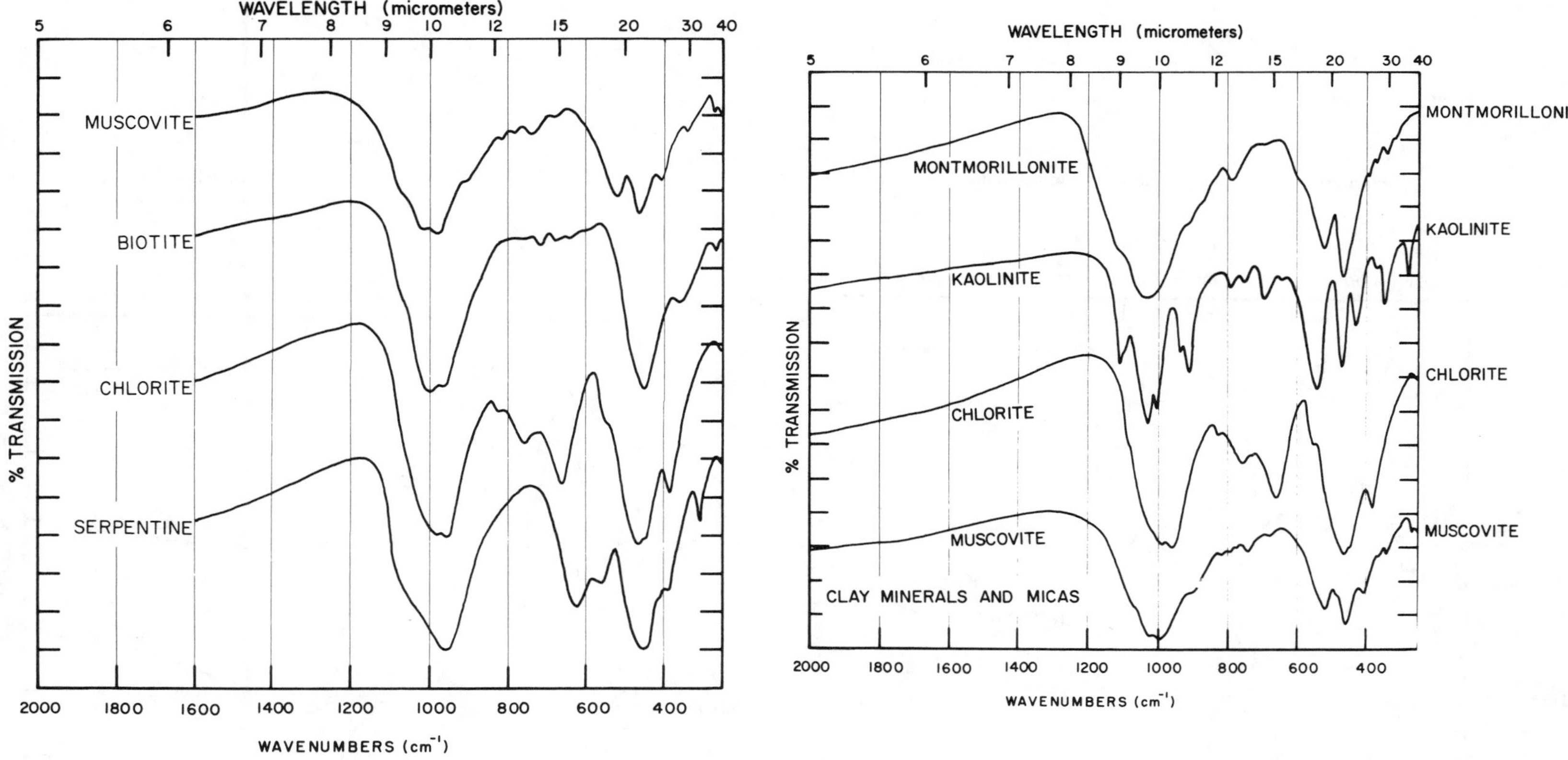

FIGURE 47. Spectra of accessory and alteration-product minerals.[12]

FIGURE 48. Spectra of clay minerals and micas.[13]

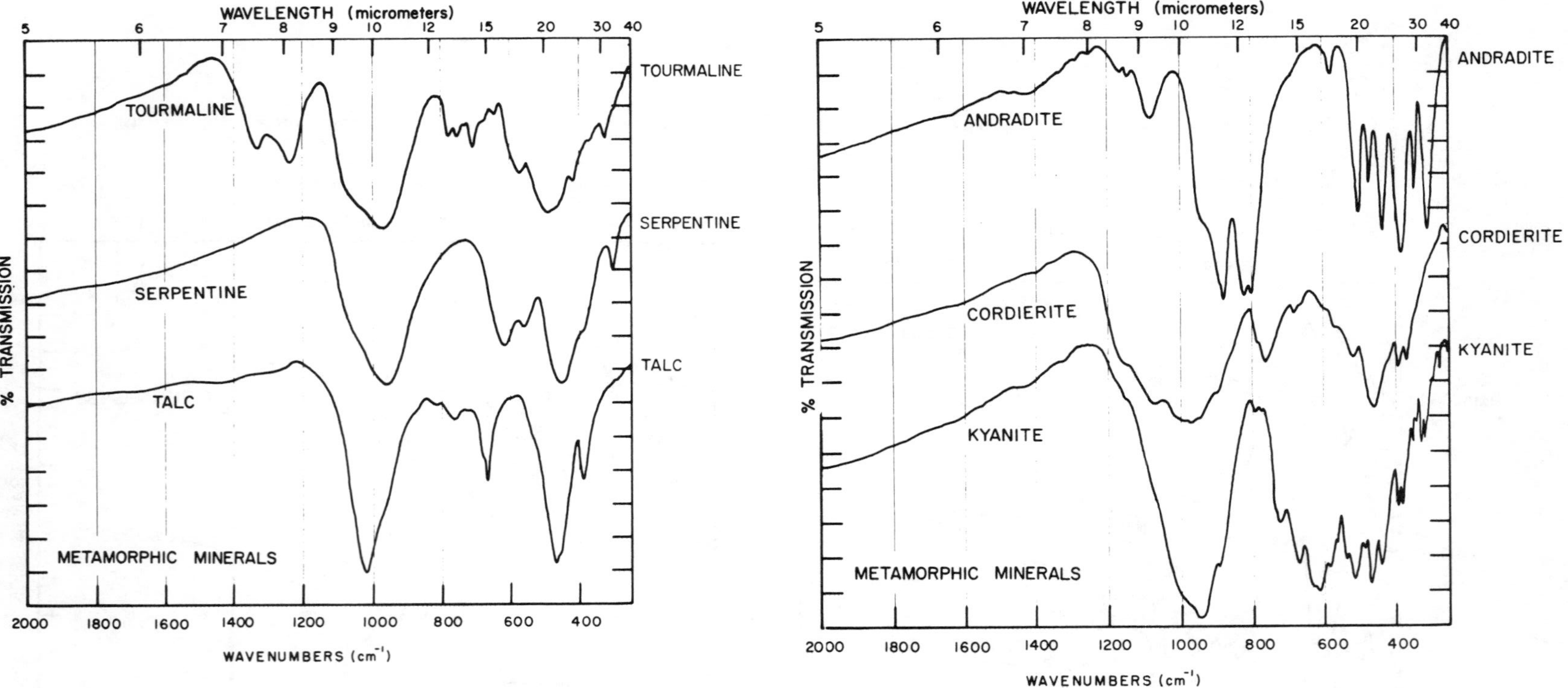

FIGURE 49. Spectra of metamorphic minerals.[13]

FIGURE 50. Spectra of metamorphic minerals.[13]

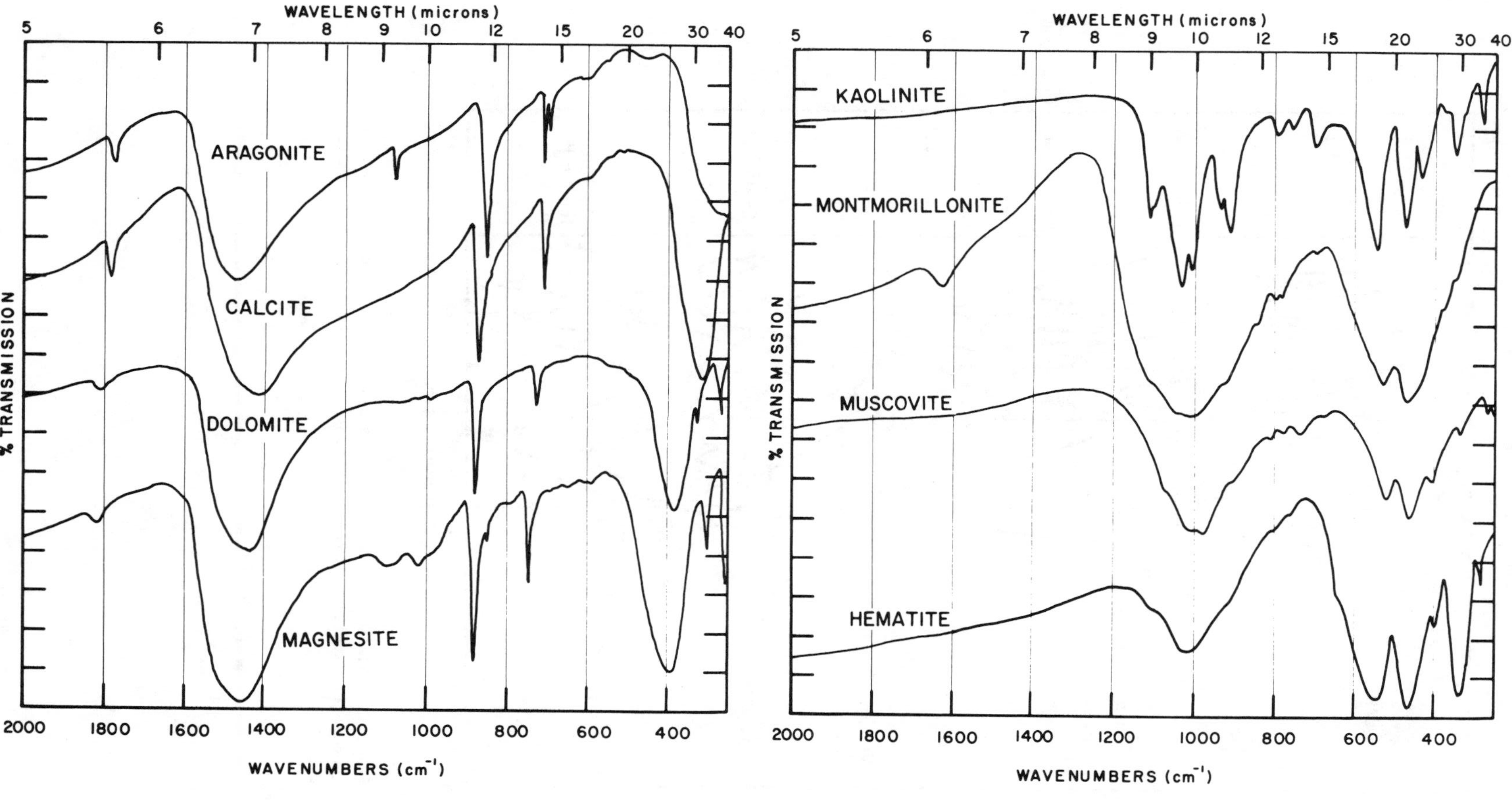

FIGURE 51. Spectra of carbonate minerals.[12]

FIGURE 52. Spectra of clay and minor accessory minerals.[13]

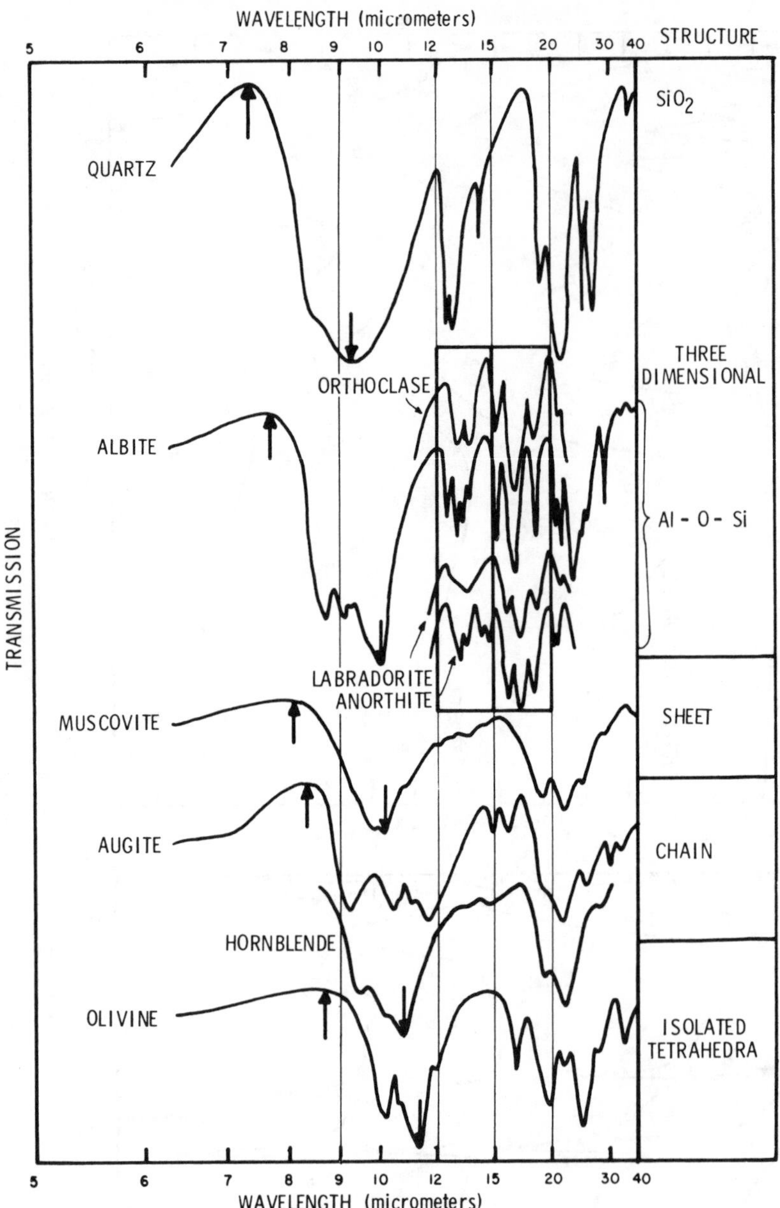

FIGURE 53. Transmission spectra of silicate minerals recorded in air. Five spectral regions are indicated and the causes of the features that occur there are discussed in the text.[9]

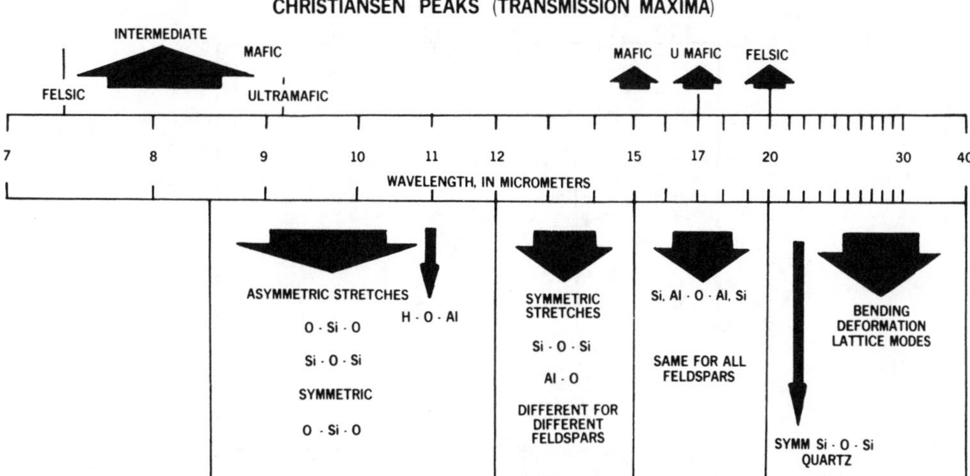

FIGURE 54. Diagrammatic representation of the location and cause of maxima and minima in the spectra of silicates in the 7 to 40 μm region.[9]

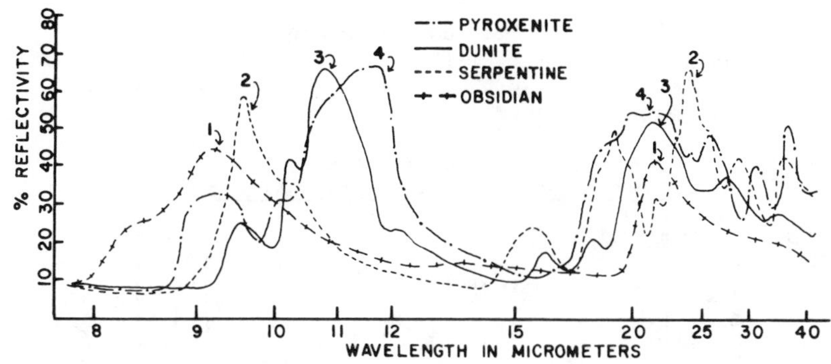

FIGURE 55. Spectral reflectance curves for four mafic minerals.[11]

Reflection from rough and particulate samples — The appearance of the spectra recorded from these samples depends upon the condition of the sample surface. Two components, a specular and a bulk, contribute to the total spectrum. For polished surfaces the spectral component dominates. For finely particulate samples, the bulk component dominates. The effect for such surfaces is illustrated in the bottom halves of Figures 56 and 57, where some features are seen to reverse polarity as the particle size is changed.

Emission Spectra

In the infrared, the emission source is the black or gray body distribution of energy established within the bulk of the sample, so that spectral information is present as a consequence of its interaction with material and boundaries as it escapes from the sample. The actual form of the emission spectrum is determined by many sample and environmental parameters which vary the intensity, location, and polarity of the features.

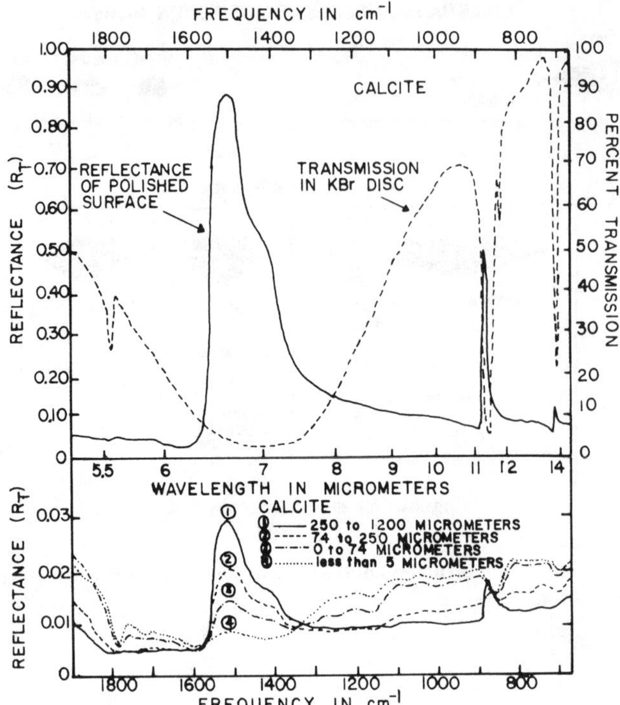

FIGURE 56. Reflectance and transmission spectra of calcite. Top: transmission spectrum of sample in KBr (dashed line) and reflection from polished sample (solid line). Bottom: reflection spectra for four particle size samples.[18]

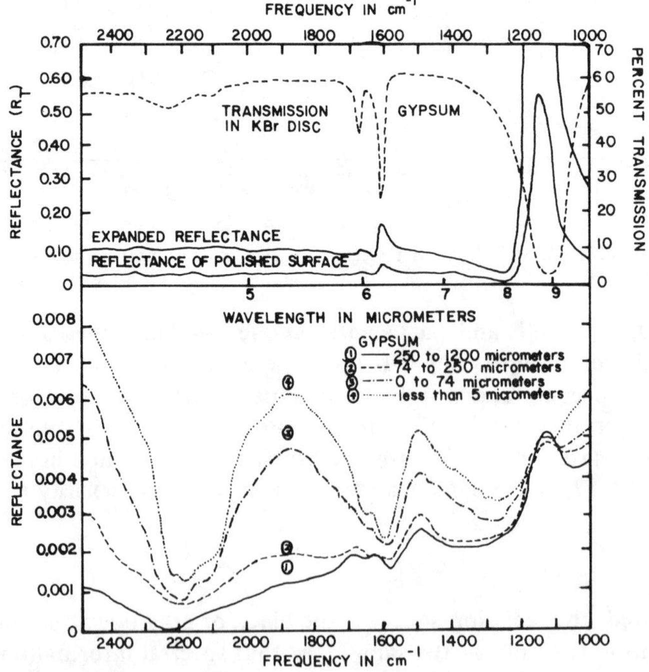

FIGURE 57. Reflectance and transmission spectra of gypsum. Top: transmission spectrum of a sample in KBr (dashed line) and reflection from polished sample (solid line). Bottom: reflection spectra for four particle size samples.[18]

Single very small particle emission — When particles are sufficiently small they emit as they absorb, very much in the manner that isolated molecules do. This is illustrated for submicro-sized quartz particles in Figure 58.

Emission from smooth plane plates — Here the emission spectrum is essentially the inverse of the reflection spectrum. The emission Eλ at wavelength λ can be expressed as

$$E\lambda = E_{BB\lambda}(1 - R\lambda)$$

where $E_{Bb\lambda}$ is the emission of a black body at the same temperature and wavelength and Rλ is the reflection of the surface. The emission spectrum of a quartz plate over a limited wavelength range is shown as the dotted curve in Figure 59.

Emission from particulate sample — Principally because of radiative and convection interaction between the individual particles making up a sample, the form of their emission spectra is principally affected by particle size, packing, background temperature, uniformity and mode of heating, pressure, and insolation angle. Figure 59 illustrates the changes to the appearance of an emission spectrum of quartz purely as a function of particle size change in the sample. Table 5 presents emission data for several particulate mineral samples to illustrate the effects on the emission features of various experimental parameters. All data were recorded under vacuum for either 0 to 5 μm or 0 to 74 μm particle size ranges; for very lightly packed (''fairy castle'') or more lightly packed samples and for a viewed background shield at 77 or 295°K.

Rock Spectra

Often the spectral features of the constituent minerals of a rock are obvious in its spectrum. This is particularly true of intense sharp features, such as those displayed by quartz. In other cases, apparently sharp features that occur close together in wavelength in more than one constituent may be broadened into single features, resulting in an anomalously bland rock spectrum. Consequently, rock spectra cannot always be simply generated by adding the constituent mineral spectra in the proportions in which they occur in rocks. Rock spectra that do and do not reflect component mineral spectra are shown in Figures 60 to 63.

These figures illustrate how features in the spectrum of a rock composed essentially of only two minerals can be readily identified, while such identification is less easy for multimineralic rocks, and that it is easier to recognize mineral features in felsic than mafic mineral containing rocks.

Igneous Rocks

Felsic rocks — Figures 64 to 67 show the collected individual spectra of granites, rhyolites, tuffs, and volcanic glasses, displaced vertically for display at the top. Beneath these are shown the same spectra (thinner lines) superimposed. At the bottom is shown a ''composite'' spectrum generated as the average of all the individual spectra. This type of display is used throughout this section, except in those figures where the composite spectra are collected, as illustrated in Figure 68 for the collected composite spectra of felsic rocks.

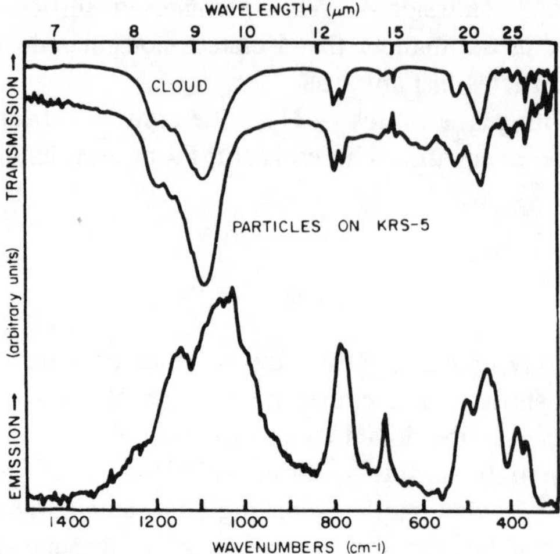

FIGURE 58. Transmission and emission spectra of individual particles of micron-sized particles of quartz. Top spectrum is the transmission of a cloud of particles, and the middle curve is of individual particles suspended on a KRS-5 plate. Bottom spectrum is the transmission of individual particles heated on a brass plate.[7]

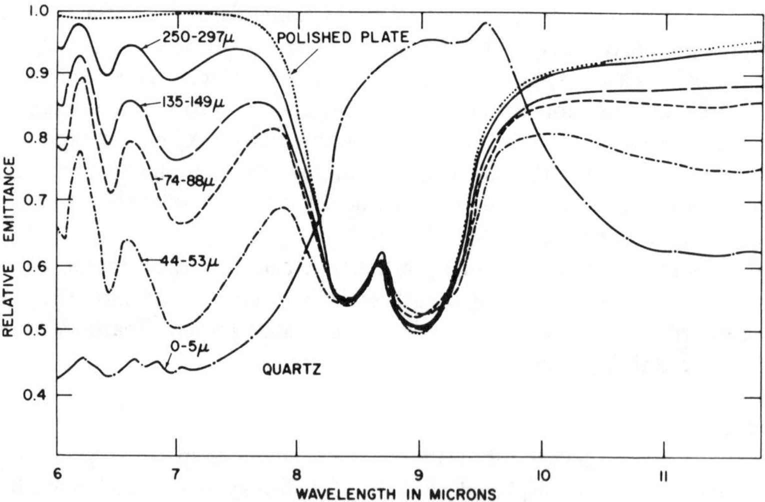

FIGURE 59. Relative emittance spectra of different, narrow size range samples of quartz particles. Spectra displaying well-defined minima were normalized at 8.5 μm.[16]

Table 5
SPECTRAL EMISSION DATA FOR SELECTED MINERAL SAMPLES[16]

Sample	Transmission maximum	Emissivity Data for 0—5 μm Particle Size							
		295°K Shield				77°K Shield			
		Fairy castle		Packed		Fairy castle		Packed	
		$\lambda_{\epsilon max}$	ϵ_{av}	$\lambda_{\epsilon max}$	ϵ_{av}	$\lambda_{\epsilon max}$	ϵ_{av}	$\lambda_{\epsilon max}$	ϵ_{av}
Quartz	7.35	7.45	0.884	7.16	0.904	7.44	0.838	7.18	—
Albite	7.63	7.83	0.885	7.49	0.946	7.80	0.855	7.47	0.937
Anorthoclase	7.78	8.30	—	—	—	8.22	0.906	7.93	0.961
Labradorite	7.91	8.12	0.882	7.97	0.957	8.03	0.873	7.88	0.944
Hornblende	8.47	8.84	0.885	8.80	0.949	8.75	0.858	8.72	0.938
Olivine	8.70	9.40	0.848	9.23	0.952	9.45	0.793	9.15	—
		Emissivity Data for 0—74 μm Particle Size							
Quartz	7.35	7.16	0.821	7.16	0.867	7.15	0.789	7.15	0.899
Albite	7.63	7.55	0.884	7.53	0.914	7.55	0.841	7.53	0.886
Anorthoclase	7.78	7.68	0.897	7.76	0.917	7.65	0.931	7.61	0.936
Labradorite	7.91	7.71	0.887	7.75	0.951	7.71	0.894	7.75	0.937
Hornblende	8.47	8.50	0.952	8.60	0.961	8.43	0.922	8.55	0.939
Olivine	8.70	8.73	0.900	8.73	0.936	8.88	0.853	—	—

Note: All data were recorded under vacuum with either a viewed background shield at 77°K or 295°K, and for lightly or lightly packed < 5 μm or > 74 μm particle sized samples.

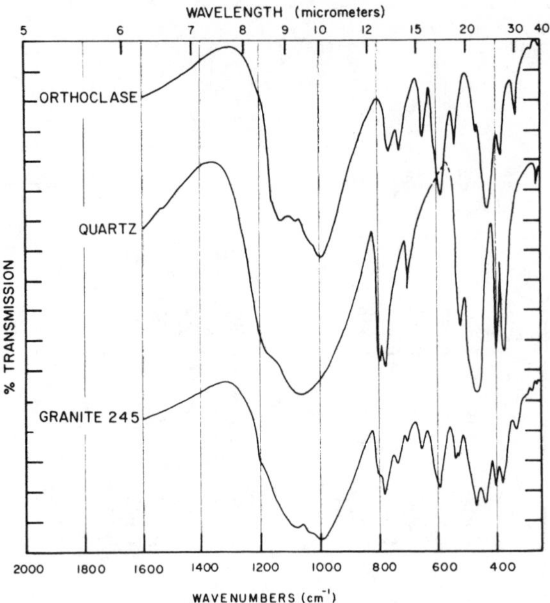

FIGURE 60. Spectra of granite and component minerals.[12]

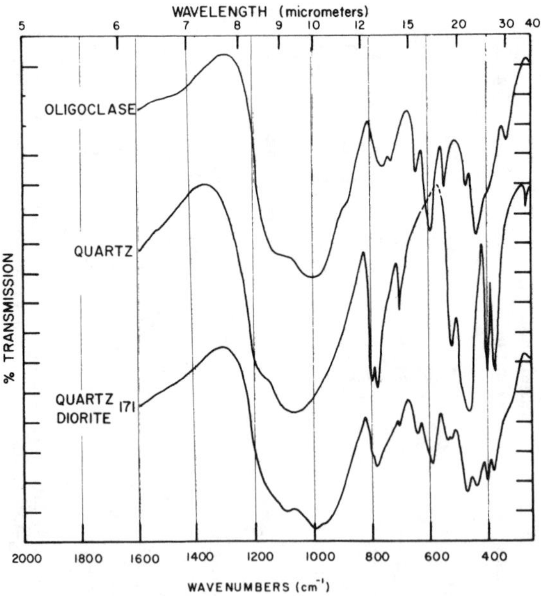

FIGURE 61. Spectra of quartz diorite and component minerals.[12]

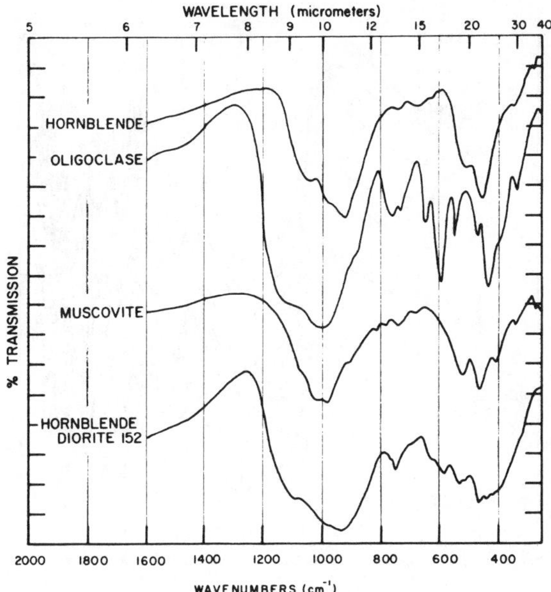

FIGURE 62. Spectra of hornblende diorite and component minerals.[12]

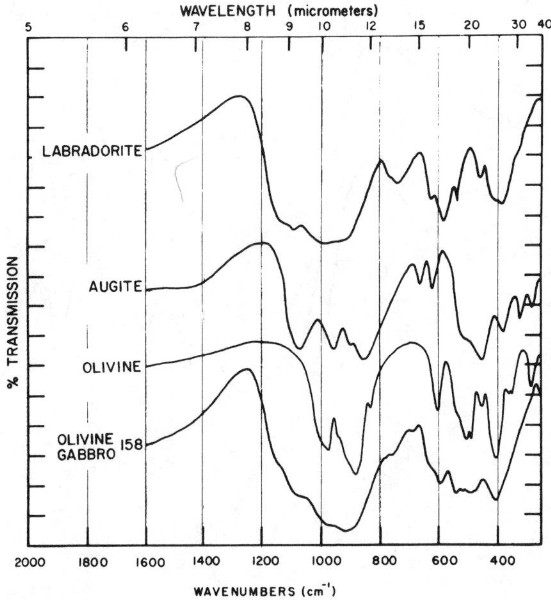

FIGURE 63. Spectra of olivine gabbro and component minerals.[12]

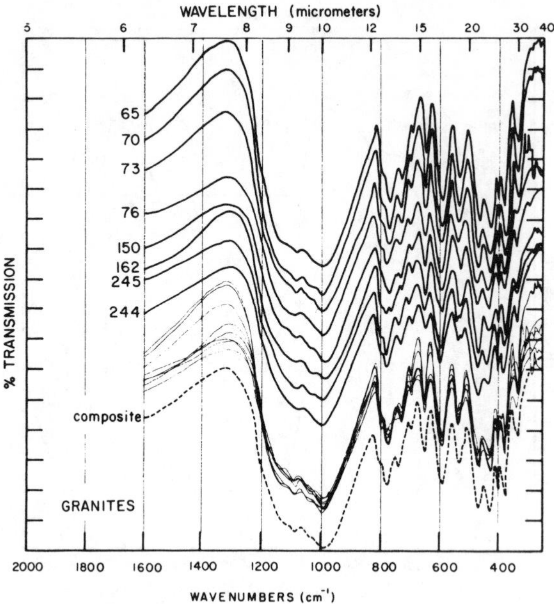

FIGURE 64. Collected spectra of granites (same samples as in Figure 19). Spectra are displaced vertically at tip. Spectra are superimposed at bottom. Composite spectrum is dashed curve.[12]

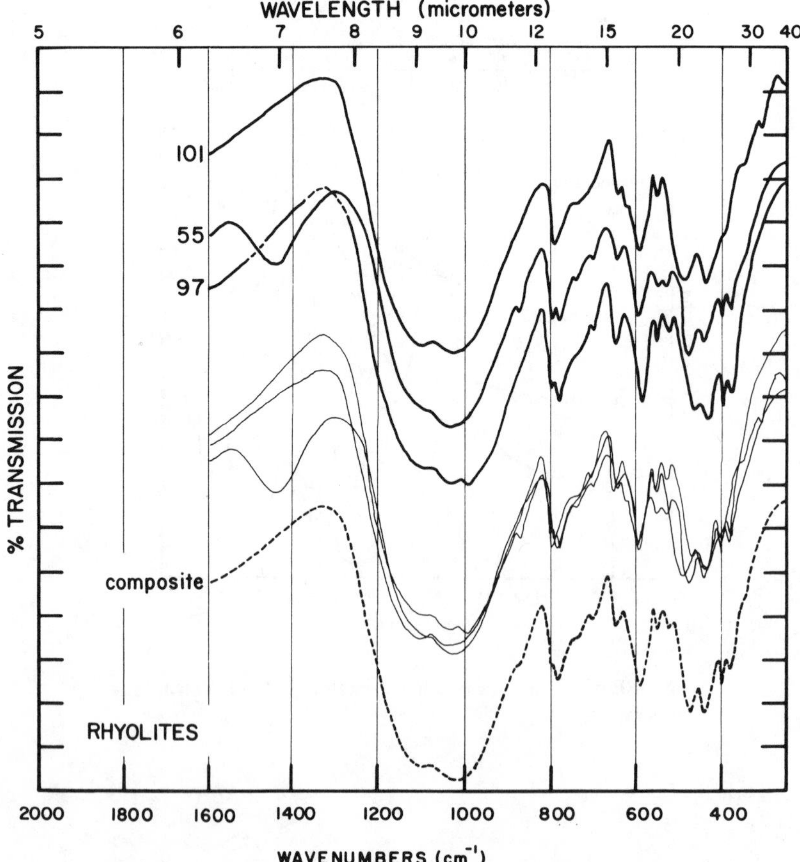

FIGURE 65. Collected spectra of rhyolites (same samples as in Figure 21).[12]

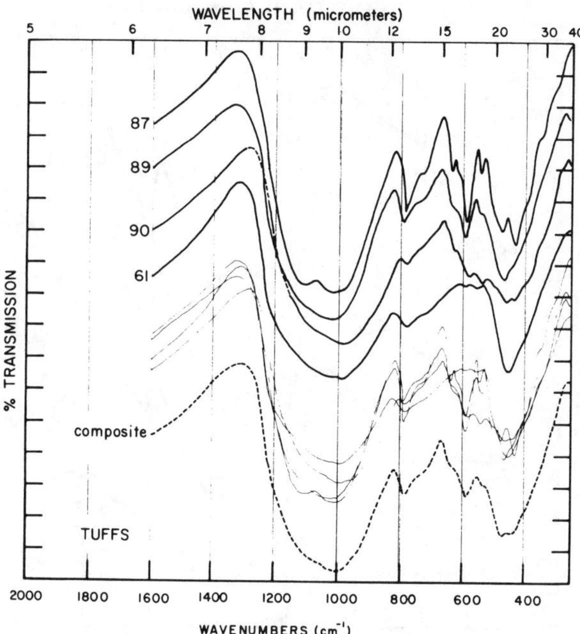

FIGURE 66. Collected spectra of tuffs (same samples as in Figure 22). 89, Butte, Mont.; 90, Lapilli, Calif.; 61, U.S.B.M. (altered).[12]

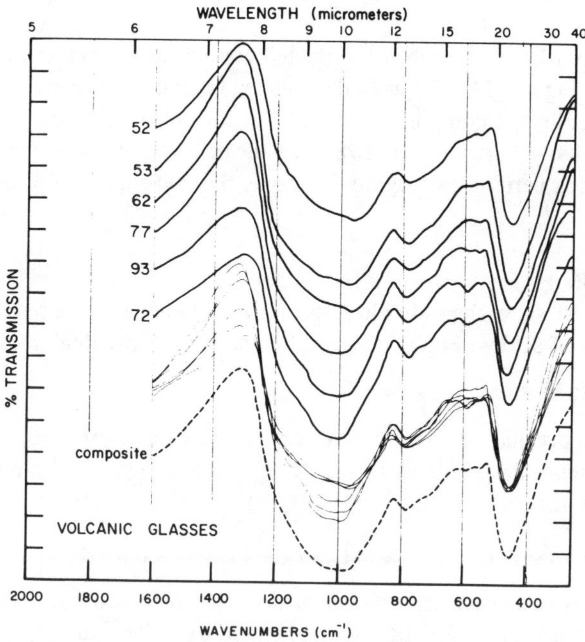

FIGURE 67. Collected spectra of volcanic glasses (same samples as in Figure 20).[12]

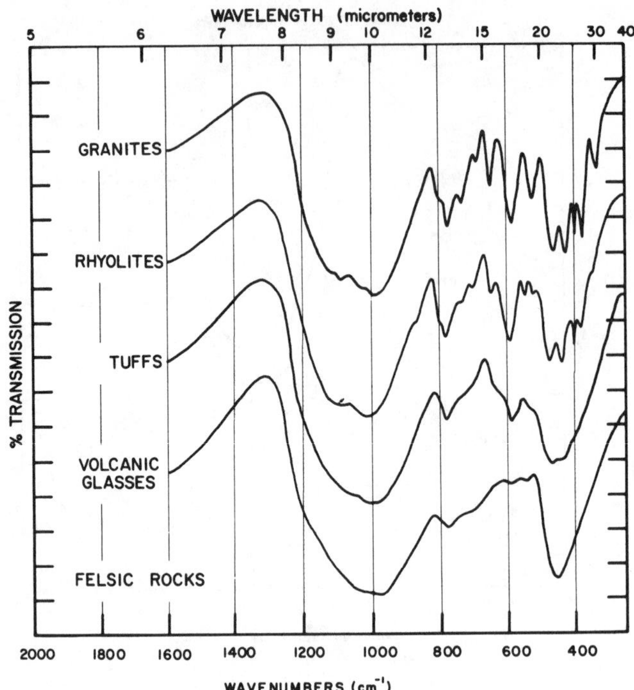

FIGURE 68. Collected composite spectra of felsic rock.[12]

Intermediate rocks — Figures 85 and 86 show the collected spectra of diordites and andesites.

In Figures 71 and 72 are shown the collected composite spectra of intermediate rocks.

Mafic rocks — Figures 73 to 75 show the collected individual spectra of gabbros, diabases, and basalts. The collected composite spectra of mafic rocks are shown in Figure 92.

Ultramafic rocks — Figures 77 and 78 show the collected individual spectra of pyroxene-rich and olivine-rich peridotites. Figure 79 shows the collected composite spectra of ultramafic rocks.

Sedimentary Rocks

Figures 80 and 81 show the individual spectra of limestones. Figures 82 and 83 show the individual spectra of shales. Figures 84 and 85 show the individual spectra of sandstones.

Metamorphic Rocks

The collected individual spectra of marbles are shown in Figure 86, quartzites in Figure 87, hornfels in Figure 88, gneisses in Figure 89 and 90, slates and phyllites in Figure 91, and schists in Figures 92 and 93.

Emission Spectra

The emission spectral data for igneous rocks were recorded for >74 μm particle size range under vacuum with a 77°K background shield over the 6 to 12 μm wavelength range (Table 6).

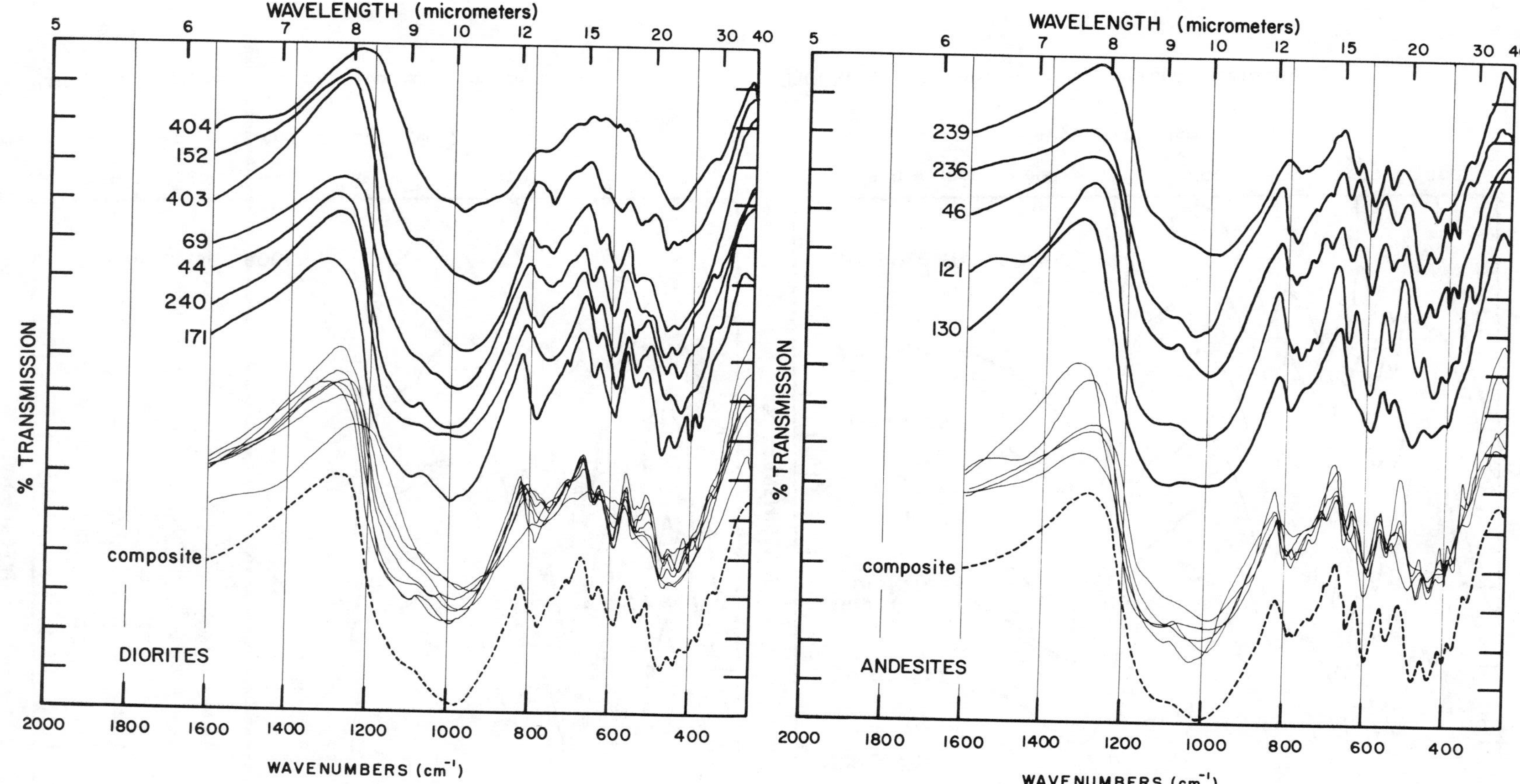

FIGURE 69. Collected spectra of diorites (same samples as in Figure 25, 69, hornblende, Salem, Mass.; 44, porphyry, Wyo.; 171, quartz, Australia.[12]

FIGURE 70. Collected spectra of andesites (same samples as in Figure 25. 46, porphyry, Colo.[12]

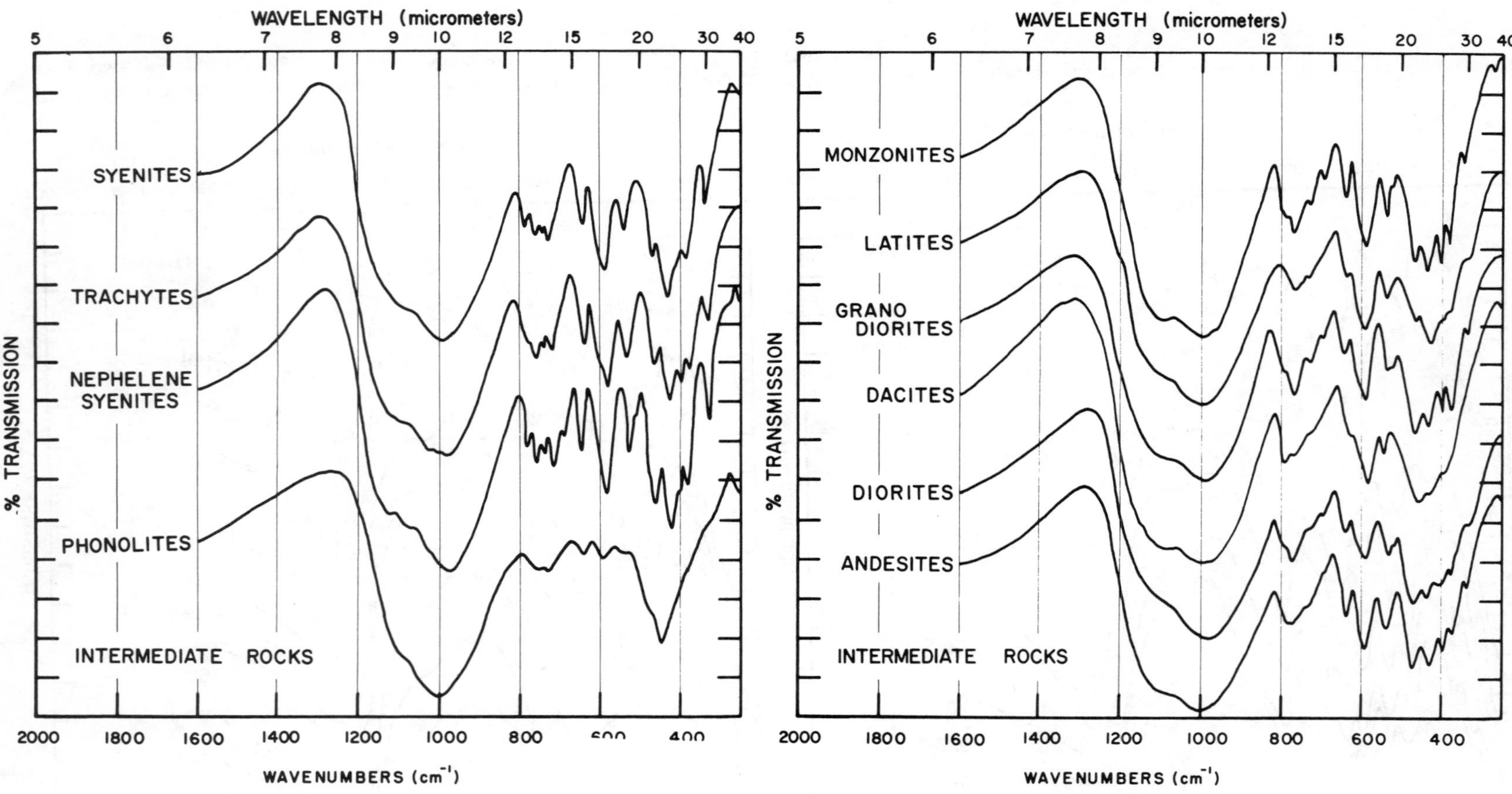

FIGURE 71. Collected composite spectra of intermediate rocks.[12]

FIGURE 72. Collected composite spectra of intermediate rocks.[12]

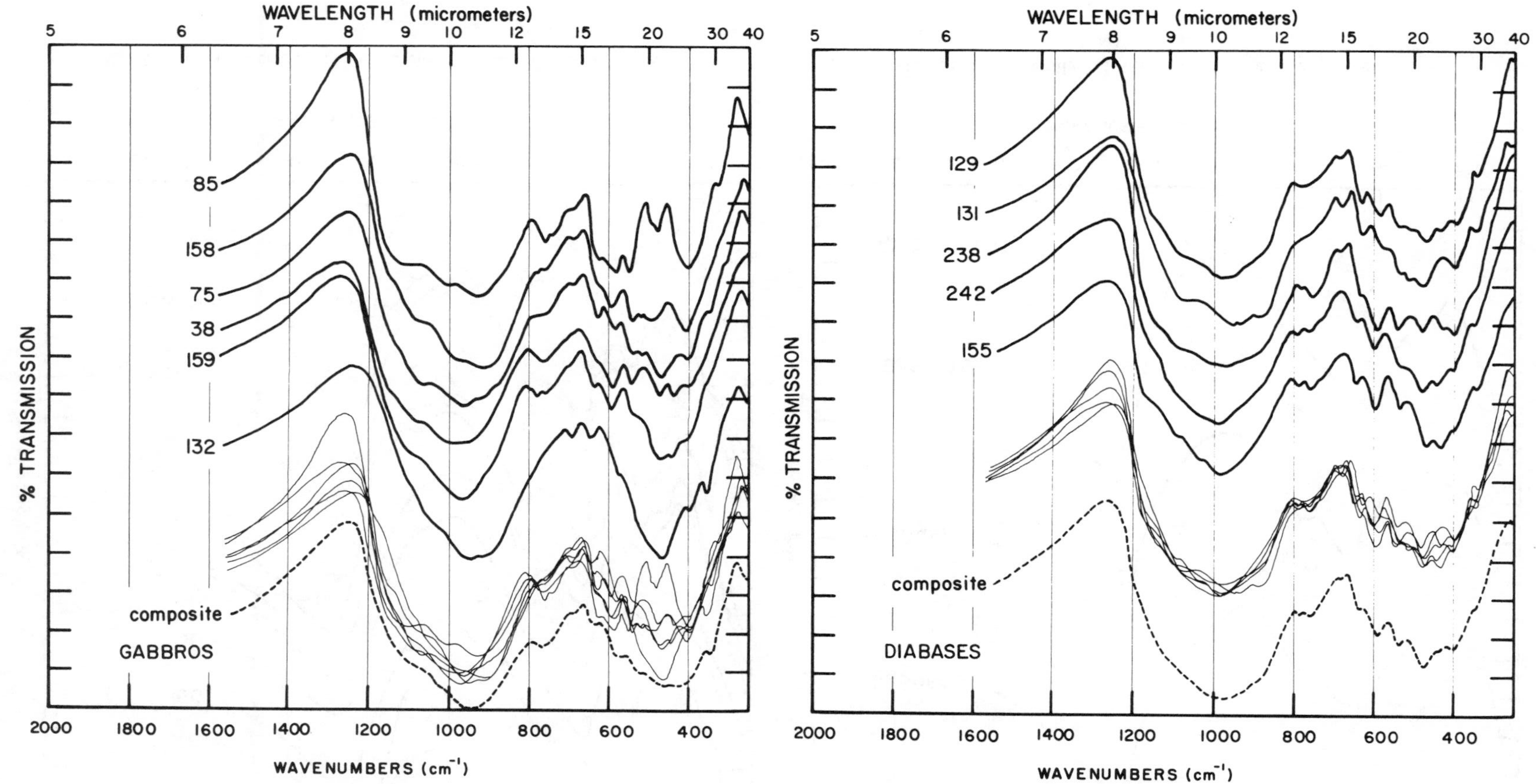

FIGURE 73. Collected spectra of gabbros (same samples as in Figure 28. 132, hornblende, N. Y.[12]

FIGURE 74. Collected spectra of diabases (same samples as in Figure 29.[12]

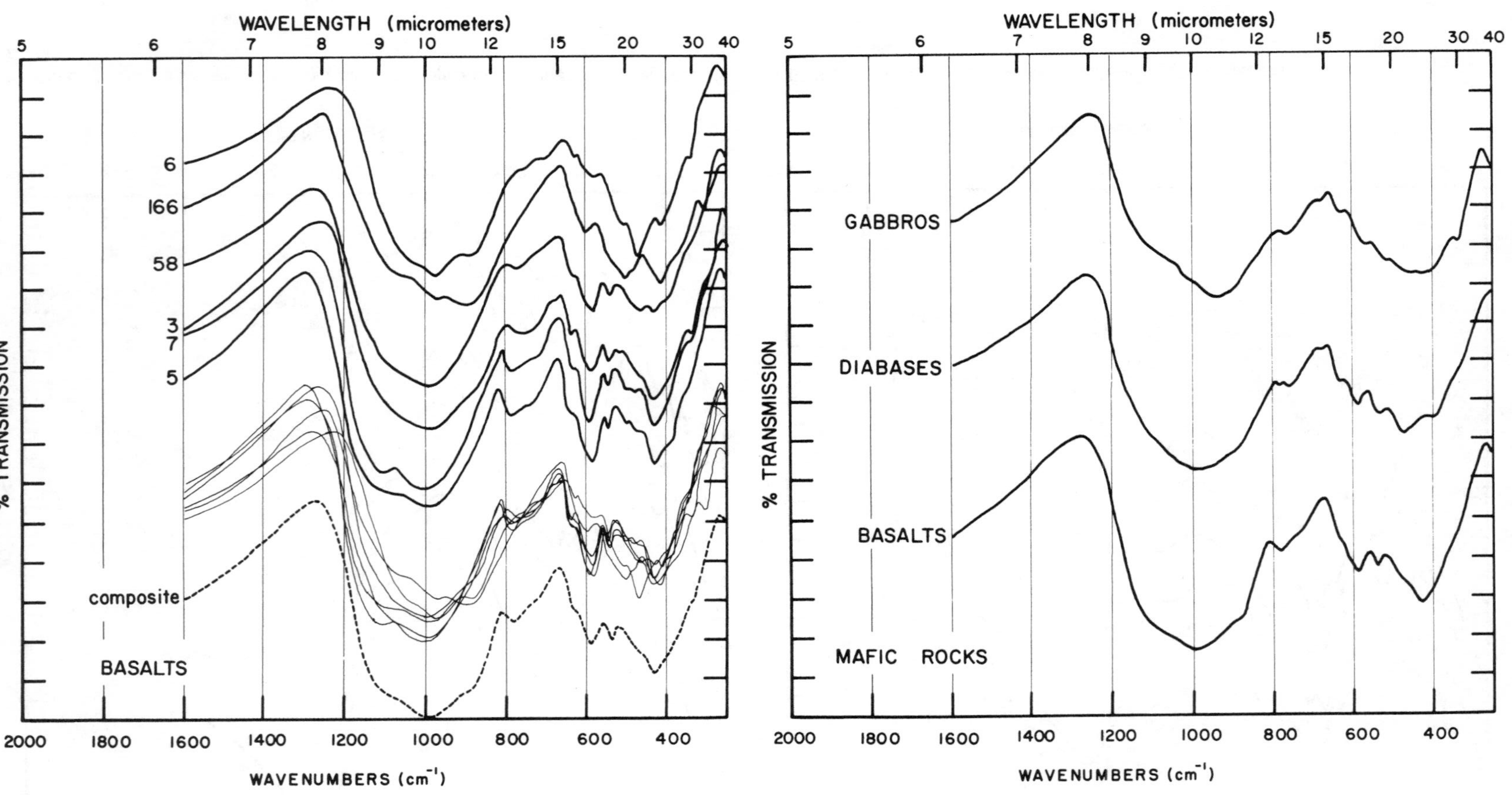

FIGURE 75. Collected spectra of basalts (same samples as in Figure 28.[12]

FIGURE 76. Collected composite spectra of mafic rocks.[12]

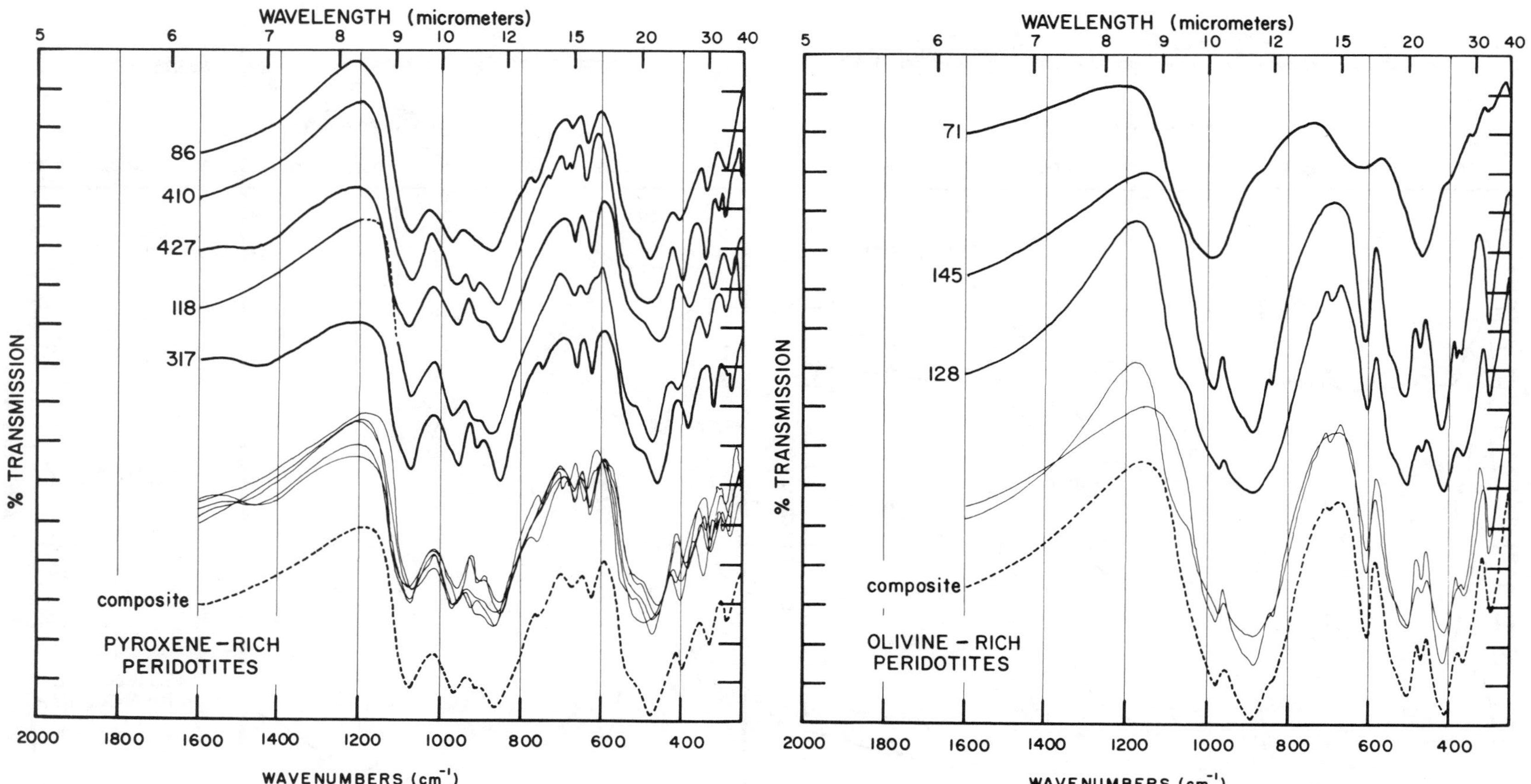

FIGURE 77. Collected spectra of pyroxene-rich peridotites (same samples as Figure 31). 118, peridotite, Helena, Mont.; 317, peridotite, Finland.[12]

FIGURE 78. Collected spectra of livine-rich peridotites (same samples as in Figure 31). 71, peridotite, Arkansas.[12]

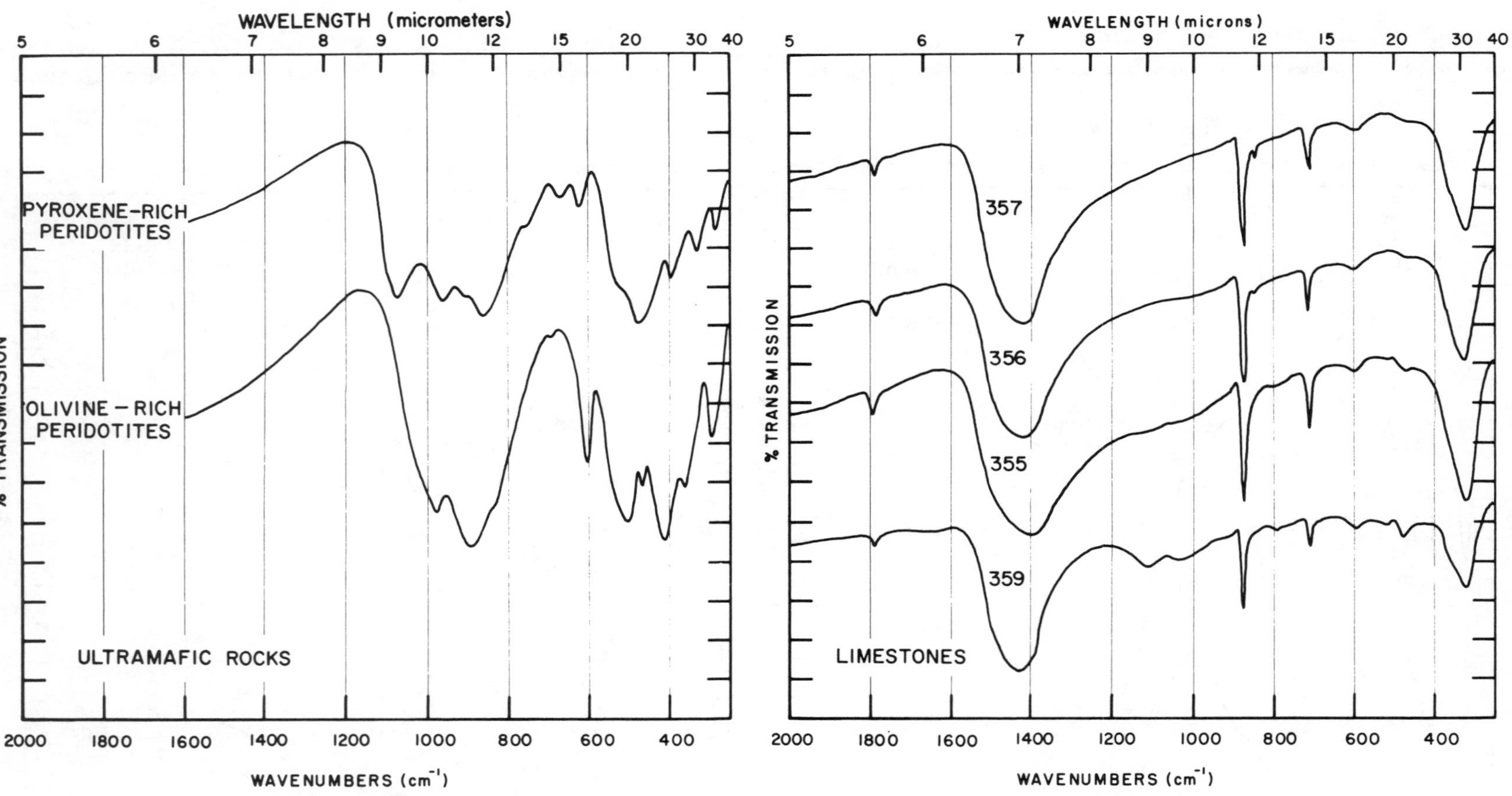

FIGURE 79. Collected composite spectra of ultramafic rocks.[12]

FIGURE 80. Collected spectra of limestones (same samples as in Figure 33.)[12]

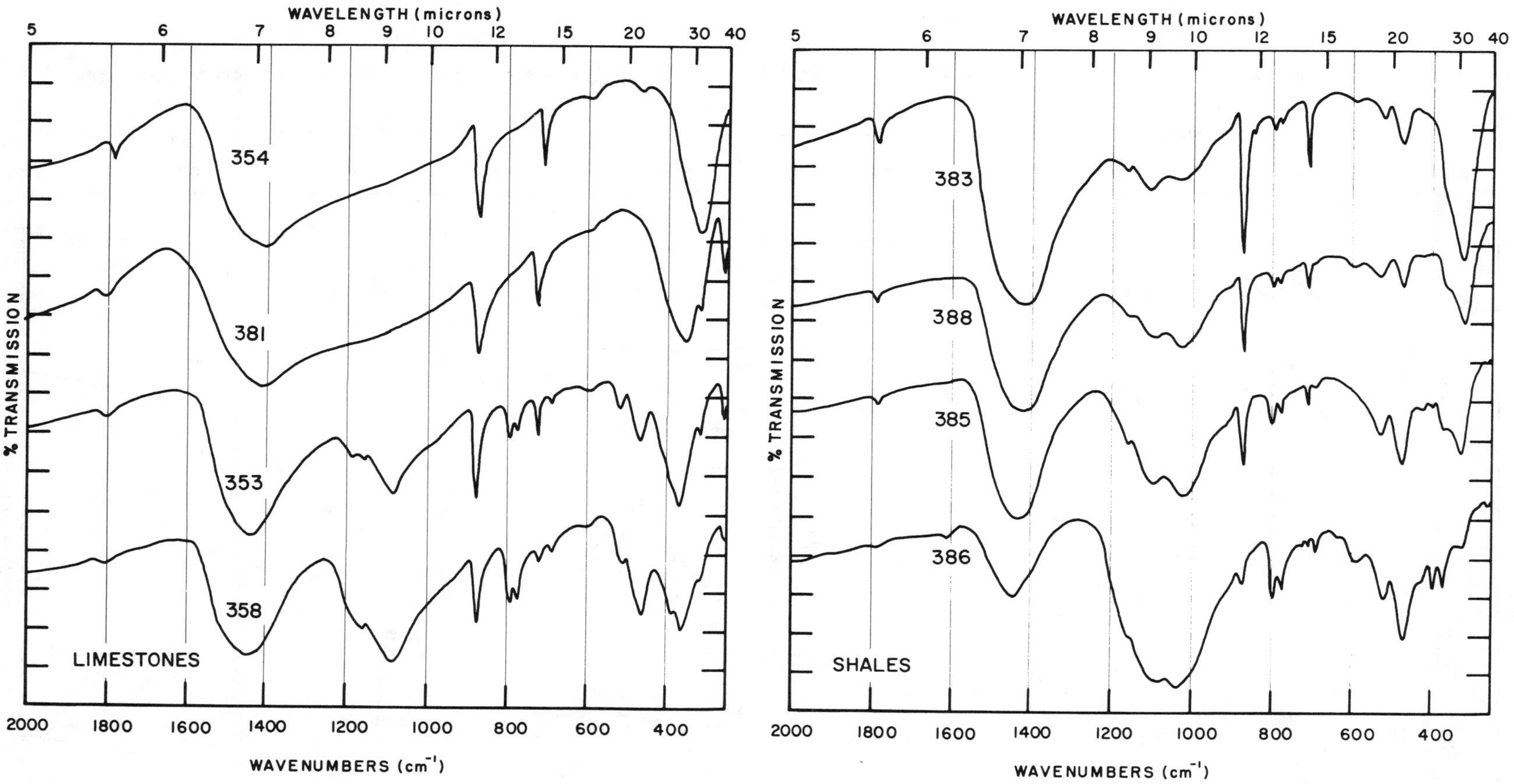

FIGURE 81. Collected spectra of limestones (same samples as in Figure 33).[12]

FIGURE 82. Collected spectra of shales (same samples as in Figure 35).[12]

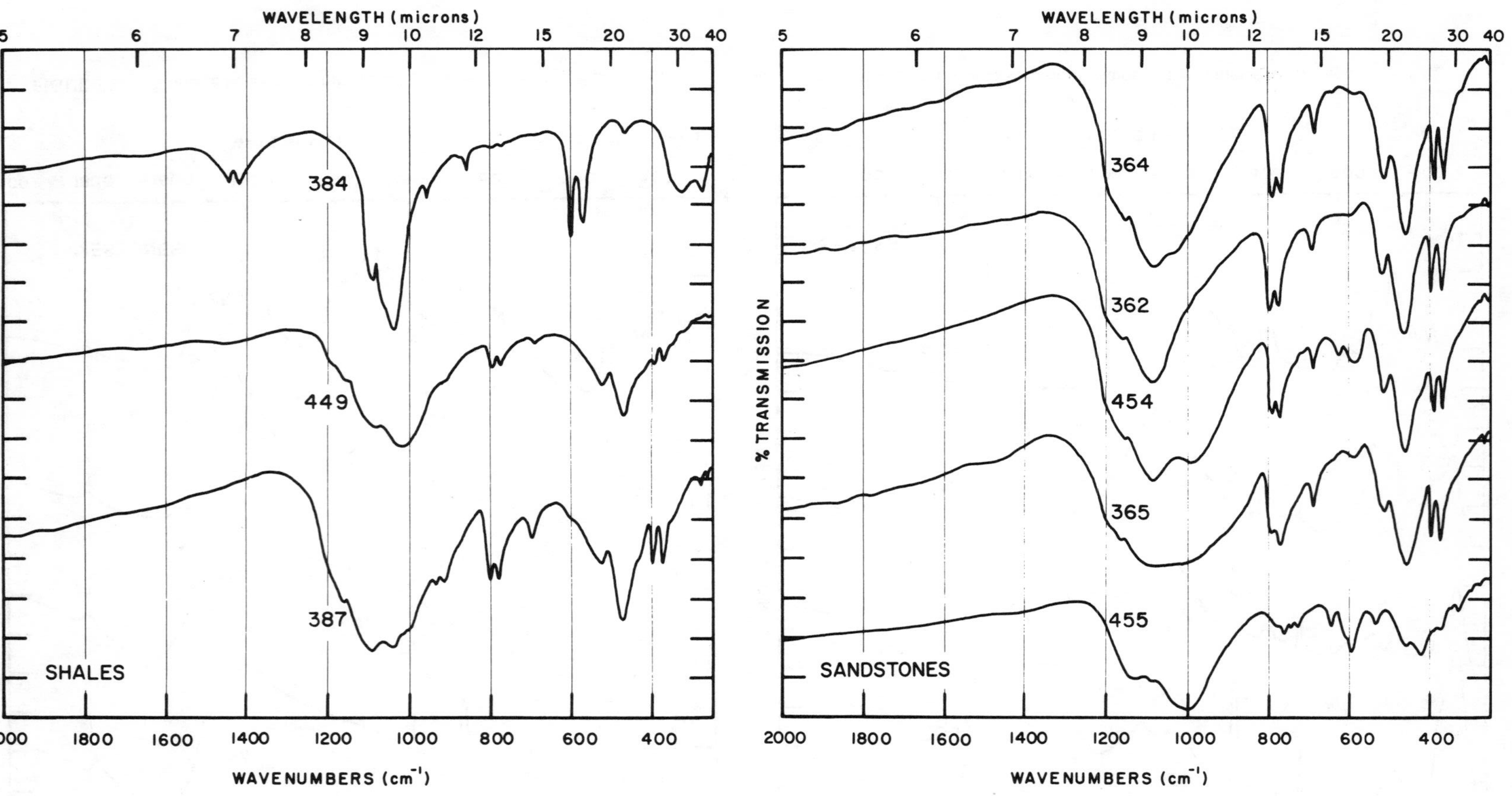

FIGURE 83. Collected spectra of shales (same samples as in Figure 36).[12]

FIGURE 84. Collected spectra of sandstones (same samples as in Figure 34). 352, dark gray siltstone, Pa.[12]

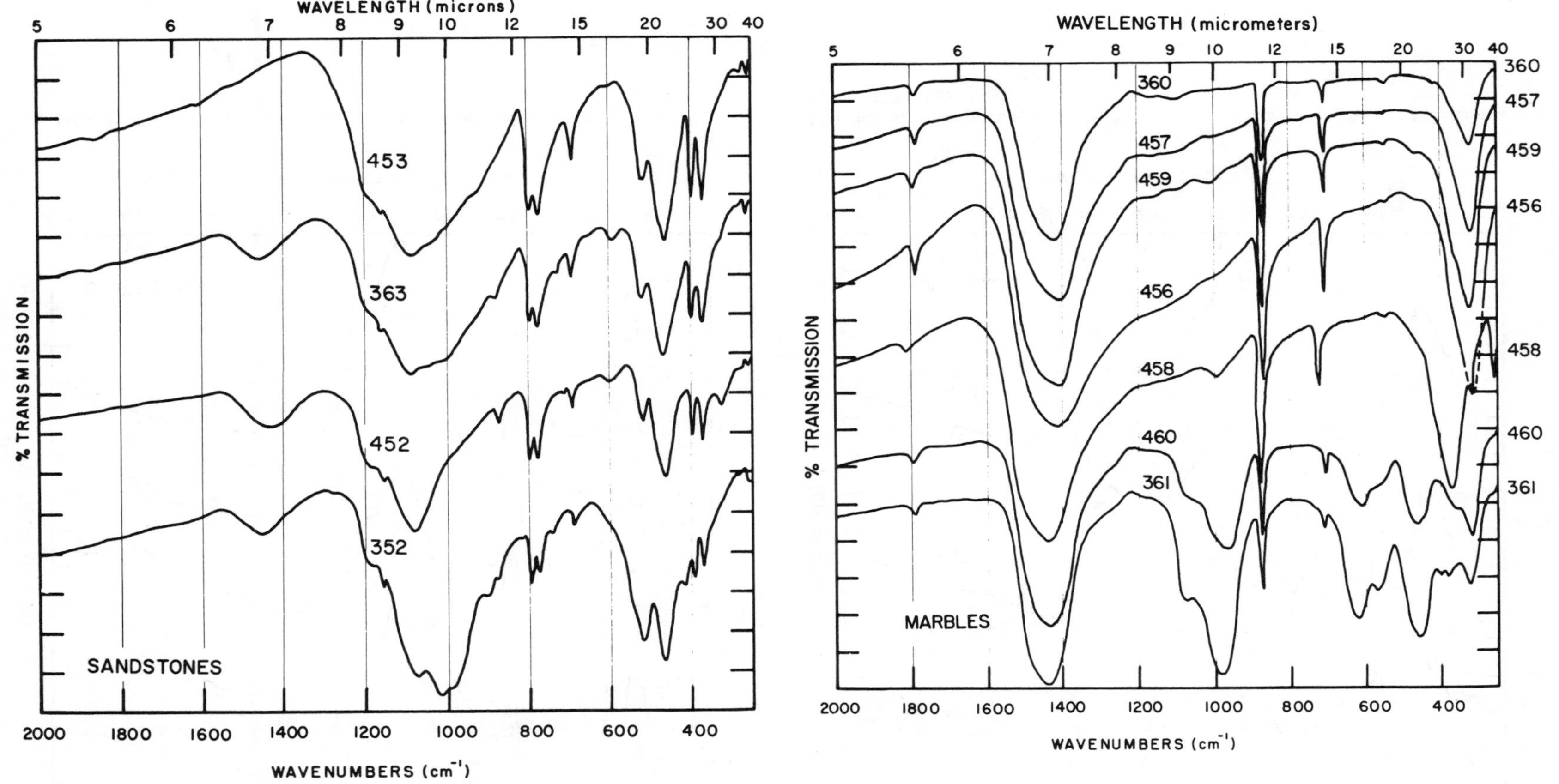

FIGURE 85. Collected spectra of sandstones (same samples as in Figure 34). 352, dark gray siltstone, Pa.[12]

FIGURE 86. Collected spectra of marbles (same samples as in Figure 36).[13]

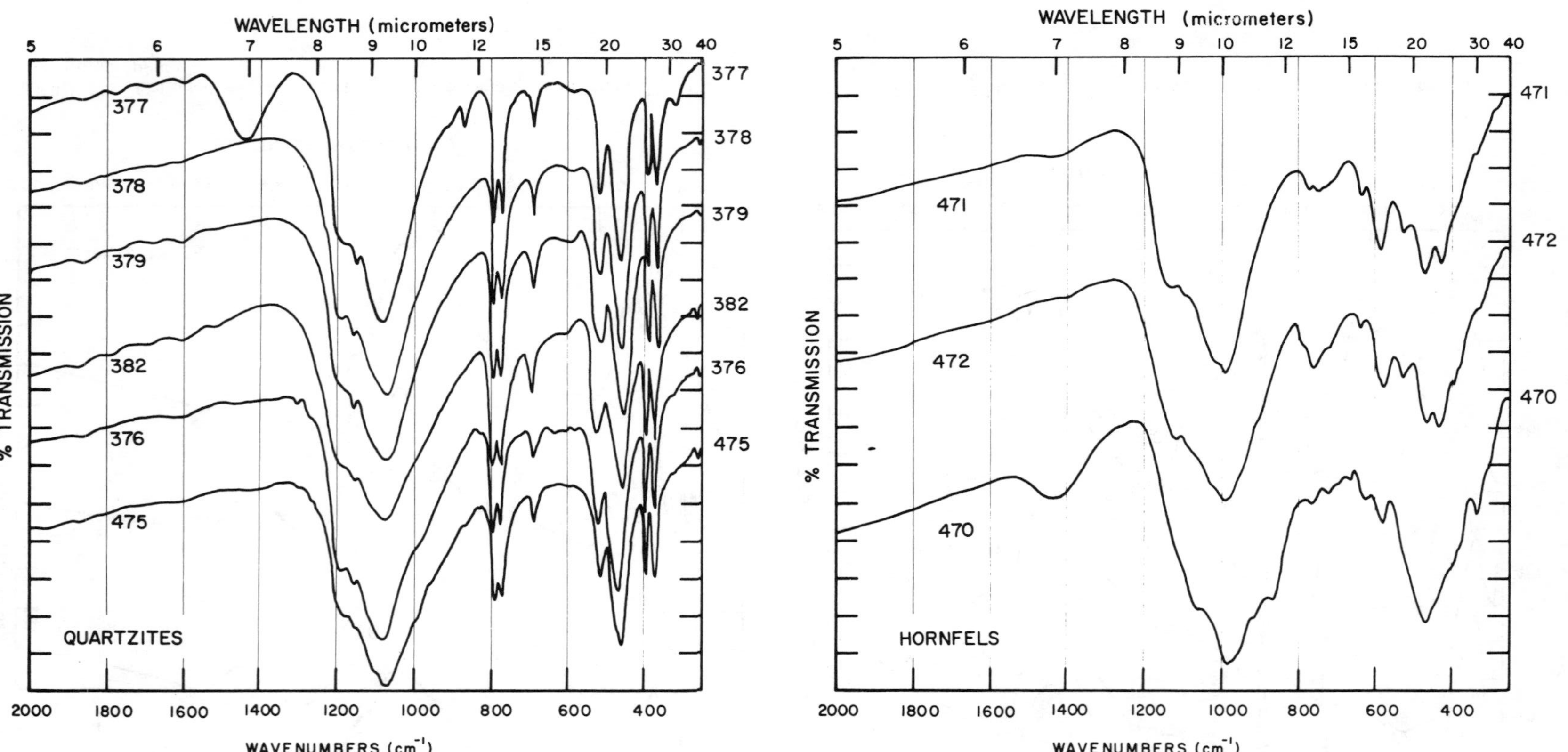

FIGURE 87. Collected spectra of quartzites (same samples as in Figure 37).[13]

FIGURE 88. Collected spectra of hornfels. 471, calcium silicate, Mont.; 472, dark brown, Mont.; 470, Mont.[13]

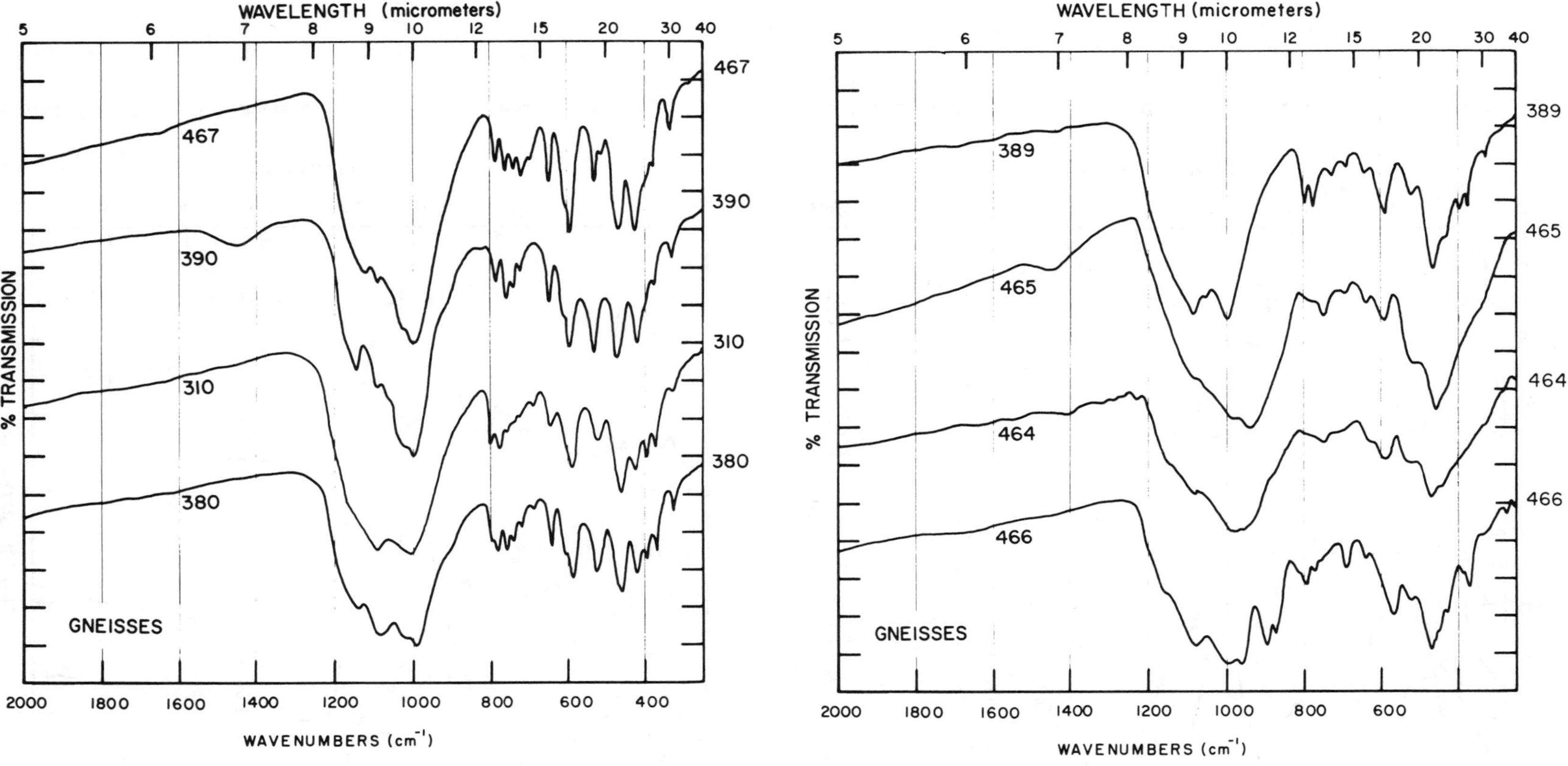

FIGURE 89. Collected spectra of gneisses (same samples as in Figure 38). 380, chloritic, S. Africa.[13]

FIGURE 90. Collected spectra gneisses (same samples as in Figure 38). 380, chloritic, S. Africa.[13]

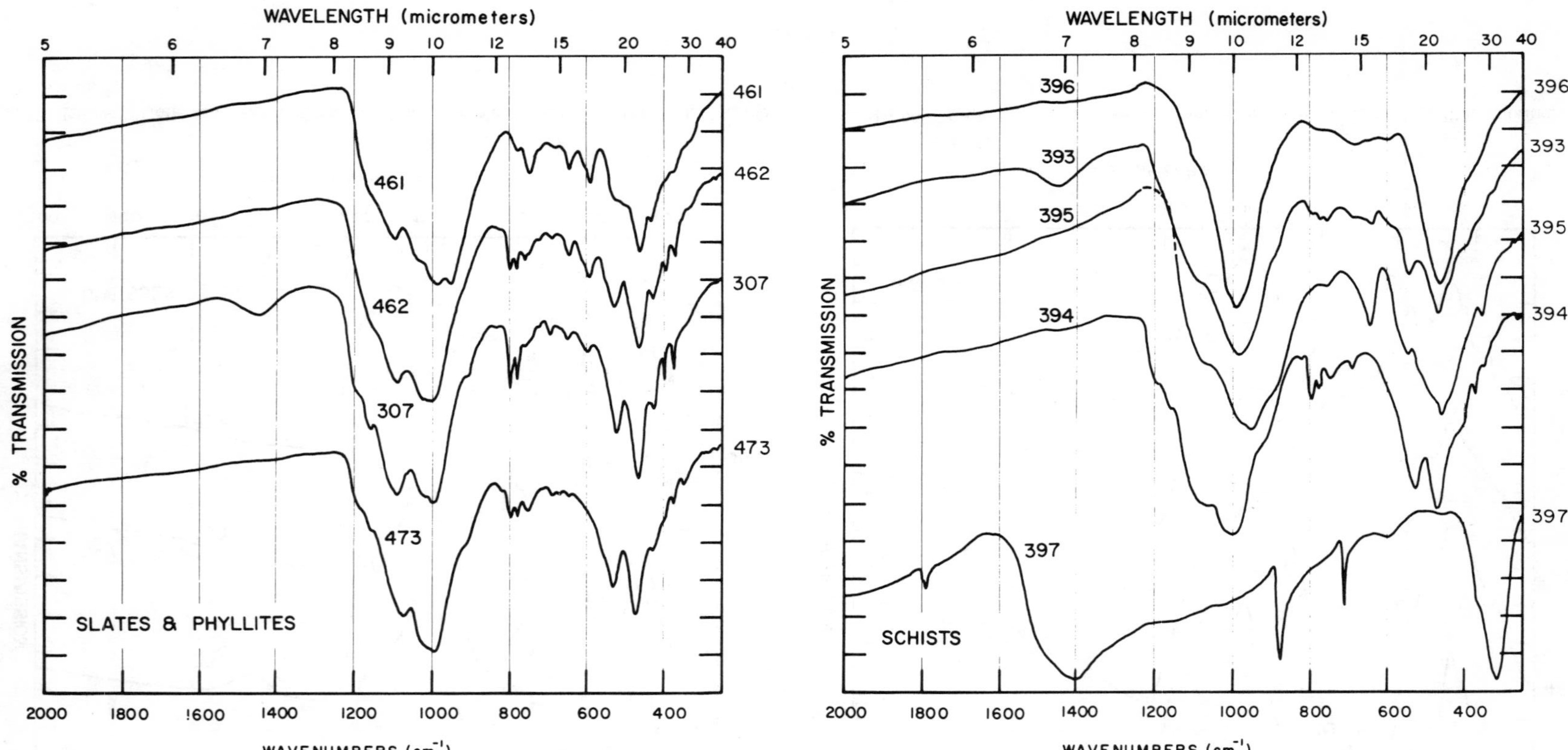

FIGURE 91. Collected spectra of slates and phyllites (same samples as in Figure 39).[13] re

FIGURE 92. Collected spectra of schistose rocks (same samples as in Figure 40). 241, hornblende, S. Dak.[13]

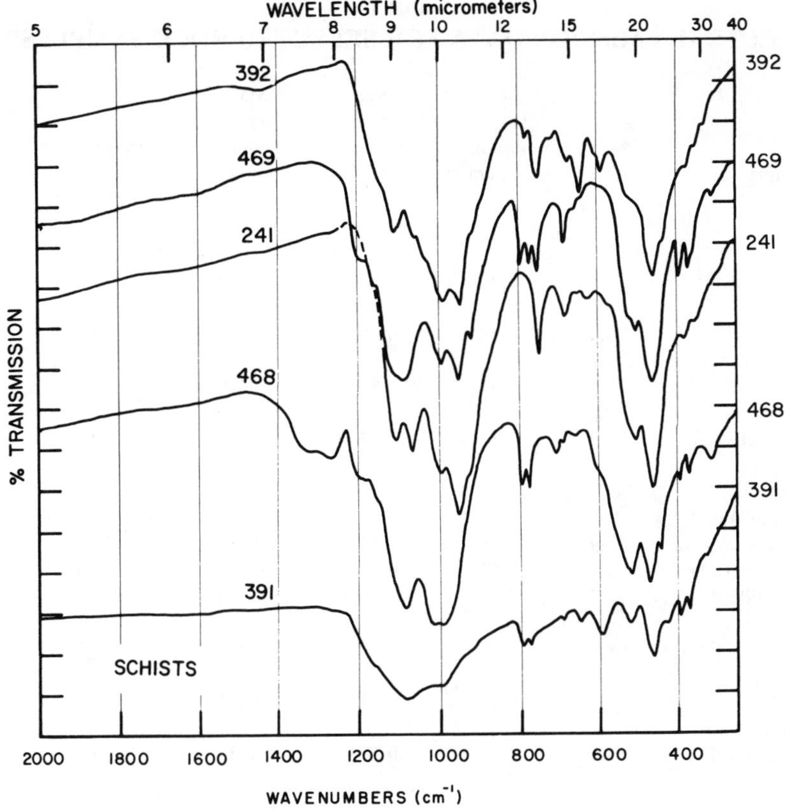

FIGURE 93. Collected spectra of schistose rocks (same samples as in Figure 40). 241, hornblende, S. Dak.[13]

Table 6
SPECTRAL EMISSION DATA FOR SELECTED ROCK SAMPLES[16]

Sample	% SiO_2	Emission max., μm	Avg. emissivity, (6—12 μm)	Max. depart. from unit emissivity	Transmission max., μm	Albedo (0.3—2.5 μm)
			Felsic			
Rhyolite, 97, Chaffee, Colo.	78.6	7.40	0.780	0.379	7.65	0.69
Graphic granite, 73, Grand Canyon, Ariz.	76.3	7.34	0.756	0.363	7.65	1.13
Granite, 150, Rockport, Mass.	75.9	7.42	0.805	0.300	7.67	0.68
Perlite, 72, Chaffee Co., Colo.,	74.9	7.43	0.770	0.345	7.68	0.79
Aplite Granite, 65, Boulder Co., Colo.	74.7	7.49	0.773	0.355	7.70	0.84
Granite, 245, Wisconsin	72.9	7.34	0.745	0.388	7.68	0.91
Rhyolite, 243, Nevada	72.9	7.52	0.735	0.391	7.54	0.95
Pumice, 50, Millard Co., Utah	71.7	7.48	0.754	0.368	7.62	0.73
Granite, 70, Pikes Peak, Colo.	68.9	7.49	0.778	0.342	7.67	0.76
Biotite Granite, 76, Rhode Island	67.5	7.45	0.789	0.332	7.70	0.81
Granite, 167, Lowell Lakes, Ont.	64.0	7.47	0.779	0.333	7.61	0.94
			Intermediate			
Quartz monzonite, 148, Westerly, R. I.	72.9	7.55	0.783	0.312	7.75	0.75
Trachyte, 99, Custer, Colo.	66.9	7.61	0.848	0.249	7.80	0.78
Syenite, 178, Wassau, Wisc.	66.4	7.65	0.829	0.264	7.75	0.61
Andesite, 130, Mt. Shasta, Calif.	66.3	7.50	0.820	0.286	7.65	0.56
Trachyte, 109, Germany	66.2	7.70	0.805	0.300	7.77	0.75
Andesite, 236, Colorado	65.9	7.62	0.895	0.174	7.71	0.42
Quartz diorite, 171, Australia	65.5	7.62	0.847	0.235	7.87	0.61
Quartz monzonite, 173	65.4	7.52	0.829	0.269	7.74	0.60
Monzonite, 154, San Juan Co., Colo.	63.2	7.86	0.875	0.202	7.76	0.59
Latite, 174, Butte, Mont.	62.9	7.66	0.863	0.218	7.72	
Quartz monzonite, 234	62.8	7.70	0.853	0.223	7.74	0.55
Quartz monzonite, 149, Chaffee Co., Colo.	61.8	7.52	0.849	0.240	7.75	0.64
Granodiorite, 64, St. Cloud, Minn.	61.8	7.68	0.875	0.198	7.75	0.61
Syenite, 170, Larvik, Norway	60.0	7.74	0.845	0.248	7.81	0.45
Andesite breccia, 88, Ouray, Colo.	60.0	7.69	0.882	0.186	7.75	0.64
Phonolite, 153, Cripple Creek, Colo.	59.8	7.72	0.870	0.210	7.87	

Table 6 (continued)
SPECTRAL EMISSION DATA FOR SELECTED ROCK SAMPLES[16]

Sample	% SiO_2	Emission max., μm	Avg. Emissivity (6—12 μm)	Max. depart. from unit emissivity	Transmission max., μm	Albedo (0.3—2.5 μm)
Nepheline syenite, 83, McClure Mt., Colo.	59.1	7.74	0.863	0.226	7.86	
Syenite, 172, Vermontville, N. Y.	58.5	7.72	0.850	0.235	7.81	0.50
Nephelene syenite, 156, Red Hill, N. H.	57.6	7.79	0.796	0.308	7.85	0.65
Diorite, 240, Texas	56.6	7.73	0.835	0.247	7.89	0.53
Syenite, 192, Litchfield, Me.	56.0	7.70	0.798	0.306	7.72	0.70
Hornblende diorite, 69, Salem, Mass.	54.2	7.86	0.920	0.128	7.93	0.41
Trachyte, 42, Cripple Creek, Colo.	55.3	7.72	0.863	0.220	7.81	
Quartz monzonite, 233, Arkansas	53.8	7.75	0.860	0.221	7.91	0.47
Syenite, 39, Victor, Colo.	52.2	7.83	0.865	0.227	7.88	0.50
Trachyte Pordhyry, 123, Ontario	50.8	7.65	0.843	0.252	7.75	0.92
Diorite, 152, Freemont Co., Colo.	48.7	7.91	0.860	0.218	8.05	0.81
Andesite, 239, Texas	54.1	7.76	0.881	0.193	7.87	0.53
			Mafic			
Basalt, 5, Chaffee, Colo.	57.9	7.80	0.904	0.151	7.84	0.41
Gabbro, 160, Salem Neck, Mass.	57.7	7.75	0.913	0.151	7.86	0.50
Diabase, 131, St. Peters	52.9	7.85	0.892	0.167	8.01	0.54
Gabbro, 159, Butte, Mont.	51.3	7.93	0.917	0.135	7.94	0.46
Gabbro, 161, Butte, Mont.	51.2	7.90	0.917	0.258	8.06	0.80
Gabbro, 38, Duluth, Minn.	50.9	7.97	0.936	0.104	7.93	
Basalt, 4, Somerset Co., N. J.	49.3	8.10	0.950	0.80	8.06	0.24
Hypersthene gabbro, 75, Ontario	48.2	7.98	0.909	0.143	8.01	0.34
Anorthositic gabbro, 85, Custer Co., Colo.	48.1	7.85	0.916	0.244	8.03	0.63
Diabase, 155, Mt. Tom, Mass.	48.1	7.95	0.918	0.141	7.92	0.42
Diabase, 238, Finland	46.9	7.85	0.889	0.178	7.95	0.43
Basalt, 3, Boulder Co., Colo.	44.3	7.94	0.912	0.148	8.00	0.44
Gabbro, 158, Wichita Mt., Okla.	44.2	7.98	0.894	0.170	8.06	0.38
Basalt, 7, Chaffee, Colo.	57.8	7.79	0.931	0.106	7.79	0.29
Anorthosite, 502, San Gabriel Mts., Calif.	53.9	7.78	0.923	0.340	7.92	0.98

Table 6 (continued)
SPECTRAL EMISSION DATA FOR SELECTED ROCK SAMPLES[16]

Sample	% SiO_2	Emission max., μm	Avg. Emissivity (6—12 μm)	Max. depart. from unit emissivity	Transmission max., μm	Albedo (0.3—2.5 μm)
Anorthosite, 501, Montana	47.2	7.80	0.920	0.302	8.02	0.88
Anorthosite, 504, Split Rock, Minn.	48.4	7.82	0.920	0.292	8.03	0.89
Anorthosite, 505, Essex Co., N. Y.	50.0	7.88	0.911	0.260	7.95	0.45
Anorthosite, 503, Lk. St. John, Quebec	50.1	7.88	0.914	0.217	8.00	0.57
Nepheline basalt	35.6	8.16	0.936	0.124		
Basalt, 6, Germany	42.7	8.22	0.938	0.100	8.20	0.32
Diabase, 129, Jersey City, N. J.	48.1	7.92	0.914	0.136	7.93	0.33
			Ultramafic			
Peridotite, 86, McClure Mt., Colo.	46.3	8.38	0.903	0.147	8.40	0.46
Peridotite, 128, Nye, Mont.	45.8	8.60	0.924	0.154	8.63	0.51
Peridotite, 71, Arkansas	38.6	8.43	0.927	0.138	8.40	0.53
Peridotite, 145, New Zealand	39.1	8.88	0.878	0.201	8.84	0.70
Pyroxenite, 118		8.52	0.864	0.202	8.62	0.77

Note: All data were recorded under vacuum conditions with a 77°K background shield temperature and for samples composed of particles < 74 μm in diameter.

REFERENCES

1. **Adams, J. B.,** Interpretation of visible and near-infrared diffuse reflectance spectra of pyroxenes and other rock-forming minerals, in *Infrared and Raman Spectroscopy of Lunar and Terrestrial Minerals,* Karr, C., Ed., Academic Press, New York, 1975, 91.
2. **Bell, P. M., Rao, H. K., and Rossman, G. R.,** Absorption spectroscopy of ionic and molecular units in crystals and glasses, in *Infrared and Raman Spectroscopy of Lunar and Terrestrial Minerals,* Karr, C., Ed., Academic Press, New York, 1975, 1.
3. **Burns, R. G. and Strens, R.,** Infrared study of the hydroxyl and in clinoamphiboles, *Science,* 153, 89, 1966.
4. **Burns, R. G.,** *Mineralogical Applications of Crystal Field Theory,* Cambridge University Press, Cambridge, 1970.
5. **Burns, R. G. and Vaughan, D. J.,** Polarized electronic spectra, in *Infrared and Raman Spectroscopy of Lunar and Terrestrial Minerals,* Karr, C., Ed., Academic Press, New York, 1975, 39.
6. **Griffith, W. P.,** Raman spectroscopy of terrestrial minerals, in *Infrared and Raman Spectroscopy of Lunar and Terrestrial Minerals,* Karr, C., Ed., Academic Press, New York, 1975, 299.
7. **Hunt, G. R.,** Infrared spectral behavior of fine particulate solids, *J. Phys. Chem.,* 80, 1195, 1976.
8. **Hunt, G. R.,** Spectral signatures of particulate minerals in the visible and near infrared, *Geophysics,* 42, 501, 1977.
9. **Hunt, G. R.,** Electromagnetic radiation: the communication link in remote sensing, in *Remote Sensing in Geology,* Gillespie and Siegal, Eds., 1978.
10. **Hunt, G. R., Salisbury, J. W., and Lenhoff, C. J.,** Visible and near infrared spectra of minerals and rocks, *Mod. Geol.,* 4, 217, 1973; 4, 237, 1973; 5, 15, 1974.
11. **Hunt, G. R. and Salisbury, J. W.,** Lunar surface features: mid-infrared spectral observations, *Science,* 146, 641, 1964.
12. **Hunt, G. R. and Salisbury, J. W.,** Mid-Infrared Spectral Behavior of Igneous Rocks, Env. Res. Paper No. 496, AFCRL-TR-74-0625, 1974; Mid-Infrared Spectral Behavior of Sedimentary Rocks, Env. Res. Paper No. 520, AFCRL-TR-75-0356, 1975; Mid-Infrared Spectral Behavior of Metamorphic Rocks, Env. Res. Paper No. 543, AFCRL-TR-76-0003, 1976.
13. **Hunt, G. R. and Salisbury, J. W.,** Visible and near infrared spectra of minerals and rocks, *Mod. Geol.,* 5, 211, 1976; 5, 219, 1976.
14. **Hunt, G. R. and Ashley, R. S.,** Altered rock spectra in the visible and infrared, in *Economic Geology,* 74, 1613, 1979.
15. **Liese, H. C.,** Selected terrestrial minerals and their infrared absorption: spectral data 4000-300 cm^{-1}, in *Infrared and Raman Spectroscopy of Lunar and Terrestrial Minerals,* Karr, C., Ed., Academic Press, New York, 1975, 197.
16. **Logan, L. M., Hunt, G. R., Salisbury, J. W., and Balsamo, S. R.,** Compositional implication of Christensen frequency maximums for infrared remote sensing applications, *J. Geophys. Res.,* 78, 4983, 1973.
17. **Logan, L. M., Hunt, G. R., and Salisbury, J. W.,** The use of mid-infrared spectroscopy in remote sensing of space targets, in *Infrared and Raman Spectroscopy of Lunar and Terrestrial Minerals,* Karr, C., Ed., Academic Press, New York, 1975, 117.
18. **Vincent, R. K. and Hunt, G. R.,** Infrared reflectance from mat surface, *Appl. Optics,* 7, 53, 1968.
19. **Hush, N. S.,** *Progr. Inorg. Chem.,* 8, 357, 1967.
20. **Faye, G. H.,** *Can. Mineral.,* 9, 403, 1968.

REFERENCES

Section X
Engineering Properties of Rock

By
Allen W. Hatheway and George A. Kiersch

INTRODUCTION

Rock is utilized as a construction medium in five broad categories: (1) underground openings or structures, (2) open cuts and excavations, (3) load-bearing surfaces, (4) as a dimensioned or structural component, and (5) as crushed mineral aggregate or stone. Due to man's originally limited means of excavating and cutting, rock has been historically used primarily as a structural component. The ancients could attack rock with only their own manpower, furthered later by gunpowder and harnessed steam. Modern man is now served by a wide variety of rock excavating and cutting machines employing petroleum products, air, electricity, jet flame, and laser beams. These kinds of energy, along with improved explosives, have vastly improved and expanded the use of rock as a construction medium. The use of both underground space and the removal of rock as an obstruction is now largely a matter of economics only.

Many factors affect the engineering design process involving rock and rock masses. However, the ultimate design is usually a reflection of the engineering properties of the structural material as defined by the reaction of the rock mass to its environment (Figure 1). The two-stage process is to first define the force field acting and then to determine the reaction of the rock material to this force. The second stage involves a knowledge of the engineering properties of the rock and the more comprehensive the knowledge, the more exact will be the design.[22]

Unfortunately, in the field of rock engineering a general lack of knowledge of engineering properties sometimes hinders correct design practice. This is due to several factors, enhanced by the high acceptance of risk in many rock engineering applications. Under such circumstances, recorded case histories and personal experience can be overriding guidelines to the exclusion of other approaches. A further difficulty lies in the variability of rock properties, often within apparent similar rock units, and in the problem of defining satisfactorily, for engineering purposes, the exact nature of adequate rock as distinct from a degree of weathered rock and/or soil.

Rock is a general geological term including all naturally occurring mineral aggregates. From the broad point of view of engineering terminology, a *rock* may be defined as a *competent* natural occurring material as distinct from a *soil,* which may be defined as an *incompetent* natural material.[22] In this context, *competence* may be taken to refer to the relative cohesion of the water-saturated rock under zero-confining stress conditions. Of course, a rock loaded beyond its failure point will cease to be competent and likewise a soil under confinement will gain in competence. Such a criterion for rocks would eliminate all loose soils and most clays. As an example, the weakest unweathered or unaltered rock therefore becomes some form of stratified shale or claystone.

A classification of the engineering properties and their relative importance is the purpose of this chapter. The requirements of any classification can best be understood by considering the approach of engineers to designing works in rock. The practical and experienced engineer invariably incorporates experience and case history data into a design while the theoretical engineer bases his design on an assumption of rock as a brittle elastic solid and adjusts the design for the anelastic feature in the rock mass.[22] Although reasonable, the latter approach assumes the more competent rocks to be essentially elastic materials. However, because time-dependent effects are invariably large, over a given time period, many structures are subject to creep or flow at subfailure stresses. Design of some structures in rock on the basis of brittle failure may therefore be incorrect.

Consequently, a principal factor in the design of an engineering structure in rock is whether the natural material will creep or flow significantly under given loading conditions. A second important consideration is the geologic environment and inherent

FIGURE 1. Engineering properties of rock and the discontinuities which separate the rock mass into discrete blocks play an important part in accommodating underground openings and design components. Shown here is the jointed Upper Cambrian to Lower Paleozoic Cambridge Argillite at the pilot tunnel for the Porter Square subway station, Cambridge, Massachusetts. (Photography courtesy of John T. Humphrey, Haley & Aldrich, Inc., 1979). For purposes of scaling, the pipe in the foreground is 7.6 cm in diameter.

conditions, such as structural features, external or internal stress, and water content. These and other conditions may markedly alter the properties of a rock mass as represented solely by laboratory or *in situ* tests.

Engineers now have the means of making underground rock openings and excavations to almost any dimension required. In their assessments of rock, in terms of design measures and ultimate costs, engineers utilize those physical characteristics of rock commonly known as *engineering properties,* as well as certain aspects of geological characteristics and structural elements.

The body of empirical knowledge relating engineering properties, geologic characteristics, and structural elements to engineering design is growing at a rapid rate. Each of these three categories of rock property data are mutually valuable and reinforcing; they are used in a variety of ways to assist the civil, mining, petroleum, and geological

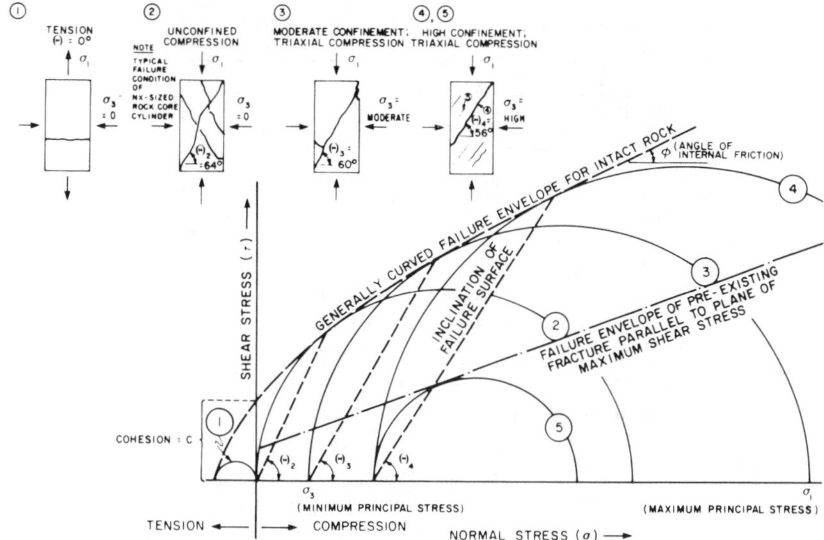

FIGURE 2. The Mohr concept of rock strength relating shear strength components, cohesion (c), and angle of internal friction (φ) to imposed stress. Shown are five typical examples of engineering property tests on NX-sized rock core; (1) tension, (2) unconfined compression, (3) triaxial compression at moderate confinement, (4) triaxial compression at high confinement, and (5) triaxial compression at high confinement with fracturing present.

engineers, as well as the geologist, hydrologist, and geophysicist, in a wide spectrum of applications. Regardless of the ultimate use of rock, engineering properties are considered the *lingua franca* among those who engineer with rock.

Direct application of rock properties is essential in planning the following types of work involving rock:

1. Slope, wall, and face retention, in open excavations
2. Anchorages
3. Size of internal support of underground openings
4. Machine excavation characteristics
5. Response to blasting
6. Countering subsidence effects
7. Waste handling and transport characteristics
8. Natural aggregate and broken stone preparation

Most rock engineering is conducted according to the Mohr-Coulomb stress-strain theory for brittle elastic materials. For computations involving deformation short of failure, many other standard strength-of-materials formulae are used. The principles of utilizing tensional, compressional, and uniaxial test data are shown on Figure 2.

MEASUREMENT OF THE PROPERTIES

Effective use of engineering properties requires careful laboratory and *in situ* measurements that must later be evaluated along with a variety of geological observations describing the rock and conditions of the rock mass (Figure 3). Standards of practice have been developed to specify the methods by which these tests are conducted. Standards are beginning to appear for conduct of field explorations and *in situ* testing. The

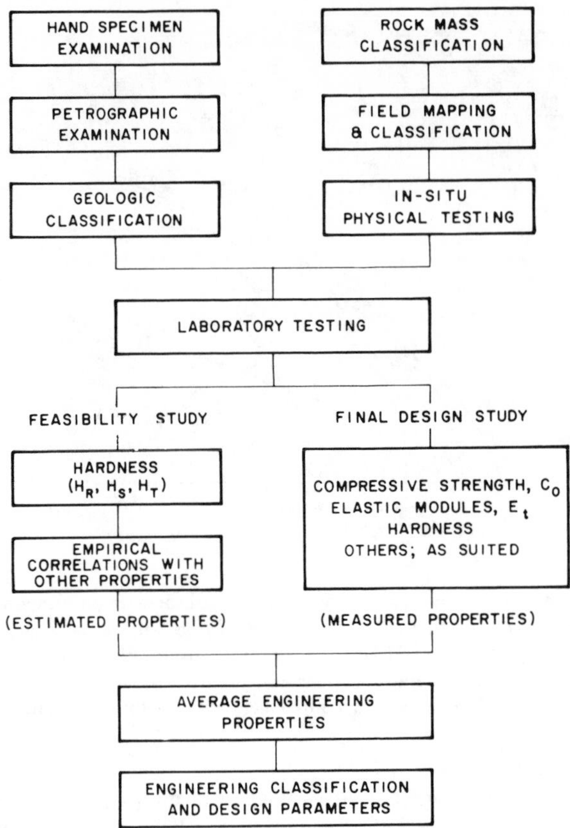

FIGURE 3. A scheme for engineering classification of rock, based on field and laboratory geologic examination and laboratory testing of engineering properties.[7]

organizations mainly responsible for the standardization are the International Society for Rock Mechanics (ISRM, Lisbon, Portugal)[33-37] and the American Society for Testing and Materials (ASTM, Philadelphia).[1-4]

A typical result of standardization is that most rock core is of the NX variety (4.5 cm diameter) or similar HQ-size and is obtained by the double- or triple-core barrel technique, which minimizes drilling damage (Figure 4). Many observations and properties are dependent on the size of core recovered, and in comparison with other sites and rock conditions are more accurate on that basis.

During the drilling process, records of the fugitive data such as penetration rate, gauge pressure on bit, and drilling fluid color and/or loss are compiled to aid in evaluating the core and laboratory data. The latter include the recovery percent and physical description of core conditions, classification, and inherent structural features.

Microscopic assessment of rock for engineering uses is common practice. An accurate petrographic analysis can classify rock lithologically so the designer or contractor can apply empirical knowledge gained from previous experience. More specifically, much can be learned concerning the rock texture and internal bonds, the presence of alteration or weathering, unstable mineralogy, strained minerals, subtle foliation or bedding characteristics, as well as the quantities and distribution of the harder (e.g., quartz) and softer (e.g., the clays) minerals, and those which may tend toward chemical dissolution, such as the carbonates.

FIGURE 4. NX-sized rock core recovered from exploratory drilling at the site of a proposed underground structure. The three core cylinders to the left have been cut and prepared for laboratory strength testing; the fourth specimen preserves the geological contact between an intrusive dike (left side) and the country rock, which is a metasedimentary mudstone.

IMPORTANCE OF GEOLOGIC CHARACTERISTICS

A variety of geologic characteristics of rock considered for engineering use may affect its engineering properties. Most of the effect tends to show up within the statistical variation that will usually be encountered in laboratory test values.

The geologic approach to planning a study of a rock mass and performing the specialized tests and analyses, as well as other forms of exploration, is outlined below. Such an investigation frequently deals with an unstable rock mass that is part of a much larger unstable element, the Earth's crust. The rock mass of an engineering site may possess extremely varied characteristics of strength and retained stresses, within short distances of inches and feet (Figure 5).

Too frequently, investigators of sites give little or no reference to the basic geologic circumstances, for example, their influence on: (1) physical properties of rock mass and (2) reaction of mass in time (days, months, years) to the changed conditions imposed by engineering works. Instead, arbitrary assumptions are commonly made based on past experiences with similar-appearing rock masses.

Geologic Factors in Appraising A Rock Mass

In order to reduce the number of assumptions, some of the interrelated geologic factors that can be supplied for either surface or subsurface sites are

1. The principal geologic conditions and features of rock units, e.g., fracture systems and other inherent structural weaknesses (Figure 6), physical properties and any changes since origin, processes responsible, and groundwater level and permeability of each rock unit

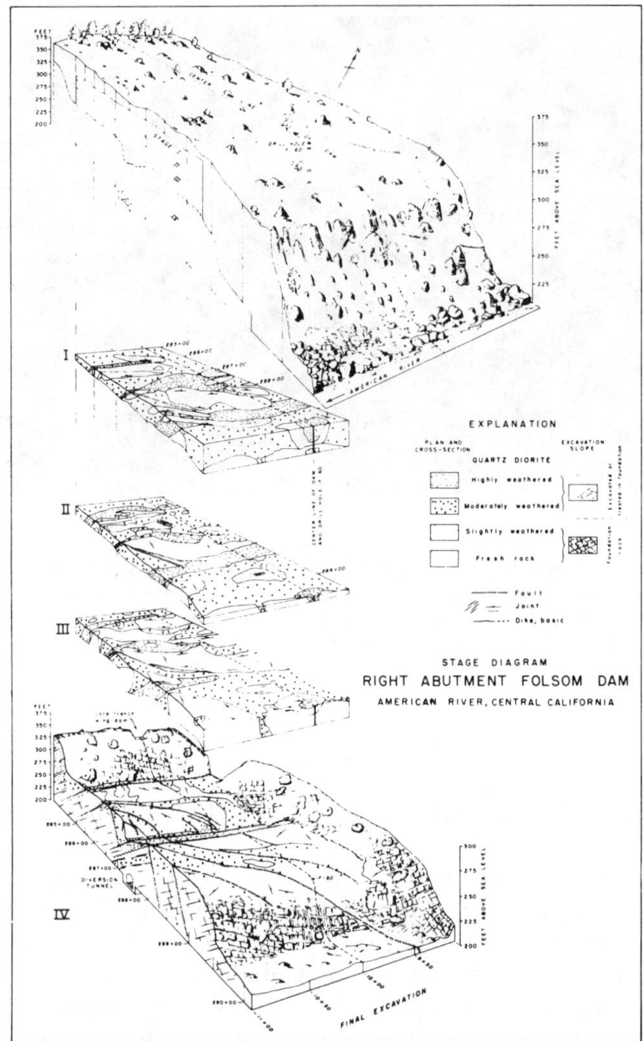

FIGURE 5. Stage block diagram used by geologists to portray variations in expected engineering characteristics of foundation rock due to degree of weathering in country rock. Rock types are differentiated on the basis of visual geologic observations from outcrops and rock core and verified or modified by laboratory test results. (From Kiersch, G. A. and Treasher, R. C., 1955.)

2. The age of rock mass and historical events that affected strength in some manner — adverse or beneficial, e.g., multiple magma injections, metamorphic cycles, differentiation, morphological changes to surface features, rate of unloading by erosion or ice, former stress distribution from character of folds and faults (Figure 7), rebound phenomena with relief joints and whether now active or in equilibrium (stable), fluctuations of groundwater level, any dissolving action or precipitation and cementation, and geochemical changes in rocks (Figure 8), increased uplift pressures from clay and glauconite
3. Effect of tectonic history on mass, e.g., unaffected, long-time stable element (tectonic stresses largely dispersed), one or more periods of deformation, some tectonic stress retained, area under active tectonic stresses with periodic seismic events (shallow to deep in origin), isostatic adjustments, and active gravitational creep from near-surface rebound phenomena

FIGURE 6. Rock discontinuities; left-hand NX core depicts joints which have opened up in the drilling process along relict bedding planes; center core illustrates a tectonically induced joint lying at an oblique angle to the relict bedding and now coated with a thin film of calcite (whitish streaks); right-hand NX core shows a presently healed sedimentary slump structure (dipping to the right side of the core and bounded on the bottom by the dark, truncated band) formed shortly after the parent mud and silt were deposited.

FIGURE 7. NX rock core showing the brecciated condition formed as a result of ancient shearing between two rock masses and commonly described as a fault or shear zone. Many such zones become healed or recemented over long periods of time and may have greater strength than otherwise unaffected country rock.

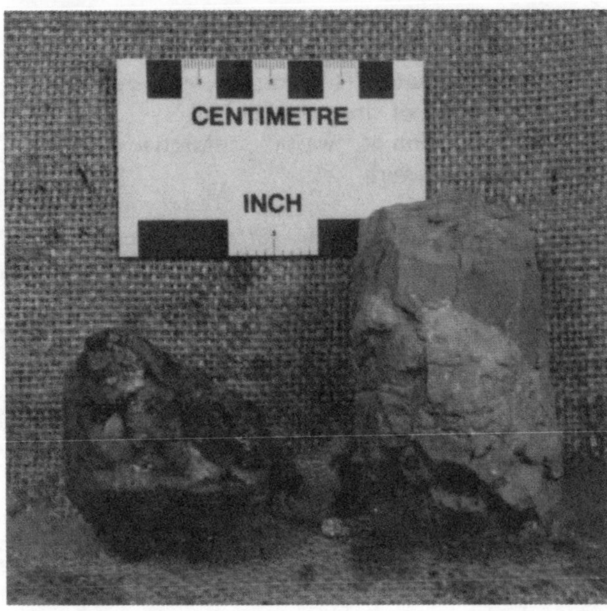

FIGURE 8. NX rock core showing the debilitating effects of chemical alteration on rock strength. Here the alteration has produced a high percentage of clay minerals from the otherwise medium-strength rock.

4. Nonhomogeneity and tendency for partial retention of stress within mass, e.g., potential for retention with a forecast of the magnitude of the three stress components, whether retained stress is most likely to be dispersed throughout mass or concentrated in zones adjacent to certain structural features

Strong active stresses cause discing of core from borings. Changes in stresses over geologic time can be signified by indicator minerals and their derivatives.

1. Effects of time (short interval) — changes imposed by engineering works
   - A. Immediate deformation effects due to loading, e.g., elastic deformation or viscoelastic "creep" initiated
   - B. Progressive deformations due to structures, e.g., elastic deformation, viscoelastic deformation, viscoplastic deformation, percolation of pore waters and chemical alteration
   - C. Stress changes induced on rock mass
2. Effects of time (future) — changes forthcoming due to geologic actions in motion
   - A. Geologic reaction, tectonic development (action); may or may not be due to project-imposed conditions
   - B. Effect of tectonic development, inherent to rock
   - C. New tectonic deformation due to this stress change (unloading, tectonic structures of rebound)
   - D. Hydrological, chemical, or consolidation changes due to natural physical and chemical conditions (volume, water content, etc.)
   - E. Effects due to changes of climate

3. Effects of time (future) — changes forthcoming due to the conditions induced by engineering works
   A. Long-time deformation (settlement, creep, etc.)
   B. Long-time change of stress (relaxation)
   C. Leaching or reaction of "waste" radioactive components with wall rock of a chamber or cavern
   D. Consolidation
   E. Swelling of slopes induced by surface or subsurface clayey constituents (claystone, gypsum, etc.)
   F. Drainage
   G. Relaxation of rock anchors
   H. Chemical influence of the engineered structure on the rock mass, such as infiltration of chemical waters from engineering works into rock and reaction and/or deterioration
   I. Statical changes

Obviously, many of these features cannot be adequately assessed in quantitative terms by the initial geologic analysis. However, recognition of their potential by the geologic approach improves on the plan-of-attack for the next step, the specialized techniques of rock testings. Furthermore, a comprehensive geologic evaluation of the potential conditions and causes is the most factual basis for interpreting the specialized data from rock testing techniques. Normally, test data represent an analysis of a very small part of the rock mass. If the test data are extrapolated without recognition of the geologic inhomogeneity and why it occurs, conclusions can be more misleading than helpful. Consequently, extensive geologic investigations and understanding as a basis for evaluating test data combined with a judgment factor are more reliable for assessing the reaction of a rock mass than extensive engineering test data with minimal geological reasoning and judgment for arriving at an evaluation (Figure 9).

Some of the indicators of discontinuities that are used to evaluate the in situ character of a rock mass are those observed in core extracted from exploratory borings. Standards of core logging, such as those of the American Society of Civil Engineers (1972), British Association of Engineering Geologists (1971), and South African Association of Engineering Geologists (1978) call for identification of discontinuities, description of the nature of their surfaces, and the reporting of such features on the boring logs. From these data, the frequency of joint occurrence can be determined, as well as the computation of the Rock Quality Designation (RQD),[14] now frequently used in geological and geotechnical engineering practice. The RQD represents the percentage of rock core in each drill run that exceeds 4 in. in length without the presence of a discontinuity, discounting any mechanical fractures or breaks as follows:

Rock quality	RQD (%)
Very poor	0—25
Poor	25—50
Fair	50—75
Good	75—90
Excellent	90—100

In addition, recovery is also reported in terms of the length of rock attempted for coring vs. that length of rock actually retrieved. If, for example, 1.52 m (60 in.) is attempted and 1.27 m (50 in.) recovered, the core recovery would equal 84%. Recovery and RQD are analyzed together and sometimes can be utilized to further evaluate effective engineering properties and to devise bulk moduli for rock masses. Conversely,

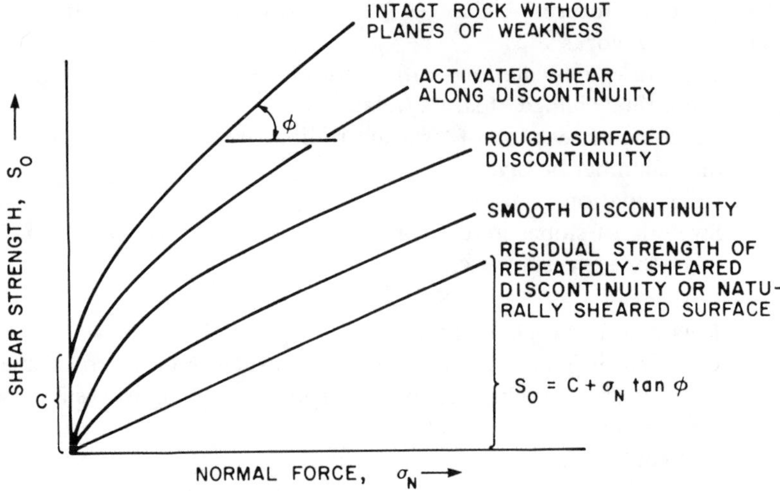

FIGURE 9. Idealized shear strength (So) envelopes for a given rock lithology, showing effects of intact condition and various types of discontinuities. C represents cohesion and ϕ represents angle of internal friction.[12]

the statistical approach to evaluating a rock mass is not always appropriate. The needs of engineering design are for an *adequate rock* (material or mass) only and accordingly many jointed rocks are satisfactory even though they have a low RQD.

DEVELOPMENT OF A ROCK STRENGTH ASSESSMENT

Optimal design takes into consideration that properly managed exploration can continue to provide information that can be used to refine design. The goal of engineers and geologists of a design team is to produce an assessment of the rock mass that defines the natural material in terms of specific geologic subdivisions of a rock mass, and then to relate such units to the element of design. This is a much different evaluation from the traditional time-rock or rock-stratigraphic description of classical geological mapping. Properly completed, this effort should provide a three-dimensional geologic fabric of the construction site. The fabric is the result of careful surface geologic mapping of the site area, to whatever radius is required to understand the structural and stratigraphic relationships observed. Subsurface explorations by direct techniques of borings, trenches, test pits, shafts, and adits all provide means for extending surficial observations to depth as do the indirect techniques such as geophysical measurements and logging, sensing, imagery, and photo interpretation. The key elements of geologic fabrics are

1. Thickness, strike, and dip of bedding
2. Attitudes of discontinuities
3. Attitudes of intrusive bodies
4. Location and attitudes of contacts; physical integrity (healed, brecciated)
5. Presence of faults and shear zones
6. Groundwater/pore pressure
7. Depth of weathering and "top of adequate rock"
8. Alteration zones and characteristics
9. Joint sets, statistics
10. Foliation/cleavage planes

All of these factors modify the effect of measured engineering properties. Engineering properties are best applied after the geologic characteristics have been compiled and analyzed. Once the rock mass has been delimited in terms of similar factors of geologic fabric, measured engineering properties are viewed as a statistical array and reduced in value to the degree that they are affected by the various geologic conditions which surround each engineering geologic unit.

COMPONENTS OF DESIGN IN ROCK

If all rock were essentially free of defects, engineering design would utilize a geometry for each structure that would minimize stress concentrations calculated to exceed the available strength of such rock, on the basis of measured engineering properties. However, rock condition varies widely and the defects that are present are acted upon by gravitational, residual, and structural (facility) stresses and hydrostatic pressure. Primary concern is given to displacements that may occur within rock masses, from minor rock slabbing in small, shallow tunnels to instantaneous and explosive rock bursts in deep, underground openings and large rockfalls from excavated slopes at the ground surface.

When stress exceeds available strength, in compression, tension, or shear, the rock mass begins to fail and displace, however slightly (Figure 2). Some means of support must be applied, or a compensating change must be made in geometry of the structure. Since dimensions and alignments are difficult to alter once construction has begun, the changes are generally applied in the form of strengthening or reinforcing members. Such components of design as rock bolts, steel support rings, poured-in-place and prestressed concrete liners, shotcrete, pore pressure-relief boreholes, and various types of buttressing are used. Although these measures are meant largely to counter defects in the rock mass, they are generally applied on an estimated basis, utilizing engineering properties as a guide to lengths, depths, and spacings.

THE PROPERTIES AND THEIR MEASUREMENT

Engineering properties of rock fall into six general categories, according to the nature and engineering application of the data. Details of each of the properties, the method in which it has been standardized for measurement, its use and limitations of use for engineered construction, and appropriate standard references are given in Table 1. In summary, the six categories of tests are

Identification	Unit weight
	Moisture content
Water-related	Porosity
	Coefficient of permeability
Strength	Compressive (Figure 10, Table 2)
	Induced tensile
	Unconfined shear
	Triaxial shear (Figure 11)
Hardness	Abrasivity
	Hardness
	Shore scleroscope value
	Schmidt rebound number
	Taber hardness
	Total hardness
Velocity	Sonic velocity
	Compressive wave velocity
	Shear wave velocity

Deformational	Poisson's ratio or Poisson's number
	Modulus of rigidity
	Bulk modulus
	Tangent modulus of elasticity
	Secant modulus of elasticity

Tables 3 to 5 are an overall compilation of the various engineering properties from published and other sources for igneous, metamorphic, and sedimentary rocks. The relative number of each type of value represents somewhat the time of general acceptance of that property; those which have been in common engineering usage for many years tend to have appeared more repeatedly in the literature. Although a substantial amount of rock testing is being done for design-level investigations associated with underground openings and large open rock cuts, a relatively small amount of the data is being released into the literature today.

Effect of Pore Water Pressure

Where present and under hydrostatic head, water exerts two profound influences on the strength of a rock mass. First, water reduces the surface tension along microfractures and hence reduces the amount of cohesion present as a part of shear strength. As hydrostatic pressure builds, point or surface contacts are reduced in area and the angle of internal friction (Ø) is reduced through application of pore water (neutral; u) stress. The available or *effective* stress between particles and surfaces (σ') is the pore pressure-reduced remnant of the applied or gravitational stress in the rock body ($\sigma' = \sigma - u$). Since the angle of internal friction (Ø) of rock is the prime component of shear strength (τ), the overall strength of the rock mass is significantly reduced, especially along rock mass boundary fractures (Figure 11).

Pore pressure can be effective in reducing rock mass strength even when permeability is low and the resulting inflow of water at a face is less than a few liters per minute.

Table 1
ENGINEERING PROPERTIES OF ROCKS AND THEIR USES IN ENGINEERED CONSTRUCTION

Engineering property	Symbol	Nature: field (F) laboratory (L) derived (D)	Method of measurement	Units of measure[a]	Use in engineered construction	Limitations	Ref.
Unit weight	γ	L	Volumetric displacement in water; weight per unit volume; usually weighed as oven dried but may be specified on several other bases	kg/m^3	Weight, per unit volume, of entire rock; primary term in many computations; useful in computing *in situ* stress	Expected statistical variances in the more porous rocks	1 35 26
Moisture content	w	L	Oven drying	%	Effect of moisture variations on other physical properties considered in data evaluation and assessment	Varies seasonally and under influences of construction activities: of little use to refer to tabulations by other except as a condition of specific property tests.	35 26
Porosity	n	L	20.106 mm^3 sample placed in porosimeter device.	%	Indication of ability to retain fluids or gases	Often highly depth dependent	45 26 21 35
Coefficient of permeability	k	L/F	Cylinder of rock placed in permeameter (L) or pressured water flow into borehole (F)	cm/sec	Yield of water for dewatering purposes; determination of pore water pressure distribution for support design of slope stability analyses, planning for grouting	Field tests require careful isolation (packing) of borehole segment or interest; avoid high-pressure hydrotracturing; greatly influenced by discontinuities	21
Compressive strength	c_o	L	Uniaxial or Triaxial conditions, in universal test machine; strain gages used for moduli determinations	N/m^2	Index classification test; load bearing capacity; other properties according to Mohr failure concept; slope stability; mine pillar	Avoid unrepresentative anisotropic fabric elements or discontinuities; select representative sample; consider data scatter;	T-15 13 30 T-24 T-25 34

Table 1 (continued)
ENGINEERING PROPERTIES OF ROCKS AND THEIR USES IN ENGINEERED CONSTRUCTION

Engineering property	Symbol	Nature: field(F) laboratory (L) derived (D)	Method of measurement	Units of measure	Use in engineered construction	Limitations	Ref.
					stress; subsidence; excavation, blasting, drilling and mole boring performance	L/D ratio is quite important (standard is 2.1); peak strength is obtained	28
Induced tensile strength	T_o	L	Point-bearing compression of tabular cylindrical specimen (Brazilian test); more classical tensile pull test is not generally utilized	N/m^2	Load-bearing capacity under tension	Introduces a contributory compressive stress, therefore not a true determination of tensile strength; magnifies effect of microdefects; theoretically of greater T_o than direct-test value	T-15 56 20 13 T-23 34 26
Unconfined shear strength	S_o	L/F	Direct shear box, utilizing natural discontinuities or planes or weakness; also triaxial compression test which initiates shear failure on basis of sample geometry or inherent weaknesses	N/m^2; Ø(deg); $c(N/m^2)$	Index classification test; load-bearing capacity; slope stability; mine pillar stress; subsidence	Slope stability; residual friction angle (ϕ_r); strong effect by anisotropy	T-15 32 25 38 16 28
Abrasivity	g	L	400g crushed rock rotated in paddle-and-drum test device	Dimensionless	Resistance to abrasion as indicator of percussion drillability index	Must be carefully related to petrographic nature of rock lithology and fabric in order to provide for meaningful assessment of effect	9 52

Hardness			Small laboratory test holding devices for impact. Height or rebound of small diamond-tipped device	Dimensionless	Indicator or relative hardness; useful in tunnel boring rate estimates		
Shore scleroscope	H_r	L			As above	Multiple readings are required to overcome effect of sharp point impact on very small area	59 58 63
Schmidt rebound	H_s	L		Dimensionless	As above	Softer rock breaks on impact; must use Type L device of minimal energy	41
Taber	H_a	L		Dimensionless	As above		T-23
Total	H_t	L		Dimensionless			T-26 T-27
Sonic velocity	v	L	Core sample, held in accoustical bench; subjected to pulsation of electrically-generated mechanical vibration	Km/sec	Computational input for calculation of E, G, and K	Difficult to accommodate fabric anisotropies; difficult to know volume tested *in situ;* does not introduce time-dependent strain variations in derived properties	19 44 3
Seismic velocity Compressional Shear	 v_p v_s	F	Mechanical or explosive energy wave arrival sensed by geophone, timer, and recorder; measured on ground surface or in borehole configurations	Km/sec	As above, indicator of overall mature of rock due to averaging effect on wave travel paths, depending on geophone/energy source array	Degree of saturation important; test does not introduce nonlinear, time-dependent strain-variation in derived properties; most valid in homogeneous and isotropic rock	49 3 28
Poisson's ratio	u	D	Calculated from sonic/seismic velocity tests or by use of electrical resistance strain gages on compression tests	Dimensionless; in range 0 to 0.5	Computational input for calculation of stress distribution patterns and of predicted strain in elastic media; required for Finite Element Modeling	Difficult to extroplate laboratory measurement to field conditions; often estimated without testing; best approximation is from triaxial compression test at confinement	47 3 28

Table 1 (continued)
ENGINEERING PROPERTIES OF ROCKS AND THEIR USES IN ENGINEERED CONSTRUCTION

Engineering property	Symbol	Nature: field(F) laboratory (L) derived (D)	Method of measurement	Units of measure	Use in engineered construction	Limitations	Ref.
						equivalent to *in situ* conditions	
Modulus of rigidity	G	D	As above	N/m²	Indicator of seismic design stiffness	Strain-related	49
Bulk modulus	K	D	As above	N/m²		Strain-related	49
Tangent modulus of elasticity (Young's modulus, or modulus of deformation)	E_t	L	Triaxial compression in universal test machine; electrical resistance strain gages	N/m²	The fundamental stress-strain relationship; input for static displacement computations and for dynamic, seismic analyses	Requires accommodation of any anisotropy of rock fabric and model *in situ* conditions	T-6 4 28
Secant modulus of elasticity	E_s	L	As above	N/m²	Alternate expression of the fundamental stress-strain relationship	As above	3
Swelling slake durability	I_s I_d	L L	Rotational cyclic immersion in water; measure percent loss by weight; measure swelling displacement (d) and initial weight of specimen L (I_s = d/L × 100); measure slaking weight percentages lost in wetting cycles	Dimensionless (0—100)	Indication of relative resistance to weathering on exposure to elements	Semi-quantitative, in use, swell potential is proportional to confining pressure; slake potential relates to moisture exposure and water chemistry at site	36 26

[a] Conform to I.S.R.M. standards, 1970.

ENGINEERING CLASSIFICATION OF INTACT ROCK[T-6]

Based on compressive strength (C_O)[a]

Class	Type	Metric system ($N/m^2 \times 10^8$)	English system ($psi \times 10^3$)
A	Very high strength	>2.21	>32
B	High strength	1.10—2.21	16—32
C	Medium strength	0.55—1.10	8—16
D	Low strength	0.28—0.55	4—8
E	Very low strength	<0.28	<4

Based on compressibility, Tangent modulus of elasticity (E_t)[b]

Class	Type	Metric system ($N/m^2 \times 10^{10}$)	English system ($psi \times 10^6$)
1	Low compressibility	>8.3	>12
2	Medium compressibility	4.1—8.3	6—12
3	High compressibility	<4.1	<6

[a] 10^5 N/m^2 = 14.5 lb/in.^{2} = 0.1 MPa.

[b] Tangent modulus generally measured at the 50% point of strain, as noted in strain-gauge monitored compression tests, uniaxial (unconfined) or triaxial.

Table 3
ENGINEERING PROPERTIES OF IGNEOUS ROCKS

Rock type; geologic unit	Location	γ (N/m³)	C_o (N/m²)[a]	H[b]	V_p (Km/sec)	u	G (N/m²)[a]		E_t (N/m²)[a]		E_s (N/m²)[a]		Ref.
Amphibolite	McLeese Lake, B. C.,	2920	2.65			0.24			1.75	10			7
Amphibolite, fine	Oorgaum, Mysore State India	3070	4.23	92	5.79		4.58	10	1.04	11			20
Andesite, hypersthene	Palisades Dam, ID	2570	1.29—1.32			0.18					5.45	10	1
Andesite, amygdaloidal	Painesdale, MH	2810	1.83	63	4.79		2.72	10	6.46	10			20
Anorthosite, Labradorite, C.	Ukranian Shield, USSR	2770	2.27			0.36	3.41	10	9.28	10			2
Basalt, Lower Granite	Pullman, WA	2727	2.27—3.55	57:412:116	5.27				5.02	10			6
Basalt, Little Goose	Walla Walla, WA	2820	2.96						7.76	10			6
Basalt, John Day	Arlington, OR	2868	3.55						8.38	10			6
Basalt, McCartys	McCartys, NM	2430	2.27—7.03			0.10—0.35			1.83	11			9
Basalt, Pisgah	Pisgah, CA	2250	2.21			0.19			6.24	10			9
Basalt, Olivine, dense	Nevada Test Site, NV	2720							2.47	10			14
Basalt, Olivine, sl. vesicular		2660							2.86	10			14
Basalt, Olivine, mod. vesicular		2560				0.19			2.50	10			14
Basalt, Olivine, vesicular		2450				0.17			2.76	10			14
Basalt, Olivine, Western Cascade	Medford OR	2730	1.69—2.20			0.25					4.21	10	1
Basalt	Painesdale, MI	2850	2.30	69	4.63		2.68	10	6.15	10			20
	Ahmeek, MI	2940	2.58—3.59	79	5.15		3.17	10	7.79	10			20

Basalt, amygdular, H. altered	Ahmeek, MI	2040	1.24		1.77		3.31	9	6.27	9	21
		2700	3.45	49	3.87	0.09	1.85	10	4.07	10	21
		2800	3.42	85		0.15			6.60	10	21
Basalt, subaqueous	Eniwetok, PTT	2860	1.94	71		0.18			6.93	10	4
Basalt, vesicular	Bergstrom, TX	2362	7.78		4.20	0.24	1.61	10	3.45	10	19
		2552	8.00			0.25			5.41	10	19
		2413	5.63			0.22			4.07	10	19
		2495	5.52			0.19			2.90	10	19
		2441			4.00	0.18			1.96	10	19
		2523			4.28	0.16			4.86	10	19
		2523	4.63		4.72	0.22			5.17	10	19
		2330	2.28		4.47	0.21					19
		2430	1.03		4.20	0.22			4.47	10	19
Basalt, vesicular	Bergstrom, TX	2550	7.44		4.65	0.13			3.74	10	19
		2580	8.34		5.04	0.19			4.05	10	19
Basalt, dense		2593	1.13			0.20			5.21	10	19
		2761	1.32		5.56	0.17			7.65	10	19
		2752	1.25		4.70				5.79	10	19
Charnokite (hypersthene granite)	Ukrainian Shield, USSR	2730	2.47			0.22	2.75	10	6.73	10	2
Diabase; Medford	Cambridge, MA	2882	1.77	44:2.13:60							8
Diabase											10
											10
Diabase; Palisades	W. Nyack, NY	2932	2.41	59					8.19	10	8
Diabase; Coggins	Culpepper, WV	3044	3.21	57					9.74	10	8
Diabase; French Creek	St. Peters, PA	3060	3.01	58					9.94	10	8
Diabase, altered	Clinton Co., NY	2940	3.21	92	5.70		3.73	10	9.58	10	20
Diabase, c.	Ahmeek, MI	2900	2.28—2.74	78	5.11		2.95	10	1.14	11	20
Diabase, f.		2930	2.39—3.18	80	5.16		3.05	10	7.77	10	20
Diabase, amygdular, altered	Painesdale, MH	2810	1.56	63	5.03	0.25	2.83	10	7.03	10	21
Diorite; Kennsington	Washington, D.C.	2820	8.09(7)—2.76[a]	50:8.7:150							8
Diorite, hornblende; Kennsington		2852	1.12—1.49	51:4.35:127							8

Table 3 (continued)
ENGINEERING PROPERTIES OF IGNEOUS ROCKS

Rock type; geologic unit	Location	γ (N/m³)	C_o (N/m²)[a]	H[b]	V_p (Km/sec)	u	G (N/m²)[a]	Ref.	E_t (N/m²)[a]	Ref.	E_s (N/m²)	Ref.	Ref.
Diorite, quartz; Kennsington		2772	2.21	52:4.4:110									8
Diorite, gneissic	Mineville, NY	3030	1.86	90	4.27		2.78	10	5.53	10			20
Diorite, augite, fresh	Keetley, UT	2740	3.33	82	5.55	0.25	3.37	10	8.41	10			21
Diorite, augite, sl. altered		2720	2.79	83	5.43	0.26	3.18	10	8.00	10			21
Diorite, Augite, altered		2720	2.15	71	4.94	0.30	2.56	10	6.64	10			21
Diorite, biotite, porph., sl. altered		2690	2.28	77	4.97	0.27	2.83	10	6.68	10			21
Diorite, biotite, porph., s. altered		2660	1.80	67	4.75	0.22	2.45	10	6.01	10			21
Diorite, hornblende	Ishpeming, MI	3010	2.74	84	6.00	0.29	4.22	10	1.07	11			4
Gabbro; Salem	Beverly, MA	3060	1.33—1.49	52:6.47:129					8.76	10			8
Gabbro, altered	Clinton Co., NY	2930	2.77	82	5.36		3.36	10	8.48	10			20
Gabbro/diabase	Ukrainian Shield, USSR	3000	3.09			0.33	4.41	10	1.19	11			2
Gabbro/diabase	Karelian SSR, USSR	3190	3.14						1.17	11			2
Granite, f.	Grand Coulee, WA	2571	1.94	53:10.5:172	4.64				5.48	10			5
Granite, c.		2627	1.61	52:9.5:161	4.08				5.24	10			5
Granite; Cape Ann	Beverly, MA	2637	1.18—2.22	55:9.56:162					7.38	10			8
Granite; Pikes Peak	Colorado Springs, CO	2675	1.57	58					7.06	10			8
Granite; Barre	Barre, VT	2643	1.94	53					6.15	10			15
Granite, sl. altered; Colville	Grand Coulee, WA	2610	6.48			0.06					6.90	9	1
		2610	5.69			0.04					5.52	9	1
Granite; Pre-Cambrian	Loveland, CO	2630	7.21			0.14					2.69	10	1
Granite, pegmatitic; pre-Cambrian		2620	5.06								1.72	10	1

Granite; Barre	Barre VT	2660	2.29	95	3.38		1.68	10	3.04	10	20
Granite	Woodstock, MD	2650	2.51	98	4.51		2.54	10	5.46	10	20
	Tem Piute Dist., NV	2630	2.72	100	4.42		2.25	10	5.13	10	20
	Mt. Airy, NC	2600	2.10	90	2.44		1.02	10	1.57	10	20
Granite, biotite, m.-c.,	Ukrainian Shield, USSR	2660	2.60			0.20	2.36	10	5.92	10	2
Granite, aplitic, f.		2650	3.53			0.26	3.28	10	8.06	10	2
Granite, biotite, m.	Tok, Ukrainian Shield, USSR	2650	2.69						7.08	10	2
Granite, biotite, trachyoidal	Ukrainian Shield, USSR	2650	2.40			0.19	2.27	10	5.38	10	2
Granite, red, f.		2640	2.80			0.19	2.58	10	6.33	10	2
Granite, biotite, m.	Karelian SSE, USSR	2700	2.39			0.25	2.41	10	6.93	10	2
Granite; pre-Cambrian	Valencia Co., NM	2800		80	5.23	0.27	5.23	10	8.21	10	3
Granite, gneissic; Lithonia	Lithonia, GA	2640	1.93	85	2.71	−0.19	1.18	10	1.91	10	3
		2640	2.13	85	2.50	−0.23	1.09	10	1.64	10	3
		2660	2.09	89	2.62	0.02	8.96	9	1.86	10	3
		2620	2.05	85	1.08	−0.28	7.10	9	1.04	10	3
Granite, f.-m; unaweep	Grand Junction, CO	2670	1.74	59	3.17	−0.19	1.68	10	2.72	10	4
		2710	1.59	53	3.75	0.00	1.91	10	3.82	10	4
Granite, par. to foliation; unaweep		2730	1.61	44	3.93	0.12	1.90	10	4.23	10	4
Granite; unaweep		2660	1.74	37	3.17	−0.13	1.55	10	2.72	10	4
Granite, pink	Bergstrom, TX	2710			6.47	0.29	6.84	10	8.57	10	19
Granite, pink	Bergstrom, TX	2620			5.78	0.25	5.02	10	7.54	10	19
		2640			5.79	0.22	4.99	10	7.94	10	19
		2690			6.52	0.29	6.97	10	8.64	10	19
		2630	1.07		5.20				4.92	10	19
		2650	1.18		5.12				4.92	10	19
		2650	1.05		5.16				5.06	10	19
Granite, weathered		2650			5.83	0.29	4.72	10	5.75	10	19
		2620			5.33	0.30	4.65	10	5.36	10	19
Granodiorite, biotite, m.	Oenitea, Ukraine, USSR	2740	2.52			0.24	2.80	10	6.86	10	2

Table 3 (continued)
ENGINEERING PROPERTIES OF IGNEOUS ROCKS

Rock type; geologic unit	Location	γ (N/m³)	C_o (N/m²)[a]	H[b]	V_p (Km/sec)	u	G (N/m²)[a]	Ref.	E_t (N/m²)[a]	Ref.	E_s (N/m³)	Ref.	Ref.
Granodiorite	Bergstrom, TX	2689	4.07		5.72	0.70			5.84	10			19
		2703	1.39		5.89	0.22			7.99	10			19
		2699	8.51		5.80	0.17			6.87	10			19
		2702	1.29		5.72	0.19			7.3	10			19
		2700	1.15		5.93	0.19			7.10	10			19
Magnetite, ore	Mineville, NY	4230	1.41	72	2.72		1.86	10	3.14	10			20
Monzonite, porphyritic; Colville	Grand Coulee, WA	2575	1.49			0.18					4.14	10	1
		2575	1.71			0.15					4.21	10	1
Pegmatite	Star Lake, NY	2590	2.14	87	4.88		2.28	10	6.16	10			20
Pyroxenite	Clinton, Co., NY	3450	1.70	70	1.98		1.03	10	1.31	10			20
Pyroxenite, fresh	Star Lake, NY	3430	1.82	60	6.03		5.03	10	1.24	11			20
Pyroxenite, moderately altered		3310	1.22	49	5.82		4.06	10	1.13	11			20
Pyroxenite, heavily altered		2530	5.86	28	2.96		7.58	9	2.20	10			20
Quartz, diorite	Mountain Home, ID		8.74			0.05					2.14	10	1
Quartz, monzonite	Bergstrom, TX	2669	1.48			0.70			6.74	10			19
		2680	1.55			0.22			7.24	10			19
		2670	1.30			0.17			6.68	10			19
		2673	1.29			0.19			7.65	10			19
		2667	1.39			0.19			7.72	10			19
Rapakivi (granite)	Ukrainian Shield, USSR	2640	2.72			0.20	2.43	10	5.81	10			2
Shonkinite (dark syenite)	Clinton, Co., NY	3350	1.85	78	3.23		1.94	10	3.54	10			20

Syenite	**Kirkland Lake, ONT**	2820	3.03	5.12	2.83	10	7.38	10	20
Syenite, porphyritic		2700	4.34	5.12	3.03	10	7.10	10	20

[a] Data are followed by exponent.

[b] Hardness values are in order, Schmidt rebound (Hs); Taber (Ha); Total (Ht).

Table 4
ENGINEERING PROPERTIES OF METAMORPHIC ROCKS

Rock type; geologic unit	Location	γ (N/m³)	C_o (N/m²)[a]		H[b]	V_p (Km/sec)	u	G (N/m²)[a]		E_t (N/m²)[a]		E_s (N/m²)	Ref.
													10
Argillite, Cambridge	Dorchester, MA	2810	1.36	8						8.41	10		8
	Cambridge, MA	2642	6.61	7									8
		2510	3.15	7	15:0.45:10								8
		2759	1.55	8	26:1.2					4.83	10		8
		2715	1.55	8						3.86	10		8
Gneiss, quartz diorite	Bethesda, MD	2775	9.60	7	64:5.17:139					7.24	10		8
Gneiss, schistose		2796	5.73	7	51:5.24:78					5.52	10		8
Gneiss, schistose; Wissahickon	Washington, D.C.	2980	7.01	7	46:2.90:79								8
Gneiss, Wissahickon		3140	1.06	8	54:8.30:129								8
Gneiss, diorite; Wissahickon		2739	6.43	7	44:7.40:133								8
Gneiss, Dworssak	Orofino, ID	2804	1.62	8	48:					5.36	10		8
Gneiss, diorite; Idaho Springs	Montezuma Quad., Co.	2865	8.41	7			0.06			6.41	10		18
Gneiss, granite	Mineville, NY	27.50	2.12	8	99	3.63		1.96	10	3.85	10		20
Gneiss, quartz		3180	2.25	8	97	3.42		2.05	10	3.70	10		20
Gneiss, granite, pegmatitic	Star Lake, NY	3040	1.53	8	75	4.66		2.88	10	6.67	10		20
Gneiss, pegmatitic		2650	1.96	8	81	4.11		2.12	10	4.46	10		20
Gneiss, pyroxene		2710	2.23	8	85	4.45		2.58	10	5.52	10		20
Gneiss, biotite		2640	2.51	8	89	4.27		2.34	10	5.00	10		20
Gneiss, hornblende		2750	2.30	8	82	4.02		2.25	10	4.49	10		20
Gneiss, augite	Hackettstown, NJ	3360	2.19	8	74	5.55	0.27	4.07	10	1.03	11		21
Gneiss, biotite		2910	1.61	8	74	4.79	0.24	2.71	10	6.72	10		21
Gneiss	Bergstrom, TX	2710				4.58	0.15	2.59	10	5.38	10		19
		2810				6.28	0.29	6.74	10	8.32	10		19
Greenstone	Mt. Weather, VA	3020	2.69	8	81	5.85		4.21	10	1.05	11		20
		2960	3.05	8	80	5.21		3.86	10	8.07	10		20

Greenstone, amygdaloidal	Catoctin, PA	3040	2.01	8	64	3.99	−0.21	3.07	10	4.90	10	3
Greenstone, albitic		3070	1.80	8	60	2.71	−0.35	2.22	10	2.35	10	3
Hematite, ore	Soudan, MN	5070	6.07	8	74	6.28		7.79	10	2.00	11	20
Hematite, ore; par. bedding	Bessemer, AL	3780	1.19	8	51	4.30		2.69	10	6.69	10	20
		3670	1.39	8	50	4.30		2.70	10	6.73	10	20
Hornfels	Tem Piute Dist., NV	3190	5.33	8		5.49		4.09	10	9.58	10	20
Marble, Cherokee	Tate, GA	2707	6.69	7	36					5.59	10	8
Marble, taconic	Rutland, VT	2707	6.21	7	31					4.79	10	8
Marble, perp. bedding	Cockeysville, MD	2870	2.12	8	56	4.18		2.61	10	4.93	10	20
Marble, par. bedding		2870	2.23	8	27			2.83	10	6.74	10	20
Marble,white	Tem Piute, Dist., NV	3200	2.38	8		5.06		3.46	10	8.21	10	20
Marble	Star Lake, NY	2720	1.27	8	49	4.42		2.31	10	5.41	10	20
Marble, paleozoic	Ural Mtns., USSR	2710	1.49	8						7.67	10	2
Marble, dolomitic, f.	Karelian SSR, USSR	2820	2.74	8			0.26	3.00	10	8.94	10	2
Marble, Oro Grande	Oro Grande, CA	2720	1.65	8	56	5.40	0.30	3.03	10	7.86	10	3
		2680	5.52	7	42	4.90	0.16	2.80	10	6.52	10	3
Marble, magnesian; Oro Grande		2720	5.72	7	43	4.63	0.17	2.52	10	6.10	10	3
Metarhyolite	Soudan, MN	2840	1.25	8	47	5.06		3.16	10	7.86	10	70
Quartzite, Wissahickon	Washington, D.C.	2804	4.71	7	38:2.78:63							8
Quartzite, phyllite lenses	Raven, Yugoslavia	2590				8.22				1.27	9	13
Quarzite, altered		2590				2.5				1.21	10	13
Quartzite, ferruginous	Kursk, USSR	3510	3.43	8						1.71	11	2
Quartzite, pre-Cambrian & Paleozoic	Urals, Ukraine, and Karelia, USSR	2650	3.74	8			0.13	3.08	10	7.00	10	2

Table 4 (continued)
ENGINEERING PROPERTIES OF METAMORPHIC ROCKS

Rock type; geologic unit	Location	γ (N/m³)	C_o (N/m²)[a]		H[b]	V_p (Km/sec)	u	G (N/m²)[a]		E_s (N/m²)[a]		E_d (N/m²)		Ref.
Quartzite, Biwabik	Babbitt, MN	2750	6.29	8		5.55	0.10	3.86	10	8.48	10			3
Quartzite, hematitic	Ishpeming, MI	4070	2.93	8	71	5.21	0.20	4.06	10	9.79	10			4
Phyllite, sericite	Eldorado Co., CA	2340	9.79	6								1.79	10	1
Phyllite, quartzose		2180	9.38	6								7.58	9	1
Phyllite, graphitic	El Dorado Co., CA	2350	6.69	6								9.65	9	1
Phyllite, pink	Raven, Yugoslavia	2450				7.10				8.30	8			13
Phyllite, green	Ishpeming, MI	3240	1.26	8	40	4.85		3.28	10	7.65	10			20
Schist, chlorite	Bethesda, MD	2813	2.53	7	37:1.89:51					3.10	10			8
Schist, chlorite; Wissahickon	Washington, D.C.	3028	8.195	7	57:6.67:140									8
Schist, hornblende; Wissahickon		3028	1.39	8	50:4.35:96									8
Schist, biotite; Idaho Springs	Montezuma Quad., CO	2720	2.09	7								2.48	10	1
Schist, clayey, altered	Raven, Yugoslavia	2490				7.20				9.81	8	3.65	8	13
Schist, banded		2670				2.63				1.37	10			13
Schist, sericite	Superior, AZ	2700	1.62	8	82	4.72		2.62	10	6.00	10			20
Skarn, garnet-pyroxene	Star Lake, NY	3280	1.30	8	61	5.12		3.48	10	8.62	10			20
Slate, par. bedding, calcareous	Bangor, PA	2740	1.83	8	56					8.88	10			20
Tactite, epidote	Ophir, UT	2870	2.66	8	65	4.60	0.11	2.77	10	6.14	10			21

[a] Data are followed by exponent.

[b] Hardness values are in order: Schmidt rebound (H_s), Taber (H_a), total (H_t).

Table 5
ENGINEERING PROPERTIES OF SEDIMENTARY ROCKS

Rock type; geologic unit	Location	γ (N/m³)	C_o (N/m²)[a]		H[b]	V_p (km/sec)	u	G (N/m²)[a]		E_t (N/m²)[a]		E_s		Ref.
Borax, ore; Ricardo	Boron, CA	2140	4.41	7	22							4.21	9	4
Chert, chalcedonic; Boone	Picker, OK	2560	3.60	8	96		0.09			5.34	10			3
Chert, dolomitic; Fort Payne	Smithville, TN	2630	2.10	8	74	3.35	0.00	1.65	10	3.54	10			4
		2670	2.02	8	67	4.48	0.14	2.37	10	5.62	10			4
Conglomerate; Roxbury	Boston, MA	2679	8.28	7	41:6.37:102									8
Conglomerate	Kirkland Lake, ONT	2670	1.65	8		5.40		3.24	10	7.79	10			20
Dolomite, Lockport	Rochester, NY	2765	2.12	8	50:3.24:86					4.48	10			8
	Niagara Falls, NY	2579	9.10	7	44:					5.10	10			8
Dolomite, Akron	Buffalo, NY	2664	1.98	8	37:0.96:36					5.56	10			8
Dolomite, Oneota	Kasota, MN	2451	8.69	7	43:1.00:43	4.97				4.40	10			6
Dolomite, Bonne Terre	Bonne Terre, MD	2673	1.52	8	49:					6.63	10			8
Dolomite	Jefferson City, TN	2760	3.59	8	69:	5.30		3.17	10	7.79	10			20
Dolomite, m.	Jefferson City, TN	2800	3.28	8	71:	5.24		3.38	10	7.72	10			20
Dolomite, siliceous	Jefferson City, TN	2770	2.45	8	66:	5.18		3.19	10	7.52	10			20
Dolomite, Niagara	Unk, OH	2400	8.96	7	42:		−0.09	1.03	10	1.93	10			21
		2600	1.58	8	51:		0.05	2.21	10	4.62	10			21
Dolomite, Clinton/Niagara	Unk., OH	2600	1.03	8	53:		0.03	1.38	10	2.83	10			21
		2600	1.31	8	58:		0.18	2.00	10	4.76	10			21
		2400	7.58	7	39:		0.07	1.03	10	2.21	10			21
Dolomite, Beekmantown	Wood Co., WV	2833					0.22	2.95	10	7.23	10			17
		3004					0.22	3.05	10	7.50	10			17
		2783					0.26	3.75	10	9.50	10			17
		2832					0.19	3.64	10	8.65	10			17
Dolomite, DeCew	Niagara Falls, ONT	2740					0.46			4.58	10			11
Dolomite, Maple Mill	Omaha, NB		3.47	7			0.36			4.79	10			18
		2827	4.32	7			0.05			2.74	10			18
		2818	1.13	8			0.12			6.13	10			18
		2528	4.45	7			0.51			4.39	10			18
		2507	7.08	7			0.09			2.14	10			18
		2531	6.09	7			0.40			8.62	10			18

Table 5 (continued)
ENGINEERING PROPERTIES OF SEDIMENTARY ROCKS

Rock type; geologic unit	Location		C_o			V_p		G		E_t				
		(N/m³)	(N/m²)[a]		H[b]	(km/sec)	u	(N/m²)[a]		(N/m²)[a]		E_s		Ref.
Dolomite, jointed; Jurassic	Gojak, Yugoslavia	2800				2.51				1.27	10			13
Dolomite	Mascot, TN	2840	3.22	8	74	5.46		3.52	10	8.48	10			20
Graywacke, m.; Chico	Monticello Dam, CA	2440	4.88	7			0.03					1.24	10	1
		2490	5.07	7			0.02					9.65	9	1
Graywacke, f.; Chico		2410	4.83	7			0.04					1.10	10	1
Gypsum	Buffalo, NY	2262	1.25	7	18									8
Jaspillite, ferrugtinous, siliceous sandstone	Ishpeming, MI	3390	3.42	8	85	5.55		4.83	10	1.03	11			20
Limestone	Bedford, IN	2206	5.10	7	33:0.43:20	3.91				2.85	10			6
Limestone, Solenhofen	Bavaria, FGR	2621	2.45	8	54:1.75:72	5.78				6.38	10			6
Limestone, Irondequoit	Rochester, NY	2781	8.66	7	44:1.77:56					3.65	10			8
Limestone, Reynales	Rochester, NY	2853	1.66	7	54:4.89:115					6.48	10			8
Limestone, Onondaga	Buffalo, NY	2720	1.56	8	45:7.94:126					7.93	10			8
Limestone, Ozark tavern	Carthage, MO	2659	9.79	7	49					5.59	10			8
Limestone, f.; redwall	Lee's Ferry, AZ	2710	8.04	7			0.25					6.69	10	1
Limestone, m.; redwall	Lee's Ferry, AZ	2680	1.27	8			0.17					3.38	8	1
Limestone, porous; redwall	Lee's Ferry, AZ	2440	1.33	8			0.18					1.65	10	1
Limestone, cherry; redwall	Lee's Ferry, AZ	2600	1.07	8			0.18					5.56	10	1
Limestone, colitic; redwall	Lee's Ferry, AZ	2670	9.94	7			0.18					4.55	10	1
Limestone, reef	Eniwetok, PTT	2300	3.42	7			0.16					3.79	10	1
Limestone, porous; reef	Eniewtok, PTT	1820	5.93	6			0.10					6.90	9	1
Limestone, reef head	Eniwetok, PTT	1790	2.12	7			0.25					2.00	10	1
Limestone, stylotic; redwall	Lee's Ferry, AZ	2730	7.95	7			0.11					3.86	10	1
Limestone, fossiliferous	Bedford, IN	2370	7.52	7	27	3.78		1.42	10	3.34	10			20
Limestone, fossiliferous, par. bed.	Bedford, IN	2370	6.85	7	27			1.56	10	3.91	10			20
Limestone	Barberton, OH	2690	1.97	8	58	4.69		251	10	5.50	10			20
Limestone, c.	Bessemer, AL	2830	1.65	8	66	4.33		2.42	10	5.27	10			20
Limestone, limonitic	Bessemer, AL	2920	1.72	8	61	4.75		2.82	10	4.54	10			20
Limestone, Marly	Rifle, CO	2250	1.10	8	56	2.38		6.90	9	1.25	10			20
Limestone, marly; par bed.	Rifle, CO	2180				3.11		6.76	9	2.14	10			20

Limestone	Ophir, UT	2780	1.93	8	52	4.85	0.20	2.71	10	6.50	10	21
Limestone, Martinsburg	Martinsburg, WV	2680	1.59	8	61	5.00	0.21	2.73	10	6.59	10	21
Limestone, dolomitic	Unk., OH	2500	8.96	7	30		0.19	1.79	10	4.21	10	21
Limestone	Unk., OH	2500	8.27	7	36		0.23	1.93	10	4.69	10	21
Limestone, Brassfield	Unk., OH	2800	1.79	8	55		0.16	2.83	10	6.62	10	21
		2600	5.52	7	33		0.06	1.38	10	2.90	10	21
Limestone, Black River	Trenton, WV	2688					0.16	2.45	10	5.70	10	17
Limestone, dolomitic		2701					0.01	2.20	10	4.40	10	17
		2693					0.32	2.20	10	5.80	10	17
Limestone		2690					0.16	2.50	10	5.90	10	17
		2694					0.30	3.05	10	7.90	10	17
		2692					0.30	3.00	10	7.75	10	17
		2693					0.26	2.95	10	7.40	10	17
Limestone, dolomitic; Mesozoic	Turkmenian SSR, USSR	2700	2.10	8						7.62	10	2
Limestone, dimension; Sarmatian		1730	1.18	7						1.08	10	2
Limestone, L. Paleozoic	Estonian SSR, USSR	2490	1.04	8			0.21	1.53	10	3.67	10	2
Limestone, detrital	Moscow Syncline, USSR	2160	5.20	7						2.90	10	2
Limestone, fossiliferous		2300	8.83	7			0.25	1.73	10	3.67	10	2
Limestone, dolomitic; U. Carboniferous	Samarskaia Luka, USSR	2480	1.32	8			0.28	1.93	10	4.95	10	2
Limestone, Gasport	Niagara Falls, ONT	2720					0.21			6.70	10	11
		2670					0.19			5.61	10	11
Limestone, Irondequoit	Niagara Falls, ONT	2680					0.22			5.34	10	11
		2660					0.13			5.02	10	11
Limestone, Gasport	Niagara Falls, ONT	2540					0.23			4.70	10	11
Limestone, Reynales	Niagara Falls, ONT	2690					0.11			4.10	10	11
		2670					0.26			4.26	10	11
Limestone, St. Louis	Prairie du Rocher, IL	2680	1.54	8	52	5.03	0.28	2.65	10	6.81	10	3
Limestone, dolomitic, keragenous; mahogony	Rifle, CO	2100	6.90	7	46	1.86	−0.06	4.21	9	8.34	9	3
Limestone, chalky; Smokey Hill	Pickstown, SD	1410	8.27	6	10	1.34	0.30	1.59	9	2.90	9	3
		1710	1.65	7	13	1.74	−0.13	2.55	9	4.48	9	3
Limestone, chalky; Fort Hayes	Pickstown, SD	1810	2.55	7	16	1.92	−0.13	3.93	9	6.76	9	3
		1890	2.90	7	16	2.44	0.02	5.38	9	1.11	10	3
		2000	1.24	7	13	1.65	−0.11	2.90	9	5.17	9	3
		2150	1.03	7	8	1.92	−0.11	4.62	9	8.27	9	3

Table 5 (continued)
ENGINEERING PROPERTIES OF SEDIMENTARY ROCKS

Rock type; geologic unit	Location	γ	C_o		H^b	V_p	u	G		E_t		E_s		Ref.
		(N/m³)	(N/m²)[a]			(km/sec)		(N/m²)[a]		(N/m²)[a]				
Limestone, dolomitic; Bonne Terre	Bonne Terre, MO	2660	1.75	8	51	5.09	0.22	2.85	10	6.96	10			3
		2780	1.98	8	59	5.88	0.29	3.76	10	9.72	10			3
		2710	1.96	8	49		0.05			1.99	10			3
		2690	1.96	8	33	5.36	0.22	3.13	10	7.65	10			3
		2670	1.46	8	48	3.78	−0.07	2.10	10	3.87	10			3
Limestone, dolomitic, sandy; Bonne Terre	Bonne Terre, MO	2680	2.05	8	54		−0.05			3.92	10			3
Limestone	Picker, OK	2670	1.30	8	59	4.11	0.24	1.83	10	4.47	10			3
Limestone, fossiliferous; St. Louis	St. Genevieve, MO	2670	1.64	8	48	5.00	0.24	2.68	10	6.67	10			4
Limestone, fossiliferous; spergen		2650	1.43	8	46	5.18	0.29	2.74	10	7.08	10			4
Limestone, oolite, fossiliferous; St. Louis		2560	1.16	8	41		0.20					5.48	10	4
Limestone, fossiliferous; Maxville	E. Fultonham, OH	2730	1.47	8	54	4.94	0.24	2.19	10	655	10			4
Limestone, fossiliferous, par. bed.	E. Fultonham, OH	2810	1.80	8	52		0.20					6.03	10	4
		2690	1.41	8	48		0.20					6.14	10	4
Limestone, fossiliferous; Maxville	E. Fultonham, OH	2690	1.49	8	56		0.25					7.40	10	4
Limestone, Sandy; Maxville	E. Fultonham, OH	2590	1.59	8	46	4.15	0.09	2.03	10	4.53	10			4
Limestone, f.; Maxville	E. Fultonham, OH	2410	1.09	8	34	3.96	0.14	1.66	10	3.77	10			4
Limestone; Wyandotte	Omaha, NB	2546	1.15	7			0.24			2.11	9			18
		2605	4.90	7			0.64			1.61	10			18
Limestone, Winterset	Omaha, NB	2493	3.68	7			0.02			7.34	9			18
Limestone, LaBette	Omaha, NB	2558	2.62	7						7.68	9			18
Limestone, Meramec-Osage	Omaha, NB	2571	9.00	7			0.05			1.12	10			18
Limestone, Independence	Omaha, NB	2647	8.60	7			0.02			3.00	10			18
Limestone, dolomitic; Wapsipinicon	Omaha, NB	2760	6.87	7			0.04			4.31	10			18
Limestone, silurian	Omaha, NB	2352	9.60	7			0.19			3.07	10			18
Limestone, dolomitic; Stewartville-Prosser	Omaha, NB	2595	4.64	7			0.03			8.41	9			18
		2494	6.65	7			0.05			1.31	10			18

Limestone, Chickamauga	Smithville, TN	2740	1.73	8	53	4.39	0.14	2.33	10	5.30	10			4
		2730	1.73	8	52	3.08	0.22	1.17	10	2.72	10			4
Limestone, recrystallized	Eniwetok, PTT	2511	1.35	8	54	5.00	0.12	2.83	10	6.33	10			4
Limestone, foramiferal	Eniwetok, PTT	2390	9.72	7	52	4.60	0.13	2.29	10	5.18	10			4
Limestone, well cemented	Eniwetok, PTT	2530	1.22	8	52	5.33	0.01	3.65	10	7.39	10			4
Limestone, dolomitic, f.-m.	Pondera Co., MT	2700	1.52	8			0.11					1.78	10	16
		2640	1.10	8			0.11					1.92	10	16
		2770	2.22	8										16
Limestone, dolomitic, well-cemented	Pondera Co., MT	2710	1.68	8			0.31					7.65	10	16
Limestone, jointed; jurassic	Gojak, Yugoslavia	2700				1.92				9.16	9			13
Marlstone	Rifle, CO	2310	1.51	8	56	3.20	0.11	1.11	10	2.49	10			21
Marlstone, keragenaceous	Rifle, CO	2240	8.96	7	47	7.87	0.18	7.79	9	1.86	10			21
Marlstone	Rifle, CO	2310	1.49	8	62	3.41	0.21	1.16	10	2.71	10			21
Marlstone, keragenaceous	Rifle, CO	2020	8.62	7	47	2.32	0.02	5.79	9	1.30	10			21
Marlstone	Rifle, CO	2450	1.94	8	59	4.54	0.28	1.94	10	4.86	10			21
		2260	1.60	8	57	3.90	0.28	1.34	10	3.54	10			21
Marlstone, keragenaceous	Rifle,CO	2190	6.62	7	46	3.75		9.79	9	3.08	10			21
		2250	9.24	7	44	3.41		6.90	9	2.64	10			21
		2080	7.17	7	47	3.02		6.90	9	1.98	10			21
Marlstone, dolomitic; keragenaceous	Rifle, CO	2100	6.90	7	46	1.86	−0.06	4.21	9	8.34	9			3
Marlstone, mahagony	Rifle, CO	2220	8.14	7	49	3.20	0.17	1.02	10	2.41	10			3
Marlstone, par. bed.; mahagony	Rifle, CO	2360	1.72	8	61	4.18	0.33	1.53	10	4.10	10			3
Marlstone, Maxville	E. Fultonham, OH	2190	5.59	7	23	3.38	0.13	1.10	10	2.50	10			4
Oil Shale, Parachute Creek	Rio Blanco, CO	2044	8.28	7			0.33					6.24	9	12
		2220	1.10	8			0.37					1.12	10	12
		2190	1.81	8			0.30					1.08	10	12
		2124	9.35	7			0.24					7.03	9	12
Quartzite, Baraboo	Baraboo, WI	2627	3.21	8	59					8.84	10			8
Quarzite	Bergstrom, TX	2610	6.45	7						2.76	6			19
		2570	1.26	8						3.56	10			19
		2610	1.75	8						5.91	10			19
		264	2.23	8						6.36	10			19
		2570	1.64	8						5.44	10			19
Salt; diamond crystal	Jefferson Island, LA	2163	2.14	7	23					4.90	9			8

Table 5 (continued) ENGINEERING PROPERTIES OF SEDIMENTARY ROCKS

Rock type; geologic unit	Location	γ (N/m³)	C_o (N/m²)[a]		H[b]	V_p (km/sec)	u	G (N/m²)[a]		E_t (N/m²)[a]		E_s		Ref.
Salt	Bergstrom, TX	2167	1.81	7		3.76				6.14	9			19
		2168	1.89	7		3.37	0.06			3.45	9			19
		2167	2.85	7		4.08				3.45	10			19
		2298	2.20	7		4.07	0.189			2.05	10			19
		2317	3.07	7			0.03			3.28	10			19
Sandstone, Navajo	Page, AZ	2015	4.35	7	30:0.04:6	2.52						1.53	10	6
Sandstone, Thorold	Rochester, NY	2640	1.79	8	50:9.52:154					6.90	10			8
Sandstone, Grimsby		2653	1.53	8	48:5.0:107					2.76	10			8
Sandstone, Cambridge	Cambridge, MA	2582	3.85	7	37:1.84:51					1.03	10			8
Sandstone, Arkosic; Triassic	New Haven, CT	2558	2.48	7										8
Sandstone, Grimsby	Rochester, NY	2462	5.96	7	24:0.26:12									8
Sandstone, Cambridge	Cambridge, MA	2647	4.93	7	27:0.44:18									8
Sandstone, Crab orchard	Crossville, TN	2531	2.14	8	47					3.92	10			6
Sandstone, f.; tensleep	Casper, WO	2325	7.25	7			0.06					1.31	10	1
Sandstone	Amherst, OH	2060	7.17	7	31	1.71		3.17	9	6.00	9			20
Sandstone, par. bed.	Amherst, OH	2060	5.41	7	31			3.79	9	7.75	9	3.79	9	20
Sandstone, c.	Amherst, OH	2170	4.21	7	20	1.20		4.00	9	7.10	9			20
Sandstone, c., par. bed.	Amherst, OH	2170	3.55	7	20			4.65	9	1.09	10			20
Sandstone, ferruginous	Bessemer, AL	3140	1.69	8	58	3.11		1.84	10	3.07	10			20
Sandstone, fossiliferous, red	Bassemer, AL	3260	1.54	8	50	3.72		2.25	10	4.45	10			20
Sandstone, ferruginous	Bessemer, AL	2930	2.35	8	65	4.05		2.42	10	4.96	10			20
	Monogalia Co., WV	2600	1.32	8	53	3.42	0.22	1.51	10	3.83	10			21
Sandstone	Huntington, UT	2200	1.07	8		2.44	−0.10	7.03	9	1.31	10			21
		2170	7.93	7		2.56	0.04	7.03	9	1.45	10			21
		2140	9.79	7		2.19	0.04	4.83	9	1.01	10			21
		2350	2.23	8		2.96	−0.11	1.17	10	2.07	10			21
		2330	1.91	8		2.87	−0.07	1.02	10	1.86	10			21
Sandstone; carboniferous	Donets Basin, USSR	2650	2.56	8			0.14	2.43	10	5.55	10			21
Sandstone, calcareous; L. Cretaceous	Turkmenian SSR, USSR	2600	2.10	8						5.46	10			2
Sandstone, Thorold	Niagara Falls, ONT	2460					−0.12			2.13	10			11
		2510					−0.18			3.31	9			11
Sandstone, calcareous, nonesuch	White Pine, MT	2600	1.58	8	62	4.63	0.16	2.39	10	5.53	10			3

Sandstone, Homewood	Franklin, PA	2160	7.65	7	23	1.80	0.09	3.93	9	7.31	9	3
		2150	7.65	7	33	2.23	0.05	5.55	9	1.08	10	3
		2490	1.23	8	50	2.50	0.23	9.52	9	1.48	10	3
		2430	1.02	8	55	2.38	0.20	8.69	9	1.38	10	3
Sandstone, cemented; Navajo	Huntington, VT	2150	8.69	7	45	2.26	−0.09	6.14	9	1.04	10	3
Sandstone, uncemented; Navajo		2220	5.93	7	29	2.77	−0.04	8.96	9	1.72	10	3
Sandstone, cemented, Navajo	Huntington, UT	2880	1.24	8	50	2.77	−0.07	9.45	9	1.75	10	3
Sandstone, sl. cemented; Navajo	Huntington, UT	2290	9.52	7	42	2.87	−0.06	1.01	10	1.90	10	3
Sandstone, cemented; Navajo	Huntington, UT	2310	9.03	7	44	3.08	−0.03	1.12	10	2.17	10	3
Sandstone, cemented; obl. bed; Navajo	Huntington, UT	2370	3.38	7	54	3.38	0.05	1.41	10	2.71	10	3
Sandstone, uncemented; obl. bed.; Navajo	Huntington, UT	2130	5.59	7	32	2.29	−0.05	5.86	9	1.12	10	3
Sandstone, uncemented, par. bed.; Navajo	Huntington, UT	2130	3.31	7	36	2.10	−0.04	4.96	9	9.58	9	3
Sandstone, carboniferous	Woodrow, PA	2150	6.69	7	21	1.80	0.01	3.52	9	6.90	9	3
Sandstone, par. bed.; carboniferous	Woodrow, PA	2130	6.69	7	21	1.95	0.10	3.72	9	8.21	9	3
Sandstone; Dakota	Grants, NM	2120			25	1.59	−0.02	2.79	9	5.38	10	3
Sandstone, Graywacke; Kanawha	DeHue, WV	2600	1.41	8	55	2.93	−0.17	1.34	10	2.23	10	3
Sandstone, argilaceous; Kanawha	DeHue, WV	2800	1.05	8	42	3.32	0.05	1.48	10	3.11	10	3
Sandstone, f., Allegheny	Bakerton, PA	2700	1.59	8	56	2.93		1.98	10	2.46	10	4
Sandstone, m.; Morrison/Bushy Basin	Lone Park, CO	2530				2.35	−0.16	8.27	9	1.39	10	4
Sandstone, m.; Morrison/Salt Wash	Long Park, CO	2680				3.23	0.36	1.01	10	2.78	10	4
Sandstone, f.; Morrison/Bushy Basin	Long Park, CO	2540				2.62	−0.04	9.10	9	1.76	10	4
		2260				2.29	−0.31	8.62	9	1.18	10	4
Sandstone, f.; Morrison/Salt Wash	Long Park, CO	2200				2.32	−0.36	9.10	9	1.17	10	4
		2290				1.71	−0.47	6.27	9	6.62	9	4
		2250				1.83	−0.45	6.90	9	7.58	9	4
		2280				1.86	−0.51	8.27	9	8.14	9	4
Sandstone; Cherokee	Omaha, NB	2347	3.29	6			0.10			3.99	8	18
Sandstone, Shaly; St. Peter	Omaha, NB	2344	3.73	7			0.05			7.19	9	18
		2450	3.46	7			0.06			1.25	10	18

Table 5 (continued)
ENGINEERING PROPERTIES OF SEDIMENTARY ROCKS

Rock type; geologic unit	Location	γ (N/m³)	C_o (N/m²)[a]		H[b]	V_p (km/sec)	u	G (N/m²)[a]		E_t (N/m²)[a]		E_s	Ref.
Sandstone, shaly	Bergstrom, TX	2530	9.79	6						9.39	9		19
		2580	2.76	7						6.00	9		19
		2570	1.09	7						5.72	9		19
		2600	4.14	7						1.23	10		19
		2610	4.92	7						1.35	10		19
		2660	2.78	7						9.52	9		19
		2580	1.32	7						1.93	9		19
		2540	2.78	7						4.76	9		19
Sandstone, silty; Seminole	Tulsa, OK	2500	7.45	7	31	2.87		1.08	10	2.19	9		4
Sandstone; L. Connoquenessing	Franklin, PA	2450	1.08	8	51	2.93	−0.06	1.12	10	2.10	10		4
Sandstone; Homewood	Franklin, PA	2200	8.69	7	43	1.92	−0.11	4.69	9	8.27	9		4
Sandstone, Homewood	Franklin, PA	2210	8.67	7	39	1.59	−0.12	3.17	9	5.59	9		4
Sandstone, Berea	Amherst, OH	2182	7.38	7	42:0.47:29	2.64				1.93	10		6
Shale, Rochester	Rochester, NY	2738	1.22	8	45:0.73:39					3.79	10		8
Shale, Williamson	Rochester, NY	2712	8.38	6	37:0.75:26					2.14	10		8
Shale, Sodus	Rochester, NY	2749	7.15	7	30:0.64:23					3.65	10		8
Shale, Maplewood	Rochester, NY	2697	7.35	7	24:0.67:19					4.34	10		8
Shale, Camulus	Buffalo, NY	2689	9.27	7	40:2.0:40					4.00	10		8
Shale, Brunswick	Highland Park, NJ	2631	8.29	7	38:0.70:31					1.38	10		8
Shale, Queenston	Rochester, NY	2680	1.12	8	47:0.92:45					2.48	10		8
Shale, Rochester	Rochester, NY	2760	1.13	8	48:0.76:39					3.79	10		8
Shale, Bertie	Buffalo, NY	2712	1.97	8	42:1.92:59					5.03	10		8
Shale, calcareous; mauv	Lee's Ferry, AZ	2670	3.60	7			0.02			1.59	10		2
Shale, quartzose; mauv	Lee's Ferry, AZ	2690	1.23	8			0.08			1.65	10		1
Shale	Monongalia Co., UV	2600	8.00	7		4.15		1.25	10	4.64	10		21
	Ophir, UT	2810	2.16	8	58	4.54	0.09	2.66	10	5.82	10		21
Shale, siliceous	Ophir, UT	2800	2.31	8	71	4.94	0.12	3.05	10	6.81	10		21
	White Pine, MI	2730	1.96	8	51		0.20			4.79	10		3
		2780	1.97	8	62		0.15			5.17	10		3
Shale, siderite, banded; Kanawha	DeHue, WV	2760	1.12	8	38	2.16	−0.43	1.17	10	1.33	10		3
Shale, micaceous; Maxville	E. Fultonham, OH	2560	7.51	7	31	2.07	−0.29	7.93	9	1.11	10		4
Shale, calcareous; Wyandotte	Omaha, NB	2177	1.19	7			0.32			1.97	9		18

Shale, Bourbon	Omaha, NB	2408	7.72	6			0.34			9.42	8			18
Shale, Bandera	Omaha, NB	2430	9.59	6			0.02			5.85	8			18
Shale, La Bette	Omaha, NB	2429	2.49	6			0.03			3.97	8			18
Shale, Cherokee	Omaha, NB	2007	1.49	7			0.06			2.43	9			18
		2411	6.26	6			0.25			2.30	9			18
Shale, Sl. weathered; Cherokee	Omaha, NB	2496	8.34	6			0.15			1.67	9			18
Shale, Calcareous; Sheffield	Omaha, NB	2602	5.98	6			0.14			3.09	10			18
Shale, Maqueketa	Omaha, NB	2618	4.25	7			0.01			7.32	9			18
Shale, Lon	Omaha, NB	2687	5.37	7			0.04			9.73	9			18
Shale, Specht's Ferry	Omaha, NB	2478	3.79	7			0.05			6.18	9			18
Shale	Bergstrom, TX	2600	2.34	7						1.10	10			19
Shale	Bergstrom, TX	2650	1.65	7						1.01	10			19
		2660	8.48	6						1.03	10			19
		2610	1.11	7						6.90	9			19
Shale, carbonaceous; Chattanooga	Smithville, TN	2300	1.12	8	50	2.38	00.00	6.55	9	1.39	10			4
		2300	1.10	8	48	2.38	−0.02	7.10	9	1.34	10			4
Shale, silty; Chattanooga	Smithville, TN	2530	8.34	7	42	2.59		5.17	9	1.73	10			4
Siltstone; Hackensack	Hackensack, NJ	2595	1.23	8	47:154:58	3.99				2.63	10			6
Siltstone; Chico	Monticello Dam, CA	2500	2.41	7			0.05			1.3	10			1
	Bessemer, AL	2760	2.56	8	71	4.82		2.53	10	5.32	10			20
Siltstone, par. bedding; Maxville	E. Fultonham, OH	2660	3.65	7	20		0.13					4.81	10	4
		2680	3.45	7	19		0.26					8.68	10	4
Siltstone, poorly cemented; Bandera	Omaha, NB	2304	3.54	6			0.35			1.25	8			18

[a] Data are followed by exponent.

[b] Hardness values are in order: Schmidt rebound (H_s), Taber (H_a), total (H_t).

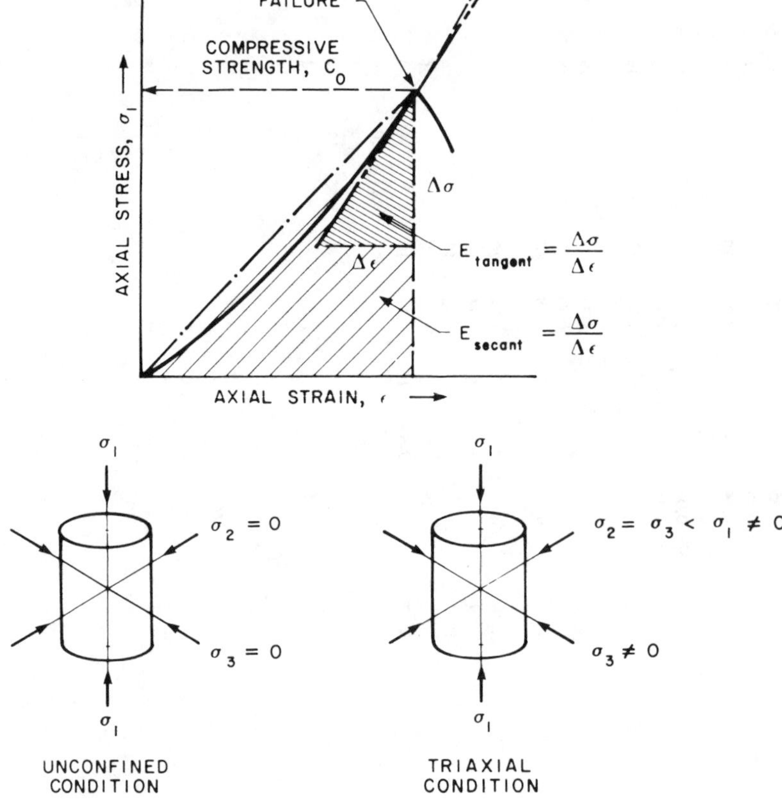

FIGURE 10. The compressive strength test and elastic moduli of a nonlinear, isotropic rock. Moduli may be determined at any strain state at or below failure; failure case is illustrated.

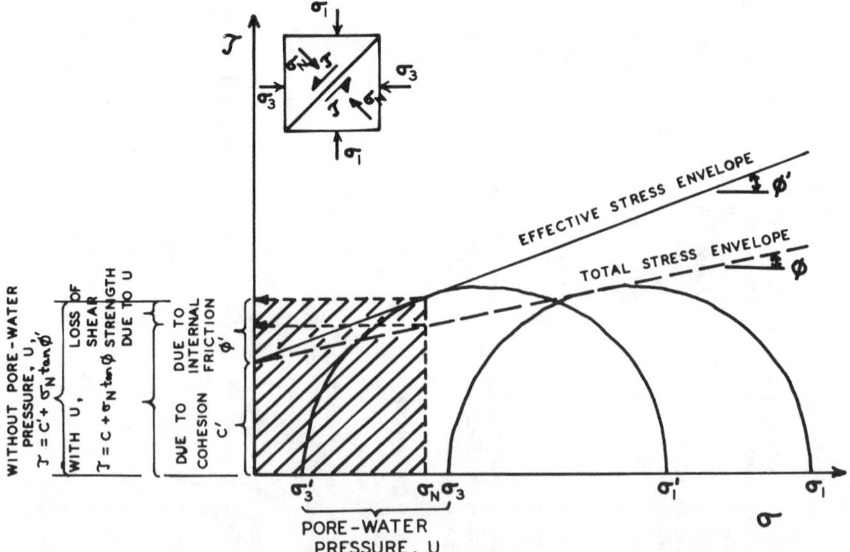

FIGURE 11. Effect of pore-water pressure (u) in reducing shear strength (τ) available as the total of cohesion (c) and coefficient of internal friction (tangent Ø) along an existing fracture. (σ) is normal strength (stress). Shaded box shows typical available shear strength for a given state of normal strss. (σ_n) bearing on fracture or two surfaces in contact; determined by construction.

SUMMARY

Engineering properties of rock are used in conjunction with geologic observations which deal with the geometry and spacing of natural discontinuities and the boundaries of geologic units of similar engineering properties and characteristics. The goal is to accommodate the stresses generated by the presence of engineered structures or the processes underway within or on such structures. The use of engineering properties of rock is always warranted in the design of structures to be placed on or in rock. However, the value of such design is highly dependent upon the care with which representative samples of rock are obtained and tested, the rock mass evaluated, and the manner in which related assumptions of values or ranges of values are utilized in the engineering design.

Carefully measured engineering properties are of value to geologists, engineers, and contractors in all aspects of engineered construction in rock. Much of this value comes from the experiences of each individual working in rock engineering and rock mechanics, for it is by way of experience that the final impact of engineering properties of rock is best assessed.

GLOSSARY OF ROCK ENGINEERING TERMS*

Angle of internal friction (angle of shear resistance) Ø (degrees) — Angle between the axis of normal stress and the tangent to the Mohr envelope at a point representing a given failure-stress condition for solid material.

Coefficient of permeability (permeability), (hydraulic conductivity) k (LT^{-1}) — The rate of discharge of water under laminar flow conditions through a unit cross-sectional area of a porous medium under a unit hydraulic gradient and standard temperature conditions (usually 20°C).

Compressive strength (unconfined or uniaxial compressive strength), C_o (FL^{-2}) — The load per unit area at which an unconfined cylindrical specimen of soil or rock will fail in a simple compression test. Commonly the failure load is the maximum that the specimen can withstand in the test.

Controlled-strain test — A test in which the load is so applied that a controlled rate of strain results.

Controlled-stress test — A test in which the stress to which a specimen is subjected is applied at a controlled rate.

Deviator stress, Δ, σ (FL^{-2}) — The difference between the major and minor principal stresses in a triaxial test.

Direct shear test — A shear test in which soil or rock under an applied normal load is stressed to failure by moving one section of the sample or sample container (shear box) relative to the other section.

Fault — A fracture or fracture zone along which there has been displacement of the two sides relative to one another parallel to the fracture (this displacement may be a few centimeters or many kilometers). (See definitions of joint set and joint system for definitions of fault set and fault system.)

Fault gouge — A clay-like material occurring between the walls of a fault as a result of the movement along the fault surfaces.

Filling — Generally the material occupying the space between joint surfaces, faults, and other rock discontinuities. The filling material may be clay, gouge, various natural cementing agents, or alteration products of the adjacent rock.

Fold — A bend in the strata or other planar structures within the rock mass.

* Selected from ASTM Draft Standard D653, 1977. Definitions include those also adopted by ISRM.

Foliation — The somewhat laminated structure resulting from segregation of different minerals into layers parallel to the schistosity.

Fracture — The general term for any mechanical discontinuity in the rock; it therefore is the collective term for joints, faults, cracks, etc.

Fragmentation — The breaking of rock in such a way that the bulk of the material is of convenient size for handling it.

Internal friction (shear resistance), s (FL^{-2}) — The portion of the shearing strength of a soil or rock indicated by the terms p tan Ø in Coulomb's equation $s_o = c + p$ tan Ø. It is usually considered to be due to the interlocking of the soil or rock grains and the resistance to sliding between the grains.

Intrinsic shear strength, S_o (FL^{-2}) — The shear strength of a rock indicated by Coulomb's equation when p tan Ø (shear resistance or internal friction) vanishes. Corresponds to cohesion, c, in soil mechanics.

Joint — A break of geological origin in the continuity of a body of rock occurring either singly, or more frequently in a set or system, but not attended by a visible movement parallel to the surface of discontinuity.

Lineation — The parallel orientation of structural features that are lines rather than planes; some examples are parellel orientation of the long dimensions of minerals, long axes of pebbles, striae on slickensides, and cleavage-bedding plane intersections.

Normal force — A force directed normal to the surface element across which it acts.

Peak shear strength — Maximum shear strength along a failure surface.

Plane stress (strain) — A state of stress (strain) in a solid body in which all stress (strain) components normal to a certain plane are zero.

Primary state of stress — The stress in a geological formation before it is disturbed by man-made works.

Principal stress (strain) — The stress (strain) normal to one of three mutually perpendicular planes on which shear stresses (strains) at a point in the body are zero.

Residual stress — Stress remaining in a solid under zero external stress after some process that causes the dimensions of the various parts of the solid to be incompatible under zero stress, e.g., (1) deformation under the action of external stress when some parts of the body suffer permanent strain; (2) heating or cooling of a body in which the thermal expansion coefficient is not uniform throughout the body.

Rock mass — Rock as it occurs *in situ,* including its structural discontinuities.

Rock mechanics — Theoretical and applied science of the mechanical behavior of rock.

Rupture — That stage in the development of a fracture where instability occurs. It is not recommended that the term be used in rock mechanics as a synonym for fracture.

Shear failure (failure by rupture) — Failure in which movement caused by shearing stresses in a soil or rock mass is of sufficient magnitude to destroy or seriously endanger a structure.

General shear failure — Failure in which the ultimate strength of the soil or rock is mobilized along the entire potential surface of sliding before the structure supported by the soil or rock is impaired by excessive movement.

Local shear failure — Failure in which the ultimate shearing strength of the soil or rock is mobilized only locally along the potential surface of sliding at the time the structure supported by the soil or rock is impaired by excessive movement.

Shear strain — The change in shape, expressed by the relative change of the right angles at the corner of what was in the undeformed state an infinitesimally small rectangle or cube.

Stability — The condition of a structure or a mass of material when it is able to support the applied stress for a long time without suffering any significant deformation or movement that is not reversed by the release of stress.

Strength — Maximum stress which a material can resist without failing for any given type of loading.

Stress (strain) field — The ensemble of stress (strain) states defined at all points of an elastic solid.

Structure — One of the larger features of a rock mass, such as bedding, foliation, jointing, cleavage, or brecciation; also the sum total of such features as contrasted with texture. Also, in a broader sense, it refers to the structural features of an area such as anticlines or synclines.

Tensile strength (unconfined or uniaxial tensile strength) T_o (FL^{-2}) — The load per unit area at which an unconfined cylindrical specimen will fail in a simple tensile (pull) test.

Tensile stress — Normal stress tending to lengthen the body in the direction in which it acts.

Triaxial compression — Compression caused by the application of normal stresses in three perpendicular directions.

Triaxial shear test (triaxial compression test) — A test in which a cylindrical specimen of soil or rock encased in an impervious membrane is subjected to a confining pressure and then loaded axially to failure.

Triaxial state of stress — State of stress in which none of the three principal stresses is zero.

Ultimate bearing capacity, q_e, q_{ult} (FL^{-2}) — The average load per unit of area required to produce failure by rupture of a supporting soil or rock mass.

Dry unit weight (unit dry weight), γ, (FL^{-3}) — The weight of soil or rock solids per unit of total volume of soil or rock mass.

Conversion factors — Factors for conversion of metric to standard English units previously common to rock engineering works are listed below, as selected from American Society for Testing and Materials (ASTM) Standard Designation E380-70, *Metric Practice Guide.* Factors are shown to three significant digits, followed by the appropriate power of 10.

To convert from	to	Multiply by
	Mass/Volume	
kg/m³	pound-mass/ft³ (pcf)	6.243×10^2
	Stress (force/area)	
N/m²	bar	1.00×10^5
N/m²	kips/in.² (ksi)	1.450×10^7
N/m²	lb/in.² (psi)	1.450×10^4
N/m²	tons/ft² (tsf)	1.044×10^5
	Velocity	
m/sec	ft/sec (fps)	3.281

Properties relating to the deformability in the range of nonpermanent or recoverable strain are termed *elastic properties.* Such properties relate to the compressibility of rock under pressure. The properties commonly measured for engineering purposes are the various moduli and Poisson's ratio (Tables 1 and 3 to 5). The properties are interrelated on the basis of any two measurable constants of the same category.[6] These conversions assume isotropic (nondirectional) engineering properties. Elastic properties are used to model or predict three-dimensional response of underground structures and behavior of rock to blasting and wave transmission.

Elastic properties reported in Tables 3 to 5 represent high values for E and H and an average of values for C_o.[6]

ACKNOWLEDGMENTS

In addition to rock property data contained in the authors' files, recognition is given herewith to the literature collection of William R. Judd, School of Civil Engineering, Purdue University, available on microfiche at a nominal charge from CINDAS (Center for Information, Numerical Data Analysis and Synthesis.)

REFERENCES

1. Standard Test Methods for Apparent Porosity, Water Absorption, Apparent Specific Gravity, and Bulk Density of Burned Refractory Bricks, Standard C 20-46 (Part 13), American Society for Testing and Materials, Philadelphia, 1967.
2. Standard Definitions of Terms and Symbols Relating to Rock Mechanics, Standard D653-78, American Society for Testing and Materials, Philadelphia, 1972.
3. Standard Test Method for Laboratory Determination of Pulse Velocities and Ultrasonic Elastic Constants of Rock, (Part 19), Standard D2845-69, American Society for Testing and Materials, Philadelphia, 1980.
4. Standard Method of Test for Elastic Moduli of Rock Specimens in Uniaxial Compression, Standard D3148-72, American Society for Testing and Materials, Philadelphia, 1974.
5. **Attewell, P. B. and Farmer, I. W.,** *Principles of Engineering Geology,* John Wiley & Sons, New York, 1976.
6. **Birch, F.,** Compressibility; elastic constants, in *Handbook of Physical Constants,* Clark, S. P., Ed., Geological Society of America, Boulder, Colo., 1966.
7. **Beverly, B. E., Schoenwolf, D. A., and Brierley, G. S.,** Correlations of rock index values with engineering properties and classification of intact rock, in 58th Annu. Meet. Transportation Research Board, Washington, D.C., 1979.
8. **Boyum, B. H.,** Subsidence case histories in Michigan mines, *Min. Ind. Stn. Bull.,* 76, 19, 1961.
9. **Burbank, R. B.,** Measuring the relative abrasiveness of rocks and ores, *Pit Quarry,* 114, 117, 1955.
10. **Coates, D. F.,** Classification of rock for rock mechanics, *Int. J. Rock Mech. Min. Sci.,* 1, 421, 1964.
11. **Coates, D. F.,** Rock Mechanics Principles, Mines Branch Monogr. 894, Department of Mines and Technical Surveys, Ottawa, 1970.
12. **Coulson, J. H.,** The Effects of Surface Roughness on the Shear Strength of Joints in Rock, Tech. Rep. No. MRD-2-80, AD 714-244, Missouri River Division, U.S. Army Corps of Engineers, Omaha, 1970.
13. **D'Andrea, D. V., Fisher, R. L., and Fogelson, D. F.,** Prediction of Compressive Strength from Other Rock Properties, Invest. Rep. No. 6702, U.S. Bureau of Mines, Washington, D. C., 1965.
14. **Deere, D. U.,** Technical description of rock cores for engineering purposes, *Rock Mech. Eng. Geol.,* 1, 18, 1963.
15. **Deere, D. U., Merritt, A. H., and Coon, R. F.,** Engineering Classification of In Situ Rock, Rep. No. AFWL-T-67-144, U.S. Air Force Weapons Laboratory, Kirtland Air Force Base, New Mexico, 1969.
16. **Dodds, R. K.,** Suggested method for test for in situ shear strength of rock, *Am. Soc. Test. Mater. Spec. Tech. Publ.,* 479, 618, 1970.
17. **East, H. H., Jr. and Gardner, F. D.,** Oil Shale Mining, Rifle, Colorado, 1954-56, Bull. 611, U.S. Bureau of Mines, Washington, D.C., 1964.
18. **Everell, M. D., Herget, G., Sage, R., and Coates, D. F.,** Mechanical properties of rocks and rock masses, in Proc. 3rd Congr. Int. Soc. Rock Mechanics, National Academy of Sciences, Washington, D.C., 1974.
19. **Fairhurst, C.,** Laboratory measurements of some physical properties of rock, in *Proc. 4th Symp. Rock Mechanics,* Pennsylvania State University, University Park, 1961.
20. **Fairhurst, C.,** On the validity of the "Brazilian" test for brittle materials, *Int. J. Rock Mech. Min. Sci.,* 4, 535, 1964.
21. **Fancher, G. H.,** Porosity and permeability in clastic rocks, in *Subsurface Geology,* 4th ed., LeRoy, L. W., LeRoy, D. O., and Raese, L. W., Eds., Colorado School of Mines, Golden, 1977.
22. **Farmer, I. W.,** *Engineering Properties of Rocks,* E & F. N. Spon Ltd., London, 1968.
23. **Franklin, J. A. and Chandra, R.,** The slake-durability test, *Int. J. Rock Mech. Min. Sci.,* 9, 325, 1972.

24. **Gamble, J. C.**, Plasticity Classification of Shales and Other Argillaceous Rocks, Ph.D. Thesis, University of Illinois, Champaign, 1971.
25. **Goodman, R.**, The mechanics and properties of joints, in Proc. 3rd Congr. Int. Soc. Rock Mechanics, National Academy of Sciences, Washington, D.C., 1974.
26. **Gyenge, M.**, Laboratory classification tests rock, in *Pit Slope Manual,* Rep. No. 75-Z5 (Suppl.3-1), Canada Centre for Mineral and Energy Technology, Ottawa, 1977.
27. **Gyenge, M. and Herget, G.**, Mechanical Properties (rock), in *Pit Slope Manual,* Rep. No. 77-12, Canada Centre for Mineral and Energy Technology, Ottawa, 1977.
28. **Gyenge, M. and Herget, G.**, Laboratory tests for design parameters (rock), in *Pit Slope Manual,* Rep. No. 77-26, Canada Centre for Mineral and Energy Technology, Ottawa, 1977.
29. **Hall, W. J., Newmark, N. M., and Hendron, A. J.**, Classification, Engineering Properties and Field Explorations of Soils, Intact Rock and In Situ Rock Masses, Rep. No. WASH 1301-UC-11, U.S. Atomic Energy Commission, Washington, D.C., 1974.
30. **Heck, W. J.**, Suggested method of test for triaxial compressive strength of undrained rock core specimens with induced pore pressure measurements, *Am. Soc. Test. Mater. Spec. Tech. Publ.,* 479, 604, 1970.
31. **Heuze, F. E.**, The Design of Room and Pillar Structures in Competent Jointed Rock—the Crestmore Mine, California, Ph.D. Thesis, University of California, Berkeley, 1970.
32. **Hoek, E. and Bray, J. W.**, *Rock Slope Engineering,* Institute of Mining and Metallurgy, London, 1974.
33. List of Symbols (Rock Mechanics), Commission on Terminology, Symbols and Graphic Representation, International Society for Rock Mechanics, Lisbon, 1970.
34. Suggested Methods for Determining the Uniaxial Compressive Strength of Rock Materials and the Point Load Index, Doc. No. 1, International Society for Rock Mechanics, Lisbon, 1972.
35. Suggested Methods for Determining Water Content, Porosity, Density, Absorption and Related Properties and Swelling and Slake-Durability Index Properties, Doc. No. 2, International Society for Rock Mechanics, Lisbon, 1972.
36. Suggested Methods for Determining Swelling and Slake-Durability Index Properties, International Society for Rock Mechanics, Lisbon, 1972.
37. Suggested Methods for Determining Shear Strength, International Society for Rock Mechanics, Lisbon, 1972.
38. **Kenty, J. D.**, Suggested method of test for direct shear strength of rock core specimens, *Am. Soc. Test. Mater. Spec. Tech. Publ.,* 479, 613, 1970.
39. **Kiersch, G. A. and Treasher, R. C.**, Investigations, areal and engineering geology: Folsom dam project, Central California, *Econ. Geol.,* 50, 271, 1955.
40. **King, R. U.**, A study of geological structure at Climax in relation to mining and block caving, *Am. Inst. Min. Metall. Pet. Eng.,* 163, 145, 1945.
41. **Knill, J. L. and Jones, K. S.**, The recording and interpretation of geological conditions in the foundations of the Roseires, Kariba, and Latiyan dams, *Geotechnique,* 1, 94, 1965.
42. **Knill, J. L.**, The application of engineering geology to the construction of dams in the United Kingdom, in *La Geologie de L'Ingenieur,* Calembert, L. E., Ed., Society of Geology Belgique, Liege, 1974, 113.
43. **Kraatz, P.**, Rockwell Hardness as an Index Property of Rock, M. S. thesis, University of Illinois, Champaign, 1964.
44. **LeComte, P.**, Methods for measuring the dynamic properties of rocks, in Proc. Rock Mechanics Symp., Queens University, Ottawa, 1963.
45. **Lewis, W. E. and Tandanand, S., Eds.**, Bureau of Mines Test Procedures for Rock, Circ. No. 8628, U.S. Bureau of Mines, Washington, D.C., 1974.
46. **Long, A. E. and Obert, L.**, Block caving in limestone at the Crestmore Mine, Riverside Cement Co., Riverside, California, Invest. Circ. No. 7838, U.S. Bureau of Mines, Washington, D.C., 1958.
47. **Merrill, R. H.**, Design of Underground Openings, Oil-Shale Mine, Rifle, Colorado, Invest. Rep. No. 5089, U.S. Bureau of Mines, Washington, D.C., 1954.
48. **Miller, R. P.**, Engineering Classification and Index Properties for Intact Rock, Ph.D. thesis, University of Illinois, Champaign, 1965.
49. **Obert, L. and Duvall, W. I.**, Design and stability of excavations in rock — subsurface, in *Mining Engineering Handbook,* Society of Mining Engineers, New York, 1973.
50. **Obert, L. and Long, A. E.**, Underground Borate Mining, Kern County, California, Invest. Rep. No. 6110, U.S. Bureau of Mines, Washington, D.C., 1962.
51. **Onodera, T. F.**, Dynamic investigation of foundation rocks in situ, in *Proc. 5th Symp. Rock Mechanics,* Pergamon Press, New York, 1963.
52. **Paone, J., Madson, D., and Bruce, W. E.**, Drillability Studies, Laboratory Percussive Drilling, Invest. Rep. No. 7300, U.S. Bureau of Mines, Washington, D.C., 1969.

53. **Parker, J.,** Mining in a lateral stress field at White Pine, *Can. Inst. Min. Metall.*, 64, 1966.
54. **Piteau, D. R.,** *Rock Slope Engineering: Planning, Design, Construction and Maintenance of Rock Slopes for Highways and Railways,* Federal Highway Administration, U.S. Department of Transportation, Washington, D.C., 1978.
55. **Protokyakanov, M. M.,** Mechanical properties and drillability of rocks, in *Proc. 5th Symp. Rock Mechanics,* Pergamon Press, New York, 1963, 103.
56. **Reichmuth, D. R.,** Correlations of force-displacement data with physical properties of rock for percussive drilling systems, in *Proc. 5th Symp. Rock Mechanics,* Pergamon Press, New York, 1963.
57. **Rogiers, J. C., Crawford, A. W., McKay, D. A., and McLennon, J. C.,** *Rock Mechanics, Laboratory Manual,* University of Toronto, Toronto, 1975.
58. **Shepard, R.,** Physical properties and durability of mine rock, *Colliery Eng. (London),* 1953.
59. **Shore, A. F.,** Report on hardness testing: relation between ball hardness and scleroscope hardness, *J. Iron Steel Inst. London,* 98(2), 59, 1918.
60. Embankment and Excavation, Chatfield Dam and Reservoir, South Platte River, Colorado, Design Memorandum, No. PC-24, U.S. Army Engineer District, Omaha, Neb., 1968.
61. **Vanderwilt, J. W.,** Ground movement adjacent to a caving block in the Climax Molybdenum Mine, *Am. Inst. Min. Metall. Pet. Eng. Trans.,* 181, 360, 1949.
62. **Wrightman, R. H.,** A new caving procedure at the Crestmore Limestone Mine, *Am. Inst. Min. Metall. Pet. Eng. Trans.,* 163, 215, 1945.
63. **Wuerker, R. G.,** The status of testing strength of rocks, *Min. Eng. N.Y.,* 1108, 1953.

REFERENCES FOR TABLES 3-5

1. **Balmer, G. G.,** Physical Properties of Some Typical Foundatio n Rocks, Concrete Lab. Rep. No. SP-39, U.S. Bureau of Reclamation, Denver, 1953.
2. **Belikow, B. P.,** Elastic properties of rock, *Stud. Geophys. Geol.,* 6, 75, 1962.
3. **Blair, B. E.,** Physical Properties of Mine Rock, Part 3, Inve st. Rep. No. 5130, U.S. Bureau of Mines, Washington, D.C., 1955.
4. **Blair, B. E.,** Physical Properties of Mine Rock, Part 4, Invest. Rep. No. 5244, U.S. Bureau of Mines, Washington, D.C., 1956.
5. **Coulson, J. H.,** Shear strength of flat surfaces in rock, in Proc. 13th Symp. Rock Mechanics, American Society of Civil Engineers, New York, 1971, 77.
6. **Deere, D. U. and Miller, R. P.,** Engineering Classification and Index Properties for Intact Rock, Rep. No. AFWL-TR-65-116, U.S. Air Force Weapons Laboratory, Kirtland Air Force Base, New Mexico, 1966, 324.
7. **Gyenge, M. and Herget, G.,** Mechanical properties (rock), in *Pit and Slope Manual,* Rep. No. 72-12, Canada Centre for Mineral and Energy Technology, Ottawa, 1977.
8. **Brierley, G. S. and Beverly, B. E., Eds.,** ROTEDA Computer File of Rock Properties, Haley & Aldrich, Inc., Cambridge, Mass., 1980.
9. **Hatheway, A. W.,** Lava Tubes and Collapse Depressions, Ph.D. thesis, University of Arizona, Tucson, 1971.
10. **Hatheway, A. W. and Paris, W. C., Jr.,** Geologic conditions and considerations for underground construction in rock, Boston, Massachusetts, in *Engineering Geology in New England,* Hatheway, A. W., Ed., Preprint 3602, American Society of Civil Engineers, New York, 1979.
11. **Hogg, A. D.,** Some engineering studies of rock movement in the Niagara area (Canada), in *Engineering Geology Case Histories* No. 3, Geological Society of America, Boulder, Colo., 1959, 1.
12. **Horino, F. G. and Hooker, V. E.,** Mechanical Properties of Cores Obtained from the Unleached Saline Zone, Piceance Creek Basin, Rio Blanco County, Colorado, Invest. Rep. No. 8297, U.S. Bureau of Mines, Washington, D.C., 1978, 21.
13. **Kunundzic, B. and Colic, B.,** Determination of the Elasticity Modulus of Rock and the Depth of the Loose Zone in Hydraulic Tunnels by Seismic Refraction Method, Radovi, (Proceedings), Water Resources Engineering Institute, (OTS 60-21644), Sarajevo, Yugoslavia, 1961, 7.
14. **Lutton, R. J., Girucky, F. E., and Hunt, R. W.,** Project Pre-Schooner; Geologic and Engineering Properties Investigations, Rep. No. PNE-50SF, Waterways Experiment Station, U.S. Army Corps of Engineers, Vicksburg, Miss., 1967.
15. **Obert, L., Windes, S. L., and Duvall, W. I.,** Standardized Tests for Determining the Physical Properties of Mine Rock, Invest. Rep. No. 3891, U.S. Bureau of Mines, Washington, D.C., 1946.
16. **Ortel, W. J.,** Laboratory Investigations for Foundation Rock, Swift Damsite-Pondera County Canal and Reservoir Company, MT, Rep. No. C-1153, U.S. Bureau of Reclamation, Concrete and Structural Branch, Denver, Colo., 1965.

17. **Robertson, E. C.**, Physical Properties of Limestone and Dolomite Cores from the Sandhill Well, Wood County, W. Va., Invest. Rep. No. 18, West Virginia Geological Survey, Charleston, 1959, 113.
18. Subsurface Investigation Report, Headquarters, SAC Combat Operations Center, Offutt AFB, U.S. Army Engineer District. Omaha, 1961.
19. Report of Data, Rock Property Test/Program, Bergstrom area (near Austin, Tex.), Waterways Experiment Station, Concrete Division, U.S. Army, Vicksburg, Miss., letters of 30 July and 11 August, 1969. (N.B.: Actual locations may vary; not strictly identified.)
20. **Windes, S. L.**, Physical Properties of Mine Rock, Part 1, Invest. Rep. No. 4459, U.S. Bureau of Mines, Washington, D.C., 1949.
21. **Windes, S. L.**, Physical Properties of Mine Rock, Part 2, Invest. Rep. No. 4727, U.S. Bureau of Mines, Washington, D.C., 1950.
22. Standard Test Method for Direct Tensile Strength of Intact Rock Core Specimens, Standard 2936-78, American Society for Testing and Materials, Philadelphia, 1980.
23. Standard Test Method for Resistance of Transparent Plastic Materials to Abrasion, Standard 1044-78, American Society for Testing and Materials, Philadelphia, 1980.
24. Standard Test for Triaxial Compressive Strength of Undrained Rock Core Specimens Without Pore Pressure Measurements, Standard D2664-67, American Society for Testing and Materials, Philadelphia, 1974.
25. Standard Test Method for Unconfined Compressive Strength of Intact Rock Core Specimens, Standard D2938-79, American Society for Testing and Materials, Philadelphia, 1974.
26. **Tarkoy, P. J.**, Rock Hardness Index Properties and Geotechnical Parameters for Predicting TBM (Tunnel Boring Machine) Performance, Ph.D. thesis, University of Illinois, Champaign, 1975.
27. **Tarkoy, P. J. and Hendron, A. J., Jr.**, Rock Hardness Index Properties and Geotechnical Parameters for Predicting TBM (Tunnel Boring Machine) Performance, Rep. No. 246293, National Science Foundation, Washington, D.C., 1975.

Index

INDEX

A

B

C

H

I

J

K

L

M

P

Q

R

T

Z